# SCIENTIFIC FOUNDATIONS OF VACUUM TECHNIQUE

# SCIENTIFIC FOUNDATIONS OF VACUUM TECHNIQUE

## SECOND EDITION


SAUL DUSHMAN

Late Assistant Director, Research Laboratory
General Electric Company
Schenectady, New York

Revised by Members of the Research Staff
General Electric Research Laboratory

J. M. LAFFERTY, EDITOR



John Wiley & Sons, Inc., New York, London, Sydney

Replacing # 982

7 8 9 10



**Library of Congress Catalog Card Number: 61-17361**
**Printed in the United States of America**

ISBN 0 471 22803 6

# Preface to the Second Edition

The first edition of this book has enjoyed unprecedented success in its field. This is attested to by the yearly sales record, which remained essentially constant up to the time the book went out of print, even though it was ten years old at that time and badly in need of revision. The revision has presented a difficult problem. The book is classic and represents a lifetime of research and study by Saul Dushman on the basic sciences involved in high-vacuum practice. It is questionable whether any one person today could revise such a comprehensive treatise. It therefore seemed appropriate for this task to select, from the General Electric Research Laboratory where Dr. Dushman and his early contemporaries made many important contributions to vacuum technology, a number of scientists with specialized knowledge in the disciplines involved. An attempt has been made to retain, wherever possible, the original plan of the book and to cover essentially the same subject matter introduced by Dr. Dushman. The inclusion of new developments in the field has forced the removal of some older and possibly less important material in order to keep the cost and size of the revised edition within reasonable limits.

Vacuum technology has continued to grow in scientific and industrial importance since the publication of the first edition in 1949. An indication of the rate of growth is the attendance at the annual symposia of the American Vacuum Society and its forerunner, the Committee on Vacuum Techniques. About 300 attended the first meeting, held in Asbury Park in 1954. By 1957, when the American Vacuum Society was formed, the attendance had increased to 570, and in 1960 the attendance at the Cleveland meeting had risen to 930. Since 1949 we have seen the formation of national vacuum societies not only in America, but in many other countries as well. This trend has culminated in the formation of an international vacuum society, the International Organization for Vacuum Science and Technology (I.O.V.S.T.), organized in Brussels in 1958.

Ultrahigh-vacuum technology, involving the production and measurement of pressures below $10^{-9}$ Torr, has developed only since 1947. These new techniques have already had a profound influence on work in surface physics, surface chemistry, and gaseous electronics. They have also been important in the study of semiconductor surfaces and will certainly play a large part in the future development of active electronic circuit elements based on vacuum-evaporated films.

In Chapter 1 the treatment of kinetic theory has been retained as it appeared in the first edition with only minor changes to improve clarity.

An introductory section has been added in Chapter 2 to indicate the theoretical methods of flow analysis and the reasons why they are used. Flow conductance is discussed in a separate section, since this concept is used almost universally in vacuum system work and appears in the text in all subsequent discussions on flow. In the section on viscous flow, material has been added on the range of validity of the Poiseuille equation, which is based on the theory of incompressible flow. If this range of validity is exceeded, as it may be if high-speed pumps are used in conjunction with low-conductance tubes, flow calculations may be revised by procedures thoroughly discussed in the treatise cited in the text. New material has been added concerning corrections to be applied for short tubes and for junctions between tubes of different radii in the incompressible flow range. Additional material has also been presented on molecular flow through short tubes and tubes with complex geometries. The section on flow through porous material has been removed, since this subject is now comprehensively treated elsewhere and has become too specialized for adequate coverage in the limited space available.

The material on mechanical and vapor pumps has been completely revised and condensed to a single chapter in this edition. Much of the catalog-type material has been removed, and the subject is presented in a more critical manner. Chapter 3 describes the major classes of vacuum pumps and their modes of operation. Particular emphasis is given to the overall similarities between these devices. Discussion of the operation of vapor-stream pumps has been expanded. A section on cryogenic pumping has been added because of the current importance of this subject in connection with the production of ultrahigh vacua. Chapter 4 is mainly concerned with vacuum-pump applications. System design and use of traps and baffles are discussed from the standpoint of proper pump applications. Use of the thermodynamic properties of pump work fluids is emphasized in their selection for particular applications.

The chapter on manometers for low gas pressures that appeared in the first edition gave a complete survey of the many different types of gauges used before 1949. Since this chapter represented one of the few comprehensive surveys on this subject, it seemed desirable to retain most of its material on mechanical, viscosity, and radiometer-type gauges, even though the modern trend is towards electronic methods of measuring pressure.

Major revisions in Chapter 5 have been made in the sections on ionization-type gauges and leak detectors. A considerable amount of new material has been added on ultrahigh-vacuum ionization gauges, spectrometers, and a new leak-detection method. Much of the older material on ionization gauge circuits has been removed.

Considerable progress has been made since the publication of the first edition in understanding the interactions between gases and solids. This is reflected in a complete revision of Chapter 6 and the addition of considerable data to Chapter 7. The earlier macroscopic phenomenological description of gas-solid interactions in terms of sorption isotherms has been supplemented by increased understanding of the atomic and molecular properties that govern these phenomena. The thorough treatment of macroscopic description is retained from the first edition, and new interpretations, based upon recent research, have been added.

Chapter 7 has been revised to include new data on the sorptive properties of materials used in vacuum systems. Recent studies of the behavior of glass and silicates have been described, and a new section on the interaction of gases with polymers has been added. Many of the data on sorption of gases by charcoal, cellulose, and silicates, contained in the first edition, have been retained and brought up to date.

Chapter 8 has been revised to include new experimental work in diffusion studies using nuclear magnetic resonance, internal fraction measurements, and radioactive and stable isotope tracers. A new section has been added on the behavior of rare gases injected into metals by ion bombardment and nuclear processes.

Several new additions have been made to Chapter 9. The processes of gas sorption are now understood for many materials. These processes are discussed, and extensive data are given for rates and total quantities of gas that can be sorbed under vacuum tube conditions. The performances of several new getter materials are discussed and compared with older types. A new technique for measuring gas sorption rates is compared with more conventional methods. The section on ultrahigh vacuum has been greatly expanded. Ultrahigh-vacuum-system components are described, and the mechanism of ion pumping is emphasized.

In Chapter 10 the data on vapor pressures and evaporation rates have been brought up to date. The tables have been completely recalculated on the basis of the most recent available determinations of these quantities. Because of the current interest in modern electronic components that utilize thin films, new material on the evaporation of thin films has been added to the section on vacuum distillation of metals and deposition of films.

The section on oxidation rates in Chapter 11 has been removed from the new edition because this subject is peripheral to the main theme of the book. Monographs on metal oxidation are now available, which present a more adequate survey of the field than could be done in the limited space available. Because of increasing interest in the properties of materials at high temperatures, data have been added on nonelementary modes of dissociation of oxides and on the reduction of oxides by metals. Other data in this chapter have been consolidated and brought up to date. The material on the oxides of iron and the nitrides of chrome has been corrected and clarified.

The revisers are indebted to the following for supplying illustrations and information used in various sections of the book: Dr. V. O. Altemose, Corning Glass Works; Mr. W. E. Briggs, Instrument Department, General Electric Company; Dr. B. B. Dayton, Consolidated Vacuum Corporation; Dr. P. della Porta, S.A.E.S.; Prof. W. B. Nottingham, Massachusetts Institute of Technology; Dr. P. A. Redhead, National Research Council; Dr. G. J. Schulz, Westinghouse Research Laboratories; Dr. N. W. Spencer, University of Michigan; Dr. H. A. Steinherz, National Research Corporation; Dr. J. S. Wagener, Kemet Company, Division of Union Carbide Corporation; and Mr. J. L. Winter, Consolidated Electrodynamics, Inc. The revisers also appreciate the criticisms, suggestions, and information which they have received from a number of their associates in the General Electric Research Laboratory.

Finally, the revisers wish to express their appreciation to Mrs. Josephine F. Morecroft for calculations used in the tables in Chapter 11 and to Miss Marilyn Smith for her secretarial help during the preparation of the manuscript and its processing before publication.

J. M. LAFFERTY

*Schenectady, N.Y.*
*October, 1961*

# Preface to the First Edition

In 1922, the writer published a small volume entitled *The Production and Measurement of High Vacua.* This work consisted of a revision of a series of articles which had appeared in the *General Electric Review* during 1920–1921. The following quotation is taken from the Preface to that volume:

"The subject of high vacua has become of such technical and scientific importance that it is hardly necessary to apologize for attempting to incorporate the information available on this topic into a convenient form for reference.

"Naturally the author has drawn to a large extent upon the results of numerous investigations carried out in this laboratory by Dr. Langmuir and his associates. To Dr. Langmuir, especially, he wishes to express his sincere gratitude for constant encouragement and helpful suggestions.

"He also feels greatly indebted to Dr. Whitney for constant interest in the present undertaking, and for placing at his disposal the facilities of the laboratory for pursuing various lines of purely scientific work."

To Dr. W. D. Coolidge, the former Director of the Research Laboratory, and Dr. C. G. Suits, the present Director, the writer feels likewise indebted for encouraging him in an extensive revision of the earlier work and for their continued interest in the present undertaking.

During the quarter century that has intervened, the subject of vacuum technique has increased in importance to such an extent as to necessitate not only a complete revision of the material in the previous volume but also the addition of a number of new topics.

The industrial and scientific importance of the investigations in this field that have been published during the past twenty-five years, and which are being increased almost daily, was emphasized by the large attendance at the Symposium on Vacuum Technique held at Cambridge, Mass., October 30–31, 1947, under the sponsorship of the National Research Corporation and Division of Industrial and Engineering Chemistry of the American Chemical Society.

The interest in vacuum technique which was initiated by the development of the incandescent-filament vacuum lamp and received an additional impetus by the subsequent developments in the study and production of electronic devices was accelerated by certain extremely important problems which arose during the second world war.

The importance of vacuum technique both in purely scientific investigations and in industry is evident from a brief summary of some of the most striking applications. In the electronic field we have the vast range of devices from the smallest receiving tube to the 100-kw transmitting tube; in the field now designated nucleonics we have high-voltage devices such as the cyclotron, betatron, and synchrotron, as well as 2-million-volt X-ray tubes. Moreover, the industrial utilization of vacuum distillation, vacuum dehydration, and vacuum fusion of metals has increased to a significant extent. In short, vacuum technique has emerged from the purely scientific environment of the research laboratory and has entered the engineering stage on a scale that could not have been conceived even as recently as two decades ago.

In this treatise the writer has not dealt specially with experimental procedures for using vacuum technique in the same manner as in Strong's book, *Procedures in Experimental Physics*, or the more recent work by Bachman, *Techniques in Experimental Electronics*. He has attempted to present a survey of fundamental ideas in physics, chemistry, and (to a smaller extent) metallurgy, which will be found useful to both scientists and engineers in dealing with problems in the production and measurement of high vacua.

The writer appreciates the criticisms, suggestions, and information which he has received from a number of his associates in the Research Laboratory and from friends engaged in the industrial applications of vacuum technology. Every effort has been made to give credit to those individuals and organizations, in connection with the specific topics.

Finally, the writer wishes to express his appreciation of the help in typing the manuscript, and for subsequent details which had to be worked out while the volume was going through the press, to Miss Constance Edwards, his former secretary, Miss Gloria Feola, his present secretary, Mrs Rose Hunton, and Miss Jean Oldham.

# Contents

# List of important symbols

## A TABLE OF SYMBOLS FREQUENTLY USED IN THE TEXT WITH THE PAGE NUMBER WHERE EACH IS INTRODUCED

### Symbols Common throughout the Book

|  |  |
|---|---|
| $k$ | Boltzmann's constant |
| $K$ | Clausing's factor |
| $M$ | molecular mass in grams |
| $N_a$ | Avogadro's constant |
| $P$ | pressure ($P_b$, $P_{atm}$, $P_{\mu b}$, $P_{mm}$, and $P_\mu$ designate pressure in bars, atmospheres, microbars, millimeters of mercury, and microns, respectively) |
| $R_0$ | universal gas constant |
| $T$ | absolute temperature in degrees Kelvin |
| $V$ | volume of gas |
| $W$ | mass in grams |

### Chapter 2

| Symbol | Meaning | Page |
|---|---|---|
| $a$ | characteristic dimension of channel, tube, etc. | 81 |
| $A$ | area of channel, tube, etc. | 87 |
| $b$ | characteristic dimension of channel, tube, etc. | 86 |
| $b$ | factor defined by equation 2.65 | 112 |
| $c$ | molecular mean free path at 1 microbar | 112 |
| $C$ | factor in equation 2.46 | 99 |
| $d$ | diameter of tube | 93 |
| $f$ | transfer ratio of momentum | 105 |
| $F$ | flow conductance | 82 |
| $H$ | perimeter of tube, channel, etc. | 87 |
| $k$ | factor defined by equation 2.66 | 112 |
| $K'$ | Dushman's approximation for $K$ | 91 |
| $l$ | length | 82 |
| $L$ | molecular mean free path | 81 |
| $m$ | mass of molecules | 84 |
| $M$ | Mach number | 83 |
| $n$ | number density of gas | 90 |
| $N$ | number of molecules | 82 |
| $Q$ | flow rate | 81 |
| $r$ | ratio of conductances | 91 |
| $R$ | Reynolds number | 84 |

| | | |
|---|---|---|
| $\Delta t_0$ | time constant for exhaust | 113 |
| $U_f$ | flow velocity | 83 |
| $v_a$ | mean molecular speed | 87 |
| $Z$ | flow impedance or resistance | 82 |
| $Z$ | factor defined by equation 2.55 | 106 |
| $\eta$ | viscosity of gas (see p. 26) | 82 |
| $\rho$ | gas density (see p. 2) | 84 |
| $\xi$ | coefficient of slip | 105 |

## Chapter 3

| | | |
|---|---|---|
| $B$ | speed factor $= k\alpha$ | 157 |
| $E$ | actual speed of exhaust | 121 |
| $F$ | system conductance | 120 |
| $\Delta H^v$ | latent heat of vaporization | 163 |
| $k$ | extracting power | 144 |
| $p$ | pressure | 137 |
| $P_s$ | ultimate pressure | 120 |
| $Q$ | rate of gas evolution from vacuum system | 123 |
| $S$ | pump speed | 120 |
| $\alpha$ | distance reduced to mean free path units | 144 |
| $\alpha(p)$ | local velocity of sound at gas pressure $p$ | 146 |
| $\gamma$ | ratio of specific heat at constant pressure to specific heat at constant volume | 138 |
| $\eta$ | gas viscosity | 132 |
| $\mu$ | $(\gamma + 1)/(\gamma - 1)$ | 146 |
| $\rho$ | gas density | 138 |

## Chapter 5

| | | |
|---|---|---|
| $i_e$ | electron emission current | 301 |
| $i_p$ | positive-ion current | 301 |
| $n$ | number of molecules per unit volume | 261 |
| $N$ | number of ion pairs produced by electron in traveling distance $L$ | 303 |
| $P_i$ | probability of ionization by electron impact | 302 |
| $r$ | $\dfrac{\text{gauge sensitivity for gas}}{\text{gauge sensitivity for argon}}$ | 323 |
| $S$ | ionization-gauge sensitivity | 301 |
| $V_c$ | voltage on ionization-gauge positive-ion collector | 306 |
| $V_g$ | voltage on ionization-gauge electron collector | 306 |
| $V_i$ | ionization potential | 301 |
| $\eta$ | viscosity of gas (see p. 26) | 244 |
| $v_a$ | average velocity of gas molecules (see p. 9) | 244 |
| $\rho$ | gas density (see p. 2) | 244 |

## Chapter 6

| | | |
|---|---|---|
| $N_M$ | number of moles adsorbed per gram | 381 |
| $N_s$ | number of molecules per square centimeter in monolayer | 381 |
| $P_0$ | saturation vapor pressure | 397 |
| $Q_s$ | heat of adsorption | 379 |
| $S$ | area covered by one molecule | 420 |
| $V_m$ | volume of monolayer (BET equation) | 397 |
| $V_s$ | volume of monolayer (Langmuir equation) | 391 |
| $\delta$ | molecular diameter | 420 |

| | | |
|---|---|---|
| $\theta$ | fraction of surface covered by adsorbate | 390 |
| $\nu$ | number of molecules striking unit area of surface per second | 390 |
| $\rho$ | density ($\rho_s$, density of solid; $\rho_\ell$, density of liquid) | 420 |

## Chapter 7

| | | |
|---|---|---|
| $A_m$ | surface area of sorbent | 469 |
| $D$ | diffusion constant | 500 |
| $K$ | permeability ($cm^3$ per sec per $cm^2$ per mm thickness, for 1 cm pressure difference) | 491 |
| $P_0$ | saturation vapor pressure | 447 |
| $Q_s$ | heat of adsorption | 443 |
| $Q_{\mu l}$ | permeation rate (micron-liters per hr per $cm^2$ per mm thickness, for 76 cm pressure difference) | 492 |
| $t$ | time | 487 |
| $t$ | temperature, degrees Centigrade | 440 |
| $V_m$ | volume of monolayer (BET equation) | 441 |
| $x$ | amount adsorbed (mg per gram adsorbed) | 506 |
| $\theta$ | contact angle in Kelvin equation | 460 |
| $\sigma$ | liquid surface tension | 460 |

## Chapter 8

| | | |
|---|---|---|
| $C$ | concentration of gas [$cm^3$ of gas (STP) per $cm^3$ of metal] | 570 |
| $C_0$ | initial uniform gas concentration | 583 |
| $D$ | diffusion constant | 570 |
| $E_0$ | heat of diffusion (gram-calories per mole) | 572 |
| $f$ | fraction of total gas content which has diffused | 581 |
| $G$ | rate of gas evolution (atm · $cm^3$ per $cm^2$ per sec) | 587 |
| $P_w$ | per cent by weight of gas in metal | 518 |
| $q$ | gas diffusion rate (atm · $cm^3$ per $cm^2$ per sec) | 571 |
| $Q$ | total amount of gas diffused into the metal | 583 |
| $Q_s$ | heat adsorbed by solubility process (calories) | 519 |
| $Q_{\mu l}$ | gas permeation rate [micron · liters (0°C) per min per $cm^2$ area per mm thickness for 1 atm pressure difference] | 572 |
| $s$ | solubility of gas [$cm^3$ (STP) per 100 grams of metal] | 518 |
| $s_0$ | saturation solubility | 554 |
| STP | standard temperature and pressure (0° C, 760 mm Hg) | 518 |
| $t_m$ | melting point (degrees Centigrade) | 568 |
| $V_g$ | volume of gas (STP) per unit volume of metal | 518 |
| $v_0$ | solubility of gas [$cm^3$ (STP) per gram of metal] | 518 |
| $x$ | milligrams of gas per 100 grams of metal | 518 |
| $z$ | $\pi^2 Dt/4a^2$ | 581 |
| $\theta$ | fraction of surface covered by adsorbed gas | 575 |
| $\mu m$ | micromoles of gas per 100 grams of metal | 518 |
| $\rho$ | density of metal (grams per $cm^3$) | 518 |

## Chapter 10

| | | |
|---|---|---|
| $a$ | activity | 710 |
| $A$ | constant in vapor-pressure equation | 693 |
| $B$ | constant in vapor-pressure equation | 693 |

| | | |
|---|---|---|
| $C$ | constant in rate-of-evaporation equation | 693 |
| $f$ | activity coefficient | 714 |
| $\Delta F$ | change in free energy | 692 |
| $L_0$ | latent heat of evaporation | 693 |
| $t$ | temperature (degrees Centigrade) | 696 |
| $t_B$ | boiling point (degrees Centigrade) | 696 |
| $T_B$ | boiling point (degrees Kelvin) | 695 |
| $t_m$ | melting point (degrees Centigrade) | 696 |
| $W$ | rate of evaporation | 693 |
| $x$ | mole fraction | 710 |

## Chapter 11

| | | |
|---|---|---|
| $\Delta F$ | change in free energy | 739 |
| $\Delta F^\circ$ | standard free energy change (calories per mole) | 740 |
| $\Delta F_v$ | molar free energy of vaporization (calories per mole) | 743 |
| $\Delta H$ | change in enthalpy or heat content | 738 |
| $K$ | equilibrium constant (if stoichiometry of reaction does not lead to cancellation of units, partial pressures must be expressed in atmospheres) | 740 |

# 1 Kinetic theory of gases[1]

## 1.1. FUNDAMENTAL POSTULATES OF THE KINETIC THEORY OF GASES

For a proper understanding of phenomena in gases, more especially at low pressures, it is essential to consider these phenomena from the point of view of the kinetic theory of gases. This theory rests essentially upon two fundamental assumptions. The first of these postulates is that matter is made up of extremely small particles or molecules, and that the molecules of the same chemical substance are exactly alike as regards size, shape, mass, and so forth. The second postulate is that the molecules of a gas are in constant motion, and this motion is intimately related to the temperature. In fact, the temperature of a gas is a manifestation of the amount or intensity of molecular motion.

In the case of monatomic molecules (such as those of the rare gases and vapors of most metals) the effect of increased temperature is evidenced by increased translational (kinetic) energy of the molecules. In the case of diatomic and polyatomic molecules increase in temperature also increases rotational energy of the molecule about one or more axes, as well as vibrational energy of the constituent atoms with respect to mean positions of equilibrium. However, in the following discussion only the effect on translational energy will be considered.

## 1.2. IDEAL GAS LAWS

According to the kinetic theory a gas exerts pressure on the enclosing walls because of the impact of molecules on these walls. Since the gas suffers no loss of energy through exerting pressure on the solid wall of its enclosure, it follows that each molecule is thrown back from the wall with the same speed as that with which it impinges, but in the reverse direction; that is, the impacts are perfectly elastic.

Suppose a molecule of mass $m$ to approach the wall with velocity $v$. Since the molecule rebounds with the same speed, the change of momentum per impact is $2mv$. If $\nu$ molecules strike unit area in unit time with an average velocity $v$, the total impulse exerted on the unit area per unit time is $2mvv$. But the pressure is defined as the rate at which momentum is imparted to a unit area of surface. Hence,

$$2mvv = P. \tag{1.1}$$

It now remains to calculate $v$. Let $n$ denote the number of molecules per unit volume. It is evident that at any instant we can consider the molecules as moving in six directions corresponding to the six faces of a cube. Since the velocity of the molecules is $v$, it follows that, on the average, $(n/6)v$ molecules will cross unit area in unit time.

Equation 1.1 therefore becomes

$$P = \tfrac{1}{3}mnv^2. \tag{1.2}$$

Since

$$mn = \rho, \tag{1.3}$$

where $\rho$ denotes the density, equation 1.2 can be expressed in the form

$$P = \tfrac{1}{3}\rho v^2, \tag{1.4}$$

which shows that, at constant temperature, the pressure varies directly as the density, or inversely as the volume. This is known as Boyle's law.

Also, from equation 1.2 it follows that the total kinetic energy of the molecules in volume $V$ is

$$\tfrac{1}{2}mnv^2V = \tfrac{3}{2}PV. \tag{1.5}$$

Now it is a fact that no change in temperature occurs if two different gases, originally at the same temperature, are mixed. This result is valid independently of the relative volumes. Consequently, the average kinetic energy of the molecules must be the same for all gases at any given temperature, and the rate of increase with temperature must be the same for all gases. We may therefore define temperature in terms of the average kinetic energy of the molecules, and this suggestion leads to the relation

$$\tfrac{1}{2}mv^2 = \tfrac{3}{2}kT, \tag{1.6}$$

where $T$ is the absolute temperature (degrees Kelvin), defined by the relation $T = 273.16 + t$ ($t$ = degrees Centigrade), and $k$ is a *universal constant*, known as the Boltzmann constant.

Combining equation 1.6 with 1.5 it follows that

$$P = nkT. \tag{1.7}$$

This is an extremely useful relation, as will be shown subsequently, for the determination of $n$ for given values of $P$ and $T$. Also, from equation 1.3 and equation 1.7, it follows that

$$P = \frac{k}{m}\rho T, \tag{1.8}$$

that is,

$$\frac{P}{\rho} = \frac{k}{m}T, \tag{1.9}$$

which is known as Charles' law or Gay-Lussac's law.

Lastly, let us consider *equal* volumes of any two different gases at the same values of $P$ and $T$. Since $P$ and $V$ are respectively the same for each gas, and $\frac{1}{2}mv^2$ is constant at constant value of $T$, it follows from equations 1.5 and 1.7 that $n$ must be the same for both gases. That is, *equal volumes of all gases at any given values of temperature and pressure contain an equal number of molecules.* This was enunciated as a fundamental principle by Avogadro in 1811, but about 50 years was required for chemists to understand its full significance.

On the basis of Avogadro's law the molecular mass, $M$, of any gas or vapor is defined as that *mass in grams,* calculated for an ideal gas, which occupies, at 0° C and 1 atmosphere, a volume[2]

$$V_0 = 22{,}414.6 \text{ cm}^3.$$

This is therefore designated the *molar volume*, and the *equation of state for an ideal gas* can be written in the form

$$PV = \frac{W}{M} \cdot R_0 T, \tag{1.10}$$

where $W$ = mass in grams,
$M$ = molecular mass in grams,
$R_0$ = a *universal constant,*
$V$ = volume of $W$ grams at pressure $P$ and absolute temperature $T$.

It is convenient to express equation 1.10 in the form

$$PV = n_M \cdot R_0 T, \tag{1.11}$$

where $n_M$ denotes the *number of moles* (corresponding to $M$ in grams) in the volume $V$ under the given conditions of temperature and pressure.

Obviously the exact value to be used for $R_0$ must depend upon the units in which $P$, $V$, $T$, and $W$ are expressed.

In the cgs system the unit of pressure is 1 dyne per cm². This is designated 1 *microbar* (= $10^{-6}$ bar).[3]

The standard atmosphere ($A_0$) is defined as the pressure exerted per square centimeter by a mercury column 760 mm in height, at 0° C. Since the density of mercury at 0° C is 13.595 g · cm⁻³, it follows that

$$1\ \text{A}_0 = 76 \times 13.595 = 1033.22 \text{ g} \cdot \text{cm}^{-2}$$

$$= 1033.2 \times 980.67 = 1.01325 \cdot 10^6 \textit{ microbars.}$$

Hence, $10^6$ microbars = 1 bar = 750.06 mm Hg, and $10^3$ microbars = 1 *millibar* = 0.75 mm Hg.

In vacuum technology the units of pressure commonly used are the following:[4]

$$1 \text{ mm Hg} = 1333.22 \text{ microbars} = 1.333 \text{ millibars.}$$

$$1 \text{ micron} = 1\ \mu = 10^{-3} \text{ mm Hg}$$
$$= 1.3332 \text{ microbars.}$$

$$10^{-3} \text{ micron} = 1\ \text{m}\mu = 1.333 \cdot 10^{-3} \text{ microbar.}$$

Hence, 1 microbar = 0.75 micron. Also, 1 Torr (or 1 Tor) is used to designate 1 mm Hg pressure.

The unit of volume in the cgs system is 1 cm$^3$, and that of mass, 1 g. In industrial operations, these are commonly replaced by the liter, cubic foot (cu ft), and ounce (or pound) avoirdupois.

For temperature we have the absolute Centigrade or Kelvin scale ($T = t°\,\text{C} + 273.16$) or the absolute Fahrenheit scale[5] ($T_F = t°\,\text{F} + 459.7$).

It should be observed, as a matter of considerable practical importance, that (to within 0.12 per cent) there is approximately the same relation between liters and cubic feet as between grams and ounces avoirdupois; that is,

$$1 \text{ liter} = 0.035316 \text{ cu ft} \quad (\text{reciprocal} = 28.316),$$

and

$$1 \text{ gram} = 0.035274 \text{ oz av} \quad (\text{reciprocal} = 28.3495).$$
$$\text{Ratio} = 1.0012.$$

By means of these conversion factors we obtain the following values for $R_0$:

$$R_0 = \frac{760 \times 22.415}{273.16} = 62.364 \text{ mm} \cdot \text{liter} \cdot \text{deg}^{-1}\,\text{K} \cdot \text{g-mole}^{-1}$$

$$= \frac{1.01325 \times 10^6 \times 22{,}415}{273.16} = 8.3146 \cdot 10^7 \text{ ergs} \cdot \text{deg}^{-1}\,\text{K} \cdot \text{g-mole}^{-1}$$

$$= 62.364 \times 0.035316 = 2.2024 \text{ mm} \cdot \text{cu ft} \cdot \text{deg}^{-1}\,\text{K} \cdot \text{g-mole}^{-1}$$

$$= 2.2024 \times 28.3495 = 62.439 \text{ mm} \cdot \text{cu ft} \cdot \text{deg}^{-1}\,\text{K} \cdot \text{oz-mole}^{-1}$$

$$= 62.439 \times 16 = 999.024 \text{ mm} \cdot \text{cu ft} \cdot \text{deg}^{-1}\,\text{K} \cdot \text{lb-mole}^{-1}$$

$$= \frac{760 \times 22.415 \times 1.0012}{(459.7 + 32)} = 34.687 \text{ mm} \cdot \text{cu ft} \cdot \text{deg}^{-1}\,T_F \cdot \text{oz-mole}^{-1}.$$

It will be observed that the last three relations refer to the molecular weight expressed in ounces avoirdupois or pounds avoirdupois.

In dealing with gases at low pressures, it is convenient to express the *quantity of gas* ($Q = PV$) in the following units:

1 micron · liter = quantity of gas in 1 liter at 1 micron.

1 mm · cu ft = quantity in 1 cubic foot at 1 mm.

1 mm · liter = $10^3$ micron · liters.

Since the amount (or mass) of gas in a given volume depends upon the temperature as well as the pressure, we shall assume, unless otherwise stated, that we are dealing with an ideal gas at room temperature, that is, 25° C.[6]

Let $Q_l$ and $Q_c$ designate the number of *micron · liters* and *millimeter · cubic feet*, respectively. Then, at 25° C,

$$Q_l = \frac{10^{-3}}{62.364 \times 298.16} = 5.3779 \cdot 10^{-8}\, Q_l \text{ g-mole}$$
$$= 5.3779 \cdot 10^{-8}\, Q_l M \text{ g},$$

$$Q_c = \frac{1}{2.2024 \times 298.16} = 1.5228 \cdot 10^{-3}\, Q_c \text{ g-mole}$$
$$= 1.5228 \cdot 10^{-3}\, Q_c M \text{ g},$$

$$Q_c = \frac{1}{62.439 \times 298.16} = 5.3715 \cdot 10^{-5}\, Q_c M' \text{ oz av},$$

where $M'$ = molecular weight in ounces avoirdupois ($M'$ for $O_2 = 32.0$).

Introducing the mechanical equivalent of heat,

$$J_{15} = 4.1855 \cdot 10^7 \text{ ergs} \cdot \text{cal}_{15}^{-1},$$

we obtain the relation

$$J_{15}R_0 = \frac{8.3146}{4.1855}$$
$$= 1.98647 \text{ cal}_{15} \cdot \text{deg}^{-1} \cdot \text{g-mole}^{-1}.$$

From the values given above for $R_0$ the following relations are derived for the density:

$$\rho = mn = \frac{M}{V_0} = \frac{MP_{\mu b}}{R_0 T}$$
$$= 1.2027 \cdot 10^{-8} \frac{MP_{\mu b}}{T} \text{ g} \cdot \text{cm}^{-3}$$
$$= 1.6035 \cdot 10^{-5} \frac{MP_{mm}}{T} \text{ g} \cdot \text{cm}^{-3} \tag{1.12}$$

where $P_{\mu b}$ = pressure in microbars, and $P_{mm}$ = pressure in millimeters of mercury.

Let $\rho_1{}^0$ = density at 1 microbar and 0° C

$$= 4.403 \cdot 10^{-11}\, M,$$

Values of $\rho_1{}^0$ for a number of gases are given in the fourth column of Table 1.3.

In terms of *cubic feet* and *millimeters of mercury*,

$$\rho = \frac{MP_{mm}}{2.2024T} = 0.45405MP_{mm}/T \quad \text{g/cu ft}$$

$$= \frac{M'P_{mm}}{34.687T_F} = 0.02883M'P_{mm}/T_F \quad \text{oz/cu ft.}$$

## 1.3. AVOGADRO NUMBER

Avogadro's law states that the number of molecules per gram-molecular mass is a constant, which is designated $N_A$. Although a number of different methods have been used for the determination of this constant, the most accurate method depends upon the determination of both the Faraday ($F$) and the charge of the electron ($e$).

For the deposition of one gram-equivalent the most accurate value is

$$F = 96{,}488 \text{ absolute coulombs.}$$

Assuming that each univalent ion has a charge equal in magnitude to that of one electron,

$$N_A = \frac{F}{e}.$$

The value of the charge of the electron is

$$e = 4.8025 \cdot 10^{-10} \text{ absolute esu}$$

$$= 10\frac{e}{c} \text{ absolute coulombs,}$$

where $c$ = velocity of light = $2.99776 \cdot 10^{10}$ cm · sec$^{-1}$. Hence $e = 1.60203 \cdot 10^{-19}$ absolute coulomb and $N_A = 6.0228 \cdot 10^{23}$ mole$^{-1}$.

From equations 1.7 and 1.11 it follows that for the molar volume

$$P = \frac{N_A kT}{V}$$

and hence

$$k = \frac{R_0}{N_A}$$
$$= 1.3805 \cdot 10^{-16} \text{ erg} \cdot \text{deg}^{-1} \text{ K}. \quad (1.13)$$

From equation 1.6 it also follows that the *average kinetic energy per molecule* is given by

$$\tfrac{1}{2}mv^2 = \tfrac{3}{2}kT = 2.0707 \cdot 10^{-16} T \text{ erg}$$
$$= 5.6565 \cdot 10^{-14} \text{ erg at } 0° \text{ C}. \quad (1.14)$$

The mass per molecule is evidently

$$m = \frac{M}{N_A}$$
$$= 1.66035 \cdot 10^{-24} M \text{ g}. \quad (1.15)$$

The second and third columns in Table 1.3 give, respectively, values of $M$ and $m$ for a number of gases.

From equation 1.7 it follows that

$$n = \text{number of molecules per cubic centimeter}$$
$$= 7.244 \cdot 10^{15} \frac{P_{\mu b}}{T}, \text{ where } P \text{ is given in microbars} \quad (1.16)$$
$$= 9.656 \cdot 10^{18} \frac{P_{mm}}{T}, \text{ where } P \text{ is given in millimeters.} \quad (1.17)$$

Table 1.1 gives values of $n$ for a series of values of $T$, $P_{\mu b}$, and $P_{mm}$. The value $n = 2.687 \cdot 10^{19}$ is known as the Loschmidt number.

**TABLE 1.1**

**Number of Molecules per Cubic Centimeter**

| $T$ (° K) | $P_{\mu b}$ | $P_{mm}$ | $n$ |
|---|---|---|---|
| 273.16 | $1.0133 \cdot 10^6$ | 760 | $2.687 \cdot 10^{19}$ |
| 298.16 | $1.0133 \cdot 10^6$ | 760 | $2.462 \cdot 10^{19}$ |
| 273.16 | $1.333 \cdot 10^3$ | 1 | $3.536 \cdot 10^{16}$ |
| 298.16 | $1.333 \cdot 10^3$ | 1 | $3.240 \cdot 10^{16}$ |
| 273.16 | 1.000 | $7.50 \cdot 10^{-4}$ | $2.653 \cdot 10^{13}$ |
| 298.16 | 1.000 | $7.50 \cdot 10^{-4}$ | $2.430 \cdot 10^{13}$ |

## 1.4. MOLECULAR VELOCITIES; MAXWELL-BOLTZMANN DISTRIBUTION LAWS

It is evident that there must be a nonuniform distribution of velocities among all the molecules in a given volume because of the constant occurrence of collisions. J. C. Maxwell and L. Boltzmann showed that it is possible to determine the law according to which the molecular velocities are distributed at any given temperature.

The form of the distribution law differs according to the particular type of velocity distribution of interest. If we let $\dot{x}$, $\dot{y}$, and $\dot{z}$ designate the components, along the three coordinate axes, of the randomly directed velocity $v$, then

$$v^2 = \dot{x}^2 + \dot{y}^2 + \dot{z}^2, \tag{1.18}$$

and the distribution function, $f_{\dot{x}}$, with respect to, say, $\dot{x}$ is given by the relation

$$\frac{1}{N}\frac{dN}{d\dot{x}} = f_{\dot{x}} = \left(\frac{m}{2\pi kT}\right)^{1/2} \epsilon^{-m\dot{x}^2/(2kT)}, \tag{1.19}$$

where $N$ = number of molecules in the volume under consideration.

The distribution function for all three components of $v$ has the form

$$\frac{1}{N}\frac{dN}{d\dot{x}\,d\dot{y}\,d\dot{z}} = f_{\dot{x}\dot{y}\dot{z}} = \left(\frac{m}{2\pi kT}\right)^{3/2} \epsilon^{-m(\dot{x}^2+\dot{y}^2+\dot{z}^2)/(2kT)} \tag{1.20}$$

With respect to the polar coordinates $\theta$ and $\phi$,

$$\frac{1}{N}\frac{dN}{dv\,d\theta\,d\phi} = f_{v\theta\phi} = \left(\frac{m}{2\pi kT}\right)^{3/2} \epsilon^{-mv^2/(2kT)} \sin\theta. \tag{1.21}$$

The most important distribution function is that with respect to $v$ *in a random direction*, which is given by the relation

$$\frac{1}{N}\frac{dN}{dv} = f_v = \frac{4}{\pi^{1/2}}\left(\frac{m}{2kT}\right)^{3/2} v^2 \epsilon^{-mv^2/(2kT)}. \tag{1.22}$$

Differentiating $f_v$ with respect to $v$, it is observed that the maximum value occurs for

$$v = \alpha = \left(\frac{2kT}{m}\right)^{1/2}. \tag{1.23}$$

Hence $\alpha$ corresponds to the value of the *most probable velocity*, which is given by the relation

$$\alpha = \left(\frac{2R_0T}{M}\right)^{1/2} = 12{,}895\left(\frac{T}{M}\right)^{1/2} \quad \text{cm} \cdot \text{sec}^{-1}. \tag{1.24}$$

In terms of $c = v/\alpha$, we can express equation 1.22 in the form

$$\frac{dy}{dc} = \frac{1}{N}\frac{dN}{dc} = f_c = \frac{4c^2}{\pi^{1/2}}\epsilon^{-c^2}, \tag{1.25}$$

where $c$ varies from 0 to $\infty$, and $dy$ ($= f_c\,dc$) corresponds to the fraction of the total number of molecules which have values of $c$ ranging between $c$ and $c + dc$. Hence

$$\int_0^1 dy = \int_0^\infty f_c\,dc = 1. \tag{1.26}$$

From equation 1.26 we derive the value of the *arithmetical average velocity*, $v_a = \alpha c_a$, where

$$c_a = \frac{\int_0^\infty c f_c\,dc}{\int_0^\infty f_c\,dc} = \frac{2}{\pi^{1/2}} = 1.1284,$$

and

$$v_a = \left(\frac{8R_0T}{\pi M}\right)^{1/2} \text{ from equation 1.24}$$

$$= 14{,}551\left(\frac{T}{M}\right)^{1/2} \text{cm} \cdot \text{sec}^{-1}. \tag{1.27}$$

The *root-mean-square velocity*, $v_r$, corresponds to the square root of the average value of $v^2$ as derived from equation 1.6 and is therefore given by the relation

$$\begin{aligned} v_r &= \left(\frac{3R_0T}{M}\right)^{1/2} \\ &= 15{,}794\left(\frac{T}{M}\right)^{1/2} \text{cm} \cdot \text{sec}^{-1} \\ &= c_r\alpha \end{aligned} \tag{1.28}$$

and

$$\begin{aligned} c_r &= (\tfrac{3}{2})^{1/2} = 1.2247 \\ &= 1.0854c_a. \end{aligned}$$

The second column in Table 1.2 gives values of $f_c$ for a series of values of $c$, and Fig. 1.1 shows a plot of these data, on which the values of $f_c$ are indicated for the values $c = 1$, 1.1284, and 1.2247, respectively.

$$y = \frac{N_c}{N} = \int_0^c f_c\,dc \tag{1.29}$$

gives the *fraction of the total number of molecules* which have a random

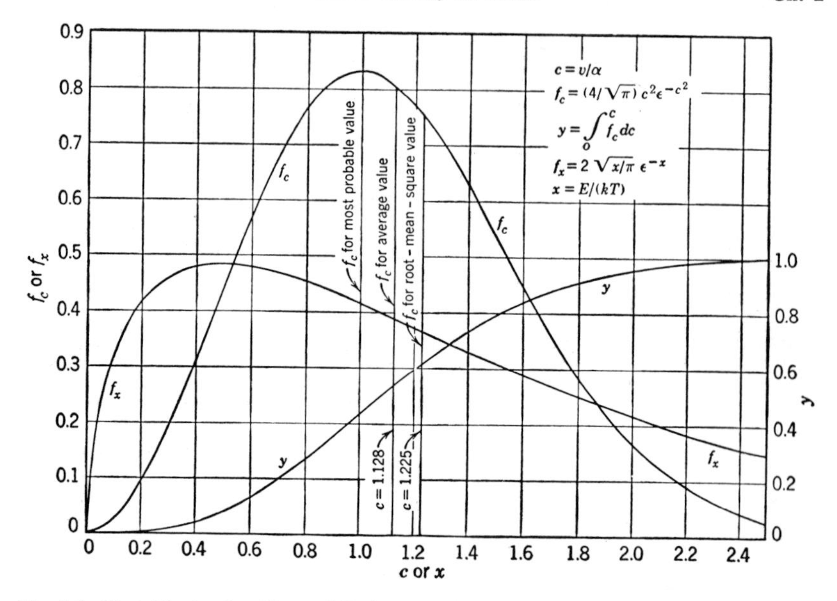


Fig. 1.1. Plots illustrating Maxwell-Boltzmann distribution laws. Plot $f_c$ shows distribution function for random velocity, $c$ expressed in terms of the most probable velocity $\alpha$; plot $f_x$ shows distribution function for energy, $E$, in terms of $x = E/(kT)$; $y$ corresponds to the fraction of the total number of molecules for which the random velocity (expressed in terms of $\alpha$) is less than or equal to a given value $c$.

velocity *equal to or less than* that corresponding to the value $c$, or to $v = \alpha c$. The third and fourth columns in Table 1.2 show values of $y$ and of $\Delta y$, where $\Delta y$ gives the fraction of the total number which have velocities (in terms of $\alpha$ as a unit) ranging between $c$ and the immediately preceding value of $c$. Thus, 8.35 per cent of the molecules have velocities between $c = 1$ and $c = 1.1$, and 42.76 per cent have velocities equal to or less than the most probable value. The values in parentheses are those of $(1 - y)$. A plot of $y$ versus $c$ is shown in Fig. 1.1. It is evident that $y$ corresponds to the area under the curve for $f_c$ from the origin to the given value of $c$.

From equation 1.22 the distribution formula for translational energy ($E$) can be derived. It has the form

$$\frac{1}{N}\frac{dN}{dE} = f_E = \frac{2}{\pi^{1/2}}\frac{1}{kT}\left(\frac{E}{kT}\right)^{1/2} \epsilon^{-E/(kT)}. \tag{1.30}$$

Substituting the variable $x = E/(kT)$, equation 1.30 becomes

$$\frac{1}{N}\frac{dN}{dx} = f_x = \left(\frac{4x}{\pi}\right)^{1/2} \epsilon^{-x}. \tag{1.31}$$

TABLE 1.2

Values of $f_c$, $y$, and $f_x$, Illustrating Application of Distribution Laws

| $c$ | $f_c$ | $y$ | $\Delta y$ | $x$ | $f_x$ |
|---|---|---|---|---|---|
| 0 | 0 | 0 | | 0 | 0 |
| 0.1 | 0.0223 | 0.0008 | 0.0008 | 0.05 | 0.2401 |
| 0.2 | 0.0867 | 0.0059 | 0.0051 | 0.1 | 0.3229 |
| 0.3 | 0.1856 | 0.0193 | 0.0134 | 0.2 | 0.4131 |
| 0.4 | 0.3077 | 0.0438 | 0.0245 | 0.3 | 0.4578 |
| 0.5 | 0.4393 | 0.0812 | 0.0374 | 0.4 | 0.4785 |
| 0.6 | 0.5668 | 0.1316 | 0.0504 | 0.5 | 0.4839 |
| 0.7 | 0.6775 | 0.1939 | 0.0623 | 0.6 | 0.4797 |
| 0.8 | 0.7613 | 0.2663 | 0.0724 | 0.7 | 0.4688 |
| 0.9 | 0.8129 | 0.3453 | 0.0790 | 0.8 | 0.4535 |
| 1.0 | 0.8302 | 0.4276 | 0.0823 | 0.9 | 0.4352 |
| 1.1 | 0.8142 | 0.5101 | 0.0835 | 1.0 | 0.4152 |
| 1.2 | 0.7697 | 0.5896 | 0.0795 | 1.2 | 0.3722 |
| 1.3 | 0.7036 | 0.6634 | 0.0738 | 1.4 | 0.3294 |
| 1.4 | 0.6232 | 0.7286 | 0.0642 | 1.6 | 0.2882 |
| 1.5 | 0.5350 | 0.7878 | 0.0602 | 1.8 | 0.2502 |
| 1.6 | 0.4464 | 0.8369 | 0.0491 | 2.0 | 0.2160 |
| 1.7 | 0.3624 | 0.8772 | 0.0403 | 2.2 | 0.1855 |
| 1.8 | 0.2862 | 0.9096 | 0.0324 | 2.5 | 0.1464 |
| 1.9 | 0.2204 | 0.9348 | 0.0252 | 3.0 | 0.0973 |
| 2.0 | 0.1652 | 0.9540 | 0.0192 | 3.5 | 0.0637 |
| 2.2 | 0.0864 | 0.9784 | 0.0244 | 4.0 | 0.0413 |
| 2.5 | 0.0272 | 0.9941 | 0.0157 | 4.5 | 0.0266 |
| 3.0 | 0.0024 | $(4.2 \cdot 10^{-4})$ | 0.0055 | 5.0 | 0.0170 |
| 4.0 | $4.1 \cdot 10^{-6}$ | $(5.1 \cdot 10^{-7})$ | | 6.0 | 0.0069 |
| 5.0 | $7.8 \cdot 10^{-10}$ | $(7.9 \cdot 10^{-11})$ | | 7.0 | 0.0027 |
| 6.0 | $1.9 \cdot 10^{-14}$ | $(4.4 \cdot 10^{-16})$ | | 8.0 | 0.0011 |

The last two columns in Table 1.2 give values of $f_x$ as a function of $x$, and Fig. 1.1 shows a plot of this function. By differentiating $f_x$ with respect to $x$ and equating the result to zero, it is readily shown that $f_x$ has a maximum value for $x = 0.5$; that is, $f_E$ has a maximum value for $E = \frac{1}{2}kT$. On the other hand, as stated in equation 1.6, $E_{av} = \frac{3}{2}kT$. Since

$$x = \frac{mv^2}{\alpha^2} = mc^2, \tag{1.32}$$

it is possible, from the plot for $y$ in Fig. 1.1 to determine the fraction of the total number of molecules which have an energy equal to or less than that corresponding to a given value of $E$.

It follows from the equations above that the value of $v$ for which $f_v$ is a maximum increases with $T^{1/2}$, while that of $E$ for which $f_E$ is a maximum increases with $T$.

Values of $v_a$, at 0° C and 25° C, for a number of gases and vapors, are given in Table 1.3 (p. 16).

### Relation between Molecular Velocities and Velocity of Sound

It is of interest to note that the relations for $\alpha$, $v_a$, and $v_r$ can also be expressed in terms of the velocity of sound, which we shall designate by *u*.

Since
$$\rho = \frac{MP}{R_0 T} = P\rho_1,$$

where $\rho_1$ = density at 1 microbar, and $P = P_{\mu b}$, we can write the relations for molecular velocities in the forms

$$\alpha = \left(2\frac{P}{\rho}\right)^{1/2} = \left(\frac{2}{\rho_1}\right)^{1/2}, \tag{1.33}$$

$$v_a = \left(\frac{8}{\pi}\cdot\frac{P}{\rho}\right)^{1/2} = 4\left(\frac{1}{2\pi\rho_1}\right)^{1/2}, \tag{1.34}$$

$$v_r = \left(3\frac{P}{\rho}\right)^{1/2} = \left(\frac{3}{\rho_1}\right)^{1/2}. \tag{1.35}$$

On the other hand,

$$u = \left(\gamma\frac{P}{\rho}\right)^{1/2} = \left(\frac{\gamma}{\rho_1}\right)^{1/2}, \tag{1.36}$$

where $\gamma = C_p/C_v$ = ratio of specific heats (per gram-mole) at constant pressure and constant volume. Hence

$$\frac{v_a}{u} = \left(\frac{8}{\pi\gamma}\right)^{1/2}, \tag{1.37}$$

and

$$\frac{\alpha}{u} = \left(\frac{2}{\gamma}\right)^{1/2}. \tag{1.38}$$

For mercury and other monatomic gases, $\gamma = 1.667$; for diatomic gases (such as $H_2$, $N_2$, and $O_2$), $\gamma = 1.40$ (approximately). Hence,

$$\frac{v_a}{u} = 1.236 \text{ for monatomic gases}$$
$$= 1.349 \text{ for diatomic gases,}$$

and

$$\frac{u}{\alpha} = 0.9124 \text{ for monatomic gas.}$$

Thus the velocity of sound in a gas approaches molecular velocities very closely.

### Determination of Avogadro Constant from Distribution of Particles in Brownian Motion

Under high magnification, all suspensions of very fine particles in gases or liquids exhibit "Brownian" motions. Einstein (1905) suggested that the motion of these particles is essentially that to be expected, on the basis of the kinetic theory of gases, of "large molecules," and therefore subject *to the same laws as gas molecules.* That is, the average energy per particle at any given temperature $T$ is $\frac{3}{2}kT$, and the average velocity of the particles is given by the relation

$$v_a = [8kT/(\pi m)]^{1/2}, \tag{1.39}$$

where $m$ = mass of particle.

Application of the Boltzmann-Maxwell laws leads to the following relation for the distribution of particles at different levels in a gravitational field:

$$\frac{n}{n_0} = \epsilon^{-m'gh/(kT)}, \tag{1.40}$$

where $n_0$ = number of particles per cubic centimeter at $h = 0$,
$n$ = number of particles per cubic centimeter at height $h$ (centimeters),
$g$ = 981 dynes,
$m'$ = apparent mass of particles, which is different from the actual mass because of the buoyancy of the medium.

Let $m$ = actual mass of particle. Then

$$m' = m\left(1 - \frac{\rho'}{\rho}\right), \tag{1.41}$$

where $\rho'$ = density of the medium and $\rho$ = density of the particles.

Actually $m'$ is determined from the rate of settling of the particles, by application of Stokes' law.

Thus it is possible to determine $k$ (and consequently the value of $N_A = R_0/k$) from observations on the value of $m'$ and the relation between $n/n_0$ and $h$. Using a fine suspension of gum arabic in water, Perrin obtained the value $N_A = 6.8 \cdot 10^{23}$.

Equation 1.40 has been applied to the determination of the variation with altitude of the density of the atmosphere.

Assuming an average temperature of $T = 230°$ K at higher altitudes,

$$\frac{m'}{k} = \frac{M}{R_0} = \frac{29 \cdot 10^{-7}}{8.315},$$

and hence

$$\log P_{mm} = \log 760 - \frac{29 \cdot 981 \cdot 10^{-5} H}{2.303 \cdot 8.315 \cdot 230}$$

$$= 2.8808 - 6.452 \cdot 10^{-5} H, \tag{1.42}$$

where $P_{mm}$ is the pressure at the altitude $H$ (in meters) above sea level.

Figure 1.2 shows plots of $P_{mm}$ versus the height in kilometers ($AA'$) as calculated by Warfield.[7] Plot $A$ is satisfactorily represented over the range 20–40 kilometers by equation 1.42. According to this relation the pressure should decrease to one-tenth its value at any given altitude for an increase in altitude of 15,500 meters.

Plot $B$ represents observations made by Best, Durand, Gale, and Havens.[8] Evidently these determinations lie below those calculated by Warfield (plot $A'$). Linear plot $C$ corresponds to the extension of equation 1.42 to altitudes above 40 kilometers.

## 1.5. RATE AT WHICH MOLECULES STRIKE A SURFACE

It was shown by O. E. Meyer that the number of molecules of a gas at rest as a whole that strike unit area per unit time is given by the relation

$$\nu = \tfrac{1}{4} n v_a. \tag{1.43}$$

Substituting for $n$ and $v_a$, from equations 1.16, 1.17, and 1.27, we obtain the relations

$$\nu = 2.635 \cdot 10^{19} \frac{P_{\mu b}}{(MT)^{1/2}} \text{ cm}^{-2} \cdot \text{sec}^{-1} \tag{1.44}$$

$$= 3.513 \cdot 10^{22} \frac{P_{mm}}{(MT)^{1/2}} \text{ cm}^{-2} \cdot \text{sec}^{-1}. \tag{1.45}$$

$G$ = mass of gas incident on unit area per unit time

$$= m\nu = \tfrac{1}{4} \rho v_a \tag{1.46}$$

$$= 1.6604 \cdot 10^{-24} M \nu \tag{1.47}$$

$$= P_{\mu b} \left( \frac{M}{2\pi R_0 T} \right)^{1/2} = P_{\mu b} \left( \frac{\rho_1}{2\pi} \right)^{1/2} \tag{1.48}$$

$$= 4.375 \cdot 10^{-5} P_{\mu b} \left( \frac{M}{T} \right)^{1/2} \text{ g} \cdot \text{cm}^{-2} \cdot \text{sec}^{-1} \tag{1.49}$$

$$= 5.833 \cdot 10^{-2} P_{mm} \left( \frac{M}{T} \right)^{1/2} \text{ g} \cdot \text{cm}^{-2} \cdot \text{sec}^{-1}. \tag{1.50}$$

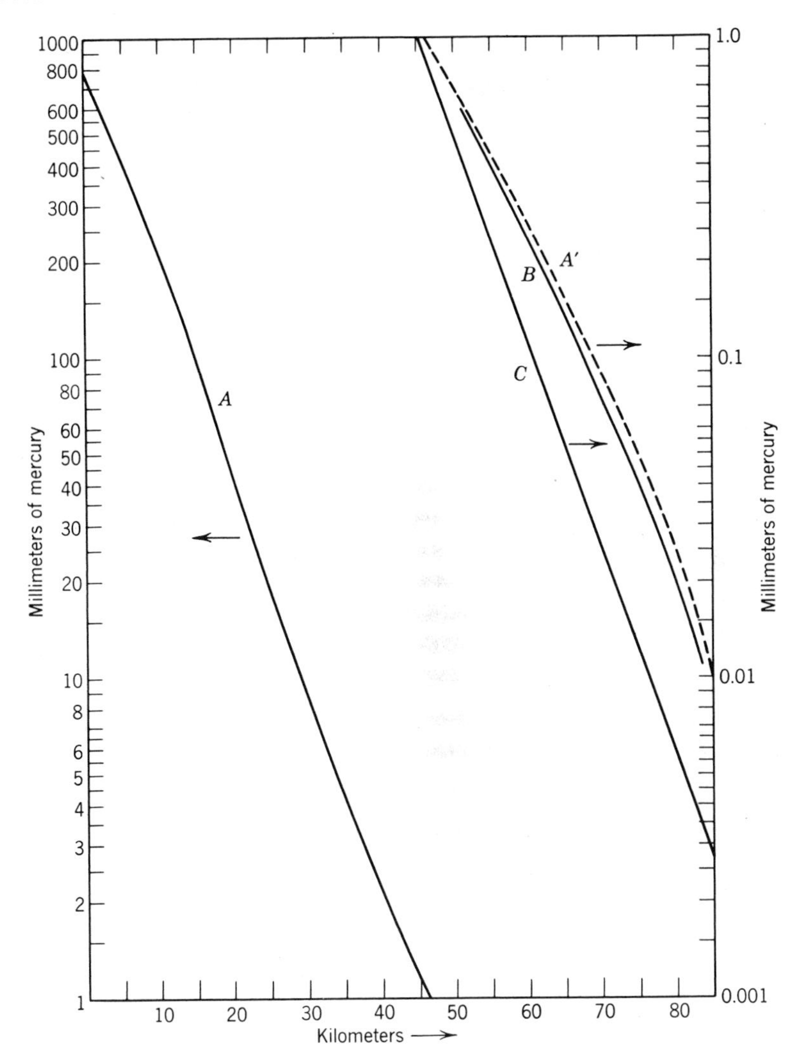


Fig. 1.2. Variation in pressure of atmosphere with height. (1 kilometer = 0.6214 statute mile = 3281 feet.) Plots $A$ and $A'$ from data by NACA (1947); plot $B$ from data in *Phys. Rev.*, 1946; plot $C$, pressure calculated according to equation 1.42. Appropriate scales of pressure are indicated by arrows.

The *volume* which strikes *unit area per unit time* is given by

$$F = \frac{\nu}{n} = \frac{v_a}{4} = 3638\left(\frac{T}{M}\right)^{1/2} \text{cm}^3 \cdot \text{sec}^{-1} \cdot \text{cm}^{-2} \tag{1.51}$$

and is therefore a constant at all pressures, but varies with $(T/M)^{1/2}$.

## TABLE 1.3

**Masses, Velocities, and Rates of Incidence of Molecules**

Note: $\nu_1$ = rate of incidence of molecules per square centimeter per second, at 0° C and 1 microbar.

$\nu_1'$ = rate of incidence of molecules per square centimeter per second, at 0° C and 1 mm.

$G_1$ = mass of gas corresponding to $\nu_1$ ($g \cdot cm^{-2} \cdot sec^{-1}$).

$G_1'$ = mass of gas corresponding to $\nu_1'$ ($g \cdot cm^{-2} \cdot sec^{-1}$).

$m$ = mass of molecule (g); $\rho_1^0$ = density of gas at 0° C and 1 microbar ($g \cdot cm^{-3}$).

$v_a$ = average velocity ($cm \cdot sec^{-1}$).

| Gas or Vapor | $M$ | $10^{23}m$ | $10^{10}\rho_1^0$ | $10^{-4} \cdot v_a$ 0° C | $10^{-4} \cdot v_a$ 25° C | $10^{-17}\nu_1$ | $10^{-20}\nu_1'$ | $10^5 G_1$ | $10^2 G_1'$ |
|---|---|---|---|---|---|---|---|---|---|
| $H_2$ | 2.016 | 0.3347 | 0.8878 | 16.93 | 17.70 | 11.23 | 14.97 | 0.3759 | 0.5012 |
| He | 4.003 | 0.6646 | 1.7631 | 12.01 | 12.56 | 7.969 | 10.63 | 0.5297 | 0.7062 |
| $CH_4$ | 16.04 | 2.663 | 7.063 | 6.005 | 6.273 | 3.981 | 5.308 | 1.060 | 1.414 |
| $NH_3$ | 17.03 | 2.827 | 7.498 | 5.829 | 6.089 | 3.865 | 5.152 | 1.092 | 1.456 |
| $H_2O$ | 18.02 | 2.992 | 7.936 | 5.665 | 5.919 | 3.756 | 5.007 | 1.124 | 1.498 |
| Ne | 20.18 | 3.351 | 8.886 | 5.355 | 5.594 | 3.550 | 4.733 | 1.190 | 1.586 |
| CO | 28.01 | 4.651 | 12.34 | 4.543 | 4.746 | 3.012 | 4.016 | 1.402 | 1.868 |
| $N_2$ | 28.02 | 4.652 | 12.34 | 4.542 | 4.745 | 3.011 | 4.015 | 1.402 | 1.868 |
| Air | 28.98* | 4.811 | 12.77 | 4.468 | 4.668 | 2.962 | 3.950 | 1.425 | 1.900 |
| $O_2$ | 32.00 | 5.313 | 14.09 | 4.252 | 4.442 | 2.819 | 3.758 | 1.497 | 1.996 |
| Ar | 39.94 | 6.631 | 17.59 | 3.805 | 3.976 | 2.523 | 3.363 | 1.675 | 2.230 |
| $CO_2$ | 44.01 | 7.308 | 19.38 | 3.624 | 3.787 | 2.403 | 3.204 | 1.756 | 2.342 |
| $CH_3Cl$ | 50.49 | 8.383 | 22.23 | 3.385 | 3.356 | 2.244 | 2.991 | 1.881 | 2.508 |
| $SO_2$ | 64.06 | 10.64 | 28.21 | 3.004 | 3.139 | 1.992 | 2.656 | 2.118 | 2.825 |
| $Cl_2$ | 70.91 | 11.77 | 31.23 | 2.856 | 2.984 | 1.893 | 2.524 | 2.229 | 2.973 |
| Kr | 83.7 | 13.90 | 36.85 | 2.629 | 2.747 | 1.743 | 2.324 | 2.422 | 3.229 |
| $C_7H_{16}$ | 100.2 | 16.63 | 44.12 | 2.403 | 2.510 | 1.593 | 2.123 | 2.650 | 3.533 |
| Xe | 131.3 | 21.80 | 57.82 | 2.099 | 2.193 | 1.392 | 1.856 | 3.034 | 4.044 |
| $CCl_4$ | 153.8 | 25.54 | 67.72 | 1.939 | 2.026 | 1.286 | 1.714 | 3.283 | 4.377 |
| Hg† | 200.6 | 33.31 | (88.33) | 1.698 | 1.774 | (1.126 | 1.501 | 3.750 | 4.998) |

* Calculated from the value $\rho = 1.293 \cdot 10^{-3}$ at 0° C and 760 mm.

† Since the vapor pressure of mercury at 0° C is $1.85 \cdot 10^{-4}$ mm (= 0.247 microbar), the values given in parentheses have no physical significance. Actual values at 0° C, corresponding to saturation pressure, are as follows: $\rho = 21.79 \cdot 10^{-10}$; $\nu = 2.777 \cdot 10^{16}$; $G = 9.249 \cdot 10^{-6}$.

In the literature, especially that originating in Germany, relation 1.51 is expressed in the form

$$F = \frac{1}{4}\left(\frac{8R_0T}{\pi M}\right)^{1/2} = \frac{1}{(2\pi\rho_1)^{1/2}}. \tag{1.52}$$

The last four columns of Table 1.3 give values of $\nu_1$ and $G_1$, the values calculated for a pressure of 1 microbar, and of $\nu_1'$ and $G_1'$, the values calculated for a pressure of 1 mm—all at 0° C.

The equations given above for $\nu$ and $G$ are also applicable to the effusion of gases *at low pressures* through small holes in *very thin plates.* The requisite condition for the application of Meyer's relation to effusion is that the diameter of the opening should be small compared with the mean free path.

A comparison of relative values of $\nu$ or $G$ for different gases or vapors streaming through such a hole makes it possible to obtain relative values of $M$, since, for constant values of $P$ and $T$, $\nu$ varies *inversely* as $M^{1/2}$, and $G$ directly as $M^{1/2}$.

A good check on the above equations was obtained by Knudsen[9] in some experiments in which hydrogen, oxygen, and carbon dioxide, at pressures ranging from 10 to 0.001 cm of Hg, were made to flow into a vacuum through a 0.025-mm hole in a 0.0025-mm-thick platinum strip.

Equation 1.51 shows that the volume per unit area per unit time, measured at the pressure $P$, is always the same. It follows that $FP_{\mu b}$ *corresponds to the volume at* 1 *microbar.* Hence, if $P_{\mu b1}$ and $P_{\mu b2}$ ($>P_{\mu b1}$) denote the pressures on the two sides of a very thin-walled orifice of area $A$, the net *quantity* of gas ($Q$) flowing through the orifice per unit time is given by

$$\begin{aligned} FA(P_{\mu b2} - P_{\mu b1}) &= Q_2 - Q_1 \\ &= Q, \end{aligned} \tag{1.53}$$

where $Q$ = volume in cubic centimeters per second, measured at 1 microbar. That is, $Q$ denotes microbars × cubic centimeters per unit time.

Similarly, if the pressure is measured in microns (denoted in this volume by $P_\mu$), then $Q$ denotes microns × cubic centimeters per unit time, and $10^{-3}Q$ denotes the number of micron · liters per unit time.

## 1.6. RATE OF EVAPORATION AND VAPOR PRESSURE

An interesting application of equation 1.48 was first made by Langmuir[10] to the determination of vapor pressures from rates of evaporation in high vacua. Since this subject is discussed more fully in Chapter 10 it will be sufficient in the present connection to quote from Langmuir's original paper, "The Vapor Pressure of Metallic Tungsten."

Let us consider a surface of metal in equilibrium with its saturated vapor. According to the kinetic theory we look upon the equilibrium *as a balance between the rate of evaporation and rate of condensation.* That is, we conceive of these two processes going on simultaneously *at equal rates.*
At temperatures so low that the vapor pressure of a substance does not exceed a millimeter, we may consider that the actual rate of evaporation of a substance is independent of the presence of vapor around it. That is, the rate of evaporation in a high vacuum is the same as the rate of evaporation in presence of saturated vapor. Similarly we may consider that the rate of condensation is determined only by the pressure of the vapor.

The rate at which molecules will, in general, condense on a surface is given by $\alpha\nu$, where $\alpha$ is known as the condensation coefficient or sticking coefficient. It represents the ratio between the rate at which molecules actually condense on the surface and the rate at which they strike the surface. If we let $\mu$ denote the rate at which molecules evaporate from the surface, then, at equilibrium,

$$\mu = \alpha\nu. \tag{1.54}$$

Langmuir[11] has shown that for metal atoms condensing on the surface of a metal the value of $\alpha$ may be assumed to be equal to 1. In a later paper on the vapor pressures of high-boiling point organic liquids, Verhoek and Marshall[12] showed that the same assumption is justified in respect to these liquids. Hence, we may, in practically all cases of evaporation, express the relation for rate of evaporation in the form

$$\mu = \nu = \tfrac{1}{4}nv_a. \tag{1.55}$$

For the purpose of calculating the vapor pressure of a metal from a determination of loss of weight per unit area per unit time it is convenient to express equations 1.49 and 1.50 in the forms

$$P_{\mu b} = 2.286 \cdot 10^4 G\left(\frac{T}{M}\right)^{1/2} \tag{1.56}$$

and

$$P_{mm} = 17.14G\left(\frac{T}{M}\right)^{1/2}, \tag{1.57}$$

where $G$ = rate of evaporation in grams per square centimeter per second.
As an illustration of the application of these equations, Table 1.4 gives values of $G$ for tungsten[13] and tantalum[14] at a series of temperatures (degrees K) together with calculated values of $P_{\mu b}$.

TABLE 1.4

Rates of Evaporation and Vapor Pressures of Tungsten and Tantalum

| Metal | $T°$ K | $G$ | $P_{\mu b}$ |
|---|---|---|---|
| Tungsten | 2600 | $8.41 \cdot 10^{-9}$ | $7.23 \cdot 10^{-4}$ |
| $M = 183.92$ | 2800 | $1.10 \cdot 10^{-7}$ | $9.81 \cdot 10^{-3}$ |
| | 3000 | $9.95 \cdot 10^{-7}$ | $9.18 \cdot 10^{-2}$ |
| | 3200 | $6.38 \cdot 10^{-6}$ | $6.08 \cdot 10^{-1}$ |
| | 3400 | $3.47 \cdot 10^{-5}$ | 3.41 |
| Tantalum | 2400 | $3.04 \cdot 10^{-9}$ | $2.58 \cdot 10^{-4}$ |
| $M = 180.88$ | 2600 | $5.54 \cdot 10^{-8}$ | $4.90 \cdot 10^{-3}$ |
| | 2800 | $6.61 \cdot 10^{-7}$ | $6.07 \cdot 10^{-2}$ |
| | 3000 | $5.79 \cdot 10^{-6}$ | $5.40 \cdot 10^{-1}$ |
| | 3200 | $3.82 \cdot 10^{-5}$ | 3.77 |

For the evaporation from a wire of diameter $d'$ (in *mils*) the *loss in weight per second per centimeter length* is given by

$$G_l = 2.54 \cdot 10^{-3}\pi d'G.$$

Hence equations 1.56 and 1.57 assume the forms

$$P_{\mu b} = 2.865 \cdot 10^6 \frac{G_l}{d'}\left(\frac{T}{M}\right)^{1/2}, \tag{1.58}$$

$$P_{mm} = 2.148 \cdot 10^3 \frac{G_l}{d'}\left(\frac{T}{M}\right)^{1/2}. \tag{1.59}$$

Equations 1.56 and 1.57 have also been applied by Knudsen and subsequent investigators to the determination of vapor pressures from *rates of effusion through a small orifice.*

Thus let us consider the case in which molecules evaporating from a hot surface pass through a small orifice into another chamber in which they are condensed. If the pressure of residual gas in this "cool" compartment is extremely low and the *radius of the opening is less than L*, the mean free path of the evaporating molecules in the "hot" compartment, then the rate at which molecules pass through the hole is equal to the rate at which they strike this opening. Consequently, the vapor pressure for any given temperature will be given by equation 1.56 or 1.57, where $G$ represents the weight passing through the orifice per *unit area, per unit time.*

These equations are, however, strictly applicable only if the thickness ($l$) of the wall, in which the orifice of area $\pi a^2$ is located, is vanishingly small compared to $a$. If the orifice consists of a short tube for which

$l/a$ is appreciable, then a correction factor has to be applied, and instead of equation 1.46 we have the relation

$$G = K(\tfrac{1}{4}\rho v_a) = KP_{\mu b}\left(\frac{\rho_1}{2\pi}\right)^{1/2}, \tag{1.60}$$

where $K$ is a function of $l/a$ which is less than 1 for $l/a > 0$. The manner in which the value of $K$ varies with $l/a$ is discussed subsequently in Chapter 2. Hence, if $G'$ denotes the actual loss in weight, at temperature $T$, of material of molecular mass $M$, through an opening of area $A$, over a period of $t$ seconds, then

$$P_{\mu b} = 2.286 \cdot 10^4 \frac{G'}{KAt}\left(\frac{T}{M}\right)^{1/2}, \tag{1.61}$$

$$P_{mm} = 17.14 \frac{G'}{KAt}\left(\frac{T}{M}\right)^{1/2}. \tag{1.62}$$

These equations have been applied by a number of investigators for the determination of vapor pressure at low temperatures, where the values are of the order of a fraction of a millimeter of mercury. The method has been used, for instance, by Egerton for such metals as zinc, cadmium, mercury,[15] and lead.[16]

As an illustration let us consider one such determination made for mercury vapor. In this case the area of the opening was $A = 0.0335$ cm$^2$. At 33.7° C, the loss of mercury through this orifice was 0.7867 g over a period of 2370 min. To correct for the fact that $l/a$ was not negligible, the value of $K$ was found (by means of the relations given in Chapter 2) to be 0.93.

Hence the corrected value of $G$ is given by

$$G = \frac{0.7867}{2370 \times 60 \times 0.0335 \times 0.93}$$

$$= 1.651 \cdot 10^{-4}\ \text{g} \cdot \text{cm}^{-2} \cdot \text{sec}^{-1}.$$

Since $T = 306.9$, and $M = 200.6$, it follows from equation 1.57 that $P_{mm} = 3.77 \cdot 10^{-3}$.

Another interesting application of the above relations, and one which is of increasing importance in industrial chemistry, is provided by the development of *high-vacuum distillation* for the separation of certain organic compounds in the pure state from naturally occurring oils. The great advantage of this process arises from the fact that these organic compounds are unstable at higher temperatures and therefore they can be distilled only at lower temperatures, at which the vapor pressures are in the range of $10^{-4}$ to $10^{-6}$ atm.[17]

In this operation, evaporation takes place from a very thin film of liquid, which is renewed continuously, and condensation occurs on an adjacent cooled surface. In a sufficiently high vacuum (pressure of residual gas less than 1 micron) the rate of transfer of distilland is in accordance with equations 1.49 and 1.50. As the pressure of residual gas is increased, however, the rate of distillation is decreased because of

**TABLE 1.5**

**Variation of Rate of Distillation with Pressure***
(Hickman)

| Pressure ($\mu$) of Residual Gas (Air) | $W$ ($g \cdot sec^{-1} \cdot m^{-2}$) | | |
|---|---|---|---|
| | $P_\mu = 1$, $T = 368°$ K | $P_\mu = 3$, $T = 383°$ K | $P_\mu = 10$, $T = 393°$ K |
| 0.3 | 0.6 | 1.85 | 6.4 |
| 4.0 | 0.46 | 1.59 | 5.7 |
| 7.0 | 0.38 | 1.37 | 5.2 |
| 10.0 | 0.32 | 1.18 | 4.6 |
| 15.0 | 0.25 | 0.95 | 3.8 |
| 25.0 | 0.21 | 0.70 | 2.1 |
| 50.0 | 0.12 | 0.40 | 1.67 |

* The values given for $T$ were taken from a plot of log $P_{mm}$ versus $1/T$ and are therefore only approximate, which accounts for the fact that values of $W$ calculated by means of equation 1.63 for extremely low pressure are slightly less for 3 and 10 microns than those given in the first row.

collisions between the molecules of the distilland and those of the gas. This is illustrated by the data shown in Table 1.5, taken from Hickman's paper.

The distilland used was Octoil, which has the chemical formula $C_6H_4(COOC_8H_{17})_2$, and molecular weight $M = 390.3$. From equation 1.50, it follows that the rate of evaporation, $W$, in grams per second *per square meter* at pressure $P_\mu$, is

$$W = 0.5833 P_\mu \left(\frac{390.3}{T}\right)^{1/2}, \tag{1.63}$$

where $T$ is the absolute temperature corresponding to the vapor pressure of $P_\mu$.

It is evident from these observations, as well as observations of a similar nature mentioned in the next section, that molecules leaving the surface of the distilland are prevented from reaching the surface of condensation because of collisions with the molecules of the residual gas. As a result of such collisions many of the molecules leaving the hot film are driven back, the number of such molecules increasing with the magnitude of residual pressure. We are thus led to the concept of free path as the distance that a molecule can travel without suffering a collision.

## 1.7. FREE PATHS OF MOLECULES

Although the individual molecules in a gas at rest possess very high velocities, as shown previously, it is a matter of ordinary observation that gases diffuse into one another very slowly. This is explained on the kinetic point of view by assuming that the molecules do not travel continuously in straight lines, but undergo frequent collisions. The term "collision" naturally leads to the notion of *free path*. This may be defined as the distance traversed by a molecule between successive collisions. Since, manifestly, the magnitude of this distance is a function of the velocities of the molecules, we are further led to use the expression "mean free path" (denoted by $L$), which is defined as the average distance traversed by all the molecules between successive collisions.

However, this definition assumes that the molecules actually collide like billiard balls; that is, the molecules are assumed to be rigid elastic spheres possessing definite dimensions and exerting no attractive or repulsive forces on one another. But this concept can certainly not be in accord with the facts. We have every reason to believe that the structure of atoms and molecules is exceedingly complex. It is probably impossible to state definitely the diameter of a hydrogen atom or molecule, much less that of a polyatomic molecule. Also there is no doubt that the molecules exert attractive forces on one another for certain distances and repulsive forces when they approach exceptionally close. Otherwise how could we explain surface tension, discrepancies from Boyle's law, and a host of related phenomena? To speak of collisions among molecules, such as these, is impossible. What meaning, therefore, shall we assign to the free path under these conditions?

Let us consider at $t = 0$ a group of $N_0$ "tagged" molecules moving in a given direction. As time goes on, these molecules will suffer random collisions and a number will disappear from the original group. Let $N$ denote the number which, after a period $t$, are still identified with the original group, and let $\omega$ denote the collision frequency. Then

$$-dN = N\omega\, dt. \tag{1.64}$$

Integrating this equation, we obtain the result,

$$N = N_0 \epsilon^{-\omega t}. \tag{1.65}$$

If we let $l$ ($= v_a t$) designate the path that has been traversed by a molecule without suffering collision during the interval $t$, then equation 1.65 can be written in the form

$$N = N_0 \epsilon^{-\omega l / v_a}. \tag{1.66}$$

Furthermore we can write

$$\omega = \frac{v_a}{L}, \tag{1.67}$$

where $L$ is a distance covered between collisions. Then it follows that equation 1.66 assumes the form

$$\phi(l) = \frac{N}{N_0} = \epsilon^{-l/L}. \tag{1.68}$$

That is, $\phi(l)$ is the *fraction* of the original group of molecules that are still traveling without having suffered a collision in the distance $l$.

Furthermore it follows from equation 1.66 that

$$\psi(l)\, dl = \frac{dN}{N_0} = \frac{\omega}{v_a} \epsilon^{-\omega l / v_a}\, dl = \frac{1}{L} \epsilon^{-l/L}\, dl \tag{1.69}$$

represents the fraction of all the free paths that have a length between $l$ and $l + dl$. (Hence the omission of the negative sign in the differentiation.)

It follows from relation 1.69 that the *average* value of the free path is

$$l_a = \int_0^\infty l \psi(l)\, dl = \frac{1}{L} \int_0^\infty l \epsilon^{-l/L}\, dl = L. \tag{1.70}$$

For $l = L$, $\phi(l) = \epsilon^{-1} = 0.3679$. This result shows that 63.21 per cent of the molecules collide with other molecules in a distance equal to or less than $L$. Furthermore it is seen from equation 1.65 that this 63.21 per cent of collisions occur in the interval $\tau = 1/\omega$. Thus $1/\omega$ is a constant of the same nature as the "decay" constant in radioactive disintegrations, while $1/L$ may be regarded as an "absorption" coefficient similar to the coefficient that measures the decrease in intensity of a beam of light in passing through a medium.

Equation 1.68 indicates an experimental method for the determination of $L$ which has been used by Born[18] and Bielz,[19] and which is described by Fraser.[20] A beam of silver atoms is sent into nitrogen or air, and a determination is made of the amount of silver deposited by the beam in a given time $t$ on a surface distant $l$ from the source. If we let $I_0$ designate the intensity of the beam at the source, then the intensity at the collector is

$$I = I_0 \epsilon^{-l/L_P}, \tag{1.71}$$

where $L_P$ is the mean free path of silver atoms in the gas at the pressure in the collecting chamber.

Measurements of the mean free path of potassium in nitrogen have also been reported by Weigle and Plesset.[21]

As Fraser points out,

> With noncondensable gases, it is not possible to measure $I_0$ directly. We assign therefore to $I_0$ a different meaning: namely, the intensity which the beam would have if it were not, as is actually the case, weakened through scattering by the alien molecules present in the collimator chamber. Now, clearly $I_0$ is directly proportional to the quantity of gas issuing from the source slit per second; but so also is the pressure $P$ in the collimator chamber, if a constant pump speed is assumed. We can therefore set $I_0 = c \cdot P$. On the other hand, $L_P$ is inversely proportional to $P$; that is, $L_P = L/P$, where if $P$ is measured say in microns of mercury, $L$ is the mean free path at a pressure of 1 micron. $I$ can therefore be expressed as a function of the pressure $P$; thus if $l$ is the distance between source slit and image slit,
>
> $$I = c \cdot P \cdot e^{-P \cdot l/L},$$
>
> it being assumed that the pressure in the observation chamber is negligibly small. $I$ is a maximum for that value of $P$ which makes $P \cdot l/L = 1$. To make a measurement, the intensity is plotted as a function of the pressure in the collimator chamber, and the value of the latter at the maximum intensity is observed. Then $L_P = L/P = l$.
>
> At this value of $P$, $I = 0.3679 I_0$.

It is of interest to observe, as Fraser emphasizes, that the requisite condition for obtaining a directional effect of the molecules passing through the slit is that *$L_P$ must not be less than $d$*, the width of the slit.

The determination of mean free path for hydrogen has been carried out by Knauer and Stern.[22] The value they obtained, however, is only about 0.44 times that derived from viscosity relations (see discussion in the following section). The reason, as Fraser points out, is that

> the standard methods require an intimate encounter in order that the molecules may exchange energy and momentum in amounts capable of affecting the viscosity or heat conductivity of the gas. The molecular ray method on the other hand counts as a collision an approach of two molecules sufficiently close to deflect them very slightly out of their paths; with narrow slits angular deflections of less than $10^{-4}$ are detectable.

In this connection the reader will find an interesting description of the many uses of molecular beams in a paper by Taylor.[23] As he states, "Molecular beams, narrow rays of molecules formed by a slit system and moving in one direction in an evacuated apparatus, may be used to advantage in many types of research." Among these are determinations of molecular velocities (involving experimental tests of the validity of the Maxwell-Boltzmann distribution law), mean free paths, vapor pressures, accommodation coefficients, and mechanism of chemical reactions and of adsorption.

Evidently the mean free path must depend upon the molecular diameter, and simple considerations indicate that the length of the mean free path must vary inversely as the total cross-sectional area of the molecules per unit volume. Again, the magnitudes of the coefficients of viscosity, heat conductivity, and diffusivity of gases are intimately bound up with the length of the free path; whether it be transference of momentum from one layer to another as in viscosity, or transference of increased kinetic energy of the molecules as in heat conductivity, the rate of this transference must depend upon the number of collisions which each molecule experiences as it passes from point to point. It is therefore to be expected that there should exist very similar relations between the values of the mean free path and those of the coefficients of viscosity, heat conductivity, and diffusion. However, in attempting to deduce such relations, the theoretical physicist has found himself confronted with the problem regarding the laws governing the variation with distance of attractive and repulsive forces between molecules. As a result of successive attacks on this problem, by a number of investigators, the exact forms of these relations have been modified from time to time. The reader is referred to Chapman and Cowling (Ref. 1*d*) for a detailed discussion of the whole problem.

## 1.8. RELATION BETWEEN COEFFICIENT OF VISCOSITY, MEAN FREE PATH, AND MOLECULAR DIAMETER

A gas streaming through a narrow-bore tube experiences a resistance to flow, so that the velocity of this flow decreases uniformly from the center outwards until it reaches zero at the walls. Each layer of gas parallel to the direction of flow exerts a tangential force on the adjacent layer, tending to decrease the velocity of the faster-moving and to increase that of the slower-moving layers. The property of a gas (or liquid) by virtue of which it exhibits this phenomenon is known as *internal viscosity*.

As a simple working hypothesis we may assume, as Newton did, that the internal viscosity is directly proportional to the velocity gradient in the gas. Furthermore, the viscosity must depend upon the nature of the fluid, so that in a more viscous fluid the tangential force between adjacent layers, for constant velocity gradient, will be greater than in a less viscous fluid. We thus arrive at the following definition of the *coefficient of viscosity*:

*The coefficient of viscosity is the tangential force per unit area for unit rate of decrease of velocity with distance* (that is, per unit velocity gradient).

With this definition we are in a position to deduce the approximate form of the relation between the coefficient of viscosity and the free path.

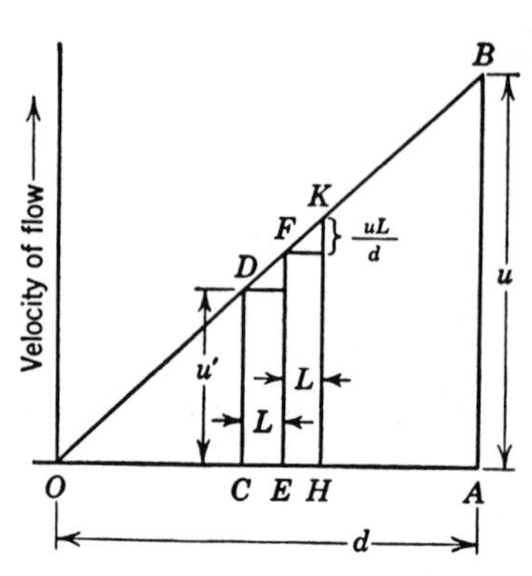


Fig. 1.3. Diagram illustrating the derivation of simple relation between the coefficient of viscosity ($\eta$) of gas and the molecular mean free path ($L$).

Let $u$ denote the velocity of flow of the gas at a distance $d$ from a stationary surface. In uniform flow along a surface, the velocity will decrease uniformly to zero as the surface is approached. We can therefore represent (see Fig. 1.3) the velocity at distance $OA = d$ by the ordinate $AB = u$ and velocities at intermediate distances by the corresponding ordinates of the line $OB$.

We shall imagine the gas divided into layers parallel to the surface, each having a depth equal to the free path, $L$.

Let us denote the tangential force per unit area between adjacent layers by $B$. By definition:

$$B = \eta \times \text{velocity gradient}$$

$$= \eta \times \frac{u}{d}, \tag{1.72}$$

where $\eta$ denotes the coefficient of internal viscosity.

But, according to the kinetic theory, the tangential force per unit area is measured by the rate at which momentum is transferred per unit area between adjacent layers. Because of the relative motion of the layers, the molecules moving from a faster- into a slower-moving layer possess more momentum in the direction of flow than those moving in the opposite direction.

Let us consider any layer, $CE$ or $EH$, of thickness equal to $L$. We have chosen this particular value of the thickness so that we may be justified, as a first approximation, in assuming that the molecules starting at either of the planes $CD$ and $HK$ reach the plane $EF$ without suffering collision, that is, without change of momentum.

The momentum, parallel to the surface, of any molecule reaching the plane $EF$ from the plane $CD$ is $m(u' + v)$, where $u'$ denotes the velocity of flow at the plane $CD$ and $v$ is the mean velocity of the molecules.

The momentum, parallel to the surface, of a molecule reaching the plane *EF* from the plane *HK* is

$$m\left(u' + v + 2\frac{uL}{d}\right).$$

The number of molecules that cross unit area per unit time in any direction in a gas at rest is equal to $\frac{1}{6}nv$, and this must be the same for the molecules traveling in a direction perpendicular to the plane *EF*, for the velocity of flow is assumed to be so small that the density remains constant throughout the different layers.

Hence the net rate of transference of momentum across unit area of the plane *EF* is equal to

$$B = \frac{1}{3} mnv \frac{uL}{d}. \tag{1.73}$$

From equations 1.72 and 1.73 it follows that

$$\eta = \tfrac{1}{3}mnvL = \tfrac{1}{3}\rho vL. \tag{1.74}$$

The dimensions of $\eta$ are evidently $ml^{-1}t^{-1}$, and in the cgs system the unit of viscosity is 1 poise = $1 \text{ g} \cdot \text{cm}^{-1} \cdot \text{sec}^{-1} = 1 \text{ dyne} \cdot \text{sec} \cdot \text{cm}^{-2}$. This is the unit of coefficient of viscosity used in this volume.

In deducing equation 1.74 it has been assumed that all the molecules possess the same velocity $v$ and the same free path $L$. Introducing the law of distribution of velocities, Boltzmann (1881) deduced the relation

$$\eta = 0.3502\rho v_a L_B, \tag{1.75}$$

where $v_a$ = average velocity, and $L_B$ is defined as the average free path.

O. E. Meyer, in his *Kinetic Theory of Gases*, used a different method of calculation and derived a relation of the form

$$\eta = 0.3097\rho v_a L_M \tag{1.76}$$

where $L_M$ is also defined as the average free path.

From these equations an interesting conclusion may be deduced regarding the dependence of viscosity on pressure. As has been mentioned, it is evident from very simple considerations that $L$ must vary inversely as the number of molecules present per unit volume. Consequently the product $\rho L$ is constant and independent of the pressure. The velocity, $v$, depends only upon the temperature and molecular weight. It therefore follows that, for any gas at constant temperature, the *viscosity is independent of the pressure* and *must increase with the temperature*. The confirmation of these two deductions has been justly regarded as one of the most signal triumphs of the kinetic theory of gases. As is well known, the viscosity of all ordinary liquids *decreases* with increase in temperature. That the viscosity of gases must increase with temperature was therefore regarded as a remarkable conclusion.

At both extremely low pressures and very high pressures, the conclusion that the viscosity is independent of the pressure is not in accord with the observations, but this is due to the fact that the same derivation as has been presented above is not valid under those conditions where either attractive forces between the molecules come into play or the pressure is so low that a molecule can travel over the whole distance between the walls of the enclosure without suffering collision.

Both relations 1.75 and 1.76 have been used by physicists, until comparatively recently, for deriving values of $L$ from $\eta$. However, the work of S. Chapman and D. Enskog, since about 1911, has led to the following relations, which are discussed by Chapman and Cowling[1d] in their treatise.

To derive a more exact relation between $\eta$ and $L$ it is necessary to introduce the relation between $L$ and $\delta$, the molecular diameter, which has been shown to be of the form

$$L = \frac{1}{2^{1/2} \cdot \pi n \delta^2}, \tag{1.77}$$

where $n$ = number of molecules per cubic centimeter.

For smooth rigid elastic spherical molecules, it is shown that

$$\eta = \frac{5}{16\delta^2}\left(\frac{kmT}{\pi}\right)^{1/2}. \tag{1.78}$$

Substituting in this equation the relations $v_a = (8kT/\pi m)^{1/2}$ and $\rho = mn$, it follows that

$$\eta = \frac{0.491\rho v_a}{2^{1/2} \cdot \pi n \delta^2}.$$

A further approximation leads to the conclusion that the right-hand side of the last equation should be multiplied by the factor 1.016, and consequently

$$\eta = \frac{0.499\rho v_a}{2^{1/2} \cdot \pi n \delta^2}. \tag{1.79}$$

Combining this with equation 1.77, the result is the relation, used in the following discussion,

$$\eta = 0.499\rho v_a L. \tag{1.80}$$

It follows directly from equation 1.78 that, for two gases having approximately equal values of $\delta$, the viscosities should be in the same ratio as the square roots of the molecular masses. This conclusion has been confirmed by observations on the relative viscosities of $H_2$ and $D_2$ (deuterium).[24] At room temperature $\eta_{D_2}/\eta_{H_2} = 1.39$, which is approximately equal to $2^{1/2}$.

The difference between equation 1.80 and the equations of Boltzmann and Meyer arises from the fact that the two latter investigators failed to take into account the existence of forces of attraction and repulsion between molecules. The accurate form of the relation between $\eta$ and $\delta$ has been a topic of considerable discussion by theoretical physicists.[25]

One interesting contribution to this subject, to which reference is made in a subsequent section, is the model suggested by W. Sutherland.[26] Let us assume that the molecules are "smooth rigid elastic spheres surrounded by fields of attractive force."

The relation between coefficient of viscosity and molecular diameter then has the form

$$\eta = \frac{(5/16\delta_m^{\ 2})(kmT/\pi)^{1/2}}{1 + C/T}, \tag{1.81}$$

where the constant $C$ is a measure of the strength of the attractive forces between the molecules. In this equation the notation $\delta_m$ has been introduced to indicate that the value of the molecular diameter derived in this manner is different from the value $\delta$ used in equation 1.78. From a comparison with equation 1.78 it follows that equation 1.81 is equivalent to the relation

$$\eta = \frac{0.499\,\rho v_a}{2^{1/2}\cdot\pi n\delta_m^{\ 2}(1 + C/T)} \tag{1.82}$$

$$= \frac{0.499\rho v_a L_m}{1 + C/T}, \tag{1.83}$$

where

$$\delta_m^{\ 2} = \frac{\delta^2}{1 + C/T} \tag{1.84}$$

and

$$L_m = \frac{1}{2^{1/2}\cdot\pi n\delta_m^{\ 2}} = L\left(1 + \frac{C}{T}\right). \tag{1.85}$$

Equation 1.81 leads to the relation, used for calculating $C$, which has the form

$$\frac{\eta_T}{\eta_0} = \left(\frac{T}{T_0}\right)^{1/2}\left(\frac{1 + C/T_0}{1 + C/T}\right), \tag{1.86}$$

where $\eta_T$ and $\eta_0$ are the values of the coefficient of viscosity measured at $T$ and $T_0$ respectively. Equation 1.86 is also often expressed in the form

$$\eta_T = \frac{KT^{3/2}}{C + T}, \tag{1.87}$$

where $K = \eta_0(C + T_0)\cdot T_0^{-3/2}$.

By combining equations 1.87 and 1.82 it is seen that the value of $\delta_m$ thus deduced is independent of $T$ and actually corresponds to the value of the molecular diameter at infinitely high temperature, whereas, as pointed out below, the value of $\delta$ decreases with increase in $T$. For this reason equations 1.82 and 1.83 were used formerly by many writers on this subject. However, following the procedure of Chapman and Cowling, and of Kennard, equations 1.79 and 1.80 are used in the following discussion for the calculations of the values of $\delta$ and $L$, respectively. The conversion to values deduced by application of Sutherland's theory may then be made by means of equations 1.84 and 1.85.

From equation 1.80 the following numerical relations are derived:

$$L = 1.1451 \cdot 10^4 \frac{\eta}{P_{\mu b}} \left(\frac{T}{M}\right)^{1/2} \text{ cm} \tag{1.88}$$

$$= 8.589 \cdot \frac{\eta}{P_{mm}} \left(\frac{T}{M}\right)^{1/2} \text{ cm,} \tag{1.89}$$

where $P_{\mu b}$ and $P_{mm}$ denote the pressure in microbars and millimeters of mercury, respectively, and $\eta$ is expressed in poises. These relations have been used to calculate values of $L$ and, from these, values of $\delta$ given in Tables 1.6 and 1.7 (p. 32).

From equation 1.79 we obtain the following relations for $\delta^2$ and $S_c = \pi\delta^2$, the latter being defined as *the mean equivalent cross section for viscosity*:[27]

$$\delta^2 = 2.714 \cdot 10^{-21} \frac{(MT)^{1/2}}{\eta_T} \text{ cm}^2. \tag{1.90}$$

$$\pi\delta^2 = 8.524 \cdot 10^{-21} \frac{(MT)^{1/2}}{\eta_T} \text{ cm}^2. \tag{1.91}$$

Where $\delta$ has been determined by some other method,[28] an approximate calculation of $L$ may be made by means of the relations

$$L = \frac{3.107 \cdot 10^{-17} T}{P_{\mu b}\delta^2} \text{ cm} \tag{1.92}$$

$$= \frac{2.331 \cdot 10^{-20} T}{P_{mm}\delta^2} \text{ cm.} \tag{1.93}$$

Another useful magnitude is the *collision frequency* per molecule, which is given by the relation

$$\omega = \frac{v_a}{L} \tag{1.94}$$

$$= 1.271 \frac{P_{\mu b}}{\eta} \text{ sec}^{-1} \tag{1.95}$$

$$= 1.694 . 10^3 \frac{P_{mm}}{\eta} \text{ sec}^{-1}. \tag{1.96}$$

Values of $\omega$ are shown in the last row in Table 1.6.

Although Sutherland's equation (1.87) has been used most frequently to express the temperature variation of viscosity, several other relations have been derived in the literature. The simplest of these is the *exponential relation*

$$\eta_T = aT^x, \tag{1.97}$$

where $a$ and $x$ are constants characteristic of each gas. This relation has been recommended especially for relatively small ranges of temperature.

Equations 1.87 and 1.97, as well as some other relations, have been tested by Licht and Stechert[29] for a number of typical gases and vapors, using for this purpose data published in the Landolt-Bornstein tables.[30]

According to these investigators, "For twenty-four representative gases and vapors at atmospheric pressure Sutherland's equation has been found to fit extensive experimental data with an average error of less than 1 per cent." Values of the constants $C$ and $K$ in equation 1.87 and of $a$ and $x$ in equation 1.97, taken from the original discussion, are shown in Table 1.8 (p. 33). These constants were used to derive values of $\eta_{25}$ shown in the fifth and eighth columns of the table.

The values of $C$ given in Table 1.6 are those deduced by Schuil,[31] and it is of interest to compare them with those given in Table 1.8.

In deriving the values of $L$ and $\delta$ shown in Tables 1.6 and 1.7, the values of $\eta$ used are those given by Kennard[32] for 15° C. Values for 0° C and 25° C ($\eta_0$ and $\eta_{25}$, respectively) were derived from the values for 15° C by means of equation 1.97, using the values of $x$, taken from Kennard's table, which are given in the second row of Table 1.6. The values of $\delta$ were derived from those of $\eta_0$ and therefore apply strictly only at 0° C.

Combining equation 1.97 with equation 1.78, it follows that at room temperature $\delta$ varies as $T^{-0.5(x-0.5)}$. Since $x$ is greater than 0.5, it also follows that the calculated value of $\delta$ must decrease with increase in temperature.

An interesting method of calculating values of $L$ for air as a function of the temperature has been used by Tsien.[33] From equations 1.80 and 1.36 it follows that

$$L = 1.255 \frac{\eta}{\rho} \frac{\gamma^{1/2}}{u}.$$

Thus $L$ may be expressed in terms of the "kinematic viscosity," $\eta/\rho$, and the velocity of sound, $u$. By means of this relation, Tsien has calculated values of the mean free path for air for the range 0–500° C. A few of the values thus deduced (in centimeters) are as follows:

| $t$° C: | 0 | 20 | 40 | 100 | 200 | 300 | 400 | 500 |
|---|---|---|---|---|---|---|---|---|
| $10^6 L(P/P_0)$: | 5.89 | 6.48 | 7.06 | 8.78 | 11.73 | 14.86 | 17.78 | 20.83 |

In the lower row, $P_0$ denotes the standard pressure, 1 atmosphere, and $P$ any other value of the pressure in atmospheres. The value 5.89 for 0° C is to be compared with the value 5.98 given for $10^6 L_0^{760}$ in Table 1.6.

**TABLE 1.6**

**Mean Free Paths, Molecular Diameters, and Related Data for a Number of Gases**

| Gas: | $H_2$ | He | Ne | Air | $O_2$ | Ar | $CO_2$ | Kr | Xe |
|---|---|---|---|---|---|---|---|---|---|
| $10^7\eta_{15}$ | 871 | 1943 | 3095 | 1796 | 2003 | 2196 | 1448 | 2431 | 2236 |
| $x$ | 0.69 | 0.64 | 0.67 | 0.79 | 0.81 | 0.86 | 0.95 | 0.85 | 0.92 |
| $10^7\eta_0$ | 839 | 1878 | 2986 | 1722 | 1918 | 2097 | 1377 | 2372 | 2129 |
| $10^7\eta_{25}$ | 892 | 1986 | 3166 | 1845 | 2059 | 2261 | 1496 | 2502 | 2308 |
| $10^3L_0^1$ | 8.39 | 13.32 | 9.44 | 4.54 | 4.81 | 4.71 | 2.95 | 3.69 | 2.64 |
| $10^6L_0^{760}$ | 11.04 | 17.53 | 12.42 | 5.98 | 6.33 | 6.20 | 3.88 | 4.85 | 3.47 |
| $10^3L_{25}^1$ | 9.31 | 14.72 | 10.45 | 5.09 | 5.40 | 5.31 | 3.34 | 4.06 | 2.98 |
| $10^6L_{25}^{760}$ | 12.26 | 19.36 | 13.75 | 6.69 | 7.10 | 6.67 | 4.40 | 5.34 | 3.93 |
| $10^8\delta$ | 2.75 | 2.18 | 2.60 | 3.74 | 3.64 | 3.67 | 4.65 | 4.15 | 4.91 |
| $C$ | 84.4 | 80 | 56 | 112 | 125 | 142 | 254 | 188 | 252 |
| $10^{-14}N_S$ | 15.22 | 24.16 | 17.12 | 8.24 | 8.71 | 8.54 | 5.34 | 6.69 | 4.78 |
| $10^{-9}\omega$ | 14.45 | 7.16 | 1.68 | 6.98 | 6.26 | 5.70 | 8.61 | 6.48 | 5.71 |

**TABLE 1.7**

**Mean Free Paths, Molecular Diameters, and Related Data for Water and Mercury Vapors**

| | $t°$ C | $P_{mm}$ | $10^5\eta$ | $10^3L_t^1$ | $L_t^P$ | $10^8\delta_t$ | $10^{-14}N_S$ |
|---|---|---|---|---|---|---|---|
| $H_2O$ | 0 | 4.58 | 8.69 | 2.90 | $6.34 \cdot 10^{-4}$ | 4.68 | 5.27 |
| | 15 | 12.79 | 9.26 | | | | |
| | 25 | 23.76 | 9.64 | 3.37 | $1.42 \cdot 10^{-4}$ | | |
| Hg | 219.4 | 31.57 | 46.66 | 6.28 | $1.99 \cdot 10^{-4}$ | 4.27 | 6.32 |
| | 150.0 | 2.807 | 39.04 | 4.87 | $1.74 \cdot 10^{-3}$ | 4.50 | 5.70 |
| | 100.0 | 0.2729 | 33.56 | 3.93 | $1.44 \cdot 10^{-2}$ | 4.70 | 5.22 |
| | 25.0 | 0.0018 | 25.40 | 2.66 | 1.45 | 5.11 | 4.42 |
| | 0.0 | | 16.2(J) | | | 6.26(J) | |

$P_{mm}$ = vapor pressure at $t°$ C.

In Tables 1.6 and 1.7 the following notation has been used:

$L_0^1$ = mean free path in centimeters at 0° C and 1 mm of Hg.
$L_{25}^1$ = mean free path in centimeters at 25° C and 1 mm of Hg.
$L_t^P$ = mean free path in centimeters at $t°$ C and $P$ mm of Hg.
$\omega$ = collision frequency (per second) at 25° C and 760 mm of Hg.

In the case of $H_2O$, for which values of $L$ and $\delta$ for a series of temperatures are given in Table 1.7, the Sutherland relation was used with $C = 650$ and $\eta_{15} = 9.26 \cdot 10^{-5}$ cgs unit.

In the case of Hg (see Table 1.7), the values of $\eta$ used were based on that given by Braune, Basch, and Wentzel[34] for $t = 219.4°$ C. Values at other temperatures were derived by means of Sutherland's relation, with $C = 942.2$.[35] It should be observed that in their publication these authors used equation 1.82 to calculate values of $\delta_m$.

**TABLE 1.8**

**Characteristic Constants for Viscosity-Temperature Functions**

| Substance | Temperature Range (° C) | Sutherland Equation | | | Exponential Equation | | |
|---|---|---|---|---|---|---|---|
| | | $C$ | $10^6K$ | $10^5\eta_{25}$ | $10^6a$ | $x$ | $10^5\eta_{25}$ |
| Ammonia | −77–441 | 472 | 15.42 | 10.30 | 0.274 | 1.041 | 10.30 |
| Argon | −183–827 | 133 | 19.00 | 22.67 | 2.782 | 0.766 | 21.83 |
| Benzene | 0–313 | 403 | 10.33 | 7.58 | 0.299 | 0.974 | 7.71 |
| Carbon dioxide | −98–1052 | 233 | 15.52 | 15.03 | 1.057 | 0.868 | 14.86 |
| Helium | −258–817 | 97.6 | 15.13 | 19.66 | 4.894 | 0.653 | 20.18 |
| Hydrogen | −258–825 | 70.6 | 6.48 | 9.04 | 1.860 | 0.678 | 8.85 |
| Mercury | 218–610 | 996 | 63.00 | 25.03 | 0.573 | 1.082 | 27.20 |
| Methane | 18–499 | 155 | 9.82 | 11.14 | 1.360 | 0.770 | 10.92 |
| Nitrogen | −191–825 | 102 | 13.85 | 17.80 | 3.213 | 0.702 | 17.60 |
| Oxygen | −191–829 | 110 | 16.49 | 20.78 | 3.355 | 0.721 | 20.42 |
| Water vapor | 0–407 | 659 | 18.31 | 9.84 | 0.170 | 1.116 | 9.82 |

The value of $\eta$ for 0° C is quoted by Jeans (from Kaye and Laby, *Physical Constants*, 1936 edition) in his book (p. 183).

In addition, Tables 1.6 and 1.7 give values of $N_S$, the number of molecules per square centimeter to form a monomolecular layer at 0° C. On the assumption that the spacing is that of a close-packed (face-centered) lattice,

$$N_S = \frac{2}{3^{1/2}\delta^2} = \frac{1.154}{\delta^2}. \tag{1.98}$$

Formulas for the viscosity of mixed gases are given by Kennard.[36] As he points out, the viscosity of a binary mixture does not necessarily lie between the values for the pure components; it may be below or above both these values.

The relations for collision frequency per unit volume between molecules are of importance in many problems of interaction between molecules.

Let $Z_{AA}$ and $Z_{AB}$ denote the number of collisions between *like* and *unlike* molecules, respectively, per cubic centimeter, per second. Then,

$$Z_{AA} = n^2\delta^2\left(\frac{4\pi R_0 T}{M}\right)^{1/2}$$

$$= 3.232 \cdot 10^4 n^2\delta^2\left(\frac{T}{M}\right)^{1/2} \text{ cm}^{-3} \cdot \text{sec}^{-1}, \tag{1.99}$$

$$Z_{AB} = n_A n_B \delta_{AB}{}^2 \left[8\pi R_0 T\left(\frac{1}{M_A} + \frac{1}{M_B}\right)\right]^{1/2}$$

$$= 4.5712 \cdot 10^4 n_A n_B \delta_{AB}{}^2 \left[T\left(\frac{1}{M_A} + \frac{1}{M_B}\right)\right]^{1/2} \text{ cm}^{-3} \cdot \text{sec}^{-1} \tag{1.100}$$

$$= 2.398 \cdot 10^{36} P_A P_B \delta_{AB}{}^2 \cdot T^{-3/2}\left[\frac{1}{M_A} + \frac{1}{M_B}\right]^{1/2}, \tag{1.101}$$

where $n_A$ = number of molecules per cubic centimeter of $A$ at pressure $P_A$ (microbars), with similar definition for $n_B$, and $\delta_{AB} = \frac{1}{2}(\delta_A + \delta_B)$.

For instance, for nitrogen ($M = 28.02$, $\delta = 3.62 \cdot 10^{-8}$) at $T = 298.2°$ K,

$$Z = 8.157 \cdot 10^{16} \text{ cm}^{-3} \cdot \text{sec}^{-1} \text{ at 1 microbar}$$

$$= 8.357 \cdot 10^{28} \text{ cm}^{-3} \cdot \text{sec}^{-1} \text{ at 1 atmosphere.}$$

For the *mean free paths* of the molecules $A$ and $B$ in a mixture, the following general relations have been derived:

$$\frac{1}{L_A} = 2^{1/2}\pi n_A \delta_A{}^2 + \pi n_B \delta_{AB}{}^2\left(1 + \frac{v_B{}^2}{v_A{}^2}\right)^{1/2}, \tag{1.102}$$

$$\frac{1}{L_B} = 2^{1/2}\pi n_B \delta_B{}^2 + \pi n_A \delta_{AB}{}^2\left(1 + \frac{v_A{}^2}{v_B{}^2}\right)^{1/2}, \tag{1.103}$$

where $v_A$ and $v_B$ refer to the *average velocity* of each type of molecule.

For $T_A = T_B$, equation 1.102 becomes

$$\frac{1}{L_A} = 2^{1/2}\pi n_A \delta_A{}^2 + \pi n_B \delta_{AB}{}^2\left(1 + \frac{m_A}{m_B}\right)^{1/2}, \tag{1.104}$$

and similarly for $1/L_B$.

For $\delta_A$ very much smaller than $\delta_B$, $m_A \ll m_B$, and $n_A \ll n_B$,

$$\frac{1}{L_A} = 2^{1/2}\pi n_A \delta_A{}^2 + \pi n_B \delta_{AB}{}^2. \tag{1.105}$$

For $\delta_A$ very much smaller than $\delta_B$,

$$\frac{1}{L_A} = \pi n_B \delta_{AB}{}^2 = \tfrac{1}{4}\pi n_B \delta_B{}^2. \tag{1.106}$$

For $n_A$ very much less than $n_B$, and $T_A$ *not identical with* $T_B$,

$$\frac{1}{L_A} = \pi n_B \delta_{AB}^2 \left[1 + \left(\frac{v_B}{v_A}\right)^2\right]^{1/2} \tag{1.107}$$

$$= \pi n_B \delta_{AB}^2 \left[1 + \frac{T_B}{T_A}\frac{M_A}{M_B}\right]^{1/2}.$$

The last equation has been used by Gaede[37] to calculate the mean free path of nitrogen ($A$) in the blast of a mercury-vapor pump. In this case we may assume: $T_A = 300°$ K, and the temperature of the mercury vapor $T_B = 400°$ K. At this temperature the pressure of mercury vapor is about 1 mm. Hence, $n_B = 2.414 \cdot 10^{16}$.

From the data in Tables 1.6 and 1.7,

$$10^8 \cdot \delta_{AB} = \tfrac{1}{2}(3.78 + 4.70) = 4.24.$$

Substituting these values and those for $M_A$ and $M_B$, in equation 1.107, the result is $L_A = 6.73 \cdot 10^{-3}$ cm, whereas, from the data for $L_t^1$ in Table 1.7, the value derived for the mean free path of mercury molecules in the saturated vapor at 400° K is $L = 4.4 \cdot 10^{-3}$ cm.

### Viscosity at Low Pressures

As mentioned previously, the coefficient of viscosity is not independent of pressure when the pressure decreases to a low value. Under those conditions it was observed by Kundt and Warburg (1875) that the damping of a vibrating surface by the surrounding gas is decreased, as if the gas were slipping over the surface. If one surface is at rest, and another surface, at a distance $d$, is moving parallel to the first surface with a uniform velocity, $u$, the viscous drag upon each surface at normal pressures is given in accordance with the definition of $\eta$ by the relation

$$B = \frac{\eta u}{d}. \tag{1.108}$$

At low pressures, however, the observed value of the tangential force $B$ is less than that given by equation 1.108, corresponding to an increase in the value of $d$. That is, the equation assumes the form

$$B = \frac{\eta u}{d + 2\zeta}, \tag{1.109}$$

where $\zeta$ is known as the *coefficient of slip.*

Considerations based on the kinetic theory of gases lead to the relation

$$\zeta = \left(\frac{2-f}{f}\right)\frac{\eta}{P_{\mu b}}\left(\frac{\pi R_0 T}{2M}\right)^{1/2}, \tag{1.110}$$

which, combined with equation 1.80, leads to the relation

$$\zeta = 2 \cdot 0.499 \frac{(2-f)}{f} L, \tag{1.111}$$

where $f$ is a numerical coefficient which has a maximum value of 1. It was introduced by Maxwell with an interpretation given by Kennard as follows:[38]

> The value of $f$, the transfer ratio for momentum, will presumably depend upon the character of the interaction between the gas molecules and the surface; it may vary with the temperature. We can imagine a surface that is absolutely smooth and reflects the molecules "specularly" with no change in their tangential velocities; in such a case $f = 0$ and $\zeta = \infty$, viscosity being unable to get a grip upon the wall at all. On the other hand, we can imagine the molecules to be reflected without regard to their directions of incidence and therefore with complete loss of their initial average tangential velocity.

In this case $f = 1$, and $\zeta = L$.[39]

More generally, $\zeta = \beta L$, where $\beta$ is of the order of unity, and at very low pressures, where $L \gg d$, equation 1.109 becomes

$$B = \frac{\eta u}{2\beta L} = \frac{u}{\beta} \cdot \frac{\rho v_a}{4} \tag{1.112}$$

$$= \frac{u}{\beta} \cdot P_{\mu b} \left(\frac{M}{2\pi R_0 T}\right)^{1/2}. \tag{1.113}$$

Thus, at very low pressures the rate of transference of momentum from a moving surface to another surface adjacent and parallel is *directly proportional to the pressure* and to the velocity of the moving surface. This conclusion was applied by Langmuir[40] to the design of a *molecular gauge* for the measurement of low pressures, which is described in section 5.4.

The quantity $\rho v_a/4$, which as noted previously corresponds to the mass of gas incident on unit area per unit time, has been designated by Kennard as the *free-molecule* viscosity of the gas between the plates.

The concept of coefficient of slip has also been used to interpret observations on the flow of gases at very low pressures, through capillaries. This topic will be discussed in Chapter 2.

## Molecular Diameters

Mention has been made in the previous section of the two relations for deducing values of $\delta$ and of $\delta_m$ from viscosity measurements. These relations are as follows:

$$\delta^2 = 2.714 \cdot 10^{-21} \frac{(MT)^{1/2}}{\eta} \tag{1.90}$$

and

$$\delta_m{}^2 = \frac{\delta^2 T}{C + T}. \tag{1.84}$$

As illustrated in Table 1.7 by the values of $\delta$ for mercury, these values exhibit a considerable decrease with increase in temperature, and in this case it is found that $\delta_m = 2.50 \cdot 10^{-8}$. The variation with $T$ is obviously greater for those molecules for which the Sutherland constant, $C$, has a large value.

In spite of the fact that the values of $\delta$ thus deduced exhibit a variation with $T$, Chapman and Cowling[1d] have been followed in this discussion in choosing equation 1.90 rather than 1.84. While, as pointed out by Chapman and Cowling, such a variation in the value of $\delta$ with $T$ "receives a simple explanation on the hypothesis that the molecules are centers of repulsive forces, not hard spheres," it is of interest to compare the results obtained by means of equation 1.90 with those obtained by means of 1.84 and also by other methods.

There are several such methods, and only a few of the more important ones can be mentioned briefly, together with some of the results deduced.

### Application of the van der Waals Equation

Near the critical temperature and pressure the behavior of gases can be described very satisfactorily by a modified form of equation 1.10, deduced by van der Waals, which is as follows:

$$\left(P + \frac{a}{V^2}\right)(V - b) = R_0 T. \tag{1.114}$$

In this equation $V$ is the volume per mole, the term $a/V^2$ is a correction term which takes into account the attractive forces between the molecules, and the constant $b$ is a measure of the actual volume of the total number of molecules in accordance with the relation

$$\frac{b}{4} = N_A \cdot \frac{4}{3} \cdot \frac{\pi\delta^3}{8}. \tag{1.115}$$

Hence,

$$\delta^3 = \frac{3b}{2\pi N_A} = 7.929 \cdot 10^{-25} b. \tag{1.116}$$

The value[41] of the constant $b$ may be determined for any given gas from the values of the critical temperature ($T_c$) and critical pressure ($P_c$) by means of the relation

$$b = \frac{R_0 T_c}{8 P_c}. \tag{1.117}$$

### From the Density of the Solid or Liquid

Assuming that the molecules are closely packed, as in a face-centered cubic lattice, the projected area per molecule is given by the relation[42]

$$\frac{1}{N_S} = \sigma = \frac{4(3)^{1/2}}{2}\left[\frac{m}{4(2)^{1/2}\rho}\right]^{2/3} = \frac{10^{-15}}{6.537}\left(\frac{M}{\rho}\right)^{2/3}, \tag{1.118}$$

where $m$ = mass of molecule = $M/N_A$,
$\rho$ = density of *condensed* phase.

But, according to equation 1.98,

$$\sigma = \frac{3^{1/2}}{2}\,\delta^2 = 0.866\delta^2. \tag{1.119}$$

Hence,

$$\delta = 1.122\left(\frac{m}{\rho}\right)^{1/3} = 1.329 \cdot 10^{-8}\left(\frac{M}{\rho}\right)^{1/3}. \tag{1.120}$$

Equation 1.118 has been used to calculate the number of molecules per unit area required to form a unimolecular layer (or monolayer). (See last column of Table 1.9.)

### Cross Section for Collision with Electrons[43]

A cathode-ray beam of initial intensity $I_0$ is decreased to intensity $I$, after passing through a layer of the gas of thickness $x$, in accordance with the relation

$$I = I_0\epsilon^{-\alpha x}, \tag{1.121}$$

where, as shown previously in equation 1.68, $\alpha$ is a measure of $1/L_e$, where $L_e$ = mean free path for electrons = $4(2)^{1/2}L$. Hence it follows from equation 1.77 that

$$\alpha = \frac{n\pi\delta^2}{4}. \tag{1.122}$$

Thus the collision cross section is given by $\alpha/n$. However, as has been observed experimentally, the value of $\alpha$ varies in a rather complex manner with the potential used to accelerate the electrons, with the result that it is actually impossible to assign a definite value to $\delta$ as derived from electron-collision measurements.

The structure of molecules has also been determined from measurements of dipole moments and from electron-diffraction experiments, all of which are discussed at length by Stuart[44] in his book.

Table 1.9 gives, for comparison, values of $\delta$ for a number of gases and vapors as deduced by at least four different relations. The second and third columns give values of $\delta_0$ and $\delta_m$ as deduced by means of equations

TABLE 1.9

Values of Molecular Diameter (cm · $10^{-8}$)

| Gas | From $\eta$ | | From $b$ | From $\rho$ | Electron Collision | $10^{-14} \cdot N_S$ from $\rho$ |
|---|---|---|---|---|---|---|
| | $\delta_0$ | $\delta_m$ | | | | |
| $H_2$ | 2.75 | 2.10 | 2.76 | 4.19 | 2.2 | 6.58 |
| He | 2.18 | 1.69 | 2.66 | 4.21 | 1.70 | 6.49 |
| Ne | 2.60 | 2.16 | 2.38 | 3.40 | 2.2 | 9.98 |
| Ar | 3.67 | 2.42 | 2.94 | 4.15 | 3.6 | 6.71 |
| $O_2$ | 3.64 | 2.50 | 2.93 | 3.73 | 3.4 | 8.30 |
| Hg | 6.26 | 2.50 | 2.38 | 3.26 | | 10.86 |
| $CO_2$ | 4.65 | 3.32 | 3.24 | 4.05 | 4.4 | 7.04 |
| $H_2O$ | 4.68 | 2.45 | 2.89 | 3.48 | 3.8 | 9.53 |
| $C_6H_6$ | 7.65 | 4.71 | 4.51 | 5.89 | | 3.33 |
| $CH_4$ | 4.19 | 3.31 | 3.24 | 4.49 | | 5.73 |
| $C_2H_6$ | 5.37 | 3.87 | 3.70 | 5.01 | | 4.60 |
| $C_3H_8$ | 6.32 | 4.45 | 4.06 | 5.61 | | 3.67 |
| $n$-$C_4H_{10}$ | 7.06 | 4.84 | 4.60 | 6.10 | | 3.18 |
| $n$-$C_5H_{12}$ | 7.82 | 5.05 | 4.89 | 6.45 | | 2.78 |
| $n$-$C_6H_{14}$ | 8.42 | 5.22 | 5.16 | 6.74 | | 2.54 |

1.90 and 1.84, respectively, from the values of $\eta$ (extrapolated to 0° C) and the values of $C$ given by Schuil.[45] The fourth column gives values of $\delta$ calculated by means of equation 1.116 from the values of the van der Waals constant, $b$; the fifth column gives values deduced by means of 1.120 from values of $\rho$ at extremely low temperatures (in general for the solid state). The sixth column gives values, derived from observed values of $\alpha$ by means of equation 1.122, for 36-volt electrons,[46] and the last column gives values of $N_S$ calculated by means of equation 1.119 from the values of $\delta$ in the fifth column.[47] The values thus derived are to be compared with those deduced in Tables 1.6 and 1.7 from kinetic-theory values of $\delta$.

## 1.9. HEAT CONDUCTIVITY OF GASES

The kinetic theory of gases achieved a great triumph when it led to the conclusion that the viscosity is independent of the pressure. It led to still further important results when it predicted the existence of simple relations between the properties of viscosity, heat conductivity, and diffusivity.

From the kinetic point of view it is the same whether the molecules transfer momentum or translational energy from one layer to another. The equations are quite analogous.

As in the case of viscosity, we consider any two layers $CE$, $EH$ (Fig. 1.3), each of thickness $L$, between two plates whose temperatures are $T_1$ and $T_2$ and distance apart $d$. Let $c_v$ denote the heat capacity *per unit mass* at constant volume. The relative temperature drop between the planes $CD$ and $HK$ is equal to

$$2(T_1 - T_2)\frac{L}{d}.$$

Hence the heat transferred per unit area is

$$\begin{aligned} E &= \frac{1}{6}\,nv \cdot 2mc_v \frac{(T_1 - T_2)L}{d} \qquad (1.123) \\ &= \frac{1}{3}\,\rho v c_v \cdot L \frac{(T_1 - T_2)}{d}. \end{aligned}$$

Therefore the coefficient of heat conductivity

$$\lambda = \tfrac{1}{3}\rho v_a L c_v. \qquad (1.124)$$

If $c_v$ is expressed in calories per gram, the unit of $\lambda$ is 1 $\text{cal} \cdot \text{cm}^{-1} \cdot \text{sec}^{-1} \cdot \text{deg}^{-1}$.

Comparing the last equation with 1.74, it follows that

$$\lambda = \eta c_v. \qquad (1.125)$$

As in the case of the relation for $\eta$, a more careful consideration of the mechanism of energy transfer by means of the molecules leads to the relation

$$\lambda = \epsilon \eta c_v, \qquad (1.126)$$

where, according to Eucken,[48]

$$\epsilon = \frac{9\gamma - 5}{4} \qquad (1.127)$$

and $\gamma$ = ratio of specific heat at constant pressure to that at constant volume.

For monatomic gases $\gamma = \frac{5}{3}$, and for polyatomic gases $\gamma$ tends to approach the value 1, with increase in total number of atoms per molecule. Hence

$$1 \leqslant \epsilon \leqslant 2.5.$$

Table 1.10 gives data published by Kannuluik and Martin[49] on values of $\lambda_0$ (conductivity at 0° C) and of $\epsilon$, as derived from observation by means of equation 1.126 and as calculated by means of equation 1.127. Similar data for these gases and a number of others are given in the treatise by Chapman and Cowling.[1d]

One important conclusion that follows from equation 1.126 is that the thermal conductivity of a gas is *independent of pressure*, which is valid as long as the pressure is higher than the range in which molecular flow occurs. (See the following section.)

With regard to the variation in $\lambda$ with $T$ the following remarks may be made. To a first approximation the variation in value of $\lambda$ follows that in the value of $\eta$, since $\epsilon$ exhibits only a slight variation with $T$.

**TABLE 1.10***

**Values of Heat Conductivity Compared with Coefficients of Viscosity**

| Gas | $10^5\lambda_0$ | $10^5\eta_0$ | $c_v(\text{cal} \cdot \text{g}^{-1})$ | $\epsilon$ (obs) | $\epsilon$ (calc) |
|---|---|---|---|---|---|
| He | 34.3 | 18.76 | 0.746 | 2.45 | 2.44 |
| Ne | 11.12 | 29.81 | 0.150 | 2.50 | 2.44 |
| Ar | 3.82 | 21.02 | 0.0745 | 2.44 | 2.44 |
| $H_2$ | 41.3 | 8.50 | 2.43 | 2.00 | 1.90 |
| Air | 5.76 | 17.22 | 0.171 | 1.96 | 1.91 |
| $O_2$ | 5.83 | 19.31 | 0.157 | 1.92 | 1.95 |
| CO | 5.37 | 16.65 | 0.178 | 1.81 | 1.91 |
| $CO_2$ | 3.43 | 13.74 | 0.153 | 1.64 | 1.72 |
| $N_2O$ | 3.61 | 13.66 | 0.155 | 1.71 | 1.73 |

* To convert values of $\lambda_0$ from $\text{cal} \cdot \text{cm}^{-1} \cdot \text{sec}^{-1} \cdot \text{deg}^{-1}$ to *watts* $\cdot \text{cm}^{-1} \cdot \text{deg}^{-1}$ multiply values in Table 1.10 by 4.186.

However, for larger ranges of $T$, account must be taken of the increase with $T$ in the value of $c_v$. Denoting the *molecular* specific heat, at *constant* volume, by $C_v$, Partington and Shilling[50] give the following relations for different gases:

| Gas | $\gamma = C_p/C_v$ | $C_v$ |
|---|---|---|
| Air | 1.4034 | $4.924 + 1.7 \cdot 10^{-4}T + 3.1 \cdot 10^{-7}T^2$ |
| $N_2$ | 1.405 | $4.924 + 1.7 \cdot 10^{-4}T + 3.1 \cdot 10^{-7}T^2$ |
| $O_2$ | 1.396 | $4.924 + 1.7 \cdot 10^{-4}T + 3.1 \cdot 10^{-7}T^2$ |
| CO | 1.404 | $4.924 + 1.7 \cdot 10^{-4}T + 3.1 \cdot 10^{-7}T^2$ |
| $H_2$ | 1.408 | $4.659 + 7.0 \cdot 10^{-4}T$ |
| $CO_2$ | 1.302 | $5.547 + 4.5 \cdot 10^{-3}T - 1.02 \cdot 10^{-6}T^2$ |
| $H_2O$ | ... | $6.901 - 1.19 \cdot 10^{-3}T + 2.34 \cdot 10^{-6}T^2$ |
| Hg, Ar, etc. | 1.667 | 2.990 |

Since the variation in $\eta$ with $T$ is given by the Sutherland relation, equation 1.87, it follows that, in terms of $\lambda_0$ in calories per centimeter per second per degree,

$$\lambda_T = 4.186\lambda_0 \cdot \frac{(C + 273.2)}{(273.2)^{3/2}} \cdot \frac{T^{1/2}(1 + \alpha T + \beta T^2)}{1 + C/T}, \qquad (1.128)$$

where $\alpha$ and $\beta$ are determined from the expression for $C_v$ as a function of $T$, and $\lambda_T$ is expressed in *watts* per centimeter per degree. For *hydrogen*,

$$\lambda_T = 1.369 \cdot 10^{-4}\, T^{1/2} \left(\frac{1 + 1.50 \cdot 10^{-4} T}{1 + 84.4/T}\right);$$

for *argon*,

$$\lambda_T = 1.470 \cdot 10^{-5} \frac{T^{1/2}}{1 + 142/T};$$

for *nitrogen*,

$$\lambda_T = 2.009 \cdot 10^{-5}\, T^{1/2} \left(\frac{1 + 3.45 \cdot 10^{-5} T + 6.30 \cdot 10^{-8} T^2}{1 + 104/T}\right).$$

Similar expressions for gases for which $\lambda$ has not been determined can be derived, as is evident, from determinations of $\eta$, using the observed values of $C$ and values of $\epsilon$ calculated from those of $\gamma$ by means of equation 1.127.

It will be observed that the heat conductivities of hydrogen and helium are much greater than those of heavier gases, such as oxygen and carbon dioxide.

For the case of a wire of radius $a$ and length $l$, suspended along the axis of a cylinder of radius $r$, the *energy loss per unit time* due to thermal conduction by the gas is[51]

$$E = \frac{2\pi a l \lambda_m (T - T_0)}{a \ln (r/a)} = \frac{2\pi \lambda_m l (T - T_0)}{\ln (r/a)}, \qquad (1.129)$$

where $T - T_0$ is the difference in temperature and $r/a$ is not "excessively" large,[52] while $\lambda_m$ is the average conductivity over the temperature range $T - T_0$.

The units in which $E$ is usually expressed are watts per square centimeter. The *energy loss per unit area* per unit time is

$$E_0 = \frac{\lambda_m (T - T_0)}{a \ln (r/a)}, \qquad (1.130)$$

and the *energy loss per unit length of wire* per unit time is

$$E_l = \frac{2\pi \lambda_m (T - T_0)}{\ln (r/a)}. \qquad (1.131)$$

Hence

$$E_l = 2\pi a E_0. \tag{1.132}$$

Since the total energy loss from a heated wire is the sum of that lost by radiation (which varies as $T^4 - T_0^4$) and that lost by conduction of the gas, the former has to be subtracted from the total energy loss in order to obtain the amount due to the latter. Furthermore, in the case of short wires especially, a correction has to be made for the loss by conduction at the ends.

For wires of low emissivity, such as platinum, operating at a temperature below about 500° C, the loss due to radiation is negligible compared to that due to conduction.

Since the thermal-conduction loss in the case of mixtures varies with both the nature of the gas and the composition, this fact has been applied to the analysis of gases.[53] An instrument devised for this purpose by Shakespear,[54] known as a *katharometer*, consists of a platinum spiral filament in a copper block. This instrument has been used extensively by English investigators for determinations of rates in gaseous reactions and for experiments on thermal transpiration (which is discussed in section 1.12).[55]

## 1.10. THERMAL CONDUCTIVITY AT LOW PRESSURES

As mentioned in the previous section, the heat conductivity of gases should theoretically be independent of pressure. That this is at least approximately confirmed by observation is illustrated by the data shown in Table 1.11, obtained by Dickins.[56]

The values under $W$ represent the total heat loss (in calories) by conduction, from a platinum wire ($a = 3.765 \cdot 10^{-3}$ cm, $l = 20.09$ cm) suspended along the axis of a Pyrex glass tube (inside radius, $r = 0.3346$ cm). The tube was maintained at about 0° C by external cooling, and $\Delta t$ is the temperature differential between the wire and the wall. The pressure of the gas in millimeters of mercury is given in the third column, and the value indicated by $W_\infty$ represents the heat loss extrapolated for $P = \infty$. The values under $\lambda_t$ give the thermal heat conductivity at the mean temperature indicated in the last column. The values of $\lambda_t$ (in calories) were derived by means of the relation, deduced from equation 1.129,

$$W_\infty = \frac{2\pi a l \cdot \lambda_t \Delta t}{a \ln (r/a)}. \tag{1.133}$$

It will be observed that over a range of pressures the values of $W$ did not exhibit any considerable decrease.

That the heat conductivity is practically constant over a large range of

TABLE 1.11

Variation in Thermal Conduction with Pressure

(Dickins)

| Gas | $\Delta t$ | $P_{mm}$ | $W$ | $W_\infty$ | $10^5\lambda_t$ | $t°$ C |
|---|---|---|---|---|---|---|
| $H_2$ | 17.866 | 625.0 | 0.21323 | 0.21494 | 42.77 | 9.04 |
| | | 442.0 | 0.21258 | | | |
| | | 302.3 | 0.21145 | | | |
| | | 229.9 | 0.21038 | | | |
| | | 168.2 | 0.20875 | | | |
| | | 129.8 | 0.20692 | | | |
| Air | 23.832 | 91.7 | 0.04021 | 0.04052 | 6.044 | 11.94 |
| | | 52.1 | 0.03997 | | | |
| | | 31.3 | 0.03961 | | | |
| | | 22.3 | 0.03926 | | | |
| | | 17.1 | 0.03891 | | | |
| | | 11.9 | 0.03825 | | | |
| $CO_2$ | 23.80 | 83.3 | 0.02461 | 0.02473 | 3.694 | 11.91 |
| | | 40.5 | 0.02447 | | | |
| | | 19.2 | 0.02417 | | | |
| | | 11.1 | 0.02379 | | | |

pressures is also shown by the plots in Fig. 1.4, of the energy loss (in watts) at constant temperature (about 99° C) from a platinum filament. The filament was a 14-cm length of 3-mil wire located along the axis of a glass tube 2.54 cm in diameter. The wall temperature was 0° C.

As will be observed, the heat loss in hydrogen was about 10 times that in argon. The loss at a pressure of about 1 micron was 0.006 watt. A comparison of the relative losses in the three gases at 76 cm yields values which agree, within a few per cent, with the values of $\lambda$ given in Table 1.10.

Thus, over the range 10–76 cm Hg, the heat loss in hydrogen increased only about 7 per cent, whereas in the range below 5 cm the decrease was nearly 100 per cent. Careful measurements show that at very low pressures the thermal conductivity decreases linearly with the pressure.

The theory of heat conduction at these pressures has been developed from two different points of view. The first of these, due to Knudsen,[57] involves a consideration of the mechanism of energy transfer by individual molecules incident on the hot surface.

The second point of view, due to Smoluchowski,[58] is based upon the concept of a temperature discontinuity which is the thermal analogue of the phenomenon of "slip" discussed in section 1.8.

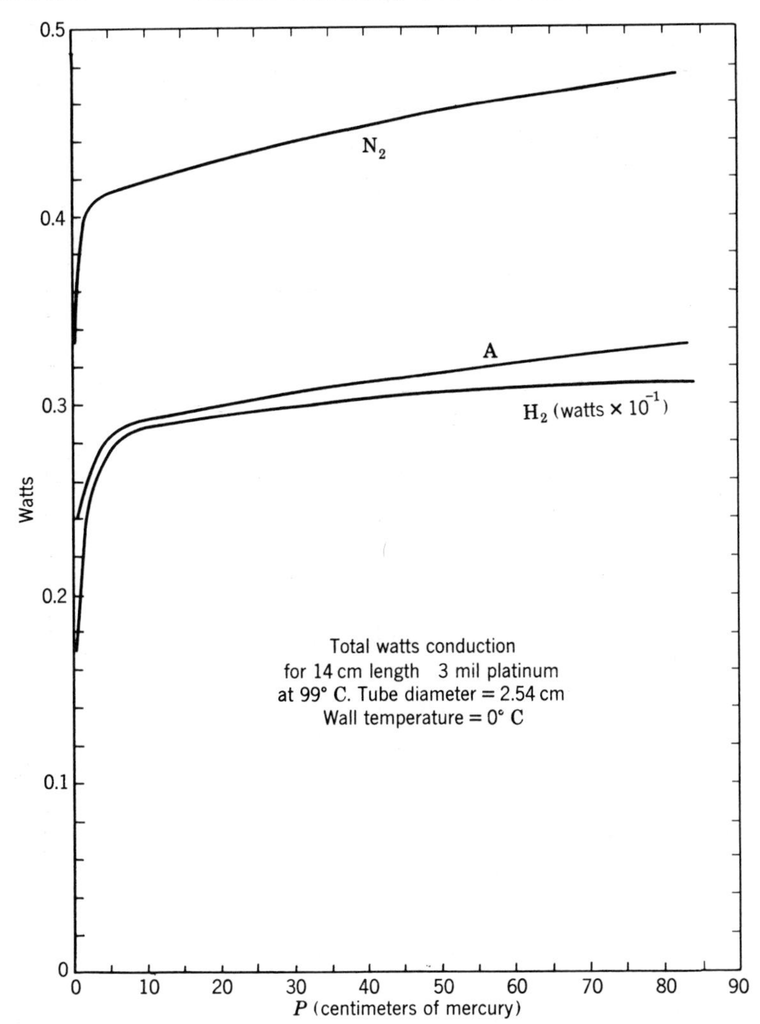


Fig. 1.4. Plots illustrating the variation in thermal conductivity with pressure, for nitrogen, argon, and hydrogen. Ordinates give values of total watts conducted from a platinum filament located along the axis of a cylindrical glass tube. Scale of watts for hydrogen should be multiplied by 10. Abscissas give pressures in centimeters of mercury.

## Free-Molecule Conductivity (Knudsen)

When molecules originally at a temperature $T_i$ strike a hot surface at temperature $T_s$ ($> T_i$), complete interchange of energy does not occur at the first collision. In fact it may often require many collisions for this to occur. Therefore Knudsen introduced a constant, known as the *accommodation coefficient*, designated by $\alpha$, which "can be defined as standing for the fractional extent to which those molecules that fall on the surface and are reflected or re-emitted from it, have their mean energy adjusted or 'accommodated' toward what it would be if the returning molecules were issuing as a stream out of a mass of gas at the temperature of the wall."[59]

The molecules re-emitted or reflected from the hot surface consequently possess a mean energy which corresponds to a temperature lower than $T_s$, which we shall designate as $T_r$, and the accommodation coefficient is defined by the relation

$$\alpha = \frac{T_r - T_i}{T_s - T_i}. \tag{1.134}$$

For $\alpha = 1$, $T_r = T_s$; for $\alpha < 1$, $T_s > T_r > T_i$. It should be noted that the temperature $T_r$ is not clearly defined, as Blodgett and Langmuir[60] have pointed out, unless the molecules leaving the surface have a Maxwellian distribution of velocities.

In any treatise on kinetic theory of gases it is shown that the energy transferred from a surface at temperature $T$ is given by $E = 2kT$ (instead of $\frac{3}{2}kT$, which is the average energy of the molecules in a volume).

Let us now consider heat transfer in a monatomic gas at low pressure.

$E_0$ = energy transfer from hot to cold surface (at temperature $T_i$) per square centimeter of hot surface per second

$$\begin{aligned} &= \nu \cdot 2k(T_r - T_i) \\ &= \tfrac{1}{4} n v_i \cdot 2k(T_r - T_i) \\ &= \frac{1}{2} \cdot \frac{P v_i}{T_i}(T_r - T_i) \\ &= \frac{\alpha}{2} \cdot \frac{P v_i}{T_i}(T_s - T_i), \end{aligned} \tag{1.135}$$

where $v_i$ = average velocity at $T_i$.

Thus the rate of energy transfer at low pressure is proportional to the pressure and the temperature difference.

For diatomic and polyatomic gases, the molecules striking the hot surface acquire not only increased translational energy but also increased amounts of both rotational and vibrational energy. The amount of the vibrational energy possessed by molecules as compared with the amount of translational energy is measured by the value of $\gamma$. A detailed calculation leads, in these cases, to the relation

$$E_0 = \frac{\alpha}{8}\left(\frac{\gamma + 1}{\gamma - 1}\right)\frac{P v_i}{T_i}(T_s - T_i), \tag{1.136}$$

which for $\gamma = \frac{5}{3}$ (case of monatomic gases) becomes identical with equation 1.135.

In this equation, $\alpha$ has the value[61]

$$\alpha = \frac{\alpha_1 \alpha_2}{\alpha_1 + \alpha_2 - \alpha_1 \alpha_2}, \tag{1.137}$$

where $\alpha_1$ and $\alpha_2$ are the values of the accommodation coefficient for the two surfaces.

Substituting for $v_i$ from the relation in terms of $T_i$ and $M$, equation 1.136 assumes the form

$$E_0 = \frac{\alpha}{2}\left(\frac{\gamma + 1}{\gamma - 1}\right)\left[\frac{R_0}{2\pi M(273.2)}\right]^{1/2} \cdot P_{\mu b}\left(\frac{273.2}{T_i}\right)^{1/2}(T_s - T_i)$$

$$= \alpha \Lambda_0 P_{\mu b}\left(\frac{273.2}{T_i}\right)^{1/2} \cdot (T_s - T_i), \tag{1.138}$$

where $\Lambda_0$ = *free-molecule heat conductivity* at 0° C

$$= \frac{1}{2}\left(\frac{\gamma + 1}{\gamma - 1}\right)\left(\frac{R_0}{2\pi}\right)^{1/2} \cdot \frac{1}{[M(273.2)]^{1/2}}$$

$$= \frac{1.819 \cdot 10^{-4}}{[M(273.2)]^{1/2}}\left(\frac{\gamma + 1}{\gamma - 1}\right) \quad \text{watts} \cdot \text{cm}^{-2} \cdot \text{deg}^{-1} \cdot \textit{microbar}^{-1} \tag{1.139}$$

$$= \frac{1.468 \cdot 10^{-5}}{M^{1/2}}\left(\frac{\gamma + 1}{\gamma - 1}\right) \quad \text{watts} \cdot \text{cm}^{-2} \cdot \text{deg}^{-1} \cdot \textit{micron}^{-1}. \tag{1.140}$$

(In terms of calories per second, $\Lambda_0$ should be multiplied by 0.2389.)

Table 1.12 gives values of $\Lambda_0$ in *watts per square centimeter per degree per micron* for a number of gases and vapors,[62] and, for comparison, values of $4.186\lambda_0$ (derived from the values in Table 1.10 and from other sources) which correspond to the conductivity at 0° C in terms of *watts per centimeter per degree.*

In terms of calories per mole

$$R_0 = \frac{8.3144 \cdot 10^7}{4.8155 \cdot 10^7} = 1.9865 \quad \text{cal} \cdot \text{deg}^{-1} \cdot \text{mole}^{-1}.$$

Since $\gamma + 1 = (2C_v + 1.9865)/C_v$, and $\gamma - 1 = 1.9865/C_v$, where $C_v$ = *molar* specific heat in calories, at constant volume,

$$\frac{\gamma + 1}{\gamma - 1} = 1 + \frac{2C_v}{1.9865}.$$

TABLE 1.12

Values of Molecular Heat Conductivity

| Gas | $M$ | $\gamma$ | $10^6\Lambda_0$ | $4.186 \cdot 10^4\lambda_0$ |
|---|---|---|---|---|
| $H_2$ | 2.016 | 1.41 | 60.72 | 17.30 |
| He | 4.003 | 1.67 | 29.35 | 14.36 |
| $H_2O$ | 18.016 | 1.30 | 26.49 | ... |
| Ne | 20.18 | 1.67 | 13.07 | 4.66 |
| $N_2$ | 28.02 | 1.40 | 16.63 | 2.38 |
| $O_2$ | 32.00 | 1.40 | 15.57 | 2.44 |
| Ar | 39.94 | 1.67 | 9.29 | 1.60 |
| $CO_2$ | 44.01 | 1.30 | 16.96 | 1.44 |
| Hg | 200.6 | 1.67 | 4.15 | ... |

Hence equation 1.139 can be expressed in the form

$$\Lambda_0 = \left(\frac{C_v}{1.9865} + \frac{1}{2}\right)\left[\frac{R_0}{2\pi M(273.2)}\right]^{1/2}. \tag{1.141}$$

For *coaxial cylinders* of radii $a$ and $r$ ($r > a$), the rate of energy transfer *per unit area* from the inner cylinder or wire, at temperature $T_s$ and at *low pressures*, is given by the relation

$$E_0 = \alpha_r \cdot \Lambda_0 P_{\mu b}\left(\frac{273.2}{T_i}\right)^{1/2} \cdot (T_s - T_i), \tag{1.142}$$

where

$$\alpha_r = \frac{\alpha}{1 - (1 - \alpha)(a/r)}, \tag{1.143}$$

and $\alpha$ is assumed to be the same for the two surfaces.[63]

It is of interest to compare the rate of energy loss in hydrogen at atmospheric pressure with that at low pressures. At 1 atm, $E_0 = 1.730 \cdot 10^{-3}$ watt $\cdot$ cm$^{-2}$ $\cdot$ deg$^{-1}$, whereas, at $P_\mu$ and $T_i = 273.2$, $E_0 = 6.07 \cdot 10^{-5} P_\mu$ watt $\cdot$ cm$^{-2}$ $\cdot$ deg$^{-1}$.

Jeans has pointed out[64] that, according to general dynamical considerations, the constant $\alpha$ should be determined by a relation of the form

$$\alpha = \frac{4mm'}{(m + m')^2} = 1 - \left(\frac{m - m'}{m + m'}\right)^2, \tag{1.144}$$

where $m'$ = mass for molecules striking the surface for which the molecular mass is $m$.

For completely roughened surfaces $\alpha = 1$. Values of $\alpha$ which have been observed for different cases are given in Table 1.13.[65]

TABLE 1.13

**Values of the Accommodation Coefficient for Several Gases**

| Surface | $H_2$ | | $O_2$ | | $CO_2$ |
|---|---|---|---|---|---|
| Polished Pt | 0.358 | | 0.835 | | 0.868 |
| Pt slightly coated with black | 0.556 | | 0.927 | | 0.945 |
| Pt heavily coated with black | 0.712 | | 0.956 | | 0.975 |
| | $H_2$ | $N_2$ | Ar | Hg | Air |
| Tungsten | 0.20 | 0.57 | 0.85 | 0.95 | |
| Ordinary Pt | 0.36 | 0.89 | 0.89 | | 0.90 |

**Temperature Discontinuity (Smoluchowski)**

In section 1.8 mention was made of the phenomenon of "slip" which occurs in a gas at moderately low pressures. There was introduced in equation 1.109 a coefficient of slip $\zeta = \beta L$, to account for the apparent decrease in viscosity at low pressures.

An analogous phenomenon was observed by Smoluchowski in the case of thermal conduction at low pressures. For two parallel plane surfaces separated by a distance $d$, the *heat loss per unit area per unit time* may be represented at these pressures by a relation of the form

$$E_0 = \frac{\lambda_m(T_1 - T_0)}{d + 2g}, \tag{1.145}$$

where $\lambda_m$ is the mean conductivity over the range $T_1 - T_0$.

In this equation, $g$ is a coefficient defined by the relation

$$\Delta T = -g\frac{dT}{dx}, \tag{1.146}$$

where $\Delta T$ represents the temperature discontinuity at any one surface, and $dT/dx$ is the temperature gradient normal to the surface.

As shown by Kennard,

$$\begin{aligned} g &= \frac{2-\alpha}{\alpha}\cdot\frac{2}{\gamma+1}\cdot\frac{\lambda}{\eta c_v}\cdot L \\ &= \frac{2-\alpha}{\alpha}\cdot\frac{2\epsilon}{\gamma+1}\cdot L = \beta' L, \end{aligned} \tag{1.147}$$

where the numerical constant $\epsilon$ has a value, according to equation 1.127, which varies between 1 and 2.5, and $\beta'$ is a numerical constant of the order of unity.

Hence equation 1.145 may be expressed in the form

$$E_{0p} = \frac{\lambda_m(T_1 - T_0)}{d + 2\beta' L} = \frac{\lambda_m(T_1 - T_0)}{d + 2\beta' c/P}, \tag{1.148}$$

where $\alpha$ is assumed to be the same for both surfaces, also $L = c/P$, and the symbol $E_{0p}$ indicates that the conductivity per unit area is a function of $P$.

For a hot wire of radius $a$ situated along the axis of a cylindrical tube of radius $r$, equation 1.148 must be replaced by the relation

$$E_{0p} = \frac{\lambda_m(T_1 - T_0)}{a \ln(r/a) + (\beta' c/P)(a/r + 1)}, \tag{1.149}$$

where it is assumed that $\alpha$ is the same for both the wire and inside surface of the cylinder.

For very low values of $P$, equation 1.148 becomes

$$E_{0p} = \frac{\lambda_m(T_1 - T_0)}{2\beta' c} \cdot P, \tag{1.150}$$

and equation 1.149 becomes

$$E_{0p} = \frac{\lambda_m(T_1 - T_0) \cdot P}{\beta' c(1 + a/r)}. \tag{1.151}$$

That is, at very low pressures $E_{0p}$ is independent of $d$ and *varies linearly with P*. This conclusion is identical with that deduced above by Knudsen for the case of free-molecule flow. That equation 1.150 is identical with 1.138 may be shown as follows.

Substituting for $\beta'$ from equation 1.147, for $\lambda_m$ from 1.126, and utilizing the relation $\eta = 0.5\rho v_a L$, we obtain the result

$$\begin{aligned}
\frac{\lambda_m}{2\beta' c} &= \frac{\alpha\epsilon\eta c_v(\gamma + 1)}{(2 - \alpha)4\epsilon PL} \\
&= \left(\frac{\alpha}{2 - \alpha}\right)\frac{1}{4}(\gamma + 1)\frac{C_v}{M}\frac{1}{2}\frac{\rho v_a L}{PL} \\
&= \left(\frac{\alpha}{2 - \alpha}\right)\frac{1}{8}\left(\frac{\gamma + 1}{\gamma - 1}\right)\frac{R_0 MP}{PMR_0 T_i}\left(\frac{8R_0 T_i}{\pi M}\right)^{1/2} \\
&= \left(\frac{\alpha}{2 - \alpha}\right)\frac{1}{2}\left(\frac{\gamma + 1}{\gamma - 1}\right)\left(\frac{R_0}{2\pi M T_i}\right)^{1/2} \\
&= \left(\frac{\alpha}{2 - \alpha}\right)\Lambda_0\left(\frac{273.2}{T_i}\right)^{1/2},
\end{aligned}$$

where $\alpha/(2 - \alpha)$ takes the place of $\alpha$ used in equation 1.138.

Equations 1.148 and 1.149 can be expressed in a form which is very convenient for calculation. We shall consider especially equation 1.149, which is more important in practical measurements of heat conductivity. This equation may obviously be written in the form

$$\frac{1}{E_{0p}} = \frac{a \ln r/a}{\lambda_m(T_1 - T_0)} + \frac{X}{P}. \tag{1.152}$$

That is, if $1/E_{0p}$ is plotted against $1/P$, a *straight line* is obtained. From the intercept for $1/P = 0$, which we shall designate by $1/E_{0\infty}$, the value of $\lambda_m$ may be obtained, while the slope is given by the relation

$$\frac{\Delta(1/E_0)}{\Delta(1/P)} = X,$$

where $X$ is a constant determined by the relation

$$X = \frac{\beta' c(1 + a/r)}{E_{0\infty}\, a \ln (r/a)}. \tag{1.153}$$

Equation 1.152 thus applies to the transition range of pressures in which $E_0$ changes from being independent of pressure (at higher pressures) to varying linearly with the pressure (at very low pressures).

The validity of this conclusion can be illustrated by its application to two sets of data, of which the first is that given in Table 1.11. A plot of $1/W = 1/(2\pi l a E_0)$ against $1/P$ yields, as Dickins has shown, a straight line for each gas, and the values of $\lambda_t$ given in the table were deduced from the values of $1/W_\infty$ which correspond to the intercepts on the ordinate axis for $1/P = 0$. (See equation 1.133.)

The other set of data is that obtained by Knudsen[66] on the rate of energy transfer from a very fine tungsten wire in hydrogen. In this case, $a = 2.2 \cdot 10^{-4}$ cm, $l = 9.65$, and $r = 0.535$. Hence $2\pi a l = 0.01334$ cm$^2$. For the range $P_{\mu b} = 10^6$ to $P_{\mu b} = 6000$, $\Delta T = 2.98°$ C; from 6000 to 400 microbars, $\Delta T = 32.73$; and for pressures below 400 microbars, $\Delta T = 104.2°$ C.

Table 1.14 gives the values of $P_{\mu b}$ and $E_0/\Delta T$ in calories per square centimeter per degree per second, along with the reciprocals of each of these quantities.

A plot of $\Delta T/E_0$ versus $10^5/P_{\mu b}$ (see Fig. 1.5) yields a straight line, for the range $P_{\mu b} = 3.2 \cdot 10^3$ to $P_{\mu b} = 1.016 \cdot 10^6$ (2.4–762 mm) with the values

$$\frac{d(\Delta T/E_0)}{d(1/P_{\mu b})} = 2.60 \cdot 10^5; \quad \left(\frac{\Delta T}{E_0}\right)_\infty = 4.4; \quad \left(\frac{E_0}{\Delta T}\right)_\infty = 0.227.$$

**TABLE 1.14**

**Data (According to Knudsen) Used in Plot of Fig. 1.5**

| $P_{\mu b}$ | $10^3/P_{\mu b}$ | $10^5 E_0/\Delta T$ | $10^{-3}\Delta T/E_0$ | $10^{-3}P_{\mu b}$ | $10^5/P_{\mu b}$ | $10^3 E_0/\Delta T$ | $\Delta T/E_0$ |
|---|---|---|---|---|---|---|---|
| 107.3 | 9.33 | 40.77 | 2.453 | 3.200 | 31.3 | 11.65 | 85.8 |
| 214.8 | 4.66 | 81.56 | 1.226 | 6.355 | 15.74 | 22.05 | 45.4 |
| 322.1 | 3.11 | 122.1 | 0.819 | 13.45 | 7.435 | 42.12 | 23.75 |
| 428.6 | 2.33 | 162.1 | 0.617 | 35.0 | 2.857 | 83.9 | 11.92 |
| 534.8 | 1.87 | 201.5 | 0.496 | 66.8 | 1.495 | 121.9 | 8.20 |
| 640.0 | 1.56 | 241.4 | 0.414 | 129.0 | 0.775 | 190.9 | 5.24 |
| 744.6 | 1.34 | 280.5 | 0.357 | 505.0 | 0.198 | 211.1 | 4.74 |
| 848.6 | 1.18 | 319.2 | 0.313 | 1016.0 | 0.0984 | 223.9 | 4.47 |
| 950.9 | 1.05 | 357.6 | 0.280 | $\infty$ | 0 | (227) | (4.4) |
| 1053.3 | 0.949 | 395.4 | 0.253 | | | | |
| 1075.0 | 0.930 | 400.9 | 0.249 | | | | |

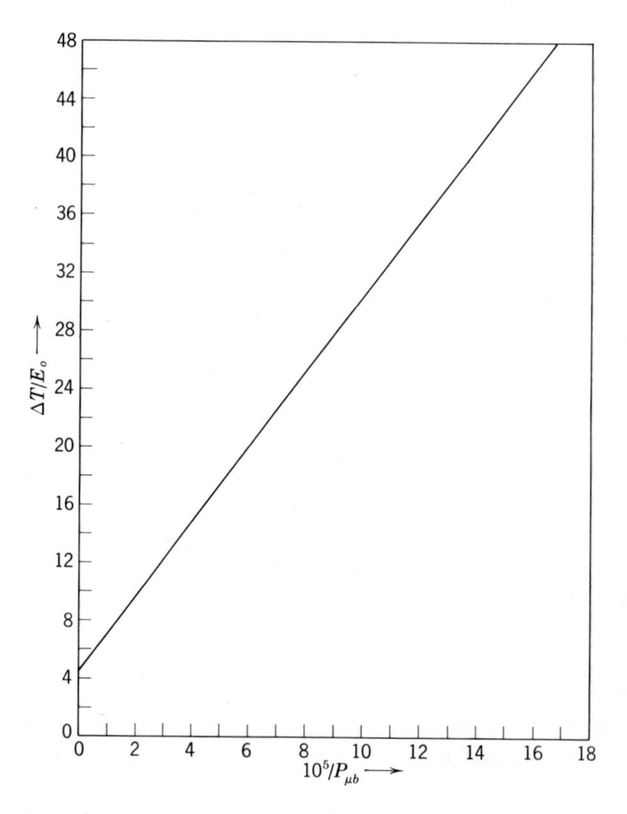


Fig. 1.5. Plot of $\Delta T/E_0$ versus $10^5/P_{\mu b}$ of Knudsen's data in Table 1.14. Ordinates give values of deg C/(cal · cm$^{-2}$ · sec$^{-1}$).

In the range $P_{\mu b} = 1053.3$ to $P_{\mu b} = 107$,

$$\frac{E_0}{\Delta T} = 3.72 \cdot 10^{-6} P_{\mu b}.$$

That is, the energy loss varies linearly with $P$.

From the value

$$\left(\frac{E_0}{\Delta T}\right)_\infty = 0.227$$

$$= \frac{\lambda_m}{a \ln (r/a)} = \frac{\lambda_m}{2.2 \cdot 10^{-4} \ln (535/0.22)}$$

it follows that $\lambda_m = 38.9 \cdot 10^{-5}$ cal · cm$^{-1}$ · deg$^{-1}$ · sec$^{-1}$, which is in good agreement with the value given for $\lambda_0$ in Table 1.10.

On the other hand, it follows from the values in Table 1.12 that, at low pressures and 0° C,

$$\frac{E_0}{\Delta T} = \frac{60.72 \cdot 10^{-6}}{4.186} \cdot \frac{3}{4} \alpha P_{\mu b} = 1.09 \cdot 10^{-5} \alpha_r P_{\mu b} \quad \text{cal} \cdot \text{deg}^{-1} \cdot \text{cm}^{-2} \cdot \text{sec}^{-1}.$$

Hence $\alpha_r = \dfrac{3.72 \cdot 10^{-6}}{10.9 \cdot 10^{-6}} = 0.34$, and $\alpha = 0.34$, since $a/r$ is negligibly small. That is, the accommodation coefficient for hydrogen on tungsten, as deduced from Knudsen's measurements, is 0.34.

It follows from the relations given above for $E_{0p}$ that, in order to obtain the minimum dependence on pressure, $a$ should be made as large as possible and $r$ nearly equal to $a$.

## 1.11. LANGMUIR'S FILM THEORY OF HEAT CONDUCTION

In a very interesting paper published in 1912,[67] Langmuir formulated a theory of the mechanism of heat conduction from wires (as well as plane surfaces), which deserves a great deal more attention than it has received.

Observations on the heat losses in hydrogen from tungsten wires heated to incandescence showed that whether the wire was in a horizontal or vertical position made a difference of only a few per cent (at constant temperature) in the heat loss. "This," states Langmuir, "was strong indication that the heat loss was dependent practically only on heat conduction very close to the filament and that the convection currents had practically no effect except to carry the heat away after it passed out through the film of adhering gas."

Let us consider, first, a *plane* surface in a gas at normal pressures. Assuming that the *whole temperature drop occurs in a film of thickness B*, adjacent to the surface, the rate of energy loss (due to conduction) per unit area is given by the relation

$$W = \frac{\lambda \cdot \Delta T}{B} \quad \text{cal} \cdot \text{cm}^{-2} \cdot \text{sec}^{-1}, \tag{1.154}$$

where $\Delta T$ is a relatively small temperature drop. Since, as shown in equation 1.128, $\lambda$ is a function of $T$, it follows that for a large difference in temperature, $T_2 - T_1$,

$$\begin{aligned} W &= \frac{4.186}{B} \int_{T_1}^{T_2} \lambda \, dT \quad \text{watts} \cdot \text{cm}^{-2} \\ &= \frac{4.186}{B} \left\{ \int_0^{T_2} \lambda \, dt - \int_0^{T_1} \lambda \, dT \right\} \\ &= \frac{4.186}{B} (\phi_2 - \phi_1), \end{aligned} \tag{1.155}$$

where $\phi = \int_0^T \lambda \, dT$.

Log-log plots of the values of $4.186\phi$ versus $T$, as calculated by Langmuir for both hydrogen and air, show that the data may be represented, to within an error not exceeding 2.5 per cent, in the range $1300 \geqslant T \geqslant 300$, by the following equations:[68]

$$\text{For } H_2\text{:} \quad \log(4.186\phi) = -4.965 + 1.78 \log T. \tag{1.156}$$

$$\text{For air:} \quad \log(4.186\phi) = -5.877 + 1.81 \log T. \tag{1.157}$$

In the case of a wire of radius $a$, the energy loss will occur through a film of adherent gas, the outer edge of which is at a distance $b$ from the axis of the wire. As shown by Langmuir, the value of $b$ satisfies the relation

$$b \ln \frac{b}{a} = B, \tag{1.158}$$

and the energy loss *per centimeter length* of wire is given by the relation

$$W = 4.186 \frac{2\pi(\phi_2 - \phi_1)}{\ln(b/a)} \quad \text{watts} \tag{1.159}$$

$$= 4.186 s(\phi_2 - \phi_1) \quad \text{watts}, \tag{1.160}$$

where

$$s = \frac{2\pi}{\ln(b/a)}. \tag{1.161}$$

Introducing into equation 1.161 the relation expressed by equation 1.158, it follows that

$$\frac{a}{B} = \frac{s\epsilon^{-2\pi/s}}{\pi}. \quad (1.162)$$

Table 1.15 (taken from Langmuir's paper) gives corresponding values of $s$ and $a/B$ as derived by means of equation 1.162. Figure 1.6 shows a plot of these data on semilog scale.

**TABLE 1.15**

**Values of $s$ as a Function of $a/B$**

(Langmuir)

| $s$ | $a/B$ | $s$ | $a/B$ | $s$ | $a/B$ | $s$ | $a/B$ |
|---|---|---|---|---|---|---|---|
| 0.0 | 0.0 | 5.0 | 0.453 | 10 | 1.696 | 30 | 7.738 |
| 0.5 | $0.735 \cdot 10^{-6}$ | 5.5 | 0.558 | 12 | 2.263 | 32 | 8.370 |
| 1.0 | $0.594 \cdot 10^{-3}$ | 6.0 | 0.671 | 14 | 2.844 | 34 | 8.995 |
| 1.5 | $0.725 \cdot 10^{-2}$ | 6.5 | 0.788 | 16 | 3.438 | 36 | 9.622 |
| 2.0 | $2.752 \cdot 10^{-2}$ | 7.0 | 0.908 | 18 | 4.040 | 38 | 10.25 |
| 2.5 | 0.0644 | 7.5 | 1.032 | 20 | 4.645 | 40 | 10.87 |
| 3.0 | 0.1176 | 8.0 | 1.160 | 22 | 5.263 | 42 | 11.50 |
| 3.5 | 0.185 | 8.5 | 1.291 | 24 | 5.877 | 44 | 12.14 |
| 4.0 | 0.265 | 9.0 | 1.424 | 26 | 6.505 | 46 | 12.77 |
| 4.5 | 0.354 | 9.5 | 1.561 | 28 | 7.122 | 48 | 13.40 |
| 5.0 | 0.453 | 10.0 | 1.696 | 30 | 7.738 | 50 | 14.03 |

From measurements of the energy losses from platinum wires in air,[69] Langmuir derived the conclusions that, for air at 20° C and 760 mm Hg, $B = 0.043$ cm and is "surprisingly independent" of $T$. He also concluded that $B$ should be proportional to the viscosity and inversely proportional to the density, "for it is the viscosity that causes the existence of the film and it is the difference of density between hot and cold gas (proportional to the density itself) that keeps the film from becoming indefinitely large." That is,

$$B = \text{const}\,\frac{\eta}{\rho}$$

$$= \text{const}\; v_a L = \text{const}\; L\left(\frac{T}{M}\right)^{1/2}, \quad (1.163)$$

where $v_a$ and $L$ correspond to values at $T_1$ (the lower temperature).

Assuming that, for air at 1 atm and 20° C, $B = 0.43$ cm, it follows from equation 1.163 that, for $H_2$ at 1 atm and 20° C, $B = 3.0$ cm. On the other hand, observations on the heat losses from tungsten wires in $H_2$ and $N_2$,[70] in which the values of $s$ were determined from the observed

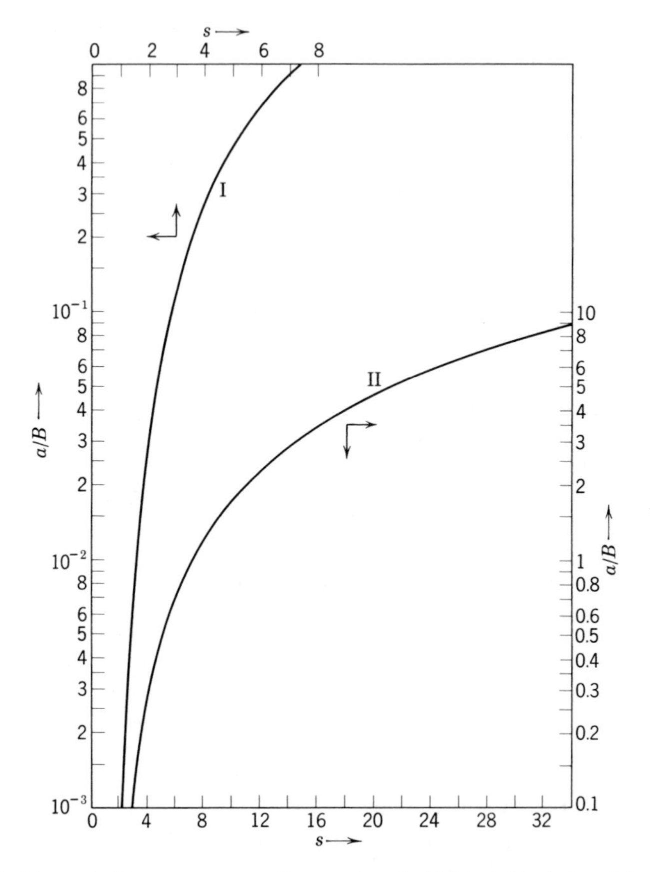


Fig. 1.6. Plots of $a/B$ versus $s$ according to data in Table 1.15, deduced for thermal conduction, at normal pressures, through a film surrounding a wire. Appropriate scales for curves I and II are indicated by the arrows.

values of $W$ and the values of $(\phi_2 - \phi_1)$ calculated according to equation 1.160, lead to the values $B = 1.35$ for $H_2$ and 0.68 for $N_2$ at 750 mm pressure.

As the pressure is decreased, the value of $b$ increases, and the following table[71] gives values deduced from experimentally determined values of $s$ for the heat loss from a tungsten wire in hydrogen ($a = 0.00353$), at

| $P_{mm}$ | $s$ | $b$ (cm) | $B$ (cm) |
|---|---|---|---|
| 750 | 1.25 | 0.54 | 1.35 |
| 200 | 0.88 | 9.0 | 32.1 |
| 100 | 0.72 | 45.0 | 196.3 |
| 50 | 0.56 | 530.0 | |

20° C. The last column gives values of $B$ calculated from those of $b$ by means of equation 1.158. It is evident that, for pressures below atmospheric, $B$ becomes greater than the distance from wire to wall. "This means," as Langmuir states,[72] "that the heat loss from the filament is actually less than if there were no convection currents and the laws of heat conduction could be applied." The decrease in value of $s$ at the lower pressures is actually due, as Langmuir points out, "to a temperature discontinuity at the surface of the wire, which becomes appreciable at pressures even as high as 200 mm, and increases with decrease in pressure."

The problem of heat conduction at low pressures from a wire of radius $a$, suspended coaxially in a glass tube of radius $r$, where $r$ is very much greater than $a$, has been discussed from the point of view of Langmuir's film theory by Jones.[73] Following the relations developed by Smoluchowski, which have been discussed above, Jones substitutes for equation 1.159, in the case of low pressures, a relation of the form

$$W = \frac{4.186 \cdot 2\pi(\phi_2 - \phi_1)}{\ln (b/a) + y(1/a + 1/b)}, \tag{1.164}$$

where $W$ = watts lost by conduction per unit length of wire, $b$ = radius of the outside edge of the film, and $y$ varies as $L$. Since $b$ is very large compared to $a$, equation 1.164 may be replaced by

$$W = \frac{4.186 \cdot 2\pi(\phi_2 - \phi_1)}{\ln (b/a) + y/a}. \tag{1.165}$$

Jones has applied this equation to calculate the heat loss from iron wires in hydrogen at 100 mm of Hg, and at the two temperatures, 840° K and 1045° K, respectively, between which ballasting action is obtained. (This is the temperature range in which the current is maintained practically constant when the lamp is used in series with a load on a variable voltage source.)

The value of $b$ has been deduced by Jones by means of equation 1.158 from the following semiempirical relation for $B$, deduced by Rice[74] for hydrogen:

$$B = \frac{0.5442a^{0.19}}{P_{mm}^{0.54}} \left\{ \frac{(T + T_0)^{5/2}}{154 + T + T_0(T - T_0)^{1/2}} \right\}^{0.54} \text{cm} \tag{1.166}$$

where $T$ = temperature of the wire in degrees K,

$T_0$ = temperature of the wall in degrees K.

The values of $B$ and $b$ thus calculated for different values of $a$ are given in the respective columns in Table 1.16.

According to Smoluchowski's views, $y$ is proportional to the mean free path $L$, and Jones derives the values given in the last row of the table. According to the data in Table 1.16, $L = 9.32 \cdot 10^{-5}$ cm, for $H_2$ at 25° C and 100 mm. It follows from equations 1.88, 1.89, and 1.86 that, for any other temperature, $T$,

$$\frac{L_T}{L_{298}} = \frac{T}{298.2}\left(\frac{1 + 84.4/298.2}{1 + 84.4/T}\right).$$

TABLE 1.16

**Film Thickness, Energy Loss, and Related Data for Iron Filaments in Hydrogen at 100 mm Hg and at Two Temperatures**

| | $T = 840°$ K | | | | | $T = 1045°$ K | | | | |
|---|---|---|---|---|---|---|---|---|---|---|
| $a$ | $B$ | $\ln b/a$ | $b$ | $y/a$ | $W$ | $B$ | $\ln b/a$ | $b$ | $y/a$ | $W$ |
| 0.00127 | 2.66 | 5.86 | 0.454 | 0.939 | 1.359 | 2.57 | 5.85 | 0.439 | 0.988 | 2.1 |
| 0.0127 | 2.88 | 4.025 | 0.715 | 0.0939 | 2.248 | 2.76 | 4.0 | 0.690 | 0.0988 | 3.5 |
| 0.0635 | 2.82 | 2.772 | 1.017 | 0.0188 | 3.31 | 2.70 | 2.74 | 0.985 | 0.0198 | 5.19 |
| 0.127 | 2.75 | 2.265 | 1.213 | 0.0094 | 4.065 | 2.65 | 2.25 | 1.177 | 0.0099 | 6.35 |
| $L_T = 3.06 \cdot 10^{-4}$, $y = 11.92 \cdot 10^{-4}$ cm | | | | | | $L_T = 3.88 \cdot 10^{-4}$, $y = 12.55 \cdot 10^{-4}$ cm | | | | |

The values of $L_T$ thus deduced are also given in the last line in Table 1.16. Thus, at 840° K, $y = 3.89L_T$, and at 1045° K, $y = 3.23L_T$.

Finally, the values of the function $4.186(\phi_2 - \phi_1)$ used in equation 1.165, as deduced from equation 1.156, are 1.468 watts $\cdot$ cm$^{-2}$ and 2.294 watts $\cdot$ cm$^{-2}$ for $T = 840°$ K and $T = 1045°$ K respectively, and $T_0 =$ 293° K.

The values under $W$ give the *watts lost, by conduction, per unit length of wire.*

## 1.12. THERMAL TRANSPIRATION (THERMOMOLECULAR FLOW)

From equation 1.46,

$$G = mv = \frac{1}{4}\rho v_a = \rho\left(\frac{R_0 T}{2\pi M}\right)^{1/2},$$

it follows that the rate of efflux of a given gas through a small opening varies as $\rho(T)^{1/2}$. If we have two chambers $A$ and $B$, separated by a small orifice or a porous plug, at two different temperatures, $T_A$ and $T_B$, transpiration will occur until a steady state is established at which

$$\rho_A(T_A)^{1/2} = \rho_B(T_B)^{1/2} \tag{1.167}$$

and, since

$$\rho = \frac{MP_{\mu b}}{R_0 T},$$

$$\frac{P_A}{P_B} = \left(\frac{T_A}{T_B}\right)^{1/2}. \tag{1.168}$$

This is of importance in observations at low pressures. For instance, if the chamber $A$ is a part of a system at liquid-air temperature ($T_A = 90$) and the pressure is measured by means of a gauge at room temperature ($T_B = 300$), then the real value of $P_A$ is given by the relation

$$P_A = P_B\left(\frac{90}{300}\right)^{1/2} = 0.55P_B.$$

Equation 1.167 or 1.168 is valid for two vessels *connected by small-bore tubing* provided that $2a$, the diameter, is small compared to $L$, the mean free path. When, however, $L$ is very much smaller than $2a$, so that collision between molecules become predominant over collisions against the walls, then the condition of equilibrium is $P_A = P_B$, in consequence of which

$$\frac{\rho_A}{\rho_B} = \frac{T_B}{T_A}. \tag{1.169}$$

There exists therefore a range of pressures in which the pressure relation changes from that given by equation 1.169 to that given by equation 1.168. The smaller the diameter of the connecting tube, the higher the pressure at which equation 1.168 is valid.

The theory of the phenomenon was first discussed by Maxwell (1879) and subsequently by Knudsen,[75] who derived equations for the pressure gradient along the connecting tube as a function of the tube diameter and pressure in the hot chamber. Rigorous mathematical analyses of the problem have yielded results which are not useful in practice,[76] and, at present, thermal transpiration corrections are best handled by semiempirical methods. Bennett and Tompkins[77] have recommended that an equation proposed by Liang[78] be used. The equation is supported by a kinetic argument, and Bennett and Tompkins found it to fit experimental data over a wide range. It should be noted, however, that these workers used glass systems; Los and Fergusson[79] observed that the nature of the wall affects the magnitude of the correction, as Liang had predicted.[80]

The form of Liang's equation given by Bennett and Tompkins is

$$\frac{P_A}{P_B} = R = \frac{\alpha_{\mathrm{He}}(f\phi_g X)^2 + \beta_{\mathrm{He}}(f\phi_g X) + R_m}{\alpha_{\mathrm{He}}(f\phi_g X)^2 + \beta_{\mathrm{He}}(f\phi_g X) + 1}, \tag{1.170}$$

where $R$ = the ratio of the pressures in the two regions at different temperatures, $T_A$ and $T_B$,

$R_m = (T_A/T_B)^{1/2}$,

$f = 1$ for connecting tubes of internal diameter, $d < 1$ cm

$= 1.22$ for connecting tubes of internal diameter, $d > 1$ cm,

$X = P_B d$,

$\alpha_{\mathrm{He}} = 3.70(1.70 - 2.6 \cdot 10^{-3}\,\Delta T)^{-2} \quad (\Delta T = T_B - T_A)$,

$\beta_{\mathrm{He}} = 7.88(1 - R_m)$ for $R_m \leqslant 1$,

and values of $\phi_g$ are given in Table 1.17.

Podgurski and Davis[106] have also done work on thermal transpiration.

**TABLE 1.17**

**Experimental $\phi_g$ Values for Glass Systems**
(Bennett and Tompkins)

| Gas: | He | Ne | Ar | Kr | Xe | $H_2$ | $O_2$ | $N_2$ | CO | $CO_2$ | $C_2H_4$ |
|---|---|---|---|---|---|---|---|---|---|---|---|
| $\phi_g$: | 1.00 | 1.30 | 2.70 | 3.90 | 6.41 | 1.44 | 2.87 | 3.53 | 3.31 | 4.52 | 6.72 |

Bennett and Tompkins also point out that $\phi_g$ values may be calculated from the collision diameters, $r_0$, given by Hirschfelder, Bird, and Spotz,[81] by making use of the equation

$$\log r_0 = 0.43 + 0.24 \log \phi_g. \tag{1.171}$$

## 1.13. THERMAL DIFFUSION

The thermal-diffusion effect has been described by Ibbs[82] as follows:

> If a temperature gradient is applied to a mixture of two gases of uniform concentration there is a tendency for the heavier and large molecules (mass $m_1$, diameter $\delta_1$) to move to the cold side, and for the lighter and smaller molecules (mass $m_2$ and diameter $\delta_2$) to move to the hot side. The separating effect of thermal diffusion (coefficient $D_T$) is ultimately balanced by the mixing effect of ordinary diffusion (coefficient $D_{12}$) so that finally a steady state is reached and a concentration gradient is associated with the temperature gradient. The amount of thermal separation thus produced by a given difference in temperature depends upon the ratios $m_1/m_2$ and $\delta_1/\delta_2$, and upon the proportion by volume of the heavier gas $f_1$ and of the lighter gas $f_2$ (where $f_1 + f_2 = 1$) and also upon the nature of the field of force operating between the unlike molecules.

The effect was predicted by Enskog (1917) and, independently, by Chapman and Dootson.[83]

Following the earlier experimental investigations of Ibbs and others, Clusius and Dickel[84] devised a continuous method for separating mixtures of gases and isotopes which has been applied extensively by subsequent investigators.

In this section the discussion will be limited largely to the experimental data, since the mathematical theory is quite complex and beyond the scope of the objectives of this chapter.

The *coefficient of thermal separation* is defined by the relation

$$k_T = \frac{D_T}{D_{12}}, \tag{1.172}$$

where $D_T$ and $D_{12}$ have been defined above. Now

$$k_T = \frac{-df_1}{d \ln T} = \frac{df_2}{d \ln T}. \tag{1.173}$$

If $k_T$ is a constant, then we obtain from equation 1.173 the relation

$$\Delta f = k_T \ln \frac{T_1}{T_2} \tag{1.174}$$

$$= \text{amount of separation},$$

where $T_1$ and $T_2$ denote the hot and cold temperatures, respectively.

Figures 1.7 and 1.8 illustrate results obtained by Ibbs[85] with mixtures

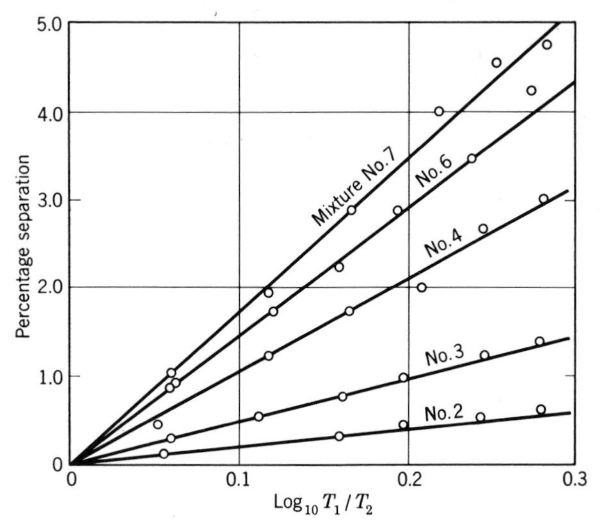


Fig. 1.7. Relation between degree of separation by thermal diffusion and log $(T_1/T_2)$ for mixtures of $H_2$ and $N_2$.

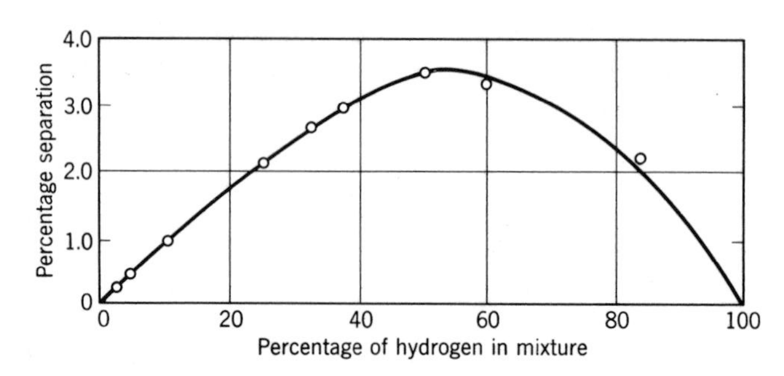


Fig. 1.8. Relation between composition of $H_2$-$N_2$ mixtures and degree of separation for $\log (T_1/T_2) = 0.2$.

of $H_2$ and $N_2$. The concentrations of $H_2$ in the different mixtures varied from 4.7 per cent for No. 2 to 50.5 per cent for No. 7. It will be observed that these observations are in accord with equation 1.174, since $\Delta f$ is found to be a linear function of $\log (T_1/T_2)$. Figure 1.8 shows the variation in value of $\Delta f$ with composition for a constant value of the ratio $T_1/T_2$.

Chapman has shown that $k_T$ can be expressed in the form[86]

$$k_T = \frac{Af_1 + Bf_2}{1 + Cf_1/f_2 + Df_2/f_1}, \tag{1.175}$$

where the coefficients $A$ to $D$ are functions of the molecular weights and of the ratio $\delta_1/\delta_2$. These can be calculated from the force constants $s_{12}$ in the relation $F = cr^{-s_{12}}$, which expresses the force of repulsion between a pair of unlike molecules as a function of $r = \frac{1}{2}(\delta_1 + \delta_2)$.

The separation in a mixture of $H_2$ and $CO_2$ was investigated by Ibbs and Wakeman[87] over a range of 600° C. For any given mixture it was observed that the amount of separation was in accordance with equation 1.174. For a constant difference in temperature, $\Delta f$ increased from 0 for pure $H_2$ to a maximum value of 0.0365 (for $\Delta T = 600°$), for a mixture containing 61 per cent $H_2$.

According to Chapman, $k_T$ in this case should be given by a relation of the form

$$k_T = \frac{5}{2}\left(\frac{0.216f_1 + 0.139f_2}{1.275 + 1.409f_1/f_2 + 0.232f_2/f_1}\right), \tag{1.176}$$

where $f_1$ refers to $CO_2$, and $f_2$ to $H_2$. Since this relation is deduced on the assumption that the molecules behave as rigid elastic spheres with a force relation in which $s$ has a definite value for the interaction between the two types of molecules, it was expected that the observed values of $k_T$ would differ from those calculated. The following table gives the results of a series of measurements. The last column gives values of a constant defined by the relation

$$R_T = \frac{k_T \text{ observed}}{k_T \text{ calculated for rigid elastic spheres}}. \tag{1.177}$$

| $f_1$ | Temperature Range (° C) | $k_T$ (calc) | $k_T$ (obs) | $R_T$ |
|---|---|---|---|---|
| 0.469 | Below 144 | 0.1575 | 0.0665 | 0.422 |
| | Above 144 | 0.1575 | 0.0939 | 0.596 |
| | At 470 | 0.1575 | 0.1045 | 0.663 |
| 0.34 | Below 145 | 0.1685 | 0.0695 | 0.415 |
| | Above 145 | 0.1685 | 0.0929 | 0.552 |

For $CO_2$-$N_2$ mixture, Chapman's expression for $k_T$ is of the form

$$k_T = \frac{5}{2}\left(\frac{0.163f_1 + 0.167f_2}{3.38 + 2.154f_1/f_2 + 1.480f_2/f_1}\right), \tag{1.178}$$

where $f_1$ again refers to the *heavier* gas. (In all the following equations $f_1$ has the same significance.)

For $f_1 = 0.494$, the value of $k_T$ calculated $= 0.0564$, whereas the value of $R_T$ observed varied from 0.247 for temperatures below 144° C to 0.441 for temperatures above 144° C.

For the mixtures Ar-He, Ne-$H_2$, and $N_2$-He, relations similar to 1.176 and 1.178 for $k_T$ have been published by Ibbs and Grew.[88] All the observed values of $R_T$ showed a decrease with decrease in temperature down to −190° C.

According to Chapman and Cowling,

$$R_T = \frac{13}{10}\left(\frac{s_{12} - 5}{s_{12} - 1}\right). \tag{1.179}$$

Hence observations on the values of $R_T$ should yield values of $s_{12}$. In fact the observations on *thermal diffusion* of gases should reveal interesting information on the forces between molecules. Table 1.18 gives values of the repulsive force constant $s_{12}$, as determined from observed values of $R_T$, and, for comparison, values of $s_1$ and $s_2$ for the individual gases, as determined from other phenomena.

In 1938, as mentioned previously, Clusius and Dickel[84] devised an arrangement of hot and cold surfaces that could be utilized practically for the separation of mixtures of different gases and of isotopes. They used a long vertical tube with a hot wire along the axis. Because of thermal diffusion, the relative concentration of the heavier molecules is

TABLE 1.18

Values of "Force" Constants for Interaction of Molecules

| Gases | $R_T$ | $s_{12}$ | $s_1$ | $s_2$ |
|---|---|---|---|---|
| $O_2$-$H_2$ | 0.48 | 7.3 | 7.6 | 11.3 |
| $N_2$-Ar | 0.47 | 7.2 | 8.8 | 7.35 |
| $H_2$-$N_2$ | 0.58 | 8.2 | 11.3 | 8.8 |
| He-Ar | 0.65 | 9.0 | 14.6 | 7.35 |
| $H_2$-Ne | 0.74 | 11.4 | 11.3 | 14.5 |
| $H_2$-$CO_2$ | 0.47 | 7.2 | 11.3 | 5.6 |
| Ne-Ar | 0.54 | 7.9 | 14.5 | 7.35 |
| He-Ne | 0.80 | 11.4 | 14.6 | 14.5 |

greater at the cold wall. Convection causes the gas at the hot surface to rise to the top, where it is deflected to the cold wall. There the gas sinks to the bottom and the cycle is then repeated. As a result, the *heavier component concentrates at the bottom*, and the lighter at the top.

For instance, in a chamber 1 meter in height, for a temperature difference of 600° C, it was possible to separate a mixture containing 40 per cent (by volume) $CO_2$ and 60 per cent $H_2$, with a yield of practically 100 per cent $CO_2$ at the bottom and 100 per cent $H_2$ at the top. In a chamber 2.9 meters in height, for a temperature difference of 600° C, air (21 per cent $O_2$, 78 per cent $N_2$) could be separated, yielding 85 per cent $O_2$ at the bottom. Finally, from a mixture of 23 per cent $HCl^{37}$ and 77 per cent $HCl^{35}$, a mixture containing 40 per cent of the heavier isotope was obtained at the bottom of the diffusion chamber.

Brewer and Bramley[89] used a heated inner tube 1 cm in diameter, and an outer cooled concentric tube 2 cm in diameter, each 1 meter long. With a 350° C difference in temperature, He and $Br_2$ could be separated, so that after 15 min no $Br_2$ could be detected at the top. Under similar conditions a 50-50 mixture of $CH_4$ and $NH_3$ was enriched 25 per cent in $NH_3$ at the bottom. Brewer and Bramley deduced the following relation: "If $l$ denotes the cylinder length, $r$ the radius of the outer tube, and $d$ the difference in radii, then, to a first approximation, the rate of separation varies as $rd$ and the purity as $l/d$."

Figure 1.9 shows a thermal-diffusion column developed by Nier[90] for the concentration of $C^{13}H_4$. A Nichrome wire was inserted along the axis of a steel tube $\frac{3}{4}$ in. in outside diameter, which was concentric with a steel tube $1\frac{3}{4}$ in. in outside diameter having a wall thickness of 0.035 in. This tube constituted the inner wall of the annular space containing the gas, and the outer wall consisted of a 2-in.-outside-diameter brass tube, which was water cooled as shown in the figure. The column was 24 ft in

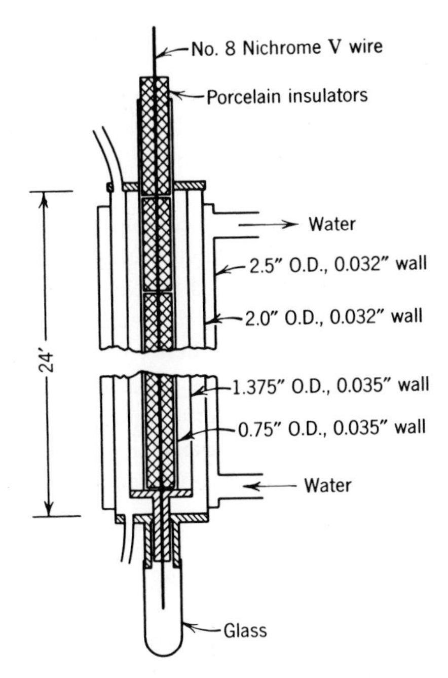


Fig. 1.9. Cross-sectional view of column used for thermal-diffusion experiments by Nier. The column is used as a return lead for the current flowing through the Nichrome wire.

length. At the lower end, the two steel tubes were brazed concentrically to a circular steel plate. The average temperature of the outer wall was 27° C, and that of the inner approximately 375° C. The samples of methane taken at the bottom and top were analyzed by means of a mass spectrometer for the ratio of $C^{13}H_4$ to $C^{12}H_4$. At a pressure of 656 mm, and after operating for 23.5 hours, the values of this ratio at the bottom and top were 0.0215 and 0.0054 respectively, so that the ratio of $C^{13}H_4$ at the bottom to that at the top was 3.99.

The results obtained were found to be in agreement with conclusions deduced from theoretical considerations by Furry, Jones, and Onsager.[91] They derived the relation

$$\frac{C_1^L}{C_1^U} = \epsilon^x, \tag{1.180}$$

where

$$x = \frac{a/p^2}{1 + b/p^4}, \tag{1.181}$$

$C_1^L$ and $C_1^U$ represent equilibrium concentrations of $C^{13}H_4$ in the lower and upper ends of the column, respectively, $a$ and $b$ are constants for the gas used and the dimensions of the column, and $p$ denotes the pressure. Thus $\epsilon^x$ corresponds to the separation factor. In the investigation under consideration, this factor was observed to pass through a maximum at $p = 0.6$ atm, approximately.

The separation of $C^{13}H_4$ and $C^{12}H_4$ has also been investigated, in a column 40 ft in length, by Taylor and Glockler,[92] who found that the separation factor could be represented, in agreement with the theory, by the above equations.

The rate at which equilibrium is approached in such a column has been discussed by Bardeen,[93] and the conclusions reached have been found to be in good agreement with experimental results obtained by Nier.

Although the above discussion has dealt largely with the separation of gases which differ in value of $M$, it follows from the theory of thermal diffusion that it should be possible by this method to effect a certain degree of separation in a mixture of gases which have the same value of $M$ but differ with respect to the magnitude of the molecular diameter. Results in agreement with this prediction have been obtained by Wall and Holley.[94]

## 1.14. THEORY OF DIFFUSION OF GASES

As in the case of viscosity and heat conduction, approximate kinetic-theory considerations lead to the conclusion that the *coefficient of self-diffusion*, $D$, is given by a relation of the form

$$D = \tfrac{1}{3} v_a L, \tag{1.182}$$

which indicates that, at constant temperature, $D$ varies directly as the mean free path or inversely as the pressure.

Combining equation 1.182 with the relation

$$\eta = \tfrac{1}{3} \rho v_a L,$$

it follows that

$$D = \frac{\eta}{\rho}.$$

Introducing the correction for persistence of velocities and that for Maxwellian distribution of velocities, it is found that the more accurate relation is

$$D = 1.342 \frac{\eta}{\rho}. \tag{1.183}$$

According to Chapman and Enskog,[95] the coefficient in equation 1.183 should have a value lying between 1.200 for hard spheres and 1.543 for inverse-fifth-power repulsion. Actually observed values of $D$ yield the following values of the coefficient:[96]

$CO$, 1.30; $H_2$, 1.37; $O_2$, 1.40; $CO_2$, 1.58.

For the derivation of a relation for the coefficient of interdiffusion, $D_{12}$ (corresponding to the fact that molecules 1 and 2 are involved), we shall follow the Stefan-Maxwell derivation,[97] which, as will be evident, is based on the same considerations as those used in deriving equation 1.182 above.

The relation for the mean free path, $L_1$, in a mixture of two gases, designated 1 and 2, according to equation 1.104, is

$$\frac{1}{L_1} = 2^{1/2} \cdot \pi n_1 \delta_1^{\,2} + \pi n_2 \delta_{12}^{\,2}\left(1 + \frac{m_1}{m_2}\right)^{1/2}.$$

However, as Stefan and Maxwell have pointed out, the collisions between like molecules do not influence diffusion, and hence for this case the mean free paths are given by the relations

$$\frac{1}{L_1} = \pi n_2 \delta_{12}^{\,2}\left(1 + \frac{m_1}{m_2}\right)^{1/2} \qquad (1.184)$$

and

$$\frac{1}{L_2} = \pi n_1 \delta_{12}^{\,2}\left(1 + \frac{m_2}{m_1}\right)^{1/2}. \qquad (1.185)$$

On this basis it is found that the relation for the coefficient of interdiffusion is

$$D_{12} = \frac{(v_1^{\,2} + v_2^{\,2})^{1/2}}{3\pi n \delta_{12}^{\,2}}, \qquad (1.186)$$

where $v_1$ and $v_2$ refer to the value of $v_a$, and $n = n_1 + n_2$.

It is evident that for like molecules this relation becomes

$$D_{11} = \frac{2}{3}\,\frac{v_a}{2^{1/2}\pi n \delta^2}. \qquad (1.187)$$

Combining this with the relation

$$\eta = \frac{0.499 \rho v_a}{2^{1/2}\pi n \delta^2},$$

the result is

$$D_{11} = 1.336\,\frac{\eta}{\rho}, \qquad (1.188)$$

which is in substantial agreement with equation 1.183.

Table 1.19[98] gives the values of $D_{12}$ ($cm^2 \cdot sec^{-1}$) observed for several pairs of gases, at 0° C and 1 atm, also values of $\delta_{12}$ calculated from the values of $D_{12}$ by means of equation 1.186, and, for comparison, values of $\delta_{12}$ calculated from viscosity measurements (see Table 1.6) for each of the two gases.

For the first six pairs, which are constituted of "hard" molecules (with a value for the repulsive force constant greater than 7.4), the agreement is very satisfactory. Even in the case of "softer" molecules the agreement is fair, as shown by the values for the second set of five pairs.

**TABLE 1.19**

**Coefficients of Interdiffusion and Average Molecular Diameters**

| Gases | $D_{12}$ (obs) | $10^8\delta_{12}$ (calc from $D_{12}$) | $10^8\delta_{12}$ (calc from $\eta$) |
|---|---|---|---|
| $H_2$-air | 0.661 | 3.23 | 3.23 |
| $H_2$-$O_2$ | 0.679 | 3.18 | 3.17 |
| $O_2$-air | 0.1775 | 3.69 | 3.68 |
| $O_2$-$N_2$ | 0.174 | 3.74 | 3.70 |
| CO-$H_2$ | 0.642 | 3.28 | 3.25 |
| CO-$O_2$ | 0.183 | 3.65 | 3.70 |
| $CO_2$-$H_2$ | 0.538 | 3.56 | 3.69 |
| $CO_2$-air | 0.138 | 4.03 | 4.20 |
| $CO_2$-CO | 0.136 | 4.09 | 4.22 |
| $N_2O$-$H_2$ | 0.535 | 3.57 | 3.69 |
| $N_2O$-$CO_2$ | 0.0983 | 4.53 | 4.66 |

Summerhays[99] has used the katharometer (see section 1.9) to measure the diffusion coefficient of water vapor in air. The value observed was $D = 0.282\ \text{cm}^2 \cdot \text{sec}^{-1}$ at 16.1° C.

As will be noted, equation 1.186 indicates that the interdiffusion coefficient should be independent of composition. However, a series of experiments carried out at Halle[100] to test this conclusion show that there is a variation of not more than a few per cent with variation in the ratio $n_1/n_2$.

Mention should be made in this connection of a relation for the coefficient of interdiffusion, given by Loeb,[101] which is of the form

$$D_{12} = \beta\left(\frac{n_1}{n} \cdot \frac{\eta_2}{\rho_2} + \frac{n_2}{n} \cdot \frac{\eta_1}{\rho_1}\right), \tag{1.189}$$

in which $\beta$ is a numerical factor, the value of which is between 1.00 and 1.50, and $n = n_1 + n_2$.

From the equations for $D$ it follows that to a first approximation $D$ should vary as $T^{3/2}$ and as $P^{-1}$. Since, however, $\eta$ varies with $T$ in accordance with Sutherland's relation, equation 1.86, it is expected that the exponent of $T$ would exceed $\frac{3}{2}$. Lonius[100] gives the following relation for $D$ as a function of $T$:

$$D_T = D\left(\frac{T}{288.2}\right)^x \frac{760}{P_{mm}}, \tag{1.190}$$

where $D_T$ = value at $T°$ K and $P_{mm}$,
$D$ = value at 15° C and 760 mm.

The values of $x$ used by Lonius are as follows:

| | |
|---|---|
| $H_2$-$O_2$ | 1.754 |
| $N_2$-$O_2$ | 1.795 |
| $H_2$-$CO_2$ | 1.74 |
| $H_2$-$N_2$ | 1.707 |
| He-Ar | 1.75 |

### Maxwell-Loschmidt Method for Determination of Diffusion Coefficients

In view of the fact that in vacuum technique a problem arises occasionally regarding the time required for gases to mix, it is of interest to review briefly one method which has been used for the determination of $D$ for gases.

A tube of length $l$ is divided by means of a wide-bore stopcock into two parts of equal length. This is mounted in a vertical position; the *heavier* gas ($A$) is put in the *lower* part, and the *lighter* gas ($B$) is put in the *upper* part. At time $t = 0$ the stopcock is opened and diffusion is allowed to occur for a definite period, $t$. The stopcock is then closed, and the proportions of $A$ and $B$ are measured in each compartment.

Let $U$ = amount of $A$ expressed as the fraction of the total number of moles in the *lower* compartment,
$Q$ = amount of $A$ in the *upper* compartment, expressed similarly.

Then, on the basis of the equations for diffusion,

$$U - Q = \frac{8}{\pi^2}\left\{\epsilon^{-x} + \frac{1}{9}\epsilon^{-9x} + \frac{1}{25}\epsilon^{-25x} + \cdots\right\}, \tag{1.191}$$

where

$$x = \frac{\pi^2}{l^2} Dt. \tag{1.192}$$

For instance, in the case of $H_2$-$O_2$ mixture, for $l = 99.93$ cm and $t = 1800$ sec, $U = 0.5982$, $Q = 0.4018$, at 14.8° C and 749.3 mm. Hence, at 15° C and 760 mm, $D = 0.788$.

Table 1.20 gives values of $U - Q$ for two different values of $l$ as a function of $Dt$, also values of $t$, for $l = 100$ cm, assuming $D = 0.2$ $cm^2 \cdot sec^{-1}$.

**TABLE 1.20**

**Illustrating the Application of Equation 1.191**

| $Dt$ | $l = 100$ cm $U - Q$ | $l = 25$ cm $U - Q$ | $t$ (min) for $D = 0.2\ cm^2 \cdot sec^{-1}$ |
|---|---|---|---|
| 100 | 0.9550 | 0.2063 | 8.33 |
| 200 | 0.8401 | 0.0425 | 16.67 |
| 300 | 0.7514 | 0.0088 | 25.00 |
| 400 | 0.6771 | | 33.34 |
| 500 | 0.6118 | | 41.67 |
| 700 | 0.5014 | | 58.67 |
| 1000 | 0.3728 | | 83.83 |
| 1500 | 0.2276 | | 125.00 |
| 2000 | 0.1390 | | 166.67 |
| 3000 | 0.0518 | | 250.00 |
| 5000 | 0.0071 | | 416.67 |

For $l = 100$, only the first term in the series on the right-hand side of equation 1.191 is of importance for values of $Dt > 500$. For $l = 25$, only the first term is of importance for $Dt \geqslant 100$.

Equation 1.192 shows that for values of $l$ and $Dt$ for which the first term only is sufficiently accurate $t$ varies as $l^2$ for constant value of $(U - Q)$.

## Effect of Pressure of Gas on Rates of Evaporation of Metals

In section 1.6, equations were derived for rates of evaporation in a *vacuum*. These equations have been applied, as illustrated in the above connection, to the calculation of vapor pressures of high-melting-point metals from observations on rate of loss in weight as a function of the temperature. However, in the presence of a gas which does not react chemically with the metal, the observed rate of evaporation is lower; as is well known, this fact was utilized by Langmuir in the invention and development of the gas-filled tungsten-filament lamp.

This phenomenon has been explained by Langmuir by assuming the existence of a film adjacent to the evaporating surface through which the atoms evaporated from the surface diffuse. Thus this theory is quite similar to that also suggested by Langmuir, and discussed in section 1.11, for the effect of varying the diameter on heat conduction from a wire.

Fonda[102] has shown that this theory accounts very satisfactorily for his observations on the rate of evaporation of tungsten in the presence of argon at various pressures, and the following remarks on the mechanism of evaporation under these conditions are quoted from his paper:

> The filament is considered to be surrounded by tungsten vapor at the same pressure as would be present in a vacuum. The atoms of this vapor, however, instead of being projected directly from the filament, as in a vacuum, are pictured as diffusing through the stationary film of gas. Once an atom reaches the outer boundary of the film, it would be carried away in the convection current of gas and would be lost to the filament; but the path within the film is so irregular that an atom may in fact return to the filament and be deposited on it, thus leading to a reduced evaporation as compared with that in a vacuum.

Let $dc/dr$ denote the concentration gradient at the surface of the wire, where $r$ is the distance from the axis, and let $D$ denote the diffusion constant. Assuming uniform distribution over the surface of the wire, the rate of diffusion *per unit length* of the wire is

$$q = 2\pi r D \frac{dc}{dr}, \tag{1.193}$$

where $D$ is given by equation 1.186.

Since the value of $n$ for tungsten atoms is negligible compared with that of $n$ for argon gas, the expression for $D$ becomes identical with equation 1.182, where $L$ and $v_a$ refer to the values of the mean free path and average velocity, respectively, of the tungsten atoms at the temperature of the filament. Thus $D$ varies inversely as the pressure, $P$, of argon, and equation 1.193 becomes

$$q = \frac{Ar}{P}\frac{dc}{dr}, \tag{1.194}$$

where $A$ is a proportionality constant.

For a wire of diameter $2a$ and gas film of diameter $2b$, it follows, from the same considerations as those that lead to equation 1.158 for the heat loss, that

$$\int_b^a \frac{dr}{r} = \ln\frac{b}{a}. \tag{1.195}$$

Let $m$ denote the rate of evaporation in grams per square centimeter per second. Then it follows from equations 1.194 and 1.195 that

$$maP\log\frac{b}{a} = \int dc = c_a - c_b = \text{constant}, \tag{1.196}$$

where $c_a$ and $c_b$ are the concentrations of the tungsten atoms at $r = a$ and $r = b$, respectively.

Table 1.21 shows results obtained by Fonda for the rate of evaporation from a filament in a mixture of argon and nitrogen, such as is used in gas-filled lamps. The value $m = 230 \cdot 10^{-9}\ \text{g} \cdot \text{cm}^{-2} \cdot \text{sec}^{-1}$ observed for

$P = 0$ is in agreement with the value $250 \cdot 10^{-9}$ observed by Langmuir, Jones, and Mackay.[103]

**TABLE 1.21**

**Rate of Evaporation ($m$) of Tungsten in 86 Per Cent Argon, 14 Per Cent Nitrogen at 2870° K**

Diameter of Filament ($2a$) = 0.00978 cm

| $P_{cm}$ | $10^9m$ | $b$ (cm) | $10^9maP_{cm}\log(b/a)$ | $10^9m/b$ |
|---|---|---|---|---|
| 0 | 230 | ... | ... | ... |
| 1 | 57.5 | 9.68 | 3.9 | ... |
| 5 | 23.5 | 2.42 | 6.3 | ... |
| 10 | 20.5 | 1.31 | 9.8 | 15.6 |
| 25 | 10.3 | 0.63 | 10.4 | 16.2 |
| 50 | 5.4 | 0.36 | 9.6 | 14.8 |
| 70 | 4.2 | 0.28 | 9.6 | 14.8 |
| 165 | 2.0 | 0.15 | 8.8 | 13.5 |

As Fonda states,

> The expression $maP_{cm}\log(b/a)$ is sufficiently constant at pressures above 10 cm to allow of credence being given the hypothesis developed above.
>
> The constancy of the expression has a further significance, for, as is evident from its derivation, it denotes a constant difference between the vapor pressure of tungsten at the surface of the filament and at the surface of the film of gas. For constant filament temperature this implies that the concentration of tungsten vapor at the surface of the stationary film of gas should be constant for all pressures above 10 cm. It is this vapor which constitutes that which effectively evaporates from the filament, for the rising convection currents of gas carry it off to be deposited on the bulb. The rate of evaporation at different pressures should be determined therefore by the area of the assumed film of gas if the concentration of vapor at its border is in fact a constant.

The constancy of the ratio $m/b$ is therefore a further confirmation of the validity of the theory. Other evidence, based on the appearance of the surface of the filament, has also been shown by Fonda to be in agreement with the views expressed above.

From equation 1.196 it also follows that at constant pressure and for filaments of different diameters the expression $ma\log(b/a)$ should be constant. The validity of this conclusion was confirmed by Fonda in a series of observations made with both pure nitrogen and a mixture of nitrogen and argon.[104] In these cases the value actually used for $B$ (the thickness of the film for a plane surface in gas at atmospheric pressure) was the same as that obtained from data on heat conduction in nitrogen (see section 1.11).

## 1.15. RANDOM MOTIONS AND FLUCTUATIONS[105]

Let us consider the case in which a group of $N$ molecules start from the plane $z = 0$, at the instant $t = 0$, and diffuse through the gas. The differential equation for the diffusion is

$$\frac{dn}{dt} = D\frac{d^2n}{dz^2}. \tag{1.197}$$

At any instant, $t$, the distribution of these molecules along the $z$ axis is given by the relation

$$dn = \frac{N}{2(\pi Dt)^{1/2}} \cdot \epsilon^{-z^2/4Dt}\, dz, \tag{1.198}$$

which satisfies the condition

$$N = \frac{N}{\pi^{1/2}} \int_{-\infty}^{\infty} \frac{1}{2(Dt)^{1/2}}\, \epsilon^{-z^2/4Dt}\, dz.$$

Hence, the average value of $z$ is

$$z_a = \frac{1}{\pi^{1/2}} \int_{-\infty}^{\infty} \frac{z}{2(Dt)^{1/2}}\, \epsilon^{-z^2/4Dt}\, dz$$

$$= 2\left(\frac{Dt}{\pi}\right)^{1/2}. \tag{1.199}$$

And similarly it can be shown that the root-mean-square value is given by the relation

$$z_r = \left(\frac{\pi}{2}\right)^{1/2} z_a = (2Dt)^{1/2}. \tag{1.200}$$

That is, the *total net displacement* in any given direction varies as $t^{1/2}$.

These equations are of special importance in the determination of diffusion coefficients from observations on the Brownian motion of small particles.

We can also express equations 1.199 and 1.200 in terms of the mean free path $L$, on the basis of the relation

$$D = \tfrac{1}{3} v_a L. \tag{1.182}$$

Substituting in the above equations, we obtain the relations

$$z_a = \left(\frac{4Lv_a t}{3\pi}\right)^{1/2} \tag{1.201}$$

and

$$z_r = \left(\frac{2Lv_a t}{3}\right)^{1/2} = \left(\frac{\pi}{2}\right)^{1/2} z_a. \tag{1.202}$$

The collision frequency per molecule is

$$\omega = \frac{v_a}{L},$$

while the total length of path actually traversed by a molecule in time $t$ is

$$l = \omega t = \frac{v_a t}{L}. \tag{1.203}$$

Hence

$$z_a^{\,2} = \left(\frac{4L^2}{3\pi}\right) l \tag{1.204}$$

or

$$l = 2.356 \frac{z_a^{\,2}}{L^2} \tag{1.205}$$

$$= 1.5 \frac{z_r^{\,2}}{L^2}. \tag{1.206}$$

That is, the *total length of path varies as the square of the net displacement* from the point of origin at $t = 0$.

As an illustration of the above equations let us consider the case of molecules in air at 25° C and atmospheric pressure. Since $L = 6.69 \cdot 10^{-6}$ and $v_a = 4.67 \cdot 10^4$, it follows from equation 1.182 that $D = 0.104$ $\text{cm}^2 \cdot \text{sec}^{-1}$, and for $t = 60$ sec, $z_r = 3.53$ cm, while $l = 6.98 \cdot 10^9 t = 4.19 \cdot 10^{11}$ cm.

That $l = 1.19 \cdot 10^{11} z_r$ is obviously due to the fact that after each collision the probability of a displacement towards lower (or more negative) values of $z$ is just as great as that of a displacement towards more positive values of $z$.

The equations for random motion have been applied to determine the value of the Avogadro constant from observations on rates of diffusion of Brownian particles. It may be demonstrated that the rate of diffusion of spherical particles is given by the relation

$$D = \frac{R_0 T}{N_A} \cdot \frac{1}{6\pi a \eta}, \tag{1.207}$$

where $a$ = radius of particle, and $\eta$ = viscosity of medium.

Also, the mean square of the displacement per second

$$\frac{z^2}{t} = \frac{R_0 T}{N_A} \cdot \frac{1}{3\pi a \eta}. \tag{1.208}$$

Values of $N_A$ obtained by application of these relations, though not nearly as accurate as those obtained from determinations of the unit electric charge (the method used by Millikan), are in good agreement with them.

## REFERENCES AND NOTES

1. In connection with this chapter, use has been made of the following treatises:

*a.* E. H. Kennard, *Kinetic Theory of Gases with an Introduction to Statistical Mechanics*, McGraw-Hill Book Company, New York, 1938.

*b.* L. B. Loeb, *Kinetic Theory of Gases*, McGraw-Hill Book Company, New York, 2nd edition, 1934.

*c.* J. H. Jeans, *An Introduction to the Kinetic Theory of Gases*, The Macmillan Company, New York, 1940.

*d.* S. Chapman and T. G. Cowling, *The Mathematical Theory of Non-Uniform Gases*, Cambridge, England, The University Press, 1939.

2. The values given for $V_0$ and other constants in this and subsequent sections are those recommended by R. T. Birge, *Reports on Progress in Physics*, Vol. VIII, 90, 1941. The original paper gives, for each constant, the probable error. While the values are stated to the same number of decimal points as by Birge, it is evident that for most calculations values rounded out to an accuracy of 0.1 per cent are sufficiently satisfactory. [A later discussion (*Am. J. Phys.*, **13**, 63, 1945) leaves the 1941 values substantially unaltered.]

As far as possible, the notation used for the constants is the same as that given by Birge. For $R_0$, the molar gas constant, writers on the kinetic theory of gases often use the notation $R_M$ or $R$. On the other hand, $R$ is also used to designate $R_0/M$.

3. Formerly the designations used were as follows: 1 bar (*ca.* 1915), and 1 barye (1920–1930), approximately); what is now designated 1 bar was termed a megabar.

4. In engineering literature the following units are frequently used:

$$\begin{aligned} 1\ \mathrm{A}_0 &= 29.921 \text{ in. Hg at } 32°\ \mathrm{F} \\ &= 14.696 \text{ lb/in.}^2 \text{ (psi).} \\ x \text{ psig} &= x \text{ lb/in.}^2 \textit{ on gauge} \\ &= x \text{ lb/in.}^2 \textit{ above} \text{ atmospheric pressure.} \end{aligned}$$

Also, in the mks system,

$$\begin{aligned} 1 \text{ newton} &= 10^5 \text{ dynes,} \\ 1 \text{ newton/cm}^2 &= 0.1 \text{ bar} = 10^5 \text{ microbars.} \end{aligned}$$

The notation $P$ will be used to designate pressure, in general; the notations $P_b$, $P_{atm}$, $P_{\mu b}$, $P_{mm}$, and $P_\mu$ will designate the pressure in bars, atmospheres, microbars, millimeters, and microns, respectively.

5. $T_F$ is designated "degrees Rankine."

6. The following additional conversion factors will be found useful:

$$\begin{aligned} 1 \text{ micron} \cdot \text{liter } (25°\ \mathrm{C}) &= 0.9162 \text{ micron} \cdot \text{liter at } 0°\ \mathrm{C} \\ &= 1.205 \text{ atm} \cdot \text{mm}^3 \text{ at } 0°\ \mathrm{C} \\ &= 1.316 \text{ atm} \cdot \text{mm}^3 \text{ at } 25°\ \mathrm{C} \\ &= 3.240 \cdot 10^{16} \text{ molecules (see section 1.3)} \\ &= 0.0353 \text{ micron} \cdot \text{cu ft at } 25°\ \mathrm{C.} \end{aligned}$$

7. C. N. Warfield, *Technical Note* 1200, National Advisory Committee for Aeronautics, 1947.

8. N. R. Best, E. Durand, D. I. Gale, and R. J. Havens, *Phys. Rev.*, **70**, 985, (1946).

9. M. Knudsen, *Ann. Physik*, **28**, 75 (1909).

10. I. Langmuir, *Phys. Rev.*, **2**, 329 (1913).

11. I. Langmuir, *Phys. Rev.*, **2**, 329 (1913), and subsequent papers.

12. F. H. Verhoek and A. L. Marshall, *J. Am. Chem. Soc.*, **61**, 2737 (1939).

13. H. A. Jones and I. Langmuir, *Gen. Elec. Rev.*, **30,** 310, 354, and 408 (1927).
14. D. B. Langmuir and L. Malter, *Phys. Rev.*, **55,** 748 (1939).
15. A. C. Egerton, *Phil. Mag.*, Series 6, **33,** 33 (1917).
16. A. C. Egerton, *Proc. Roy. Soc. London,* **A, 103,** 469 (1923).
17. A comprehensive review of this topic has been published by K. C. D. Hickman, one of the pioneers in this field, in *Chem. Revs.*, **34,** 51 (1944).
18. M. Born, *Physik. Z.*, **21,** 578 (1920).
19. F. Bielz, *Z. Physik*, **32,** 81 (1925).
20. R. G. J. Fraser, *Molecular Rays*, Cambridge University Press, 1931; also *Molecular Beams*, Methuen and Company, London, 1937.
21. J. J. Weigle and M. S. Plesset, *Phys. Rev.*, **36,** 373 (1930).
22. F. Knauer and O. Stern, *Z. Physik*, **53,** 766 (1929).
23. J. B. Taylor, *J. Ind. Eng. Chem.*, **23,** 1228 (1931). Also see W. H. Bessey and O. C. Simpson, *Chem. Revs.*, **30,** 239 (1942).
24. A. C. Torrey, *Phys. Rev.*, **47,** 644 (1935); A. B. Van Cleave and O. Maass, *Can. J. Research*, (B) **12,** 57 (1935); **13,** 384 (1935).
25. This topic is discussed at length by Chapman and Cowling (Ref. 1*d*) in their treatise on this subject. It will be observed that

$$L_M = 1.611L = 1.131L_B$$

and

$$L_B = 1.425L.$$

26. See section 10.41 of Chapman and Cowling (Ref. 1*d*).
27. See p. 147 of Kennard (Ref. 1*a*).
28. See section on molecular diameters, beginning on p. 36.
29. W. Licht, Jr., and D. G. Stechert, *J. Phys. Chem.*, **48,** 23 (1944).
30. *Physikalisch-Chemische Tabellen*, 5th edition, Julius Springer, Berlin, 1923–1935.
31. A. E. Schuil, *Phil. Mag.*, **28,** 679 (1939). This paper gives data on values of $\eta$ for gases and vapors over the range 0–250° C.
32. Ref. 1*a*, p. 149.
33. Hsue-Shen Tsien, *J. Aeronaut. Sci.*, **13,** 653 (1946).
34. H. Braune, R. Basch, and W. Wentzel, *Z. physik. Chem.*, **A, 137,** 447 (1928).
35. Value given in Ref. 34, which differs somewhat from that given in Table 1.8.
36. Ref. 1*a*, pp. 160–162.
37. W. Gaede, *Ann. Physik*, **46,** 357 (1915).
38. Ref. 1*a*, p. 297.
39. A. Timiriazeff, *Ann. Physik*, **40,** 971 (1913), assumed that $f$, which may be regarded as an accommodation coefficient for transfer of momentum, has the same value for any gas-surface combination as $\alpha$, the accommodation coefficient for heat transfer (discussed in section 1.10); but B. Baule, *Ann. Physik*, **44,** 145 (1914), disagreed with this assumption and concluded that the value of the ratio between $\zeta$ and $L$ (denoted by $\beta$ in equations 1.112 and 1.113) must be a complicated function of the diameters of the molecules in the gas and those constituting the surface. The theoretical investigations on this topic are discussed by Kennard and Loeb. Loeb quotes results obtained by R. A. Millikan, *Phys. Rev.*, **21,** 217 (1923), which lead to values of $f$ ranging from 0.87 to 1.00.
40. I. Langmuir, *Phys. Rev.*, **1,** 337 (1913).
41. Values of $b$ for a large number of gases and vapors are given in the *Handbook of Chemistry and Physics*, published by Chemical Rubber Publishing Company, Cleveland, Ohio. Since these values are expressed in terms of the volume at 0° C and 1 atm as unity, they should be multiplied by 22,415 to give values for use in equation 1.116. See

also *Handbook of Chemistry*, by N. A. Lange, Handbook Publishers, Sandusky, Ohio, 1941, p. 1307.

42. H. K. Livingston, *J. Am. Chem. Soc.*, **66,** 569 (1944); also S. Brunauer, *The Adsorption of Gases and Vapors*, Princeton University Press, Princeton, 1942, p. 287.

43. These observations are discussed very fully by H. A. Stuart, *Molekülstruktur*, Julius Springer, Berlin, 1934, pp. 49 *et seq.*

44. See Ref. 43.

45. E. Schuil, *Phil. Mag.*, **28,** 679 (1939).

46. H. A. Stuart (Ref. 43), p. 51.

47. See also P. H, Emmett and S. Brunauer, *J. Am. Chem. Soc.*, **59,** 1553 (1937).

48. A. Eucken, *Physik. Z.*, **14,** 324 (1913). See also Ref. 1*b*, pp. 234–252.

49. W. C. Kannuluik and L. K. Martin, *Proc. Roy. Soc. London*, **A, 144,** 496 (1934). See also the following references on this topic:

H. Gregory and C. T. Archer, *Proc. Roy. Soc. London*, **A, 110,** 91 (1926); **121,** 285 (1928).

H. Gregory and S. Marshall, *Proc. Roy. Soc. London*, **A, 114,** 354 (1927); **118,** 594 (1928).

B. G. Dickins, *Proc. Roy. Soc. London*, **A, 143,** 517 (1934).

H. A. Daynes, *Gas Analysis by Measurements of Thermal Conductivity*, Cambridge University Press, 1933, gives a very comprehensive table of relative thermal conductivities (air = 1) for a large number of gases, including hydrocarbons.

50. J. R. Partington and W. G. Shilling, The *Specific Heats of Gases*, 1924. See also G. N. Lewis and M. Randall, *Thermodynamics*, McGraw-Hill Book Company, New York, 1923, p. 80. *Bulletin* 30, Cornell University Engineering Experiment Station, October, 1942, gives specific heats of a number of gases over a wide range of pressures and temperatures.

51. $\ln = \log_e = 2.303 \log_{10}$.

52. The exact meaning of this requirement will appear in the subsequent discussion.

53. See A. Farkas and H. W. Melville, *Experimental Methods in Gas Reactions*, Cambridge University Press, 1939, p. 190; also A. Farkas, *Orthohydrogen, Parahydrogen, and Heavy Hydrogen*, Cambridge University Press, 1935, for illustrations of the application of this method.

54. G. A. Shakespear, *Proc. Roy. Soc. London*, **A, 97,** 273 (1920). See also the comprehensive discussion of this instrument by H. A. Daynes, *Gas Analysis by Measurement of Thermal Conductivity*, Cambridge University Press, 1933.

55. See especially a description of a modified construction by T. L. Ibbs, *Proc. Roy. Soc. London*, **A, 99,** 385 (1921), also **107,** 470 (1925). W. E. Summerhays, *Proc. Phys. Soc. London*, **42,** 218 (1930), has used the katharometer to measure the coefficient of diffusion of water vapor.

56. B. G. Dickins, *Proc. Roy. Soc. London*, **A, 143,** 517 (1934).

57. M. Knudsen, *Ann. Physik*, **31,** 205 (1910); **32,** 809 (1910); **33,** 1435 (1910); **34,** 593 (1911); and **6,** 129 (1930). See also Loeb (Ref. 1*b*), pp. 310–325, and Kennard (Ref. 1*a*), pp. 311–320.

58. M. von Smoluchowski, a number of papers published before 1911, and *Ann. Physik*, **35,** 983 (1911). Also discussed by Loeb (Ref. 1*b*) and Kennard (Ref. 1*a*).

59. Ref. 1*a*, pp. 311–312.

60. K. B. Blodgett and I. Langmuir, *Phys. Rev.*, **40,** 78 (1932).

61. Ref. 1*a*, p. 316.

62. A table of values of $\gamma$ is given by Loeb (Ref. 1*b*), p. 445. Also see J. R. Partington and W. G. Shilling (Ref. 50).

63. M. von Smoluchowski; *Ann. Physik*, **35,** 983 (1911).

64. Ref. 1*c*, p. 193.

65. The data for platinum are from Loeb (Ref. 1*b*), p. 321. The data for tungsten are from H. A. Jones and I. Langmuir, *Gen. Elec. Rev.*, **30,** 354 (1927), while those for "ordinary" platinum are from B. G. Dickins, *Proc. Roy. Soc. London*, **A, 143,** 517 (1934). That the value of the accommodation coefficient is greatly affected by the nature of the surface film has been shown by K. B. Blodgett and I. Langmuir, *Phys. Rev.*, **40,** 78 (1932). In the case of a tungsten wire in hydrogen at 0.20 mm, the value of $\alpha$ changes from 0.534 for bare tungsten to 0.094 for tungsten with an adsorbed film of oxygen. The general problem of the interaction between molecules and solid surfaces, which is obviously involved in the interpretation of $\alpha$, has received considerable attention from a number of investigators. For detailed discussion, especially with regard to the role of $\alpha$ in adsorption phenomena, see Loeb (Ref. 1*b*), p. 311, and also J. K. Roberts, *Some Problems in Adsorption*, Cambridge University Press, 1939. Values of $\alpha$ for air and a number of different metals having etched, polished, and machined surfaces have been determined by M. L. Wiedmann, *Trans. Am. Soc. Mech. Eng.*, **68,** 57 (1946). These values range from 0.87 to 0.97.

66. M. Knudsen, *Ann. Physik*, **34,** 593 (1911).

67. I. Langmuir, *Phys. Rev.*, **34,** 401 (1912). See also H. A. Jones, *Gen. Elec. Rev.*, **28,** 650 (1925).

68. Values of $4.186\,(\phi_2 - \phi_1)$ for hydrogen for the case in which the lower temperature is 80° K (boiling point of nitrogen) are given by K. B. Blodgett and I. Langmuir, *Phys. Rev.*, **40,** 78 (1932).

69. Of course, the loss due to radiation was subtracted from the total observed energy loss.

70. I. Langmuir and G. M. J. Mackay, *J. Am. Chem. Soc.*, **36,** 1708 (1914), and I. Langmuir, *J. Am. Chem. Soc.*, **37,** 417 (1915).

71. I. Langmuir, *J. Am. Chem. Soc.*, **37,** 419 (1915).

72. A further application of Langmuir's film theory to account for the effect of gas pressure on rate of evaporation is discussed in section 1.14.

73. H. A. Jones, *Gen. Elec. Rev.*, **28,** 650 (1925).

74. C. W. Rice, *Trans. A.I.E.E.*, June, 1923, also February, 1924.

75. M. Knudsen, *Ann. Physik*, **31,** 205, 633 (1910). A detailed discussion of Knudsen's theory is given in *Vacuum Practice*, by Louis Dunoyer, translated by J. H. Smith, D. Van Nostrand Company, Princeton, N.J., 1926.

76. Chapter 7 of Ref. 1*d*.

77. M. J. Bennett and F. C. Tompkins, *Trans. Faraday Soc.*, **53,** 185 (1957).

78. S. C. Liang, *Can. J. Chem.*, **33,** 279 (1955).

79. J. M. Los and R. R. Fergusson, *Trans. Faraday Soc.*, **48,** 730 (1952).

80. S. C. Liang, *J. Appl. Phys.*, **22,** 148 (1951).

81. J. O Hirschfelder, R. B. Bird, and E. L. Spotz, *J. Chem. Phys.*, **16,** 968 (1948); *Chem. Revs.*, **44,** 205 (1949).

82. T. L. Ibbs, *Physica*, **4,** 1135 (1937). This is a nonmathematical discussion of the subject which gives a list of references to previous papers.

83. S. Chapman and F. W. Dootson, *Phil. Mag.*, **33,** 248 (1917). See also Jeans (Ref. 1*c*), p. 251. A mathematical treatment of the subject is given by Chapman and Cowling (Ref. 1*d*), pp. 253–258, as well as by W. H. Furry, R. Clark Jones, and L. Onsager, *Phys. Rev.*, **55,** 1083 (1939), and R. C. Jones and W. H. Furry, *Revs. Modern Phys.*, **18,** 151 (1946). A "simple theory" has been published by L. J. Gillespie, *J. Chem. Phys.*, **7,** 530 (1939), and a much more elaborate mathematical discussion by S. Chapman, *Proc. Roy. Soc. London*, **A, 177,** 38 (1940–1941).

84. K. Clusius and G. Dickel, *Naturwissenschaften*, **26,** 546 (1938); **27,** 148 (1939).
85. T. L. Ibbs, *Proc. Roy. Soc. London*, **A, 107,** 470 (1925).
86. See L. J. Gillespie, *J. Chem. Phys.*, **7,** 530 (1939), for discussion of the derivation of this equation.
87. T. L. Ibbs and A. C. R. Wakeman, *Proc. Roy. Soc. London*, **A, 134,** 613 (1932).
88. K. E. Grew, *Proc. Phys. Soc. London*, **43,** 142 (1931).
89. A. K. Brewer and A. Bramley, *Phys. Rev.*, **55,** 590 (1939).
90. A. O. Nier, *Phys. Rev.*, **57,** 30 (1940).
91. W. Furry, R. Jones, and L. Onsager, *Phys. Rev.*, **55,** 1083 (1939).
92. T. I. Taylor and G. Glockler, *J. Chem. Phys.*, **8,** 843 (1940).
93. J. Bardeen, *Phys. Rev.*, **57,** 35 (1940).
94. F. T. Wall and C. E. Holley, Jr., *J. Chem. Phys.*, **8,** 949 (1940).
95. See Kennard (Ref. 1*a*), pp. 195–196.
96. See Jeans (Ref. 1*c*), p. 216.
97. See Jeans (Ref. 1*c*), pp. 207–210.
98. See Jeans (Ref. 1*c*), pp. 217–218.
99. W. E. Summerhays, *Proc. Phys. Soc. London*, **42,** 218 (1930).
100. The results are summarized by A. Lonius, *Ann. Physik*, **29,** 664 (1909).
101. Loeb (Ref. 1*b*), p. 272.
102. G. R. Fonda, *Phys. Rev.*, **31,** 260 (1928).
103. I. Langmuir, H. A. Jones, and G. M. J. Mackay, *Phys. Rev.*, **30,** 211 (1927).
104. G. R. Fonda, *Phys. Rev.*, **21,** 343 (1923).
105. See Kennard (Ref. 1*a*), Chapter 7; Jeans (Ref. 1*c*), pp. 218–224.
106. (*Note added in proof*). H. H. Podgurski and F. N. Davis, *J. Phys. Chem.* **65,** 1343 (1961), have re-examined the thermal transpiration effect for several gases. They find that the results for argon and xenon are satisfactorily described by Bennett and Tompkins' equation, but that the equation is not accurate for neon and hydrogen at low pressures.

# 2 Flow of gases through tubes and orifices

Revised by David G. Worden

## 2.1. INTRODUCTION

The description of gas flow in vacuum systems is generally divided into three parts, the division being specified by three ranges of values of a dimensionless parameter called the *Knudsen number*. The Knudsen number is defined as the ratio of the mean free path of a molecule to a characteristic dimension of the channel through which the gas is flowing, for example, the radius in the case of a cylindrical tube.

In the high-pressure range of vacuum-system operation, where the mean free path is small compared to the characteristic dimension of the channel, that is, the range of small Knudsen numbers, collisions between molecules occur more frequently than collisions of molecules with the channel walls. Consequently, intermolecular collisions predominate in determining the characteristics of the flow. In conventional flow in this range, the properties of the gas (temperature, density, flow velocity) do not vary appreciably in one mean free path and the gas can be considered to be a continuous medium, that is, a viscous fluid. The flow is therefore described and analyzed hydrodynamically and is called *viscous flow*. The coefficient of viscosity, which appears in all viscous-flow equations, reflects the influence of intermolecular interactions.

At low pressures, the mean free path is large compared to the characteristic dimension and the flow of gas is limited by molecular collisions with the walls of the channel. The analysis of the flow is primarily a geometrical problem of determining the restrictive effect of the walls on the free flight of a molecule. Since there are comparatively few intermolecular collisions, each molecule acts independently of the others. Flow at large Knudsen numbers is therefore called *free-molecule flow*, or simply *molecular flow*.

The transition from viscous flow to molecular flow occurs at intermediate values of the Knudsen number where both types of collisions are influential in determining the flow characteristics. There are no general derivations of flow equations in this range which are based on first principles and applicable to vacuum technique. It is therefore necessary to describe the flow by semiempirical equations.

Let $L_a$ be the mean free path evaluated at the average pressure in the channel, and let $a$ be the characteristic dimension. It is shown in section 2.5 that a practical assignment of limits to the Knudsen numbers which delineate the ranges of the high- and low-pressure descriptions, and the transition range between them, is as follows:

When $L_a/a < 0.01$, the flow is viscous.
When $L_a/a > 1.00$, the flow is molecular.
When $0.01 < L_a/a < 1.00$, the flow is in the transition range.

Since the mean free path in *centimeters* for air at 25° C is related to the pressure, $P_\mu$, in *microns of Hg* by (see section 1.8)

$$L_a = \frac{5.09}{P_\mu} \approx \frac{5}{P_\mu},$$

the division can also be stated:

When $aP_\mu > 500$, the flow is viscous.
When $aP_\mu < 5$, the flow is molecular.
When $5 < aP_\mu < 500$, the flow is in the transition range.

The dimension $a$ is measured in *centimeters* in the last inequalities.

The flow equations for the viscous, molecular, and transition ranges are discussed in sections 2.3, 2.4, and 2.5, respectively. Derivations of the equations are not given, since there are a number of textbooks where the basic theories are discussed extensively. Several references are listed in each section for the reader interested in studying the subject in more detail.

All equations are written, at least once in the text, in a form which will permit the user to employ any consistent set of units. Where numerical coefficients are given, the units used in the calculation are stated.

It should be noted that it is assumed throughout the chapter that the flow is isothermal.

## 2.2. FLOW CONDUCTANCE AND IMPEDANCE

A flow rate $Q$ will be used in the following sections, defined as the product of the volumetric flow rate $dV/dt$ across a plane, and the pressure $P$

at which it is measured, that is,

$$Q = P\frac{dV}{dt}. \tag{2.1}$$

Convenient units for $Q$ are micron · liters per second. Using the ideal gas law, it follows that

$$Q = P\frac{dV}{dt} = kT\frac{dN}{dt}, \tag{2.2}$$

where $k$ is the Boltzmann constant, $T$ is the absolute temperature, and $dN/dt$ is the rate at which molecules cross the plane (number per unit time). Thus $Q$ is directly proportional to a molecular current.

It is sometimes advantageous to use a quantity, $F$, defined by

$$F = \frac{Q}{P_2 - P_1}, \tag{2.3}$$

where $P_2$ is the upstream pressure (measured at the entrance to a channel) and $P_1$ is the downstream pressure (measured at the exit). This quantity corresponds to the rate of flow per unit difference of pressure between the ends of the channel and is analogous to the quantity which in the conduction of electricity is known as *conductance*. That is, the pressure is analogous to electric potential, and $Q$, to the current. In consequence of this, $1/F$ may be treated as an *impedance* or *resistance* which will be denoted by $Z$. If two or more tubes of different dimensions are connected in series, the total impedance may be approximated by adding the values of $Z$ for each tube, as in dealing with electrical impedances. Similarly, if there are two or more tubes in parallel, the total conductance may be approximated by adding the values of $F$ for each tube.

## 2.3. VISCOUS FLOW

### The Poiseuille Equation

The most familiar of the viscous-flow equations is the Poiseuille equation[1] for the flow through a straight tube of circular cross section:

$$Q = \frac{\pi a^4}{8\eta l} P_a(P_2 - P_1), \tag{2.4}$$

where $a$ is the tube radius; $l$, the tube length; $\eta$, the viscosity of the gas; and $P_a$, the arithmetic mean of $P_1$ and $P_2$. In the range where equation 2.4 is applicable, it has been substantially verified experimentally. The viscous conductance of a tube, by definition 2.3, is

$$F = \frac{\pi a^4}{8\eta l} P_a. \tag{2.5}$$

Explicit in the derivation of the Poiseuille equation are four assumptions: (1) the gas is incompressible; (2) the flow is fully developed—that is, the flow-velocity profile is constant throughout the length $l$; (3) there is no turbulent motion of the gas; and (4) the flow velocity at the tube walls is zero. The first three assumptions will be examined in this section. A discussion of slip flow (where the flow velocity at the walls is not zero) will be given in section 2.5 on flow in the transition range.

In order to establish the consequences of the first assumption, it must be determined under what conditions compressibility will be an important factor. It can be shown[2] that compressibility of the gas can be neglected in the analysis if

$$\tfrac{1}{2}M^2 \ll 1, \tag{2.6}$$

where $M$ is the Mach number of the flow, defined as the ratio of the flow velocity, $U_f$, to the velocity of sound in the gas, which is $3.5 \cdot 10^4$ cm · sec$^{-1}$ in air[3] at 25° C. Following Schlichting,[2] $M$ is taken equal to $\frac{1}{3}$ as a practical upper limit. The average flow velocity across a plane in the tube where the pressure is $P$ is defined by

$$U_{fa} = \frac{Q}{\pi a^2 P}. \tag{2.7}$$

Consider a point in a tube of radius 2.5 cm where the pressure is 100 microns of Hg. $M = \frac{1}{3}$ for the average flow velocity when the flow rate is

$$Q = \tfrac{1}{3}(3.5 \cdot 10^4)\,(3.14)\,(2.5)^2\,(100) \text{ micron} \cdot \text{cm}^3 \text{ sec}^{-1}$$

or

$$Q = 2.3 \cdot 10^4 \text{ micron} \cdot \text{liters sec}^{-1}.$$

Since this flow rate is not outside the range encountered in vacuum-system operation, the application of the Poiseuille equation may be invalid in some instances. On the basis of the limit $M = \frac{1}{3}$, it can be stated that for air at 25° C, when

$$Q < 37a^2P_\mu \text{ micron} \cdot \text{liters} \cdot \text{sec}^{-1}, \tag{2.8}$$

where $a$ is in centimeters and $P_\mu$ is in microns, the Poiseuille equation can be applied. The tube must be long, however, because of assumption 2, that the flow is fully developed. The meaning of "long" will become apparent in the following paragraphs.

When gas flows into the entrance of a tube from a large volume, the flow velocity is approximately uniform over the entrance area. As the gas proceeds down the tube, the frictional forces encountered at the walls tend to retard the flow adjacent to the walls relative to the flow in the central core near the tube axis. The flow is thus continually modified until, at a distance from the entrance, a velocity profile is established which then remains unchanged throughout the remainder of the passage through the tube. When the profile becomes constant, the flow is said to be fully developed.

Calculations of the effects of this entrance transition on the flow have been made by a number of authors. Summaries are given by Goldstein[4] and by Shapiro, Seigel, and Kline.[5] The calculation of Langhaar,[6] for the case of incompressible flow, is utilized here.

According to Langhaar, the flow will be fully developed in a distance, $l_e$, from the entrance of a tube which is given by

$$l_e = 0.227aR, \tag{2.9}$$

where $R$ is the Reynolds number of the flow. $R$ is a dimensionless quantity defined by

$$R = \frac{a\rho U_{fa}}{\eta}, \tag{2.10}$$

where $\rho = mP/kT$, the density of an ideal gas. $R$ and $l_e$ can be expressed in terms of $Q$ and the radius $a$ by substituting definition 2.7 for $U_{fa}$ in equation 2.10. Thus,

$$R = \left(\frac{m}{\pi\eta kT}\right)\frac{Q}{a}, \tag{2.11}$$

which, for air at 25° C, with $a$ in centimeters and $Q$ in micron · liters per second, is

$$R = 2.69 \cdot 10^{-3}\frac{Q}{a}. \tag{2.12}$$

The transition length is therefore

$$l_e = 6.11 \cdot 10^{-4}Q \text{ cm} \tag{2.13}$$

For a flow rate of $10^4$ micron · liters sec$^{-1}$, the transition length is about 6 cm. Hence, the effects of the transition to fully developed flow will be most pronounced when short tubulation is used.

The net effect of the transition is to lower the flow rate for a given pressure difference between the ends of the tube, or, alternately, increase the pressure drop required for a given flow rate, when compared with the Poiseuille case. Langhaar showed that for

$$l \geqslant 0.304aR, \tag{2.14}$$

or, using 2.12,

$$l \geqslant 8.18 \cdot 10^{-4}Q \text{ cm}, \tag{2.15}$$

the total pressure drop between the ends of the tube can be written

$$P_2 - P_1 = \frac{8\eta l}{a^2}U_{fa} + 1.14\rho U_{fa}^{\ 2}. \tag{2.16}$$

Using equation 2.7 for $U_{fa}$ at the arithmetic mean pressure, $P_a$, and the ideal gas law, equation 2.16 becomes

$$Q = \frac{\pi a^4 P_a(P_2 - P_1)}{8\eta l\left[1 + 1.14\left(\frac{m}{8\pi\eta kT}\right)\frac{Q}{l}\right]}. \tag{2.17}$$

This is the Poiseuille equation again, with an additional factor in the denominator. For $Q$ in micron liters per second, and $l$ in centimeters,

$$1.14\left(\frac{m}{8\pi\eta kT}\right) = 3.83 \cdot 10^{-4} \tag{2.18}$$

for air at 25° C. As an example, consider a flow rate of $10^4$ micron · liters $\text{sec}^{-1}$. According to 2.15, equation 2.17 can be used in this case for $l > 8.2$ cm. Using $l = 8.2$ cm, the value of the factor is 1.47. Thus, the pressure difference would be 47 per cent higher than that calculated with the unmodified Poiseuille formula. For short tubulation, therefore, this is not a negligible effect.

The preceding remarks are valid only for viscous laminar flow (assumption 3). When the Reynolds number of the flow exceeds a critical value, the flow becomes turbulent and the pressure difference between the ends of the tube exceeds that calculated from the formulas presented. The critical value has been shown to be dependent upon the entrance conditions and the roughness of the walls of the tube. In general, for smooth tubes with "well-rounded" entrances, the critical value[7] is about 1000. From equation 2.12 the inequality

$$(2.69 \cdot 10^{-3})\frac{Q}{a} < 1000,$$

or

$$Q < 3.72 \cdot 10^5 a \text{ micron} \cdot \text{liters} \cdot \text{sec}^{-1} \tag{2.19}$$

is obtained as the condition for the flow to be laminar. For a tube with a radius of 1 cm, $Q$ would have to be greater than $3.72 \cdot 10^5$ micron · liters · $\text{sec}^{-1}$ before turbulence would be expected. This is not an exceedingly severe restriction. If a flow rate of this magnitude is anticipated a tube of larger radius will usually be employed.

If conditions 2.8 and 2.19 are satisfied,[8] the Poiseuille equation 2.4 or, in modified form, equation 2.17 should be applicable. For air at 25° C ($\eta = 1.845 \cdot 10^{-4}$ poise), with $a$ and $l$ in centimeters and pressure in microns of Hg, the equations for the flow conductances are:

$$F = 2.84\frac{a^4}{l}P_{\mu a} \text{ liters} \cdot \text{sec}^{-1} \tag{2.20}$$

and

$$F = 2.84\frac{a^4}{l}\frac{P_{\mu a}}{[1 + 3.83 \cdot 10^{-4}(Q/l)]} \text{ liters} \cdot \text{sec}^{-1}. \quad (2.21)$$

The concept of treating tubes or channels as circuit elements with simple conductances, as discussed in section 2.2, is nullified to some extent in the case of Poiseuille flow. In order to determine the conductance of a tube, it is necessary to estimate the mean pressure at which the flow will occur, and also, in the case of equation 2.21, the flow rate.

There will be an additional increase in the total pressure difference between the ends of a tube if there are abrupt changes in the tube radius. This may be important at high flow rates and should be considered in utilizing the rule stated in section 2.2 on the additivity of flow impedances for tubes of different dimensions connected in series. It can be shown[9] that the pressure increase, $\Delta P$, is of the order of

$$\tfrac{1}{2}\rho U_{fa}^{2},$$

which is

$$\frac{mP_a}{2kT} U_{fa}^{2}.$$

For air at 25° C, and an average flow velocity of $3.5 \cdot 10^3$ cm · sec$^{-1}$ ($M = 0.1$),

$$\Delta P = 5.85 \cdot 10^{-10} P_a U_{fa}^{2},$$
$$= 7.2 \cdot 10^{-3} P_a,$$

that is, about seven-tenths of 1 per cent of the average pressure at the point where the radius changes.

**Flow through Channels with Cross Sections Other than Circular**

The formulas presented here are derived on the same basis as the Poiseuille equation and hence the general remarks made in the preceding paragraphs apply here also. Calculations of entrance effects for non-circular cross sections, applicable to vacuum technique, have not appeared in the literature.

Consider a channel of annular section, with outer radius $b$ and inner radius $a$. The conductance is given by[1]

$$F = \frac{\pi P_a}{8\eta l}\left[b^4 - a^4 - \frac{(b^2 - a^2)^2}{\log_e (b/a)}\right]. \quad (2.22)$$

For air at 25° C,

$$F = 2.84\frac{P_{\mu a}}{l}\left[b^4 - a^4 - \frac{(b^2 - a^2)^2}{\log_e (b/a)}\right] \text{ liters} \cdot \text{sec}^{-1}. \quad (2.22a)$$

For a channel of elliptic section[1] with semiaxes $a$ and $b$,

$$F = \frac{\pi P_a}{4\eta l}\left(\frac{a^3b^3}{a^2 + b^2}\right). \tag{2.23}$$

For air at 25° C,

$$F = 5.68\,\frac{P_{\mu a}}{l}\left(\frac{a^3b^3}{a^2 + b^2}\right) \text{ liters} \cdot \text{sec}^{-1}. \tag{2.23a}$$

Loevinger[10] offers the following formula for a rectangular duct, for air at 25° C:

$$F = 0.26Y\,\frac{a^2b^2}{l}\,P_{\mu a} \text{ liters} \cdot \text{sec}^{-1}. \tag{2.24}$$

Here $a$ and $b$ are the dimensions of the sides of the duct, and $Y$ is given in the following table:

| $a/b$ | 1.0 | 0.9 | 0.8 | 0.7 | 0.6 | 0.5 | 0.4 | 0.3 | 0.2 | 0.1 |
|---|---|---|---|---|---|---|---|---|---|---|
| $Y$ | 1.00 | 0.99 | 0.98 | 0.95 | 0.90 | 0.82 | 0.71 | 0.58 | 0.42 | 0.23 |

In formulas 2.22*a*, 2.23*a*, and 2.24, all dimensions are in centimeters and pressures in microns of Hg.

## 2.4. MOLECULAR FLOW

### Molecular Flow through Long Tubes and Channels

At pressures corresponding to $aP_\mu < 5$ micron · cm, where the mean free path is greater than the characteristic dimension, the rate of flow is limited, not by collisions between molecules, but by collisions of molecules with the walls. The manner of attacking the problem of flow in this range theoretically and the first experimental verifications of the theory are due to Knudsen.[11] Knudsen's work and the results of more recent investigations are discussed in several texts.[12,13,14,15]

For a long tube[16] of length $l$, varying cross section $A$, and perimeter $H$, the fundamental relation deduced by Knudsen is

$$Q = \frac{4}{3}\,\frac{v_a}{\displaystyle\int_0^l \frac{H}{A^2}\,dl}\,(P_2 - P_1), \tag{2.25}$$

where $v_a$ is the mean molecular speed (see section 1.4) given by

$$v_a = \left(\frac{8kT}{\pi m}\right)^{1/2} = \left(\frac{8R_0T}{\pi M}\right)^{1/2}, \tag{2.26}$$

$R_0$ is the molar gas constant, and $M$ is the molecular weight of the gas. It follows that the flow conductance,

$$F = \frac{4}{3} \frac{v_a}{\int_0^l \frac{H}{A^2} dl}, \tag{2.27}$$

is independent of the pressure, and the concept of conductance, therefore, has more utility than in the region of viscous flow. For a given gas, flowing at a given temperature, $F$ is dependent only on the geometry of the tube.

From equations 2.27 and 2.26, a number of useful relations can be derived. In formulas 2.28*a*–2.33*a*, $T$ is measured in degrees Kelvin, $M$ in grams per mole, and the dimensions in centimeters. They apply to long tubes, where the length of the tube is about 100 times the largest tranverse dimension.

In general, for a long tube of uniform cross section (see also equation 2.48),

$$F = \frac{4}{3} \frac{A^2}{Hl} v_a \tag{2.28}$$

$$= 19.40 \frac{A^2}{Hl} \left(\frac{T}{M}\right)^{1/2} \text{liters} \cdot \text{sec}^{-1}. \tag{2.28a}$$

Thus, for a cylindrical tube of radius $a$,

$$F = \frac{2}{3} \pi \frac{a^3}{l} v_a \tag{2.29}$$

$$= 30.48 \frac{a^3}{l} \left(\frac{T}{M}\right)^{1/2} \text{liters} \cdot \text{sec}^{-1}. \tag{2.29a}$$

For a duct of uniform rectangular cross section (lengths of sides $a$ and $b$),

$$F = \frac{2}{3} \frac{a^2 b^2}{(a + b)l} v_a \tag{2.30}$$

$$= 9.70 \frac{a^2 b^2}{(a + b)l} \left(\frac{T}{M}\right)^{1/2} \text{liters} \cdot \text{sec}^{-1}. \tag{2.30a}$$

For the case $a = b$ in 2.30,

$$F = \frac{1}{3} \frac{a^3}{l} v_a \tag{2.31}$$

$$= 4.85 \frac{a^3}{l} \left(\frac{T}{M}\right)^{1/2} \text{liters} \cdot \text{sec}^{-1}. \tag{2.31a}$$

In the case of an ellipse, $A = \pi ab$, and $H = 2\pi[(a^2 + b^2)/2]^{1/2}$ approximately, where $a$ and $b$ designate the lengths of the two semiaxes. Hence, for uniform elliptical cross section,

$$F = \frac{2\pi}{3l} \frac{a^2b^2}{[(a^2 + b^2)/2]^{1/2}} v_a \tag{2.32}$$

$$= 43.10 \frac{a^2b^2}{l(a^2 + b^2)^{1/2}} \left(\frac{T}{M}\right)^{1/2} \text{liters} \cdot \text{sec}^{-1}. \tag{2.32a}$$

Another interesting application of equation 2.27 is that of calculating the conductance of a long tapered round tube having the radius $a_0$ at one end and $a_l$ $(= a_0 + bl)$ at the other end. The result of the integration is

$$F = \frac{2}{3}\pi \frac{a_0^2 a_l^2}{a_{l/2} l} v_a \tag{2.33}$$

$$= 30.48 \frac{a_0^2 a_l^2}{a_{l/2} l} \left(\frac{T}{M}\right)^{1/2} \text{liters} \cdot \text{sec}^{-1}, \tag{2.33a}$$

where $a_{l/2}$ is the value of the radius halfway along the tube. It is evident that, for $a = a_0 = a_{l/2} = a_l$, equation 2.33$a$ becomes identical with equation 2.29$a$.

That the conductance for a cylindrical tube at low pressures is considerably less than that at high pressures is evident from the following example.

For the flow of air at 25° C through a tube, for which $l = 100$ cm and $a = 0.5$ cm, it follows from equation 2.20 that, at $P_a = 3.80 \cdot 10^5$ microns (0.5 atm), the viscous flow conductance is

$$F_{viscous} = 675 \text{ liters} \cdot \text{sec}^{-1}.$$

$F_{viscous}$ decreases linearly with decrease in $P_a$. On the other hand, at low pressures, $F_{molecular}$ is independent of the pressure, and according to equation 2.29$a$, using $(T/M)^{1/2} = 3.207$, for air at 25° C,

$$F_{molecular} = 0.122 \text{ liter} \cdot \text{sec}^{-1},$$

for the case under consideration. In section 2.5 the relation governing $F$ for the transition region between viscous and molecular flow will be considered at further length.

Table 2.1 gives the values of the conductances of a cylindrical tube ($F_t$, column 3), an orifice of area $A$ ($F_o$, column 4), and a circular orifice of radius $a$ ($F_o$, column 5), for a few important gases at $T = 298°$ K. The orifice conductances are discussed in the next section.

A useful approximation for a cylindrical tube, pointed out by Yarwood,[17] is

$$F_t = 100 \frac{a^3}{l} \text{ liters} \cdot \text{sec}^{-1}$$

for air at 25° C, where $a$ and $l$ are in centimeters.

TABLE 2.1

Values of the Conductance of Cylindrical Tubes and Openings at 25° C

| Gas | $M$ | $F_t(l/a^3)$ (liters · sec$^{-1}$) $= 526.2/M^{1/2}$ | $F_oA^{-1}$ (liters · sec$^{-1}$) $= 62.81/M^{1/2}$ | $F_oa^{-2}$ (liters · sec$^{-1}$) $= 197.3/M^{1/2}$ |
|---|---|---|---|---|
| $H_2$ | 2.016 | 370.6 | 44.24 | 139.0 |
| He | 4.003 | 263.0 | 31.39 | 98.58 |
| Air | 28.98 | 97.75 | 11.67 | 36.66 |
| Ar | 39.94 | 83.28 | 9.94 | 31.23 |

From the above equations for $F$ it follows that for different gases $F$ varies inversely as $M^{1/2}$. This indicates a method for the separation of gases of different molecular weights which has been applied very successfully in certain cases. For instance, by means of porous walls, Hertz[18] and his associates have effected a partial separation of the isotopes of helium, carbon, and nitrogen, while Farkas[19] has been able to separate the isotopes of hydrogen ($H_2$ and $D_2$).

## Molecular Flow through Orifices and Short Tubes

Consider an isothermal vessel containing a gas at a low pressure $P_2$. Let one wall of the vessel contain a small[20] orifice through which the gas can effuse into an adjacent vessel, where the pressure is $P_1$ ($P_1 < P_2$) at the same temperature. From the discussion in section 1.5, the total rate of flow, in molecules per second, is given by

$$\frac{dN}{dt} = \tfrac{1}{4}v_a A(n_2 - n_1), \tag{2.34}$$

where $n_2$ is the number density of the gas corresponding to the pressure $P_2$, $n_1$ is the number density corresponding to $P_1$, and $A$ is the area of the orifice. From the ideal gas law and equation 2.2, the flow rate through the orifice is

$$Q = \tfrac{1}{4}v_a A(P_2 - P_1). \tag{2.35}$$

Therefore the conductance of an orifice is

$$F = \tfrac{1}{4}v_a A. \tag{2.36}$$

It should be emphasized that this relation is strictly applicable, as shown by Knudsen, only under those conditions in which the diameter is one-tenth, or less, of the mean free path for the gas molecules at the pressure $P_2$. For practical purposes the relation may be applied to pressures which are ten times greater, but it is no longer valid, even approximately, if the area of the orifice is so large, or the pressure so high, that mass motion of the gas occurs. Also, the wall of the vessel in the region of the orifice should be vanishingly small, so that the orifice has no "length," ideally.

Introducing expression 2.26 for $v_a$,

$$F = A\left(\frac{R_0 T}{2\pi M}\right)^{1/2} \tag{2.37}$$

$$= 3.64A\left(\frac{T}{M}\right)^{1/2} \quad \text{liters} \cdot \text{sec}^{-1}, \tag{2.37a}$$

where $A$ is measured in square centimeters, $T$ in degrees Kelvin, and $M$ in grams per mole.

From the relations given above, it follows that, at very low pressures, the ratio, $r$, of the conductance, $F_o$, of a circular orifice of radius $a$, to that for a cylindrical tube of the same radius, $F_t$, is

$$r = \frac{F_o}{F_t} = \frac{\frac{1}{4}\pi a^2 v_a}{\frac{2}{3}(\pi a^3/l)v_a}$$

$$= \frac{3}{8}\frac{l}{a}.$$

For $l/a = 50$, $r = 18.75$; that is, the resistance to flow offered by the walls is 18.75 times that offered by the open end. This leads to the conclusion that, for small values of $l/a$, the relations for the conductances of long tubes would give values which are too high. Reasoning from the point of view that the open end of a long cylindrical tube can be considered as a vacuum circuit element with impedance $Z_0 = 1/F_o$ in series with the tube proper with impedance $Z_t = 1/F_t$, or

$$\frac{1}{F} = \frac{1}{F_t} + \frac{1}{F_o}$$

$$= \frac{1}{F_o}\left(1 + \frac{F_o}{F_t}\right),$$

Dushman[21] arrived at an equation which approximates the total conductance of a cylindrical tube. It can be written

$$F = K'A\frac{v_a}{4} \tag{2.38}$$

$$= 3.64K'A\left(\frac{T}{M}\right)^{1/2} \quad \text{liters} \cdot \text{sec}^{-1}, \tag{2.38a}$$

where

$$K' = \frac{1}{1 + \frac{3}{8}l/a}, \tag{2.39}$$

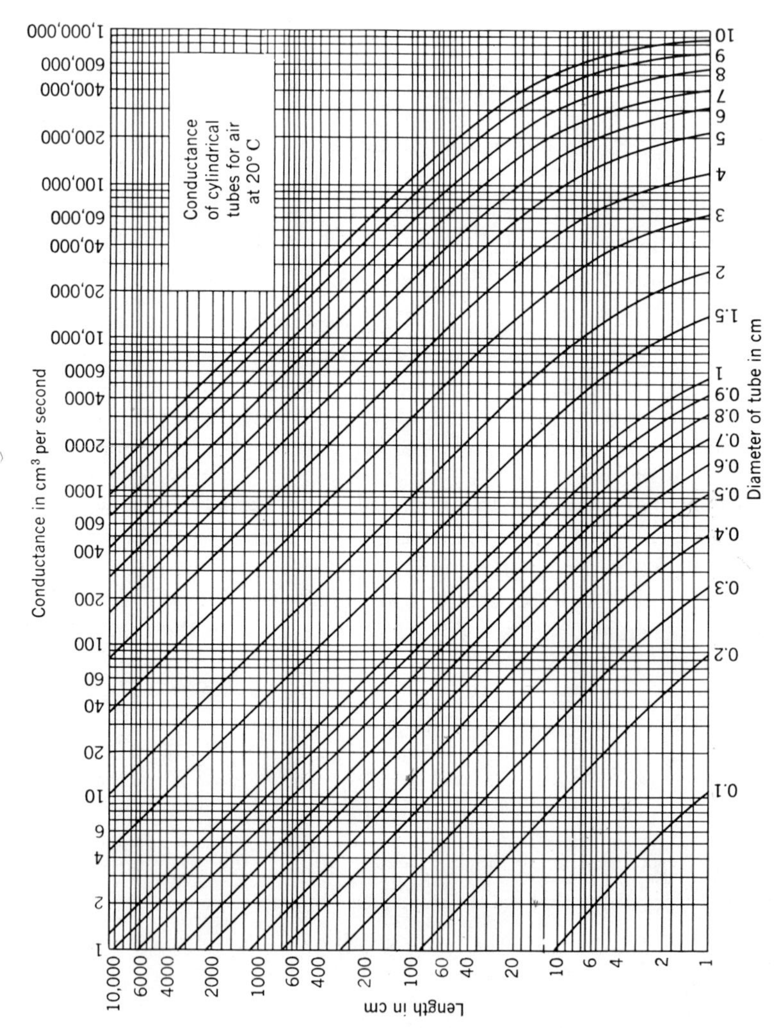


Fig. 2.1. Plots of conductance of cylindrical tubes for air at 20° C calculated according to equation 2.40.

and $A = \pi a^2$. The factor $K'$ may be regarded as representing the ratio between the rate at which gas leaves the outlet of the tube and that at which gas strikes the inlet.

Equation 2.38 can be expressed, for $F$ in cubic centimeters per second, as

$$\frac{1}{F} = \left(\frac{2.394l}{d^3} + \frac{3.184}{d^2}\right)(\rho_1)^{1/2}; \tag{2.40}$$

where $d = 2a$ = diameter of the tube, and

$$\rho_1 = \frac{M}{83.144 \cdot 10^6 T}.$$

Figure 2.1 represents plots of values of $F$, calculated, by means of equation 2.40 for air at 20° C, drawn on log-log scale. Each curve gives $F$ as a function of $l$ (in centimeters) for a constant value of $d$ (in centimeters) which is indicated at the bottom of the figure. The linear portion of each curve represents the effect of the term $l/d^3$, while the curved portion represents the added effect of the term containing $1/d^2$. For other gases the conductance, for constant values of $l$ and $d$, varies inversely as $M^{1/2}$.

As shown by P. Clausing,[22] equation 2.38, with the factor $K'$ given by 2.39, is only approximate. From a detailed investigation he deduced the relation

$$F = KA\frac{v_a}{4} = 3638KA\left(\frac{T}{M}\right)^{1/2} \quad \text{cm}^3 \cdot \text{sec}^{-1}, \tag{2.41}$$

where, as shown in Table 2.2, $K$ is a dimensionless function of $l/a$, the value of which decreases from 1 (for $l/a = 0$) to $8a/(3l)$ for $l/a$ very large. Clausing's values of $K$ are given in the second and fifth columns; the values of $K'$ deduced by means of equation 2.39 are given, for comparison, in the third and sixth columns. It will be observed that for low values of $l/a$ the difference amounts to as much as 12 per cent. For $l/a$ very large, $K'$ tends to become equal to $K$. A series of plots of $K$ versus $l/a$ are shown in Fig. 2.2. The scale for $l/a$ is indicated underneath each curve. It is important to realize that for relatively short tubes serious errors may be made if the value of $F$ is calculated by means of equation 2.29. This is evident from an inspection of the data given in Table 2.3 (p. 96).

Let $F_1$ = value calculated by means of equation 2.29 and $F_2 = KF_0 = KAv_a/4$, and let

$$r = \frac{F_1}{F_2} = \frac{8}{3K}\frac{a}{l}.$$

Then the value of $r$ denotes the ratio between the value of $F$ deduced for the condition $l/a$ very large and the correct value deduced by means of Clausing's correction factor $K$. It will be observed from the last column

TABLE 2.2

**Values of Clausing's Factor $K$ and the Approximate Correction $K'$ for a Series of Values of $l/a$**

| $l/a$ | $K$ | $K'$ | $l/a$ | $K$ | $K'$ |
|---|---|---|---|---|---|
| 0 | 1 | 1 | 3.2 | 0.4062 | 0.454 |
| 0.1 | 0.9524 | 0.965 | 3.4 | 0.3931 | 0.439 |
| 0.2 | 0.9092 | 0.931 | 3.6 | 0.3809 | 0.426 |
| 0.3 | 0.8699 | 0.899 | 3.8 | 0.3695 | 0.412 |
| 0.4 | 0.8341 | 0.870 | 4.0 | 0.3589 | 0.400 |
| 0.5 | 0.8013 | 0.842 | 5 | 0.3146 | 0.348 |
| 0.6 | 0.7711 | 0.816 | 6 | 0.2807 | 0.307 |
| 0.7 | 0.7434 | 0.792 | 7 | 0.2537 | 0.276 |
| 0.8 | 0.7177 | 0.769 | 8 | 0.2316 | 0.250 |
| 0.9 | 0.6940 | 0.747 | 9 | 0.2131 | 0.229 |
| 1.0 | 0.6720 | 0.727 | 10 | 0.1973 | 0.210 |
| 1.1 | 0.6514 | 0.708 | 12 | 0.1719 | 0.182 |
| 1.2 | 0.6320 | 0.690 | 14 | 0.1523 | 0.160 |
| 1.3 | 0.6139 | 0.672 | 16 | 0.1367 | 0.143 |
| 1.4 | 0.5970 | 0.656 | 18 | 0.1240 | 0.129 |
| 1.5 | 0.5810 | 0.640 | 20 | 0.1135 | 0.117 |
| 1.6 | 0.5659 | 0.625 | 30 | 0.0797 | 0.0817 |
| 1.7 | 0.5518 | 0.611 | 40 | 0.0613 | 0.0625 |
| 1.8 | 0.5384 | 0.597 | 50 | 0.0499 | 0.0506 |
| 1.9 | 0.5256 | 0.584 | 60 | 0.0420 | 0.0425 |
| 2.0 | 0.5136 | 0.572 | 70 | 0.0363 | 0.0367 |
| 2.2 | 0.4914 | 0.548 | 80 | 0.0319 | 0.0322 |
| 2.4 | 0.4711 | 0.526 | 90 | 0.0285 | 0.0288 |
| 2.6 | 0.4527 | 0.506 | 100 | 0.0258 | 0.0260 |
| 2.8 | 0.4359 | 0.488 | 1000 | 0.002658 | 0.002660 |
| 3.0 | 0.4205 | 0.470 | | $8a/(3l)$ | $8a/(3l)$ |

in Table 2.3 that, for low values of $l/a$, $r$ is much greater than 1 and approaches unity as a limiting value only for $l/a > 100$ approximately.

DeMarcus[15] has reinvestigated the problem of flow through short cylindrical tubes. His calculations yield a vigorous upper limit for the Clausing factor $K$ and agree with the values of $K$ in Table 2.2 within 1 per cent for $l/a < 4.0$ and in the limit of $l/a = \infty$. For $l/a > 4.0$, the tabulated values are higher, becoming 3.7 per cent too high at $l/a = 20$. DeMarcus' calculations do not extend beyond $l/a = 20$. The trend indicates, however, that Clausing's calculations tend to over-estimate the conductance for higher $l/a$ values. This fact would make the ratio $r$ in Table 2.3 too low.

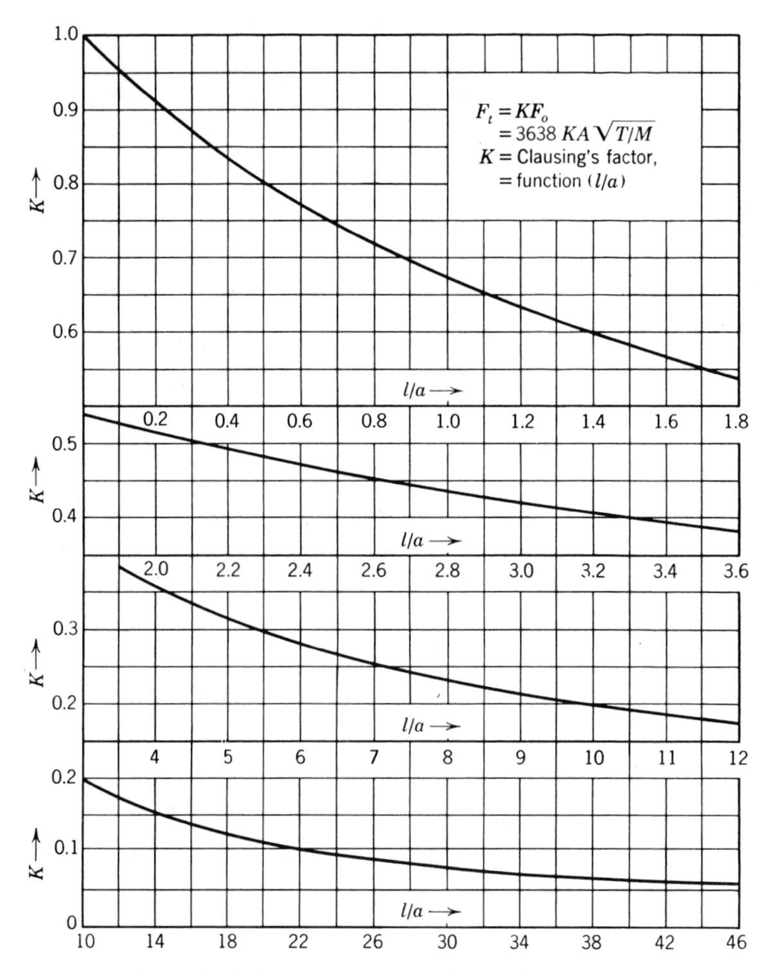


Fig. 2.2. Plot of Clausing's factor, $K$, as a function of $l/a$. Scales for $K$ and $l/a$ are indicated for each curve.

Clausing[22] has also deduced values of $K$ for tubes with rectangular section, $a \gg b$, and $a \gg l$ (as in slits for molecular ray experiments).[23]

$$F = 3638Kab\left(\frac{T}{M}\right)^{1/2} \text{cm}^3 \cdot \text{sec}^{-1}$$

$$= 3.668 \cdot 10^4 Kab \text{ cm}^3 \cdot \text{sec}^{-1}, \tag{2.42}$$

for air at 25° C, where $a$ and $b$ are expressed in centimeters. Table 2.4 gives a selection of values of $K$.

TABLE 2.3

**Illustrating the Effect of Neglecting "End Correction" in Calculating Conductances of Cylindrical Tubes**

| $l/a$ | $K$ | $r$ |
|---|---|---|
| 0.5 | 0.8013 | 6.656 |
| 1.0 | 0.6720 | 3.968 |
| 2.0 | 0.5136 | 2.597 |
| 3.0 | 0.4205 | 2.114 |
| 4.0 | 0.3589 | 1.857 |
| 5.0 | 0.3146 | 1.695 |
| 7.0 | 0.2537 | 1.502 |
| 10.0 | 0.1973 | 1.352 |
| 20.0 | 0.1135 | 1.175 |
| 30.0 | 0.0797 | 1.115 |
| 40.0 | 0.0613 | 1.087 |
| 50.0 | 0.0499 | 1.069 |
| 100 | 0.0258 | 1.034 |
| 1000 | 0.00266 | 1.0035 |

TABLE 2.4

**Values of Clausing's Factor for Tubes of Rectangular Cross Section**

| $l/b$ | $K$ | $l/b$ | $K$ |
|---|---|---|---|
| 0 | 1.000 | 1.50 | 0.6024 |
| 0.1 | 0.9525 | 2.00 | 0.5417 |
| 0.2 | 0.9096 | 3.00 | 0.4570 |
| 0.4 | 0.8362 | 4.00 | 0.3999 |
| 0.8 | 0.7266 | 5.00 | 0.3582 |
| 1.0 | 0.6848 | 10.00 | 0.2457 |
| | | $\infty$ | $\frac{b}{l}\ln\frac{l}{b}$ |

For a short tube of any form of uniform cross section, an approximate result may be obtained, as in the case of a cylindrical tube (equations 2.38 and 2.39), by means of relation

$$\frac{1}{F} = \frac{1}{F_t} + \frac{1}{F_o}, \tag{2.43}$$

where $F_t$ is the conductance deduced for a very long tube, $F_o$ denotes the conductance of one end, and $F$ is the conductance corrected for "end effects."

Hence, in terms of the cross-sectional area, $A$, in square centimeters, and $H$ and $l$ in centimeters,

$$F = \frac{F_o}{1 + \frac{3}{16}(Hl/A)} = \frac{3.638A(T/M)^{1/2}}{1 + \frac{3}{16}(Hl/A)} \text{ liters} \cdot \text{sec}^{-1}, \qquad (2.44)$$

which is identical with equation 2.38$a$ for a cylindrical tube.

As illustrations of the application of the equations for molecular flow, Table 2.5 gives values of $F_t$ (cubic centimeters per second) for a number of

**TABLE 2.5**

**Values of Conductance, $F_t$, for a Number of Typical Cases**

| $a$ | $l$ | $l/a$ | $K$ | Gas | $F_t$ ($cm^3 \cdot sec^{-1}$) | $2303/F_t$ (sec) |
|---|---|---|---|---|---|---|
| 0.51 | 3.2 | 6.273 | 0.273 | $H_2$ | $9.868 \cdot 10^3$ | 0.2333 |
| (Wide-bore stopcock) | | | | Ne | $3.120 \cdot 10^3$ | 0.7381 |
| | | | | Air | $2.620 \cdot 10^3$ | 0.8790 |
| | | | | Ar | $2.216 \cdot 10^3$ | 1.040 |
| | | | | Kr | $1.532 \cdot 10^3$ | 1.503 |
| 0.20 | 2.0 | 10.0 | 0.197 | $H_2$ | $1.088 \cdot 10^3$ | 2.116 |
| (Narrow-bore stopcock) | | | | Ne | $3.442 \cdot 10^2$ | 6.689 |
| | | | | Air | $2{\cdot}871 \cdot 10^2$ | 8.021 |
| | | | | Ar | $2.445 \cdot 10^2$ | 9.419 |
| | | | | Kr | $1.690 \cdot 10^2$ | 13.61 |
| 0.40 | 40.0 | 100.0 | 0.0258 | $H_2$ | $5.709 \cdot 10^2$ | 4.033 |
| | | | | Ne | $1.805 \cdot 10^2$ | 12.76 |
| | | | | Air | $1.505 \cdot 10^2$ | 15.31 |
| | | | | Ar | $1.282 \cdot 10^2$ | 17.97 |
| | | | | Kr | $0.886 \cdot 10^2$ | 25.99 |
| 2.0 | 40.0 | 20.0 | 0.1135 | $H_2$ | $6.319 \cdot 10^4$ | $3.645 \cdot 10^{-2}$ |
| | | | | Air | $1.665 \cdot 10^4$ | $1{\cdot}383 \cdot 10^{-1}$ |
| 5.0 | 20.0 | 4.0 | 0.3589 | $H_2$ | $1.248 \cdot 10^6$ | $1.845 \cdot 10^{-3}$ |
| | | | | Air | $3.291 \cdot 10^5$ | $6.998 \cdot 10^{-3}$ |
| 20.0 | 20.0 | 1.0 | 0.6720 | $H_2$ | $3.741 \cdot 10^7$ | $6.156 \cdot 10^{-5}$ |
| | | | | Air | $9.860 \cdot 10^6$ | $2.335 \cdot 10^{-4}$ |
| 2.5 | 20.0 | 8.0 | 0.2316 | $H_2$ | $2.014 \cdot 10^5$ | $1.144 \cdot 10^{-2}$ |
| | | | | Air | $5.309 \cdot 10^4$ | $4.338 \cdot 10^{-2}$ |
| 2.0 | 1.0 | 0.5 | 0.8013 | $H_2$ | $4.461 \cdot 10^5$ | $5.163 \cdot 10^{-3}$ |
| | | | | Air | $1.176 \cdot 10^5$ | $1.959 \cdot 10^{-2}$ |

typical cases, corrected for end effects. The temperature is assumed to be 25° C in all cases. At 25° C,

$$F_t = 1.972 \cdot 10^5 K a^2 M^{-1/2} \text{ cm}^3 \cdot \text{sec}^{-1}.$$

The last column in the table gives values of the time constant, $\Delta t_o = 2.303 \cdot 10^3 F_t^{-1}$, which, as will be shown in a subsequent section, corresponds to the interval of time (in seconds) required to reduce the pressure in a 1-liter volume to 10 per cent of its instantaneous value, through a tube of the dimensions given, assuming that the pressure at the inlet to the pump is maintained at zero.

As will be observed, the values of $F_t$ for large values of $a$ are considerably greater than for smaller values. This is, of course, due to the $a^3$ factor. An increase of 10 per cent in the value of $a$ leads to an increase of 33.1 per cent in the value of $F_t$ (for long tubes), whereas a 10 per cent decrease in the value of $l$ leads to an increase of only 11.1 per cent in the value of $F_t$.

The effect of the "end correction," which is really the significance of the factor $K$, is shown by comparing the value of $F_t$ for the 40-cm length of tubing, 2.0 cm radius, with that for the 1-cm length. If only $l$ were involved, $F_t$ for the shorter tubing should be equal to $40 \cdot 1.665 \cdot 10^4 = 6.860 \cdot 10^5$ (for air), whereas it is actually equal to $1.176 \cdot 10^5$.

For an opening of radius $a$ in a very thin plate, the value of $F_o$, in the case of air at 25° C, is given, as shown in Table 2.1, by

$$F_o = 36.66 \cdot 10^3 a^2 \text{ cm}^3 \cdot \text{sec}^{-1}$$

and

$$\frac{2303}{F_o} = 6.279 \cdot 10^{-2} a^{-2} \text{ sec.}$$

Table 2.6 gives values, for air at 25° C, of $F_o$, the conductance of an opening of radius $a$, and of $F_t$, the conductance of a tube of the same radius, for different values of $l/a$. The corresponding values of $K$ are given at the top of the table, and the value of $F_t = F_o K = 36.66 K a^2$ liters · sec$^{-1}$.

The relation derived in this section for the conductance of an orifice, $F_o$, was based on the supposition that the orifice dimensions were small compared to the dimensions of the vessels on either side (p. 90). If, however, an orifice of area $A_o$ is located in a diaphragm across the exit of a tube of area $A_t$, which is comparable to that of the orifice, Loevinger[10] has shown that the conductance must be modified. Loevinger concludes from elementary considerations that the effective conductance, $F_{eff}$, of the orifice is

$$F_{eff} = \frac{1}{1 - (A_o/A_t)} F_o. \tag{2.45}$$

TABLE 2.6

**Values of Conductances of Orifices and Tubes for a Range of Values of $a$ for Air at 25° C**

| $a$ (cm) | $F_o$ (liters · sec$^{-1}$) | $F_t$, Conductance of Tube (liters · sec$^{-1}$) for Air at 25° C | | | | | | |
|---|---|---|---|---|---|---|---|---|
| | | $l/a = 1$ | 2 | 4 | 8 | 12 | 16 | 30 |
| | | $K = 0.672$ | 0.514 | 0.359 | 0.232 | 0.172 | 0.137 | 0.080 |
| 0.1 | 0.367 | 0.246 | 0.188 | 0.132 | 0.085 | 0.063 | 0.050 | 0.029 |
| 0.2 | 1.466 | 0.986 | 0.753 | 0.527 | 0.340 | 0.252 | 0.200 | 0.117 |
| 0.3 | 3.300 | 2.217 | 1.664 | 1.184 | 0.764 | 0.567 | 0.451 | 0.263 |
| 0.4 | 5.866 | 3.943 | 3.013 | 2.106 | 1.358 | 1.008 | 0.802 | 0.468 |
| 0.5 | 9.166 | 6.160 | 4.708 | 3.291 | 2.122 | 1.575 | 1.253 | 0.731 |
| 0.6 | 13.20 | 8.872 | 6.779 | 4.739 | 3.057 | 2.269 | 1.805 | 1.052 |
| 0.7 | 17.97 | 12.08 | 9.228 | 6.449 | 4.161 | 3.088 | 2.457 | 1.432 |
| 0.8 | 23.47 | 15.77 | 12.05 | 8.424 | 5.436 | 4.033 | 3.208 | 1.871 |
| 0.9 | 29.70 | 19.96 | 15.25 | 10.66 | 6.879 | 5.105 | 4.061 | 2.368 |
| 1.0 | 36.66 | 24.64 | 18.83 | 13.16 | 8.492 | 6.302 | 5.013 | 2.922 |
| 2.0 | 146.6 | 98.56 | 75.34 | 52.65 | 33.97 | 25.21 | 20.05 | 11.69 |
| 3.0 | 330.0 | 221.7 | 166.4 | 118.4 | 76.42 | 56.71 | 45.11 | 26.30 |
| 4.0 | 586.6 | 394.3 | 301.3 | 210.6 | 135.8 | 100.8 | 80.21 | 46.77 |
| 5.0 | 916.6 | 616.0 | 470.8 | 329.1 | 212.2 | 157.5 | 125.3 | 73.10 |
| 6.0 | 1320.0 | 887.2 | 677.9 | 473.9 | 305.7 | 226.9 | 180.5 | 105.2 |
| 7.0 | 1797.0 | 1208.0 | 922.8 | 644.9 | 416.1 | 308.8 | 245.7 | 143.2 |
| 8.0 | 2347.0 | 1577.0 | 1205.0 | 842.4 | 543.6 | 403.3 | 320.8 | 187.1 |
| 9.0 | 2970.0 | 1996.0 | 1525.0 | 1066.0 | 687.9 | 510.5 | 406.1 | 236.8 |
| 10.0 | 3666.0 | 2464.0 | 1883.0 | 1316.0 | 849.2 | 630.2 | 501.3 | 292.2 |

For the range $a = 10$ to $a = 100$ cm, the values of both $F_o$ and $F_t$ are obviously 100 times those given in the table for the corresponding values in the range $a = 1$ to $a = 10$.

Keller[24] has offered a more detailed treatment of a similar problem, the case of a circular orifice located at the midpoint of a long cylindrical tube, and coaxial with the tube. Assuming that the pressure gradient is constant throughout the tube, with the exception of a discontinuous change in pressure at the diaphragm, he arrives at the relation

$$F_{eff} = \frac{C}{1 - (A_o/A_t)} F_o, \tag{2.46}$$

where the coefficient $C$ is given in Table 2.7. It is a function of the ratio of the orifice radius, $a_o$, to the tube radius, $a_t$.

TABLE 2.7

**Values of the Factor $C$ in Equation 2.46 for the Effective Conductance of a Circular Orifice**

| $a_o/a_t$ | $C$ | $a_o/a_t$ | $C$ |
|---|---|---|---|
| 0.0 | 1.000 | 0.6 | 1.074 |
| 0.1 | 1.002 | 0.7 | 1.107 |
| 0.2 | 1.007 | 0.8 | 1.152 |
| 0.3 | 1.017 | 0.9 | 1.216 |
| 0.4 | 1.030 | 1.0 | 1.333 |
| 0.5 | 1.049 | | |

Both equations 2.45 and 2.46 contain the factor $1 - (A_o/A_t)$, which gives the right sense to the correction. If $A_o = A_t$, $F_{eff}$ is infinite, that is, there is no impedance to the flow at the point formerly occupied by the diaphragm. As $A_o$ becomes small compared to $A_t$, $F_{eff}$ approaches $F_o$, for in this case the orifice is small compared to the vessel.

If an orifice is located in a diaphragm between two points in a long tube separated by a distance $l$, combining $F_{eff}$ given by 2.46 with the conductances of the segments of the tube on either side (of length $l/2$) by means of equation 2.43 yields

$$F = \frac{1}{\frac{1}{C}\left(1 - \frac{A_o}{A_t}\right) + \frac{3}{8}\frac{A_o}{A_t}\frac{l}{a_t}} F_o \tag{2.47}$$

for the total conductance between the two points. This formula has been investigated experimentally[24] (for $l/a_t = 5.1$ and 0.93) and found to predict the variation of $F$ with $A_o/A_t$ within a few per cent.

A natural extension of these results is to apply the factor $(1 - A_o/A_t)^{-1}$ as an end correction for a tube of area $A_o$ when it is in series with a tube of area $A_t$, that is, a correction for the junction. This will probably give the proper order of magnitude for the correction but should be considered as only approximate.[25]

Molecular flow rates through channels with more complex geometries than those discussed above have been calculated by Davis.[26] He has analyzed the flow through a cylindrical elbow, a cylindrical annulus, a straight pipe with restricted openings, and a straight pipe with restricted openings and a centrally located plate to prevent "line-of-sight" transmission. Results are given in tabular and graphical form of a factor $P$ which is analogous to the Clausing factor $K$ in equation 2.41, that is,

$$F = PA\frac{v_a}{4}.$$

### The Conductance of a Liquid-Air Trap

One case of importance in vacuum-tube technique is the determination of the conductance (or "speed") of a liquid-air trap.

Let us consider a trap, such as that shown in Fig. 2.3, in which $2a_2$ = the inside diameter of the outer cylinder, and $2a_1$ = the outside diameter of the inner cylinder. For the annular space between the two cylinders, of length $l$, equation 2.28 leads to the relation

$$F_c = \frac{2}{3}\frac{v_a}{l}\pi(a_2^2 - a_1^2)(a_2 - a_1), \tag{2.48}$$

that is,

$$F_c = \frac{9.70\pi}{l}(a_2 - a_1)^2(a_2 + a_1)\left(\frac{T}{M}\right)^{1/2} \text{ liters} \cdot \text{sec}^{-1} \tag{2.48a}$$

$$= \frac{97.7}{l}(a_2 - a_1)^2(a_2 + a_1) \text{ liters} \cdot \text{sec}^{-1}, \tag{2.48b}$$

for air at 25° C.

It will be observed that for $a_1 = 0$ the last relation is identical with that given in Table 2.1.

For short tubes an approximate correction may be made by adding the resistance of the orifice to that of the annular space. Hence,

$$\frac{1}{F_{c2}} = \frac{1}{F_c} + \frac{4}{v_a}\frac{1}{\pi(a_2^2 - a_1^2)}$$

$$= \frac{4}{\pi v_a}\frac{1}{(a_2^2 - a_1^2)}\left[1 + \frac{3l}{8(a_2 - a_1)}\right].$$

Assuming that the thickness of the wall is negligible compared to $a_1$, the resistance of the inside cylinder is

$$\frac{1}{F_{c1}} = \frac{1 + \frac{3}{8}(l/a_1)}{a_1^2}\frac{4}{\pi v_a}.$$

Hence, the total resistance is given by

$$\frac{1}{F} = \frac{1}{F_{c2}} + \frac{1}{F_{c1}}$$

$$= \frac{4}{\pi v_a}\left[\frac{1}{a_2^2 - a_1^2} + \frac{3l}{8(a_2 - a_1)(a_2^2 - a_1^2)} + \frac{1}{a_1^2} + \frac{3}{8}\frac{l}{a_1^3}\right]. \tag{2.49}$$

The problem of interest in designing a liquid-air trap is the calculation of the optimum value of the ratio $a_1/a_2$. J. M. Lafferty of the General Electric Research Laboratory has treated the problem in the following manner.

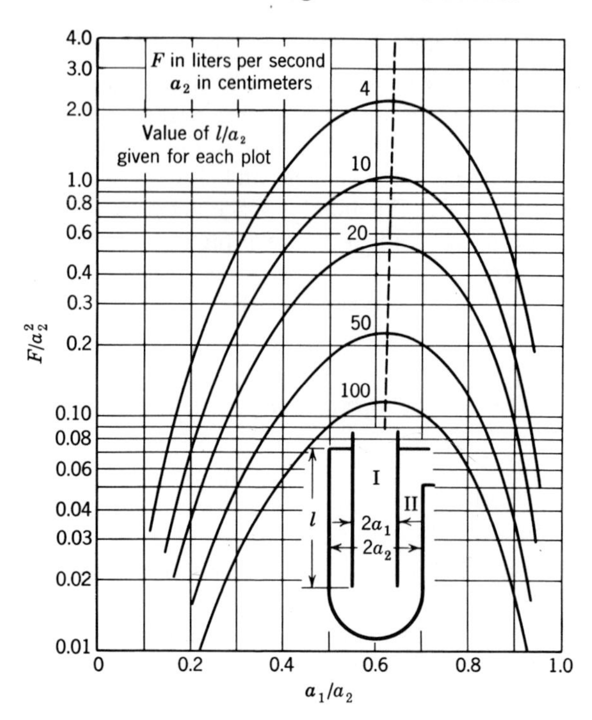


Fig. 2.3. Plots of $F/a_2^2$ versus $a_1/a_2$ for the "molecular" flow of air at 25° C through a liquid-air trap. The ordinates give values of $F/a_2^2$, where $F$ = conductance in liters per second and $a_2$ denotes the inside radius (in centimeters) of the outer tube. The abscissas give values of $a_1/a_2$, where $a_1$ denotes the outside radius (in centimeters) of the inner tube. Each curve corresponds to the indicated value of $l/a_2$, where $l$ denotes the total length (in centimeters) of the trap. For large values of $l/a_2$ the maxima (indicated by the dashed line) occur at the value $a_1 = 0.62a_2$.

Let $X = a_1/a_2$ and let $Y = l/a_2$.

Then equation 2.49 leads to an expression for $F$, the conductance of the trap, of the form

$$F = \pi a_2^2 \cdot \frac{v_a}{4} \cdot f(X, Y),$$

where

$$f(X, Y) = \frac{X^3(1 - X)(1 - X^2)}{X(1 - X) + \frac{3}{8}Y[X^3 + (1 - X)(1 - X^2)]}.$$

For air at 25° C,

$$F = 36.66a_2^2 f(X, Y).$$

In Fig. 2.3 this relation has been used to plot $F$ as a function of $X$ for a series of values of $Y$. From the approximate method used for the calculation of both $F_{c1}$ and $F_{c2}$, it is evident that the derived relation may be applied only for values of $Y$ which are greater than at least 4. The dashed curve shows the values of $X$ for maximum conductance at different values of $Y$. For large values of $Y$, the *maximum value* of $f(X, Y)$ occurs for $X^2 + X = 1$, that is, for $X = 0.618$.

Hence the *optimum conductance* of a liquid-air trap (for air at 25° C) for given values of $a_2$ and $l$ is given by the relation

$$F = (0.618)(0.382)(1.33)(36.66)a_2{}^2/Y$$
$$= 11.53a_2{}^3/l \text{ liters} \cdot \text{sec}^{-1}, \tag{2.50}$$

where $a_1 = 0.618a_2$ and $l > 5a_2$.

To the resistance thus calculated must also be added the resistance of the extension of the inner tube and also that of the side connection. Furthermore, the wall thickness of the inside tube may not be negligible compared with the inner diameter of the tube. As illustrations, the following detailed calculations give the values of $F$ for two different traps. For both, $2a_1$ is the *outside* diameter of the inner tube, $2a_2$ the inside diameter of the outer tube, and $2a$ the inside diameter of the inner tube.

*Case I. Large trap.*

1. For the *annular space*, $a_1 = 1.59$, $a_2 = 2.38$, $l = 21.6$. Hence

$$F_1 = \frac{97.83 \times 0.79 \times 3.97}{21.6/0.79 + 2.67} \text{ liters} \cdot \text{sec}^{-1}$$
$$= 10{,}220 \text{ cm}^3 \cdot \text{sec}^{-1}$$

and

$$\frac{1}{F_1} = 9.783 \cdot 10^{-5} \text{ cm}^{-3} \cdot \text{sec}.$$

2. For the *inside tube*, $a = 1.40$, $l = 24.1$ (assuming that this tube extends 2.5 cm outside the annular space). Therefore $l/a = 17.22$, $K = 0.13$.

$$F_2 = 36.66 \times (1.40)^2 \times 0.13 \text{ liters} \cdot \text{sec}^{-1}$$
$$= 9340 \text{ cm}^3 \cdot \text{sec}^{-1}.$$
$$\frac{1}{F_2} = 10.71 \cdot 10^{-5} \text{ cm}^{-3} \cdot \text{sec}.$$

3. For the *side tube*, $a = 1.40$, $l = 2.54$, $l/a = 1.81$, $K = 0.537$.

$$F_3 = 36.66 \times (1.40)^2 \times 0.537 \text{ liters} \cdot \text{sec}^{-1}$$
$$= 38{,}590 \text{ cm}^3 \cdot \text{sec}^{-1}.$$
$$\frac{1}{F_3} = 2.592 \cdot 10^{-5} \text{ cm}^{-3} \cdot \text{sec}.$$

Hence,

$$\frac{1}{F} = \frac{1}{F_1} + \frac{1}{F_2} + \frac{1}{F_3} = 23.09 \cdot 10^{-5}\ \text{cm}^{-3} \cdot \text{sec},$$

and

$$F = 4331\ \text{cm}^3 \cdot \text{sec}^{-1}.$$

In this calculation it is assumed that the side tube is only 1 in. long, and that the inner tube extends only 1 in. outside the ring seal at the top. If the inner tube extends 20 in. (50.8 cm), then $F_2 = 3682\ \text{cm}^3 \cdot \text{sec}^{-1}$, and $1/F_2 = 27.16 \cdot 10^{-5}$, with the result that

$$\frac{1}{F} = 39.54 \cdot 10^{-5} \quad \text{and} \quad F = 2529\ \text{cm}^3 \cdot \text{sec}^{-1}.$$

*Case II. Small trap.*

1. For *annular space*, $a_1 = 0.95$, $a_2 = 1.35$, $l = 19.05$. Hence

$$F_1 = 1812\ \text{cm}^3 \cdot \text{sec}^{-1}.$$

$$\frac{1}{F_1} = 5.518 \cdot 10^{-4}.$$

2. For the *inside tube*, $a = 0.794$, $l = 71.2$ cm,

$$F_2 = 687.3\ \text{cm}^3 \cdot \text{sec}^{-1}.$$

$$\frac{1}{F_2} = 14.55 \cdot 10^{-4}.$$

3. For the *side tube*, $a = 0.794$, $l = 8$, $K = 0.172$,

$$F_3 = 3968\ \text{cm}^3 \cdot \text{sec}^{-1}.$$

$$\frac{1}{F_3} = 2.520 \cdot 10^{-4}.$$

Hence $1/F = 1/F_1 + 1/F_2 + 1/F_3 = 22.59 \cdot 10^{-4}$, and

$$F = 442.7\ \text{cm}^3 \cdot \text{sec}^{-1}.$$

In this case, also, the major part of the resistance is contributed by the long length of the inside tube. Since $1/F$ varies as $l/a^3$ it is obvious that, whenever it is necessary to use a long length of tube, the diameter should be made as large as practicable.

It should be observed that at the temperature of liquid air ($T = 90°$ K approximately) the conductance for any gas will be $(90/298)^{1/2} = 0.55$ times that for the same gas at 25° C. (See section 1.12.)

## 2.5. FLOW IN THE TRANSITION RANGE

According to equation 2.5, the conductance of a tube in the viscous range is directly proportional to the average pressure, and therefore a plot of $F$ versus $P_a$ should have zero intercept at $P_a = 0$. As the pressure in a tube is gradually reduced, however, it is found that an extrapolation of the conductance measured in the range where $aP_\mu \approx 500$ (the high-pressure end of the transition range) indicates a finite conductance at zero pressure. This additional conductance, which can be included in the equations for viscous conductance as an added constant term, is attributed to the "slipping" of the gas over the walls of the tube; that is, the flow velocity at the walls is not zero. This was mentioned earlier in the discussion of the Poiseuille equation. At pressures below the viscous limit indicated by the inequality stated in the introduction, the slip correction becomes an appreciable contribution to the total conductance.

With further reduction in pressure, the dependence of the conductance on pressure becomes more complex. The flow characteristics begin a progressive change from those of viscous slip flow to those of molecular flow, where the conductance becomes independent of the pressure. The complete transition from viscous to molecular flow takes place over roughly two orders of magnitude change in pressure.

Strictly speaking, the theory of slip is applicable only in the range of pressure where the flow of gas still has viscous characteristics in regions several mean free paths from the walls. There is no quantitative kinetic-theory analysis for the remainder of the transition region at the present time. Nevertheless, the complete transition region is frequently identified with slip for the following reason.

The viscous conductance corrected for slip is[13]

$$F = \frac{\pi a^4}{8\eta l} P_a \left(1 + \frac{4\zeta}{a}\right), \tag{2.51}$$

where $\zeta$ is the coefficient of slip given by equation 1.110; that is

$$\zeta = \frac{2-f}{f} \frac{\eta}{P_a} \left(\frac{\pi R_0 T}{2M}\right)^{1/2}. \tag{2.52}$$

It will be recalled from the discussion in Chapter 1 that $f$ is the transfer ratio of momentum and represents the fraction of the molecular collisions with the walls which result in diffuse scattering of the molecules.[27] Equation 2.51 predicts the slip conductance correctly[28] if $f$ is assigned a constant value between about 0.8 and 1.0. If 2.52 is substituted in 2.51, it can be seen, after some manipulation, that 2.51 can also be written in the form

$$F = F_v + \frac{3\pi}{16} \frac{2-f}{f} F_t. \tag{2.53}$$

Here $F_v$ is the "slip-free" viscous conductance given by equation 2.5 and $F_t$ is the molecular flow conductance of a long tube given by equation 2.29.

This form of the equation evidently shows characteristics of both viscous and molecular ranges. The tendency is therefore to endow it with more generality than is implied in the derivation, and to associate flow in the transition range with slip flow.

Equation 2.53 does not predict the pressure dependence observed experimentally. For, as shown below, the ratio $F/F_t$ determined experimentally for long tubes exhibits a minimum when plotted as a function of pressure, whereas 2.53 indicates a linear dependence. For a description of the flow in the transition range, it is therefore necessary to rely primarily on an empirical relation formulated by Knudsen. As a result of a series of measurements, Knudsen deduced that the conductance of a long cylindrical tube can be written

$$F = F_v + ZF_t, \tag{2.54}$$

where $F_v$ and $F_t$ are defined above, and $Z$ is a function of the mean pressure in the tube,[29] the temperature, the tube radius, and the viscosity of the gas. Knudsen's expression for $Z$ is

$$Z = \frac{1 + \dfrac{2a}{\eta}\left(\dfrac{M}{R_0T}\right)^{1/2} P_a}{1 + 2.47\dfrac{a}{\eta}\left(\dfrac{M}{R_0T}\right)^{1/2} P_a}. \tag{2.55}$$

Using the kinetic-theory relation for the viscosity,

$$\eta = 0.499\rho v_a L,$$

given in section 1.8, equation 2.55 can be written

$$Z = \frac{1 + 2.507(a/L_a)}{1 + 3.095(a/L_a)}. \tag{2.56}$$

Here $L_a$ is the mean free path corresponding to the average pressure $P_a$. When the ratio $a/L_a$ is large, $Z$ approaches the constant value 0.810, and the molecular-flow term in 2.54 becomes a small addition to the viscous term. On the other hand, when $a/L_a$ is small, $Z$ approaches unity, and the conductance $F$ approaches the molecular-flow conductance, since $F_v$ becomes small, being proportional to $P_a$.

Equation 2.54 can be written in a form which may be more convenient for comparison:

$$F = F_t\left(\frac{F_v}{F_t} + Z\right) \tag{2.57}$$

or

$$F = F_t\left(0.1472\frac{a}{L_a} + Z\right). \tag{2.57a}$$

In equation 2.57*a*, $F_v/F_t$ has been reduced to the term in $a/L_a$ by using equation 2.26 for $v_a$, the ideal gas law, and the equation for the viscosity given above. Table 2.8 gives values of $0.1472(a/L_a)$, $Z$, and $F/F_t$, for a range of values of $a/L_a$.

**TABLE 2.8**

**Ratio of Conductance of Cylindrical Tube ($F$) to That for Molecular Flow ($F_t$) as a Function of $a/L_a$**

| $a/L_a$ | $0.1472\ (a/L_a)$ | $Z$ | $F/F_t$ |
|---|---|---|---|
| $10^4$ | 1472.0 | 0.810 | 1472.81 |
| $10^3$ | 147.2 | 0.810 | 148.01 |
| $10^2$ | 14.72 | 0.810 | 15.53 |
| 10 | 1.472 | 0.816 | 2.288 |
| 5 | 0.736 | 0.822 | 1.558 |
| 1 | 0.147 | 0.857 | 1.004 |
| 0.5 | 0.074 | 0.885 | 0.959 |
| 0.323 | 0.047 | 0.905 | 0.952 (min) |
| 0.2 | 0.029 | 0.933 | 0.962 |
| 0.1 | 0.015 | 0.955 | 0.970 |
| 0.05 | 0.007 | 0.974 | 0.981 |
| 0.01 | 0.002 | 0.995 | 0.997 |
| 0 | 0 | 1.000 | 1.000 |

Replacing $L_a$ by $L_1/P_{\mu a}$, where $P_{\mu a}$ is the mean pressure in the tube, in microns, and $L_1$ is the mean free path at 1 micron, the data in the table show that, for values of $aP_{\mu a}$ greater than $100L_1$, the flow is almost completely viscous, whereas for values of $aP_{\mu a}$ less than $L_1$, the flow is over 95 per cent molecular. We also note that $F/F_t$ has a minimum value at $a/L_a = aP_{\mu a}/L_1 = 0.323$. Figure 2.4 shows a semilog plot of $F/F_t$, as a function of $aP_{\mu a}/L_1$.

For instance, for air at 25° C, $L_1 = 5.09$ cm. Hence, the Poiseuille law is applicable for values of the average pressure above $509/a$ microns, while the relations for molecular flow are applicable in the range below $5.09/a$ microns. This is the justification for the inequalities stated in the introduction to this chapter. Furthermore, the minimum value of $F$ occurs at $P_{\mu a} = 1.644/a$ microns, and it should be carefully noted that, in these relations, $a$ = radius of tube in centimeters.

As an illustration of the application of the data in Table 2.8, it is of special interest to calculate the rate of leak of air through a "pinhole" in an evacuated device at atmospheric pressure and 25° C.

In this case, $P_{\mu 2} = 7.6 \cdot 10^5$, $P_{\mu 1} = 0$, and $P_{\mu a} = 3.8 \cdot 10^5$. For air at 25° C, $L_1 = 5.09$ cm.

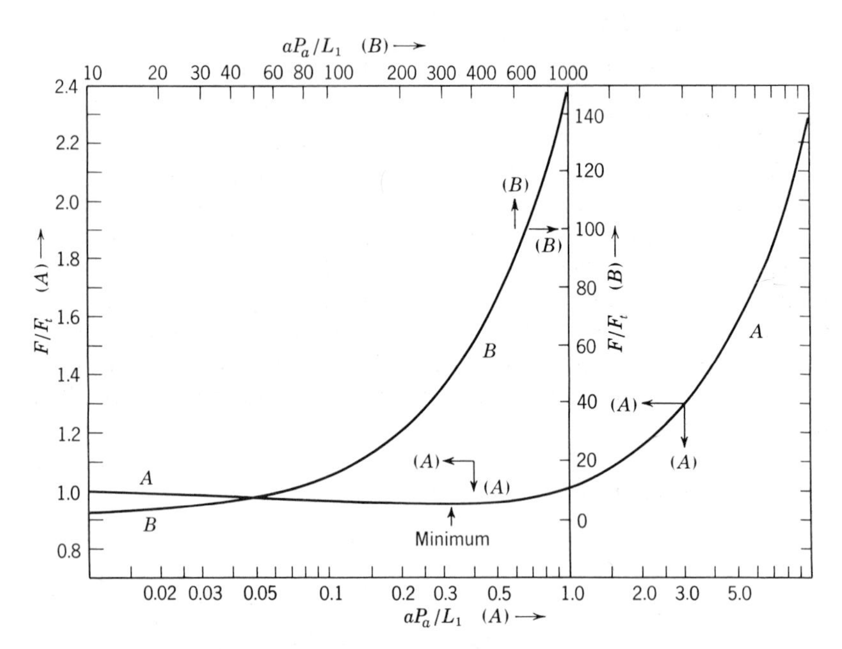


Fig. 2.4. Plot of ratio $F/F_t$ versus $aP_a/L_1$ for cylindrical tube, where $P_a$ = average pressure (microns), $a$ = radius of tube (centimeters), and $L_1$ = mean free path (centimeters) at 1 micron. $F$ = conductance for $P_a$, $F_t$ = conductance for "molecular" flow. For curve $A$ use scales ($A$), and for curve $B$, scales ($B$), as indicated by directions of arrows.

Now let us assume two sizes of holes:

1. $l = 0.1$ cm, $a = 5 \cdot 10^{-5}$ cm (0.04-mil diameter),
2. $l = 0.1$ cm, $a = 5 \cdot 10^{-4}$ cm (0.4-mil diameter).

Hence we derive the following results:

1. $a/L_a = 3.73$, $F/F_t = 1.4$, and $Q = FP_{\mu 2} = 3.75$ micron · liters per *hour*.
2. $a/L_a = 37.3$, $F/F_t = 6.3$, and $Q = 4.48$ micron · liters per *second*.

As will be observed from the values of $F/F_t$, the ratio of viscous to molecular flow is greater for the larger-diameter capillary. Also, case 1 serves to emphasize the fact that, even for a very fine hole, the rate of leak may be quite considerable.

For air at 25° C, equation 2.57 can be written in the form

$$F = F_t\left(0.0290aP_{\mu a} + \frac{1 + 0.493aP_{\mu a}}{1 + 0.608aP_{\mu a}}\right) \quad \text{liters} \cdot \text{sec}^{-1}, \qquad (2.58)$$

where $F_t$ is expressed in liters per second, $P_\mu$ in microns of Hg, and $a$ in centimeters.

Also, the rate of flow,

$$Q_{\mu l} = F(P_{\mu 2} - P_{\mu 1}) \text{ micron} \cdot \text{liters} \cdot \text{sec}^{-1}. \tag{2.59}$$

From equations 2.58 and 2.59 it follows that, in the range of molecular flow, $Q$ varies linearly with the value of $(P_{\mu 2} - P_{\mu 1})$, whereas, in the range of viscous flow, $Q$ varies with the value of $(P_{\mu 2}{}^2 - P_{\mu 1}{}^2)$.[30]

Figure 2.5 shows a plot of log $Q$ versus log $P_{\mu 2}$ for a capillary 100 cm long with 0.51-cm radius. At low pressures, the slope is unity. With increase in pressure the slope increases, and at the higher pressures it becomes 2. The plot shows the transition region of pressures in which the rate of flow changes character. At very low pressures (less than 1 micron of Hg), $F_t = 0.163$ liter · sec$^{-1}$. The value $Z = 0.9$ in the parentheses is used as an average over the range of pressures in the transition region.

In equation 2.57 and the data of Table 2.8, $F$ is expressed in terms of $F_t$. An equivalent expression in terms of $F_v$ would be

$$F = F_v\left(1 + Z\frac{F_t}{F_v}\right) \tag{2.60}$$

$$= F_v\left[1 + 6.793\frac{L_a}{a}\left(\frac{1 + 2.507(a/L_a)}{1 + 3.095(a/L_a)}\right)\right]. \tag{2.60$a$}$$

Brown and his associates[31] have chosen to write equation 2.53 for slip flow in the form

$$F = F_v\left(1 + \frac{3\pi}{16}\frac{2-f}{f}\frac{F_t}{F_v}\right), \tag{2.61}$$

which becomes

$$F = F_v\left[1 + 4\left(\frac{2}{f} - 1\right)\frac{L_a}{a}\right], \tag{2.61$a$}$$

when $F_t/F_v$ is reduced to the ratio $L_a/a$. Comparing 2.61*a* with 2.60*a*, it is found that

$$4\left(\frac{2}{f} - 1\right) = 6.793\frac{1 + 2.507(a/L_a)}{1 + 3.095(a/L_a)},$$

where now $f$ is to be regarded as an empirical function of $a/L_a$. For small $a/L_a$ (region of molecular flow), this leads to the value $f = 0.74$. Brown and his associates compared equation 2.61*a* with values of $F$ actually observed with glass capillaries by previous investigators, and found the best fit with $f = 0.77$. For higher pressures, a similar comparison with experiment yielded $f = 0.84$, which is in agreement with the high-pressure limit obtained from the relation above. Data obtained on copper pipes in the same pressure range also agreed with $f = 0.84$. However, deviations were noted in the flow of air through iron pipes, $F$ being as much as 30 per cent below the data on copper pipes. Knudsen's formula was found to

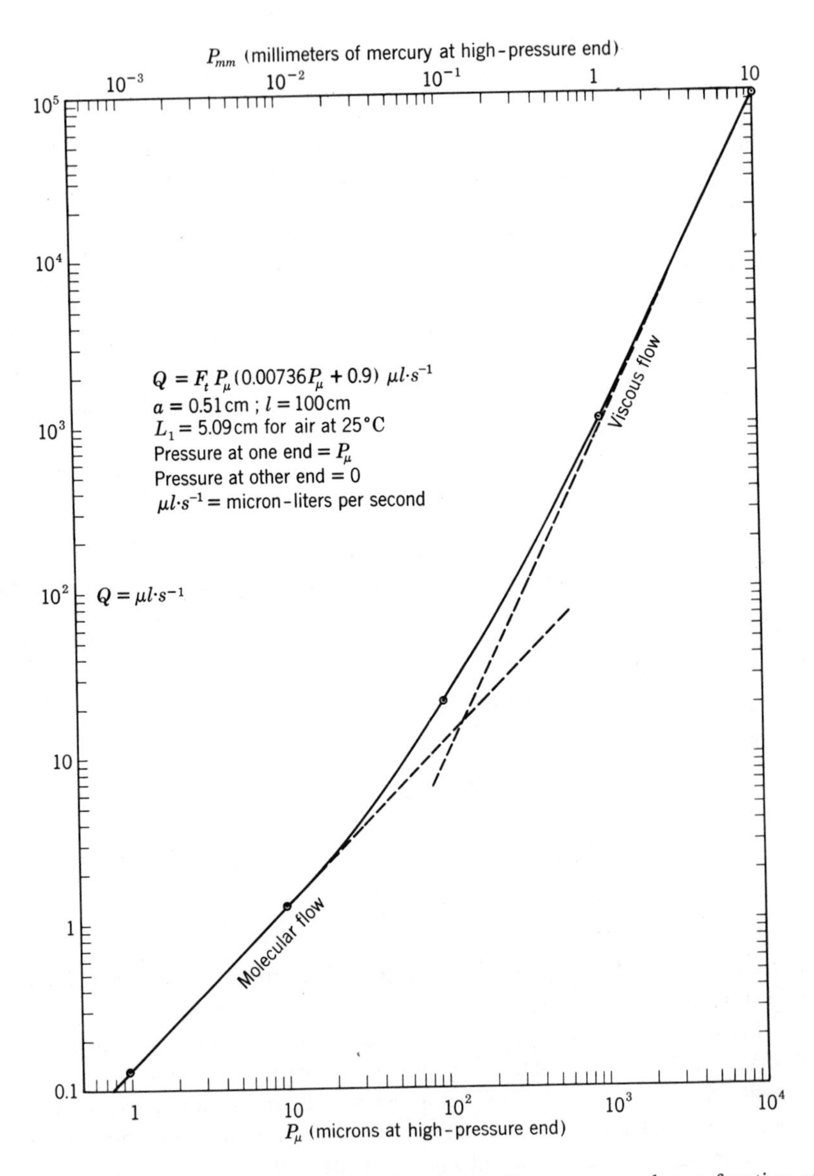


Fig. 2.5. Rate of flow of air at 25° C in micron · liters per second as a function of pressure $P_\mu$ at inlet and $P = 0$ at other end.

predict $F$ within 5 per cent over the entire transition range, with the exception of the flow through iron pipes.

The conductance minimum observed by Knudsen, appearing at $a/L_a = 0.323$ in Table 2.8, may not always occur at this value of $a/L_a$. Moreover it may be absent entirely. Pollard and Present[32] have offered a qualitative explanation for the minimum and have shown that it should depend on the length of the tube. When the pressure is sufficiently low so that the mean free path is much larger than both the length and the radius of the tube, the flow of gas is controlled solely by wall collisions. As the pressure is raised, the mean free path becomes smaller, and the increased number of intermolecular collisions tends to impede the transport of a molecule through the length of the tube. At the same time, the effect of increasing the number of intermolecular collisions is also to initiate a drift or flow velocity which will tend to increase the flow. Pollard and Present reason that, since the development of a drift velocity will be approximately proportional to $a/L_a$, when the mean free path is less than the length of the tube but still greater than the radius, the decrease in the flow caused by the shortened mean free path will outweigh the increase in the flow due to the small initiation of a drift velocity. Therefore, the conductance of a long tube should decrease from the molecular-flow value with initial increase in pressure. From this point of view, when $l$ is comparable to $a$, the two effects may counterbalance each other, and a minimum in conductance would then not be observed. The absence of a minimum in the "conductance" of porous material[28] at low pressures lends credence to this argument.

## 2.6. RATE OF EXHAUST THROUGH A TUBE OR ORIFICE

In vacuum work the problem often occurs of determining the effect on rate of exhaust of a given conductance $F$ between the volume, $V$, that is being exhausted, and the pump. It will be assumed in the following discussion that the speed of the pumps is sufficiently high to maintain the pressure $P = 0$ at the pump inlet. That is, it is assumed that the whole drop in pressure occurs in the conductance $F$.

Let $P$ designate the pressure in the volume $V$ at any instant during the exhaust. Then the rate of exhaust, that is, the volume of gas removed per second, is given by the relation

$$-V\frac{dP}{dt} = FP. \qquad (2.62)$$

For the whole range of pressures, from atmospheric to very low, $F$ varies with $P$ in the manner shown in equation 2.54. For the purpose of solving the differential equation given above, the validity of the approximate relation, derived from equation 2.57*a*,

$$F = 0.0736 \frac{a}{c} F_t P + F_t, \tag{2.63}$$

is assumed. Here $P = 2P_a$, and $c$ = mean free path at 1 microbar.

As is evident from the data in Table 2.8, the error involved in using the assumption $Z = 1$ does not exceed 10 per cent in the range $10 > a/L_a > 0.5$ and is less than this in the higher-pressure and very low-pressure regions.

Introducing equation 2.63 into 2.62, the latter assumes the form

$$-\frac{dP}{dt} = bP^2 + kP, \tag{2.64}$$

where

$$b = 0.0736 \frac{aF_t}{cV} \tag{2.65}$$

and

$$k = \frac{F_t}{V}. \tag{2.66}$$

Hence

$$-\frac{dP}{P} + \frac{dP}{(P + k/b)} = k\,dt.$$

Integrating equation 2.66, we obtain the result

$$\ln\left(\frac{P_1}{P_2}\right) - \ln\left(\frac{P_1 + k/b}{P_2 + k/b}\right) = k(t_2 - t_1), \tag{2.67}$$

where $P_1$ = pressure at $t_1$, $P_2$ = pressure at $t_2$, and $t_2$ is greater than $t_1$, so that $P_1 > P_2$. The symbol "ln" is used to indicate logarithms to base $\epsilon$.

Converting to ordinary logs, equation 2.67 becomes

$$\log\left(\frac{P_1}{P_2}\right) - \log\left(\frac{P_1 + k/b}{P_2 + k/b}\right) = \frac{k}{2.303}(t_2 - t_1). \tag{2.68}$$

There are two limiting cases for which equation 2.67 or 2.68 assumes a simpler form.

*Case I. $P_2$ is very much greater than $k/b$.* This is valid, in general, for $P_2 > 100$ microbars (75 microns).

Since

$$\ln\left(P + \frac{k}{b}\right) - \ln P = \ln\left(1 + \frac{k}{bP}\right) = \frac{k}{bP},$$

equation 2.67 becomes

$$\frac{1}{b}\left(\frac{1}{P_2} - \frac{1}{P_1}\right) = t_2 - t_1. \tag{2.69}$$

That is, $1/P$ varies linearly with $t$.

*Case II. P is very much less than k/b.* This is valid in general in the range $P_1 < 1$ microbar. Equation 2.68 then becomes

$$\log\left(\frac{P_1}{P_2}\right) = \frac{k}{2.303}(t_2 - t_1)$$

or

$$t_2 - t_1 = \frac{2.303}{k}\log\left(\frac{P_1}{P_2}\right) \tag{2.70}$$

$$= \frac{2.303V}{F_t}\log\left(\frac{P_1}{P_2}\right). \tag{2.71}$$

For $P_1/P_2 = 10$,

$$t_2 - t_1 = \Delta t_0 = \frac{2.303V}{F_t}. \tag{2.72}$$

Let $\Delta t_0$ designate the interval of time required to reduce the pressure in $V$ to one-tenth its instantaneous value, as the time constant of the conductance $F = F_t$ for the volume $V$. Evidently $\Delta t_0$ varies linearly with $V$.

Although it is customary in such an equation as 2.70 to designate, as the time constant, that value of $(t_2 - t_1)$ which reduces the pressure to $1/\epsilon$th (36.79 per cent) of its value at any given instant, the definition given in equation 2.72 is evidently much more convenient for practical applications in high-vacuum technique. It is obvious that, if $P$ is plotted against $t$ on semilog paper, $\Delta t_0$ may be obtained directly from the straight-line plot. Furthermore $2\Delta t_0$ is the period required to reduce the pressure to 1 per cent, $3\Delta t_0$ is the period to reduce the pressure to 0.1 per cent of its value at any instant, and so forth.

For the range of pressures intermediate between cases I and II, equation 2.68 must be used.

Two illustrations of the application of the above relations will now be considered. In both cases, the calculation will be made for air at 25° C, for which $c = 6.78$ cm at 1 microbar.

*Case I.* $V = 5000\ \text{cm}^3$, $l = 100$ cm, $a = 0.5$ cm. For this tube, as shown in section 2.4, $F_t = 122.2\ \text{cm}^3 \cdot \text{sec}^{-1}$. Hence

$$b = 1.326 \cdot 10^{-4}$$

$$k = 2.444 \cdot 10^{-2}$$

$$\frac{2.303}{k} = 94.2\ \text{sec} = \Delta t_0$$

$$\frac{k}{b} = 184.2\ \text{microbars} = 138\ \text{microns}.$$

Table 2.9 gives values of $t_2 - t_1$ and of the total time in seconds for the reduction of pressure in the system. The values of $P_1$ and $P_2$ are given in microbars. It will be observed that the interval of time required to reduce the pressure from $10^6$ microbars (= 750 mm) to 1 microbar (= 0.75 micron) is just slightly greater than that required to reduce the pressure from 1 to 0.01 microbar.

*Case II.* $V = 100$ liters, $l = 20$ cm, $a = 10$ cm. Hence $l/a = 2$, $K = 0.5136$, and $F_t = KF_o = 1883$ liters · sec$^{-1}$.

**TABLE 2.9**

**Rate of Exhaust of Air through a Tube for Which $l$ = 100 cm and $a$ = 0.5 cm**

| $P_1$ | $P_2$ | $t_2 - t_1$ | Total Time (sec) |
|---|---|---|---|
| $10^6$ | $10^5$ | 0.0678 | ... |
| $10^5$ | $10^4$ | 0.678 | 0.746 |
| $10^4$ | $10^3$ | 6.16 | 6.91 |
| $10^3$ | $10^2$ | 35.8 | 42.7 |
| $10^2$ | 10 | 78.6 | 121.3 |
| 10 | 1 | 92.3 | 213.6 |
| 1 | 0.1 | 94.2 | 307.8 |
| 0.1 | 0.01 | 94.2 | 402.0 |

In this case, the interval of time required to reduce the pressure to that at which molecular flow occurs is so extremely low that the only value of interest is that of $\Delta t_o$. From the preceding relations it follows that

$$\Delta t_0 = \frac{2.303V}{KF_o} = \frac{2.303 \times 100}{1883}$$

$$= 0.123 \text{ sec.}$$

A comparison of equations 2.69 and 2.72 shows that the interval of time required to reduce the pressure in a system from atmospheric to that at which molecular flow begins is relatively short when compared with the interval required for further reduction of the pressure to the order of $10^{-3}$ microbar (or micron). This follows from the fact that, in the viscous-flow range, the conductance varies as $a^4$, whereas in the molecular-flow range it varies as $a^3$. Hence, for large values of $a$, the first period is negligible in comparison with the second. This may be illustrated quantitatively by the following example.

From equation 2.69 it follows that, for $P_2 = 0.1P_1$,

$$t_2 - t_1 = \frac{9}{bP_1} = \frac{122.3cV}{P_1 a F_t}.$$

Hence it follows from equation 2.72 that

$$\frac{t_2 - t_1}{\Delta t_0} = \frac{53.1c}{P_1 a}. \tag{2.73}$$

For air at 25° C, $c = 6.78$ cm at 1 microbar. Let us assume $P_1 = 10^4$ microbars. Then the ratio

$$\frac{t_2 - t_1}{\Delta t_0} = \frac{3.60 \cdot 10^{-2}}{a}.$$

For $a = 0.5$, the ratio has the value 0.072, which is in agreement with the value 6.78/94.2 shown in Table 2.9; for $a = 10$, the ratio has the value 0.0036. That is, for the reduction of the pressure to one-tenth its value at any instant, the time required in the viscous-flow range of pressures is 0.36 per cent of that required in the molecular-flow range.

In section 2.4, Table 2.5 gives values of $F_t$ for a range of sizes of tubes. The last column in this table gives values of $2303/F_t$, that is, of $\Delta t_0$ for $V = 1000$ cm³. ($F_t$ and $V$ should be given in the same units of volume.)

Since the total impedance of two or more impedances in series is given by the relation

$$\frac{1}{F} = \frac{1}{F_1} + \frac{1}{F_2} + \cdots, \tag{2.74}$$

it follows that $\Delta t_0 =$ the time constant for the total impedance

$$= (\Delta t_0)_1 + (\Delta t_0)_2 + \cdots. \tag{2.75}$$

That is, the time constant of a series of tubes is equal to the sum of the time constants for each element of the series. For instance, the time constant of a tube in series with a stopcock and another tube is equal to the sum of the time constants for each of the sections.

It is also important to observe that, whereas $\Delta t_0$ is a constant which has the same value independently of the unit of pressure (whether microbars, microns, or millimeters of mercury), the value of $b$ depends upon the unit of pressure. Let $b_{mm}$ denote the value for $P_{mm}$, and $b_{\mu b}$ the value for $P_{\mu b}$. Then it follows that

$$b_{mm} = 133 b_{\mu b}, \tag{2.76}$$

$$\frac{k}{b_{mm}} = \frac{1}{1333}\frac{k}{b_{\mu b}}. \tag{2.77}$$

It should be noted that in deducing equation 2.70 it was assumed that the values of $P_1$ and $P_2$ are large compared with that of the ultimate pressure $P_s$ observed at the low-pressure end of the tube (of conductance $F_t$). If, however, this condition is not satisfied, then $P_1$ and $P_2$ in equation 2.70 should be replaced by $P_1 - P_s$ and $P_2 - P_s$, respectively. Obviously $P_s$ should be expressed in the same units as $P_1$ and $P_2$.

## REFERENCES AND NOTES

1. H. Lamb, *Hydrodynamics*, Dover Publications, New York, 1st American edition, 1945, Chapter 11.

2. H. Schlichting, *Boundary Layer Theory*, translated by J. Kestin, McGraw-Hill Book Company, New York, 1955, Chapter 1.

3. *American Institute of Physics Handbook*, McGraw-Hill Book Company, New York, 1957, p. 2–216.

4. S. Goldstein, *Modern Developments in Fluid Dynamics*, The Clarendon Press, Oxford, 1938, Volume 1, Chapter 7.

5. A. H. Shapiro, R. Siegel, and S. J. Kline, *Proceedings of the 2nd U.S. National Congress of Applied Mechanics*, American Society of Mechanical Engineers, New York, 1954, pp. 733–741.

6. H. L. Langhaar, *J. Appl. Mechanics*, **9**, A-55 (1942).

7. Ref. 3, section 2y.

8. For an extended discussion of flow in ducts at Mach numbers exceeding the limit (2.8), the reader is referred to Chapter 6 of the text by A. H. Shapiro, *The Dynamics and Thermodynamics of Compressible Fluid Flow*, The Ronald Press Company, New York, 1953.

9. F. J. Bayley, *An Introduction to Fluid Dynamics*, Interscience Publishers, New York, 1958, Chapter 7.

10. See A. Guthrie and R. K. Wakerling, *Vacuum Equipment and Techniques*, McGraw-Hill Book Company, New York, 1949, Chapter 1.

11. M. Knudsen, *Ann. Physik*, **28**, 75, 999 (1909); **35**, 389 (1911).

12. L. B. Loeb, *The Kinetic Theory of Gases*, McGraw-Hill Book Company, New York, 2nd edition, 1934, Chapter 7.

13. E. H. Kennard, *Kinetic Theory of Gases*, McGraw-Hill Book Company, New York, 1938, Chapter 8.

14. R. D. Present, *Kinetic Theory of Gases*, McGraw-Hill Book Company, New York, 1958, Chapter 4.

15. E. C. DeMarcus, *U.S. Atomic Energy Comm. Report K*-1302, Parts I and II (1956); Parts III and IV (1957).

16. That is, one for which the conductance of the tube is small compared with that of the ends. This is discussed in more detail in subsequent remarks.

17. J. Yarwood, *High Vacuum Techniques*, John Wiley & Sons, New York, 2nd edition, 1945.

18. G. Hertz, *Z. Physik*, **79**, 108 (1932); H. Harmsen, *Z. Physik*, **82**, 589 (1933); H. Harmsen, G. Hertz, and W. Schutze, *Z. Physik*, **90**, 703 (1934).

19. A. Farkas, *Light and Heavy Hydrogen*, Cambridge University Press, London, 1935, p. 120.

20. That is, (1) small in dimension compared to the mean free path, as discussed here, and (2) small in area compared to the area of the vessels, as discussed on p. 98.

21. S. Dushman, "The Production and Measurement of High Vacua," *Gen. Elec. Rev.*, 1922.

22. P. Clausing, *Ann. Physik*, **12,** 961 (1932).

23. See also Loeb (Ref. 12), and I. Estermann, *Revs. Modern Phys.*, **18,** 301 (1946), for more detailed data.

24. A. J. Bureau, L. J. Laslett, and J. M. Keller, *Rev. Sci. Instr.*, **23,** 683 (1952).

25. B. B. Dayton, *American Vacuum Society Transactions*, Pergamon Press, New York, 1958.

26. D. H. Davis, *J. Appl. Phys.*, **31,** 1169 (1960).

27. It should be noted that all the formulas in section 2.4 are based on diffuse surface scattering.

28. P. C. Carman, *Flow of Gases through Porous Media*, Academic Press, New York, 1956, Chapter 3.

29. $Z$ here is not a flow impedance. In this section it is used only to designate the function defined by equation 2.55.

30. Since $P_{\mu a} = \frac{1}{2}(P_{\mu 2} + P_{\mu 1})$, $P_{\mu a}(P_{\mu 2} - P_{\mu 1}) = \frac{1}{2}(P_{\mu 2}^2 - P_{\mu 1}^2)$.

31. G. P. Brown, A. DiNardo, G. K. Cheng, and T. K. Sherwood, *J. Appl. Phys.*, **17**, 802 (1946).

32. W. G. Pollard and R. D. Present, *Phys. Rev.*, **73,** 762 (1948).

# 3 Vacuum pumps

Revised by Peter Cannon

## 3.1. HISTORICAL REVIEW

Before the advent of the carbon-filament lamp (1879), interest in the properties of gases at low pressures was manifested by relatively few investigators. Hittorf in Germany and Crookes in England made important observations on electric discharges in gases at pressures ranging as low as a fraction of a millimeter of mercury, and the Geissler discharge tube was a familiar showpiece in lectures on physics. To obtain the pressures required for such electric discharge tubes and other experiments at low pressures, piston pumps were generally used, which reduced the pressure to about 0.25 mm of mercury. For still lower pressures recourse was had to the hand-operated Toepler pump.

With the introduction of incandescent-filament lamps and the consequent need for vacuum pumping equipment, rotary oil pumps were adopted, by which vacua of the order of $10^{-1}$ to $10^{-2}$ mm could be obtained. These pumps, towards the development of which W. Gaede made important contributions, exhausted into a "rough" vacuum of about 1–2 cm of mercury which, in turn, was obtained by means of a large rotary oil pump, operating against atmospheric pressure. Incidentally, it should be mentioned that, in order to obtain the much lower pressures required for efficient operation and long life of lamps, "getters" were applied to the filaments. After seal-off these materials were flashed off the filament and "cleaned up" the residual gases.

In 1905 Gaede introduced his rotary mercury pump, and a little later the rotary oil pump was applied widely in the manufacture of incandescent lamps, as mentioned above. Although with the mercury pump, backed up by the oil pump, it was possible to attain pressures as low as $10^{-1}$ micron, the speed obtained was quite low. A radically new departure in the design of vacuum pumps was represented by Gaede's invention, in 1913, of the "molecular" pump, by which much lower pressures could be obtained with much higher speeds of exhaust. In fact, this type of pump was used for 2 or 3 years in the Research Laboratory of the General Electric Company in the commercial production of Coolidge X-ray tubes.

A still more important advance in vacuum technique occurred in 1915 when Gaede published an account of his "diffusion" pump, which was the direct incentive for the Langmuir "condensation" pump in 1916. By means of these mercury-vapor pumps it was possible to obtain pressures as low as $10^{-4}$ micron and thus meet the demands of the rapidly expanding radio and communication techniques for electronic devices, which, as well known, require extremely low pressures for efficient operation. In this manner vacuum technology, which had hitherto found commercial application only in the incandescent-filament lamp industry, now entered into a completely new field which, since about 1920, has developed into one of importance, not only for communication and broadcasting, but also for industrial control operations, television, radar, and many other applications.

In 1928, C. R. Burch, of the Metropolitan-Vickers Electrical Company of England, found that certain high-boiling petroleum derivatives could be used effectively in condensation pumps to replace the mercury. This led K. C. D. Hickman and his associates at the Eastman Kodak Company Research Laboratories (and subsequently in the laboratory of the Distillation Products, Inc.) to investigate the practical possibilities of synthetic phthalates and sebacates for use in vapor pumps. Since these compounds and certain petroleum distillation products possess very low vapor pressures (less than about $10^{-6}$ mm) at room temperatures, it is possible to eliminate a cooling trap between the system to be exhausted and the pump and thus utilize pumps of very high speeds.

Such pumps have been employed by the Distillation Products, Inc., for "molecular" distillation and the fractionation of natural organic oils for the isolation of vitamins. Other applications, such as the vacuum-furnace distillation of metals and the dehydration of penicillin and plasma, have been developed extensively by the National Research Corporation.

As a result there has been developed, since about 1940, a new industry —*vacuum technology*. The production of vacua of the order of $10^{-3}$ micron and even lower has been transformed from a laboratory curiosity to an industrial operation carried out on a scale that would at one time, have been considered utterly fantastic. The vacuum engineer and vacuum technologist have taken their places in industrial activities along with the engineers and technologists trained in the older fields. Two chapters,

3 and 4, will deal with the most essential element in this new industry—the vacuum pump and the technique associated with its use.

## 3.2. GENERAL REMARKS ON THE OPERATION OF PUMPS

All vacuum pumps work in one of two ways. To create a vacuum, the total momentum of the gas in the evacuable space must be reduced. This is achieved either by providing sufficient momentum transfer to the gas at some point in the system so that it will be forced to travel out of the system through some kind of no-return path, or by removing the momentum of the gas by chemical reaction or condensation. In the former case the momentum transfer is accomplished by a working substance (for example, a sliding piston or a high-velocity fluid stream), and in the second case either by the controllable presence of a species which will react with the gas to give a condensed phase (for example, a reactive metal vapor) or by the provision of access to a heat sink (for example, a refrigerated trap).

In considering the various designs of working systems obtainable by extension of these principles, it is necessary to define certain parameters which will afford useful comparisons between the particular classes of design. The following defined parameters are frequently used in this connection, particularly with respect to mechanical pumps.

### Exhaust Pressure

This is the upper pressure limit of the range in which the device is operable. Since pumps based on mechanical momentum transfer achieve only a finite compression ratio, it is sometimes necessary to use compound pumping systems to bring the exhaust up to the pressure at the discharge port. The exhaust pressure of a pump cannot of course be less than the pressure in the next higher pressure portion of the system.

### Attainable or Ultimate Vacuum

This is the lower pressure limit of the operating range of the device in a closed system. For mechanical pumps the limit is usually set by leakage back through the pump, while for vapor-stream pumps there is no theoretical limit to the attainable vacuum.

### Speed of the Pump

The speed of a pump is defined by $S$ in the relation

$$-\frac{dP}{dt} = \frac{S}{C}(P - P_s), \tag{3.1}$$

where $P_s$ is the attainable vacuum and $C$ the volume to be exhausted.

$S$ has the units of conductance and is quoted frequently in liters per second. It is important to note that the usual values quoted are liters per second at a given pressure: thus 1 liter $\cdot$ sec$^{-1}$ at 1 mm pressure is an actual throughput of 1000/760 standard cm$^3$ $\cdot$ sec$^{-1}$.

When $S$ is independent of $P$, a plot of log $(P - P_s)$ versus $t$ gives a line of slope

$$\frac{\Delta \log (P - P_s)}{\Delta t} = \frac{S}{2.303C}. \tag{3.2}$$

We may also define $E = S(1 - P_s/P)$, so that

$$-\frac{dP}{dt} = \frac{EP}{C} \tag{3.3}$$

and $E$ is the actual speed of exhaust at any instant. Thus $E$ varies from values approximately equal to $S$ at large values of $P$ to zero at $P = P_s$. The value of $S$ in equation 3.1 then drops with time to a zero value when the attainable vacuum has been reached. We should note that there is no theoretical limit to $P_s$ for a vapor-stream pump.

The observed speed of exhaust depends not only on these factors but also on the conductance of the exhausted system. Let $F$ equal this conductance and assume $P_s = 0$: let $P$ and $P_p$ represent the pressures in the exhausted system (volume $C$) and at the pump entrance, respectively. Then $Q$, the rate of flow in the system at any instant, is given by $Q = F(P - P_p) = S_pP_p = SP$, where $S$ is the observed speed of the pump. Then

$$\frac{1}{S} = \frac{1}{S_p} + \frac{1}{F}, \tag{3.4}$$

which signifies that $S$ and $S_p$ are also to be regarded as conductances.

## 3.3. TYPES OF MECHANICAL PUMPS

The early forms of exhaust pumps were of the piston type. Since they have been superseded in modern practice, and since they are described in many elementary textbooks,[1] no detailed description of them need be made here.

The use of a water-jet suction pump is a familiar practice in laboratory work, especially in filtering operations. The *Sprengel* mercury pump operates on the same principle, and Kahlbaum described a modification capable of exhausting to 0.003 micron.[2]

### Geissler-Toepler Pump[3]

The principle of this pump is fundamentally the same as that applied by Torricelli in his famous experiment. In this type (Fig. 3.1) mercury

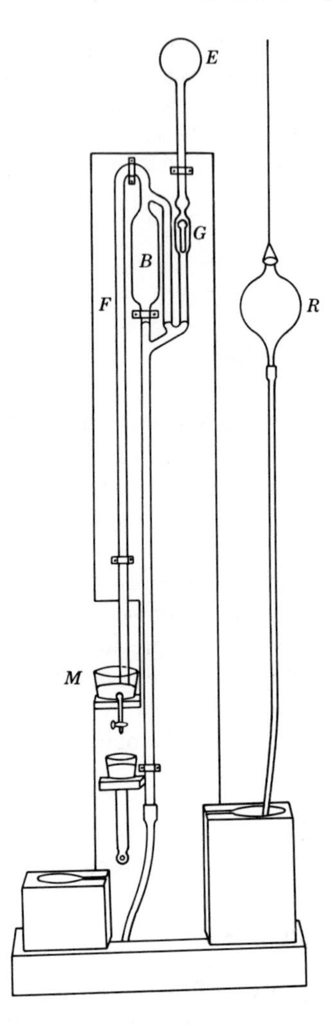


Fig. 3.1. Toepler pump.

forces the piston and also opens and closes certain ports, so that no valves are needed except one rough glass valve $G$ to prevent the mercury from entering the vessel $E$ which is being exhausted. The essential parts of the pump are made of glass, and the air from $E$ is exhausted by alternately raising and lowering the mercury reservoir $R$, which is connected to the tube of barometric length below $B$. At each upward "stroke," the gas in $B$ is closed from $E$ and forced through the tube $F$ into the atmosphere at $M$. Then, on the downward stroke, the pressure in $E$ is lowered by expansion of the gas into $B$. Bessel-Hagen[4] has described a modified form of Toepler pump with which he claims to have obtained pressures of residual gas as low as 0.012 micron.[5] Both the Sprengel and Toepler pumps have rendered very valuable service in high-vacuum investigations, and there is no doubt that with care it is possible by their use to obtain pressures as low as 0.02–0.01 micron.

The great disadvantages of these pumps are, however, twofold. First, they require constant personal attention during the exhaust; and, second, the speed of exhaust is extremely slow, since it depends upon the rate at which the mercury can be raised and lowered alternately. It is of interest to note in this connection the results obtained by Scheel and Heuse[6] in their investigation of the degree of vacuum attainable with different types of pump. They used a 6-liter bulb and measured the speed of exhaust by means of a very sensitive McLeod gauge. In the experiments with a Toepler pump, each stroke actually required 2 min, and 2 more min were allowed between strokes for equalization of pressure. Table 3.1 shows the pressures at the end of different intervals of time.

**TABLE 3.1**

**Speed of Exhaust with Toepler Pump**

| $t$ (min) | $P_{mm}$ | $E = \frac{2.3V}{60t} \log \frac{P_1}{P_2}$ |
|---|---|---|
| 0 | 0.0645 | 0.40 |
| 2 | 0.0399 | 0.38 |
| 24 | 0.0254 | |
| 48 | 0.0107 | 0.35 |
| 60 | 0.00709 | |
| 108 | 0.00141 | 0.35 |
| 120 | 0.00093 | |
| 180 | 0.00024 | 0.39 |
| 192 | 0.00015 | |
| 240 | 0.000053 | 0.28 |
| 252 | 0.000038 | |
| 264 | 0.000032 | 0.06 |
| 300 | 0.000025 | |

The last column gives the calculated speed of exhaust. Compared with the speed of even 100 $cm^3 \cdot sec^{-1}$ obtained by a Gaede rotary mercury or an ordinary oil pump, the speeds in Table 3.1 are manifestly very low. Considering, furthermore, that, in the case where gas is continually evolved from the walls, the minimum attainable pressure is given by the ratio $S/Q$, where $Q$ denotes the rate of gas evolution, it is seen that in actual practice it would be very difficult to obtain pressures below about 0.01 micron by means of a Toepler pump.

Similar results were obtained by Scheel and Heuse in investigating the rate of exhaust of a 6-liter bulb by means of a Sprengel pump (improved by L. Zehnder[7]).

## Gaede Rotary Mercury Pump

Obviously, exhausting even a relatively small volume by means of a Toepler pump is an extremely tedious operation. The invention by Kaufmann[8] of a rotary mercury pump in 1905 was therefore a welcome development in this field. However, it was very soon superseded by a rotary mercury pump, designed by W. Gaede[9] at about the same time, which was used extensively in the commercial exhaust of incandescent lamps and of Roentgen tubes. The pump consists of an iron casing (with glass front) partially filled with mercury, in which a porcelain drum is made to rotate. A *rough pump* producing a vacuum of 10–20 mm is used as *fore pump*. Figure 3.2 shows a vertical section of the pump, and Fig. 3.3 a

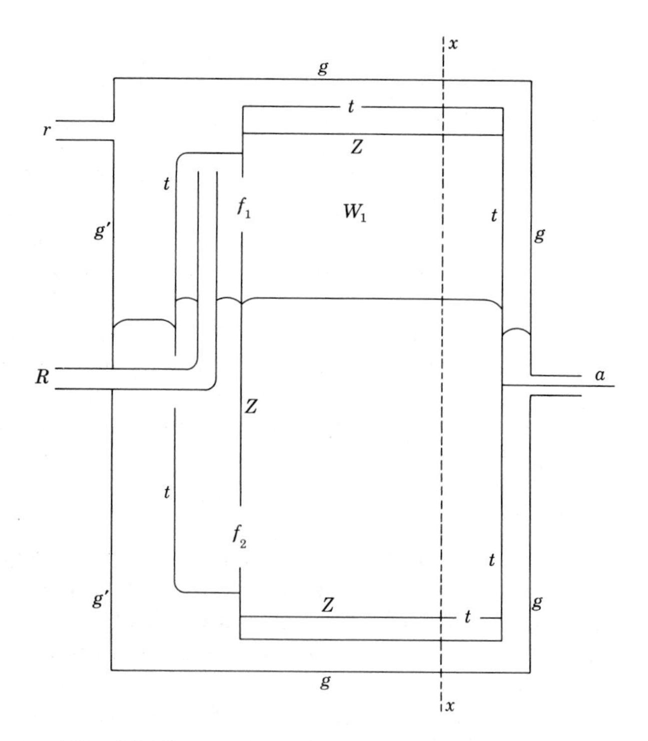


Fig. 3.2. Gaede rotary mercury pump (*vertical section*).

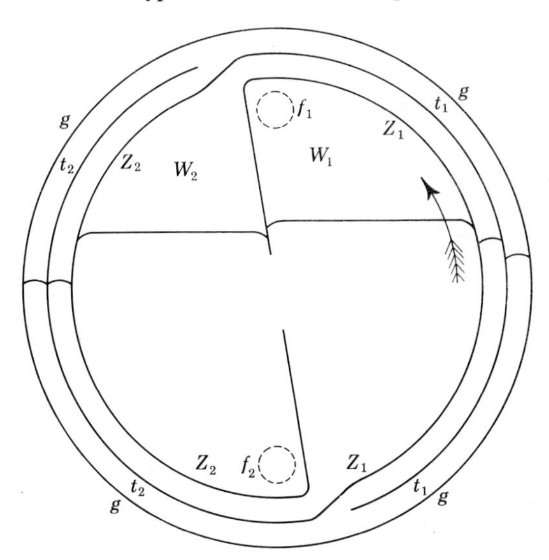


Fig. 3.3. Gaede rotary mercury pump. Section perpendicular to axis of rotation, through plane shown at $x$ in Fig. 3.2.

front view. The iron case is shown at $g$, and $g'$ is a heavy glass plate through which pass the tubes $R$ and $r$ which connect to the vessel to be exhausted and to the fore pump, respectively. The porcelain drum $t$ is built up of two (or more) sections as shown in Fig. 3.2 and rotates on the axis $a$. As the drum rotates in the direction of the arrow (see Fig. 3.3), the compartment $W_1$ is at first increased in volume and thus takes in the gas at the opening $f_1$, from the vessel to be exhausted. During the second part of the revolution, the opening $f_1$ becomes covered with mercury, as shown at $f_2$, and the gas is then forced out under pressure from the compartment $W_2$ into the space between the walls $Z_1$ and $t_1$ and into the rough pump connection at $r$.

The speed of exhaust observed with a volume of 6250 cm³ varied from about $95\ \text{cm}^3 \cdot \text{sec}^{-1}$ at $P_\mu = 30$ to about $7\ \text{cm}^3 \cdot \text{sec}^{-1}$ at $P_\mu = 0.07$.

### Gaede Rotary Oil Pump[10]

Figure 3.4 shows the construction of a pump of this type designed by Gaede primarily for the purpose of functioning as a fore pump to the rotary mercury pump just described. The pump consists of a steel cylinder $A$ which rotates eccentrically inside a steel casing $G$. The projections at $s$ are held tightly against the inner wall by means of springs, so that as the cylinder rotates the air enters at $C$ and is forced out through the valve $D$ into the atmosphere at $J$. The oil serves also as a lubricant and helps to

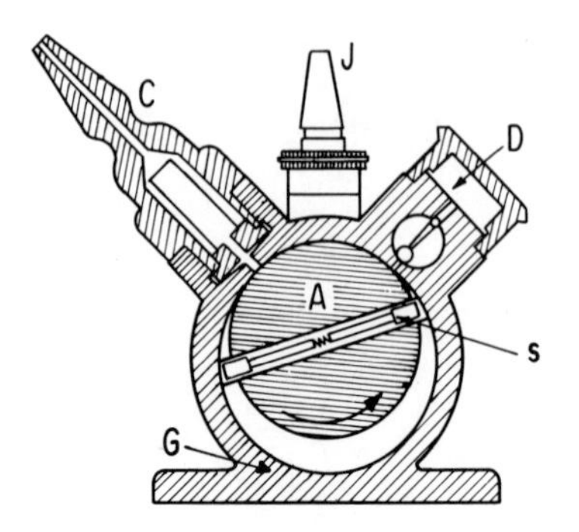


Fig. 3.4. Gaede rotary oil pump (*section*).

prevent air from leaking back into the fine-pump side, by forming a film between the rotating and the stationary members.

With a "rough" vacuum of about 10 mm mercury, such a pump could reduce the pressure to about 1 micron, with a speed of exhaust of 100–150 $cm^3 \cdot sec^{-1}$. Speed versus pressure curves for typical commercial models are shown in Fig. 3.5.

The Gaede rotary oil pump is the laboratory workhorse for work down to 1 micron of pressure: the only significant changes in modern practice

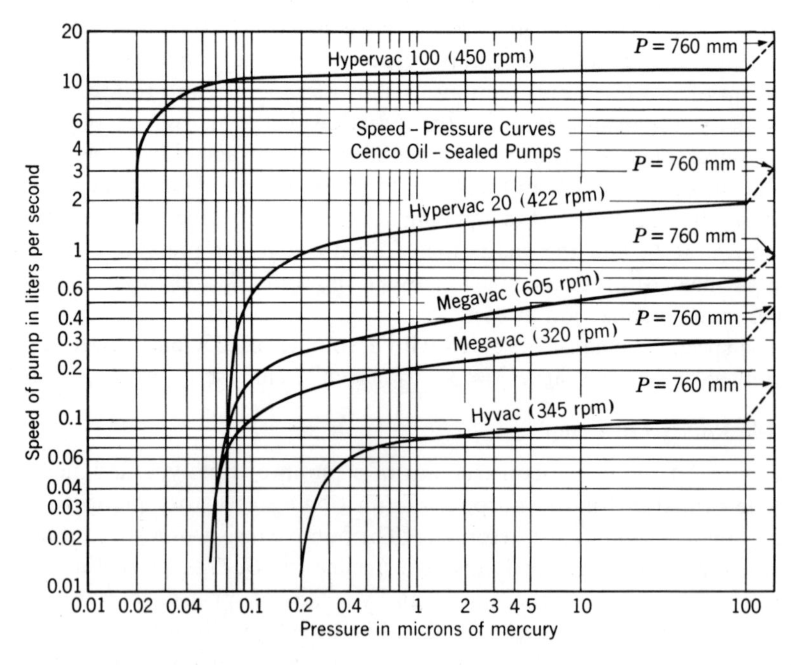


Fig. 3.5. Log-log plots of speed versus pressure for some rotary oil pumps (Cenco).

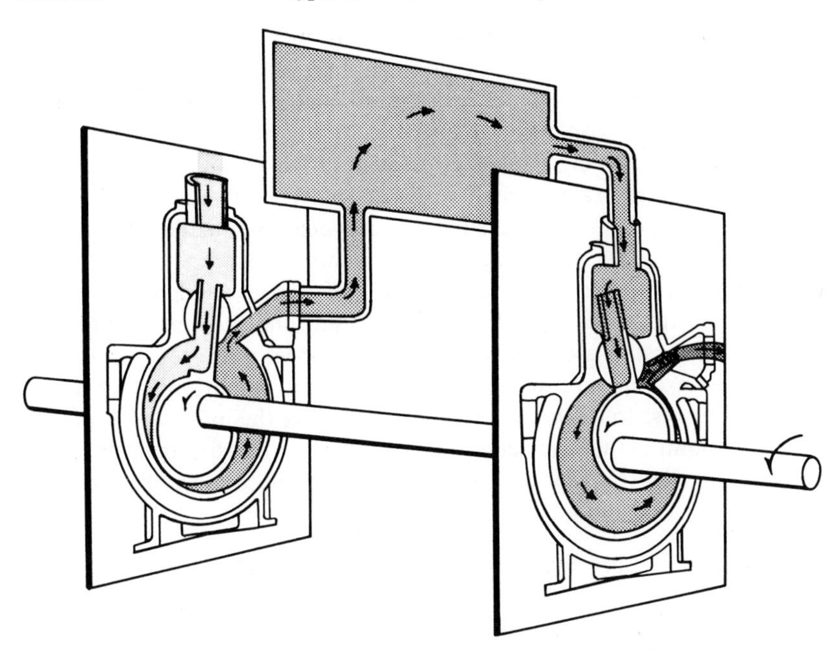

Fig. 3.6. Two-stage eccentric piston and slide mechanical pump (Kinney model KC).

are the use of rotary eccentric pistons rather than vanes and multistage or compound pumps for higher capacities (greater than 10 liters · $sec^{-1}$) (see Fig. 3.6), and the introduction of "gas ballasting" for pumps handling condensable or soluble vapors. This technique, which permits the oil in the pump chamber to purge itself of dissolved vapors, reduces slightly the minimum pressure the pump can achieve.

Condensation or dissolution of vapors in the pump oil is a consequence of the high compression ratios rotary oil pumps have to achieve in order to discharge the exhaust against atmospheric pressure: these ratios lie in the range $10^3$–$10^6$. The presence of vapors in the oil causes a rapid reduction in the attainable vacuum of the device. Methods to combat this reduction include oil separation, heating, centrifuging, and removal, but the gas ballast, introduced by Gaede, seems to be the most convenient. The ballast device is usually incorporated in the body of a normal rotary pump and may be started or stopped at will: it is easily confused in its action with air scavenging of the oil, but is considerably more efficient. The usual vapor to be stripped is water, the partial pressure of which in most systems during first evacuation can be much greater than that of the permanent gases. The essential feature of the technique is that air is drawn into the pump chamber after the chamber has been shut off from the intake side of the pump (Fig. 3.7). This air (gas ballast) lowers the effective compression ratio of the pump and prevents the condensation of vapors before they and the air are exhausted to the atmosphere. By this means, the compression ratio is reduced to 10 : 1, and such a pump should be capable of removing vapors up to 30 mm Hg partial pressure. This figure includes saturated water vapor in most climates, but if higher condensable vapor partial pressures are expected, an auxiliary condenser is fitted on the intake side. A gas-ballast pump needs traps on both sides of the pump chamber to prevent the flow of vapors condensing outside the chamber into the oil, if the pump is used under severe conditions; normal care must also be taken that the vapor does not react with the oil, to form, for example, resins. Under more normal conditions, such as are found in the research laboratory, a single trap on the intake side suffices. A further point is that the action of gas ballasting transports oil from the chamber, so that frequent make-up of the level is necessary. The effect of

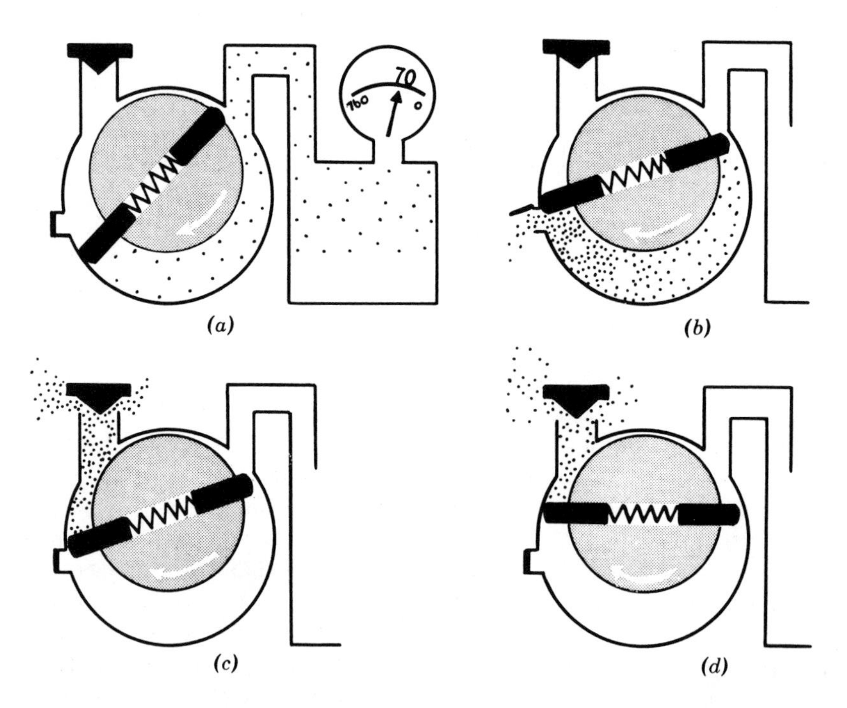


Fig. 3.7. Schematic of gas-ballast operation. (*a*) Without ballast. (*b*) The pump chamber is shut off and the ballast valve opens, admitting air to the pump chamber. (*c*) The exhaust valve opens; air and uncondensed vapor are released. (The exhaust valve opens before it ordinarily would.) (*d*) The pump continues to eject air and vapor; ordinarily the exhaust would not open till (*d*).

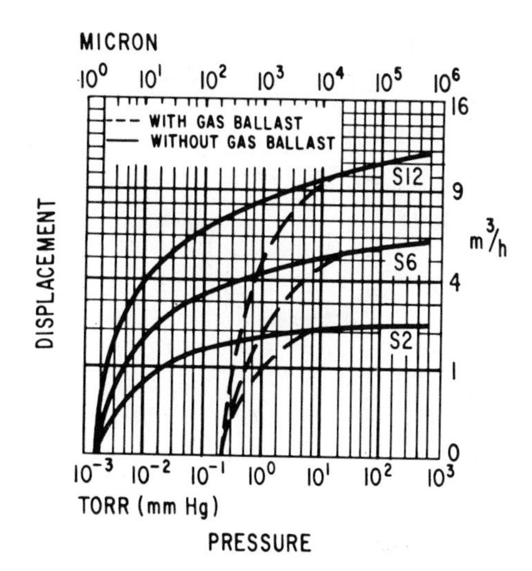


Fig. 3.8. Variation in speed of rotary pumps with gas-ballast device (Leybold pumps).

gas ballasting on the speed curve is shown in Fig. 3.8. When the vapor-contamination level is not too high (only a few millimeters partial pressure), it is possible to air-scavenge the pump oil to strip the vapor, rather than use a full gas-ballast device. This practice is, however, generally restricted to small-laboratory-scale rotary pumps running under predictable conditions.

## 3.4. ROOTS-TYPE PUMPS

The Roots principle, first proposed about 1857, has been successfully used in the manufacture of compression blowers for nearly 100 years. It utilizes two figure-eight-shaped rotors which counterrotate in a chamber without touching each other or the chamber wall. By machining these rotors to closer tolerances and increasing the speed of rotation, it was found that the pump, backed by an oil-sealed mechanical pump, could evacuate a chamber efficiently down to pressures of $10^{-5}$ mm Hg or less. The accurate shape of the rotors, together with the close tolerances, allows this counterrotation to be carried out without friction and without substantial gas backflow at pressures below 20 mm Hg. The rotor surfaces do not need oil lubrication even though their speed of rotation is high.

Figure 3.9 indicates the mode of operation. The rotors counterrotate as indicated and are synchronized by a pair of timing gears mounted on the rotor shafts. One motor per pump is used, except on the largest models which, because of the size and weight of their rotors, use two motors.

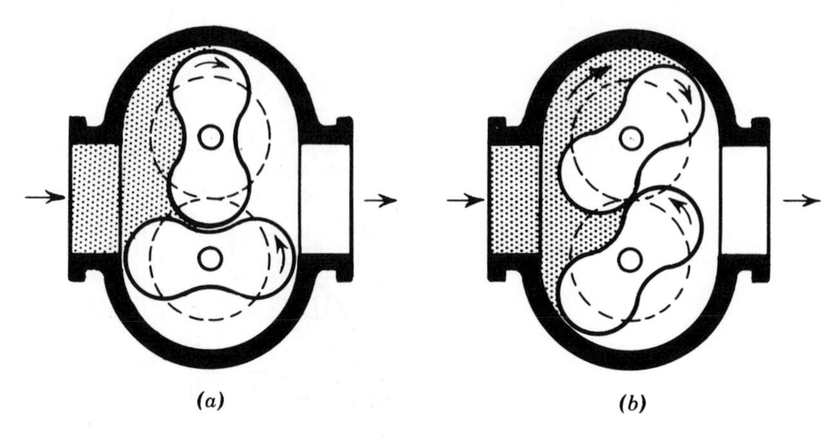


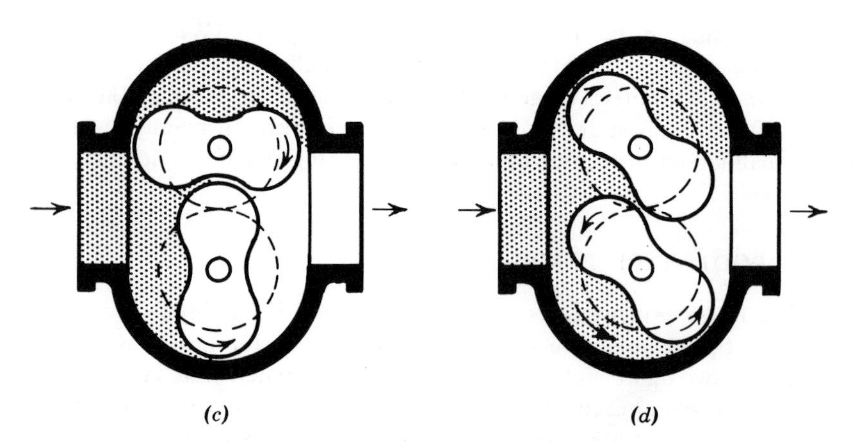


Fig. 3.9. Operational diagram of Roots pump.

The air or other gas in the chamber to be evacuated enters the Roots pump on the inlet side and fills the area which is shaded in Fig. 3.9(*a*). As the rotors advance, this gas flows into the area, shown in (*b*), between the rotors and the chamber wall, until a point is reached where a definite volume of air is trapped (*c*). With a further advance of the rotors, shown in (*d*), the trapped volume of air is exposed to the outlet or fore-pressure side of the pump. When this occurs, air from the outlet side, at higher pressure, rushes in until the pressure differential is overcome. The volume trapped between the rotor and the casing is then discharged into the outlet by the rotor. For each complete revolution of the driveshaft, this process is carried out four times.

Roots pumps can discharge directly to the atmosphere without backing pumps and have been used for short periods of time in this manner to produce a rough vacuum, but such a procedure is not recommended. The addition of a backing pump makes it possible to operate the Roots pump at lower inlet pressures. In addition, it greatly reduces the power required to drive the pump motor, and consequently the heating is reduced. The importance of the latter is apparent when one realizes that prolonged dry operation at low pressures without a backing pump will quickly produce enough heat to cause close-tolerance internal pump parts to seize.

The speed and operating pressure range of the pump depend in part upon the performance characteristics of the backing pump used. A low ultimate pressure of the mechanical backing pump results in a low ultimate pressure for the combination. The higher the speed of the mechanical pump, the higher the speed of the combination. It is thus apparent that the performance of each Roots pump can, within limits, be adapted to specific operating conditions by the proper choice of backing pump. The manufacturer's recommended backing pumps are chosen to produce the best balance between speed and ultimate pressure for general use. These specific backing pumps can be replaced by other pumps for special applications.

## 3.5. MOLECULAR PUMPS

The Gaede molecular pump undoubtedly marked a distinct advance in the design of pumps for the production of high vacua. The difference between this pump and the types previously constructed is as follows. The gas is drawn from the vessel to be exhausted into the fore vacuum by means of a cylinder rotating with high velocity inside a hermetically sealed casing. In the case of the molecular pump, therefore, there is no mechanical separation between the fine and the rough vacua.[11] The pump thus represents a logical development and application of the laws of flow of gases at very low pressures as investigated by Knudsen, Smoluchowski, and Gaede himself.

The fundamental principle of the pump may be illustrated by means of Fig. 3.10. The cylinder $A$ rotates on an axis $a$ (in the direction of the arrow) inside the airtight stator $B$, and drags the gas from the opening $n$ towards the opening $m$, so that a pressure difference is built up in the manometer $M$, as shown by the mercury levels at $o$ and $p$. Between $m$ and $n$ there is a slot in the case $B$, as shown in the diagram, while at every other point

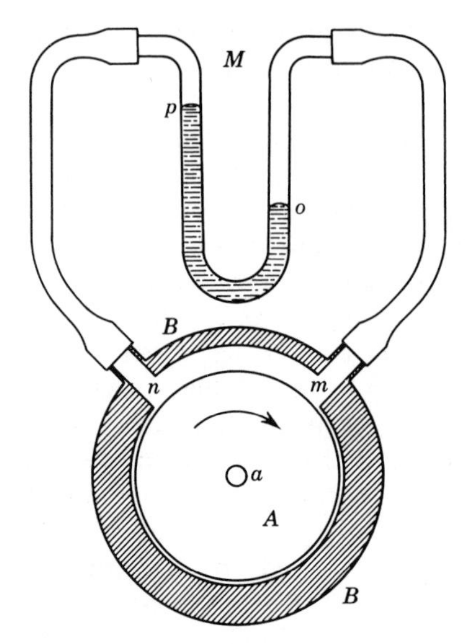


Fig. 3.10. Diagram illustrating principle of operation of Gaede molecular pump.

*A* and *B* are very close together. Now, at ordinary pressures the viscosity is independent of the pressure. Under these conditions, as Gaede showed, the difference in pressure at *o* and *p* depends only on the speed of rotation *u* of the cylinder, the coefficient of viscosity of the gas $\eta$, the length of the slot *l*, and the depth measured radially *h*, according to the following relation:

$$P_1 - P_2 = \frac{6lu\eta}{h^2}, \tag{3.5}$$

where $P_1$ is the pressure on the high side.

At *low pressures*, however, the number of collisions between gas molecules becomes relatively small compared with the number of collisions between the gas molecules and the walls. Under these conditions the molecules therefore tend to take up the same direction of motion as the surface against which they strike, if that surface is in motion. This conclusion is based upon the investigations of Knudsen on the laws of molecular flow, which have been discussed in Chapter 2. The relation previously deduced is therefore found to be no longer applicable, and, instead of the pressure difference remaining constant at constant speed of rotation, the *pressure ratio is now constant* and independent of the pressure in the fore vacuum. Gaede shows that, at very low pressures,

$$\frac{P_1}{P_2} = \epsilon^{bu}, \tag{3.6}$$

where *b* is a constant whose value depends upon the nature of the gas and the dimensions of the slot in the casing $B$ of the pump, so that, at constant speed of rotation $u$, the ratio between the pressures on the two sides of the pump is constant.

The construction of the actual pump based on these principles is illustrated in Figs. 3.11 and 3.12. The rotating cylinder $A$ has twelve parallel slots around the circumference, into which project the extensions $C$ from the outer casing. If $A$ rotates clockwise, the pressure at $m$ is greater than that at $n$, and, in order to increase this pressure, different sections are connected in series. The distance between the outer edge of the cylinder $A$ and the inside of the shell $B$ is about 0.01 cm. The overall radius of $A$ is 5 cm, and the depth of the slots varies from 0.15 cm in the outer section to 0.6 cm in the inner ones. With the cylinder rotating clockwise as indicated, the vessel to be exhausted is connected at $S$, while the opening $T$ is connected to an ordinary mercury or oil pump capable of exhausting to a pressure of less than 0.05 mm Hg. Since the speed of rotation of the

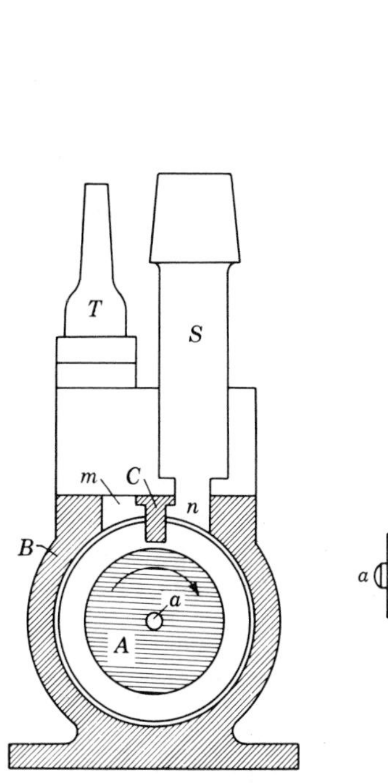


Fig. 3.11. Construction of Gaede molecular pump (*front view*).

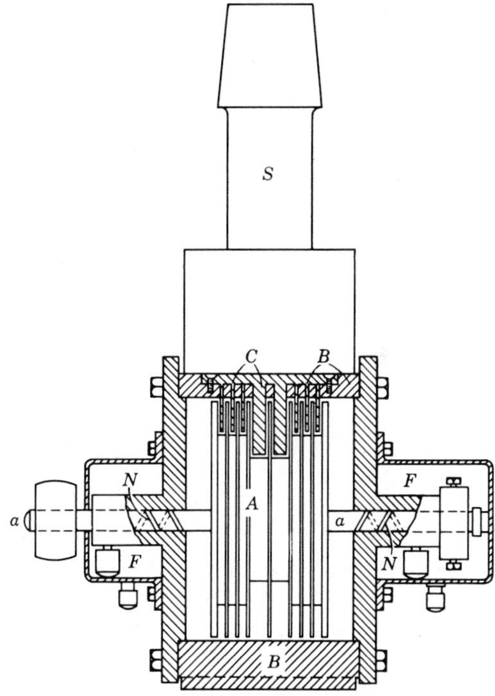


Fig. 3.12. Construction of Gaede molecular pump (*side view*).

cylinder is very high (about 8000 rpm) oil cups are provided at *F* (Fig. 3.12), and the shaft *N* is so designed that the oil in the spiral slot is driven outward by the centrifugal action. The slots in the rotor are so arranged that the lowest pressure is in the center, and the pressure increases uniformly outward until the ends, where it is equal to that produced by the rough pump, into which the outlet *T* exhausts.

The effect of variation in the speed of rotation and in the fore-pump pressure on the degree of vacuum produced by the molecular pump is shown in Table 3.2.

**TABLE 3.2**

**Effect of Speed of Rotation on Degree of Vacuum Obtained with Gaede Molecular Pump**

| Speed of Rotation (rpm) | Rough-Pump Pressure, $P_{mm}$ | Pressure on Fine Side, $P_{mm}$ |
|---|---|---|
| 12,000 | 0.05 | 0.0000003 |
| 12,000 | 1 | 0.000005 |
| 12,000 | 10 | 0.00003 |
| 12,000 | 20 | 0.0003 |
| 6,000 | 0.05 | 0.00002 |
| 2,500 | 0.05 | 0.0003 |
| 8,200 | 0.1 | Not measurable |
| 8,200 | 1 | 0.00002 |
| 8,200 | 10 | 0.0005 |
| 6,200 | 0.1 | 0.00001 |
| 6,200 | 1.0 | 0.00005 |
| 4,000 | 1.1 | 0.00003 |
| 4,000 | 1 | 0.0003 |

The pressures on the fine side were measured with an extremely sensitive McLeod gauge, except that the first result given in the table was estimated. Dushman's experiments[12] with the Gaede molecular pump at 8000 rpm showed that with a rough-pump pressure of 20 mm the fine-side pressure was 0.0004 mm, so that the ratio of the pressures was 50,000—a result in accord with figures given by Gaede.

The speed of the pump as defined by equation 3.1 was found by Gaede to vary with the magnitude of the rough-pump pressure. The curve *A* in Fig. 3.13 shows that the maximum speed was about 1400 $cm^3 \cdot sec^{-1}$ with a fore vacuum of 0.01 mm. For comparison, curve *B* also shows the speed characteristic for the Gaede rotary mercury pump, which had a speed of about 130 $cm^3 \cdot sec^{-1}$ at the maximum.

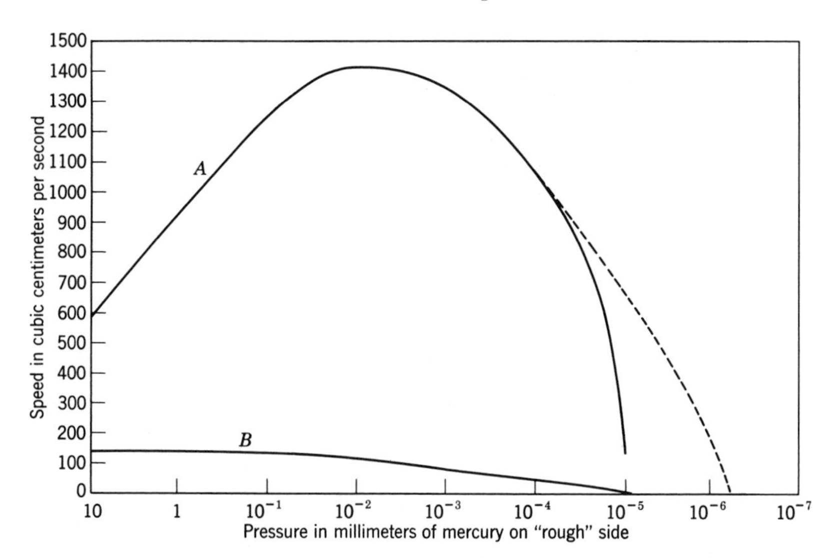


Fig. 3.13. Plot (curve *A*) showing effect on speed of molecular pump of variation in fore-pump pressure. Curve *B* shows speed characteristic of rotary Gaede mercury pump.

A modified design of the molecular pump has been described by Holweck.[13] For this design, pressure ratios as great as $10^7$–$10^8$ were observed with rotor-stator clearances of 1 mil (0.025 mm). It is also possible to construct molecular pumps with single disks, which spin inside shields in which are cut spiral grooves. The initial designs of this type are due to Siegbahn.[14] In practice, the grooves are deeper the further they are located from the pump axis. The pump shaft is supported by ball bearings located in the fore-vacuum region, so that conventional lubricants may be used, and is driven at approximately 10,000 rpm through a flexible coupling.

Figure 3.14 shows the construction of a large pump of this type. There are three spiral grooves (I, II, III) in each shield, instead of one, each of them starting at the periphery and ending near the center. The pump may thus be described as three separate pumps connected in parallel inside a common case. The groove width is 22 mm, and the depth varies between 22 and 1 mm. The high-vacuum ends are joined by passages (4) to wide metal tubes (9) which have a common outlet (10). This outlet can be provided with a flange for the use of a rubber gasket seal between the pump and the apparatus to be evacuated. Another feature in this design is the use of ball bearings to support the pulley (7) and the addition of another flexible coupling (5) at the outer end of the axis.

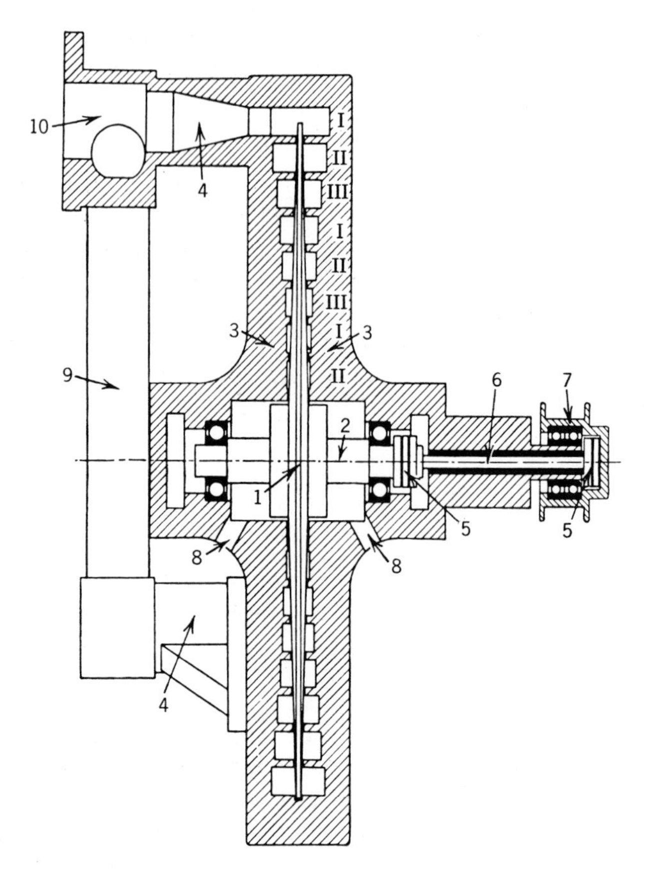


Fig. 3.14. Schematic diagram of a Siegbahn molecular pump with high pumping speed. Diameter of disk = 54 cm.

It is stated that the pump will operate even against a pressure of several millimeters, and the pumping speed at 3700 rpm is 73 liters $\cdot$ $sec^{-1}$ at $10^{-3}$ mm. The highest vacuum observed is $6 \cdot 10^{-7}$ mm, as measured by means of an ionization gauge. No refrigerant or drying agent is used. This lower pressure limit is apparently set by the vapor pressure of Apiezon grease, which, according to Seydel,[15] is about $5 \cdot 10^{-7}$ mm (see also Chapter 10).

These pumps have certain advantages, according to von Friesen.[14] They have instantaneous action and high speed for all gases and vapors, especially for heavier species. Consequently, they have found considerable acceptance as alternatives to vapor-stream pumps, particularly in Europe.

According to Eklund,[16] the fine-pump pressure $P$ obtainable with a pump of the Seigbahn design is expressed as a function of the fore-pump pressure $p$ by the empirical relation

$$P = P_0 + cp \tag{3.7}$$

where $P_0$ is the limiting pressure and $c$ is a pressure-ratio constant the value of which is of the order of $10^{-6}$ (for $P$ and $p$ in millimeters of mercury). Furthermore the speed of exhaust increases linearly with the speed of rotation, and for pumps with disk diameters between 22.5 and 27.5 cm the highest speeds (about 14 liters $\cdot$ sec$^{-1}$) were obtained at about 9000 rpm.

With a disk diameter of 54 cm, speeds of 60–80 liters $\cdot$ sec$^{-1}$ were obtained at about 8300 rpm, independently of the high-vacuum pressure (which ranged from $10^{-2}$ to $2 \cdot 10^{-6}$ mm Hg).

## 3.6. VAPOR-STREAM PUMPS

### General Remarks

Just as rotating metal parts were used to transfer momentum to the pumped gas in the preceding examples, so a high-velocity gas or vapor stream can be employed for the same purpose. Thus, an entire class of pumps is based on the use of work fluids, in the general manner described below.

When a stream of vapor enters an empty chamber, it expands and converts its pressure energy into velocity energy in a flow, the direction of which is dictated by the nozzle and chamber geometries and the pressure drop across the entrance throat. The stream of high-velocity vapor, traveling at high speed (always supersonic with respect to the velocity of sound in the vapor, at the normal boiling temperature of the corresponding liquid), creates a turbulent layer between itself and the gas in the pump chamber, leading to a transfer of momentum to the pumped gas, so that there is a net component of velocity in the gas in the direction of flow of the work fluid.

The total mass then moves into a part of the chamber where velocity energy is recovered as pressure energy (the "diffuser"), thus permitting exhaust against a higher pressure. The pumped gas is thus compressed in such a device: the only possible difference between ejector pumps, which operate at higher pressures, just as described above, and "diffusion-condensation" pumps, which operate at lower pressures, concerns the existence of a turbulent boundary layer between the work fluid and the pumped gas. In the latter case, the pumped gas may enter the high-velocity work-fluid stream by molecular diffusion, rather than by a turbulent viscous-flow mechanism involving aerodynamic shear relations, although we shall see that in this case this distinction is at best vague and frequently it is nonexistent.

As might be expected, the primary factors affecting the performance of such devices are mass flux through the chamber (which depends on the thermodynamic properties of the work fluid), the stability of the work-fluid stream (which depends on nozzle design), and the diffuser back pressure, which in its turn demands control of pump-chamber design. An important secondary factor affecting performance is the condition of the wall boundary layer, formed where the work fluid and the entrained gases strike the chamber wall, which, if too thick, can permit rapid diffusion of the pumped gas from the higher-pressure diffuser area back into the low-pressure region (backstreaming). A specific factor influencing all these properties is the chemical stability of the work fluid (especially of oils and esters). It is noteworthy that recent advances and general interest in high-speed aerodynamics cause understanding of the mechanism of vapor-stream pumps to be much less of a mental hurdle than previously.

### Ejector Pumps

The first application of the ejector principle using a steam jet to the production of low pressure appears to have been made by Leblanc, who, according to Dunoyer,[17] showed that it is possible by means of such an ejector to obtain a pressure as low as 1 mm of mercury.[18]

The diagrammatic sketch in Fig. 3.15 shows the standard terminology for the different parts of a steam ejector.[19] In entering the nozzle from the steam chest the steam expands at the throat to a pressure $P_t$, which is about 55 per cent of the pressure in the chest; in the nozzle there is a further isentropic expansion to the value $P_k$, which is the same as that at the suction opening. As shown in Fig. 3.15, the steam jet diverges in the region near the nozzle, in which the gas or vapor entering from the system to be evacuated is entrained in the steam by virtue of the shear gradient between the steam and the pumped gas. The mixture of gas or vapor and steam filling the cross section of the diffuser is then compressed isentropically until it reaches the pressure $P_D$ at the discharge opening. The relation between the kinetic energy of the steam just beyond the throat, and the pressures $P_k$ and $P_D$, is of the form:

$$\frac{v^2}{2} = \frac{\gamma}{(\gamma - 1)} \frac{P_K}{\rho_k}\left[\left(\frac{P_D}{P_k}\right)^{(\gamma-1)/\gamma} - 1\right], \tag{3.8}$$

where $\gamma$ = ratio of specific heat at constant pressure to specific heat at constant volume (1.32 for steam); $\rho_k$ = density in grams per cubic centimeter at pressure $P_k$; $P_D$ and $P_k$ denote pressures in microbars; and $v$ = velocity of jet in centimeters per second.

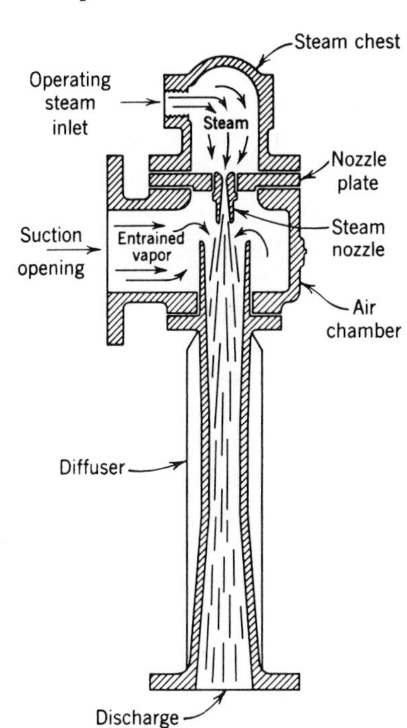


Fig. 3.15. Terminology for parts of steam-jet ejector.

Usually single-stage ejectors are operated at about the value of the ratio $P_D/P_k = 7.5$. Assuming that $P_D = 10^6$ microbars (750 mm Hg), the corresponding value of $P_k = 1.333 \cdot 10^{-5}$ microbars (100 mm Hg), and $\rho_k = 8.97 \cdot 10^{-5}\ \mathrm{g \cdot cm^{-3}}$. Also, $(7.5)^{(\gamma-1)/\gamma} = (7.5)^{0.2424} = 1.629$. Hence $v = 8.782 \cdot 10^4\ \mathrm{cm \cdot sec^{-1}}$.

At the given pressure $P_k$ and density $\rho_k$, the velocity of sound is $u = 4.43 \cdot 10^4\ \mathrm{cm \cdot sec^{-1}}$. Thus the Mach number of the jet is 1.983 (Mach number $= v/u$, where $u$ is the velocity of sound under the local conditions).

In general, the value of the Mach number for single-stage ejectors, discharging against atmospheric back pressure, ranges as high as 3. In booster stages, that is, in those stages where the back pressure is less than atmospheric, particularly in the first stages of a four- or five-stage ejector, the Mach number may be as high as 9–11, depending on the initial steam pressure and the air-chamber pressure.

Owing to this high velocity of the stream there is practically no lateral diffusion of steam or entrained gas, and no back diffusion of entrained gas from the discharge opening. Because of this, it is possible by means of a steam ejector to obtain pressures as low as 100 microns, or even less. An important consequence of operating the device at pressures below 4 mm is that the isentropic expansion produces vapor temperatures at which the solid phase is stable, so that auxiliary heat input is necessary to prevent freezing.

In these devices, the operating exhaust pressure is an important factor in the stability of the jet and the capacity or pumping speed. Thus, Fig. 3.16(*a*) shows the operating line for stability of the jet with respect to operating exhaust pressure, and Fig. 3.16(*b*) the stability line with respect to the attained vacuum. The performance of the device is such that the attained vacuum ("suction pressure") cannot be varied by changing the steam throughout: this can only be used to vary the pumping speed, since the attained vacuum depends on diffuser throat area and exhaust pressure. The latter is usually fixed by the mechanical design, and hence provision for variable pumping speed must be made by artificially loading the stage with steam or air (preferably the former), or by using a number of stages in parallel that are cut in or out as required.

As a consequence of these facts, the shape and size of the diffuser (the device by which velocity energy is converted into pressure energy) are the major factors in design and performance of ejector pumps. The position of the throat in the axial direction is critical and is adjusted by

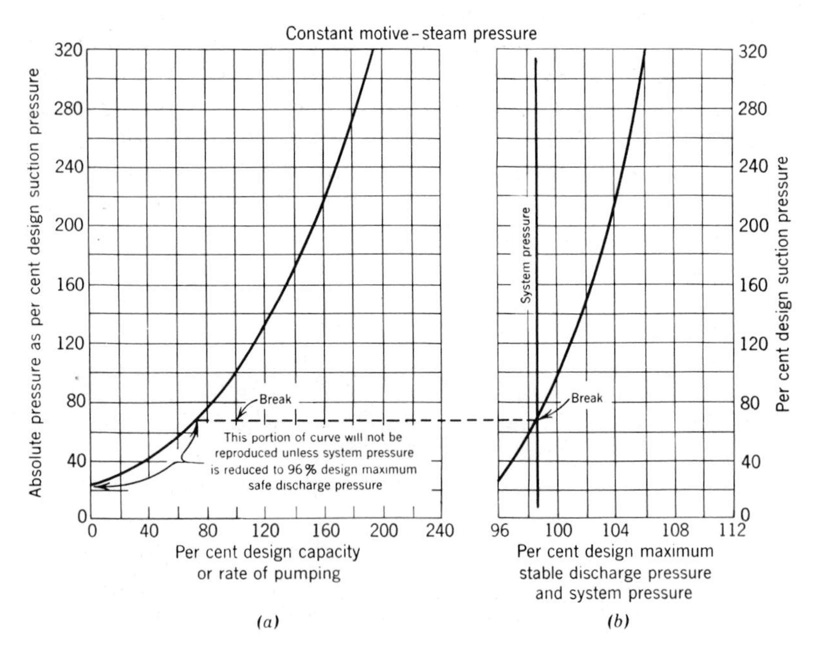


Fig. 3.16. Typical ejector performance curve for constant motive steam pressure, showing stable range and break point, for suction and discharge pressures expressed as percentage of design value (Freneau[18]).

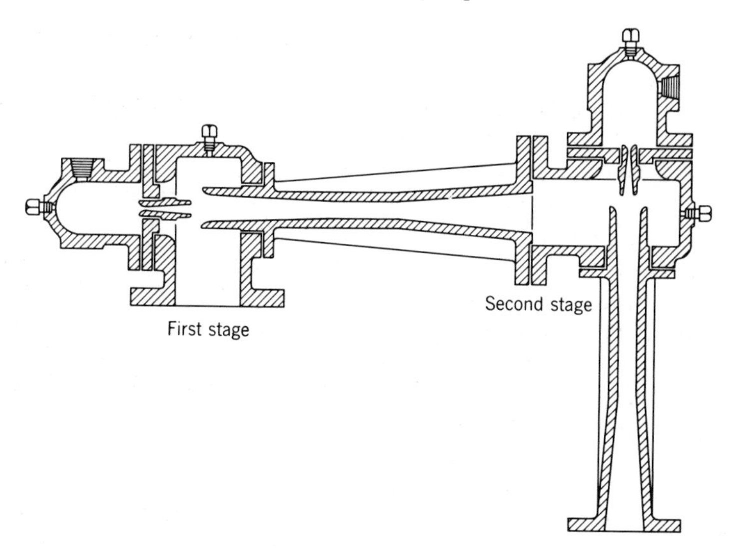


Fig. 3.17. Sectional view of two-stage noncondensing type of ejector.

including thin shims of metal between the flanges at the join of the steam chest and the expansion throat; the final adjustment is by trial and error.

As mentioned previously, it is necessary, in order to obtain low pressures, to operate an ejector stage in series with one or more booster stages, and Fig. 3.17 shows a sectional view of a two-stage noncondensing type of ejector. The number of stages actually employed to obtain a desired low pressure depends upon the capacity required. Thus it is possible with a five-stage ejector to maintain 50 microns absolute pressure with quite high capacity.

## "Diffusion" Pumps

Gaede[20] invented and developed a diffusion pump which employs a vapor stream in the following way. A blast of condensable vapor passes in the direction $AB$ past a porous diaphragm $C$ (Fig. 3.18). The vessel

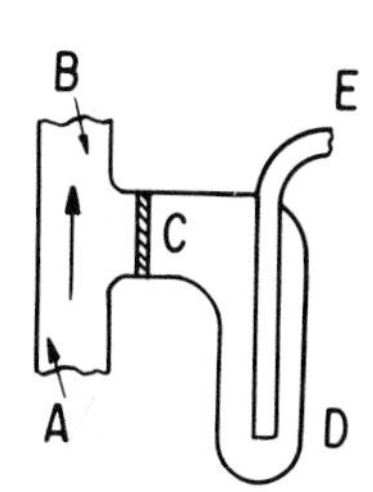


Fig. 3.18. Method of operation of Gaede diffusion pump.

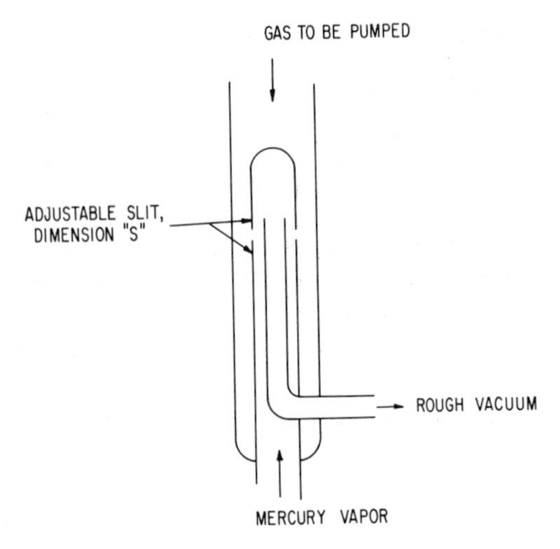


Fig. 3.19. Original Gaede diffusion pump.

to be exhausted is connected at *E*, and the trap *D* is cooled. Vapor diffuses across the diaphragm into *D* and is there condensed, preventing its passage into *E*. Gas diffuses through the diaphragm in the reverse direction and is carried away by the vapor blast. Gaede's simple concept of this device demanded an infinite speed if the diaphragm were of zero thickness, and he consequently developed another explanation, based on kinetic energy, to circumvent this difficulty. He considered the rate at which one gas diffused back through a small hole in a thin plate from which a second gas stream was issuing, and found this rate to be a maximum when the hole diameter was approximately the mean free path of the second gas. This led to the construction of a device like that in Fig. 3.19, where the slit dimension *S* was made equal to the mean free path of the molecules in the vapor stream. It will be readily appreciated that the speed of such a device is inherently low, and Langmuir[21] conceived a pump in which the primary process of introduction of the gas into the vapor stream was made easier; the second process, that of removal of the admixed stream, was already effective, and in this latter respect, the Gaede diffusion pump is very similar to the ejector pump.

Langmuir reasoned that a relatively low-pressure vapor stream, such as he considered existed in a vapor-diffusion pump, would spread in all directions and fill the expansion chamber when passed through an ejector-like nozzle. He believed that by cooling the walls of the chamber he could promote the condensation of those vapor molecules which had had appreciable lateral components of momentum and thereby permit the continued, less interrupted passage of the other, downward-directed molecules, creating a more efficient pump (Fig. 3.20). Such a pump was built and worked well, and Langmuir did some more experiments to support his contention that condensation was essential to the operation of the pump. It is indeed important, and such pumps are very popular today, but the primary effect of condensation as propounded by Langmuir is a little misleading. He considered that there was no pressure reduction at all in the vapor stream as it entered the pump chamber, but this is not the case. The vapor stream has many characteristics of a high-density stream. Subsequent experimental and theoretical researches have shown that some of the other consequences of his original condensation concepts are not necessarily true, particularly in regard to the cooling of the expansion nozzle and diffuser of an ejector and the shape of the nozzle in

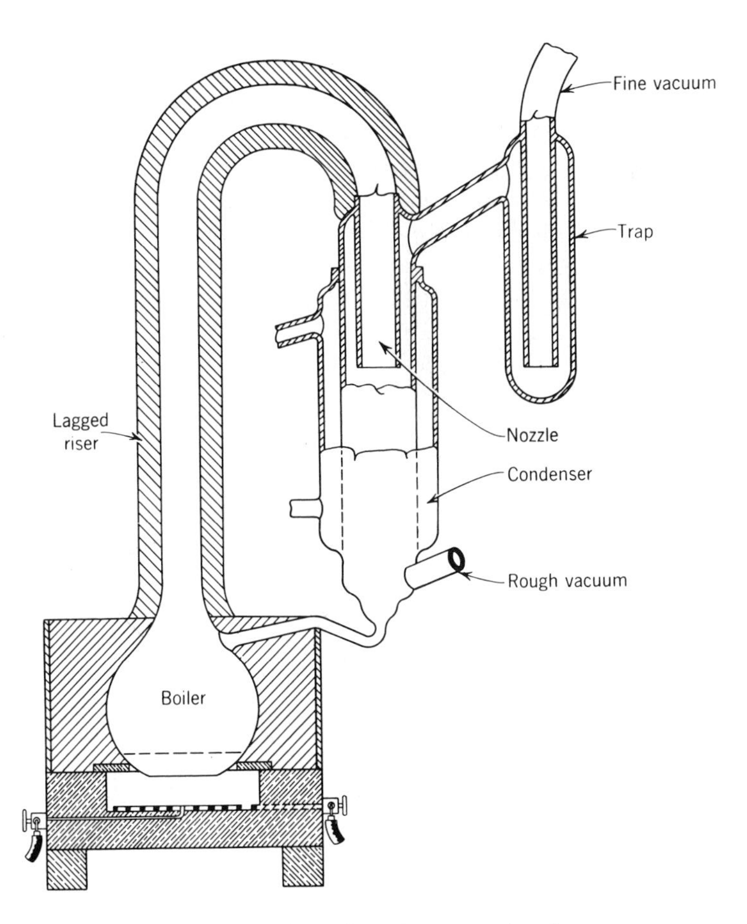


Fig. 3.20. Langmuir condensation pump, glass form.

the diffusion pump, neither of which should have any effect, according to his original paper.

It is important in a true diffusion pump to maintain the pressure (and therefore the temperature) of the vapor stream constant so that the mean free path of the vapor molecules is approximately equal to the slit dimension. It is, however, found experimentally that the condensation pump has an outstanding advantage in that careful temperature control is not necessary. The reason for this fact will become apparent.

### General Considerations Regarding the Speed of Diffusion Pumps

To a first approximation, the speed of exhaust for a diffusion vapor-stream pump is given by

$$S = \tfrac{2}{3} v_a \, \pi a^3 / l \tag{3.9}$$

where $v_a$ = average velocity of pumped molecules,
$a$ = radius of inlet port,
$l$ = length of inlet port,

provided $a$ is comparable with mean free path of the gas molecules. This expression is identical with that for the conductance $F_t$ of a cylindrical tube of radius $a$ and length $l$, with $l/a$ very large. However, $F_t$ is very small under these conditions; but, if we attempt to reduce $l/a$, then the equation is no longer valid. It is also apparent that, if the inlet is made up of a series of fine holes, the equation will be valid at higher pressures. Thus, Gaede considered the speed $S$ as

$$S = \frac{\alpha \pi a^2 v_a}{4}, \tag{3.10}$$

where $\alpha$ is a function of radius reduced to mean-free-path units $(a/L)$, where $\alpha = 0$ for $a = 0$ and nearly unity for $a/L = 0.5$, thereafter increasing very slowly to the limit $\alpha = 1$ when $a/L = 1$. This equation also implies that the speed of the pump varies as the inverse square root of the molecular weight of the gas, and that the speed is independent of the pressure for "hard-sphere" ideal gases only; it also contains an assumption that each gas molecule is removed totally and immediately from the diffusion area. The latter condition is not true, and $\alpha$ should be replaced by $k\alpha$, where $k$ can be called the "extracting power" of the vapor and can vary from zero to unity. It is a rather inaccessible number, whose *a priori* evaluation demands a complete theory of nonuniform gases.

The speed is also reduced as a result of back diffusion of gas from the fore pump. Just below the orifice the pressure of gas, $P_0$, is maintained at an extremely low value because gas is removed very rapidly by the stream of vapor. Since this pressure is lower than $P_f$, the pressure at the opening to the fore pump, there is a back diffusion of gas, and it follows that the net speed of exhaust through the orifice is given by

$$S = \frac{Av_a}{4}\left(k\alpha - \frac{P_0}{P}\right), \tag{3.11}$$

where $P$ denotes the pressure in the volume that is being exhausted, and $A$ replaces $\pi a^2$. It is then possible to stall a diffusion pump if

$$P = \frac{P_0}{k\alpha}, \tag{3.12}$$

or even run gas backwards through the device if $P_0 > k\alpha P$. For air in mercury vapor, the ratio $P_0/P_F$ decreases very rapidly with temperature and the dimension $a$, and at a sufficiently high vapor velocity

$$S = \frac{k\alpha A}{(2\pi\rho_1)^{1/2}}, \tag{3.13}$$

where $\rho_1$ is the standard gas density of the pumped gas (for air at 25° C, $S = 11.67\, k\alpha A$ liters · sec$^{-1}$ · cm$^{-2}$).

In the next paragraphs, we shall consider the magnitude of the vapor velocity and the direction of flow of the work fluid, since the quantity $v_a$ can effectively exceed the average molecular velocity at the nascent vapor temperature by a large margin, thereby influencing $k$ and $\alpha$.

## Free Expansion of a Gas into a Vacuum

It is important to realize that in a freely expanding gas the three components of molecular momentum normally associated with pressure energy are substantially replaced by one single larger component, which is associated with the flow into the void. The magnitude of this quantity is a measure of the local temperature of the moving molecule; and, although the average velocity of the molecule is at all times the local sound velocity, the molecules move at velocities supersonic with respect to the velocity of sound in the vapor under vapor-equilibrium conditions. This fact is highly significant to the operation of vapor-stream pumps at lower pressures and results in many similarities between the operation of "ejector" and "diffusion" pumps.

The velocity (and hence the momentum the work fluid will attain) is calculable as follows. We consider the molecular velocity of the particles of a freely expanding gas or vapor. The expansion of the vapor will impart a velocity $u$ to the molecules, where $\rho_0$ is the final, and $\rho$ the initial, density:

$$u = \int_{\rho}^{\rho_0} \left(\frac{\partial p}{\partial \rho}\right)^{1/2} \cdot \frac{d\rho}{\rho}. \tag{3.14}$$

For an adiabatic expansion, $p/p_0 = (\rho/\rho_0)^\gamma$, where $p_0$ and $\rho_0$ refer to the unexpanded material, and $\gamma$ has its usual significance as the ratio of specific heats of the vapor at constant pressure and constant volume, respectively, so

$$u = \frac{2\alpha_0}{\gamma - 1}\left[1 - \left(\frac{p}{p_0}\right)^{\frac{\gamma - 1}{2\gamma}}\right]. \tag{3.15}$$

From this flow velocity there must be subtracted the velocity with which vapor molecules arrive at the expansion nozzle, at constant initial pressure $p$, which is the local sound speed $\alpha(p)$.

$$\alpha(p) = \left(\frac{dp}{d\rho}\right)^{1/2} = \alpha_0\left(\frac{p}{p_0}\right)^{(\gamma-1)/2\gamma}. \tag{3.16}$$

Thus, the net velocity imparted to the molecule through free expansion will be

$$s(p) = u - \alpha = \alpha_a\left[\frac{2}{\gamma - 1} - \frac{\gamma + 1}{\gamma - 1}\left(\frac{p}{p_0}\right)^{(\gamma-1)/2\gamma}\right]. \tag{3.17}$$

Writing $\mu$ for $(\gamma + 1)/(\gamma - 1)$,

$$s = \alpha_0\left[(\mu - 1) - \mu\left(\frac{p}{p_0}\right)^{1/(\mu+1)}\right]. \tag{3.18}$$

For air, $\mu = 6$ and

$$s = \alpha_0\left[5 - 6\left(\frac{p}{p_0}\right)^{1/7}\right]. \tag{3.19}$$

The Mach number, $u/\alpha$, is greater than, equal to, or less than 1 depending on whether $(u - \alpha)$ is respectively greater than, equal to, or less than 0, and hence the flow is supersonic for air if $5 > 6(p/p_0)^{1/7}$.

For mercury, $\mu = 4$ and the flow is supersonic if $3 > 4(p_0/p)^{1/5}$, a less stringent condition in terms of required pressure drop than that for air. If the vapor molecules are expanding into a second gas or vapor, then the pressure of such a second gas must be less than $p_0$ in the above inequalities if supersonic flow is to occur.

The ratios of pressures corresponding to given values of $\gamma$, required for supersonic expansion in the above manner, are shown in Table 3.3, Since the work fluid will not transfer a net downward component of momentum to the pumped gas unless the velocity of the work fluid is supersonic, the lowest ratio $p_0/p$ for supersonic flow corresponds to the maximum back pressure permissible if a net pumping effect is to be seen. Here, $p_0$ is the saturation vapor pressure of the fluid at the temperature at which it arrives at the expansion nozzle (that is, the pump boiler temperature), and $p$ is the back pressure.

It is perhaps necessary to elaborate a little on the statement that downward momentum will not be transferred to the pumped gas unless the work-fluid flow is supersonic. Were the fluid moving at its own equilibrium

TABLE 3.3

**Pressure Drop Required for Supersonic Expansion, Varying with the Specific Heat Ratio of the Expanding Gas**

| $\frac{C_p}{C_v} = \gamma$ | $\mu = \frac{\gamma + 1}{\gamma - 1}$ | $p_0/p$ (Maximum Ratio of Final Pressure to Initial Pressure for Supersonic Flow) |
|---|---|---|
| 1.1 | 21.0 | 0.34 |
| 1.2 | 11.0 | 0.32 |
| 1.3 | 7.1 | 0.30 |
| 1.4 | 6.0 | 0.28 |
| 1.5 | 5.0 | 0.26 |
| 1.6 | 4.3 | 0.25 |
| 1.7 | 3.9 | 0.23 |
| 1.8 | 3.5 | 0.22 |
| 1.9 | 3.2 | 0.21 |
| 2.0 | 3.0 | 0.20 |

average molecular velocity (the sound speed corresponding with the temperature of the boiling liquid), the molecules of the jet would be bound to spread laterally into the expansion chamber, and no pumping of gas could occur, except that caused by an initial partial pressure gradient and maintained by the removal of a certain number of molecules of work fluid per unit time through the discharge orifice. A homely analogy is with the purging of air from the space above boiling water in a kettle: if the kettle is subsequently sealed and cooled, a partial vacuum is created within it, and the kettle probably collapses. All those with an inclination to read this book will undoubtedly have done this experiment in their early youth: they were running a true free diffusion "pump," a very slow affair indeed.

However, when the flow is supersonic, with the directional selection effect of the nozzle, only a few molecules will have enough initial lateral momentum to move out of the main flow, that is, during the first very few millimeters of jet travel. Alexander[22] calculated from the Maxwell-Boltzmann law that the proportion of molecules spreading laterally is only 4.4 per cent at Mach 1.2, and less at higher velocities. The maximum velocity usually attained by the work fluid in a single stage is around Mach 3, and it is for these reasons that we think of the work-fluid jet as being a rather compact entity. The low extent of the spread of the jet was confirmed by Alexander, who measured the deposition of cadmium vapor at different directions from the principal axis of flow of such a jet. In addition to his own calculations and experiments, Alexander also considered that, since many vapor-stream pumps begin to operate at 100 microns gas pressure, where $L$ is much smaller than the intake dimension, it is not likely that the pure diffusion process proposed by Gaede does in fact govern their operation.

The events that occur after the first efflux of the vapor are also germane to this question. Thus, further below the nozzle, the vapor gets less dense and its driving effect on the gas diminishes and finally ceases. The streaming velocity of the gas accordingly decreases, and its density increases. The density of the gas will hence be at a minimum in a region just below the nozzle and will gradually increase below this region, reaching a maximum near the backing outlet. The vapor stream has to be dense enough to prevent the gas from streaming backwards from the high-density region. The wider the throat of the pump the lower the density of the vapor will be, and it will be lowest along the wall. Evidently, at a certain width of the throat the vapor stream along the wall will not be dense enough to "seal off" the compressed gas, and the gas will spread backwards along the wall into the low-pressure space.

The curves in Fig. 3.21 show the experimental confirmation[22] of this idea. The pressure readings obtained at different points below the nozzle are shown as isobars, with the pressure marked in the left-hand column of each diagram. The thick line is the isobar with a value equal to the pressure at the opening of the pump.

Plots of the speed of the pump versus back pressure show a gradual increase of speed with increase in pressure up to a maximum, followed by a rather steep drop with further increase in pressure. At the maximum, the sealing effect of the vapor starts to break down and air streams backward along the wall as the boundary layer fails in reverse shear. The critical value of the working pressure where this decline begins is higher if the heat input and the mass flux of circulating fluid are larger.

The important advantages obtained by cooling the walls of vapor-stream pumps operating at low pressures of pumped gas are now more comprehensible. The effects are not concerned with the primary action, which remains that of momentum transfer from the work fluid to the pumped gas, but rather with the secondary action of changing the wall boundary-layer structure by making it more dense and resistant to backflow, and by restricting the number of consecutive collisions of work-fluid molecules which lead to a net backwards movement of these molecules.

The pumping effect of the vapor stream is primarily as follows. First, the high-velocity vapor stream sweeps the gas away from the space in front of the nozzle, by a combination of viscous flow and true diffusion mechanisms,

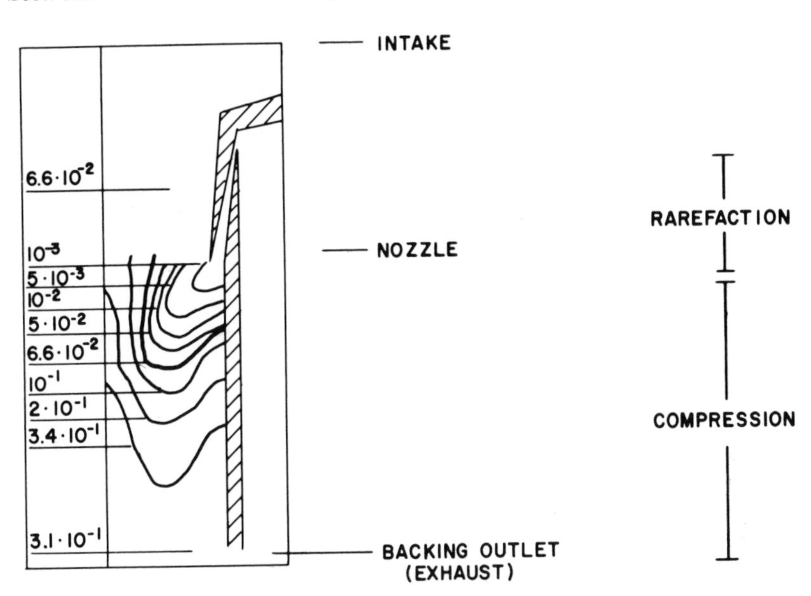


Fig. 3.21. Isobars showing distribution of pressures ($P_{mm}$) at different points below the nozzle of a mercury-vapor pump (Alexander). The thick line is the isobar with a value equal to the pressure at the opening of the pump.

such that the velocity of the pumped gas becomes higher than it would be by diffusion along a pressure gradient into the work-fluid stream. The viscous-flow characteristic dominates the pumping action. This is confirmed by the very variable efficiency of such devices, depending on geometry and cooling in the pump chamber, by the comparative absence of any sizable dependence on the nature of the pumped gas, and by the fact that most condensation pumps are insensitive to small changes in vapor temperature, none of which phenomena are expected for a true diffusion pump. The condition of efficient working of vapor-stream pumps, as shown by Alexander, is that there be a net compression of the pumped gas over a comparatively small linear dimension within the pump chamber. If this condition is met, then the speed of the pump is controlled by the rate at which pumped gas passes through the intake port, and the device will have a finite speed until the pressure of the gas to be pumped is zero, or until this pressure equals the vapor pressure of gases sorbed on and dissolved in the pump walls.

On the basis of these considerations, a high pumping speed will be reached if (1) the area of the throat is as large as possible, (2) the nozzle is directed straight into the pumping direction, (3) the density of the vapor along the outer wall is as high as possible, and (4) the walls enclosing

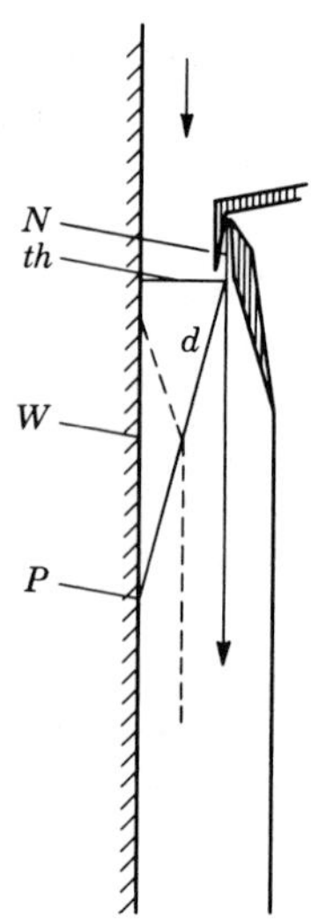


Fig. 3.22. Principle of mercury-vapor pump (Alexander).

the pumping space are cooled. In order to extend the range of pressures within which the pump works efficiently, three parameters have to be controllable, namely, the heat input, the width of the throat, and the gap of the nozzle.

Conditions 1 and 2 seem to be contradictory to condition 3, because a large area of the throat and a straight-downwards-directed nozzle mean lower density of the vapor along the wall. This difficulty is largely overcome by shaping the outer wall so that it is not parallel to the pumping direction. In Fig. 3.22 the line *th* is the width of the throat, *N* is the nozzle, and the arrow shows the pumping direction. The line *d* can be regarded as the boundary within which the densest part of the vapor is concentrated, and the density along the wall at the point *P* depends on the length of this line. By constructing the water-cooled wall *W* in such a way as indicated by the broken line, the length of *d* can be considerably deduced without reducing *th*. The consequent reduction of area will not impede the flow of gas, as the pressure of the gas in this region is comparatively high.

A pump embodying these principles was built by Alexander, with a variable nozzle orifice and throat area. Above a pressure of 50 microns, the best results were obtained with maximum heat input (mass flux) and smallest throat area, whereas at the lowest pressures the reverse was true. It is quite apparent that this pump possessed its dual function as a result not only of changing the final pressure $p_0$ in the quotient $p_0/p$ by deliberately lowering the mass flux at the lowest pressures, but also of changing the initial pressure $p$, to some extent, by restricting the throat area at the maximum mass flux, creating a higher regime of working pressures. Alexander also built another pump along the same lines that had a maximum speed of 1400 liters · sec$^{-1}$ in the pressure range 0.1–100 microns, corresponding to 4 liters · sec$^{-1}$ · cm$^{-2}$ of throat area. The performance of the first pump with respect to heat input and throat dimension is shown in Table 3.4.

**TABLE 3.4**

**Variation of Pump Speed with Changes in Throat Configuration**

| Heat Input (watts) | Throat Width (cm) | Speed/Throat Area (liters · sec$^{-1}$ · cm$^{2}$) | | | |
|---|---|---|---|---|---|
| | | $10^{-4}$ mm | $10^{-3}$ mm | $2 \cdot 10^{-3}$ mm | $10^{-2}$ mm |
| 12,000 | 1.50 | 0.9 | 1.4 | ... | 3.2 |
| 12,000 | 3.30 | 0.9 | 1.4 | ... | 2.6 |
| 12,000 | 5.00 | 0.9 | 1.4 | ... | 2.5 |
| 9,250 | 2.25 | 1.3 | 1.7 | ... | 3.3 |
| 9,250 | 3.10 | 1.3 | 1.7 | ... | 2.9 |
| 9,250 | 5.00 | 1.3 | 1.7 | ... | 2.8 |
| 2,700 | 1.50 | 2.0 | 2.2 | 2.5 | ... |
| 2,700 | 3.30 | 2.0 | 2.0 | 1.9 | ... |
| 2,700 | 5.00 | 2.0 | 1.9 | 1.6 | ... |

It is apparent that all these results are in agreement with that part of Langmuir's condensation idea which calls for the effective removal of work-fluid molecules not traveling with the useful pumping stream, and, as mentioned above, disagree with the contention that all vapor pumps function as diffusion pumps, with the speed determined by the diffusion of the pumped gas into the vapor stream.

### The Performance of Vapor Pumps—Specific Effects

*Effect of power input.* The effect of increased power is substantially that of increasing the mass flux in the pump: however, it also raises the temperature in the boiler. The resultant increased mass flux improves the performance of the pump, as evidenced by ultimate vacuum, up to a certain value, when it ceases to be further effective. The speed of the pump is also directly affected by the mass flux available for momentum transfer, provided that the fore-pump pressure is below a certain minimum value, for example, 150 microns. Reduction of fore-pump pressure below such

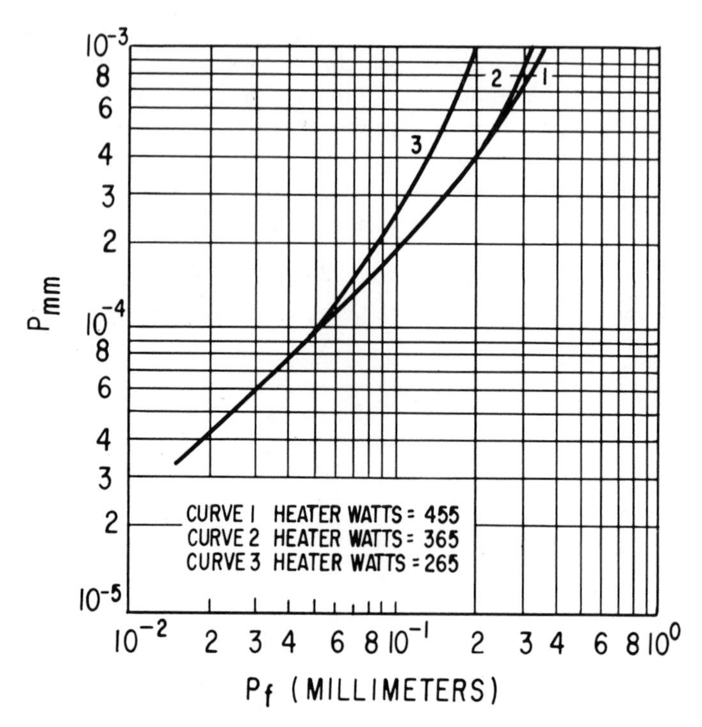


Fig. 3.23. Fine-pump pressure ($P$) versus fore-pump pressure ($P_f$) for various heater currents (gas being pumped, argon).

a limit has no further effect on speed or ultimate vacuum of the vapor pump.

*Effect of fore-pump pressure.* As mentioned above, the fore-pump pressure is a limiting factor in the operation of a vapor-stream pump. Matricon[23] has demonstrated this effect, using a simple single-stage condensing mercury pump. The consequences of varying the power input and changing the rate of cooling are shown in Figs. 3.23 and 3.24 to be related to the maximum fore-pump pressure against which the device will work.

*Effects of different nozzles.* Attention has already been drawn to the critical nature of nozzle design. In an investigation into the effect of the nozzle area on speed, Estermann and Byck[24] used hollow cone nozzles which could be varied in thickness and hence in the area of the nozzle. The cone half-angles were about 13° and their slant height 4 cm. Table 3.5 shows the effect of the clearance area (that is, the annular space between nozzle and wall) on speed (theoretically 11.67 liters · sec$^{-1}$ · cm$^{-2}$ maximum). All the speeds in this table are uncorrected for the resistance

TABLE 3.5

**Effect of Nozzle Area on Speed of Pump**

(Estermann and Byck)

| Clearance (cm) | Region Area of Diffusion (cm²) | $S$ (liters · sec$^{-1}$) | $S$/Area (liters · sec$^{-1}$ · cm$^{-2}$) |
|---|---|---|---|
| 0.1 | 1.5 | 3.0 | 2.0 |
| 0.2 | 2.9 | 5.0 | 1.72 |
| 0.33 | 4.7 | 8.5 | 1.81 |
| 0.55 | 7.4 | 13.0 | 1.76 |
| 0.63 | 8.4 | 9.5 | 1.13 |
| 0.71 | 9.2 | 8.4 | 0.91 |

to flow of the tube, from the nozzle to the measuring gauge. The corrected value gives an intrinsic pump speed of about 23 per cent of theoretical, calculated from the equation for diffusion through an orifice

$$S = \frac{Av_a}{4}\left(k\alpha - \frac{P_0}{p}\right), \tag{3.20}$$

where $Av_a/4$ corresponds to the conductance of the orifice through which the gas diffuses into the stream of vapor. However, as might be expected from the previous comments on the form of the mass flow of the vapor, $Av_a/4$ may not correspond to a quantity derived from the actual annular area of the intake, and the per cent of theoretical speed may actually be

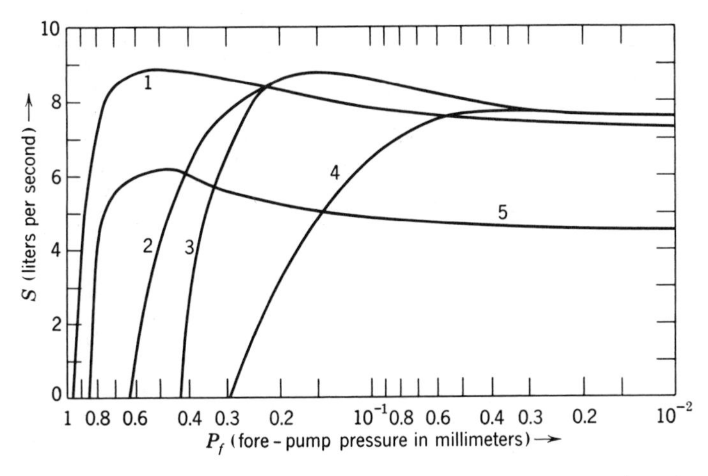


Fig. 3.24. Semilog plots of speed of pump versus fore-pump pressure (Matricon).

spuriously high. Indeed, Hickman[25] prefers to think of the entire expansion cone surface modified by the vapor admittance for the gas as a more realistic measure of the diffusion area.

The factor $k\alpha$ in equation 3.20, which depends on vapor temperature, among other things, was called the "speed factor" by Ho,[26] who investigated the effect of the number and design of nozzles upon the performance of the pump, and it has been widely used as a factor in the comparison of pumps. That it is an empirical factor was recognized by Ho himself, since he thought of its variation with temperature, the shape of the nozzle, and the vapor density distribution below the diffusion annulus. Ho found that a divergent jet gave higher values of $k\alpha$ (frequently symbolized by $B$, in this connection) than a long straight jet , and that these values also varied and could be maximized with respect to the relative physical dimensions of the jet and the chamber. He further observed that organic pump fluids gave a higher speed in the same pump than mercury. Copley *et al.*[27] also studied the speeds of pumps with divergent nozzles and found that, at the same annular width, long divergent nozzles were more efficient than short or straight single or multiple nozzles. The results of these studies are incorporated in the design of most modern glass/mercury pumps.

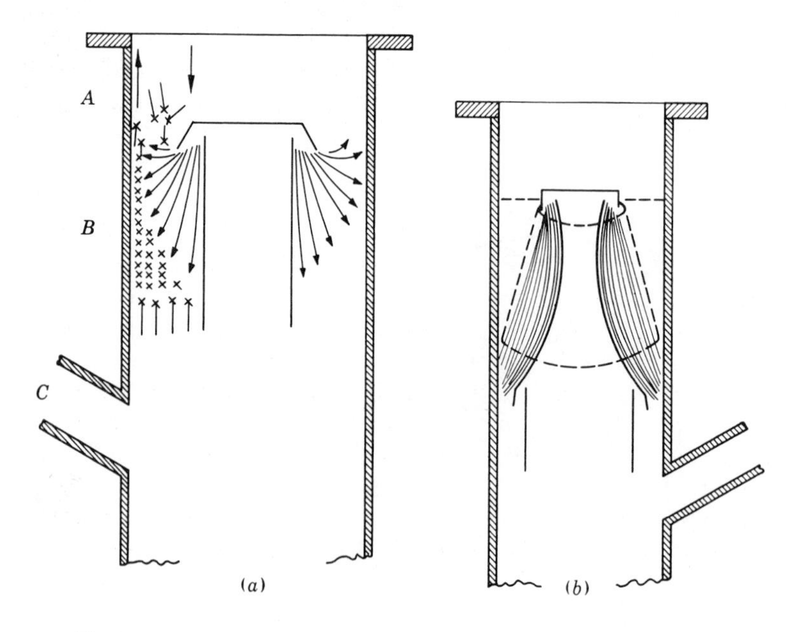


Fig. 3.25. Design of Embree jet (*b*) compared with straight-sided jet (*a*).

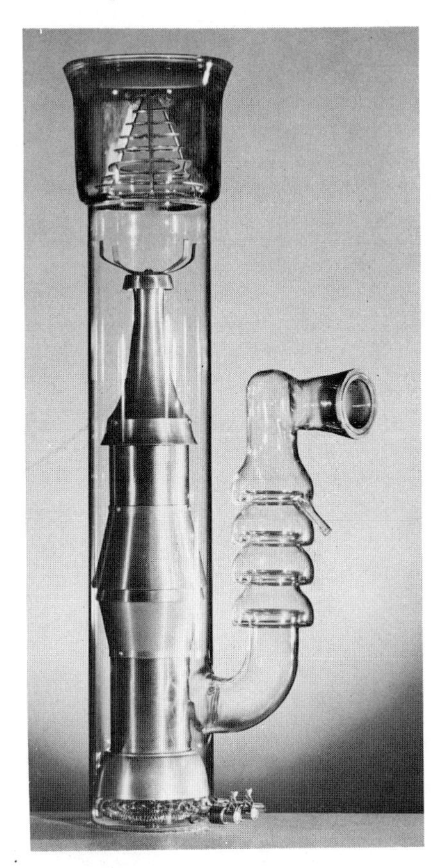

Fig. 3.26. High-speed three-stage ; oil-vapor pump with metal ii (Consolidated Vacuum Corp.)

Results similar to the above were obtained by Holtsmark *et al.*[28] for oil-vapor pumps, and Hickman[25] has considered the design of nozzles in these systems in some detail. Hickman also considered that too thorough condensation of the work fluid permitted a ready back diffusion of pumped gas, and recommended balancing the boil-up rate, the jet design, and the cooling. He emphasized the most important factor in design to be the directing of the jet straight in the principal axis of the pump, and pointed out that both the speed factor $B$ and the annular area could be increased by lowering the fore pressure, a consideration that has led to the design of multistage jet systems.

For very high-speed designs, it is necessary to reduce the drag between the inside of the expanding cone and the vapor stack in designs like those of Fig. 3.25(*a*). This has led to aerodynamic contouring of the vapor stack in the form of Fig. 3.25(*b*): the original design of this type is due to Embree.[29] An example of its application can be seen in Fig. 3.26. The contouring permits a greater subtended angle for free expansion of the pump fluid with fairly obvious advantages in the proportion of the work fluid available for useful momentum transfer. The apparent speed factor for such designs is around 30 per cent, with transients up to 50 per cent.

## Variation of Speed with the Nature of the Pumped Gas and Pump Connections; "Blank-off"

It was previously remarked that the speed of any diffusion process occurring in this class of pump should be related to the inverse square root of the molecular weight of the pumped gas. However, the ratio of the gas pressure just below the orifice to the exhaust pressure, $P_0/P_f$, is usually so small that this effect of molecular weight is not often observed (Fig. 3.27). It may, however, happen (as frequently observed) that, though a pump gives a good vacuum with air or argon, it is no longer

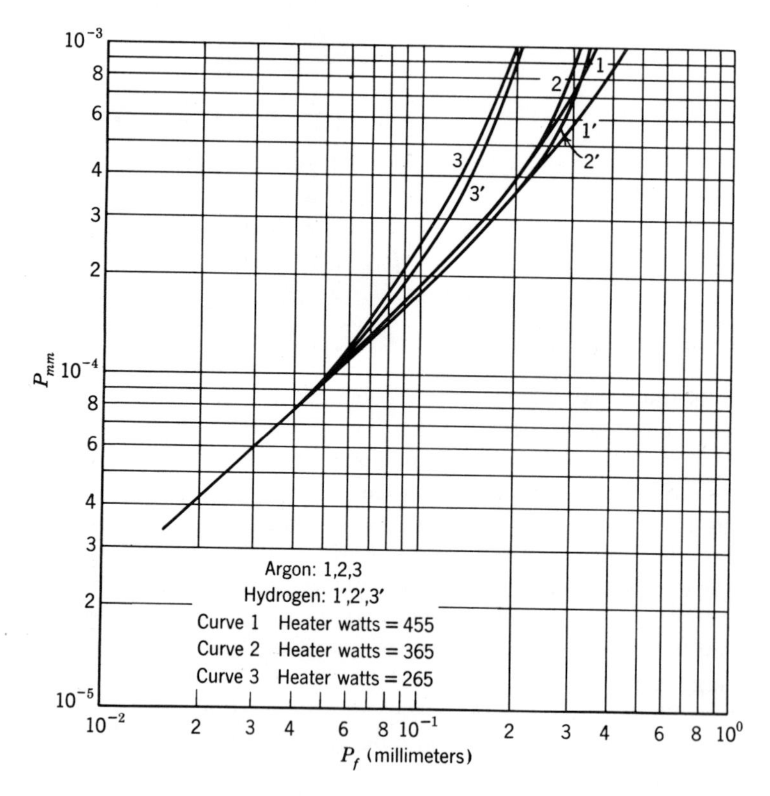


Fig. 3.27. Log-log plots of fine-pump pressure ($P$ in millimeters) versus fore-pump pressure $P_f$ (in millimeters) for argon and hydrogen (Matricon[23]).

as effective with hydrogen. Since $P_0/P_f$ decreases with increase in the velocity of the vapor, the obvious remedy for such a situation is increasing the heater power input.

One very important factor with regard to any pump is the relation between speed of exhaust and pressure $P$ in the system ($P > P_0$). In order to discuss this problem it is necessary to refer once more to equation 3.20, which can be written in the form

$$S = \frac{Av_a}{4}\left(B - \frac{P_0}{P}\right). \tag{3.21}$$

In this expression, the factor $Av_a/4$ corresponds to the conductance, $F_0$, of the orifice through which the gas diffuses into the stream of vapor. It may, therefore, be calculated for any given design of nozzle to give the theoretical maximum value of $S$. The quantity $B$ depends upon the temperature of the vapor, and it will be remembered that $B$ is $k\alpha$ or $k\alpha/L$, $a$ being the radius of the intake port, $L$ the mean free path, and $k$ a semiempirical factor ascribed to a "collision efficiency" in the nonuniform gas-vapor mixture. If the pressure becomes very small, the speed of the pump will decrease. However, when we consider that at the lowest values of pressure each gas molecule entering the vapor stream will be instantly swept away ($k \rightarrow 1$ as pressure $\rightarrow 0$), the problem of whether the speed will go to zero reduces to the question, "Can $\alpha = P_0/P$?"

Since this problem contains the question of the exact relation of the viscosity $\eta$ of a gas to its pressure $p$, an unanswered riddle, it is impossible to obtain a satisfactory answer. In the trivial case where $p$ is zero, and $p_0$ is also zero, then $\alpha$ [= constant $\times$ $(P/\eta)$] and $p_0/p$ will both be indeterminate as 0/0. In the practical case, recent observations of such pumps having a measurable speed at $10^{-10}$ mm[30] and lower[31] make it seem unlikely that any such real zero value exists, unless of course the pump chamber starts to degas, upsetting the analysis of the flow by adding to $p_0$.

Before considering the manner in which the value of the factor $B$ in equation 3.21 varies with the design of nozzle used in a pump, it is well to point out that, in any experimental determination of the value of $B$ from direct measurements of $S$ for a given pump, it is necessary to introduce certain corrections. In pumps used in practice the value of $S$ which is actually measured for the given area of orifice at the nozzle must be corrected for the effect of resistance to flow due to the length of tube above the orifice. Denoting the conductance in series with the orifice by $F$, and the intrinsic speed at the nozzle by $S_p$, the observed value of the speed is given by the relation,

$$\frac{1}{S} = \frac{1}{S_p} + \frac{1}{F}. \tag{3.22}$$

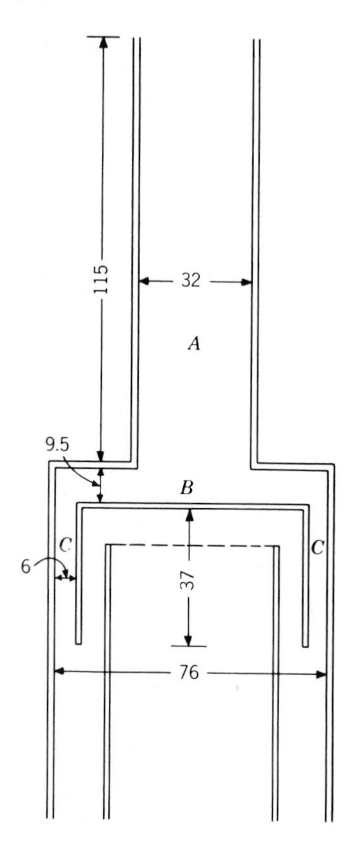


Fig. 3.28. Dimensions (in millimeters) of single-stage vapor pump of metal used by Matricon.

As an illustration of the application of this relation it is of interest to calculate the values of $S$, $S_p$, and $B$ for a pump, the dimensions of which are given in Fig. 3.28, and which was used by Matricon.[23]

From the dimensions given in Fig. 3.28 we obtain the following values of the conductances for air at 25° C (cubic centimeters per second):

Section $A$: $F_A = 2.35 \cdot 10^4$, $1/F_A = 4.36 \cdot 10^{-5}$.
Section $B$: $F_B = 2.56 \cdot 10^5$, $1/F_B = 0.39 \cdot 10^{-5}$.
Annular section $C$: $F_C = 6.67 \cdot 10^4$, $1/F_C = 1.5 \cdot 10^{-5}$.

(The value of $F_B$ in the following calculation is that given by Matricon. The other values were calculated by Dushman from the relations given in Chapter 2.) Hence,

$$\frac{1}{F} = \frac{1}{F_A} + \frac{1}{F_B} + \frac{1}{F_C} = 6.90 \cdot 10^{-5}.$$

Now the observed speed of the pump was $7.8 \cdot 10^3$, corresponding to $1/S = 12.82 \cdot 10^{-5}$. Hence the intrinsic speed of the pump is given by the relation

$$\frac{1}{S_p} = (12.82 - 6.90)10^{-5} = 0.92 \cdot 10^{-5},$$

that is, $S_p = 16{,}890 \text{ cm}^3 \cdot \text{sec}^{-1}$.

Since the area of the slit across which diffusion occurs is 13.2 cm², $Av_a/4 = 1.54 \cdot 10^5$. Hence

$$B = \frac{1.69 \cdot 10^4}{1.54 \cdot 10^5} = 0.11.$$

It will be observed that the resistance of the cylindrical section $A$ constitutes about two-thirds of the total resistance, $1/F$. By decreasing the length of this section, the value of $1/F$ would be decreased, thus yielding higher values of both $S$ and $B$. This result serves to emphasize once more the importance of making all connections between the pump and the system as short and as wide as possible if full advantage is to be taken of the maximum potential speed of a given design of pump.

Since the cross section of $A$ is 8.04 cm², the speed per unit area of opening is approximately $1000 \text{ cm}^3 \cdot \text{sec}^{-1} \cdot \text{cm}^{-2}$ compared to $11{,}670 \text{ cm}^3 \cdot \text{sec}^{-1} \cdot \text{cm}^{-2}$, which is the theoretical speed. Thus the ratio between the actually observed and the calculated speed for the pump is about 0.085. The same result has also been obtained in the General Electric Research Laboratory on metal pumps. In general, the ratio between the observed speed and that calculated for an opening of equal area from the relation $S = 11{,}670A$ varies, for mercury-vapor pumps, between 0.07 and 0.1.

## Development of Vapor-Stream Pumps

*General.* A single-stage vapor-stream pump requires a rather low fore-pump pressure for successful operation. In order to lessen this disadvantage, a large number of multistage pumps have been designed, and many have been fabricated from steel, in which an annular cap nozzle seems to be preferred. Thus Gaede[20] described a three-stage pump made entirely of steel, which could be operated against a pressure of 20 mm at a speed of 60 liters · sec$^{-1}$ for air and 100 liters · sec$^{-1}$ for hydrogen.

A two-stage pump made of Pyrex has been described by Kurth, and Fig. 3.29 shows a variation of this construction designed by W. A. Ruggles in the General Electric Research Laboratory. It has the great advantage that the glass blowing is not very difficult.

The mercury boiler shown at $M$ is heated by a standard radiant heater

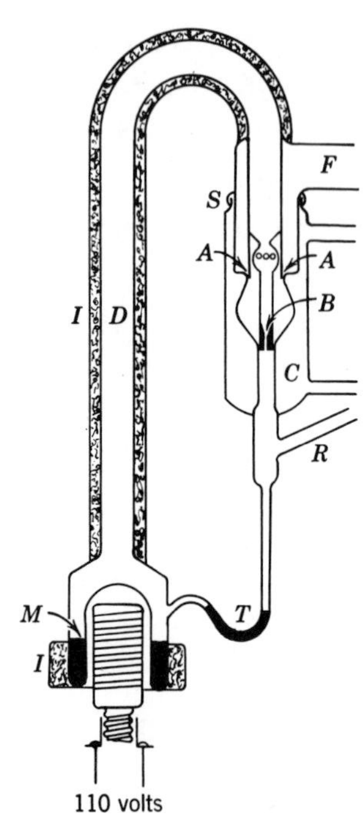


Fig. 3.29. Design of Kurth-Ruggles two-stage mercury-vapor pump, glass form.

unit as shown, and the blast of mercury passes up the tube *D*, which is covered with some heat-insulating material *I*. The mercury passing through the small holes past the opening *A* is condensed on the adjacent parts of the walls by the cooling water in the jacket *C*, while the higher-pressure stage occurs at the opening to nozzle *B*. The rough pump is connected to *R*, and the vessel to be exhausted, at *F*. Tape *S* is wound around the top of the water jacket *C* and thus provides a flexible joint there.

The speed obtained with this pump, when used in series with a liquid-air trap, is about 700 $cm^3 \cdot sec^{-1}$, and pressures as low as $5 \cdot 10^{-5}$ micron have been attained, as measured by an ionization gauge. Much lower pressures have been obtained with one-stage mercury-vapor pumps.[30,31]

Two-stage mercury-vapor pumps have also been described by Volmer,[32] Kraus,[33] and Stimson.[34] The designs of Kraus and Stimson consist essentially of two Langmuir pumps in series. The speed attained with Stimson's design is reported to be 250 $cm^3 \cdot sec^{-1}$.

Two designs of all-metal mercury-vapor pumps have been described by Backhurst and Kaye,[35] in which the first stage utilizes the principle of the ejector. The speed obtained in the second, more compact design varies from 1000 to 7000 $cm^3 \cdot sec^{-1}$, "depending on the pressure in the vessel being exhausted and also on the value of the back pressure against which the pump is working. The pump will readily produce a vacuum of the order of $10^{-5}$ mm or less when working against a back pressure of 1 mm of mercury."

About 1930 a considerable amount of work was carried on in the General Electric Research Laboratory by J. H. Payne, in cooperation with the Lamp Division of the company, in order to determine the most suitable design for a high-speed, high-vacuum, multistage metal pump.[36] Figure 3.30 shows a design of a two-stage pump having the following characteristics: volume of mercury, 70 $cm^3$; power input, 550 watts; fore-pump pressure, 1–2 mm; minimum pressure, about $3 \cdot 10^{-4}$ micron; speed, 12–15 liters $\cdot$ $sec^{-1}$.

A three-stage pump of similar design, used in the exhaust of high-power transmitting tubes, gives the same degree of vacuum at a speed of 60 liters $\cdot$ $sec.^{-1}$

It is perhaps worth noting that single-stage divergent-nozzle glass-mercury pumps are now made semiautomatically in the United States,

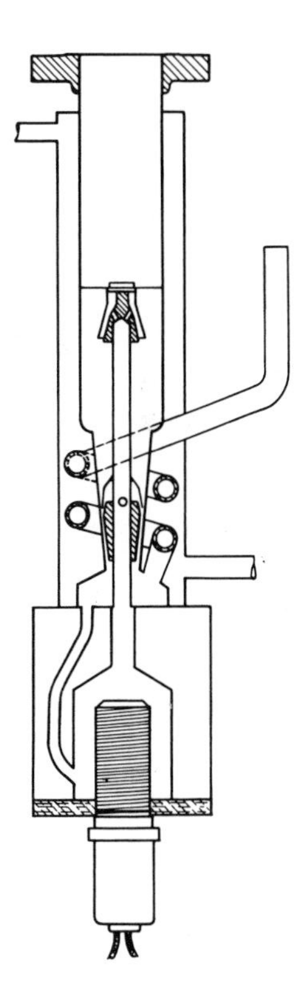

Fig. 3.30. Design of two-stage mercury-vapor pump of metal.

TABLE 3.6

Vapor-Pressure Data for Oils Used in Vapor Pumps

| Name | Ref. | $A$ | $B$ | Degrees C for $P$ (microns) = | | | | | | | $P$ (microns) at 25° C |
|---|---|---|---|---|---|---|---|---|---|---|---|
| | | | | $10^3$ | $10^2$ | 10 | 1 | $10^{-1}$ | $10^{-2}$ | $10^{-3}$ | |
| Butyl phthalate | 1 | 14.215 | 4680 | 144 | 110 | 81 | 56 | 34 | 15 | −1 | $3.3 \cdot 10^{-2}$ |
| | 2 | 15.589 | 5120 | 134 | 104 | 78 | 55 | 35 | 18 | 2 | $2.5 \cdot 10^{-2}$ |
| | 3 | 14.05 | 4625 | 145 | 111 | 81 | 57 | 37 | 20 | 4 | $2.0 \cdot 10^{-2}$ |
| | 4 | 14.50 | 4780 | 143 | 109 | 81 | 57 | 35 | 17 | 0 | $2.9 \cdot 10^{-2}$ |
| Amoil | 4 | 13.60 | 4610 | 162 | 124 | 93 | 66 | 43 | 22 | 5 | $1.3 \cdot 10^{-2}$ |
| Octoil | 1 | 15.116 | 5590 | 188 | 153 | 123 | 97 | 74 | 54 | 35 | $2.3 \cdot 10^{-4}$ |
| | 3 | 14.15 | 5205 | 194 | 155 | 123 | 95 | 72 | 53 | 35 | $2.2 \cdot 10^{-4}$ |
| | 4 | 13·00 | 4870 | 214 | 170 | 133 | 102 | 75 | 52 | 31 | $4.6 \cdot 10^{-4}$ |
| Amoil-S | 4 | 14.40 | 5190 | 182 | 146 | 114 | 87 | 64 | 43 | 25 | $1 \cdot 10^{-3}$ |
| Octoil-S | 3 | 14.26 | 5514 | 215 | 177 | 142 | 114 | 89 | 68 | 50 | $2 \cdot 10^{-5}$ |
| | 5 | 16.54 | 6434 | 205 | 171 | 141 | 116 | 94 | 74 | 56 | $9 \cdot 10^{-6}$ |
| Tri-*m*-cresyl phosphate | 2 | 15.886 | 6088 | 199 | 165 | 136 | 110 | 87 | 67 | 50 | $3 \cdot 10^{-5}$ |
| Tri-*p*-cresyl phosphate | 2 | 15.223 | 5926 | 212 | 175 | 144 | 116 | 92 | 71 | 52 | $2 \cdot 10^{-5}$ |
| Dibenzyl sebacate | 2 | 15.775 | 6320 | 222 | 186 | 155 | 128 | 104 | 82 | 64 | $4 \cdot 10^{-6}$ |
| Arochlor | 3 | | | 161 | 123 | 93 | 67 | 45 | 27 | 10 | $8 \cdot 10^{-3}$ |
| Apiezon A | 3 | | | 186 | 144 | 110 | 82 | 58 | 37 | 20 | $2 \cdot 10^{-3}$ |
| Apiezon B | 3 | | | 206 | 162 | 127 | 97 | 73 | 50 | 31 | $4 \cdot 10^{-4}$ |
| Litton oil | 3 | | | 209 | 167 | 132 | 103 | 77 | 57 | 30 | $1.4 \cdot 10^{-4}$ |
| 171 dist. (hydrocarbon) | 3 | | | 213 | 171 | 135 | 106 | 81 | 60 | 42 | $9 \cdot 10^{-5}$ |
| $(CH_3)_3SiO$-$[Si(CH_3)_2O]_x$-$Si(CH_3)_3$ | | | | | | | | | | | |
| $x = 12$ | 6 | 14.60 | 5350 | | 152 | | 93 | | 49 | | $4.4 \cdot 10^{-4}$ |
| $= 13$ | 6 | 15.04 | 5710 | | 165 | | 107 | | 62 | | $7.6 \cdot 10^{-5}$ |
| $= 14$ | 6 | 15.48 | 6070 | | 177 | | 119 | | 81 | | $1.3 \cdot 10^{-5}$ |
| $= 15$ | 6 | 15.93 | 6430 | | 189 | | 131 | | 86 | | $2.2 \cdot 10^{-6}$ |
| $= 16$ | 6 | 16.37 | 6790 | | 200 | | 142 | | 97 | | $3.8 \cdot 10^{-7}$ |
| $= 17$ | 6 | 16.81 | 7150 | | 210 | | 152 | | 107 | | $6.5 \cdot 10^{-8}$ |
| $= 18$ | 6 | 17.25 | 7510 | | 219 | | 162 | | 117 | | $1.2 \cdot 10^{-8}$ |
| Silicone 704 | 7 | 11.03 | 5570 | | 190 | | 120 | | 76 | | $4 \cdot 10^{-8}$ |

*References for Table* 3.6

1. K. C. D. Hickman, J. C. Hecker, and N. Embree, *Ind. Eng. Chem.*, Anal. Ed., **9**, 264 (1937).
2. F. H. Verhoek and A. L. Marshall, *J. Am. Chem. Soc.*, **61**, 2737 (1939).
3. Metropolitan-Vickers, personal communication, 1946.
4. Values of constants deduced by S. F. Kapff and R. B. Jacobs of Distillation Products Inc.
5. Values of constants deduced from plot sent to Dushman by Dayton of Distillation Products, Inc.
6. D. F. Wilcock, *J. Am. Chem. Soc.*, **68**, 691 (1946).
7. A. R. Huntress *et al.*, *Vacuum Symposium Transactions*, p. 104, 1959.

and are therefore quite inexpensive. The writer and his colleagues frequently use such pumps connected in series by cooled wide tubes in place of multistage pumps, to provide a significant reduction in back streaming.

*The use of organic fluids in vapor pumps.* In 1928, Burch,[37] of the Metropolitan-Vickers Electrical Company, Ltd., England, observed that mercury could be replaced in condensation pumps by certain high-boiling-point petroleum distillates. Employing one such material, of vapor pressure 1 micron at 118° C, and of thermal stability such that it could be heated to a temperature corresponding to about 100 microns, in a condensation pump, Burch obtained a vacuum of $10^{-7}$ mm without using a cold trap. This observation led to the now familiar oil vapor-stream pumps.

This work was followed up by Hickman[38] and his associates, whose contributions have been in two main areas; first, the synthesis of pure organic compounds which can be used satisfactorily in a vapor pump; and, second, the development of designs of pumps which will operate efficiently in producing the desired degree of vacuum.

An important feature of pumps using oil is that they must be permitted to purge themselves of gases and vapors, all kinds of which dissolve in the oils used, and which ruin the performance of the pump. Thus, these pumps usually have an extra chamber in which the impurities are removed, and are frequently fitted with baffles. The ultimate vacuum in such a device is not limited by the saturation vapor pressure of the clean oil: these devices, like mercury pumps, have been found to operate down in the range of $10^{-10}$ mm Hg, but their performance is limited by the stability of the pure oil, specifically the noncondensable products of oil decomposition, such as $H_2$ and CO.

A rather large number of organic fluids are satisfactory for use in vapor pumps; a partial listing is made in Table 3.6, where the parameters $A$ and $B$ are those of the vapor-pressure relation

$$\log P = A - \frac{B}{T}.$$

$B$ is also related to the latent heat of vaporization of the liquid, $\Delta H^V$ cal/mole:

$$\Delta H^V = 4.574B.$$

The given values of $P$ at 25° C were calculated from the above equation. It will be noticed that not only carbon but also silicon-based organic materials can be used in these pumps. (See also Chapter 10 for information on low-vapor-pressure organic fluids.)

A peculiarity of some of these fluids is their need for conditioning, particularly if they are unsaturated. Thus, a good vacuum is not reached until the oil has attained an equilibrium, presumably with regard to its decomposition products, and this takes time which may be measured in hours. Fresh oil almost always gives a strong pressure burst on first usage; this can easily be mistaken for a system leak.

It is also necessary to take care not to expose the boiling oil to oxidizing vapors. The various fluids vary markedly in their resistance to this treatment: the best include the silicon-based materials and phthalate esters. All these fluids are, of course, powerfully adsorbed by glass and metals surfaces, and a freshly degassed ion gauge is consequently a powerful sink for such vapors. Subsequent operation of the gauge will crack the compounds and ruin the vacuum.

A simple pump[38] embodying the fractionating feature essential to these devices is shown in Fig. 3.31. The points to note are a short, wide vapor supply, cooling restricted to the upper portions of the jacket, and an abnormally large side arm leading to the mechanical pump. The side arm is provided with catchment lobes, and its junction with the main portion is heavily lagged. The heat input is adjusted until there is perceptible refluxing in the arm, but the outflow of vapor is not allowed to exert sufficient back pressure to interfere with the jet performance.

An advantage possessed by oil pumps over mercury pumps lies in the lower room-temperature vapor pressure of the work fluid. It is possible to avoid the use of liquid-air traps, and, by providing a baffled path

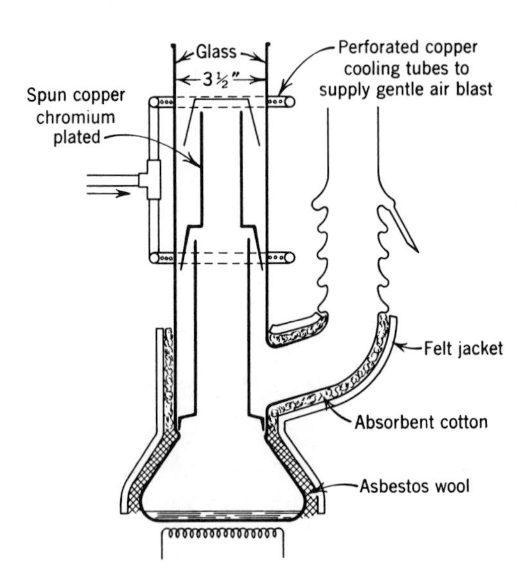


Fig. 3.31. Design of fractionating vapor pump (Hickman).

toward the high-vacuum side, to obtain high vacua in locations remote from sources of liquefied gases. These pumps have therefore found favor in small laboratories and in industrial processes. Of course, this advantage is partly offset by the oil-stability problem.

In general, the pump stages designed for use with oils differ from those intended for mercury only in those dimensions which are dictated by the different physical properties of the organic compounds. Their steep vapor-pressure curves call for a short, wide connecting tube (which must be protected against cooling) between boiler and nozzle. The ease of condensation permits a short cooling surface, and the lower vapor pressure at room temperature permits a larger clearance between nozzle and condensing wall. An annular jet is preferred over the coaxial cylindrical jet for the same reason, unless the latter can be superheated.[40] For the same area of the nozzle, pumps using oil give greater speeds of exhaust than those with mercury. In view of Ho's demonstration of the advantages of multinozzle jets in mercury pumps,[26] Zabel[41] designed a multinozzle oil pump and demonstrated the speed enhancement expected, provided that the heat input was raised as the number of nozzles (but not the total inlet area) was increased. Copley *et al.*[27] also applied their findings with mercury pumps to the design of a divergent-nozzle oil pump.

A two-stage pump has been described by Edwards[42] in which conical baffles were apparently "effective in preventing oil vapor from entering the chamber to be evacuated and at the same time in offering a minimum of impedance to the flow of gas through the pump."

A multiple-nozzle type of metal pump has been described by Amdur.[43] When this pump was operated with Apiezon B oil at a maximum forepump pressure of 30 microns, the speed obtained for hydrogen was 250 liters $\cdot$ sec$^{-1}$ at 0.1 micron. The heat input was 630 watts.

*Fractionating and other types of oil-vapor pumps.* Mention was made in the previous section of the importance of fractionating the fluid in oil-vapor pumps. Vertical, all-metal constructions, based on the self-fractionating principle, were designed by Lockenvitz, [44] Malter, and Marcuvitz,[45] and Sykes and Bancroft,[46] as well as by Hickman.[38] Frequently, no change is made in the arrangement of stages from the initial designs of Hickman. Thus, the pump of Malter and Marcuvitz differed only in the addition of fractionating cylinders (Fig. 3.32, *A* and *B*). During operation the condensate of the vapor issuing from the separate stages collects in the region surrounding the lower end of cylinder *B*. It is prevented from immediately entering into the region between *A* and *B* by the small size of the apertures in cylinder *A*. As a consequence, the lowest-boiling-point fractions are boiled off before the remaining fluid passes into the region between *A* and *B*. In the same way, further low-boiling-point fractions are boiled off in

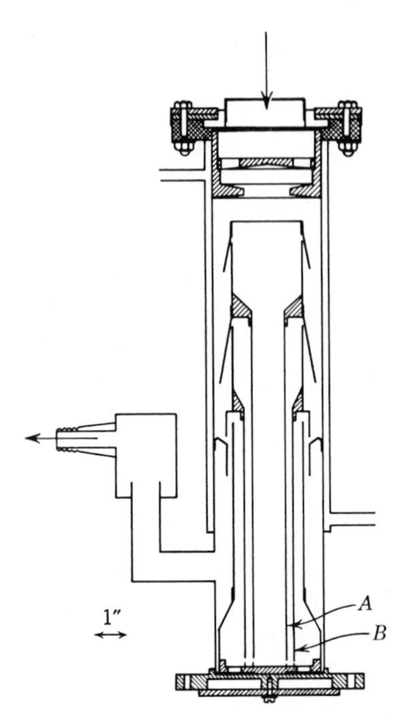


Fig. 3.32. Vertical, all-metal vapor pump of fractionating type (Malter and Marcuvitz).

the regions between cylinders *A* and *B* with the result that the fluid which finally works its way through the two sets of apertures into the interior of cylinder *A* is composed of the highest-boiling-point fractions or, in other words, the fractions of lowest vapor pressure.

Table 3.7 shows the ultimate pressures (as measured by an ionization gauge connected to the pump) obtained by these investigators, with and without the two inner fractionating cylinders. With 450 watts the pump operated against a fore pressure of 150 microns, and with 600 watts the fore pressure could be increased to 0.5 mm.

It was found necessary in this design to insert a baffle at the top of the pump, as shown in Fig. 3.32. Using Octoil-S, the speed without the baffle varied between 70 and 80 liters $\cdot$ sec$^{-1}$ over a pressure range of $10^{-3}$ to $10^{-6}$

**TABLE 3.7**

**Effect of Fractionating Chamber on Ultimate Vacuum**

| | Octoil | "Burned" Octoil | Apiezon B |
|---|---|---|---|
| Fractionating cylinders in | $8 \cdot 10^{-8}$ mm | $2.0 \cdot 10^{-7}$ mm | $2.4 \cdot 10^{-7}$ mm |
| Fractionating cylinders removed | $4 \cdot 10^{-7}$ | $9.0 \cdot 10^{-7}$ | $1.0 \cdot 10^{-6}$ |

mm of Hg. The insertion of the baffle reduced this speed to between 35 and 40 liters · sec$^{-1}$. It should also be mentioned that, according to the authors, "Octoil will stand punishment which will completely ruin Apiezon B."

Following this work, Hickman[47] devised a glass and metal fractionating pump of horizontal construction. In his paper Hickman gives reasons for favoring horizontal rather than vertical construction. Two of the main advantages of the horizontal type are that the jet diameter is independent of the boiler area and that the heat input to the separate boilers can be regulated precisely. However, it should be observed that the larger sizes of metal pumps developed by the different manufacturers are of the vertical type.

Figure 3.33 shows a photograph of a water-cooled, three-stage fractionating pump made of Pyrex; the same design has also been constructed for air cooling. A novel feature of this construction is the insertion of the heating coils in the liquid. In consequence of this feature, and of the fact that the boilers are covered with insulating material, optimum efficiency is obtained for the transfer of heat to the oil. The high-vacuum connection is at the left, through an opening of $1\frac{1}{16}$-in. inside diameter. The fore pump is connected at the right through an opening $\frac{7}{8}$ in. in inside diameter. The total length is 25 in., and the height 13 in.

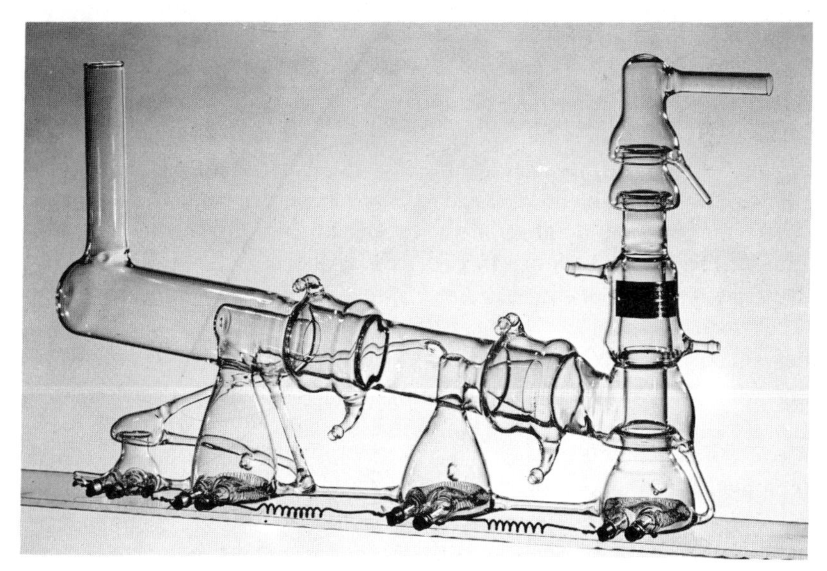

Fig. 3.33. Water-cooled, three-stage fractionating pump (Consolidated Vacuum Corp.).

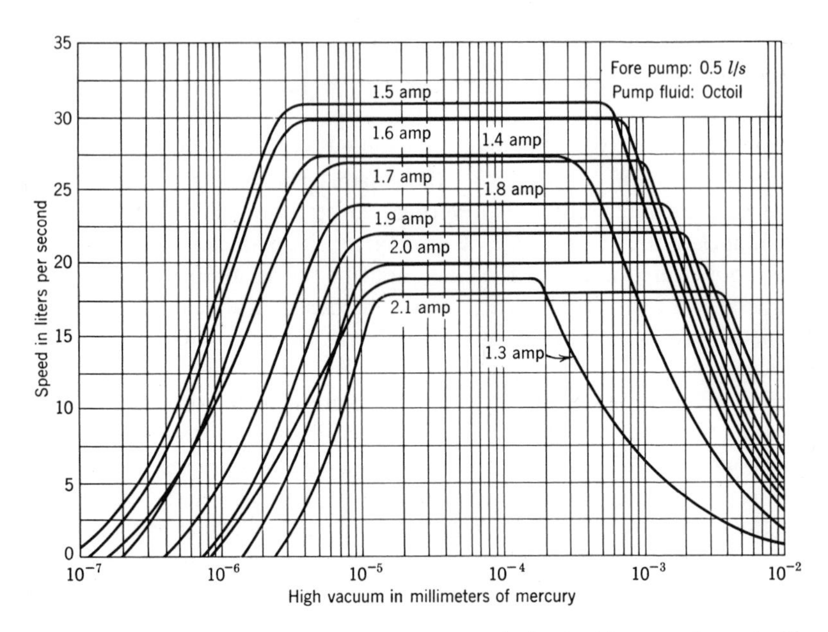


Fig. 3.34. Semilog plots of speed versus pressure obtained with pump shown in Fig. 3.33 at a series of heating currents.

The small boiler at the left functions as a sink for nonvolatiles produced during the operation of the pump. Special connecting tubes are provided between this reservoir and the space adjacent to the high vacuum as well as to the oil in the other boilers.

The pump requires 200 g of Octoil-S and operates at a fore pressure as high as 0.10 mm, with a power input varying from 90 to 250 watts. The speed is stated to be 25 liters $\cdot$ sec$^{-1}$ at $10^{-4}$ mm, and the ultimate vacuum, $5 \cdot 10^{-8}$ mm at 25° C. Values of the speed at a series of heating currents and fine pressures are shown in Fig. 3.34. Air- and water-cooled pumps of various numbers of stages are also available.

To evacuate equipment such as the electron microscope, the cyclotron, the vacuum spectrograph, and lens-coating or metal-evaporation outfits, high-speed three-stage pumps have been developed such as shown in Fig. 3.26. While the illustration shows a glass container with aluminum insert of slits, a similar design has also been incorporated in an all-metal pump which has welded on the outer surface copper coils through which water is circulated for cooling. With an opening at the top, $3\frac{7}{8}$ in. in inside diameter, the speed obtained for air is 220 liters $\cdot$ sec$^{-1}$ at $10^{-4}$ mm pressure. The ultimate vacuum, with Octoil or Amoil-S, is about $5 \cdot 10^{-6}$

mm at 25° C, with a fore-pump pressure of 0.10 mm. The speed is about 30 per cent of theoretical, and pumps of this type are provided by the major manufacturers for speeds up to 50,000 liters · sec$^{-1}$.

To provide the low fore-pump pressure ($10^{-3}$ mm or less) required by vapor pumps for obtaining very low pressures, so-called booster pumps, which operate with butyl phthalate or butyl sebacate at the higher fore-pump pressures (up to 500 microns) that are obtained from high-speed mechanical pumps, are used. These are made of metal, have a simple annular cap nozzle, and provide a certain measure of fractionation by virtue of the baffles in their stacks. In the writer's experience, such booster pumps are a very adequate means of cheaply turning a mechanical pump into a high-vacuum pumping set by simply mounting the booster on the mechanical pump base (though traps are needed on both sides of the booster to prevent oil flow through the system).

Oil-vapor ejector pumps also are employed as boosters. The usual pressures of work fluid are from 15 cm up to 70 cm Hg, and the height of the equipment is kept to a minimum, condensate being returned to the boiler by a mechanical pump. The ejector stage is usually mounted at a slight downward angle, and the pumped gas is stripped of work fluid in a baffled fore-pump line. The more volatile pump fluids are preferred in these devices; it is here that silicon-based oils have their best success. Such devices have a sharp peak in the speed curve (Fig. 3.35) and are most useful in operations where a pressure of 20–100 microns is required in the face of continued gas evolution, like distillation or vacuum smelting. (The

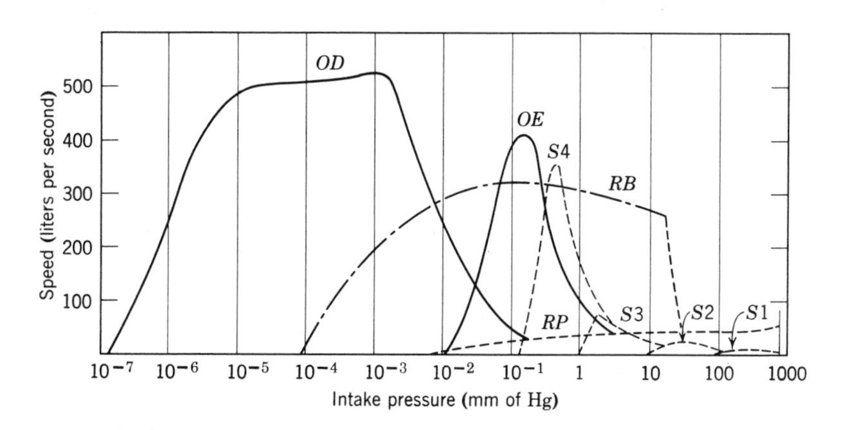


Fig. 3.35. Plots of speed against pressure for various types of pumps. *S*1, *S*2, *S*3, and *S*4 are simple steam ejector pumps, *RP* is a rotary mechanical pump, *RB* is a Roots blower, *OE* is an oil ejector pump, and *OD* is an oil vapor-stream pump.

Roots-type mechanical pump offers competition for these devices in the lower end of this pressure range.)

## 3.7. ION AND GETTER PUMPS

Before the invention of vapor-stream pumps in 1915, high vacuum was usually obtained by using a mechanical pump in conjunction with a getter, or an absorbent material, cooled to low temperatures. Recently, there has been increased interest in the latter techniques as a means of entering the ultra high-vacuum region (that is, pressures less than $10^{-9}$ mm) without raising the possibility of contamination through backstreaming. However, the speeds of such pumps are usually low, and Hnilicka[49] believes that oil contamination can still occur from the fore pumps. All pumps in this class rely on the gas-absorbing power of a nascent metal film, which can be enhanced by ionization of the pumped gases. The metal film may be formed continuously by evaporation or by sputtering (in the latter process the transfer of metal is accomplished by an electrical discharge from the metal, not necessarily at its melting point).

The action of such a device is complex, even when a pure gas is being pumped, and no complete account can be given of[50,51,52] the mechanism at this time. However, some researches have cast light on the mechanism of sorption in electrical discharges.

When a glow discharge is struck in a closed tube containing metal electrodes, the gas contained in the tube cleans up and the pressure drops.[53] This fact, known for more than 60 years, is still an annoyance today to users of thyratron tubes. If sputtering of the metal occurs, the clean-up proceeds at a rate greater than if sputtering is absent. Thus, at potential differences of more than 500 volts the effect is very marked,[54] and it has been shown that the gas-capture/metal-evaporated ratio can be 0.5 atom/atom.[52] The gases may be captured by chemical reaction with the metal or by compound formation on the cathode and subsequent evaporation, or may simply be mechanically buried beneath the relatively enormous masses of high-velocity metal striking the wall. The latter effect is by no means negligible and in some cases may dominate the clean-up process. (For example, gold cleans up oxygen better than does platinum. Gold sputters more easily than platinum but forms a less stable oxide.)

It is also important to note that an electrodeless discharge struck in a glass tube results in clean-up and in changes of the appearance of the glass. Soda glass is particularly prone to this effect, and the clean-up is presumably due to sodium sputtered from the glass.

As is fairly well known, hot-cathode devices, like ionization gauges, act as powerful pumps at pressures below 1 micron. The positive ions formed by electron impact go to the negatively charged walls of the device and can there react or be trapped[54] (the gauge walls go negative because of the inefficient secondary emission of their materials under the prevailing electrical conditions in the gauge). It is quite possible that some evaporated metal arrives on the wall and assists the process: a difficult problem to resolve in this area is the rather efficient way helium is cleaned up in such tubes, since it will diffuse readily through glass and is not soluble in metals. However, in some zeolites (complex aluminosilicates with structural elements similar to soda glass) it is possible to trap helium in the smallest interstices in a very firm manner.[55]

There is evidence that the pumping action of ion gauges is, in fact, due to two phenomena, one of which causes pumping at about five times the speed of the other. Carter[56] has shown that the potential which the glass wall assumes approximates either the cathode or the anode potential, and the action of the gauge thus depends on the immediate previous history of the device. The higher rate of pumping was found for the wall potential approximately that of the cathode (the lower value). Stabilization at the anode potential ($\sim$200 volts) could be achieved only when a gauge of the Bayard-Alpert[57] type was rather new, but the phenomenon is reproducible. The higher stabilization potential is apparently due to a balance between primary electron current and secondary emission from the walls: the latter can occur only when the walls contain gas, that is, when the device is new. These hypotheses were checked by direct measurement of the amount of the cleaned-up gas by subsequent heating of the device, and by measuring the ionic bombardment rates by a cylindrical ion collector. The speeds, 0.0003 liter $\cdot$ sec$^{-1}$ $\cdot$ ma$^{-1}$ and 0.00005 liter $\cdot$ sec$^{-1}$ $\cdot$ ma$^{-1}$, correspond to ion clean-up efficiencies of about 10 per cent and $1\frac{1}{2}$ per cent, respectively. This bistable behavior also affects the sensitivity of the device as a gauge.[56] Blears[58] has considered two gas absorption processes to be responsible for variable results obtained when ion gauges are used to measure ultimate vacua produced by oil diffusion pumps.

Pure contact getter action is preferably conducted under pressure conditions where gas-wall collisions are more likely than gas-gas or gas-getter collisions in the free space of the device. It follows then that such devices perform better the slower the getter evaporation. Clean-up can also be accomplished by dispersal gettering, in which reaction takes place during that act of arrival of the getter at the wall. Both processes are slowly being elucidated by careful studies of chemisorption of gases on evaporated metal films: it is worth noting that a clear picture of the kinetics of such a process can hardly be forthcoming when an ion gauge is operating in the system at the time of the experiment, but a considerable majority of the reports ignore this fact. Static systems give spurious results under these circumstances, though the problem may be largely circumvented by operating a flow-type experiment or by using a leak from the experiment into a mass spectrometer or equivalent device.

Getters are normally evaporated at high temperatures, which aid the diffusion of the pumped gas into the deposited material. The particle size of the evaporated film is an important factor governing the clean-up; such film getters are usually chosen from the alkaline-earth metals or from the metals

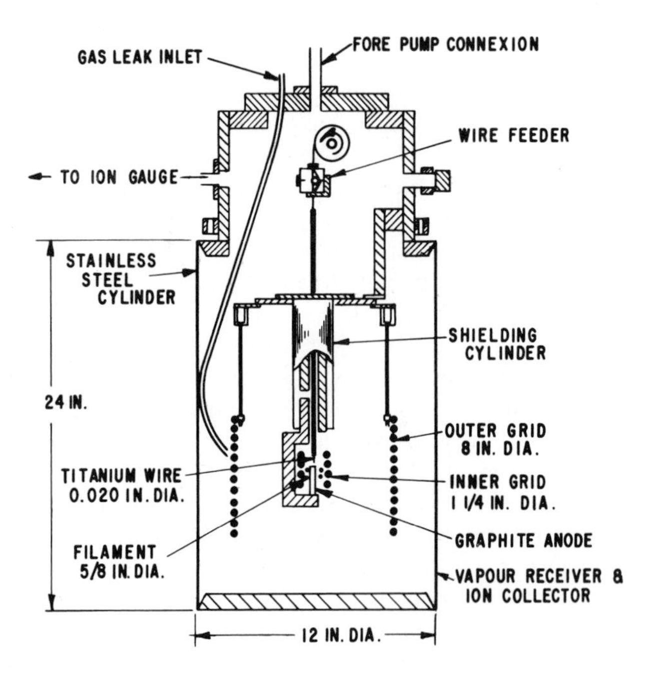


Fig. 3.36. Continuous, wire-fed evapor-ion pump (Davis and Divatia).

of groups IVa, Va, and VIa. The sorption properties and ease of evaporation of a series of metals are shown in Table 3.8, taken from Holland.[48]

There is generally concordant evidence that ionization[59] or mild electronic excitation[60] leads to enhanced and unusual chemical reactions of the activated species, and ionizing currents are therefore frequently used in conjunction with getter evaporation to accelerate pumping.

Devices which employ such a dual action are known as getter-ion, "evapor-ion," or "vac-ion" pumps. The second and third terms are also used as proprietary names and are then capitalized; the term getter-ion is preferred here for the general class of such dual-action devices. Titanium is frequently employed as the getter and may be used as a filament winding,[61] as solid blocks,[62] or in a novel manner as a continuously fed wire[63] (Fig. 3.36). It will be appreciated that the larger pumps evaporate large fractions of a gram of titanium each minute, and ion currents in the ampere range can thus be obtained. Phenomenal speeds have been obtained, up to 1000 liters · $sec^{-1}$ at $10^{-5}$ mm pressure. It must be remembered that some gases do not ionize as easily as others, and quite heavy sputtering must be resorted to in order to trap, for example, the rare gases.

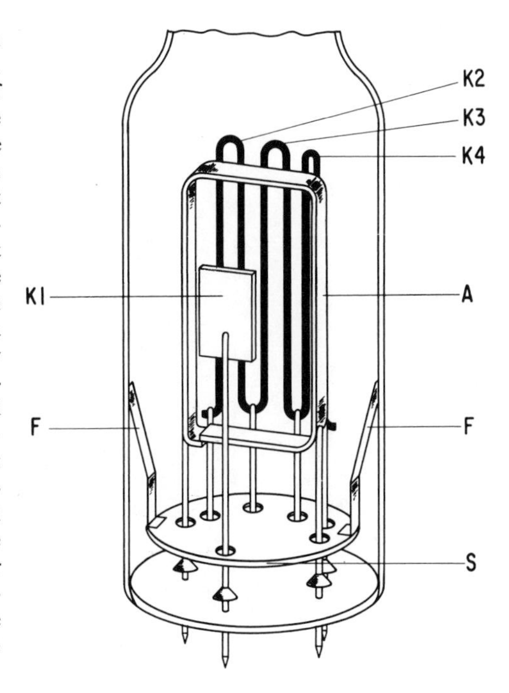


Fig. 3.37. Titanium ion pump with one cold and three hot cathodes. The cold cathode *K*1 consists of a solid plate of titanium. The anode frame is made of a massive titanium strip *A*. The getter can be evaporated by three different methods: (*a*) heating "hot" cathode *K*2, *K*3, *K*4 by passing current through them; (*b*) heating the anode by subjecting it to electron bombardment (without using a magnetic field), the necessary electrons being supplied by emission from the hot cathodes; and (*c*) sputtering from the cold cathode. At low pressures one hot cathode must be used to maintain the discharge. With method (*c*), the starting pressure can be higher than with the other methods. The shell is made conducting and is in contact with the molybdenum shield *S* via the springs *F*.

These gases can diffuse back from the wall and cause poor ultimate vacua: other sources of poor performance result from gas production by reaction of activated species, such as hydrocarbon scission and oxidation.

In some forms of getter-ion pumps, a magnetic field is applied to the device (VacIon pump, Varian Associates, Palo Alto, Calif.; Penning pump[64] Fig. 3.37). This achieves an appreciable increase in electron path length and hence results in greater ionization efficiency: these ions promote sputtering from the cathode.

In general, small getter-ion pumps permit ready access to the ultra high-vacuum region, and larger throughput devices have great promise. A further discussion of the class of phenomena employed in these devices is given in Chapter 9.

## TABLE 3.8

### Some Characteristics of Metal Films of Importance in Determining Their Usefulness as Getters

| Getter Film | Gas | Initial Adsorption Rate ($\simeq 20°$ C) ($sec^{-1} \cdot cm^{-2}$) | Film Temperature for Continuous Sorption (° C) | Sorptive Capacity (micron · liters/mg at $t$° C)† | Adsorption Rate Enhanced by Ionization | Remarks |
|---|---|---|---|---|---|---|
| Aluminum | $O_2$ | *C* | 500° | 7.5–38.5 (20) | ... | |
| mp = 660° C | $H_2$ | 0.0 | ... | 0 | ... | |
| 1291° C (vp = 100 $\mu$) | $N_2$ | 0.0 | ... | 0 | ... | |
| | $CO_2$ | <0.005 liter | ... | 0 | ... | |
| | CO | <0.005 liter | ... | ... | ... | |
| Barium | Air | *Ch* | ... | 56 (400) | ... | $H_2$ gettering increased by presence of heated cathode. Commercially available metal requires extensive degassing or distillation to remove impurities. Can be obtained in alloy form with Al. |
| mp = 717° C | $O_2$ | 0.3 liter | 40° | 57 (300) | No | |
| 730° C (vp = 100 $\mu$) | $H_2$ | 0.05 liter | 200° | 100 (400) | No | |
| | $H_2O$ | *C* | ... | 72 (300) | ... | |
| | $N_2$ | 0.003 liter | 100° | {3–25 (<100); 43–51 (<100)} | Yes | |
| | $CO_2$ | 5.0 liter | ... | 66 (400) | ... | |
| | CO | 3.5 liter | 80° | 100 (400) | No | |
| | $C_2H_2$, $C_2H_4$ | *Ch* | ... | ... | ... | |
| Calcium | $O_2$ | *C* | 425° *L* | ... | ... | Electrolytically prepared metal may contain $CaH_2$, which dissociates at the evaporation temperature. Distilled metal is commercially available and cheaper than Ba. |
| mp = 810° C | $H_2$,$N_2$,CO, $C_2H_2$ | *Ch* | ... | ... | ... | |
| 700° C (vp = 100 $\mu$) | $CO_2$,$H_2O$, $SO_2$, $NH_3$ | *Ch* | ... | ... | ... | |
| | $C_2H_4$ | *Ch* | ... | ... | ... | |
| Magnesium | $O_2$ | *C* | 450° *L* | 20–200 (20) | ... | $H_2$ gettering increased by presence of heated cathode. |
| mp = 651° C | $H_2$ | 0.0 | ... | ... | Yes | |
| 515° C (vp = 100 $\mu$) | $N_2$ | 0.0 | ... | ... | ... | |
| | $CO_2$ | <0.005 liter | ... | ... | ... | |
| | CO | <0.005 liter | ... | $\simeq$0.0 (30) | Yes | |
| Titanium | $O_2$ | *C* | ... | ... | ... | Vacuum-melted metal in form of wire and sheet is commercially obtainable, contains $\simeq$0.008% of $H_2$. |
| mp = 1660° C | $H_2$ | *Ch* | ... | ... | ... | |
| 1742° C (vp = 100 $\mu$) | $N_2$ | 3.0 liters | ... | 1.9–2.5 (30–300) | ... | |
| | $CO_2$ | 4.3 liters | ... | 4.3 (20) | ... | |
| | CO | 12.0 liters | ... | 3.4–4.2 (30–200) | ... | |
| | $SF_6$ | *Ch* | ... | ... | ... | |
| | $C_2H_2$ | *Ch* | ... | ... | ... | |
| | $C_2H_4$, $CH_4$, $CCl_2F_2$, $NH_3$ | *Ch* | ... | ... | ... | |

| | | | | | | |
|---|---|---|---|---|---|---|
| Titanium (solid) | $O_2$<br>$H_2$<br>$N_2$<br>$CO_2$<br>$H_2O$ | 2.01 $\mu$l (800°)<br>*D*<br>0.08 $\mu$l (1000°)<br>0.81 $\mu$l (1100°)<br>*D* | >650° *L*<br>20-400°<br>>700°<br>>700°<br>300–400° | 90 (800)<br>...<br>160 (1000)<br>50 (1100)<br>... | ...<br>...<br>...<br>...<br>... | $CH_4$ dissociates when striking Ti at 1200° C; C is sorbed and $H_2$ taken up at lower temperature. $H_2$ is only gas released by heating. Surface oxide reduced $H_2$ sorption below 300° C. |
| Zirconium<br>mp = 2127° C<br>2212° C (vp = 100 $\mu$) | $O_2$<br>$H_2$, $C_2H_2$, $C_2H_4$<br>$N_2$, CO | *C*<br>*Ch*<br>>2.5 liters | ...<br>...<br>... | ...<br>...<br>... | ...<br>...<br>... | |
| Zirconium (ribbon) | $O_2$<br>$H_2$<br>$N_2$<br>$CO_2$<br>CO | *C*<br>*D*<br>*C*<br>*C*<br>*C* | 885°<br>300–400°<br>1527°<br>...<br>... | 1.99 (400)<br>13.3 (350)<br>1.46 (800)<br>3.04 (800)<br>3.65 (800) | No<br>...<br>...<br>Yes<br>No | |
| Tantalum<br>mp = 2996° C<br>2820° C (vp = 1 $\mu$) | $O_2$<br>$H_2$, $C_2H_2$, $C_2H_4$<br>$N_2$<br>CO | *C*<br>*Ch*<br>>2.5 liters<br>>2.5 liters | ...<br>...<br>...<br>... | ...<br>...<br>...<br>... | No<br>...<br>...<br>No | Difficult to evaporate—low vapor pressure. |
| Molybdenum<br>mp = 2622° C<br>2533° C (vp = 10 $\mu$) | $O_2$<br>$H_2$, $C_2H_2$, $C_2H_4$<br>$N_2$<br>CO | *C*<br>*Ch*<br>2.7 liters<br>3.5 liters | ...<br>...<br>...<br>... | ...<br>...<br>1.0 (30)<br>3.0 (30–200) | ...<br>...<br>...<br>... | Difficult to evaporate thick films—low vapor pressure. May be sublimated. |
| Tungsten<br>mp = 3382° C<br>3309° C (vp = 10 $\mu$) | $O_2$<br>$H_2$, $C_2H_2$, $C_2H_4$<br>$N_2$, CO | *Ch*<br>*Ch*<br>>2.5 liters | ...<br>...<br>... | ...<br>...<br>... | ...<br>...<br>... | Difficult to evaporate thick films—low vapor pressure. May be sublimated. |
| Thorium<br>mp = 1827° C<br>2431° C (vp = 100 $\mu$) | $O_2$<br>$H_2$<br>$CO_2$ | *C*<br>*Ch*<br>*Ch* | 450° *L*<br>...<br>650° | 7.5–33.1 (20)<br>19.5–53.7 (20)<br>... | ...<br>...<br>... | Difficult to evaporate—low vapor pressure. |
| Uranium<br>mp = 1132° C<br>2098° C (vp = 100 $\mu$) | $O_2$<br><br>$H_2$ | *C*<br><br>*Ch* | 240° *L*<br><br>... | 10.6–9.3 (20)<br><br>8.9–21.5 (20) | ...<br><br>... | |
| Misch metal<br>Cerium<br>mp = 785° C<br>1439° C (vp = 100 $\mu$) | $O_2$<br>$H_2$<br>$N_2$<br>$CO_2$ | *C*<br>*Ch*<br>*Ch*<br>*Ch* | ...<br>...<br>...<br>... | 21.7–51 (20)<br>46.1–64 (20)<br>3.2–16 (20)<br>2.2–45 (20) | ...<br>...<br>...<br>... | Chiefly cerium and lanthanum. |

* Initial adsorption rates are given in terms of volumetric and mass speeds (that is, liter · $sec^{-1}$ · $cm^{-2}$ and micron · liters · $sec^{-1}$ · $cm^{-2}$) according to original source.

† Generally values are for bright films; where two values are quoted the second figure is for a black deposit.

*C* = chemical reaction. *Ch* = chemisorption. *D* = diffusion. *L* = linear law of oxidation.

## 3.8. CRYOGENIC PUMPS

In the early days of vacuum technique, a cold trap cooled by liquid air and possibly containing charcoal was very nearly the only means of obtaining good vacuum. With the advent of mechanical pumps, this type of cryogenic pump lost its popularity. However, as the demand and need for experiments and technological operations to be conducted at what are termed ultrahigh vacua (that is, pressures of the order of $10^{-9}$ mm mercury) increased, a search was made for pumping systems which were inherently simple and which did not cause any extra contamination or back-streaming to the system. Traps which operate at liquid-nitrogen or liquid-air temperature are useful as pumps for some gases and vapors, as is well known by the majority of experimenters, and traps which operate at the temperature of liquid helium are useful as pumps for almost all gases. At the temperature of liquid helium only helium and hydrogen have appreciable vapor pressures, and the vapor pressure of hydrogen itself is only of the order $10^{-6}$ mm.

The speed of a cryogenic pump, which is usually employed as a re-entrant trap in small systems, or as a normal trap in larger systems, is close to the conductance of an aperture of area equivalent to that of the cool surface times the accommodation coefficient of the trap surface for the condensing vapor. In spite of the fact that there is no substantial agreement on the precise figures for the accommodation coefficients of various gases on various surfaces, it is generally recognized that the accommodation coefficients for all gases and all vapors on all surfaces more nearly approach unity as the temperature of the surface is reduced to a very low value.

In the design of such a cryogenic pump it is necessary to reduce the radiation incident upon the trap from the walls of the vessel or experimental chamber; consequently the helium trap itself can and should be jacketed by liquid nitrogen. The jacket may reduce the pumping speed,

**TABLE 3.9**

**Reduction of Pressure in 1000-Liter Tank by Cryogenic Pumping[65]**

| Experiment | Observed Pressure without Liquid Helium (mm Hg) | Observed Pressure with Liquid Helium (mm Hg) | Calculated Pressure* with Liquid Helium (mm Hg) |
|---|---|---|---|
| 1 (trap 1) | $2.8 \cdot 10^{-7}$ | $1.5 \cdot 10^{-7}$ | $1.9 \cdot 10^{-7}$ |
| 2 (trap 2) | $1.4 \cdot 10^{-8}$ | $2.4 \cdot 10^{-9}$ | $1.4 \cdot 10^{-9}$ |
| 3 (trap 2) | $7 \cdot 10^{-9}$ | $8 \cdot 10^{-10}$ | $7 \cdot 10^{-10}$ |

* Calculated pressures assume accommodation coefficient of 1.

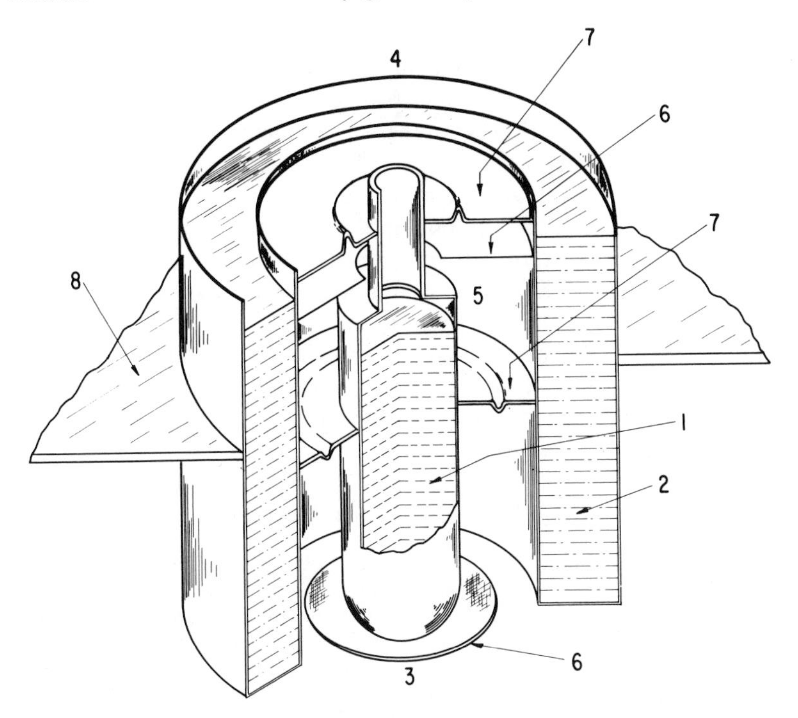


Fig. 3.38. Schematic drawing of cryogenic pump. 1, Liquid helium; 2, liquid nitrogen; 3, ultrahigh-vacuum space; 4, atmospheric pressure; 5, vacuumtight space; 6, reflecting heat shields; 7, partition; 8, chamber wall. The lower partition serves to keep the liquid helium level from the ultrahigh-vacuum region. The gases in the vacuumtight space condense, and the resulting vacuum serves as a heat barrier.

depending upon its arrangement. Figure 3.38 shows a sketch of a typical liquid-helium pump due to Farkass and Vanderschmidt.[65] The sketch is self-explanatory, and it will be readily recognized that such pumps are very simple indeed in their construction; the principal problem is the availability of liquid helium. Such pumps operate reliably and well at pressures below $1 \cdot 10^{-9}$ mm with close to their theoretical pumping speed. Table 3.9 shows the results of experiments made by Farkass and Vanderschmidt on cryogenic pumping of a 1000-liter stainless steel tank. In this case the cryopump was used as a supplement to a diffusion pump-cold trap system. The trap was filled with liquid helium, and the pressure dropped below that value which had been reached with the vapor-stream pump system itself. The table shows the results of the investigation of two traps. Trap 1 was similar to that shown in Fig. 3.38, and the second trap was an unjacketed cold finger. It is easily seen that if a cryopump is to be used it is necessary to design the system around the cryopump so as to permit the maximum utilization of liquid helium; the calculated speed of trap 1 is about the same as the speed of the diffusion-pump system; the calculated speed of traps is, however, about 10 times the calculated speed of the vapor-pump system. In the case of these experiments, it is evident that the residual gas at the time was not either hydrogen or helium. It is, of course, necessary to provide an alternate means for the removal of hydrogen and helium from the system if a cryopump is being used and these two gases are present in large quantities (that is, greater than one equivalent monolayer on the cold surface). Hydrogen may conveniently be removed by a getter-ion pump, but helium itself is much more difficult to remove. Indeed, residual pressures in systems designed for ultrahigh vacua are frequently set by the diffusion of atmospheric helium through the glass envelope, with no means being available to reduce the pressure of that gas to still lower values.

## REFERENCES AND NOTES

1. Some of these pumps are the following: (1) Geryk vacuum pump, see E. H. Barton, *An Introduction to the Mechanics of Fluids*, Longmans, Green, & Company, 1915, p. 197, and also *Encyclopædia Britannica*, 14th edition, Vol. 22, p. 126; (2) Gaede's piston pump, see E. H. Barton, *loc. cit.*, p. 188, and W. Gaede, *Physik. Z.*, **14,** 1238 (1913). A description of early types of mechanical pumps and of the Gaede and Siegbahn types, described in section 3.6, is given by R. Neumann, *Electronic Eng.*, **20** (January and February 1948). These articles also include a comprehensive bibliography.

2. See older treatises on physics. Kahlbaum's pump is described in *Wied. Ann.*, **53,** 199 (1894), and also in *Ann. Physik*, **10,** 623 (1903).

3. E. H. Barton (Ref. 1), from whose book Fig. 3.1 is taken. See also *Encyclopædia Britannica*, 14th edition, Vol. 22, p. 926.

4. E. Bessel-Hagen, *Wied. Ann.*, **12,** 425 (1881).

5. Other forms of Toepler pump are described by A. Stock, *Ber. deut. chem. Ges.*, **38,** 2182 (1905), and E. Grimsehl, *Physik. Z.*, **8,** 762 (1907).

6. K. Scheel and W. Heuse, *Z. Instrumentenk.*, **29,** 47 (1909).

7. L. Zehnder, *Ann. Physik.*, **10,** 623 (1903).

8. *Encyclopædia Britannica*, 14th edition, Vol. 22, p. 927.

9. *Verhandl. deut. physik. Ges.*, **7,** 287 (1905); *Physik. Z.*, **6,** 758 (1905); *Encyclopædia Britannica*, 14th edition, Vol. 22, p. 930.

10. *Encyclopædia Britannica*, 14th edition, Vol. 22, p. 929. Also see G. Meyer, *Verhandl. deut. physik. Ges.*, **10,** 753 (1907).

11. W. Gaede, "The Molecular Air Pump," *Ann. Physik.*, **41,** 337–380 (1913). This paper contains a complete discussion of the theory and construction of the pump. Briefer descriptions may also be found in the following: W. Gaede, *Physik. Ann.* **13,** 864–870 (1912), and *Verhandl. deut. physik. Ges.*, **14,** 775–787 (1912); K. Goes, *Physik. Z.*, **13,** 1105 (1912), and **14,** 170–172 (1913).

12. S. Dushman, *Phys. Rev.*, **5,** 224 (1915).

13. *Compt. rend.*, **177,** 43 (1923). See also *Vacuum Practice*, by Louis Dunoyer, translated by J. H. Smith, D. Van Nostrand Company, Princeton, N.J., 1926. pp. 36–38, and *Encyclopædia Britannica*, 14th edition, Vol. 22, p. 931.

14. M. Siegbahn, *Archives for Mathematics, Astronomy, and Physics*, Royal Swedish Academy, **30B,** No. 2 (1943); also S. von Friesen, *Rev. Sci. Instr.*, **11,** 362 (1940).

15. C. Seydel, *Z. tech. Physik.*, **16,** 107 (1935).

16. S. Eklund, *Archives for Mathematics, Astronomy, and Physics*, Royal Swedish Academy, **27A**, No. 21 (1940); **29A,** No. 4 (1942).

17. See Dunoyer (Ref. 13), pp. 41–42.

18. P. Freneau and J. R. Shields, private communication to S. Dushman.

19. *Bulletin* 910, Elliot Company, Jeannette, Pa.

20. W. Gaede, *Ann. Physik*, **46,** 357 (1915); *Z. tech. Physik*, **4,** 337 (1923).

21. I. Langmuir, *Gen. Elec. Rev.*, **19,** 1060 (1916), *J. Franklin Inst.*, **182,** 719 (1916).

22. P. Alexander, *J. Sci. Instr.*, **23,** 11 (1946).

23. M. Matricon, *J. Phys.*, **3,** 127 (1932).

24. I. Estermann and H. T. Byck, *Rev. Sci. Instr.*, **3,** 482 (1932).

25. K. C. D. Hickman, *J. Franklin Inst.*, **221,** 215 (1936).

26. T. L. Ho, *Physics*, **2,** 386 (1932).

27. M. J. Copley *et al.*, *Rev. Sci. Instr.*, **6,** 265 (1935).

28. J. Holtsmark *et al.*, *Rev. Sci. Instr.*, **8,** 90 (1937).

29. N. Embree, cited by K. C. D. Hickman, *Rev. Sci. Instr.*, **8,** 263 (1937).

30. G. W. Sears, *Rev. Sci. Instr.*, **20,** 458 (1949).

31. A. Venema and M. Bandinga, *Philips. Tech. Rev.*, **20,** 145 (1958).

32. M. Volmer, *Z. angew. chem.*, **34,** 149 (1921).

33. C. A. Kraus, *J. Am. Chem. Soc.*, **39,** 2183 (1917).

34. H. F. Stimson, *J. Wash. Acad. Sci.*, **7,** 477 (1917).

35. I. Backhurst and G. W. C. Kaye, *Phil. Mag.*, **47,** 918, 1016 (1924).

36. See discussion in *J. Franklin Inst.*, **211,** 689 (1931).

37. C. R. Burch, *Nature*, **122,** 729 (1928); *Proc. Roy. Soc. London*, **A, 123,** 271 (1929).

38. K. C. D. Hickman, *J. Franklin Inst.*, **221,** 383 (1936).

39. T. R. Ullman, *American Vacuum Society Symposium Transactions*, 1957, p. 95.

40. H. R. Smith, *American Vacuum Society Symposium Transactions*, 1959 (in press).

41. R. M. Zabel, *Rev. Sci. Instr.*, **6,** 54 (1935).

42. H. W. Edwards, *Rev. Sci. Instr.*, **6,** 145 (1935).

43. I. Amdur, *Rev. Sci. Instr.*, **7,** 395 (1936).

44. A. E. Lockenvitz, *Rev. Sci. Instr.*, **8,** 322 (1937).

45. L. Malter and N. Marcuvitz, *Rev. Sci. Instr.*, **9,** 92 (1938).

46. C. Sykes and F. E. Bancroft, cited by K. C. D. Hickman, *J. Appl. Phys.*, **11,** 303 (1940).

47. K. C. D. Hickman, (Ref. 46).

48. L. Holland, *J. Sci. Instr.*, **36,** 105 (1959), offers a timely review of the status of this class of pump. He describes at length the utility of various metals in these devices.

49. M. P. Hnilicka and N. Beecher, *American Vacuum Society Symposium Transactions*, 1958, p. 94.

50. J. Plücker, *Pogg. Ann.*, **103,** 91 (1889).

51. L. Vegard, *Ann Physik*, **50,** 769 (1916).

52. K. Blodgett and T. A. Vanderslice, *J. Appl. Phys.*, **31,** 1017 (1960).

53. A. Güntherschulze, *Z. Physik*, **38,** 575 (1926).

54. J. R. Young, *J. Appl. Phys.*, **27** ,926 (1956).

55. R. M. Barrer, private communication to the writer.

56. G. Carter, *Nature*, **183,** 1619 (1959); G. Carter and J. H. Leck, *Brit. J. Appl. Phys.*, **10,** 364 (1959).

57. R. T. Bayard and D. Alpert, *Rev. Sci. Instr.*, **21,** 571 (1950).

58. J. Blears, *Proc. Roy. Soc. London*, **A, 188,** 62 (1946).

59. J. H. Leck, *Chemisorption*, Butterworths Scientific Publications, London, 1956, p. 162.

60. See, for example, H. A. Dewhurst, *J. Phys. Chem.*, **63,** 1976, (1959); J. L. Weininger, *General Electric Research Laboratory Report*. RL 2374C.

61. A. J. Gale, *Committee on Vacuum Techniques, Vacuum Symposium Transactions*, Pergamon Press, London, 1956, p. 12.

62. L. D. Hall, *Rev. Sci. Instr.*, **29,** 367 (1958); L. D. Hall, *American Vacuum Society Symposium Transactions*, 1958, p. 158.

63. L. Holland, L. Laurenson, and J. H. Holden, *Nature*, **182,** 851 (1958); L. Holland and L. Laurenson, *Le Vide*, Nr. 81, 141 (1959); R. G. Herb, R. H. Davis, A. S. Divatia, and D. Saxon, *Phys. Rev.*, **89,** 897 (1953); R. H. Davis and A. S. Divatia, *Rev. Sci. Instr.*, **25,** 1193 (1934).

64. A. Klopfer and W. Ermrich, *American Vacuum Society Symposium Transactions*, 1959 (in press). The writer is grateful to them for a preprint of their paper.

65. I. Farkass and G. F. Vanderschmidt, *American Vacuum Society Symposium Transactions*, 1959 (in press). The writer is grateful to them for a preprint of their paper.

# 4 The utilization of pumps

Revised by Peter Cannon

## 4.1. GENERAL REMARKS ON SYSTEM DESIGN

In Chapter 3 the performance factor most frequently referred to was the intrinsic speed, $S_p$. In practice, the observed speed is modified by the conductance of the system to which it is attached, according to the equation

$$\frac{1}{S} = \frac{1}{S_p} + \frac{1}{F}, \tag{4.1}$$

where $S$ is the observed speed and $F$ is the system conductance, calculable from the considerations given in Chapter 2. This equation requires some remarks regarding system design before considering the evaluation of $S$.

Thus, $S$ is always smaller than $S_p$, and the value of the ratio $S/S_p$ actually obtained depends on that of $F/S_p$. More generally, let $F/S_p = n$. Then

$$\frac{S}{S_p} = \frac{n}{n+1} \tag{4.2}$$

and

$$\frac{F}{S} = n + 1. \tag{4.3}$$

The application of these relations is illustrated by the data in Table 4.1, in which $n$ is assigned a series of increasing values. That is, as the value of $F$ is increased, the value of $S$ approaches, more and more, that of $S_p$. In order to utilize to the greatest extent the speed available from a given pump, the connections between the pump and the rest of the system should be made as large as practicable.

TABLE 4.1

**Ratio between Actual Speed of Exhaust and Intrinsic Speed of Pump as a Function of Conductance of Connections**

| $n = F/S_p$ | $S/S_p$ | $S/F$ |
|---|---|---|
| 0.1 | 0.091 | 0.909 |
| 1.0 | 0.500 | 0.5 |
| 5.0 | 0.833 | 0.167 |

Dayton[1] observed that for large values of the ratio $F/S_p$ (5 or higher), equation 4.2 yields values of $S/S_p$ which are too low. The reason for this, Dayton says, is as follows:

When the speed $S_p$ of a diffusion pump has been measured by allowing air to diffuse into the mouth of the pump from all directions in a test dome whose diameter is considerably greater than that of the pump mouth, and when this speed is more than 10 per cent of the theoretical admittance of the mouth (11.67 liters · sec$^{-1}$ · cm$^{-2}$), then the measured value of $S_p$ includes an "end correction" similar to that applying to short tubes. Therefore, when a pump is joined to a short tube of about the same diameter, the net speed ($S$) will not be given correctly by formula 4.1 since the end correction is then included twice.

The procedure for calculating an approximate value of $S$, recommended by Dayton, is as follows.

We shall assume that the mouth of the pump, of radius $a$ (centimeters), is joined to the volume to be exhausted by a short length $l$ of tubing of the same radius. Using the relation

$$S_p = 36.66Ka^2, \tag{4.4}$$

where $K$ is Clausing's factor, it is now possible, from the data in Table 2.3 or from the plot of $K$ as a function of $l/a$ in Fig. 2.2, to determine a length $l_p$ equivalent in speed to that of the pump.

Now, let us assume further that the net speed desired for the added tubing and pump is $fS_p$ (where $f$ is less than 1). Hence we can now use the relation

$$fS_p = 36.66Ka^2 \tag{4.5}$$

to determine the equivalent length $l_e$ of the pump and tubing. The length of tubing to be used as a connection is then given by $l = l_e - l_p$.

For example, given a pump for which $S_p = 275$ liters · sec$^{-1}$ and $a = 5$ cm, it follows that

$$K = \frac{275}{36.66 \times 25} = 0.30.$$

From the plot in Fig. 2.2, the corresponding value of $l_p/a$ is found to be 5.5 and hence $l_p = 27.5$ cm. For $f = 0.8$,

$$K = \frac{220}{36.66 \times 25} = 0.24,$$

and the corresponding value of $l_e = 37.5$. Hence, the value of $l = 10$ cm, approximately. This is the length of tubing of radius 5 cm to be used as connection between the pump and the volume to be exhausted in order that the net speed of the system shall not be less than 80 per cent of the intrinsic speed of the pump.

In general, vacuum pumps serve one of two purposes. The first is that of producing as high a vacuum as possible in a device that will be sealed off the pump once this condition is secured. The second purpose is that of continuously maintaining an extremely low pressure in a system.

Since devices such as X-ray tubes and vacuum tubes in general are to be sealed off the pump, the most important requisite of the pump is that it shall produce such a low pressure that traces of residual gases do not affect the electrical characteristics of the device. These devices are connected to the pump through constrictions, liquid-air traps, stopcocks, and so forth. Under these conditions, the resultant conductance of the connections may vary from 100 $cm^3 \cdot sec^{-1}$ to 1000 $cm^3 \cdot sec^{-1}$. Since the value of $S_p$ for most vapor-stream pumps is at least equal to or greater than 1000 $cm^3 \cdot sec^{-1}$, the speed of exhaust is limited to a large extent by the resistance of the connections, as shown by the above equations; consequently, increasing the value of $S_p$ is of no particular advantage.

On the other hand, in the electron microscope, in the electron diffraction camera, and in distillation operations, where it is important to maintain as low a pressure as possible against low rates of leak, it is necessary that $F$ be very much greater than $S_p$ and also that $S_p$ be sufficiently large to take care of the rate at which gas may leak into the system from outside, or of the rate at which gas may be evolved in the system itself.

As illustrations of the application of these remarks we shall consider two cases.

### Case I

A certain device is connected to a pump through a tube ($l = 25$, $a = 0.5$) in series with a wide-bore stopcock ($l = 3.2$, $a = 0.51$). According to Tables 2.6 and 2.7, the values of $1/F$ for each of these parts (for air at 25° C) are $2.044 \cdot 10^{-3}$ and $3.817 \cdot 10^{-4}$ $cm^{-3} \cdot sec$, respectively. The device is sealed on by means of a constriction ($l = 15$, $a = 0.3$), for which $1/F = 5.679 \cdot 10^{-3}$. Therefore the total resistance is $8.105 \cdot 10^{-3}$, corresponding to $F_0 = 123.4$ $cm^3 \cdot sec^{-1}$.

Since $S_p$ is, in general, much greater than this value of $F$, we can use equation 4.2 to investigate the effect of varying $S_p$. Let us assume that we have two pumps available, for one of which $S_p = S_{PA} = 2F_0$, and for the other, $S_{PB} = 10F_0$. For pump $A$, $S/F_0 = \frac{2}{3}$; for pump $B$, $S/F_0 = 10/11$. That is, increasing the speed of the pump fivefold has increased the effective speed of exhaust from $0.667F_0$ to $0.909F_0$, or only about 36 per cent. Even if we used a pump having an extremely high value of $S_p$, it would not be possible to increase the effective speed beyond that of $F_0$.

### Case II

Assuming that we have available a pump for which the value of $S_p$ is known, the minimum value of $F$ may be determined, by means of equation 4.1, for which $S$ will be equal to or exceed a given value $S_0$. We shall designate this value by $F_0$, where

$$\frac{1}{F_0} \leqslant \left(\frac{1}{S_0} - \frac{1}{S_p}\right).$$

Let us assume $S_p = 200$ liters · sec$^{-1}$, and $S_0 = 100$ liters · sec$^{-1}$. Then $F_0$ must be equal to or greater than 200 liters · sec$^{-1}$. For air at 25° C, flowing through a tube,

$$F_0 = 36.66a^2K \text{ liters} \cdot \text{sec}^{-1},$$

where $K$ is given by Table 2.3. Hence

$$a^2K = \frac{200}{36.66} = 5.456.$$

It will be found that the following combinations of $l$ and $a$ will satisfy this condition:

| $a$ (cm) | $K$ | $l/a$ | $l$ (cm) |
|---|---|---|---|
| 5 | 0.218 | 9.0 | 45 |
| 4 | 0.341 | 4.4 | 17.6 |
| 3 | 0.606 | 1.35 | 4.05 |

Cases often occur in which gas is evolved from the walls or other parts of the system at such a rate that the lower limit of pressure which the pump can produce is not attained. This may also occur if the rate of leakage of gas into the system through seals or joints has an appreciable value.

Let $Q$ denote, as before, the rate at which gas is evolved or leaks into the system. A stationary condition is then attained at which the rate of gas evolution is equal to that at which this gas is removed by the pump.

Let $S_0$ denote the conductance of pump and connecting tube. At the stationary state, the pressure $P_s$ is given by the relation

$$Q = S_0 P_s. \tag{4.6}$$

Thus, the value of $Q$ may be determined from a knowledge of $S_0$, together with an observation on the value of $P_s$. For instance, Langmuir gave the following illustration:[2]

Measurements of pressure by various forms of supersensitive vacuum gauges have shown that the rate of evolution of water vapor from glass surfaces at room temperature is such that pressures below 0.2 microbar could not be attained even when using a molecular pump ($S_p = 870\ \text{cm}^3 \cdot \text{sec}^{-1}$) continuously for an hour. In this case we can calculate by equation 4.6 that $Q$ must have been 170 microbars $\cdot\ \text{cm}^3 \cdot \text{sec}^{-1}$. This corresponds to $0.00017\ \text{cm}^3 \cdot \text{sec}^{-1}$ of gas at atmospheric pressure, or $0.62\ \text{cm}^3$ of water vapor per hour. In this case the surface of glass was about $1800\ \text{cm}^2$.

It is apparent that the presence of leaks in the system can be detected through deviations from normal pump-down behavior. Thus, Kraus[3] used the equation

$$P = \frac{\phi_n{}^F}{S}(t - t_0)^{-1/n} - P_\epsilon \tag{4.7}$$

where $\phi_n$ was the "outgassing time constant," $n$ an integer, and $P_\epsilon$ the final pressure at $t = \infty$, to represent a pump down. The pressure $P$ at time $t$ is also

$$P_t = \frac{Q}{P - P_\epsilon},$$

where $Q$ is a constant "leak flow." These equations are approximately true for gas mixtures, and consequently a plot of $P$ versus $\log t$ shows a slope of $-1$ (for desorption outgassing) or $-\frac{1}{2}$ (for diffusion outgassing). This slope is also proportional to $F\phi/S$, or the rate constant of outgassing, and deviations from this value indicate leaks.

Recent developments in ultrahigh-vacuum techniques have further revealed the considerable importance of degassing effects in controlling the performance of the system. Hayashi[4] has considered the problem of fractionation by adsorption and the consequent modified flow of the degassed materials, and has remarked on the importance of realizing that, during pump down, different species are removed at different times, depending on the conditions in the system. It is an experimental fact that ultrahigh vacua cannot be obtained without paying attention to the degassing procedure, and, further, minimizing degassing is a better approach to ultrahigh vacuum than attempting to increase the pump size.

## 4.2. DETERMINATION OF SPEED OF PUMP

In this section we shall describe a number of procedures which may be used to determine the value of $S_p$, the intrinsic speed of a pump, as defined by equation 2.1.

### Constant-Pressure Method

Kaye[5] has described a simple method which is illustrated in Fig. 4.1. Into a large bulb[6] which is connected directly to the pump and also has a gauge attached, as indicated, air is admitted through a needle valve.[7] The rate at which the air is admitted is measured by the rate at which a mercury pellet travels along the calibrated capillary.

Let $Q_\mu$ denote the rate, in micron · liters per second, at which air is entering the bulb, and let $P_\mu$ denote the pressure as measured by the gauge. When the rate of motion of the pellet is constant,

$$Q_\mu = SP_\mu, \tag{4.8}$$

where $S$ is the speed of the system, in liters per second, including the connection between the bulb and the pump. If $F$ denotes the conductance of this connection, which may be calculated from the dimensions, then

$$\frac{1}{S_p} = \frac{1}{F} = \frac{1}{S}$$

and

$$S_p = \frac{S}{1 - S/F}. \tag{4.9}$$

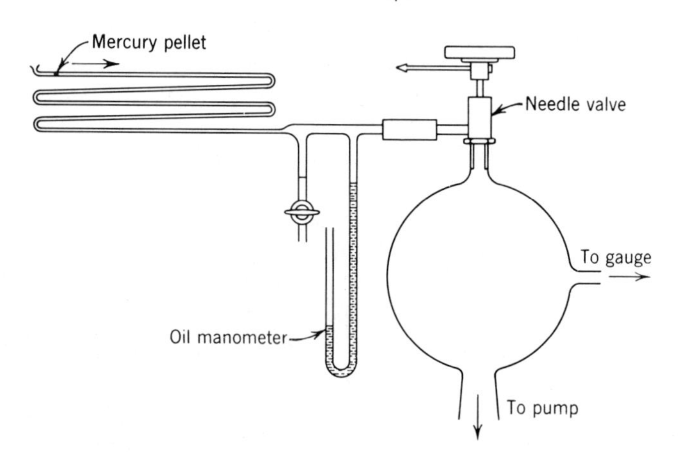


Fig. 4.1. Apparatus for measuring speed of pump (Kaye).

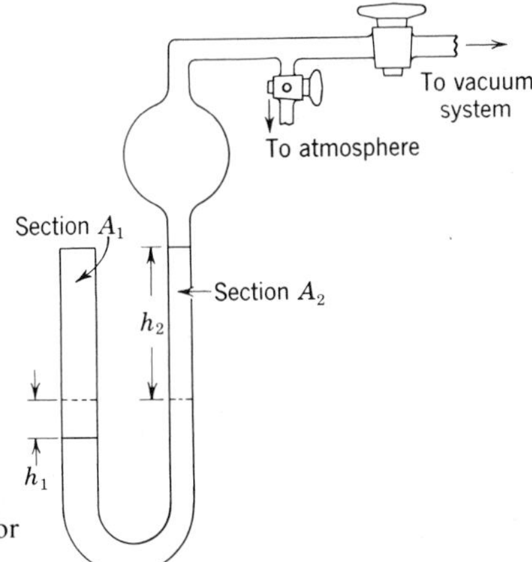


Fig. 4.2. Constant-pressure method for measuring speed of pump (Downing).

Since, in general, $S_p$, the intrinsic speed of the pump, varies with the fine-pump pressure $P$, a series of determinations should be made at different values of $Q_\mu$, and in this manner a plot is obtained of $S_p$ as a function of $P$.

Instead of measuring the rate of flow of gas by means of the rate of motion of the moving pellet, a customary procedure is to observe the rate of rise of oil in an oil manometer of known cross section, the closed end of which is connected to the vacuum system through an adjustable throttle valve, as shown in Fig. 4.2.

The theory of the method has been described by Downing[8] as follows.

Let $V_0$ = initial total free volume, and let $P_0$ = pressure when the oil is at the same height in both sides of the manometer. Also, let $A_1$ and $A_2$ denote the cross-sectional areas in the two sides.

At any subsequent time, $t$, the oil[9] in the left-hand side will have been lowered a height $h_1$, and the oil in the right-hand side will have been raised a height $h_2$. Consequently, the volume of oil displaced is given by

$$L = \int_0^{h_2} A_2\, dh = \int_0^{h_1} A_1\, dh.$$

Also,

$$V = V_0 - L,$$

$$P = P_0 - (h_1 + h_2),$$

and

$$PV = P_0V_0 - V_0(h_1 + h_2) - P_0L + (h_1 + h_2)L.$$

Therefore, the volume of air at unit pressure which enters the system, per unit time, is

$$\frac{P_0V_0 - PV}{P_0t} = \frac{L}{t}\left[1 + \frac{(h_1 + h_2)}{P_0}\left(\frac{V_0}{L} - 1\right)\right]. \tag{4.10}$$

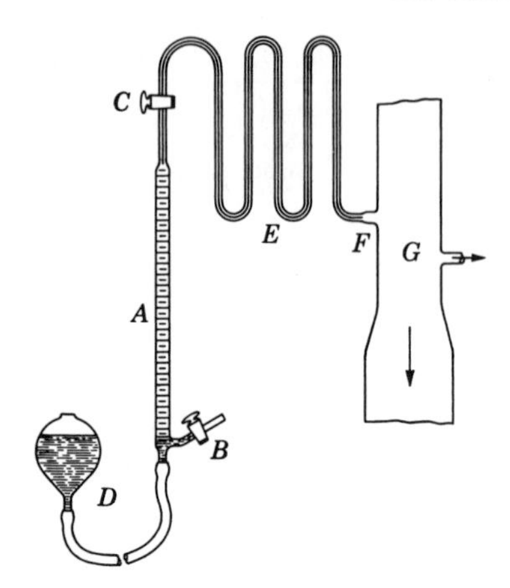


Fig. 4.3. Alternative constant-pressure method for measuring speed of pump (Howard).

As Downing points out, the most common error is caused by neglecting the term $V_0(h_1 + h_2)/(LP_0)$. As he has also shown, the effect caused by dynamic head (viscosity effect) may be neglected in comparison with that due to static head, which is given by equation 4.10.

A variation of the constant-pressure method described by Howard[10] is especially adapted for measuring speeds of exhaust up to 20 liters $\cdot$ sec$^{-1}$. The following quotation from the publication contains a description of the arrangement used, which is shown in Fig. 4.3.

> *A* is an ordinary gas burette of 50-cc capacity, calibrated to 0.1 cc, *B* a stopcock of 3-mm bore, and *C* a stopcock of 1 mm with the edge of the bore slightly nicked so as to permit adjustment of any desired leak. *D* is a mercury leveling bulb. The capillary *E* is sealed to the system at some convenient place as *F*. The pressure in the system at the point *G* is measured by a McLeod gauge.
>
> To make a determination, stopcock *C* is closed and the level of the mercury in the burette is lowered until there is free communication to the atmosphere through stopcock *B*. Stopcock *C* is then cautiously opened until the pressure at *G* reaches approximately the desired value, which in our work was 1 micron. When the pressure in the system has become constant, usually in about 30 minutes, the level of the mercury in the burette is raised until communication with the atmosphere is cut off; stopcock *B* is closed and the volume of air in the burette, the time, and the atmospheric pressure recorded. As the air is evacuated by the pump from the burette, the pressure is kept approximately atmospheric by adjusting the height of the leveling bulb. At the end of a convenient period of time, say 15 minutes, the pressure in the burette is adjusted to atmospheric and the burette reading and time recorded.

A relatively "quick method," which is a modification of the previously mentioned one, has been described by Eltenton.[11]

The formula used in the Consolidated Electrodynamics Corporation's laboratories for determining pump speed is $S = 750/(Pt \times 10^{-3})$. In this formula, $S$ is the speed of exhaust of the test device in liters per seconds, $P$ is the total pressure in mm Hg indicated by a Pirani gauge or an untrapped ion gauge calibrated for air, and $t$ is the time in seconds required for 1 cc of atmospheric air to enter the test device. At Rochester, N.Y., their plant atmospheric pressure, averaged over a year, is approximately 750 mm Hg. The variation from this average pressure is generally no greater than $\pm 25$ mm Hg, and the error introduced by using the average value in all calculations is less than 5 per cent. Errors involved in measuring the pressure inside the test device are normally greater than 5 per cent, so the use of average atmospheric pressure is justified.

*Admission of air to testing device.* The method of metering the flow of atmospheric air into the test device is extremely important. In measuring the speed of small vapor pumps, the CEC laboratories admit the air through a needle valve which is connected directly to a standard burette. The burette is inclined at an average angle of 10° with the horizontal (although this angle will vary with the bore of the burette) and is bent at the lower end to dip vertically into a beaker of butyl phthalate pump fluid. Measurements have proved that the dynamic pressure head associated with the rapid rise of fluid in the inclined burette is negligible compared to the static head, so dynamic-head corrections are not required. The fluid is always drained from the burette with a waiting time of at least 3 min between readings. Although a thin film of fluid remains in the burette, the decrease in cross section, after the waiting period, is negligible compared to the errors involved in measuring pressure within the test device.

When determining the speed of pumps which have extremely large throughputs, another measuring device used is a vertical burette 1–3 in. in diameter which is lowered into a container of fluid until the level of fluid inside the burette is the same as the outside level. Flexible tubing connects the burette to the test dome. A small hole at the top of the burette serves as an air-inlet valve and is covered with a finger while a measurement is being made. For measuring throughputs so large that significant errors are introduced by using a burette, gas flowrators and dry-gas meters are employed.

The location of the air inlet with respect to the mouth of the pump being tested has a major effect on the speeds obtained. If the admitted air should be beamed into the vapor stream in the pump, falsely high speed values would result.

In determinations of pumping speeds made in the General Electric Research Laboratory Dushman found it very convenient to use "calibrated leaks" made of partly sintered porcelain rods. These rods are coated with glass on the sides, so that flow can take place only through the ends, and

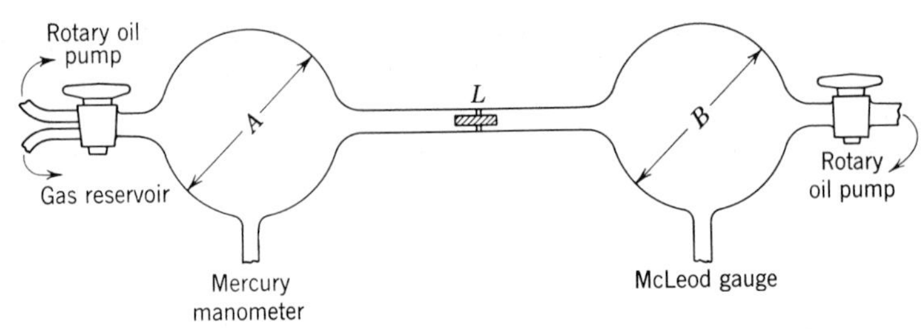


Fig. 4.4. Method of calibrating leak for use in measuring speeds of pumps.

are sealed into glass tubes, as shown in $L$ in Fig. 4.4. They are calibrated by means of the arrangement shown in the figure.

The leak is inserted between a bulb $A$ containing dried gas at a pressure $P_A$, as indicated by the attached manometer, and the bulb $B$, which is connected to a pump through a stopcock. By means of a McLeod gauge connected to this bulb the rate can be measured at which gas passes through the leak $L$.

Let $V$ denote the total volume, in liters, of the bulb $B$ and gauge. Then

$$Q_\mu = \text{rate at which gas enters bulb } B \text{, in micron} \cdot \text{liters per second,}$$

$$= V\frac{dP}{dt}, \tag{4.11}$$

where $P$ = pressure as read by the McLeod gauge at any instant, $t$ seconds, after the start of the experiment.

As shown in Chapter 2, it follows that

$$\frac{Q_\mu}{P} = A + cP, \tag{4.12}$$

where $A$ represents the free-molecule flow, and $cP$ the Poiseuille flow at pressure $P$. Hence

$$A = \frac{a}{M^{1/2}} \tag{4.13}$$

and

$$c = \frac{b}{\eta}, \tag{4.14}$$

where $a$ and $b$ are constants characteristic of each individual leak.

Experimentally it has been found possible to control temperature and time of sintering of the rods so that the value of $Q_\mu$ for air (at atmospheric pressure and room temperature) can be varied from 100 to 0.2 micron · liter min$^{-1}$ or less.

A very similar procedure is that in which a capillary of known conductance, $F$ (which may be deduced from its dimensions), is used in series with a needle valve or other form of leak. If the conductance $F$ is connected at one end to the pump, and the higher pressure as measured by a McLeod gauge is $P_{\mu 1}$, while the pressure at the pump as measured by an ionization gauge is $P_{\mu 2}$, then, at equilibrium between rate of inflow and rate of outflow,

$$SP_{\mu 2} = F(P_{\mu 1} - P_{\mu 2}). \tag{4.15}$$

### Constant-Volume Method

This is essentially an application of equation 2.4. It involves a series of measurements of the pressure in a large volume $C$ during the exhaust by the pump whose speed is to be determined.

For example, let us assume $C =$ 10 liters and $S = 0.5$ liter $\cdot$ sec$^{-1}$. Then, for $P$ large compared to $P_s$,

$$\Delta t = \frac{23.03}{0.5} \cdot \Delta \log P. \tag{4.16}$$

For $\Delta \log P = 1$, that is, for a decrease in pressure to one-tenth its initial value, $\Delta t = 46$ sec. Since $S$ varies with $P$, the most reliable method for deriving values of $S$ is to plot $\log P$ versus $t$ and then calculate the values of the slope of this curve at a series of decreasing values of $P$.

As pointed out earlier, this method actually yields values of $E$, the speed of exhaust, where

$$E = S\left(1 - \frac{P_s}{P}\right),$$

and $P_s =$ ultimate pressure attained by the pump.

The greatest source of error involved in such a series of measurements is due to gas evolution from glass or metal surfaces, which, obviously, would make the observed value of $E$ or $S$ less than the actual value for the pump itself.

Alexander[12] has described a method which, he states, eliminates difficulties inherent in both the constant-volume and constant-pressure methods. The interval of time is measured which is required to reduce the pressure in a given volume from an initial pressure $P_1$ to a final value $P_2$, where the values of these pressures are of the order of magnitude of normal atmospheric pressure. The known volume is connected in this case to the pump through a needle valve, and the speed is then determined from these data by observing the values of $P_2$ and of the pressure at the pump inlet.

### The Circulatory Method

This has been used in the General Electric Research Laboratory by Payne[13] and also by Matricon[14] and Hickman and Sanford.[15] A diagram

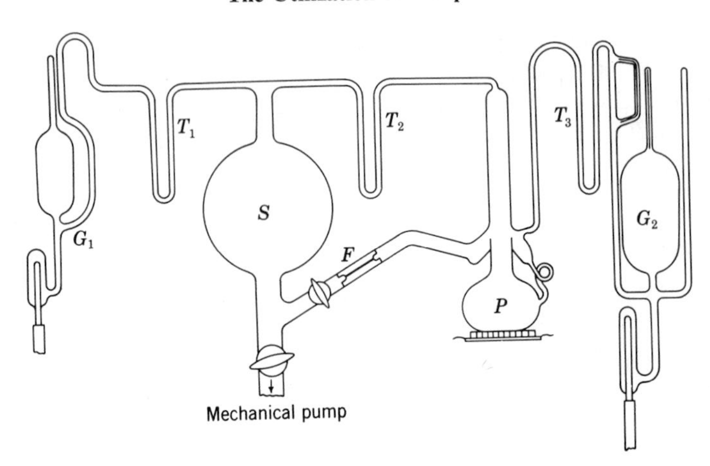


Fig. 4.5. Circulatory method for measuring speed of pump.

of the arrangement used by the last two is shown in Fig. 4.5, taken from their publication.

A calibrated leak of conductance $F$ is inserted between the pump $P$ and a reservoir $S$. The reservoir is connected through a large-bore stopcock to a mechanical pump which can exhaust to 0.1 micron. The McLeod gauges $G_2$ and $G_1$ indicate respectively the pressure $P_2$ in the jet chamber and the pressure $P_1$ in the reservoir $S$. "Traps $T_1$, $T_2$, and $T_3$ are included for the condensation of volatiles by liquid air or solid carbon dioxide. The condensation pump circulates the gas round and round against any chosen backing pressure, producing a reduction of pressure between the calibrated leak and the jet." The speed is determined by means of equation 4.15.

Summarizing the different methods for measuring speed of pumps, Dushman was of the opinion, as a result of personal experience, that the constant-pressure method, involving the use of a calibrated leak such as that described above, is extremely convenient. By varying the pressure at the inlet end of the leak (which can be measured with a manometer or coarse McLeod gauge) it is possible to obtain, in a relatively short period, values of the speed for a series of values of the equilibrium pressure at the pump inlet.

In connection with the application of equation 4.15, Dayton has pointed out that considerable variation in the measurement of $S$ may result from failure to realize the effect of location of the gauge (used for measuring the pressure $P_2$ at the pump) with respect to the gas inlet. The following remarks are based largely on Dayton's[16] excellent discussion of the procedure to be used for accurate determination of pump speeds.

When gas enters a chamber from a narrow tube under conditions of molecular flow, there is considerable beaming. The amount of this beaming has been calculated by Clausing[17] for a short tube with a length equal to its diameter. The full curve (see Fig. 4.6) which shows the distribution calculated; the dotted curve shows the cosine-law distribution predicted for free molecular flow through a hole in a thin plate. These calculations have been checked experimentally by Ellett.[18] Dayton also mentions that Korsunsky and Vekshinsky[19]

calculated the distribution to be expected when silver is evaporated on to a glass plate through narrow tubes under conditions of molecular flow. The calculations were then checked by evaporating silverthrough a narrow quartz tube using several different source temperatures and distances to the plate. In all cases the plate showed a central spot of high density and a rapid decrease in density away from the center, as measured by a microphotometer, in good agreement with the calculations. This distribution was compared with the normal spherical distribution from a point source consisting of a spherical silver drop on a tungsten wire.

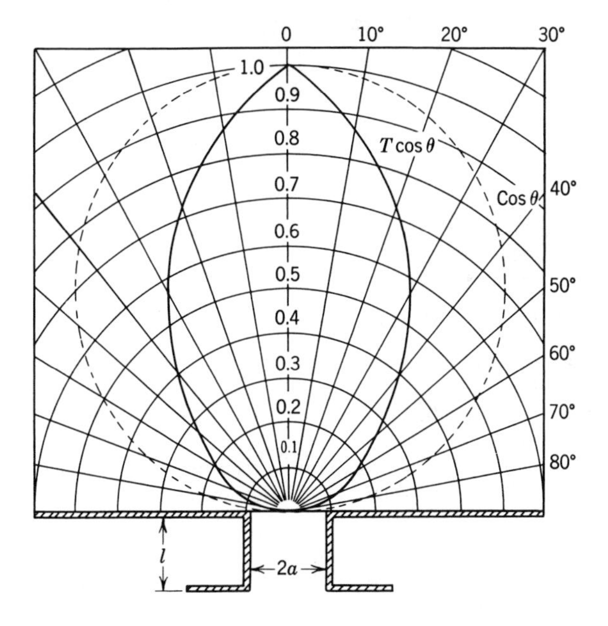


Fig. 4.6. Spatial distribution for free molecule flow through a short tube under conditions where $L$, the mean free path, is greater than $2a$, the diameter, and $l = 2a$. $T \cos \theta$, the full curve, shows the calculated distribution ($T$ = function of $\theta$), and the dotted curve shows the expected cosine law distribution (Clausing). The significance of the values on the vertical scale is illustrated by the following example: for 40°, $\cos \theta = 0.766$ and $T \cos \theta = 0.4$.

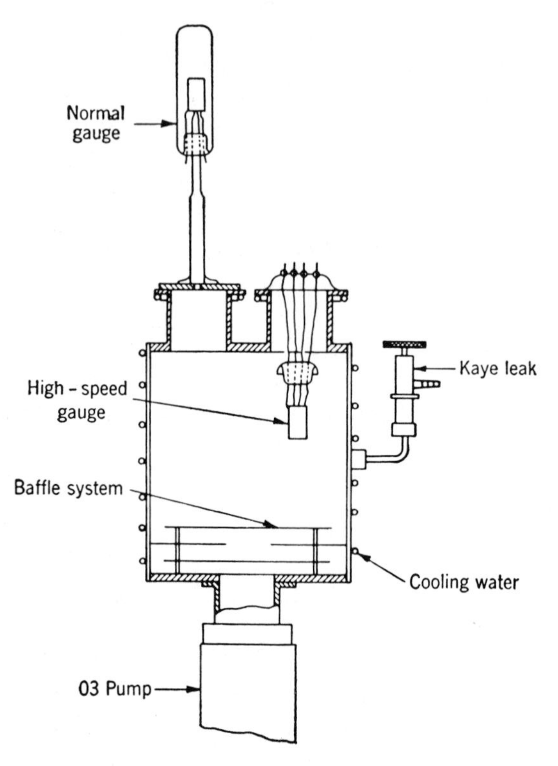


Fig. 4.7. Experimental arrangement used by Blears for measuring "ultimate" pressures obtained with oil-vapor pumps.

Consequently, as Dayton points out,

> If the air is beamed away from the mouth of the gauge, the pressure recorded will be lower than normal; on the other hand if the mouth of the gauge is located directly opposite the air inlet so that the air is beamed directly into the mouth of the gauge, the pressure reading will be higher than normal, and the measured pump speed will be less than the true speed . . . . This effect has been verified by experiments in our laboratories in which the measured speed is nearly one-half the true speed.

In order to overcome errors due to these effects, Dayton makes the following recommendation:

> We believe that the proper procedure is to build a test dome of diameter equal to, or preferably larger than, that of the pump casing, and height at least equal to the diameter. The air is then introduced from a point high on the side of the dome so that the molecules are scattered into a random distribution about the top of the dome. The molecules can then diffuse through the mouth of the pump from all directions in a way which approximates the actual flow in most applications.

In fact data obtained by Dayton show that the true speed as measured with a dome may be 50 per cent of the pseudospeed as measured with a blank-off plate with air beamed directly into the jet of an unbaffled pump, while even with a baffle located in the top of the pump casing the speed with a dome may be 25 per cent less than the value obtained with a plate.

Because a number of investigators have failed to take these precautions, the values observed by them for pump speeds are probably considerably higher than the true speeds. Furthermore, from these observed values of $S$, unrealistically high values of the Ho coefficient[20] are deduced. According to Dayton, values of this coefficient in excess of 45–50 per cent are to be regarded with suspicion.[21]

For this reason also Dayton recommended that the plane of the mouth of the gauge should be placed perpendicular to that of the pump.

An alternative suggestion for eliminating directional effects on ionization gauges is to mount the electrodes of the gauge directly *inside* the test dome, as has been done by Blears[22] and as is shown in Fig. 4.7.

It should also be mentioned in this connection that, in Part II of his paper, Dayton also treated very comprehensively the methods of measuring the gas flow which have been discussed above.

### Diffusion-Pump Speed Measurement at Very Low Pressures

Throughputs, or rates, of less than 1 micron · liter · $sec^{-1}$ are difficult to measure with the oil manometers and conventional leak valves described above. The difficulty may be better appreciated by writing a flow rate of 100 liters · $sec^{-1}$ at $10^{-7}$ mm Hg (a typically high diffusion-pump speed) in STP units, namely $\simeq 1$ $mm^3 \cdot min^{-1}$. Dayton's discussion includes consideration of errors inherent in such measurements, and Landfors and Hablanian[23] have proposed a flow measurement employing a low-conductance tube between the dome of the pump being measured and an auxiliary high-vacuum chamber. An air leak is introduced into the auxiliary chamber, and the speed of the tested pump is then

$$Q = SP_2 = C(P_1 - P_2),$$

where $Q$ is the flow through the tube, $S$ the unknown speed, $P_2$ the inlet pressure of the tested pump, $P_1$ the pressure in the auxiliary chamber, and $C$ the conductance of the "metering tube." $C$ is invariant with pressure in the testing range, since the mean free path of the molecules is larger than the apparatus dimensions: its precise value is calculated using

$$C = \frac{\pi}{3}\left(\frac{kT}{2\pi m}\right)^{1/2} \frac{D^3}{L}\,\alpha,$$

where $\alpha$ is a correction factor depending on $L/D$, and the other symbols have their usual meanings. The technique has been used for pump speeds down to $1 \cdot 10^4$ micron · liters · $sec^{-1}$. The Landfors and Hablanian paper describes suitable combinations of auxiliary chambers with unknown pumps; in general the auxiliary chamber is the dome of another diffusion pump.

## Standards for Performance Ratings of Vapor Pumps

The above discussion points up the need for general agreement between manufacturers for methods of realistically measuring the performance of their products. While there is no formalized agreement as yet, several major manufacturers use the recommendations of the American Vacuum Society's Committee on Standards and Nomenclature (chairman, B. B. Dayton). Although many of Dayton's recommendations are given above, it is worth while to quote here the deliberations of the committee.[24]

The use of a testing dome with precautions to avoid beaming the gas into the mouth of the pump has now become standard practice among vacuum equipment manufacturers. However, there is still some variation in the diameter of the dome and the location of the ionization gauge. The exact dimensions of the dome relative to the pump, and the location and method of calibration of the gauge should be standardized to obtain complete uniformity of procedure.

The test dome shall have the same inside diameter as the pump casing at the inlet port and shall have a height equal to at least 1.5 times the diameter. The gas shall be admitted from a tube projecting into the dome and bent so that its exit is located on the axis at a distance of 1.0 times the diameter from the bottom of the dome and facing the top of the dome. The mouth of the vacuum gauge shall have its plane parallel to the axis of the dome and shall be located at a distance of 0.25 times the casing diameter from the top surface of the inlet flange of the pump and project only slightly beyond the wall to avoid entry of condensed backstreaming pump fluid.

The vacuum gauge should be calibrated before and after each series of pump speed measurements. The following method of calibration is recommended to the Standards Committee for further study as a possible standard procedure. A standard impedance, consisting of a small circular orifice in a metal diaphragm placed midway between the flanges of a cylinder of the same inner diameter as the dome and of length equal to $\frac{4}{3}$ times the diameter, is introduced between the test dome and the pump. The theoretical value of the standard impedance can be calculated from the formula given in the paper entitled "The Pumping Speed of a Circular Aperture in a Diaphragm across a Circular Tube" by Bureau, Laslett, and Keller[25] or from a formula derived by W. Harries.[26]

Using the formula of Bureau, Laslett, and Keller, the pressure in the test dome will be given by

$$p = \frac{Q}{S} + \frac{Q}{K\pi R^2}\left[\frac{3L}{8R} + \left(\frac{R^2}{r^2} \cdot 1\right)\frac{1}{C}\right] \tag{1}$$

where $Q$ is the throughput, $K = (RT/2\pi M)^{1/2}$ is the effusion constant, which equals 11.7 liters · $sec^{-1}$ · $cm^{-2}$ for air at 25° C, $S$ is the speed of the pump as measured with a standard test dome directly over the inlet, $r$ is the radius of the orifice, $R$ is the common radius of the dome and of the cylinder housing the orifice, $L$ is the length of the cylinder housing the orifice, and $C$ is a dimensionless factor varying from 1 for $r/R = 0.0$ to 1.333 for $r/R = 1.0$, as shown in Table I of the article by these authors. For all values of $r/R$ less than 0.3 one can assume $C = 1$ with an error of less than 2 per cent. Since the standard impedance has $L = 8R/3$, then for air at 25° C equation 1 reduces to

$$p = \frac{Q}{S} + \frac{Q}{11.7\pi r^2} \; (r < 0.3R). \tag{2}$$

This same result is obtained using the aperture formula of W. Harries[26] omitting the entrance correction to the cylinder since the pressure is measured in a dome of equal radius. While neither the Harries formula nor that of Bureau, Laslett, and Keller is exactly correct, the error is negligible for purposes of pump speed measurement when the length $L$ is greater than $2R$.

The value of $r/R$ for the standard orifice has not been established as yet. However, it is recommended that $r$ be less than $0.3R$. If the speed of the orifice is not more than 5 per cent of the pump speed, then the term $Q/S$ in equation 2 can be neglected. This will usually be the case when $r = 0.1R$, or less. However, in this case the ultimate pressure in the test dome may be greater than $10^{-5}$ mm Hg if an ordinary synthetic rubber gasket is used between the flanges. Calibration of the gauge can then only be accurately checked at pressures above $2 \cdot 10^{-4}$ mm Hg. While gaskets with negligible outgassing rate should be used between the flanges to avoid this difficulty, it is not necessary that the speed of the orifice be so small that $Q/S$ is negligible. If $S$ has been previously measured directly with the test dome using the same gauge over the range of throughput, $Q$, which includes the values of throughput used during the calibration check, then the calibration correction factor, $k$, can be obtained by substituting in the following equation this measured value, $S'$, and the gauge reading $p'$ when the standard impedance is used between the dome and the pump:

$$k = \frac{Q}{11.7\pi r^2}\left(p' \cdot \frac{Q}{S'}\right)^{-1} (r < 0.3R). \tag{3}$$

The correct pressure is then $p = kp'$, where $p'$ is the uncorrected gauge reading. The correct speed, $S$, will be $1/k$ times the previously measured value $S'$.

If the vacuum gauge is of the ionization type and has an appreciable pumping speed for the gas being used, then the speed of the orifice should be at least twenty times the pumping speed of the ionization gauge. The thickness of the diaphragm must, of course, be negligible. The vacuum gauge should indicate total pressure, and therefore should not be trapped, unless mercury is used as the pump fluid. The speed of the orifice should be measured at several values of throughput corresponding to pressures higher than 20 times the ultimate pressure.

## 4.3. DESIGN OF HIGH-VACUUM SYSTEM

The considerations discussed in the previous sections are directly applicable to the problem of designing a high-vacuum exhaust system.

Such a system consists, in general, of the following components:

1. The device, of volume $C$ (liters), to be exhausted and maintained at a pressure $P_0$ (millimeters).

2. Connections of conductance $F_1$ (liters per second) to a "vapor" pump.

3. A vapor pump which has the speed $S_p$ at the pressure $P_0$ and is capable of operating against a backing pressure as high as $P_f$ (millimeters).

4. Connections of conductance $F_2$ (liters per second) between the vapor pump and the fore pump.

5. A fore pump, capable of operating at a reasonable speed, $S_f$ (liters per second), at the pressure $P_f$.

Let us assume that gas is leaking into the system at a rate of $Q$ (millimeter · liters per second). Hence,

$$Q = S_p'P_0, \tag{4.17}$$

where $S_p'$ is the speed actually obtained, and

$$\frac{1}{S_p'} = \frac{1}{S_p} + \frac{1}{F_1}. \tag{4.18}$$

Therefore,

$$S_p = \frac{Q}{P_0}\left(\frac{F_1}{F_1 - C/P_0}\right) \tag{4.19}$$

$$= f_1 \frac{Q}{P_0}, \tag{4.20}$$

where $f_1$ is a value greater than unity and corresponds to the value of $S_p/S$ in Table 3.1.

For the fore pump we obtain, in a similar manner, the relations

$$S_f = f_2 \frac{Q}{P_f}, \tag{4.21}$$

where

$$f_2 = \frac{F_2}{F_2 - Q/P_f}. \tag{4.22}$$

Since $Q/P_f$ is much less than $Q/P_0$, $F_2$ may be made considerably smaller than $F_1$. Like $f_1$, the factor $f_2$ is greater than unity.

In order to take care of possible contingencies it is advisable to choose values of $S_p$ and $S_f$ which will compensate adequately for even larger values of $f_1$ and $f_2$ than those calculated by means of equations 4.19, 4.20, and 4.22.

As an illustration of the application of these relations, let us consider the following case:

$$Q = 10^{-3} \text{ mm} \cdot \text{liter} \cdot \text{sec}^{-1} \quad (1 \text{ micron} \cdot \text{liter} \cdot \text{sec}^{-1}),$$

$$P_0 = 10^{-5} \text{ mm}, \quad \text{and} \quad P_f = 10^{-2} \text{ mm}.$$

Then

$$S_p = f_1 \cdot \frac{10^{-3}}{10^{-5}} = 100 f_1 \quad \text{liters} \cdot \text{sec}^{-1},$$

$$S_f = f_2 \cdot \frac{10^{-3}}{10^{-2}} = 0.1 f_2 \quad \text{liter} \cdot \text{sec}^{-1}.$$

Let us assume the values $f_1 = f_2 = 5$. Then, the pumps chosen should have the values $S_p = 500$ liters · sec$^{-1}$ and $S_f = 0.5$ liter · sec$^{-1}$.

These are, however, not the only considerations that have to be taken into account. We must also consider the periods required to pump the system from 760 to $P_f = 10^{-2}$ mm. This is given by the relation

$$t = \frac{2.303C}{f_2 S_f} \log \frac{760}{P_f}, \tag{4.23}$$

where $S_f$ now denotes the *average* value of the speed over the range of pressures.

Assuming $C = 100$ liters, and the values for $f_2$, $S_f$, and $P_f$ given above, we obtain the result

$$t = \frac{230.3 \times 4.88}{0.5} = 2830 \text{ sec}$$
$$= 47.2 \text{ min}.$$

If this is too long an interval, we should choose a larger value of $S_f$ than 0.5 liter · sec$^{-1}$. Obviously these conclusions indicate lower limits for the values of $S_p$ and $S_f$, and the actual choice of pumps will have to be governed to a large extent by the sizes available from the manufacturer.

Equation 4.23 assumes that it is possible to assign an average value of $S_f$ over the whole range of pressures. If this value varies greatly with pressure, then equation 4.23 must be replaced by a series of the form:

$$t = 2.303C \left\{ \frac{1}{S_1} \log \frac{760}{P_1} + \frac{1}{S_2} \log \frac{P_1}{P_2} + \cdots + \frac{1}{S_n} \log \frac{P_{n-1}}{P_n} \right\}, \tag{4.24}$$

where $S_1$ is the speed of exhaust for the small range 760 to $P_1$, $S_2$ is the speed for the range $P_1$ to $P_2$, and so forth. This relation is, in effect, analogous to that given previously for the speed of exhaust through a capillary tube.

For the period of exhaust by the vapor pump, we have, in the above case, the relation

$$t_1 = \frac{2.30C}{S_p} \log \frac{P}{P_0}$$

$$= \frac{2.30 \times 100 \times 3}{500} = 1.38 \quad \text{sec.}$$

Thus the rate of exhaust of the system is governed to a very large extent by the speed of the fore pump, rate of leak, and capacity of the system.

In connection with this topic it is also of interest to make some remarks about the rating of mechanical pumps. A figure frequently mentioned in manufacturers' descriptions is the *free air displacement*, designated by $D$. By this is meant the volume of air that the pump passes per unit time, at atmospheric pressure. Now, if we let $S_1$ denote the speed at 1 micron (in the same units of volume and time as $D$), then $r = S_1/D$ has been designated the "merit factor" of the pump. It is evident that, the larger the value of $r$, the greater the pumping speed in the range of 1 micron, for a given value of $D$.

## 4.4. THE USE OF TRAPS AND BAFFLES WITH VAPOR-STREAM PUMPS

### Oil-Pump Traps and Baffles

Although, as has been stated in the previous sections, it is possible with the different esters and oils mentioned above to obtain pressures below $10^{-7}$ mm without the use of extremely cold traps (such as are required for mercury-vapor pumps), it has been found advantageous to prevent, as far as possible, any transport of the oil vapors towards the high-vacuum side. In many pumps baffles are introduced for this purpose, even though such inserts cut down the speed by about 50 per cent. As a consequence of the reduction in backstreaming, it is a practical possibility to obtain pressures well below the saturation vapor pressure of the pump fluid. As mentioned in Chapter 3, it is then possible to achieve ultrahigh vacua ($p < 10^{-9}$ mm) in this way.

More, Humphreys, and Watson[27] have suggested the insertion between the pump and the system of a metal cooling chamber such as is shown in Fig. 4.8. The inner container for Dry Ice is surrounded by two helical ramps located diametrically opposite each other. These are soldered to the surface of the inner cylinder and fit snugly into the annular space between the inner and outer cylinders. These investigators stated that

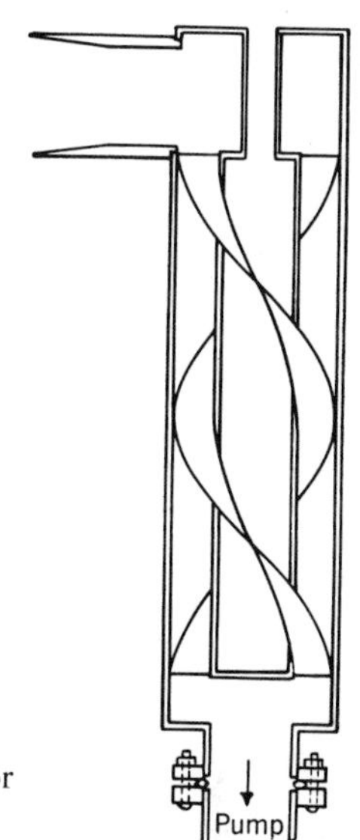


Fig. 4.8. Low-resistance cooling baffle for use with oil-vapor pumps (More *et al.*).

"such a trap offers but little resistance to the flow of air downwards while at the same time effectively preventing the passage of oil vapor upwards."

Morse[28] investigated a number of different types and found baffles may usefully be either heated or cooled. Thus, when "a grid structure, filament, or right-angle bend is introduced between the pump and system and kept hot enough to decompose all impinging oil vapor, no oil reaches the system. Regardless of the type of organic oil employed, the resulting end decomposition products are volatile gases and impure carbon. The gases are, of course, easily pumped from the system, and the residual carbon has a negligible vapor pressure." [A suggested design is illustrated in Fig. 4.9(*a*)]. "A series of electrically heated plates is introduced between the pump and the apparatus under exhaustion as shown." Under these conditions the ultimate vacuum attainable may be bettered as much as tenfold, depending upon the pumping speed available and the rate of gas evolution.

Although baffles of this type may prove advantageous under certain conditions, it would seem that better results could be obtained by mechanical baffles of suitable design, with a certain amount of simultaneous cooling.

"One of the most simple forms of efficient cooled mechanical baffles" is shown in Fig. 4.9(*b*). It is claimed that "with such an arrangement the effective pumping speed is reduced by a factor slightly greater than 2,"

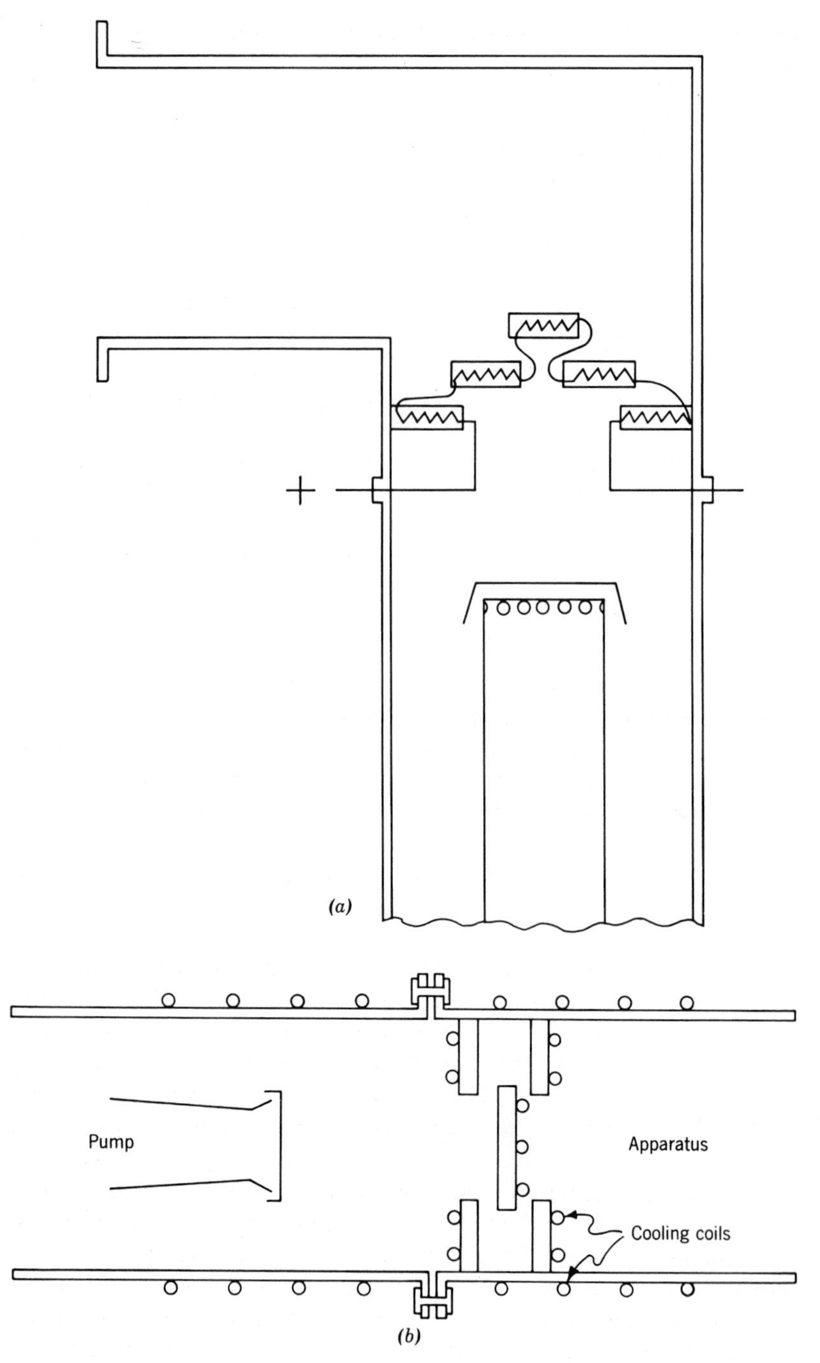


Fig. 4.9. Hot (*a*) and cold (*b*) baffles for use with oil-vapor pumps (Morse).

and at the same time oil vapor is completely prevented from entering the high-vacuum system.

A mechanical baffle of conical type has also been found effective, especially with horizontal pumps.

The efficiency of such baffles has been demonstrated by observations on the electron emissivity of a tungsten-filament cathode. After the pump is "conditioned," there is no observed decrease in emission due to back diffusion of oil vapors. In some experiments which Dushman carried out with the three-stage glass fractionating pump, a cold blast at the right-angled bend in the tubing connecting the ionization gauge to the pump was found sufficient to prevent any decrease in emission or carbonization of a tungsten filament.[29]

Becker and Jaycox demonstrated that very low pressures could be obtained with an oil-vapor pump by use of a trap containing activated charcoal. Anderson[31] has described a specially designed charcoal trap for this purpose which can be outgassed much more rapidly than the conventional[32] trap consisting of a Pyrex tube wound with a resistor. Standard activated charcoal is packed between graphite electrodes and plugs of Pyrex wool. By passage of current, the temperature during outgassing is raised to 800–900° C (instead of about 550° C as with the usual type of trap), and consequently the charcoal exhibits high absorptive capacity even while hot. Using two such traps in series with an oil-vapor pump, Anderson was able to obtain as good a vacuum as with a mercury-vapor pump and liquid-air trap.

Nowadays, vapor-stream pumps are frequently built with the baffles directly inserted in the construction, especially in the smaller sizes. However, baffled connectors which serve as cold traps or true optical baffles may be inserted in the system where desired. One such product is the series made by the Kinney Manufacturing Division of the New York Air Brake Co., Boston, Mass. The Kinney cold traps are made up specifically for use with mechanical pumps, although they can also be used with diffusion-pump systems. Their main purpose is to allow trapping out of mechanical-pump oil vapors at temperatures commonly available with mechanical refrigerators employing "Freon" 22 ($CHClF_2$) systems (−40° F) or cascade devices (−150° F). These traps are quite efficient, since they present a bypass to the flow of mechanical-pump vapors going to the high-vacuum system and, similarly, of any vapors coming from the high-vacuum system to the mechanical pump. The trap is made to be placed in line with a piping system and is thus fitted with standard flanges. The "Freon" lines and vacuum lines are permanently mounted on the end cap of the trap and need never be disturbed. Servicing and cleaning of the trap when necessary are conveniently done by unbolting the casing, which slides off and exposes the condensing surface area of the trap. The baffle plates provided are made with holes in them so arranged that the holes of one plate are not in line with the holes of the plate below, and so on. Thus, a very good baffling action is provided. Another feature of the trap is the vacuum-type isolation of the input "Freon" lines, which are brazed into stems on the end cap. The stems provide sufficient insulation to minimize chilling of the end cap.

These traps are made up for $1\frac{1}{2}$-in. lines and have a 6-in.-diameter casing. They are also available for 1-in. vacuum lines with a 4-in.-diameter casing. They are used extensively in the electronic and semiconductor fields in conjunction with baking and drying ovens, which are evacuated with mechanical pumps. They are also used in cases where it is desirable to minimize any backstreaming from a mechanical pump into a high-vacuum system, by placing these traps between the mechanical pump and the system (roughing line), or between the mechanical pumps and the diffusion pumps (fore lines).

### Mercury-Pump Traps and Baffles

Since it is necessary, in particular applications, to keep mercury vapor out of the system to be exhausted, some provision must be made for this purpose. If a pressure of the order of $10^{-3}$ micron is to be maintained in a system, mercury, which has a vapor pressure of 0.185 micron at 0° C, must be prevented from diffusing into the system. This may be accomplished by means of gold foil, sodium, or potassium with which the mercury reacts to form an amalgam. However, the usual method involves the use of a trap which is cooled either by solid carbon dioxide in acetone or by liquid air. As shown in Table 10.11, the vapor pressure of mercury at $-78.6°$ C, the temperature of sublimation of solid carbon dioxide, is about $3 \cdot 10^{-6}$ micron, and hence this temperature should be sufficiently low to achieve a satisfactory removal of mercury. But it is important to observe that, at this temperature, the vapor pressure of ice is as high as 0.5 micron; and, while the mercury-vapor pump does remove water vapor, the rate at which this vapor is eliminated from the system is not as high as when the trap is cooled in liquid air.

As mentioned previously, the insertion of such a trap decreases the effective speed of pumping, and therefore it is important to use a trap which will have the maximum practicable conductivity. Furthermore, this trap must be designed to present a maximum of condensation surface for any mercury vapor that diffuses into it from the pump.

As shown in section 2.2, the maximum conductivity for the conventional construction of trap is obtained by making the outside diameter as large as practical, and making the inside diameter of the inner tube

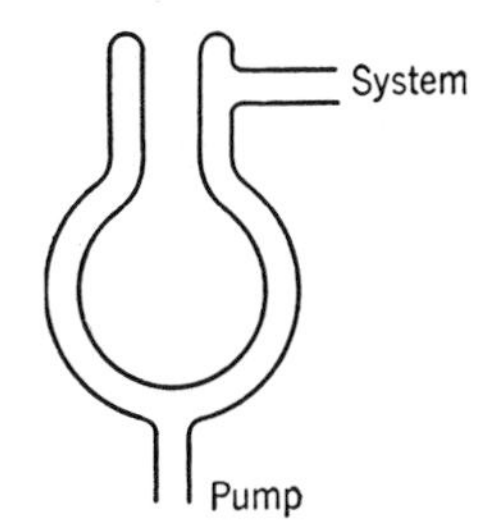


Fig. 4.10. Alternative design of liquid-air trap of glass.

0.62 of the inside diameter of the outer tube. In order to obtain as large a surface as possible for condensation of mercury, the length must not be decreased to an abnormally low value.

Another, less conventional, design[33] is shown in Fig. 4.10. Though this has a higher conductivity for gas flow, it leaves one surface exposed to room temperature. This, of course, can be taken care of by heat insulation.

The effect of the added resistance due to the trap obviously becomes more pronounced, the greater the intrinsic speed of the pump, and therefore the use of baffles is frequently favored.

An elaborate development of this idea has been described by Northrup, Van Atta, and Van Atta.[34] A gear pump is used to circulate butyl alcohol, which is cooled by solid carbon dioxide, in a coil surrounding a cylindrical tube and welded to it for good thermal conductivity. This tube, which is as wide as the opening to the mercury-vapor pump, is located between the pump and the system to be exhausted, and inserted in it is a "winding staircase" of baffle elements which may be welded to the walls. By means of this arrangement the temperature of the trap may be maintained in the range $-50$ to $-78.6°$ C. For continuous operation a standard refrigerating equipment is recommended in which "Freon" is circulated by means of a compressor unit through an expansion coil of copper tubing.

The vapor trap and refrigerant unit described above were used in conjunction with a 130-liter $\cdot$ sec$^{-1}$ mercury vapor-stream pump in evacuating a porcelain accelerating tube. In this evacuated volume of 450 liters with a measured rate of leak of $5 \cdot 10^{-6}$ liter $\cdot$ sec$^{-1}$ at 1 mm, the equilibrium pressure after 24 hr of pumping was $1 \cdot 10^{-5}$ mm Hg without liquid air and $6 \cdot 10^{-7}$ mm Hg with liquid air added to a re-entrant trap in the main volume.

However, the design of a baffle can be approached in a remarkably utilitarian manner. In the General Electric Research Laboratory, there was an operating nonferromagnetic synchrotron, with an overall diameter for the vacuum chamber of 7 ft and a chamber height of about 4 ft. This space was evacuated by a mechanically backed 20-in. oil vapor-stream pump, through a "Freon" 12 ($CCl_2F_2$) refrigerated baffle. This was made from $2 \times \frac{3}{16}$-in. angle iron, stacked in chevron style, with cooling applied to the top edge of each angle iron. The performance of this homemade baffle in reducing backstream was outstanding, while the pump speed was reduced only to about 45 per cent of its rated value.

In the commercial dehydration of products such as penicillin cold traps have been utilized extensively, and the design of a large rotary condenser type of cold trap is shown in an interesting paper by Morse.[35] The refrigerant is circulated around a chamber in which the ice is collected and then removed by means of scraper blades.

As substitutes for liquid air a number of investigators have suggested the use of traps containing sodium or potassium, which react with mercury vapor to form amalgams. Some of these amalgams possess extremely low vapor pressures at room temperature.

Poindexter[36] and Hughes and Poindexter[37] reported that, by use of potassium, pressures of mercury vapor as low as $10^{-8}$ mm were obtained, as measured by an ionization gauge. Poindexter calculated that the vapor pressure of a 1 : 1 sodium amalgam at 20° C is about $3 \cdot 10^{-8}$ mm, and he concluded that "a trap containing 10 g of sodium should hold the mercury vapor beyond the trap down to $10^{-8}$ mm for over two years."[38]

The chief objection to the use of the alkali metals to replace liquid air is that the trap has to be reactivated by distilling into it fresh alkali metal each time the system is opened to the atmosphere.

In a paper by Hunten, Woonton, and Longhurst[39] the development of a specially designed trap is reported "which is self reactivating even after admission of moist air and which, therefore, can be used with systems which are repeatedly brought to atmospheric pressure by the admission of dried or even undried air." Alloys of sodium and potassium containing 3–5 per cent potassium, which are liquid at room temperature, are much more satisfactory than either potassium or sodium alone. According to the authors, the effectiveness of these alloys "is presumably due to the ability of the small amount of liquid solution present to wander about among the crystals of sodium and transport mercury from the surface to the interior of the mass." Furthermore, "Over a period of 22 days, a vacuum system protected by this type of trap was pumped repeatedly from atmospheric pressure to $5 \cdot 10^{-7}$ mm Hg; during this time the trap required no attention."

In this connection mention should also be made of the application by Lobb and Bell[40] of liquid amalgams of sodium-bismuth and lithium-bismuth in a high-vacuum cut-off. Amalgams containing about 0.2 per cent sodium, lithium, and bismuth have vapor pressures of about $5 \cdot 10^{-8}$ mm at room temperatures and have proved satisfactory for sealing off a part of a vacuum system from the pump. The freezing points are about $-44°$ C (compared with $-39.6°$ C for mercury).

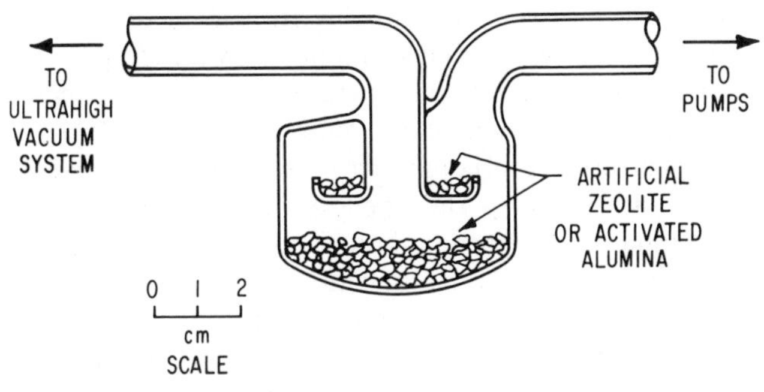


Fig. 4.11. Mineral-filled trap for ultrahigh vacua (Biondi).

At ultrahigh vacua, it is necessary to employ traps of very high conductance. Thus, Alpert[41] described and used a trap of somewhat conventional shape, which contained only a hollow, multisided piece of copper foil of approximately right prismatic form. Using such traps, Alpert[41] and others[42] were able to achieve very fine vacua with simple vapor-stream pumps. Biondi[43] has observed that the Alpert copper trap cannot conveniently be increased in size above that suitable for a system of a few liters capacity. Biondi has therefore designed nonrefrigerated high-speed traps (Fig. 4.11) containing pellets of a synthetic zeolite, which act as adsorbent traps for backstreaming pump vapors. This development bids fair to be a considerable simplification in the commercial application of pressures below $10^{-8}$ mm; however, care needs to be employed in the application of the zeolite, which can decompose quite stable organic molecules,[43] giving large amounts of hydrogen and carbon monoxide. This difficulty (which does not always arise) can be circumvented when necessary by the use of a high-specific-surface-area alumina or silica. The considerable advantages offered by these traps are claimed by Biondi to be (1) stability under bake-out conditions, with low intrinsic degassing rates, (2) effective isolation of large ultrahigh-vacuum systems from large oil pumps and, (3) the possibility of easy design scale-up. Harris[44] of the General Electric Research Laboratory has employed such traps and others of a similar design containing alumina, in the preparation of thermionic emission devices employing ultrahigh vacua, in conjunction with a simple vapor-stream pump.

The present status of carbon-filled traps is rather confusing. Active charcoal used to be employed quite successfully in refrigerated traps backed by simple piston and rotary pumps as a means of producing vacua suitable for the operation of, for example, a mass spectrometer. Today, active carbon traps would seem to offer a useful alternative to the mineral-filled designs of Biondi, and indeed Blears, Greer, and Nightingale[45] have found that such devices, of sizes ranging from 3-in. to 30 in. diameter, would improve the ultimate vacuum attainable with an oil vapor-stream pump by at least one order of magnitude. However, these authors experienced a playback of contamination from the trap, which led specifically to sulfide tarnishing of copper parts in the vacuum system. This is a very serious drawback to the application of these traps to particle accelerators containing copper, although prior treatment of the carbon with copper nitrate solution can eliminate the playback.

The nature of the residual gases in the system changes with the extent of refrigeration of the traps used. Table 4.2 shows typical gas analyses obtained by Blears *et al.* from systems evacuated by metal-oil vapor-stream pumps.

**TABLE 4.2**

**Analysis of Residual Gases with Various Fluids and Traps**

(Unit $10^{-9}$ mm Hg)

| | Mercury | | Apiezon Oils | | | | |
|---|---|---|---|---|---|---|---|
| Oil Designation | | | BW | C | BW | C | B |
| Trap Condition | −170° C | −78° C | 20° C | 20° C | −78° C | −40° C | Carbon |
| Gas | | | | | | | |
| $H_2$ | NA* | NA | 22† | 9 | 3 | NA | 1.5 |
| $H_2O$ | 6 | 2.2 | 25 | 9 | 7 | NA | 360 |
| CO | 1.1 | 1.1 | NA | 3 | 1 | NA | 10 |
| $CO_2$ | 1.1 | 1.1 | 12 | 2 | 0.5 | NA | 5 |
| Air | NA | NA | NA | 1.5 | 1.6 | NA | 7 |
| Total hydrocarbon | 1.3 | 1.2 | 1250 | 60 | 5 | NA | 30 |
| Mercury | 15 | 224 | NA | NA | NA | NA | NA |
| | 24.5 | 229.6 | 1309 | 84.5 | 18.1 | 30 | 413.5 |

* NA = datum not available.

† Probably produced by the ionization process.

## 4.5. THE CHOICE AND STABILITY OF WORK FLUIDS FOR VAPOR-STREAM PUMPS

Because of the many uses of diffusion pumps in research and engineering, need arises for a multiplicity of fluids. The pump fluid must possess properties which will allow not only the required low pressure to be attained but also the highest possible pumping speed in the working pressure range. These two requirements, however, are contradictory because a heavy pump fluid is needed to attain a low ultimate pressure, whereas a light pump fluid must be used to obtain high pumping speeds at high pressures. By means of correctly regulated distillation, however, it is possible to separate fractions which possess the most favorable properties for mastering certain tasks.

Thus, the preparation of various organic liquids for use as work fluids in vapor-stream pumps has been a major research contribution to vacuum technique. Today there exists a more or less complete line of fluids to meet all needs, with each fluid having vapor-pressure characteristics carefully balanced against stability. Great care is taken in their preparation, since an extremely small trace of more volatile contaminants can seriously degrade the ultimate pressure attainable with the fluid.

Ultimate pressure is virtually equivalent to the vapor pressure of the pump fluid at the lowest temperature to which the vapor is subjected. Thus, high-boiling-point (low-vapor-pressure) fluids produce the lowest ultimate pressures. Low-boiling-point fluids, however, operate at higher boiler pressures, which result in higher limiting fore pressures and improved throughput characteristics.

In use, organic oils have some characteristics which are totally different from those of mercury, and these oils also differ from one another. The only major disadvantage of mercury is its high vapor pressure, coupled with its toxicity. Secondary undesirable characteristics are its tendency to amalgamate with some metals and the sensitivity of a mercury vapor-stream pump to contamination by oil or grease, due probably to inefficient evaporation and condensation processes arising from the presence of grease films. To mercury's credit are its elemental character, ease of purification, and high density, the last characteristic easing dimension requirements in pump design.

Thus, mercury pumps are preferred for operation against high backing pressures, for gas handling in analytical operations where contamination must be minimized, for the preparation of rhodium mirrors, and for use in equipment containing photoemissive surfaces, as well as for the preparation of mercury-filled lamps and for handling chemical vapors which might attack organic fluids.

In general, the modern demand is for very high-speed, continuously operating pump systems, and here the use of organic fluids is favored. A number of applications exist where the choice is not so readily apparent, and we must consider the general properties of the organic oils in order to decide wisely which fluid to use.

The question of the stability of the organic liquids on heating is obviously of great importance. As mentioned previously, Hickman and his associates observed that the high-boiling-point phthalates and sebacates developed by them are extremely stable at the temperatures used in their pumps. Any tarry products that may be formed during the operation of the pump are eliminated by a small compartment which connects with both the vacuum and the oil chambers.

The stability of the oils with respect to oxidation is also of importance, since accidents may occur, during the operation of the pump, which lead to a sudden inrush of air. This problem has been investigated very fully by Jaeckel.[46] A special comparison was made between the behavior of Apiezon B and Octoil in glass and metal pumps of different designs. The pressures were measured by means of both an ionization gauge and the *Molvakumeter* devised by Gaede.[47]

The degree of vacuum attained was of the same order for both liquids and varied from $10^{-5}$ to $10^{-6}$ mm, depending upon the particular arrangement and type of diffusion pumps used. As is to be expected, the pressures indicated by the ionization gauge are higher, in general, than those indicated by Gaede's gauge, owing to the fact that the calibration of the ionization gauge varies[48] with the nature of the vapor.

Observations were made on the time intervals required to attain good vacuum after air at different pressures was allowed to leak into the system, and also on the effects of moisture and of large leaks produced suddenly.

The conclusions drawn are as follows: (1) although the organic liquids are sensitive to oxidation and thermal dissociation, the recovery after maltreatment is fairly rapid because of the fractionating chamber; (2) the pure hydrocarbons (Apiezons) are relatively more insensitive in these respects than Octoil and Octoil-S. The second conclusion, however, is not confirmed by the experience of a number of workers in this country. For instance, in connection with an investigation of equipment satisfactory for the factory exhaust of vacuum tubes, DeGroat[49] concluded that the esters are much superior in performance to other types of oils.

One interesting conclusion drawn by Jaeckel was to the effect that, though all organic liquids are dissociated at high temperatures and suffer oxidation, thermal dissociation is favored by metal surfaces. Thus glass pumps give lower ultimate pressures than metal pumps.

A series of interesting observations on the ultimate pressures of oil-vapor pumps has been reported by Blears.[50] The experimental arrangement used was as follows.

A 10-liter steel bell jar was evacuated by a standard pumping plant consisting of a rotary pump and two oil-diffusion pumps in cascade (Metrovac D.R. 1, 02, and 03). The diffusion pump fillings were Apiezon A and B oils respectively. The internal baffles of the 03 pump had been removed and replaced by a large internal baffle in the bell jar itself. Two ionization gauges having identical electrode systems were used to measure the pressure, one being mounted on a tubulation in the normal manner, and the other being removed from its glass envelope and mounted directly inside the bell jar. These gauges have been called the "normal" and the "high-speed" gauges respectively.

After the pumps were started, readings were taken on both gauges until the indicated pressures reached approximately constant values (equivalent nitrogen pressures). The ultimate pressure indicated by the normal gauge was only a few per cent of that indicated by the high-speed gauge. Experiments in which different gases were admitted led to the conclusion that the difference in pressures was due to sorption of oil vapor in the normal gauge. That the readings on both gauges were proportional to the pressure in the bell jar was shown by the fact that plots of log $P$ versus temperature of pump cooling water were found to lie on parallel straight lines for each gauge.[51]

A consideration of the sorption, evaporation, and decomposition processes in the two gauges led Blears to the conclusion that the correct ultimate pressure obtained with a given oil is that indicated by the high-speed gauge, which, according to this investigator, "is defined ideally as a gauge in which every molecule absorbed is immediately replaced and where every molecule produced by reactions is immediately removed, so that the pressure in the interelectrode space is always equal to the pressure in the rest of the vacuum system."

TABLE 4.2

**Ultimate Pressures Produced by Diffusion-Pump Oils**

(Expressed in Terms of Equivalent Nitrogen Pressures at 20° C)

| Oil | $P$ (mm · $10^{-6}$) Measured Values | | Values Quoted in the Literature |
|---|---|---|---|
| | Standard | Fractionating | |
| Octoil-S | 6.4 | 2.9 | 0.01–1.0 |
| Litton oil | 14.0 | 6.6 | 1.0 |
| Apiezon B | 17.0 | 9.2 | 0.1–5.0 |
| Apiezon A | 45.0 | 19.0 | 10.0 |
| Dibutyl phthalate | 225 | ... | 100 |
| Arochlor 1254 | 310 | 260 | 100 |

Table 4.2 gives data thus observed for the ultimate pressures obtained with a number of oils, after the pumps had been operated for 16–24 hr.

As was to be expected, the pressures obtained with a pump of fractionating type (Metro 03 was used in all cases) were lower than those obtained with the standard or nonfractionating type. The important conclusion from these observations is that the ultimate pressures obtained with oil-vapor pumps are considerably higher than those previously reported in the literature.

One very important feature about these measurements which must be given further consideration, however, is the interpretation to be assigned to the "equivalent nitrogen pressure" indications. As will be pointed out in Chapter 5, in the discussion of the calibration of an ionization gauge for different vapors, the equivalent nitrogen pressures of organic vapors are considerably higher than the absolute pressures (5–10 times larger). Consequently there may not be such a great difference between the pressures obtained by Blears and those quoted in the literature, but there is a need for considerable care when assessing the merits of various organic oils, because of this type of experimental artifact. It has become conventional to accept air-calibrated ion-gauge readings in this connection.

Organic fluids for the operation of pumps are available with vapor pressures at ordinary temperatures varying from about $10^{-4}$ to the order of $5 \cdot 10^{-8}$ mm Hg. These fluids are sufficiently inert with respect to most metals to permit a wide selection of materials to be used for pump construction, and pumps working with these fluids are insensitive to contamination in the sense that they will operate correctly even though they have been very indifferently cleaned (though the ultimate may of course be affected). Moreover, although the purification of organic fluids in bulk is much less easy and more expensive than in the case of mercury, it is possible to design the pump in such a way that, when charged with a suitable crude fluid, it will itself perform much, if not all, of the necessary purification. The major disadvantage of the fluids is that they are all complicated chemical structures and are all subject to some chemical decomposition under any one of a number of conditions normally encountered during use.

Thus, in addition to the hydrocarbon oils initially used, attention has been paid to thermally stable esters and chlorinated biphenyls, and a great deal of interest has centered around the application of silicone polymers (based on an Si—O—Si—O chain rather than C—C—C—C). Demands for thermally stable fluids for use as lubricants have led to the synthesis of many unusual classes of compounds (for example, the "Ucon" polyesters, alkylaryl hydrocarbons, and aromatic polyethers), and we may confidently expect some of these materials to appear as workable pump fluids in the next decade. Hickman has already reported[52] on the utility of the polyethers for this purpose.

Let us consider the nature of the presently popular fluids. The dibasic esters (of sebacic and similar acids) condense readily and are quite stable under normal conditions. Familiar examples are Octoil-S, excellent for obtaining the lowest ultimate pressure in glass fractionating pumps, while

Octoil is the best general-purpose pump fluid for all types of diffusion pumps. When using either pump fluid, the pump heaters should be turned off if the pressure rises above 800 microns Hg for more than a few minutes, and the fluids should be cooled below 100° C before exposure to atmosphere.

Chlorinated hydrocarbon oils produce higher limiting fore pressures and greater throughout than most ester or hydrocarbon fluids. Since they can produce high boiler pressures without danger of thermal decomposition, the performance of a given pump using them can be extended by the application of substantially more heat. There is little difference between the available examples other than in different ultimate pressures achieved with them. Thus, the choice here rests on viscosity and fore-pressure tolerances.

Hydrocarbon oils (Apiezon, Convoil) are provided over a wide property range and can be used in ejector or vapor-stream pumps. The lower the vapor pressure of the oil, the greater the preference for its application under conditions where migration through the system must be minimized.

Butyl phthalate fluid is used where the pump is required to operate against higher fore pressures (up to 400 microns) than can be tolerated by the sebacate esters. It should not be exposed to the atmosphere at temperatures greater than 100° C for any extended time.

Silicone oils are recommended for a pumping system where there is danger of sudden exposure at operating temperature to pressures above 1 mm Hg and the system is not equipped with automatic safety controls. The silicones are generally more stable than the other pump fluids,[53] but they too cannot be exposed to atmosphere for long periods (minutes) at full heater input without incurring chemical changes.

It should be noted, however, that when these silicones decompose on metal surfaces they yield a layer of an insulating material (presumably a suboxide of silicon) which can be removed only by heating above 400–500° C. As will be more fully discussed in Chapter 5, this insulating deposit formed in the course of time on the positive-ion collector of an ionization gauge gives rise to gradually decreasing positive-ion readings, and thus renders the gauge useless as a pressure indicator. This undesirable effect is believed to be accelerated by the presence of small quantities of impurities in the oil, and rigorous attempts to remove such materials are made before current silicone oils are brought to market.[54] It is convenient to summarize the vapor-pressure data of a number of pump fluids on a single graph, which may then be used to choose a fluid for a particular application (Fig. 4.12, reproduced by courtesy of the Consolidated Vacuum Corp.).

The curves represent the observed vapor pressures for organic pump

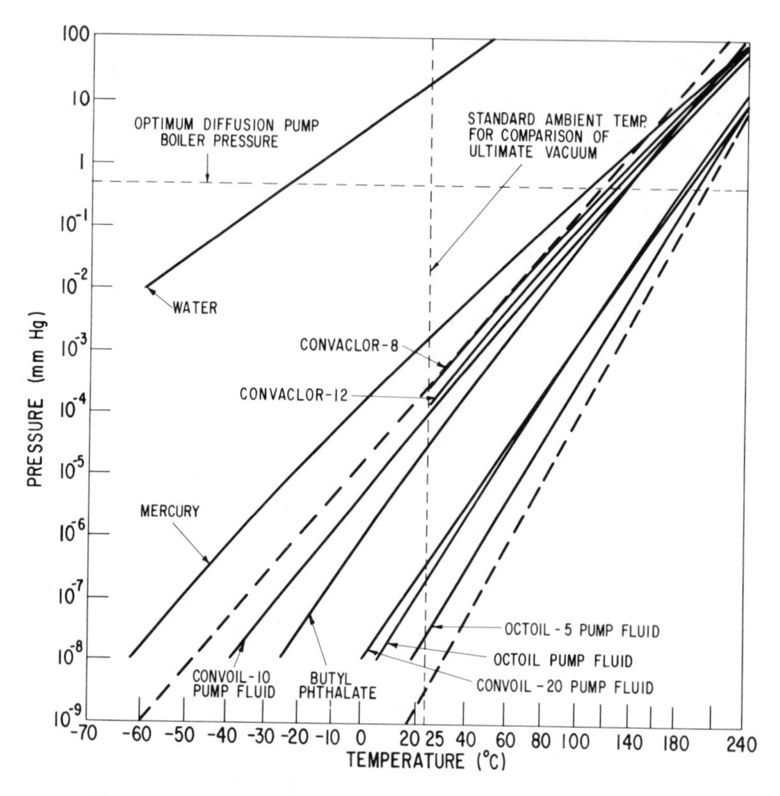


Fig. 4.12. Pressure-temperature characteristics of some pump fluids.

fluids and standard data for water and mercury. The intersections with dashed vertical line give the ultimate pressure attainable in a three-stage glass fractionating pump under standard test conditions at 25° C. With a diffusion pump under practical conditions, the ultimate pressure obtained in a leak-free outgassed system normally falls slightly above the corresponding vapor-pressure curve because of vapor from gaskets and lubricants and because of local concentrations of more volatile trace impurities and decomposition products.

If a plot of observed ultimate pressure versus ambient temperature is found not to parallel the corresponding vapor-pressure curve, a leak of noncondensable gas is usually indicated.

The lower portions of the curves with horizontal dashed line give the operating vapor temperatures in diffusion pumps using the fluids when the optimum boiler pressure of 0.5 mm Hg is maintained. The useful range of boiler pressures is from 0.1 to 1.0 mm Hg. (A thermometer below the liquid surface in the boiler may indicate from 5° to 15° C higher, depending on the ratio of nozzle throat area to boiler surface area.)

An interesting comparison of the performance of various commercial fluids has been made by Latham, Power, and Dennis.[55] They first measured the ultimate pressures in a three-stage metal fractionating pump using various fluids (Fig. 4.13). They then employed the following test:

A chamber of a few liters capacity was evacuated by a standard $2\frac{1}{4}$-in. diffusion pump (Speedivac 203) containing the fluid to be tested. After 15 minutes pumping, the pump heaters were switched off, a valve in the backing line was closed, and a second valve admitted air to the pump and the chamber. After 5 minutes of such exposure the valves were returned to their original positions, the heaters were again switched on, and pumping was resumed for a further 15 minutes, the chamber being roughed down through the hot pump. This cycle of alternating exposure and pumping continued until a cold-cathode ionization gauge mounted on the chamber indicated that the pump had stopped pumping and that breakdown of the fluid was substantially complete.

This test was not regarded as an absolute indication of merit, but rather as a useful indication of what might happen in practice. Silicones behave

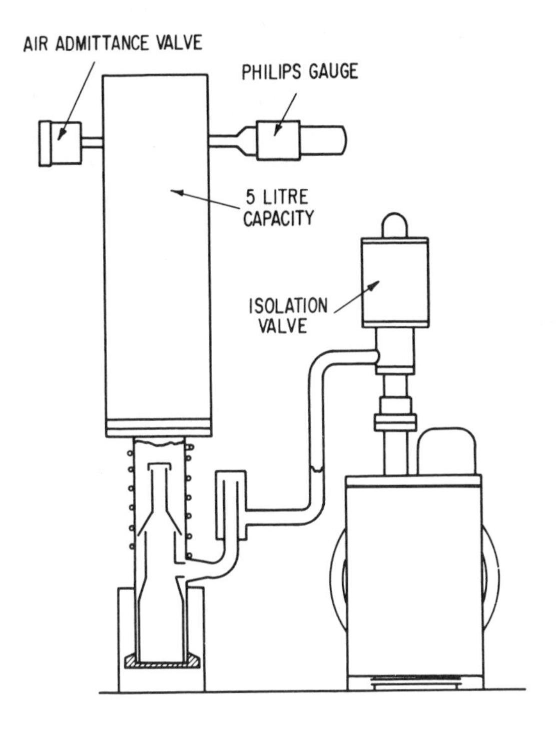


Fig. 4.13. "Breakdown" test apparatus (Latham, Power, and Dennis[55]).

well in these tests, and further research led Huntress *et al.*[54] to announce the fluid DC704, having a high stability and an improved vapor-pressure characteristic ($p$ at 25° C, $\sim 4 \cdot 10^{-8}$ mm). Brown[53] also observed superior stability characteristics for silicones over hydrocarbons.

**Physical Stability of Vapor-Stream Pumps**

Pumping speed in vapor-stream pumps is, as we have already observed, dependent upon the heat input to the pump boiler. This is especially important in leak-detection equipment, where small pressure changes must be detected, and random fluctuations in heat input upset the stability of the device. Martin and Leck[56] established that the random effects seen were in fact due to heat-input variations and not to condenser or nozzle instability, and that the stability could be maximized at a heat input corresponding to the maximum jet speed of the pump. They suggested using an oversize pump as a backing device for the lowest pressure vapor-stream pump, in order to maintain the exhaust pressure of the latter at a minimum.

Bishop[57] observed that vibration could be set up in sensitive equipment (for example, an electron microscope) by overvigorous boiling of the oil. However, it is necessary in equipment of this type to reach a good vacuum quickly (that is, to reach boiling temperature rapidly), and he suggested a modification of the power on-off temperature-dependent control to give a stepped power input in the boiling range.

In connection with unstable boiling, and the general question of providing enough vapor flux to the pump to ensure its good performance, Smith[58] has attacked the problem of boiler design and jet superheating. Thus, he found that the undesirable temperature distribution shown in Fig. 4.14 (which he measured) led to too-early condensation; by providing an axial and a top-cap heater, he could improve the performance markedly (Fig. 4.15). This research should have an important effect on future pump design.

Backstreaming of work fluid and cracked vapors also can cause fluctuations in the pump performance. Thus, in the absence of a cooled trap, the ultimate vacuum obtained with a vapor stream will not be lower than the saturation pressure of the work fluid at the lowest temperature in the pump, and will normally be higher because of components arising from sources other than those of lowest temperature. These will include cracked fluid, which can ordinarily be avoided to a great extent by a wise choice of fluid, and condensed drops of fluid on the top cap ("wet running"), which drop off into the hot cap below and give a vapor burst. It is possible, using the auxiliary heaters suggested by Smith, to avoid this latter condition, but Power and Crawley[59] found that a deliberately cooled top cap

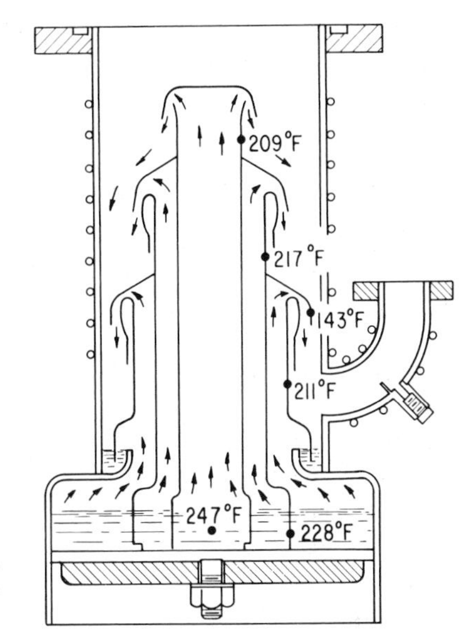


Fig. 4.14. Distribution of temperatures in a typical oil vapor-stream pump (Smith[58]).

would have the same desirable effect. The problem is essentially one of controlling the thickness of the boundary layer inside the nozzle: the thinner the better. However, it is a complex phenomenon, since Power and Crawley also found that the use of different materials in the top cap had an effect on the measured backstreaming rate. A cooled guard ring coaxial

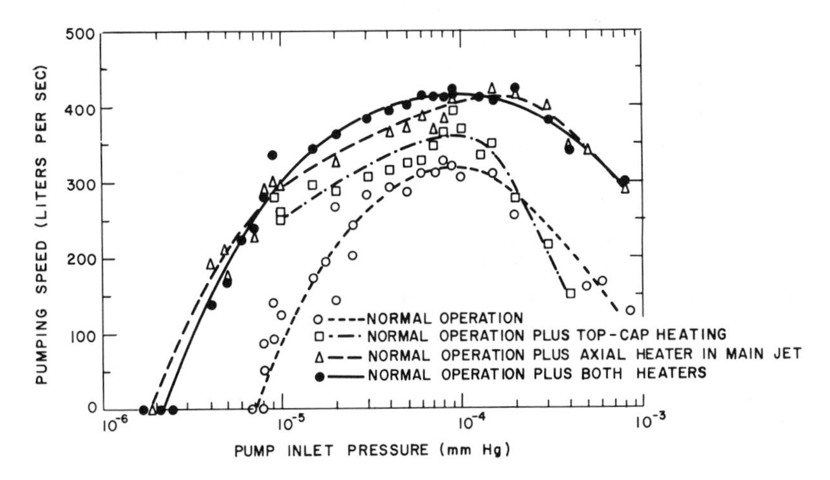


Fig. 4.15. Effect of superheaters on the speed of a 6-in. nonfractionating oil vapor-stream pump (Smith[58]).

with the jet will also serve to reduce the extent of backstreaming due to lateral movement of molecules in the pump jet, just as Langmuir suggested in his original condensation-pump paper (see Chapter 3).

## REFERENCES AND NOTES

1. B. B. Dayton, private communication to S. Dushman

2. I. Langmuir, *Gen. Elec. Rev.*, **19,** 1060 (1916).

3. T. Kraus, *American Vacuum Society Symposium Transactions*, 1959, p. 204.

4. C. Hayashi, *American Vacuum Society Symposium Transactions*, 1957, p. 13.

5. G. W. C. Kaye, *High Vacua*, p. 162; A. Farkas and H. W. Melville, *Experimental Methods in Gas Reactions*, Cambridge University Press, 1939, p. 59.

6. It is very important that gas should be able to diffuse into the pump inlet from all directions. Therefore the leak should be connected to a tube of larger diameter than that of the pump inlet. This point is discussed at further length at the end of this section.

7. J. Yarwood, *High Vacuum Technique*, John Wiley & Sons, 2nd edition, 1945, p. 23, gives an illustration of such a needle valve. An improved form of this type of valve has been described by P. Alexander, *J. Sci. Instr.*, **21,** 216 (1944). Also G. C. Eltenton, *J. Sci. Instr.*, **16,** 27 (1939), has described a "greaseless" leak, the value of which may be varied by moving a wire of gradually changing diameter along the axis of a closely fitting capillary.

8. J. R. Downing, personal communication from the National Research Corporation to S. Dushman.

9. It is evident that, in this and the following equations, $P$ and $P_0$ should be expressed in terms of the equivalent height of oil.

10. H. C. Howard, *Rev. Sci. Instr.*, **6,** 327 (1935).

11. G. C. Eltenton, *J. Sci. Instr.*, **15,** 415 (1938); also J. Yarwood (Ref. 7), p. 59.

12. P. Alexander, *J. Sci. Instr.*, **21,** 216 (1944).

13. J. H. Payne, *J. Franklin Inst.*, **211,** 689 (1931).

14. M. Matricon, *J. phys. radium*, **3,** 127 (1932).

15. K. C. D. Hickman and C. R. Sanford, *Rev. Sci. Instr.*, **1,** 140 (1930).

16. B. B. Dayton, paper presented at High-Vacuum Symposium, Oct. 30–31, 1947, Cambridge, Mass. See *Ind. Eng. Chem.*, **40,** 795 (1948).

17. P. Clausing, *Z. Physik*, **66,** 471 (1930).

18. A. Ellett, *Phys. Rev.*, **37,** 1699 (1931).

19. M. Korsunsky and S. Vekshinsky, *J. Phys. U.S.S.R.*, **9,** 399 (1945).

20. The Ho coefficient is defined as the ratio between the observed speed and the theoretical value for the rate of flow of gas through the annular area between the nozzle and the walls.

21. As an illustration Dayton cites the measurements by Copley and his associates on the speeds of the mercury-vapor jet pump shown in Fig. 4.25.

22. J. Blears, *Proc. Roy. Soc. London*, **A, 188,** 62 (1946).

23. A. A. Landfors and M. H. Hablanian, *American Vacuum Society Symposium Transactions*, 1958, p. 22.

24. *American Vacuum Society Symposium Transactions*, 1955, p. 91.

25. A. J. Bureau, L. J. Laslett, J. M. Keller, *Rev. Sci. Instr.*, **23,** 683–686 (1952).

26. W. Harries, *Z. angew. Physik*, **3,** 296–300 (1951).

27. K. R. More, R. F. Humphreys, and W. W. Watson, *Rev. Sci. Instr.*, **8,** 263 (1937).

28. R. S. Morse, *Rev. Sci. Instr.*, **11,** 277 (1940).

29. Dushman used this as a test for the presence of hydrocarbon vapor by operating the aged filament at 2000–2200° K for a period of at least 25 hr.

30. J. A. Becker and E. K. Jaycox, *Rev. Sci. Instr.*, **2,** 773 (1931).

31. P. A. Anderson, *Rev. Sci. Instr.*, **8,** 493 (1937).

32. By "conventional" is meant the type which is immersed in a Dewar flask containing the refrigerant.

33. A metal construction of this type is shown in *Bulletin* 10, "High-Vacuum Apparatus," by the Central Scientific Company. T. H. Johnson, *J. Franklin Inst.*, **205,** 99 (1928) has described a modification of the design shown in Fig. 4.33 which not only has high conductance (about 10 liters · $sec^{-1}$) and large surface area for the condensation of mercury but also reduces to a minimum the probability of back diffusion of mercury into the system.

34. D. L. Northrup, C. M. Van Atta, and L. C. Van Atta, *Rev. Sci. Instr.*, **11,** 207 (1940).

35. R. S. Morse, *J. Ind. Eng. Chem.*, **39,** 1064 (1947).

36. F. E. Poindexter, *J. Opt. Soc. Am.*, **9,** 629 (1924), also *Phys. Rev.*, **28,** 208 (1926).

37. A. L. Hughes and F. E. Poindexter, *Phil. Mag.*, **50,** 423 (1925).

38. See Chapter 10 for discussion of the vapor-pressure data for amalgams of sodium and potassium.

39. K. W. Hunten, G. A. Woonton, and E. C. Longhurst, *Rev. Sci. Instr.*, **18,** 842 (1947). Detailed information on the method used for introducing the alkali metals into the trap is given in the original publication.

40. G. W. Lobb and J. Bell, *J. Sci. Instr.*, **12,** 14 (1935).

41. D. Alpert, *Rev. Sci. Instr.*, **24,** 1004 (1953).

42. A. Venema and M. Bandringa, *Philips Tech. Rev.*, **20,** 145 (1958).

43. M. A. Biondi, *Rev. Sci. Instr.*, **30,** 831 (1959).

44. L. A. Harris, *Rev. Sci. Instr.*, **31,** 903 (1960).

45. J. Blears, E. J. Greer, C. J. Nightingale *Advances Vacuum Sci. Tech.*, **2,** 473 (1960). I am grateful to Dr. Blears for furnishing a copy of this paper before publication.

46. R. Jaeckel, *Z. tech. Physik*, **23,** 177 (1942).

47. W. Gaede, *Z. tech. Physik*, **15,** 664 (1934).

48. See Chapter 5.

49. Personal communication to S. Dushman.

50. J. Blears, *Nature*, **154,** 20 (1944); *Proc. Roy. Soc. London*, **A, 188,** 62 (1947).

51. See Chapter 5.

52. K. C. D. Hickman, *Nature*, **187,** 405 (1960).

53. G. P. Brown, *Rev. Sci. Instr.*, **16,** 316 (1945).

54. A. R. Huntress, *American Vacuum Society Symposium Transactions*, 1957, p. 104.

55. D. Latham, B. D. Power, and N. T. M. Dennis, *Vacuum*, **2,** 33 (1952).

56. C. S. Martin and J. H. Leck, *Vacuum*, **4,** 486 (1954).

57. F. W. Bishop, *Rev. Sci. Instr.*, **30,** 830 (1959).

58. H. R. Smith, *American Vacuum Society Symposium Transactions*, 1959, p. 140. *UCRL Document* 8970.

59. B. D. Power and D. J. Crawley, *Vacuum*, **4,** 415 (1954).

# 5 Manometers for low gas pressures[1]

Revised by J. M. Lafferty

## 5.1. APPLICATION OF ELECTRICAL DISCHARGE IN GAS AT LOW PRESSURES

The breakdown voltage in a gas between two electrodes (preferably plane surfaces) located a fixed distance apart is a function of the pressure. As the pressure is decreased from atmospheric, the voltage required to initiate a discharge decreases, reaching a minimum value at a pressure of about a millimeter of mercury or less, depending upon the geometrical arrangement, shape, and area of the electrodes and upon the nature of the gas. As the pressure is then decreased still further, the breakdown voltage increases rapidly until it exceeds that of the available voltage source. With a discharge tube of given design and high-voltage source it is therefore possible to obtain in this manner a semiquantitative indication of pressure, especially the value at which the discharge ceases.

The appearance of the discharge has been used by a number of observers as a criterion of the order of magnitude of the pressure. A more detailed discussion of these observations has been reported by Stintzing[2] and is also given by Yarwood.[3]

Burrows[4] has described a discharge tube suitable for metal systems. The construction is shown in Fig. 5.1. The tube is operated from a spark coil providing a spark in air about $\frac{3}{8}$–$\frac{1}{2}$ in. long. According to the author,

> The following approximate relationships exist between the discharge as observed in a discharge tube of the type shown and the corresponding air pressure:
>
> $\frac{1}{2}$-cm-diameter column of glow discharge = 10 mm Hg
> First visible striations = 1.5 mm
> Striations pitched 1 cm apart = 0.5 mm
> Green fluorescence on inside walls = 0.01 mm
> Black-out in dark (under vacuum conditions) = 0.001 mm or less

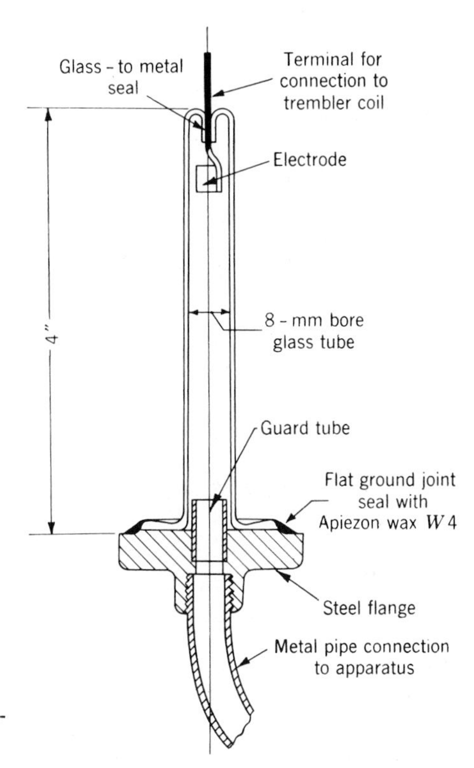


Fig. 5.1. Discharge tube for estimation of pressure (Burrows).

The discharge tube also enables an estimation to be made of the kind of gas or vapor present, thus

| Appearance | Kind of Gas |
|---|---|
| Red or pink | Air |
| Greenish gray | Decomposed oil |
| Faint (transparent) blue | Water vapor |

The gas discharge as an indicator of low pressures has also been applied in a device designated the *skanascope*.[5] One type of design consists of a cylindrical metal tube which is connected to the vacuum system and along the axis of which is located a metal rod. With a given maximum a-c voltage applied between the rod and cylinder a discharge is initiated at a certain pressure $P_i$. As the pressure is reduced the current in the discharge decreases and drops abruptly to practically zero at a certain pressure, $P_f$. The discharge excites a fluorescent screen inserted in one end of the metal cylinder, and the intensity of the luminescence serves as an indicator of the degree of vacuum. Thus, for air, the values of $P_i$ and $P_f$, corresponding to initiation of the discharge and to cessation, are 600 microns and 30 microns, respectively. The values of these critical pressures vary with the composition of the gas.

By changing the arrangement of the electrodes and replacing the fluorescent screen by a fluorescent tube in series with the discharge it is possible to decrease the value of $P_f$ for air to 5 microns.

A high-frequency spark coil is an essential device in experimental work on high vacuum. In practice a spot at which a leak is suspected may be detected more readily by wetting it with carbon tetrachloride or other volatile organic liquid. The presence of a leak is then indicated by the change in color of the glow produced by the spark coil.

For the measurement of low pressures the different types of gauges may be classified, according to the basic principles involved, as follows:

1. Manometers which use mercury or some very nonvolatile liquid.
2. Manometers which operate on the same principle as the Bourdon gauge, that is, the pressure produces a deformation in a thin wall.
3. Viscosity manometers.
4. Radiometer type of manometers, which measure the rate of transfer of momentum from a hot to a cold surface.
5. Conductivity type of manometers, involving the effect of pressure on the rate of heat transfer.
6. Ionization gauges.

## 5.2. MERCURY (AND OTHER LIQUID) MANOMETERS

Numerous attempts have been made to increase the sensitivity which may be obtained in reading the height of a barometric column or U-tube manometer.

### Rayleigh Gauge[6]

The essential parts of the Rayleigh gauge [Fig. 5.2(*a*)] are two glass bulbs, one of which communicates with a good vacuum by a tube *C*, and the other with the system in which the pressure is to be measured. Two glass pointers are sealed into the bulbs, and the bulbs are attached to a T-connection which forms the upper end of a barometric column. Mercury can be raised and lowered in the bulbs by means of a reservoir connected to the barometric column, and the level thus brought up so as to be flush with the ends of the pointers. Any difference in pressure on the mercury in the two bulbs is then measured by gradually tilting the framework on which the bulbs are fastened and observing the deflection on a mirror which is located on top of the bulbs. According to Rayleigh this gauge can be used to read pressures between 1.5 mm and $1 \cdot 10^{-3}$ mm of mercury.

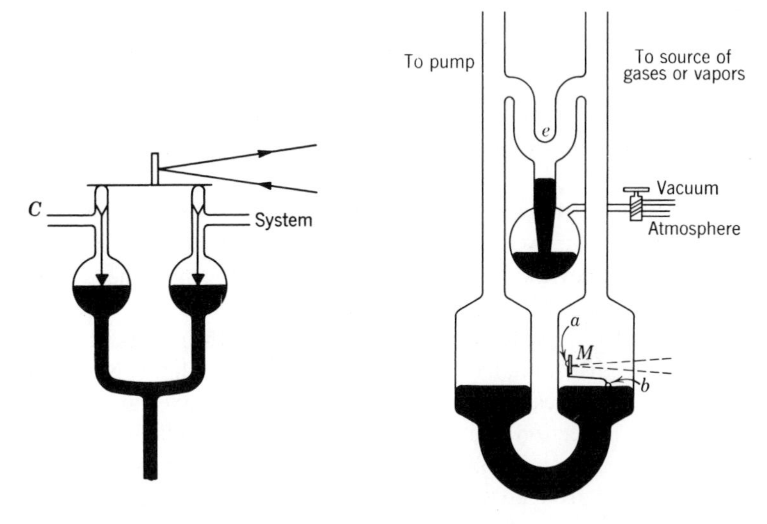


(*a*) Rayleigh form of differential manometer.

(*b*) Optical-lever manometer (Shrader and Ryder).

Fig. 5.2. Manometers.

A modified form of this gauge was used by Scheel and Heuse[7] for measuring the vapor pressure of water at temperatures below 0° C, and similar manometers have been constructed by Thiesen[8] and Hering.[9] An ingenious modification of Rayleigh's method has been used by Mündel[10] for measuring vapor pressures at very low temperatures; and a very sensitive optical method for measuring slight differences in level of two mercury surfaces, developed by Prytz,[11] has been employed extensively by different investigators in connection with Rayleigh's method.[12]

In the *optical-lever manometer*, described by Shrader and Ryder,[13] the same object is attained by a very simple construction. The following description is quoted from the original paper:

A mercury U-tube manometer [Fig. 5.2(*b*)] is formed in the usual manner, except that the surfaces of the mercury are so arranged as to be of relatively large area. Above one of the surfaces, within the tube, is arranged an optical lever as shown in the illustration. This lever is supported by two knife edges, *a*, which rest on loops of wire, which in turn are sealed into the glass walls of the tube; a glass bead *b*, fused to the end of the lever arm, acts as a float on the mercury surface, and in this way transmits the motion of the mercury surface to the lever arm. A mirror *M* attached at the position shown acts in the usual manner to reflect a beam of light from a lamp to a scale, if the gauge is to be arranged as an indicating instrument. If the gauge is to be used for recording variations in pressure, the scale may be replaced by a photographic device such as is used in oscillographic work.

The cross connection *e* provides an easy means of evacuating the whole system with one pump located as shown. With this stopcock or mercury cutoff open, a zero reading can be easily obtained, after which this connection may be closed and the gases or vapors introduced for measurement. This system provides also for the measurement of small variations in pressure, with an original pressure of any desired value, this value in no way affecting the absolute sensitivity of the gauge.

A sensitivity of $10^{-3}$ mm of mercury is claimed for the gauge, and it certainly ought to prove useful where the McLeod gauge is inapplicable.

A more elaborate modification of the optical-lever manometer has been described by Carver,[14] by which pressures as low as $10^{-4}$ mm may be measured, and a simple form, which may be used in the ranges 10 to $5 \cdot 10^{-2}$ mm, or 1 to $5 \cdot 10^{-3}$ mm, has been constructed by Hamlin.[15] Newbury and Utterback[16] have measured the vapor pressures of waxes and greases by observing the deflection of a mirror attached to an iron float on top of the mercury. The sensitivity obtained is of the order of $10^{-3}$ mm.

Instead of an iron float, Johnson and Harrison[17] have suspended "a glass float on the top of the mercury surface by means of flattened platinum wires which wrap around a roller of glass rod about 1 mm diameter. The mirror is fastened to this roller and two more flattened platinum wires also wrap round the rod and suspend it from a stout copper wire soldered to a brass ring which can slide loosely into the manometer tube." It is stated that with this construction pressure differences between 0.5 and $2 \cdot 10^{-4}$ mm of mercury can be measured.

A form of U-tube manometer sensitive to $5 \cdot 10^{-3}$ mm has been described by Pearson.[18]

A tilting differential mercury manometer for *recording* pressures between 0.01 and 5 mm, developed in several different forms by Hickman,[19] should be of interest in certain continuous operations at moderately low pressures.

In this manometer the mercury may be replaced by an organic liquid of extremely low vapor pressure, and Burrows[20] has described such a manometer in which Apiezon oil is used. Though the sensitivity, as compared with a mercury manometer, is thereby increased to as much as fifteenfold, there are two serious objections to an oil manometer. The first arises because of the solubility of the gas in the oil, and the second is due to the sluggishness of operation resulting from the high viscosity of the liquid and its tendency to stick to the glass walls.

In order to measure pressure differences of the order of $10^{-2}$ mm, in two mercury columns, Klumb and Haase[21] fasten a fine tungsten wire, about 100 cm long, along the axis of each leg of a U-tube manometer. The difference in pressure in the two sides effects a corresponding difference in the resistances of the two wires, which is determined by a Wheatstone bridge method. At pressures down to 1 mm, the difference in resistance is proportional to the pressure difference. At lower pressures (below about 0.1 mm) the indications are complicated by heat-conduction losses from the wires in the gas. However, by putting a layer of a low-vapor-pressure oil (such as *n*-dibutyl phthalate) on the surface of the mercury and using very short lengths of wire so that the wire is wholly covered by oil, the change in resistance is found to be proportional to the pressure difference, and it is claimed that pressures as low as 0.01 mm may be measured.

For measuring small pressure differences of mercury at a distance from the system, Simon and Fehér[22] have made use of a special electronic circuit in which the capacity is determined between the upper part of the mercury column and a piece of tinfoil wrapped around the outside of the glass tube.

Mention should also be made of the application by Manley[23] of the Michelson interferometer to determine pressures ranging from $10^{-4}$ to 20 mm.

### McLeod Gauge

The principle of the McLeod gauge, first described by McLeod,[24] consists in compressing a given volume $V$, of the gas whose pressure $P$ is to be measured, to a much smaller volume $v$ and observing the resultant pressure $p$, which in accordance with Boyle's law is given by the relation

$$p = \frac{PV}{v}.$$

One of the simplest forms of McLeod gauge is shown in Fig. 5.3(*a*). The bulb $B$, to which is attached a capillary tube *aa*, is connected to the low-pressure system as indicated and also to the barometric column below. In order to avoid errors due to the effect of capillarity, a tube, *bb*, of the same diameter as *aa*, is sealed on as a by-path to the larger tube $E$. To operate the gauge the reservoir $R$ is raised, thus forcing the mercury up in the barometric column $T$ until the gas in the bulb and capillary is shut off from the remainder of the system at $C$. As the mercury is raised further, the gas in $B$ is compressed until finally the mercury in the capillary *bb* is level with the top of the inside of the capillary *aa* (corresponding to the point 0 on the scale). The pressure on the gas in the capillary is then equal to the sum of the pressure in the system and that of the mercury column of length $h$, which is also the length of the capillary *aa* that contains the compressed gas.

Figure 5.3(*b*) shows a very common design of gauge in which the reservoir illustrated in Fig. 5.3(*a*) is replaced by a wide tube with a wooden (or glass) plunger which serves to raise the mercury, as shown in the figure on the right-hand side.[25]

Let $V$ denote the volume (in cubic centimeters) of the capillary and of

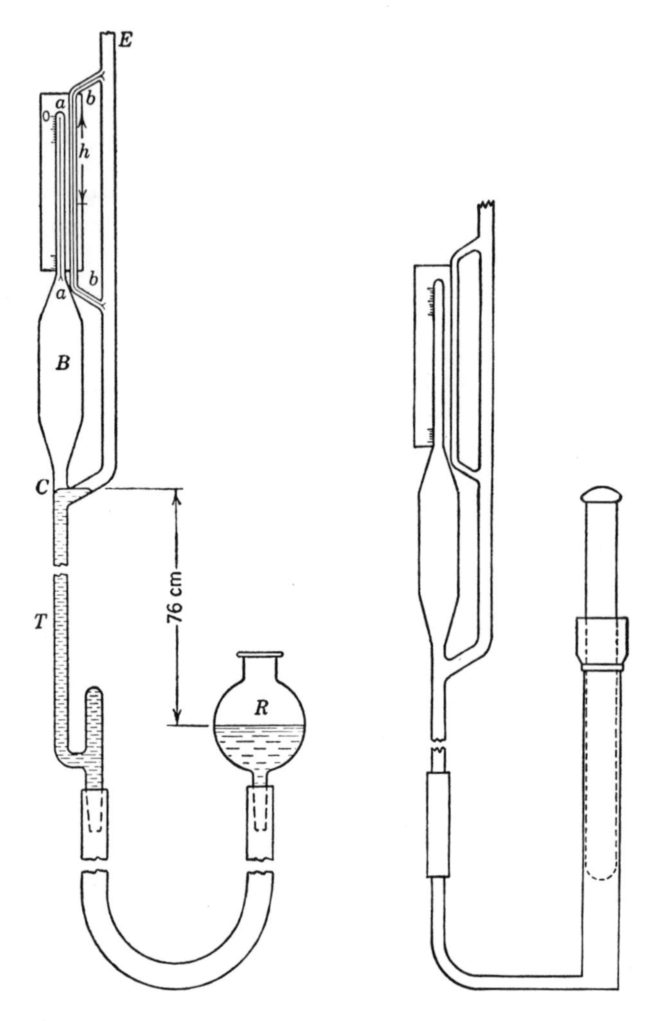


($a$) Form with reservoir that can be raised or lowered.

($b$) Form with plunger.

Fig. 5.3. McLeod gauge.

the bulb $B$ down to the level $C$ at which the gas in the bulb becomes trapped when the mercury is raised. Let $P_{mm}$ denote the pressure in the system. Also, let $b$ denote the volume, in cubic centimeters, of the capillary *per millimeter* length.

$$\text{Volume of gas trapped} = v = bh \quad \text{(cubic centimeters)},$$

$$\text{Pressure in capillary} = P_{mm} + h.$$

Hence

$$(P_{mm} + h)bh = P_{mm}V,$$

or

$$P_{mm} = \frac{bh^2}{V - bh}.$$

For gauges in which a plunger or reservoir is used to raise the mercury in the capillary, the value usually chosen for $V$ is between 75 and 100 cm$^3$, and the diameter of the capillary is between 0.7 and 2.5 mm.

Let $d =$ diameter of capillary in millimeters; then $b = 7.854 \cdot 10^{-4}d^2$ cm$^3$ per millimeter length, and

$$P_{mm} = \frac{7.854 \cdot 10^{-4}d^2h^2}{V - 7.854 \cdot 10^{-4}d^2h}.$$

For a capillary 150 mm in length (an upper limit for the length),

$$P_{mm} = \frac{17.67d^2}{V - 0.1179d^2};$$

for $d = 3$ mm, and $V = 100$,

$$P_{mm} = \frac{159.1}{100 - 1.06} = 1.607.$$

Thus, if we use a capillary 3 mm or less in bore and a bulb of 100 cm$^3$ volume, the error resulting from the omission of the $bh$ term in the denominator is less than 1 per cent.

For gauges designed for measuring maximum pressures of 1 mm or less, the more simple relation is valid, of the form

$$P_{mm} = \left(\frac{b}{V}\right)h^2 = Kh^2,$$

where $K$ is known as the "gauge constant," expressed in millimeters of mercury per millimeter length.

Expressing $h$ and $d$ in *millimeters*, and $V$ in *cubic centimeters*, the values of $K$ for gauges such as are used in ordinary laboratory practice are shown in Table 5.1. The values given in the last column, that is, $10^5K$, evidently correspond to the *pressures in microns* for $h = 10$ mm.

It will be observed that $P_{\mu b} = 1333Kh^2$.

To avoid the error which arises when the value of $h$ is small, it is often preferable to compress the gas in the capillary *aa* to a *definite* volume $v$, and then observe the height $h_1$, in millimeters, of the mercury above this level in the side capillary. In that case, since

$$P_{mm} = \frac{vh_1}{V},$$

TABLE 5.1

**Values of the McLeod Gauge Constant for Typical Dimensions of Bulb and Capillary**

| $d$ (mm) | $V$ (cm$^3$) | $10^5K$ |
|---|---|---|
| 1 | 100 | 0.785 |
| 1 | 60 | 1.309 |
| 1.5 | 100 | 1.767 |
| 1.5 | 60 | 2.945 |
| 2.5 | 100 | 4.909 |

it is seen that $h_1$ is directly proportional to $P_{mm}$, and the magnitude of this quantity can be determined by means of a cathetometer. This method, however, is not as sensitive at low pressures as that described above.

Also, if the range of pressures to be measured is large, the single capillary *aa* may be replaced by two or more capillaries of different bores, the coarser one being sealed directly to $V$, and the finer one on top of this, and so on. The serious objection to this construction is the inaccuracy involved in the readings at the junctions between the capillaries.

It is evident that the sensitivity of the gauge can be made extremely high by increasing the ratio $V/v$, that is, $V/b$. However, in the case of gauges of the construction described above, practical considerations limit the magnitude of $V$ because of the weight of mercury to be raised. Furthermore, in capillaries of very small bore (0.5–0.7 mm), the mercury tends to stick badly as the level is raised and the column breaks as the level is lowered, thus leaving a portion hanging in the capillary.

The causes of these phenomena and methods of overcoming the resulting errors in measurements have been investigated in considerable detail,[26] especially by Rosenberg[27] and Haase.[28] Haase concludes that the principal causes are as follows: (1) slight oxidation of the mercury when exposed to air, and (2) condition of the inside surface of the capillary which is due to chemical attack by water vapor or chemically corrosive vapors. Capillary tubing which has been exposed to the atmosphere for a long period should not be used for the construction of a McLeod gauge.

One method for the treatment of capillary tubing, for which satisfactory results have been claimed and which is described by Haase, is the following:

> A wire having a diameter about $\frac{2}{3}$ that of the inside diameter of the capillary is dipped into an abrasive material, such as alumina, and is then drawn carefully through the capillary. By turning the wire and successive draws through the capillary, the inner wall becomes polished until it is no longer transparent. With suitable illumination the position of the mercury meniscus can be observed quite well.

Although heating the capillary to the softening point restores the transparency, it is likely to deform the tubing; consequently as short a length as possible should be heated in forming the closed end. Obviously this treatment should be applied to the capillary before it is sealed to the bulb.

Haase has also observed that, after the capillary has been sealed off at the end and attached to the bulb, treatment with hydrofluoric acid or the vapor of the acid accomplishes the same result as the treatment with an abrasive.

Another very essential precaution is to use mercury that has been vacuum-distilled, and, if the gauge is to be used in connection with an oil-vapor pump or where water vapor is likely to be present, a liquid-air trap should be inserted between the gauge and the rest of the system.

By careful attention to these details Haase states that he has been able to construct an accurate gauge having the dimensions $V = 1400$ cm$^3$ and $d = 0.7$ mm, for which

$$P_{mm} = 3 \cdot 10^{-7} h^2,$$

where $h$ is expressed in millimeters. For $h = 180$, $P_{mm} = 10^{-2}$.

In the construction of such a gauge, the mercury reservoir is connected by means of stopcocks to the rough vacuum and the atmosphere. The mercury is raised by opening the stopcock to the atmosphere and lowered by opening the other stopcock to the vacuum. Both the reservoir and the upper bulb should be supported solidly in plaster of paris, and provision made for a box around the bottom of the reservoir to contain the mercury that would be spilled in case of breakage.

It is evident that the McLeod gauge does not indicate the pressure of mercury vapor and condensable vapors such as those of oil, water, and ammonia.

The vapor pressures of water at 15°, 20°, and 25° C are 12.79, 17.54, and 23.76 mm Hg, respectively. Hence, in the presence of water vapor condensation will occur in the capillary as $h$ is decreased, and therefore the value of $h$ will no longer correspond to the real pressure in the system. It is of interest to note in this connection the observations made by Armbruster[29] that condensation occurs in the capillary at a lower pressure than that corresponding to the vapor pressure for the temperatures at which the reading is taken.[30] This can be avoided by heating the capillary to a slightly higher temperature, so that the condensation will occur at a pressure above that of the gases in the system.[31]

Even for measuring the pressure of carbon dioxide the gauge is not reliable. In using it to measure very low pressures, such as those produced by a Gaede molecular or Langmuir condensation pump, a liquid-air trap should be inserted between the gauge and the remainder of the system in order to prevent diffusion of mercury or other vapor into the vessel to be exhausted.

Regarding the accuracy of the gauge for indicating the pressure of the so-called permanent gases (hydrogen, helium, argon, oxygen, nitrogen, and carbon monoxide), a careful investigation carried out by Scheel and Heuse[32] showed that if the bulb and tubing are carefully dried (to eliminate the presence of a film of water) the results obtained for air are certainly reliable down to pressures of 0.01 mm of mercury and are probably just as accurate at lower pressures.

Rayleigh[33] found, by means of his differential manometer, that in the range of pressures 0.001–1.5 mm Boyle's law holds accurately for nitrogen, hydrogen, and oxygen; and Scheel and Heuse[34] observed the same result with their membrane manometer. A very careful investigation on this point was carried out by Gaede[35] in connection with his work on the laws of flow of gases at low pressures. He found that, with nitrogen and hydrogen, when care is taken to dry the walls thoroughly, the McLeod gauge is quite accurate down to very low pressures (below 0.0001 mm), whereas, with oxygen, errors are likely to arise because of the formation of an oxide scum on the surface of the mercury which causes the surface to wet the glass in the capillary. However, this scum may be removed by heating the capillary carefully, and the mercury then becomes quite clean again.

For "condensable" gases, such as carbon dioxide, ammonia, sulfur dioxide, and most hydrocarbons, erratic results are frequently observed because of adsorption on the walls.[36]

If a reservoir is used with a rubber-tube connection to the barometric column, the tubing should be thoroughly cleaned and dried to remove any loose particles and also to prevent as much as possible the injurious effect of the sulfur present in rubber. Only the cleanest mercury should be used, and all glass parts of the gauge should be dried thoroughly before filling with mercury. A new McLeod gauge will give erroneous observations at the beginning until all the condensable vapors adhering to the walls have been removed by gentle heating with simultaneous exhaustion.[37]

The best procedure for calibrating the gauge is as follows. A length of capillary tubing of the desired diameter is first inspected under a microscope for uniformity of bore.[38] A drop of mercury sufficient to fill about one-third or one-fourth of the total length is then introduced into the capillary and the corresponding length measured, using a metal scale on which lengths can be measured to at least 0.5 mm (under a microscope).[39] Three or four determinations of this nature are made for different points along the capillary, and the weight of mercury used is determined to within at least $1 \cdot 10^{-3}$ g. The volume $V$ should also be obtained by weighing the bulb and capillary when filled with mercury or water to the level of the shut-off cross section.

A very important feature in connection with the construction of a

McLeod gauge is the sealing off of the top of the capillary in such a manner as to leave the inside as flat as possible, without any rounded end to the bore. The method to be used in the sealing operation in order to secure this objective has been described in detail by Clark,[40] and in the same connection there is a discussion of the precautions to be taken in calibration. Nottingham has developed an experimental procedure for determining the location of the "effective" top of the sealed capillary and a simplified method for reading pressure without setting the mercury in the open capillary level with this "effective" top. He states:[41]

Pressure measurements with a McLeod gauge depend on an application of Boyle's law for gases. The resulting equation is:

$$p = \frac{a}{V}(h' - h_0)(\Delta h). \qquad (a)$$

In this equation (see Fig. 5.4) $\Delta h$ is the difference in the mercury levels in the open and closed capillaries. This quantity is directly measurable and indicates the pressure difference between the gas compressed in the closed capillary and that in the open capillary. The capillaries must be clean and of equal and uniform cross section. The quantity $(h' - h_0)$ represents the distance, expressed in millimeters, between the mercury surface and the "effective" top of the closed capillary. Before a McLeod gauge can be used for accurate measurements, the location of the effective top of the capillary must be determined experimentally. The area of cross section of the capillaries is denoted by $a$, the total volume of gas trapped off by the closed capillary and the main bulb of the McLeod gauge by $V$. A consistent system of units is obtained if the area is expressed in square millimeters, the volume in cubic millimeters, and the distance measurements in millimeters. In that case, the pressure will be expressed in millimeters of mercury.

To determine the effective end of the closed capillary, gas pressure is introduced at some arbitrary and unknown value. The distance $h'$ is measured from an arbitrary fiducial line near the top of the closed capillary. A convenient point is the top external surface of the glass that closes this capillary. As the gas in the closed capillary is compressed, three or more readings of $\Delta h$ and the corresponding $h'$ values can be observed. This set of readings can be related by the following equation:

$$h' = h_0 + \frac{pV}{a}\left(\frac{1}{\Delta h}\right), \qquad (b)$$

which shows a linear relation between the observable quantities $h'$ and $(1/\Delta h)$. A plot of $h'$ as a function of $(1/\Delta h)$ should yield a straight line with an intercept at $h_0$. Although the choice of pressure is arbitrary for this part of the calibration, it is best to work in the range that gives $h'$ and $\Delta h$ values between 5 to 10 mm when the compression is such as to make their values equal. This procedure reduces the experimental uncertainty in determining $h_0$ by extrapolation. After $h_0$ has been determined, larger arbitrary pressures may be chosen and the linearity of the capillary examined by means of Boyle's law. If systematic differences occur, the indications are that the capillary is either dirty or nonuniform.

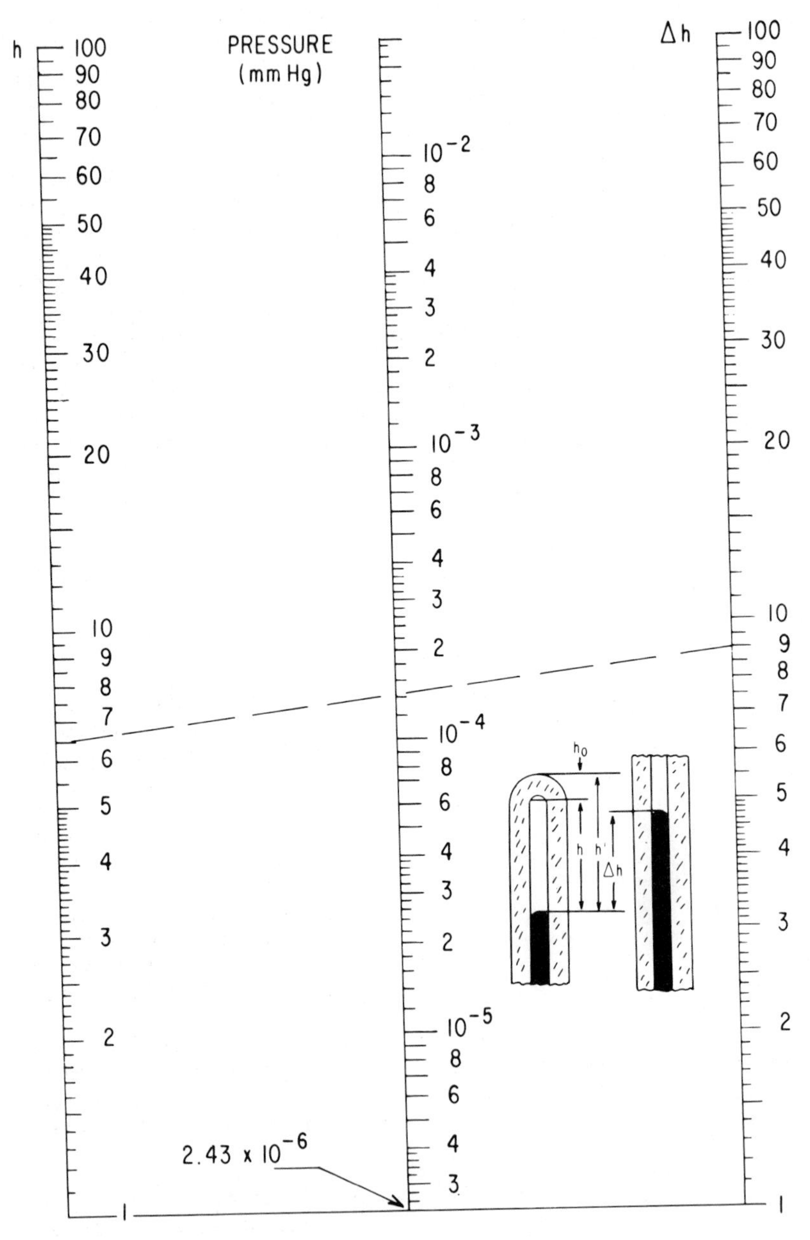


Fig. 5.4. Nomographic chart for conversion of McLeod gauge measurements to gas pressure (Nottingham).

For accuracy it is impractical to attempt to use a McLeod gauge by directly viewing the column heights against a simple ruled scale. The alternative is to use a good cathetometer with a good telescope which can be sighted with high accuracy on the top of the mercury meniscus. Even though the capillaries from which the McLeod gauge was constructed were presumably of uniform bore, a necessary preliminary test must be carried out to show that the mercury rise in the two columns is precisely the same, thus giving a $\Delta h$ of zero at all points along the capillary when the residual gas pressure is in the low-pressure range of the order of $10^{-8}$ mm Hg or lower. Experience shows that even with clean mercury and clean capillaries frictional forces between the mercury column and the glass can cause very serious random errors in the readings. These errors can be minimized by some tapping of the capillaries followed by sufficient time for a final position to be observed, after which the value of $\Delta h$ under the high-vacuum conditions will become zero at all positions or at least follow a systematic pattern of very small differences. The observer must remember that column differences under these conditions of one- or two-tenths of a millimeter may introduce important errors in the use of the McLeod gauge. The determination of $h_0$ with accuracy is not easy. If the capillaries are tapered, a systematic error may show as a reproducible nonlinearity. Repeated measurements will give an indication of the random errors that must be expected.

After $h_0$ has been determined, then the effective length of the gas-filled part of the closed capillary, which in equation $a$ is $(h' - h_0)$, may be identified by $h$ and equation $a$ rewritten as

$$p = \frac{a}{V} h\,(\Delta h). \qquad (c)$$

In actual gauge use, it is generally advisable to make the observation with $h$ approximately equal to $\Delta h$, but it is not always possible to stop the in-flow of mercury with such accuracy that these two quantities are precisely equal to within a tenth of a millimeter. The nomographic chart illustrated as Fig. 5.4 is applicable to a gauge with a value of $a/V$ of $2.43 \cdot 10^{-6}$. The method of construction involves the choice of the simple logarithmic scales identified by $h$ and $\Delta h$ in the figure. The center scale, located halfway between the two lines, has a scaling of two orders of magnitude for the same scale distance as one order of magnitude in the $h$ scales. The center scale is displaced with respect to the others, so that the straight line that joins the corresponding unit points will fall at the corresponding pressure point; in this case $2.43 \cdot 10^{-6}$. This chart is very helpful in the determination of McLeod gauge pressure from the observed $h$ and $\Delta h$.

As an example of the use of this chart, the value of $h$ might be observed as 6.5 mm and $\Delta h$ 9 mm. The observer places a straight edge across the chart with the intersection at each outside line coinciding with these values, and reads the pressure on the middle scale directly as $1.4 \cdot 10^{-4}$ mm. The dashed line illustrates this example.

There are described in the literature,[42] and in catalogs of firms selling physical apparatus, a number of variations in the construction of McLeod gauges.

Pfund[43] has combined the McLeod gauge with a hot-wire gauge (see section 5.6) in order to extend the range of the former to still lower values.

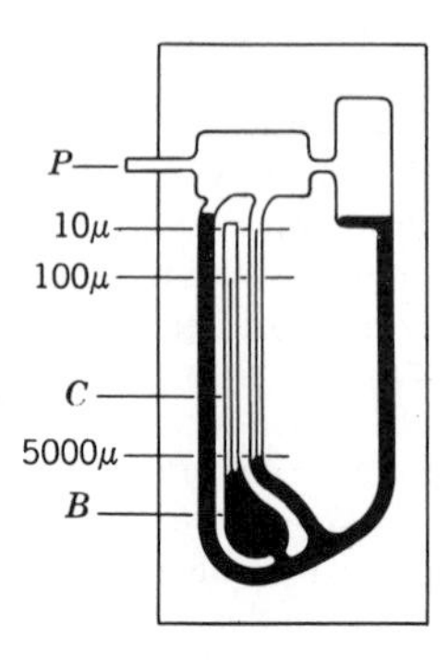


Fig. 5.5. Swivel form of McLeod gauge.

A very small filament made of fine tungsten wire is sealed into the capillary at the top. Down to a certain pressure the gauge is used as described above. At lower pressures, at which the observed value of $h$ becomes only a millimeter or so, the filament is heated to a very faint red, and the resistance at constant current is found to vary linearly with the pressure. In this range the gauge must be calibrated separately for each gas. As an illustration of the increased sensitivity thus obtained, Pfund gives the following data.

The limit of the "unaided" gauge was $5.26 \cdot 10^{-5}$ mm for $h = 1$ mm, whereas, by using the hot wire, the range was extended to $1.7 \cdot 10^{-7}$ mm. Since a portable galvanometer was employed as indicating instrument for measuring the resistance, the range could be extended to still lower values by using a sensitive wall galvanometer.

For pressures in the range such as 0–1000 microns and higher, and for factory purposes, "swivel" types of gauges are available such as that shown in Fig. 5.5, which is taken from a publication by Flosdorf.[44] The gauge is connected to the system by a rubber-tube connection at $P$, and is mounted on a framework so that turning it cuts off a definite volume of the gas in the system at pressure $P$ and causes this volume to be compressed to a smaller volume at a higher pressure. A similar design, devised by Bruner[45] and designated the *vacustat*,[46] is described by Yarwood.[47]

The Gaede vacuscope[48] resembles the last two gauges in its compactness. It consists of a central expanded part to which are sealed, on one side, one arm in the form of a U-tube, and another arm, on the other side, which functions as a McLeod gauge. When in the horizontal position the gauge is evacuated, and when tilted to one of the two other positions, which are in the vertical plane and 180° apart, it can measure pressures as low as $10^{-2}$ mm or as high as 80 mm.

Moser[49] has described a more elaborate variation of the vacustat which has three ranges: 760 to 1 mm, 1 to $10^{-2}$ mm, and $10^{-2}$ to $10^{-4}$ mm.

## 5.3. MECHANICAL MANOMETERS[50]

A number of attempts have been made to construct low-pressure manometers which depend upon the measurement of the mechanical deformation suffered by a thin wall or diaphragm under pressure. At ordinary pressures this principle has been utilized in the construction of the Bourdon spiral. Other manometers have been reported by Dibeler and Cordero[51] in which the deflection of a calibrated metallic diaphragm is used as an indication of pressure. Wallace and Tiernan[52] have developed an absolute-pressure gauge that is actuated by force on a sealed diaphragm. This device, shown in Fig. 5.6, operates over a pressure range of 0–50 mm Hg and can be read to 0.2 mm Hg. Higher ranges are available. Pressure measurement is accomplished by admitting the unknown pressure to the hermetically

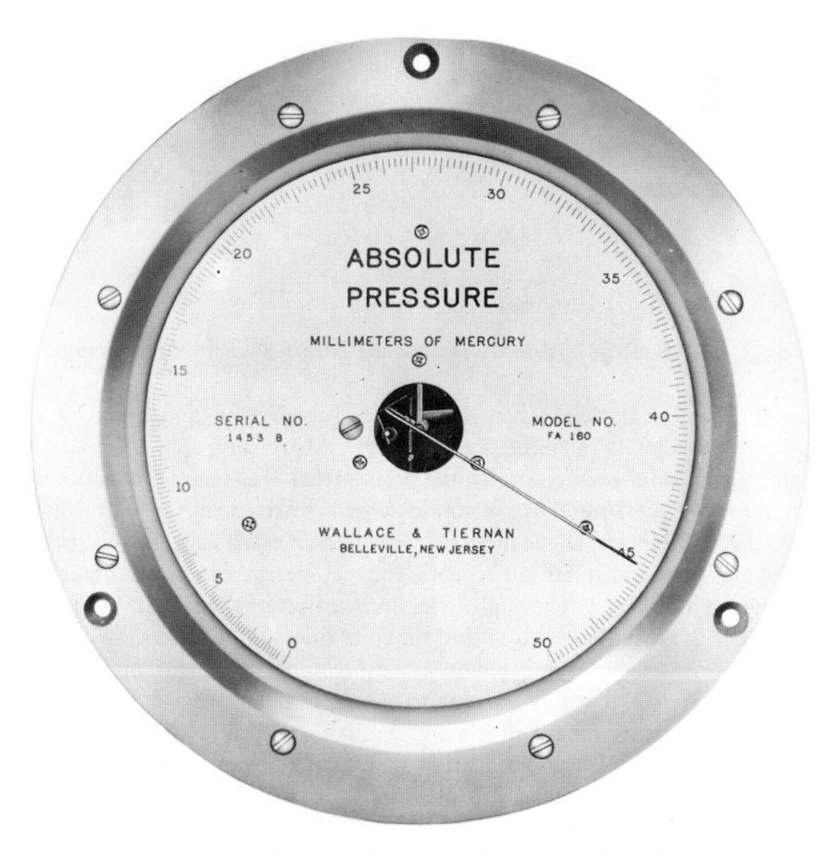


Fig. 5.6. Wallace and Tiernan absolute-pressure indicator.

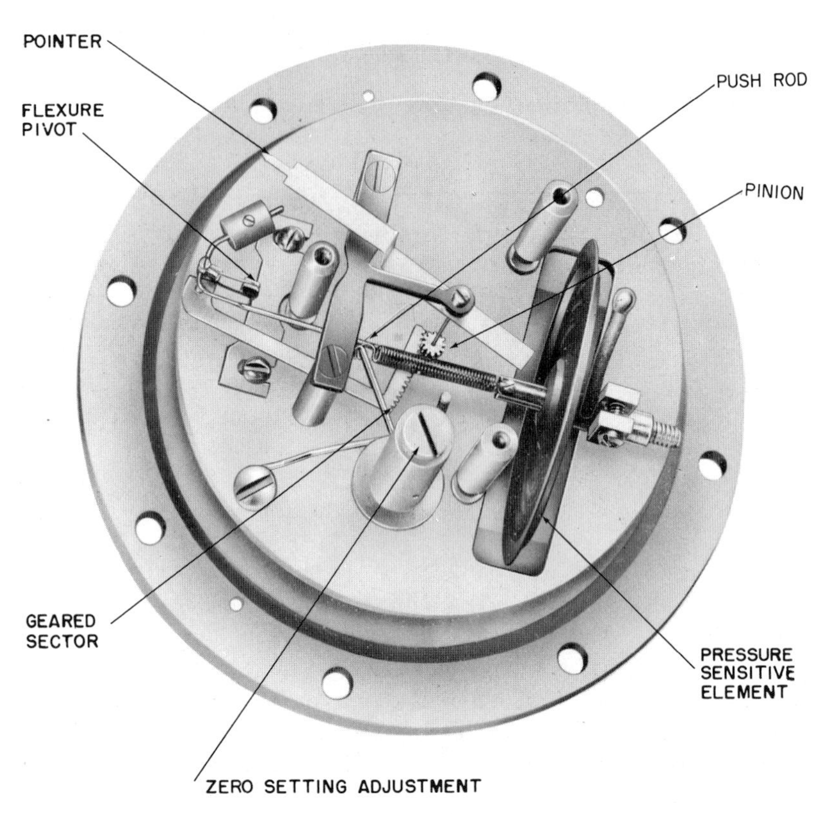


Fig. 5.7. Internal construction of the Wallace and Tiernan absolute-pressure indicator

sealed instrument case, where it exerts a pressure on a flat evacuated capsule that has been permanently sealed. Movement of the capsule is transmitted by a lever system to the pointer that registers the pressure on the dial as a pressure above absolute zero. This mechanism is shown in Fig. 5.7. The pressure-sensitive element is connected to a geared sector through a push rod and flexure pivot. The sector engages a pinion mounted on the pointer shaft. The shaft turns in bearings at the top and bottom. Backlash is prevented by a spring that provides uniform tension for all positions of the sector and pinion. The pressure-sensitive element has a built-in stop that prevents damage to the mechanism from overpressure. An excess-pressure relief valve is also available to protect the instrument from possible accidental application of pressures in excess of 1 atm. A check valve is provided at the pressure connection to release pressure slowly from the manometer, thus protecting the lever system against large pressure surges. This instrument has the advantage of simplicity, with no auxiliary electrical equipment, and reads pressure independently of the nature of the gas.

An absolute manometer that may be baked out at high temperatures for use with ultrahigh-vacuum systems has been described by Alpert, Matland, and McCoubrey.[53] This device is used for calibration or the measurement of gas pressures higher than those which can be measured with an ionization gauge. It is a null-reading instrument and measures pressures above 0.1 mm Hg with an absolute accuracy of about 0.01 mm Hg. The pressure-sensing element is a metal diaphragm that separates the ultrahigh-vacuum system from a vacuum system that includes an ordinary liquid manometer. With both sides of the diaphragm evacuated, it assumes an equilibrium position as indicated by its capacitance to an electrical probe. When a sample of pure gas is introduced into the ultra-high-vacuum system, a comparable amount of air is let into the manometer system until the diaphragm assumes its null position again as indicated by its capacitance to the probe. The liquid manometer reading then gives the pure gas pressure.

Ladenburg and Lehmann,[54] and subsequently Johnson and McIntosh,[55] have described a low-pressure gauge consisting of a flat tapered glass tube bent in the form of a spiral. The walls are usually very thin, so that the device may be sensitive to small pressure differences. A glass mirror is attached to the end of the spiral, and the spiral is sealed into another chamber in which the pressure may be varied. The system whose pressure is to be measured is connected to the spiral. In using the instrument, the pressure outside the spiral is varied until it is equal to that in the spiral, as indicated by the mirror, and the pressure outside is then measured by an ordinary mercury manometer. The device has been used for measuring the pressure of corrosive gases like chlorine and ammonium chloride vapor. A similar type of manometer has also been used by C. G. Jackson for measuring the dissociation pressure of cupric bromide.[56] These gauges, however, are not sensitive to pressures below about 100 microns. A quartz spiral gauge operated on the same principle and devised by Bodenstein and Dux[57] is shown in Fig. 5.8. As described by Farkas and Melville,[58] this gauge "consisted of a long thin spiral of flattened silica tubing carrying a mirror at the end to detect the extent of 'unwinding' of the spiral."[59]

Two methods of mounting the spirals are used: in one, a long light pointer is sealed to the closed end at 90° to the axis of the coil and the deflection is noted of the end of this pointer; in the other, a suspension, to which a galvanometer mirror is fastened, is attached to the end of the spiral and the spiral is mounted with its axis vertical and the open end (connecting to the system) at the bottom. In both methods the spiral

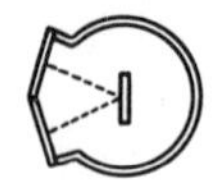

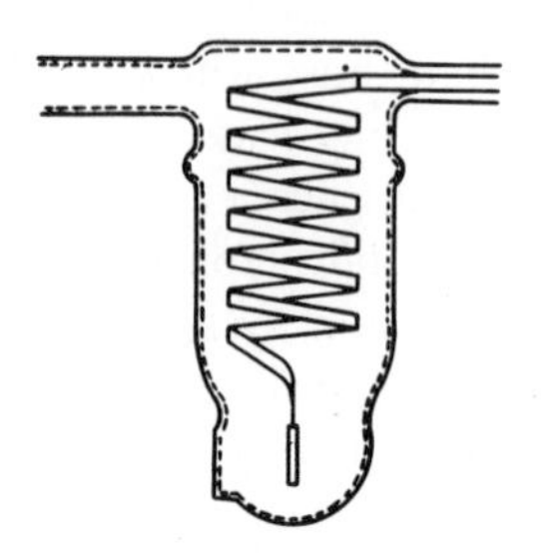

Fig. 5.8. Quartz spiral gauge.

is enclosed in a glass tube opened at one end, so that the pressure inside the spiral may be balanced, if desired, by an external pressure. It is stated that the most sensitive form of this gauge (that constructed according to the second method) is capable of measuring pressures of the order of 2 mm of mercury.

A form of Bourdon gauge, described by Foord,[60] is shown in Fig. 5.9. It consists of a collapsed thin bulb *B* to which is attached a short butt *D*. *D* is surrounded by a rigidly supported ring *C*, which limits its traverse to about $\pm 0.5$ mm. Foord states:

> By this means a gauge which would fracture at about 200 mm of mercury may be made to withstand a pressure difference of $\pm 1$ atmosphere. To compensate for the loss of sensitivity in using such a comparatively strong bulb, the deflection is multiplied by means of a simple all-glass lever system as shown, the lateral motion of the butt being transmitted by a flexible link *E* to the pointer proper *F*, which is rendered free to bend about its point of support by drawing out its lower extremity to a fine rod.

A water jacket *A* is provided, and the gauge is best operated in a vertical position. According to the author, "A gauge of this type has been constructed which gives a deflection of 0.54 mm per mm of mercury, using a pointer 20 cm in length, and there is no doubt that more sensitive gauges could be made if required, especially if not called upon to withstand such a large maximum difference of pressure." It is this requirement, of course, that puts limitations on the use of a gauge of this type. Furthermore, it obviously requires a considerable number of trials to make a bulb such as *B* which will be both thin-walled and not too fragile.

A glass membrane manometer designed by Grigorovici[61] is shown in Fig. 5.10. It consists of an evacuated glass tube, which is closed at one end by a glass plate (cemented to the tube), and inside of which is a glass membrane. This membrane is prepared by softening a thin glass bulb to the point at which one side will collapse to form a flat surface, as shown. Two wires $DD$ support the membrane against the cover plate, and the deflection of the membrane is measured by means of a small mirror, $M$, fastened at the edge, and a telescope and scale. A ground joint $S$ is provided so that the membrane can be readily replaced. The gauge may be used in the range 0.1–5.0 mm of mercury.

The membrane manometer used by Scheel and Heuse[62] for measuring the vapor pressure of ice at very low temperatures consisted of a very shallow cylindrical glass box separated into two compartments (parallel to the flat sides) by a thin copper membrane. One compartment was connected to the system; the other was connected directly to a high-vacuum pump. The deformation of the membrane, due to the slight difference in pressure on the two sides, was then measured by noting the number of interference rings produced by the pressure of the membrane against a glass plate. The instrument was found capable of measuring pressures down to about 0.001 mm of mercury, but difficulties were encountered in using it because of the continual gas evolution from the walls of the device.

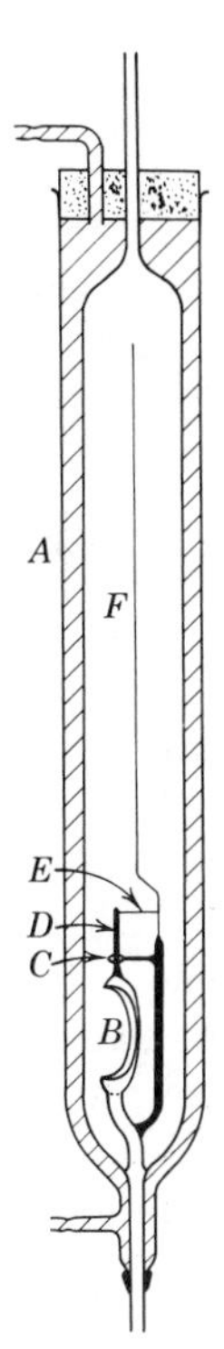


Fig. 5.9. Thin-wall type of gauge (Foord). The extent of deformation of the thin wall $B$ is observed by the deflection of the end of the pointer $F$.

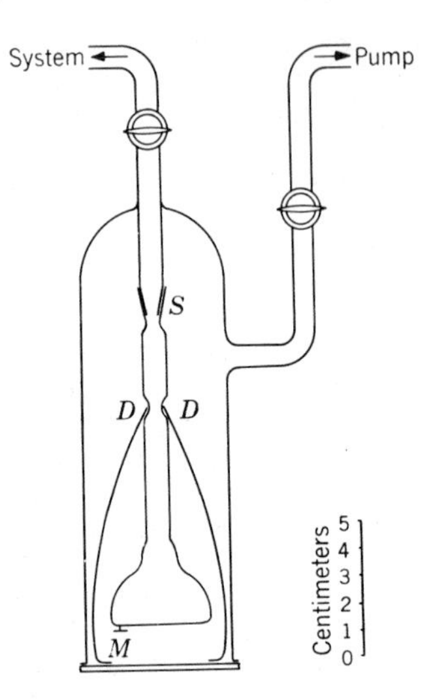


Fig. 5.10. Thin glass-wall manometer for direct reading (Grigorovici). The extent of the deformation is observed by means of the small mirror *M*.

The deflection of the membrane may, obviously, be measured by means of an optical system similar to that used in the mercury manometer devised by Shrader and Ryder, that is, by attaching a pointer to the membrane and letting this pointer deflect a mirror. Such a manometer (sensitive to $10^{-4}$ mm Hg) has been described by Stewardson.[63] In this case a thin diaphragm of mica or glass was used. This type of manometer has been applied to a considerable extent, especially where it was necessary to measure pressures of gases that react chemically with mercury.[64]

The deflection of the membrane can also be measured by observing the change in capacity with respect to a stationary membrane. This is the method used by Sommermeyer.[65] An accuracy of 0.01 mm is claimed for this particular gauge.

Although the manometer designed by Rodebush and Coons[66] is not of the same type as the other gauges described above, it has this feature in common with them, that it *measures absolute pressure directly* and is therefore independent of the molecular weight of the gas or vapor. Figure 5.11 shows an improved form of the gauge, made by Deitz,[67] for measuring the vapor pressure of potassium chloride and cesium iodide.

As Farkas and Melville[68] state, "The only disadvantage of this absolute manometer is that gas or vapor has to be continuously applied to the measuring chamber owing to leakage past the silica disk." The following description of the gauge is given by Deitz:

*L* is a disk of fused quartz about 2.5 cm in diameter, the upper surface of which is ground to seat on the ground end of the quartz tube *J*. *C* is a small needle of soft iron, enclosed in glass, and $B$ is a cantilever arm made from a quartz fiber by which the system is suspended. $B$ is about 10 in. long from the point of susp.nsion to the point of support, which is a quartz to Pyrex seal to the rest of the apparatus. The arm $B$ extends outwards for about 2 in. past the suspension support and is pulled out to a fine point, the extreme tip being about a millimeter from a plane window. A microscope was then used to measure the position of the arm and hence give the vertical position of the quartz disk. $D$ is a coil through which a current is passed to bring an electromagnetic traction on $C$, thus pulling the disk upward to seat on $J$.

The ground seat, $J$, is sealed within a transparent quartz tube, $F$, 12-mm bore, the top of which is sealed to the remainder of the apparatus by a DeKhotinsky joint at $E$, in the manner shown. Surrounding the quartz tube $F$ is another transparent quartz tube, $G$, 14 in. long, which is closed at the bottom. A short length of quartz tube of the same diameter as $F$ is sealed within $G$. The seal is made at the bottom, the top end having been previously ground to make a seat with the bottom of the inner tube $F$ at $H$. The salt crystals are placed at the bottom of the outside tube $G$. This is then slipped into place and sealed at $K$ by means of a water-cooled picein joint. The long suspension is lowered into place through the opening of the ground joint at $A$. The female part of the joint carries a test tube sealed concentrically within, which serves as a liquid-air cooled surface.

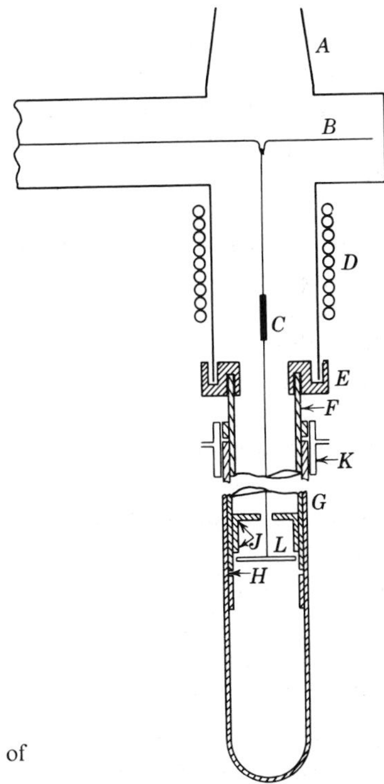


Fig. 5.11. Manometer for determination of absolute vapor pressures (Deitz).

In operation the pressure is measured by noting the value of the current at which the disk just drops from its seat on the end of the tube.

At the null point the apparatus functions as an electromagnetic balance. The forces upwards consist of the tension in the quartz cantilever support, the magnetic pull of the solenoid, and the force due to the pressure of the vapor on the quartz lid, equal to the pressure times the area. The force downward is the force of gravity. As the pressure increases, the pull necessary from the solenoid decreases.

For calibration the procedure was used as follows:

The quartz disk was replaced by an aluminum pan and sufficient weights added to make the total weight equal the weight of the quartz disk. The necessary current was then passed through the solenoid to bring the pointed end of the cantilever arm $B$ to the null-point position, previously observed with the quartz disk seated in place. By removing weights from the pan and observing the corresponding decrease in the solenoid current, a calibration was obtained.

The calibration curve was observed to be linear, and the slope gave the decrease in current per unit increase in total pressure on the disk. The pressures measured ranged from about 1 to 30 microns.

A somewhat similar method for the determination of the *vapor pressures of oils* has been described by Hickman, Hecker, and Embree,[69] and a modified design of their apparatus, used by Verhoek and Marshall,[70] is shown in Fig. 5.12.

The sample to be investigated is located in the bulb $B$, which is maintained at the desired temperature. The vapor emitted through the orifice $O$ exerts a pressure on the disk $D$, and the magnitude of the pressure is measured by turning the entire apparatus through an angle $\theta$ that is just sufficient to close the orifice.

Let $W$ = weight in grams of the assembled pendulum, and $A$ = area of orifice. Then

$$P_\mu = \frac{W \sin \theta}{0.00136A},$$

where the factor in the denominator converts the pressure from grams per square centimeter into microns. (In the device described, $W = 0.787$ g, and $A = 7.716$ cm$^2$.)

In order to eliminate possible back pressure on the disk $D$ and prevent the disk from adhering to the glass by the surface tension of any liquid formed, a condensing tube $C$, filled with liquid air, is used and a platinum resistance heater is located around the glass tube at $O$.

Further details of construction quoted from the original paper are as follows:

To measure $\theta$, the turning motion was carried through a worm gear to a disk 15.0 cm in diameter divided on its circumference into 216 divisions, each of

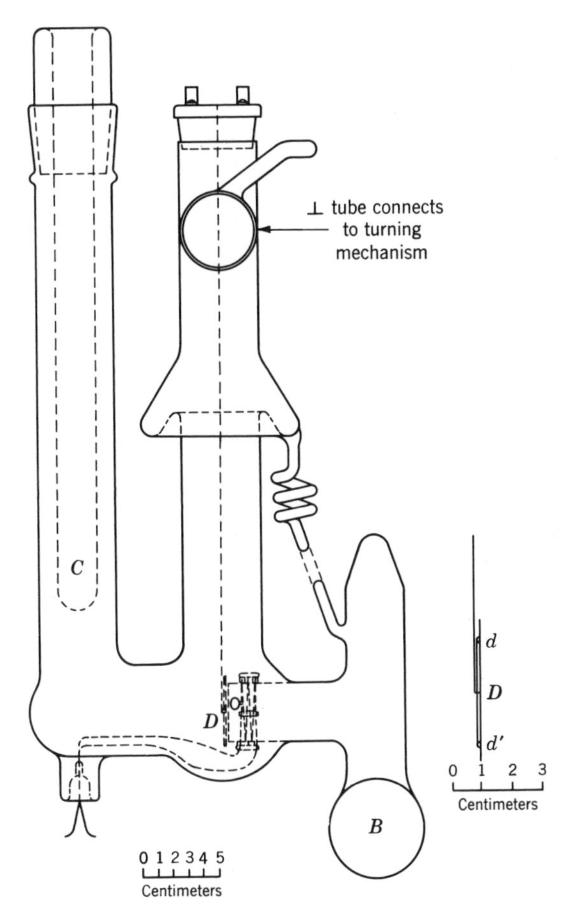


Fig. 5.12. Manometer for determination of absolute vapor pressures (Verhoek and Marshall).

which corresponded to a deflection of one minute of arc. Consecutive readings at a given temperature could easily be reproduced to within one minute of arc. The pendulum consists of a Duralumin disk *D*, 0.013 cm thick, suspended by means of Duralumin wire, 0.05 cm in diameter, from a shaft riding in jeweled bearings. The circular corrugation (*dd′*, Fig. 5.12) in the disk serves to stiffen and hold it perfectly flat.

In connection with a discussion of methods for the determination of absolute vapor pressures, it is also of interest to mention the method devised by Volmer, Heller, and Neumann[71] and applied by Neumann and Volker[72] to measure vapor pressures of potassium and mercury in the range $10^{-2}$ to $10^{-5}$ mm.

The material to be investigated is placed in a small glass vessel consisting of two bulbs joined by a short piece of rather wide tubing and provided with a tubulation at the center and with two very small holes in the bulbs which are opposite to each other. The vessel is suspended at the center, and the surrounding glass container is evacuated. The vapor streaming out of the two holes causes the suspended vessel to be rotated through an angle which is proportional to the pressure.

## 5.4. VISCOSITY MANOMETERS

As shown in section 1.8, the rate of transfer of momentum from a surface moving with velocity $u$ in the plane of the surface, to a stationary surface parallel to it at distance $d$, is given, at low pressures, by the relation

$$B = \frac{\eta u}{d + 2\zeta}. \tag{1.109}$$

In this equation $\zeta$ is the coefficient of slip, which is proportional to the mean free path $L$ and, consequently, is inversely proportional to the pressure.

Hence, equation 1.109 can be written in the form

$$B = \frac{\eta u}{d + 2bL} \tag{5.1a}$$

or

$$B = \frac{\eta u}{d + c/P}, \tag{5.1b}$$

where $b$ and $c$ are constants which vary with the nature of the gas and that of the surface. In general, the constant $b$ has a value of the order of unity.

At very low pressures, where $L \gg d$, equation 5.1$a$ becomes

$$B = \frac{\eta u}{2bL},$$

and, substituting from the relation

$$\eta = 0.5\rho v_a L,$$

we obtain the relation

$$B = K_1 \rho v_a u \tag{5.2}$$

$$= K_2 P \left(\frac{M}{R_0 T}\right)^{1/2} \cdot u, \tag{5.3}$$

where $K_1$ and $K_2$ are constants.

A relation of the same form as these equations may be derived from considerations similar to those used in deriving the laws of molecular flow.

The number of molecules crossing unit area per unit time is given by

$$\nu = \tfrac{1}{4} n v_a.$$

If the molecules are leaving a surface having a velocity $u$, the momentum transferred per molecule is $mu$. Hence,

$$\begin{aligned} B &= \nu m u \\ &= P\left(\frac{M}{2\pi R_0 T}\right)^{1/2} \cdot u. \end{aligned} \tag{5.4a}$$

Thus we can write, in general,

$$B = KuP\left(\frac{M}{T}\right)^{1/2}, \tag{5.4b}$$

where $K$ is a constant.

As will be observed, the coefficient $\eta/d$, which is independent of $P$ at normal pressures, is replaced at low pressures by the coefficient $P(M/T)^{1/2}$, which varies linearly with the pressure.

In applying the above considerations to the construction of a gauge, two different methods have been used. In the first of these, which is applied in what we may designate for reference as the "decrement" type of gauge, a surface is set in oscillation and the rate of decrease of the amplitude of oscillation is taken as a measure of the pressure. Physically, the damping may be explained as due to the gradual equalization of energy between the moving surface and the molecules of gas striking it.

In the second method, a surface is set in *continuous* rotation and the amount of twist imparted to an adjacent surface is used to measure the pressure. The molecules striking the moving surface acquire a momentum in the direction of motion which they tend in turn to impart to the other surface. If that surface is suspended and free to turn about an axis which is perpendicular to the direction of motion of the rotating surface, it will be twisted around until the force due to the incident molecules is just balanced by the torsion of the suspension. We may, therefore, designate this as the "static" type of viscosity gauge, to emphasize the fact that observations with this method are taken under stationary conditions.

## Decrement Type of Viscosity Gauge

A gauge based on this principle was first suggested by Sutherland,[73] and subsequently a very careful investigation on the same subject was carried

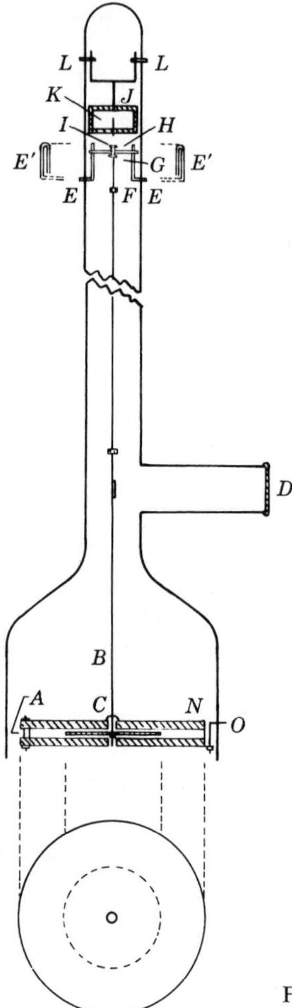


Fig. 5.13. Decrement type of viscosity gauge (Hogg).

out by Hogg.[74] The construction used by Hogg, which was essentially the same as that employed by Maxwell, and by A. Kundt and E. Warburg, in their determinations of the coefficient of viscosity, is shown in Fig. 5.13. A thin glass disk is suspended by means of a wire between two fixed horizontal plates *N*. The wire carries a mirror which may be viewed through a plate-glass window *D* by means of a telescope and scale. At the top, the wire is supported by clamps and is connected to a soft-iron armature *J* which is supported by the swivel head *K*. Turning this armature by means of an external magnet sets the center disk in oscillation, and the rate of decrease of the amplitude of these oscillations is then observed by means of the telescope pointed at the window *D*.

Let $\theta$ denote the angles of twist at any instant. Then the motion of the suspended disk is expressed by a differential equation of the form

$$I\frac{d\theta^2}{dt^2} + R\frac{d\theta}{dt} + K\theta = 0, \qquad (5.5)$$

where $I$ = moment of inertia of disk ($ml^2$),
$R$ = coefficient of damping ($ml^2t^{-1}$),
$K$ = coefficient of restoring force ($ml^2t^{-2}$).

The dimensions of each of these coefficients are given in parentheses. Hence the dimensions of each term in the equation are $ml^2t^{-2}$.

The solution of this equation is of the form

$$\theta = \theta_0\epsilon^{-Rt/2I}\cos 2\pi\nu(t - t_0), \qquad (5.6)$$

where $t_0$ may be set equal to zero, and $\nu$ is the frequency of oscillation, which is given by

$$2\pi\nu = \left(\frac{1}{IK} - \frac{R_2}{4I^2}\right)^{1/2}. \qquad (5.7)$$

That is, the maximum amplitude of $\theta$, denoted by $A$, decreases exponentially with the time. The solution expressed by equation 5.6 is valid if the damping factor is not too great. In fact, there exists a critical relation,

$$R^2 \leqslant \frac{4I}{K},$$

which must be satisfied for the occurrence of oscillations.

Let $\tau = 1/\nu$ denote the *period* of a complete oscillation. If $A_1$ and $A_2$ denote two successive values of the maximum amplitude $A$, it follows from equation 5.6 that

$$\frac{A_2}{A_1} = \epsilon^{-R\tau/2I} = \epsilon^{-\lambda}, \qquad (5.8)$$

where $\lambda$ is known as the *logarithmic decrement.*

Let $t_{0.5}$ designate the time required for the maximum amplitude to decrease to half its initial value. Then

$$t_{0.5} = \frac{\ln 2}{R/(2I)} = \frac{\tau \ln 2}{\lambda}. \qquad (5.9)$$

At normal pressures, $R$ is proportional to the product of $\eta/d$ and $l^4$, where $l$ is a length proportional to the radius of the disk, and $d$ is the distance between the disk and stationary plate. Thus, at these pressures,

$\lambda$ is independent of the pressure. However, at lower pressures, it follows, as shown above, that $R$ must vary linearly with the product $\rho v_a l^4$. Consequently, $\lambda$ should vary linearly with the pressure.

This statement, however, requires some modification. Let $\lambda_0$ denote the value of the decrement at normal pressures, and $\mu$ the decrement of the suspension itself.

From equation 5.1*b* and the above considerations regarding the variation with pressure in the value of $R$, it follows that, in the intermediate range of pressures for which $L$, the mean free path, is not very large compared to $d$,

$$\frac{\lambda_0 - \mu}{\lambda - \mu} = 1 + \frac{C}{Pd}; \tag{5.10a}$$

that is,

$$\left(\frac{\lambda_0 - \mu}{\lambda - \mu} - 1\right) P = \frac{C}{d}, \tag{5.10b}$$

where $C$ is a constant.

This is the relation used by Hogg in connection with his gauge. It is evident that this gauge is not suitable for measuring very low pressures, since the value of $\lambda$ then becomes comparable with that of $\mu$ and therefore the determination of $P$ involves large experimental errors.

A contribution to the theory of this type of gauge has also been published by Shaw.[75] He has recorded measurements down to $0.35 \cdot 10^{-3}$ mm of mercury.

*Quartz-fiber gauges.* This method was first suggested by Langmuir[76] for measuring the pressure of residual gas in a sealed-off incandescent lamp, and it has been used by Dushman in some investigations. Following Langmuir's suggestion, a gauge of this type was described by Haber and Kerschbaum[77] in a paper in which the theory of its operation was also discussed in detail.

While these investigators used the gauge to measure the pressures of vapors of water, iodine, and mercury, Henglein[78] employed the same type of gauge to measure the dissociation pressure of chlorine at high temperatures, and Scott[79] applied it to measure vapor pressures of cesium and rubidium.

The construction of the gauge described by Haber and Kerschbaum is shown in Fig. 5.14(*a*). It consists of a fine quartz fiber (diameter ranging from 0.01 to 0.005 cm) sealed into the top of a glass tube. The fiber is set in oscillation by tapping the glass bulb gently, and the rate of decrease of the amplitude of vibration is observed by means of a telescope and lamp, as shown in Fig. 5.14(*b*).

Theoretical considerations, which are discussed more fully below, lead to the following relation for the determination of the pressure.

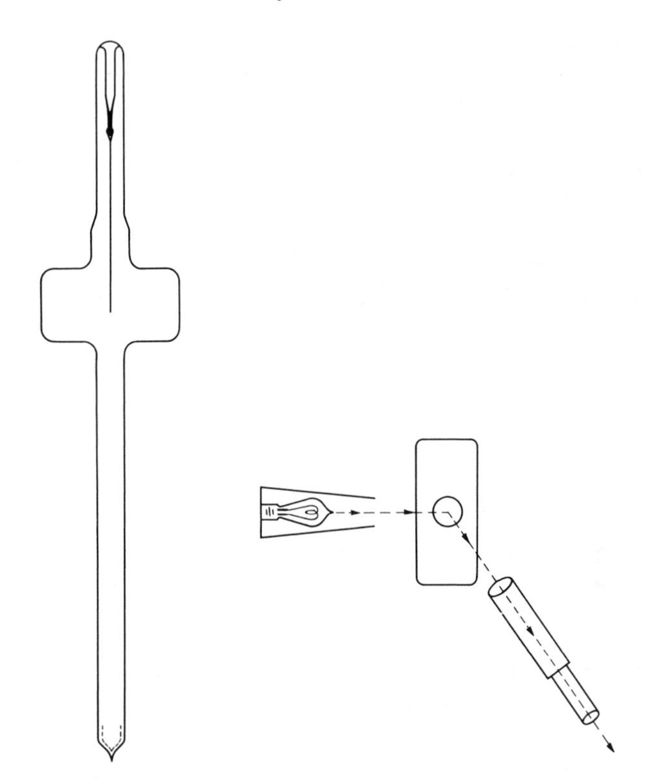

Fig. 5.14(*a*). Quartz-fiber gauge.

Fig. 5.14(*b*). Optical arrangement for quartz-fiber gauge.

Let $t_{0.5}$ denote the interval of time required for the maximum amplitude to decrease to half value. Then,

$$P(M)^{1/2} = \frac{B}{t_{0.5}} - C, \tag{5.11}$$

where $B$ and $C$ are constants characteristic of the fiber. The physical significance of these constants will be evident from the discussion of the theory of operation of the gauge.

In the case of a mixture of two or more gases, the sum of the terms $P_i(M_i)^{1/2}$ for each gas must be used instead of $P(M)^{1/2}$, where $P_i$ is the partial pressure of the gas of molecular weight $M_i$.

It is evident from the form of equation 5.11 that $B/C$ corresponds to the value $t_0$ at which the amplitude would decrease to half value in a perfect vacuum. For calibration, it is necessary to obtain only two points, corresponding to the two constants $B$ and $C$. One of these may be determined by observing the value $t_0$ in a very good vacuum; the other point may be obtained by calibrating against a McLeod gauge with some gas of definite composition.

The following data are given by Haber and Kerschbaum for a quartz fiber 7.0 cm long and 0.013 cm in diameter. Air was used for calibration.

| $P_{mm}$ | $P_{mm}(M)^{1/2}$ | $t_{0.5}$ (sec) | $B$ |
|---|---|---|---|
| 0.00302 | 0.01625 | 74 | 1.22 |
| 0.00494 | 0.02654 | 46 | 1.23 |
| 0.00775 | 0.0417 | 31 | 1.30 |
| 0.0117 | 0.0630 | 22 | 1.39 |
| 0.01880 | 0.101 | 12 | 1.23 |
| 0.0260 | 0.140 | 10 | 1.40 |
| $C = 0.0003$ | | | Avg 1.28 |

Some measurements with air taken by H. A. Huthsteiner in the General Electric Research Laboratory, using a fiber 3.8 cm long and 0.0045 cm diameter, are given for comparison.

| $P_{mm}$ | $t_{0.5}$ (sec) |
|---|---|
| 0.00058 | 105 |
| 0.00342 | 31 |
| 0.0080 | 16 |
| 0.0190 | 6.5 |

Plotting $P_{mm}$ against $1/t_{0.5}$ gives a straight line whose equation is $P_{mm} = 0.130/t_{0.5} - 6.55 \cdot 10^{-4}$. For air, $M^{1/2} = 5.38$. Hence, for this particular fiber

$$P_{mm}(M)^{1/2} = \frac{0.705}{t_{0.5}} - 0.00353.$$

Since $t_0 = 200$ in this case, it is evident that this fiber could not be used for measuring pressures below about 0.0001 mm of air. It also follows from the form of the above relation that the heavier the gas the lower the range of pressures over which the gauge may be used.

The optical arrangement suggested by Haber and Kerschbaum [Fig. 5.14(*a*)] may be varied in practice by fastening a scale to the back of the gauge and placing the lamp in such a position that the light beam passes practically parallel to this scale. The scale and tip of the quartz fiber are then sighted by means of a cathetometer.

These investigators used tubes which were more or less flattened on two sides, but ordinary cylindrical-walled tubes are more convenient and almost as satisfactory. As observed by Haber and Kerschbaum, care

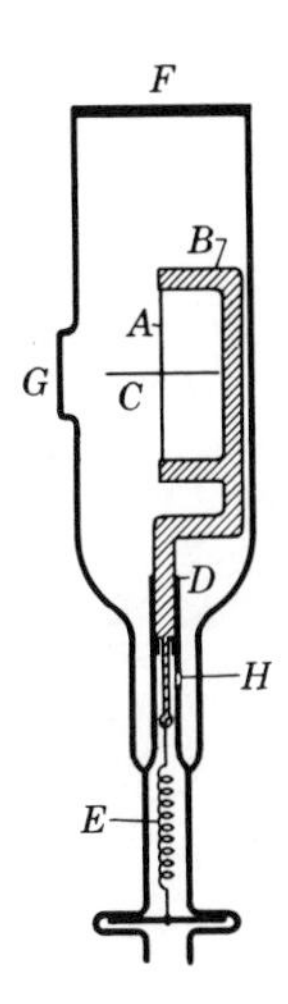


Fig. 5.15. Decrement type of viscosity gauge (King).

should be taken to tap the glass in such a manner that the fiber vibrates in the plane at right angles to the line of sight from the cathetometer. With a little experience, this can readily be accomplished. In view of the simplicity of construction and relative ease of manipulation, the quartz-fiber gauge ought to find a useful field of application in low-pressure technique, where the pressures to be measured are not below about 0.05 micron.

The difficulty encountered in getting the fiber to vibrate in a definite plane led Coolidge[80] to devise a bifilar form of gauge, in which two quartz fibers are fastened to the support 1 cm apart and fused together at the free ends, forming a *V*.

In the General Electric Research Laboratory Mrs. Andrews[81] replaced the vibration element by a strip of 0.1-mm molybdenum about 10 cm long by 2 mm wide. This was welded to a strip of 10-mil nickel, which was in turn welded to two leads in the stem of a bulb. The gauge was used to measure vapor pressures of naphthalene as low as $0.034 \cdot 10^{-3}$ mm.

A form of the decrement gauge used by King[82] is shown in Fig. 5.15. The construction is described as follows:

The essential part consists of a silica fiber *A*, between 3 and 4 $\mu$ in diameter and 4 cm long, which is fixed by a small amount of gold to the silica frame *B*, the tension, which thus does not vary with changes in temperature, being about half that required for breaking. At the center of this fiber, and at right angles to it, are fixed two 10 $\mu$ fibers 2.5 cm long, one on either side of the suspension. The ends of these remote from the frame are fused into a sphere about 50 $\mu$ in diameter, while the other ends are fixed together by a minute amount of iron, which, when activated by an electromagnet outside the case, also serves to start the oscillations of the cross-piece, the damping of which is measured. If the

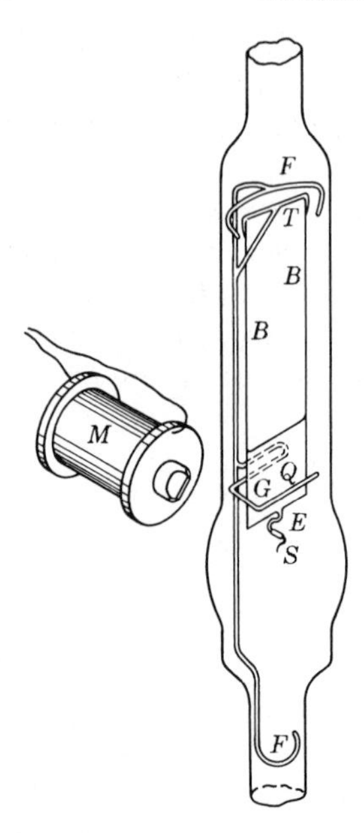


Fig. 5.16. Quartz-membrane manometer.

external case is of glass, the frame is held in the cylindrical socket *D* by the spring *E*, while if it is of silica the frame is fused directly to it. Two flat windows, *G* and *F*, are fused to the case, the former for observing and the latter for illuminating the tip of the cross-piece.

The gauge is exhausted through the opening *H*. The lower limit of the gauge for a light gas such as oxygen is stated to be $2 \cdot 10^{-6}$ mm.

A considerable increase in the sensitivity of this type of gauge was obtained by Brüche[83] by the development of a quartz "surface" manometer. The details of construction of this gauge, shown in Fig. 5.16, are as follows. To the supporting rod *T* are fastened two flat quartz fibers 75 mm long and 0.8 by 0.1 mm in cross section. These are attached to a quartz membrane 25 by 25 mm in area and 0.04 mm thick. This membrane has fastened to it, at the bottom, a very thin-walled small quartz tube in which is sealed a piece of iron wire *E*, 7 mm long and 0.5 mm in diameter. Below this is a pointer *S*, which is used to indicate the amplitude of vibration by means of a telescope *M*. A fork *G* limits the extent of the vibration. A 4-cm-diameter tube encloses the vibrating system, which is supported on the glass framework *F*.

Brüche confirmed, by measurements with hydrogen, air, and the inert gases, the validity of equation 5.11 over the range $5 \cdot 10^{-4}$ to $2 \cdot 10^{-2}$ mm, and concluded from his experience that this type of gauge should find extensive application in high-vacuum technique and for the measurement of very low pressures of chemically active vapors.

More recently, the theory and applicability of this type of gauge have been discussed very comprehensively in two papers by Wetterer.[84] The following remarks are based largely on the contents of the second of these papers.

As mentioned above, the motion of the quartz membrane and supporting wires is represented by equation 5.5, in which $\theta$ denotes the angular displacement with respect to a vertical axis. The coefficient of damping, $R$, is the sum of two coefficients, $R_a$, which represents the damping due to the gas molecules, and $R_i$, which corresponds to the intrinsic damping of the system *in vacuo*.

Consequently, we can write equation 5.9 in the form

$$t_{0.5} = \frac{2I \cdot \ln 2}{R_a + R_i} = \frac{1.3863I}{R_a + R_i}\,. \tag{5.12}$$

As stated above, $R_a$ is proportional to the product of $\rho v_a$ and a quantity which has the dimension of $l^4$. In order to calculate the magnitudes of $R_a$ and $I$, the moment of inertia, Wetterer assumes that the system of quartz fibers and membrane can be replaced by a duplex fiber and two rectangles as shown in Fig. 5.17. Using the designations for the dimensions given in this figure, Wetterer derived the following relations:

$$R = \frac{4P(M)^{1/2}}{(3R_0T)^{1/2}}\left(\frac{Dl^3}{3} + 2l^2a^2 + \frac{2}{3}\,a^4\right) + R_i, \tag{5.13}$$

$$I = \frac{m_1 l^2}{3} + \frac{m_2}{2}\left(l^2 + \frac{a^2}{3}\right), \tag{5.14}$$

where $D$ = width of fiber (centimeters),
$m_1$ = mass of each fiber (grams),
$m_2$ = mass of membrane (grams)
$= 4a^2d \cdot s$,
$d$ = thickness of membrane,
$s$ = density of quartz.

Substituting these relations in equation 5.12, it follows that

$$t_{0.5} = \frac{K_1}{K_2P(M)^{1/2} + R_i}, \tag{5.15}$$

where $K_1$, $K_2$, and $R_i$ are constants which depend upon the dimensions and density of the quartz fibers and membrane. This equation can obviously be rewritten in the form given in equation 5.11.

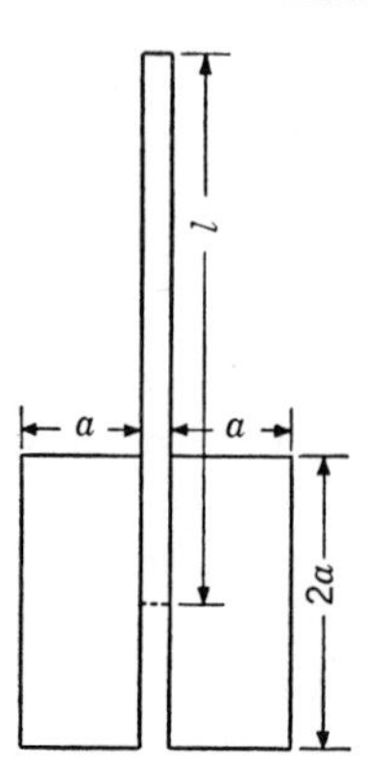


Fig. 5.17. Theory of membrane manometer.

For fine wires having negligible mass compared with that of the membrane, equation 5.15 takes the form

$$t_{0.5} = \frac{K_3 d}{P(M)^{1/2} + K_4}, \tag{5.16}$$

where $K_3$ and $K_4$ are constants. Thus the value of $t_{0.5}$ varies directly with $d$, the thickness of the membrane. At higher pressures, the value of $K_4$ is negligible compared to $P(M)^{1/2}$, and equation 5.16 then reduces to

$$P = \frac{K_3 d}{t_{0.5}(M)^{1/2}}, \tag{5.17}$$

that is, $\log P = \log (K_3 d) - \log (t_{0.5}) - 0.5 \log M$, and consequently the plot of $P$ versus $t_{0.5}$ on log-log scales should be linear with a 45° slope.

Figure 5.18 shows such plots for a series of membranes for which $a$ varies from 1 to 100 mm. As will be observed, the plots show a 45° slope at higher pressure and become flatter at lower pressures. The flatter region of each curve indicates the lower limit of pressure for practical purposes. The larger the value of $a$, the lower the pressure at which such a gauge may be operated.

For practical use, Wetterer has also developed an attachment which makes it possible to project the indicator of the quartz membrane on a matt screen.

Since values of $t_{0.5}$ greater than $10^4$ sec are obviously impractical, it would seem that the lower limit of pressure which could be measured by this type of gauge is about $10^{-4}$ micron.

A very interesting application of the considerations discussed above has been described by Beams, Young, and Moore,[85] in connection with their investigations on the production of high centrifugal fields by spinning small steel spherical rotors at extremely high speeds in a vacuum. They

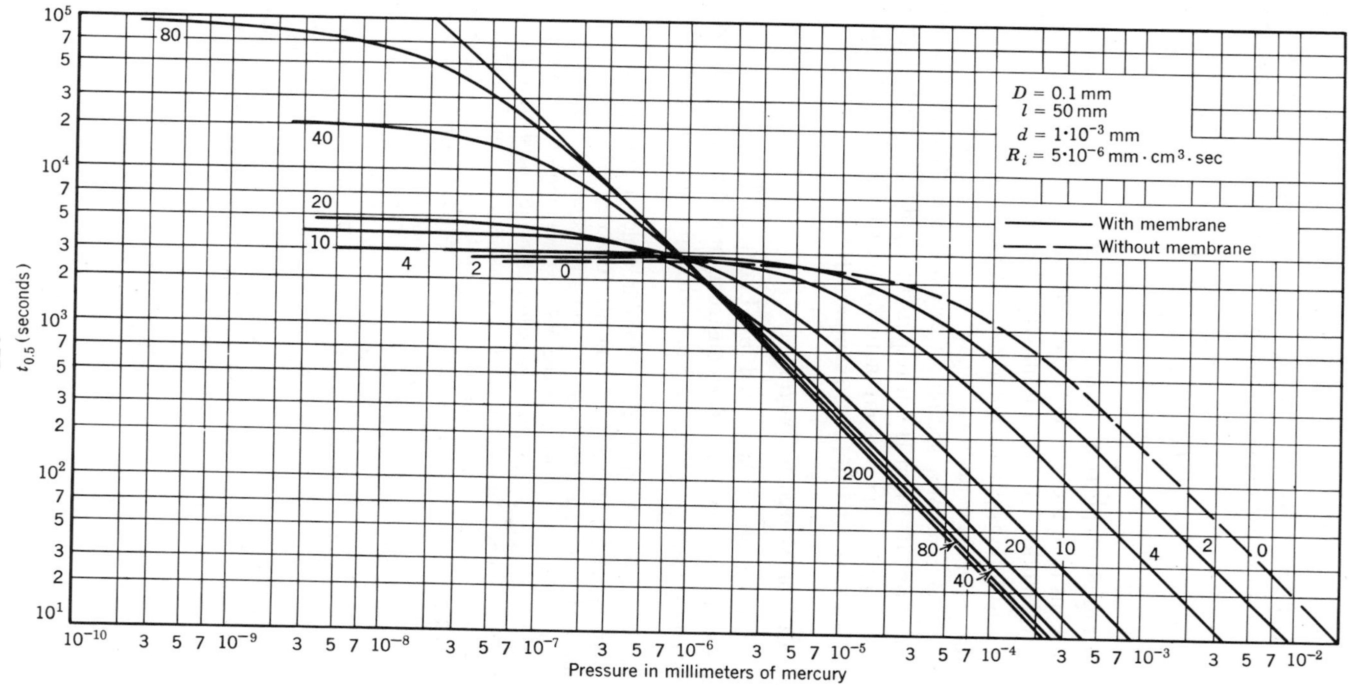


Fig. 5.18. Calibration curves for a vibrating membrane manometer (Wetterer) for a series of values of $2a$ (the length of side of a square membrane) in millimeters.

made the following observation. When the rotor, after being brought up to speed, is allowed to spin freely with driving power off, the speed decelerates at a rate that depends on the pressure in accordance with the relation

$$P_{\mu b}\left(\frac{M}{2\pi R_0 T}\right)^{1/2} = \frac{r\rho}{5t}\ln\frac{f_0}{f}, \tag{5.18}$$

where $r$ = radius of spherical rotor,
$\rho$ = density of the steel ($=7.8\ \mathrm{g \cdot cm^{-3}}$),
$f_0$ = revolutions per second initially,
$f$ = revolutions per second at time $t$ (seconds),
and $M$, $R_0$, and $T$ have the usual signification.

"With a 1.59-mm rotor spinning freely at about 120,000 rps, with a pressure of about $10^{-5}$ mm Hg, it required roughly 2 hours to lose 1 per cent of its speed." The pressure actually measured with an ionization gauge was between $10^{-5}$ and $2 \cdot 10^{-5}$ mm of Hg.

**Rotating Disk or Molecular Gauges**

The molecular gauge suggested by Langmuir[86] represents a direct application of equation 5.4*b*. The construction and results obtained with a gauge built on this principle have been described by Dushman.[87]

It consists of a glass bulb $B$ [(Fig. 5.19*a*)] in which are contained a rotating disk $A$ and, suspended above it, another disk $C$. The disk $A$ is made of thin aluminum and is attached to a steel or tungsten shaft $H$ mounted on jewel bearings and carrying a magnetic needle $NS$. Where the gauge is to be used for measuring the pressure of corrosive gases like chlorine, the shaft and disk may be made of platinum. The disk $C$ is of very thin mica, about 0.0025 cm thick and 3 cm in diameter. A small mirror $M$, about 0.5 cm square, is attached to the mica disk by a framework of thin aluminum. This framework carries a hook with square notch which fits into another hook similarly shaped, so that there is no tendency for one hook to turn on the other. The upper hook is attached to a quartz fiber, $F$, about $2 \cdot 10^{-3}$ cm diameter and 15 cm long.

The lower disk can be rotated by means of a rotating magnetic field produced outside the bulb. This field is more conveniently obtained by a Gramme ring, GG, supplied at six points with current from a commutating device rotated by a motor [Fig. 5.19(*b*)]. In this way the speed of the motor determines absolutely the speed of the disk, and the speed of the latter may thus be varied from a few revolutions per minute up to 10,000 or more.

By applying equation 5.4*b* it can be shown that the angle of torque ($\alpha$) on the upper disk is given by the equation

$$\alpha = \frac{K\tau^2 r^4}{I} \cdot P\omega\left(\frac{M}{T}\right)^{1/2}, \tag{5.19}$$

where $\tau$ = period of oscillation,
$r$ = radius of rotating disk,
$I$ = moment of inertia of disk,
$\omega$ = angular velocity of rotation,
$K$ = a constant which depends upon the nature of the gas and the accommodation coefficients for transfer of momentum.

Hence, for any one gauge, the torque on the upper disk is proportional to the product of the speed of rotation of the aluminum disk and the quantity $P(M/T)^{1/2}$. The sensitivity of the gauge can thus be increased by increasing the speed of rotation; also, by illuminating the mirror and using an arrangement similar to that employed for galvanometers, it is possible with the gauge to measure pressures of the order $10^{-3}$ to $10^{-4}$ micron.

The gauge actually used for measuring very low pressures showed a

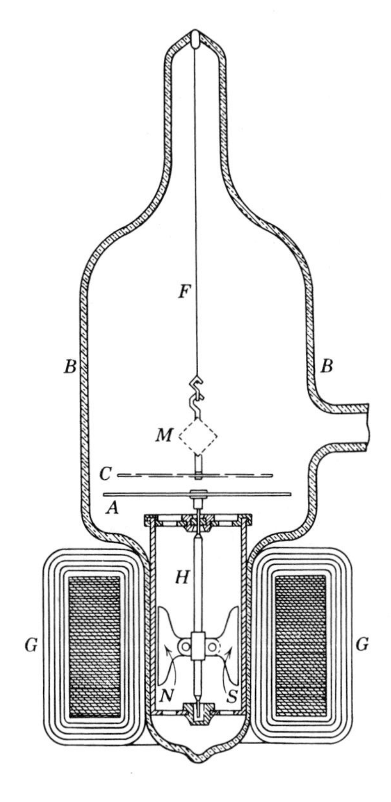

Fig. 5.19(*a*). Molecular gauge.

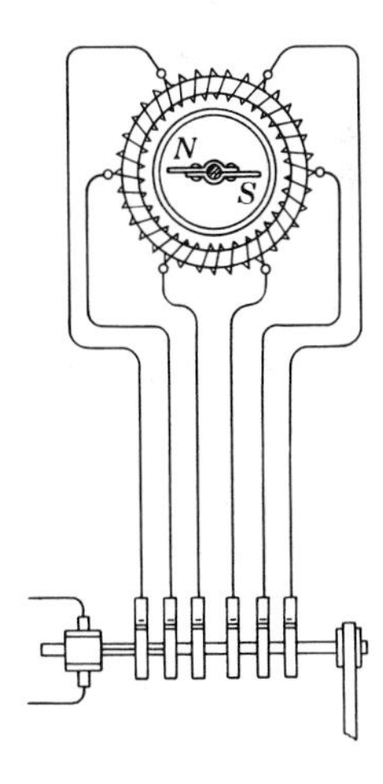


Fig. 5.19(*b*). Rotating commutator for use with molecular gauge.

deflection of 1100 mm per microbar of air, at 1000 rpm, with the scale 50 cm from the mirror. Up to a pressure at which the mean free path of the gas molecules becomes comparable with the distance between the two disks, the deflections, at constant speed of rotation, were found to be proportional to the pressure as observed by a McLeod gauge.

At extremely low pressures (below about $3 \cdot 10^{-4}$ micron) the indications of the gauge were found to be inaccurate because of two sources of error. First, the rotation of the magnetic field produced by the Gramme ring tends to induce eddy currents in the metal framework that holds the mirror; and, second, there is a tendency for the upper disk to start swinging, especially at very high speeds of rotation of the aluminum disk. As the damping at low pressures is very feeble, it is exceedingly difficult to stop this oscillation when once started.

Working independently of Langmuir, and about the same time, Timiriazeff[88] also suggested the application of equation 5.4*b* to the construction of a low-pressure gauge. Since he was primarily interested in determining the laws of slip for different gases, his actual design is not suitable for a very sensitive gauge. Instead of a rotating disk with a stationary disk situated symmetrically above it, Timiriazeff used a rotating cylinder with a stationary cylinder placed symmetrically inside it and suspended by a phosphor-bronze wire. A gauge based on the same principle has been described by Riegger.[89]

A modern version of Langmuir and Dushman's molecular gauge has been built by J. A. Roberts of the Instrument Department of the General Electric Company. This device is shown in Fig. 5.20. It consists of a vacuum chamber containing a rotating vaned cylinder driven at 3600 rpm by a synchronous motor, and a spring-restrained vaned cylinder mounted axially concentric with the rotating cylinder. The indicating pointer is attached to the spring-restrained cylinder. The field coil or stator of the synchronous motor is mounted outside the vacuum chamber and magnetically drives the rotor, which is contained inside a tubular extension of the vacuum chamber. The indicating mechanism is magnetically damped to minimize pointer oscillation. The energy received by the gas molecules from making contact with the rotating cylinder is transferred to the surface of the spring-restrained cylinder. This energy turns the spring-restrained cylinder through an angle proportional to the rate of energy transfer, which is a measure of the number of molecules present and hence the gas pressure. The pointer on the spring-restrained cylinder gives the gas pressure on a calibrated scale. Below 300 microns the gauge is essentially a viscosity gauge, but because of the vaned cylinders it performs as a windage gauge at pressures above this value. The scale is expanded on both ends and compressed around 300 microns for this reason.

Fig. 5.20. Internal construction of the General Electric direct-reading molecular vacuum gauge.

## 5.5. RADIOMETER TYPE OF GAUGES

### Crookes' Radiometer

One of the first instruments to be used for detecting low gas pressures was the radiometer devised by Sir William Crookes in 1873. The instrument, which is described in most textbooks, consists of a glass bulb in which a small vane is mounted on a vertical axis. The vane has four arms of aluminum wire on which are attached four small plates of thin mica, coated on one side with lamp black. These plates are set so that their planes are parallel to the axis. If a source of light or heat is brought near the bulb, and the rarefaction is just right, the vane rotates, but at very low pressures the rotation practically ceases.

The theory of the device was apparently not very well understood for a long time, and attempts to use it as a gauge for low pressures yielded very unsatisfactory results. Dewar[90] has stated the case for this instrument as follows:

The radiometer may be used as an efficient instrument of research for the detection of small gas pressures. For quantitative measurements the torsion balance or bifilar suspension must be employed.

Some years ago W. E. Ruder, of the General Electric Research Laboratory, developed a method of using the radiometer for the measurement of the residual gas pressure in incandescent lamps. The following account was prepared by him:

It was found that when exhausted to the degree required in an incandescent lamp the radiometer could not be made to revolve, even in the brightest sunlight. In order to get a measure of the vacuum, the radiometer vanes were revolved rapidly by shaking the lamp, and the time required to come to a complete stop was therefore a measure of the resistance offered to the vanes by the gas, together with the frictional resistance of the bearings. The latter quantity was found to be so small in most cases that a direct comparison of the rates of decay of speed of the vanes gave a satisfactory measure of the degree of evacuation. In this manner a complete set of curves was obtained which showed the change in vacuum in an incandescent lamp during its whole life and under a variety of conditions of exhaust. The chief objections to this method of measuring vacua were the difficulty in calibrating the radiometer and the difference in frictional resistance offered by different radiometers. For *comparative* results, however, the method was entirely satisfactory.

As a result of his investigations of the laws of heat transfer in gases at low pressures, Knudsen arrived at a clear explanation of the radiometer action and furthermore developed, along the same lines, an accurate gauge for the measurement of extremely low pressures.

According to Knudsen, a mechanical force is exerted between two surfaces maintained at different temperatures in a gas at low pressure. This is due to the fact that the molecules striking the hotter surface rebound with a higher average kinetic energy than those that strike the colder surface. In the case of the radiometer the blackened surfaces absorb heat from the source of light, and the molecules rebounding from the vanes are therefore at a higher temperature than those striking the walls of the bulb. Consequently a momentum is imparted to the vanes which tends to make them rotate.[91]

## Knudsen Gauge

The principle of this gauge[92] may be explained by referring to the lower part of Fig. 5.21. Let us consider two parallel strips $A$ and $B$ placed at a distance apart which is *less* than the mean free path of the molecules. Let $A$ be at the same temperature $T_0$ as the residual gas, while $B$ is maintained at a higher temperature $T_1$. On the side away from $B$, $A$ will be bombarded by molecules having a root-mean-square velocity $v_0$, corresponding to the temperature $T_0$, as given by the equation

$$v_0 = \left(\frac{3R_0T_0}{M}\right)^{1/2}.$$

These molecules will, of course, rebound from $A$ with the same velocity.

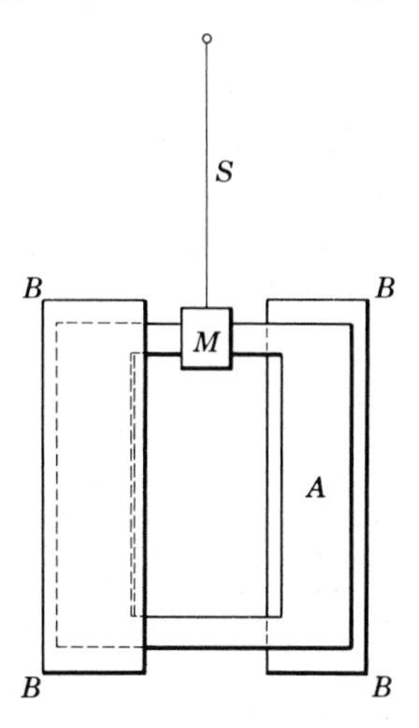


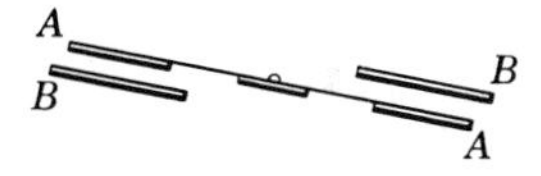


Fig. 5.21. Diagrammatic sketch of Knudsen type of gauge.

However, on the side towards *B*, *A* will be bombarded by molecules coming from *B* and having a higher velocity $v_1$ corresponding to the temperature $T_1$. Consequently *A* will receive momentum at a greater rate on the side towards *B* than on the opposite side, and will therefore be repelled from *B*.

In deriving the relation between the pressure of the gas and the force exerted on the surface *A*, we shall follow the simple derivation given by Todd.[93]

If the distance between the surfaces *A* and *B* is *less than the mean free path*, the force *F* is given by the *rate at which momentum is transferred* per unit area from *B* to *A*, that is,

$$
\begin{aligned}
F &= \tfrac{1}{6} n v_0 (m v_1 - m v_0) \\
&= \frac{1}{6} \frac{P_{\mu b}}{k T_0} \cdot 3k[(T_1 T_0)^{1/2} - T_0] \\
&= \frac{1}{2} P_{\mu b} \left[ \left( \frac{T_1}{T_0} \right)^{1/2} - 1 \right]
\end{aligned}
\tag{5.20}
$$

For small differences of temperatures, and for the purpose of pressure measurements, this equation may be written in the form

$$
P_{\mu b} = 4F \left( \frac{T_0}{T_1 - T_0} \right) . \tag{5.21}
$$

Thus, for constant value of $(T_1 - T_0)/T_0$, the force is proportional to the pressure and is *independent of the molecular weight of the gas.*

In order to measure this force of repulsion, Knudsen uses the arrangement shown diagrammatically in the upper part of Fig. 5.21. The strip $A$ is replaced by a rectangular vane, cut out in the center and suspended by means of a fiber $S$. Two strips $BB$ which can be heated are placed symmetrically on opposite sides of this vane, and the force of repulsion is then balanced by the torsion of the fiber. By means of the mirror $M$, the deflection can then be measured in the same manner as in galvanometers.

For this arrangement, equation 5.21 assumes the form

$$P_{\mu b} = \frac{4\pi^2 I D}{r A \tau^2 d} \cdot \frac{T}{T_1 - T}, \tag{5.22}$$

where $I$ = moment of inertia of the moving vane,
$r$ = mean radius of the moving vane,
$A$ = area of the vane $A$ opposite each strip $B$,
$\tau$ = period of vibration of the vane,
$D$ = scale deflection,
$d$ = scale distance.

Since all these quantities can be measured directly, it follows that the device can be used as an *absolute manometer*, without the necessity of calibrating against any other gauge. It is also evident that the indications of this gauge must be independent of the nature of the gas to be measured.

In his first paper on this subject, Knudsen mentioned several different forms of construction which might be used in making a gauge on the foregoing principle but gave very few constructional details. One form which appears to be very simple in construction is shown in Fig. 5.22. In $AA$, a glass tube about 1.4 cm in diameter, is sealed a narrow tube $BB$ which has a rectangular piece cut out at $C$, 0.41 cm wide by 2.95 cm long. A piece of mica $D$ is suspended in front of this opening by means of a fiber fastened at $E$. The tube $AA$ can be heated by means of an external water jacket $FF$. As the temperature of the water in $FF$ is raised, the mica plate is repelled by the "hot" molecules traveling through the opening $C$, and the amount of deflection can be observed by means of a microscope.

Variations of this construction were described by Knudsen in a later paper.[94] Angerer[95] described a Knudsen manometer which consisted of a silvered mica vane between two electrically heated platinum strips. He stated that pressures as low as $8 \cdot 10^{-7}$ mm of mercury could be measured with it.

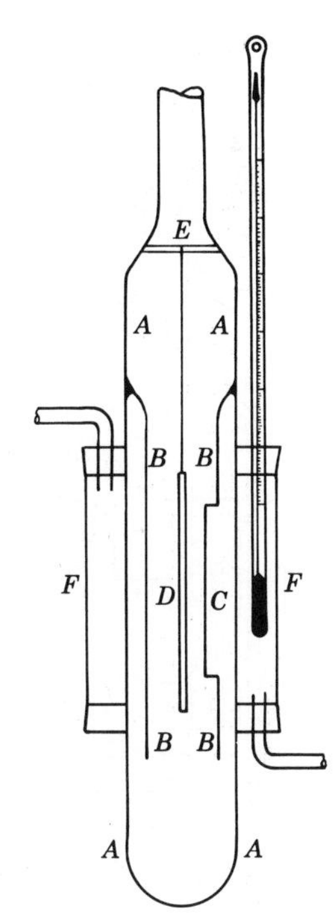


Fig. 5.22. Simple construction of Knudsen gauge.

The same type of design has also been used by Woodrow[96] and by Shrader and Sherwood.[97]

## Woodrow's Modification of the Knudsen Gauge

The following description of Woodrow's form of Knudsen gauge is quoted from the original publication:

Several different gauges were constructed varying in sensitivity so as to be used at different pressures. A typical gauge is shown in Figs. 5.23(*a*) and 5.23(*b*), and the electrical circuits are given in Fig. 5.24. The glass rods *GG* served as supports for the metallic parts of the gauge. All the internal electrical connections and adjustments, with the exception of the final leveling, were made before the outer glass walls *OO* were sealed on at *SS*. The suspension *W* was a phosphor-bronze ribbon 50 mm in length which had been obtained from W. G. Pye and Co. and was listed by them as No. 0000. The movable vane *VV* consisted of a rectangular frame of aluminum 0.076 mm in thickness, the dimensions of the outer rectangle being 30 by 36 mm and the inner 26 by 30 mm. The heating plates *PP* were platinum strips 4 mm wide, 40 mm long, and 0.025 mm thick. The deflections of the movable vane were obtained in the usual way by the reflection of a beam of light from the mirror *M*. Figure 5.23(*b*) is a cross-sectional view through the middle of Fig. 5.23(*a*).

All of the platinum connections were made by electric welding, as that was

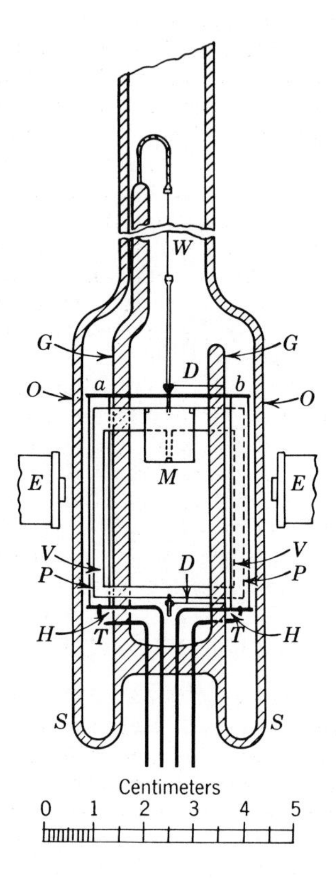


Fig. 5.23(*a*). Woodrow's modification of Knudsen gauge.

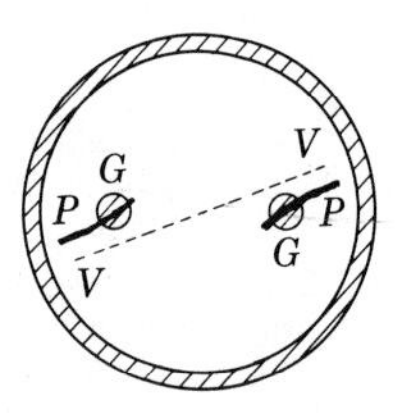


Fig. 5.23(*b*). Cross-sectional view through the middle of the gauge shown in Fig. 5.23(*a*).

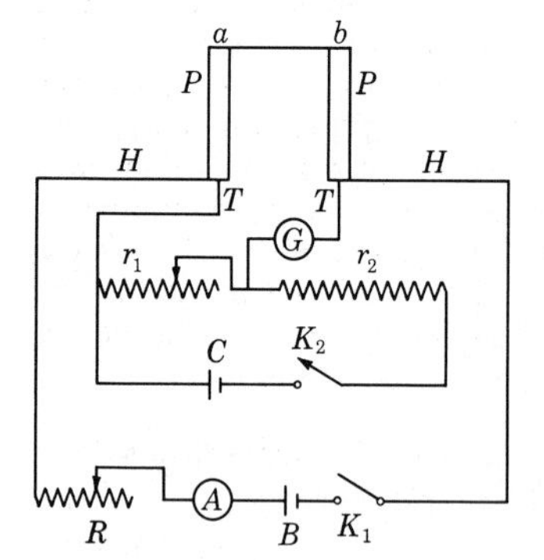


Fig. 5.24. Electrical circuit for determining the temperature of the heating strips in gauge shown in Figs. 5.23(*a*) and (*b*).

found much more satisfactory than the use of any kind of solder, especially when heated. After a little practice, it was possible to weld the thin platinum heating vanes to the heavy platinum wire so as to make a perfectly continuous contact throughout its width. The phosphor-bronze suspension was connected at both ends by threading through three small holes drilled into the flattened extremities of the platinum and aluminum wires respectively. The small loops *DD* were so placed that they supported the movable vane *V* except when the gauge was leveled for taking readings. This made the gauge readily portable and, by placing it in the inverted position when connected to the molecular pump, the danger of the breaking of the suspension by vibration was eliminated. One gauge of medium sensitivity was constructed so as to be sufficiently steady to be used when connected directly to the molecular pump. Large glass tubing was employed in all the connecting portions of the apparatus.

A small electromagnet, shown at *E* in Fig. 5.23(*a*), was employed in bringing the moving vane to rest. This was found to be quite necessary in working with the most sensitive gauges, since in a very good vacuum the damping is so small that the vane will not settle down sufficiently for the taking of readings for some time after an accidental disturbance has set it vibrating. It should be noted that the electromagnet must have either an air core or one of good, soft Norway iron, for otherwise the residual magnetism will produce a false zero if the aluminum vane is at all magnetic, as was the case with the samples of metal investigated in this laboratory. Under these conditions it is obvious that the electromagnet should be used only for damping and that the exciting current should be shut off while making observations.

Several methods were tried for determining the temperature of the heating strips, and that shown in diagram in Fig. 5.24 was finally settled upon as giving the most satisfactory results. The potentiometer leads *TT* were connected by electric welding to the very extremities of the platinum heating vanes *PP*. The heating current was regulated by the variable resistance *R* and its value was read on the ammeter *A*. The resistance $r_2$ was kept constant at 10,000 ohms and $r_1$ varied to obtain a balance of the sensitive galvanometer *G*. The potentiometer battery *C* consisted of a carefully calibrated Weston Standard Cell. This arrangement gave an accurate method of measuring the resistance of the platinum strips *PP*, plus the heavy platinum wire *ab*, the total cold resistance being 0.17 ohm. This cold resistance was determined by plotting the curve connecting resistance and heating current under a constant low pressure and extrapolating backward to the intersection with the axis of resistance. If the resistance is measured for small currents, the value at zero current, that is the cold resistance, can be determined very accurately. The temperature coefficient of resistance of the platinum, which contained a small amount of iridium, was carefully determined and was found to give a linear relation within the range of temperatures employed. The value of the coefficient was $2.35 \cdot 10^{-3}$ ohm per degree C. With this system one can determine the mean temperature of the heating strips with sufficient accuracy, the error for temperature difference of about 50° C being less than 4 per cent.

Woodrow also observed that in order to avoid electrostatic effects it was necessary to silver-coat the outside of the glass walls, which were then grounded. Similarly, the moving system was connected through the suspension to that terminal of the heating strips that was grounded.

With the gauge whose dimensions are given above, the period of a complete oscillation was 10 sec and the calculated moment of inertia of the moving vane was 0.074 g · cm². This gives, for the pressure, the relations

$$P_{\mu b} = 2.9 \cdot 10^{-5} \frac{T}{T_1 - T} d,$$

$$P_{mm} = 2.2 \cdot 10^{-8} \frac{T}{T_1 - T} d,$$

where $d$ is the deflection in millimeters on a scale at a distance of 1 meter from the mirror.

Thus, with a temperature difference of 100° C, the gauge could be used to read pressures as low as $3 \cdot 10^{-8}$ mm of mercury.

### Shrader and Sherwood's Modification of the Knudsen Gauge

The construction used by Shrader and Sherwood differed in a few details from that used by Woodrow. In view of the importance of the Knudsen gauge for low-pressure measurements, the description of this modification is worth quoting.

The gauge is shown in Fig. 5.25. It is enclosed in a hard glass tube 2 in. in diameter and 9 in. long. The heating strip *aa* is of platinum, 0.018 mm thick and 7.5 mm wide with a total length of 18 cm. It is folded at the top forming a cross piece and two parallel sides. The ends are brazed to 20-mil tungsten leading-in wires at the bottom. Fifteen-mil tungsten wires *b* sealed into the glass-rod support serve as a spring support for the platinum strip. This allows accurate adjustment of the strip and sufficient tension is secured to keep the strip taut during heating. One of these wires is carried up the glass-rod support, sealed into it at *c*, leaving a free end *d* to serve for electrical connection of the moving vane to the heating strip. Connection is made by the wire pressing under tension against the tungsten wire *e* to which the suspension of the vane is fastened. Potential leads of fine platinum wires *ff* are welded to the strip about 1 cm from the ends and are brazed to tungsten sealing-in wires.

The movable rectangular vane *g* is made of aluminum 0.0076 cm thick. A standard size adopted is 3.2 cm by 4 cm outside dimensions, the width of the vane being 0.5 cm. Because of liability of warping during heat treatment the vane is stiffened by an aluminum wire passing through slits at the top and a hole at the bottom into which the wire is hooked and fastened firmly. For portability, two copper wires *h* are sealed into the glass-rod support while the free ends form loops around the rod, these forming guides for the vane. The mirror is fastened at the bottom of the vane by leaving a small projection of the aluminum at the lower edge and cutting out small tongues from the material of the vane on either side. The mirror is laid in place and the projection and the two tongues are pressed closely over it, holding it securely.

Silver mirrors were tried, but failed to withstand the heat treatment to which the gauge and system were subjected. Mirrors made by coating microscope cover glass with china decorators' platinum solution, followed by baking at 500°, solved this difficulty.

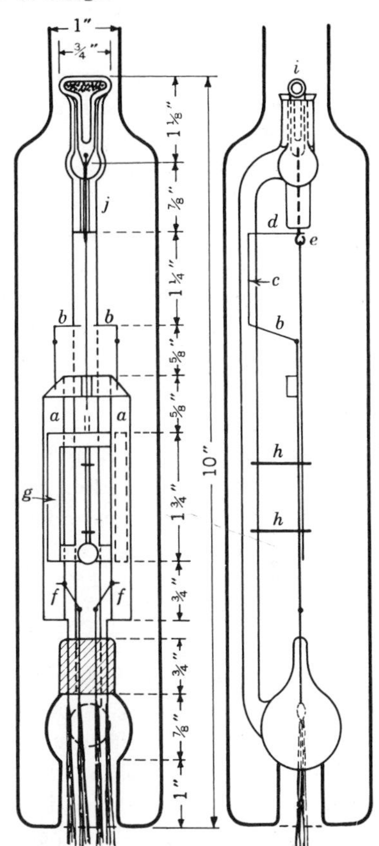


Fig. 5.25. Shrader and Sherwood's modification of Knudsen gauge.

The distance between the heating strip and the vane is adjusted from outside the case by magnetic control on a piece of soft iron *i* sealed into a glass stem to which the suspension is fastened.

The suspension is 0.0005-in. tungsten wire. This is fastened to small aluminum hooks around which the wire is wrapped several times after which the hooks are pressed firmly together. This method is not difficult and holds the wire securely. A hook on the end of a tungsten wire sealed into a glass stem, the free end passing through a capillary rod *j* on the glass support, serves to hold the suspension.

A gauge such as has been described, using a 0.0005-in. tungsten suspension from 6–7 cm long, has such a sensibility that a scale deflection of 1 mm at a meter's distance with a temperature difference of 150° C between the heating strip and the movable vane indicates pressures of $1 \cdot 10^{-8}$ to $5 \cdot 10^{-8}$ mm Hg. One gauge of other dimensions than those given above would indicate a pressure of $5 \cdot 10^{-9}$ mm Hg under the same conditions.

The temperature of the heated strips was measured by substantially the same electrical method as that used by Woodrow.

Richardson[98] has described a construction of Knudsen gauge in which the force of repulsion is balanced by means of a magnetic field, and Riegger,[99] a commercial form made in Germany at that time.

In connection with his investigations on the theory of the radiometer Fredlund[100] constructed several forms of the Knudsen gauge which are quite elaborate. In all these constructions the movable vane is a thin circular metal disk located in the center between two large metal plates that are maintained at a constant difference in temperature. Two magnet needles are attached to the suspension of the disk, and the force on the disk is balanced by passing current through two coils located in planes parallel to that of the disk. Thus the magnitude of the current required to obtain zero deflection is a direct measure of the pressure. The theoretical implications of the work of Fredlund and other investigators are discussed at the end of this section.

As DuMond and Pickels[101] have pointed out, the reason for the infrequent use of the Knudsen gauge, "in spite of its manifold advantages, lies in the elaborate designs and precautions described by the above-named workers," who really regarded the gauge itself "as the object of a research problem rather than a tool."

Figure 5.26 shows a line drawing of the gauge designed by these investigators. The movable vane (1) consists of a rectangular frame of 2.5-mil aluminum to which is fastened an axle (2) of 32-mil aluminum. This axle carries an ordinary galvanometer mirror (5), attached to a 0.5-mil tungsten wire, which is fastened to an aluminum wire held by means of a set screw (8) in the hole in the bottom of the steel taper plug (9). A vacuumtight seal is obtained by means of laboratory wax or stopcock grease. The heaters (14) are helices of Chromel resistance wire, and the current leads are provided by means of the insulated vacuumtight plugs fastened by the nuts (18). The cylindrical brass envelope (20) of the gauge with its window and water jackets (21) is slipped over the brass disks (11) and (12) and sealed at (12) by means of solder or Apiezon wax. Two Helmholtz coils outside the case are used to obtain magnetic damping.

The investigators state, "Instead of maintaining the gauge heater at constant temperature for all gas pressures in the gauge we let the heater assume whatever temperature the combined effect of radiation loss and gas conduction loss impose with a constant wattage input to the heater." Under these conditions, the relation between deflection of the mirror ($D$) and pressure is not linear, but has the form

$$D = \frac{CP}{1 + BP}, \tag{5.23}$$

where $C$ and $B$ are two constants derived from a calibration of the gauge against a McLeod gauge (with liquid-air trap inserted between the McLeod

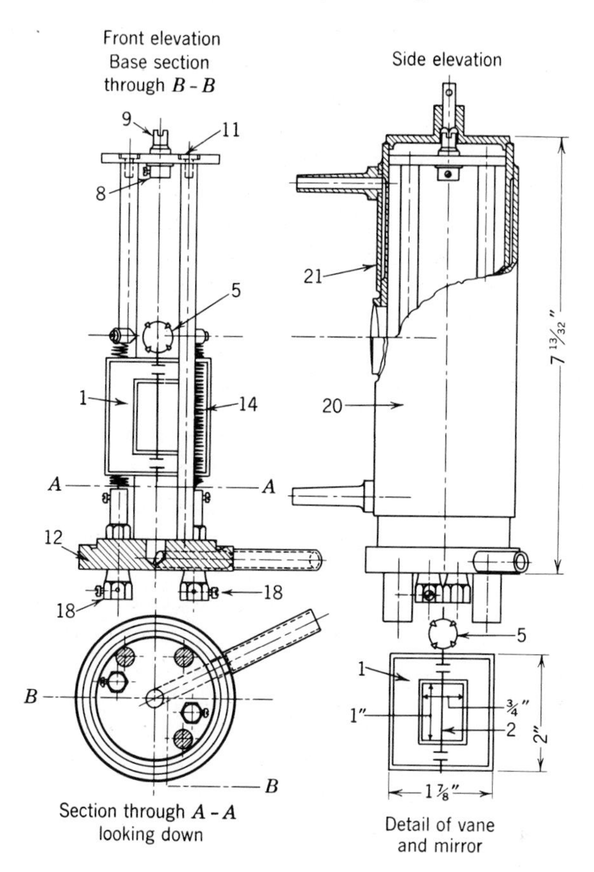


Fig. 5.26. Construction of DuMond and Pickels form of Knudsen gauge.

and the Knudsen gauge). The deflection is observed by means of a small lamp and galvanometer screen. By varying the wattage in the heater coils two scales are obtained for the pressure readings. On the more sensitive of these, pressures as low as $10^{-7}$ mm can be detected.

A more rugged form of Knudsen gauge and, consequently, one that is not as sensitive as that devised by DuMond and Pickels has been described by Lockenvitz.[102] The details of construction are shown in Fig. 5.27. As will be observed it consists, essentially, of a very thin aluminum leaf swinging freely from its upper edge between two surfaces maintained at constant temperatures. The author states,

The hot and cold plates, which are machined out of copper, are sealed to the body by means of lead washers. The glass windows for observing the deflection of the leaf are sealed in with Apiezon W wax. The 0.01-mm-thick aluminum leaf

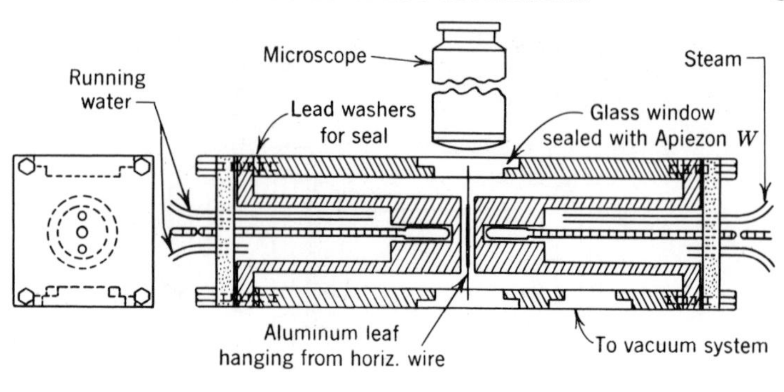


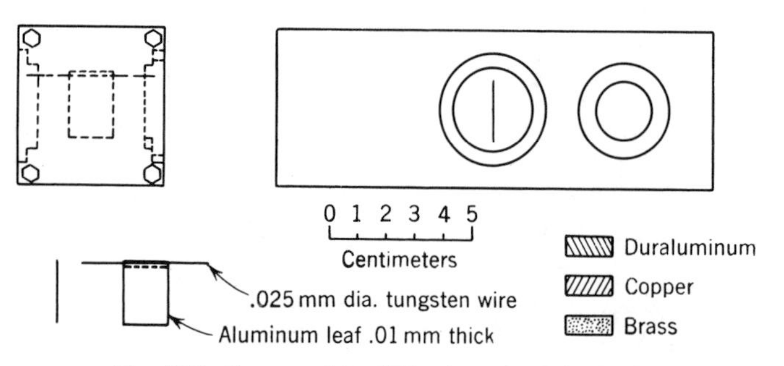


Fig. 5.27. Construction of Knudsen gauge by Lockenvitz.

is hung from a fine wire by folding the top edge of the leaf over about 0.5 mm. A microscope is fastened to the body of the gauge and is focussed on one bottom corner of the aluminum leaf so that a deflection of 0.005 mm can easily be read on the scale in the micrometer eyepiece.

The pressure is directly proportional to the deflection.

The lower limit of the gauge is stated to be about $0.75 \cdot 10^{-2}$ micron.

However, by using a thinner or a larger leaf, or both, the lower limits could easily be extended to the region of $10^{-3}$ microbar. In that case it might be advisable to construct the gauge out of glass, since gases given off by the large metal surfaces would be troublesome.

It will be observed that all such gauges in which an Apiezon wax or low-melting-point solder is used obviously cannot be given a heat treatment in order to eliminate adsorbed gases. Hence they must be maintained on the exhaust system for quite a long period before the pressure indications actually correspond to pressures in the rest of the system.

In contrast to the forms described above is the "simple Knudsen gauge" described by Hughes,[103] which is shown in Fig. 5.28. It consists of a Pyrex tube, 2.5 cm in diameter and 15 cm long. The heater, Pt, is made of a platinum strip 10 cm long, 1.3 cm wide, and $10^{-4}$ cm (0.04 mil) thick, and is connected to two 50-mil tungsten leads. The movable strip $A$ is made of aluminum leaf, 12 cm long, 0.4 cm wide, and $5 \cdot 10^{-5}$ cm (0.02 mil) thick. The strip is fastened to a 50-mil tungsten lead which is sealed in a ground-glass joint. The joint is sealed on the *outside*, as shown, by a suitable wax of very low vapor pressure. The author states,

> The aluminum leaf is connected electrically to the heater to ensure that no electrostatic effects causing undesirable deflections of the leaf can occur. The motion of the tip of the free end of the leaf is measured by means of a reading microscope with 100 divisions in its scale. Seventeen divisions correspond to 1 mm.
>
> Current to raise the temperature of this heater is provided by a storage battery and measured by an ammeter.

The sensitivity increases with increase in heating current, as shown in Table 5.2.

Calibration curves for the low-pressure and high-pressure ranges are shown in Figs. 5.29(*a*) and 5.29(*b*), respectively.

Obviously, the ground-glass joint could be replaced by a permanent seal and the aluminum foil by one of a higher-melting-point metal to permit baking out the gauge at 400–500° C.

A construction radically different from that of all the Knudsen gauges mentioned above has been described by Klumb and Schwarz.[104] Instead of a vane in a plane parallel to that of the heated surface, a cylindrical

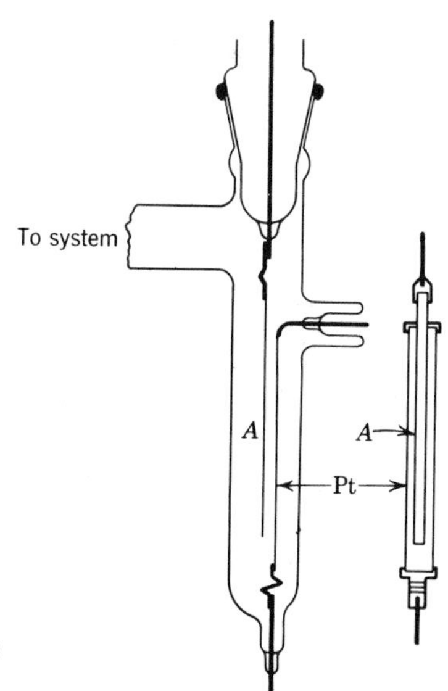


Fig. 5.28. Simple form of Knudsen gauge by Hughes.

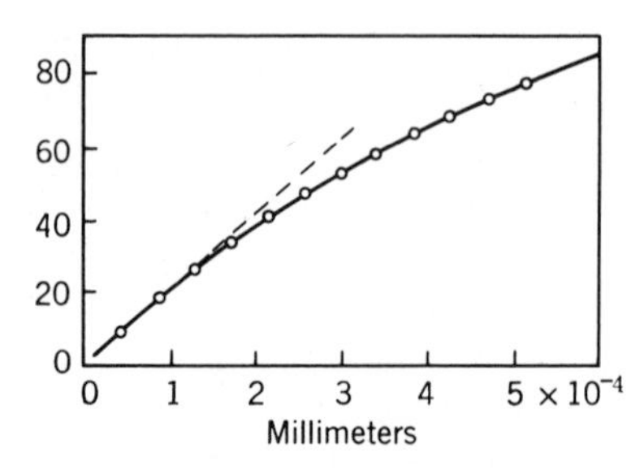


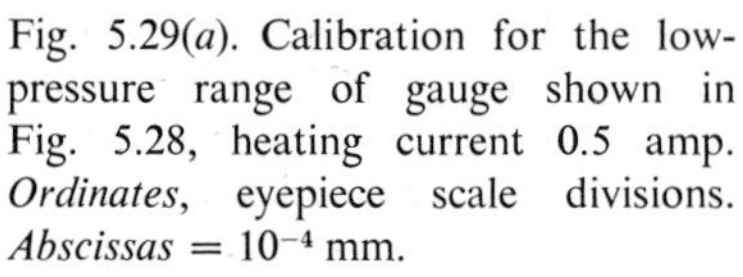
Fig. 5.29(*a*). Calibration for the low-pressure range of gauge shown in Fig. 5.28, heating current 0.5 amp. *Ordinates*, eyepiece scale divisions. *Abscissas* = $10^{-4}$ mm.

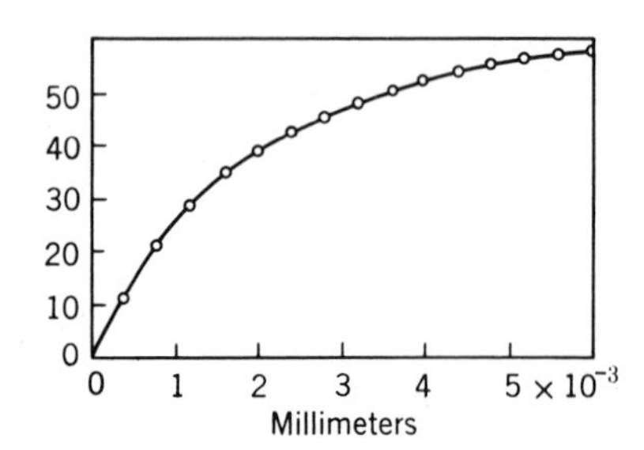


Fig. 5.29(*b*). Calibration for the high-pressure range of gauge shown in Fig. 5.28, heating current 0.2 amp. *Ordinates*, eyepiece scale divisions. *Abscissas* = $10^{-3}$ mm.

arrangement of vanes is used, as shown in Figs. 5.30(*a*) and 5.30(*b*). This system *R* is suspended in the annular space between the outer glass tube *G* and inner tube *H* [see Fig. 5.30(*b*)], and a spiral heater is inserted in *H*. The molecules striking the glass surface *H* acquire a correspondingly

**TABLE 5.2**

**Variation in Sensitivity of Knudsen Gauge with Heating Current**

(Hughes)

| Heating Current (amp) | 1 Scale Division Equals | $10^{-3}$ Micron Equals |
|---|---|---|
| 0.2 | 304 · $10^{-4}$ micron | 0.033 scale division |
| 0.5 | 46.5 | 0.22 |
| 1.0 | 12.4 | 0.81 |
| 1.5 | 6.5 | 1.54 |
| 2.0 | 4.37 | 2.28 |
| 2.5 | 3.16 | 3.17 |

higher kinetic energy; and, when these molecules strike the vertical vanes of the system *R*, *R* acquires a rotational moment which is observed by the deflection of the mirror *M* attached to the suspension. (See Fig. 5.31.) The whole assembly, including the tungsten wire suspension and cooling chamber for the glass walls opposite *R*, is shown in Fig. 5.32. The dimensions (in millimeters) are indicated on the figure.

Instead of observing the deflection directly, a compensating magnetic

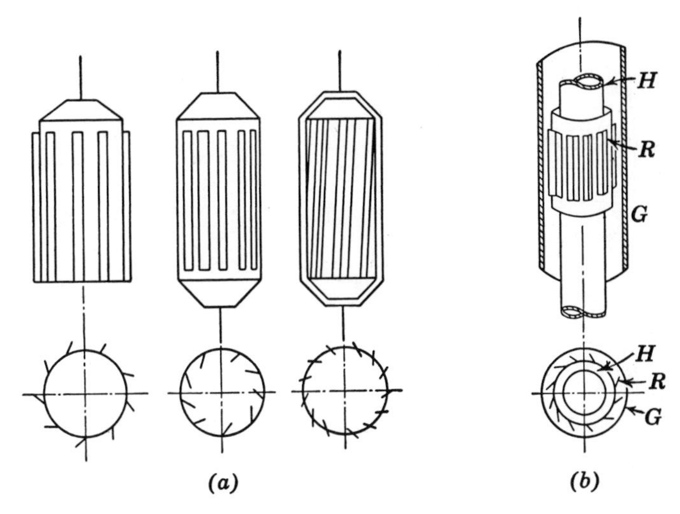


Fig. 5.30. Klumb-Schwarz modification of Knudsen gauge. (*a*) Cylindrical arrangement of vanes. (*b*) Assembly of vanes and heated inner cylinder.

field (produced by the action of the solenoids $W_1W_2$ on the iron armature $D$) may be used, as shown in Fig. 5.33.

Let $\theta$ denote the angular deflection. Then the pressure is given by the relation

$$C_0\theta = P, \quad \text{or} \quad \theta = E_0P, \tag{5.24}$$

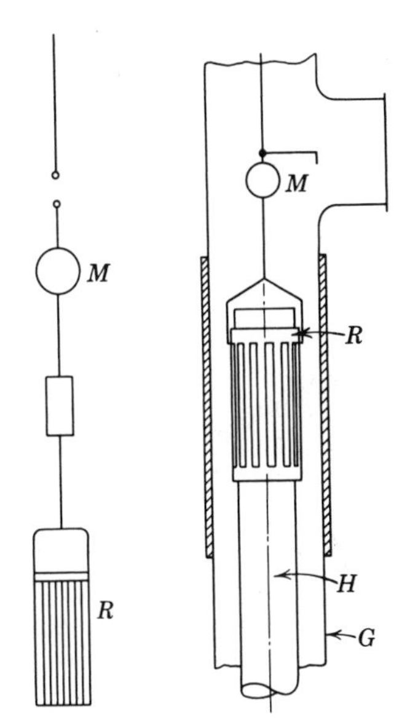


Fig. 5.31. Assembly of vanes and suspension for Klumb-Schwarz form of gauge.

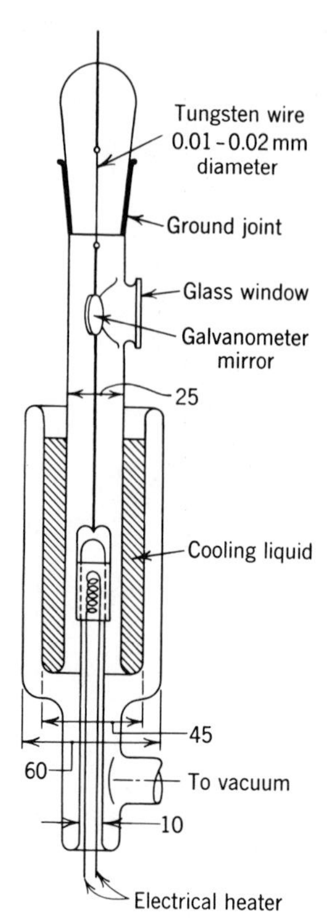


Fig. 5.32. Diagrammatic sketch of Klumb-Schwarz construction of gauge. Tube dimensions are given in millimeters.

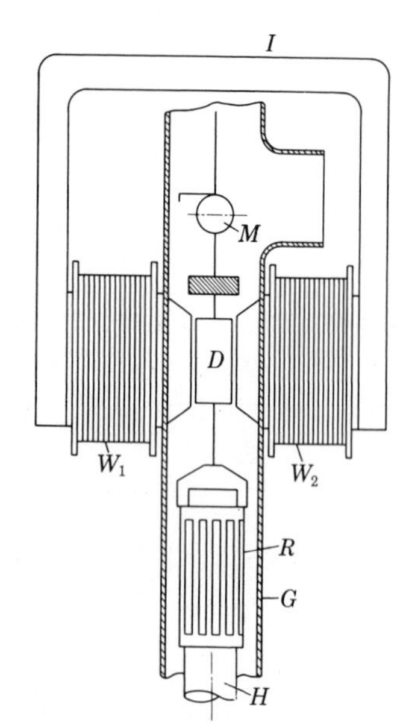


Fig. 5.33. Alternative arrangement for gauge shown in Fig. 5.32, involving use of a magnetic field.

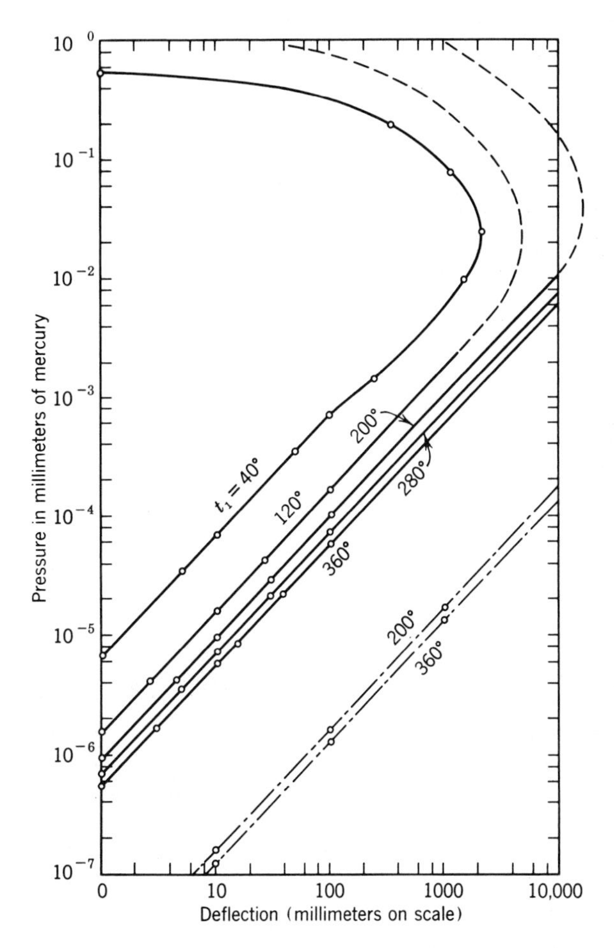


Fig. 5.34. Log-log plot of deflection versus pressure for Klumb-Schwarz gauge.

where $C_0$ is designated the "reduction factor," and $E_0$ the "gauge sensitivity." Furthermore,

$$E_0 = K\left[\left(\frac{T_1}{T_0}\right)^{1/2} - 1\right], \tag{5.25}$$

where $T_1$ is the temperature of the hot surface, $T_0$ that of the cold surface, and $K$ a constant dependent upon the geometrical arrangement of surfaces, their dimensions, and also the magnitude of the accommodation coefficient for the gas used. (See discussion below.)

Figure 5.34 gives plots of data obtained under different conditions. The continuous curves were obtained for air with a gauge having a tungsten wire suspension 0.02 mm diameter and a cold temperature of 13° C

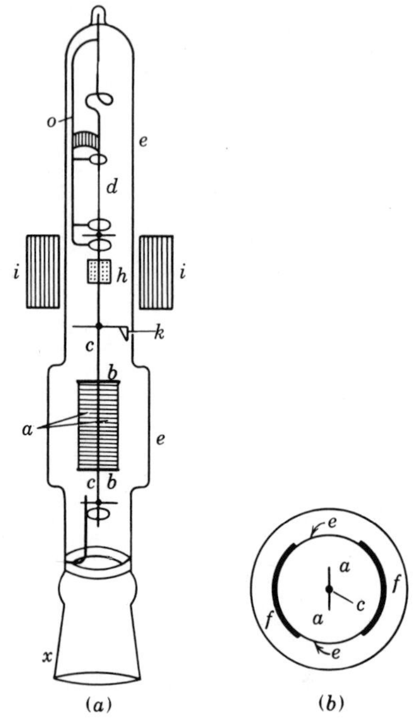


Fig. 5.35. "Molvakumeter" (Gaede).

(cooling water). The dotted curves were obtained for air, with a tungsten suspension 0.01 mm in diameter and an external temperature of $-180°$ C (liquid air). The temperatures of the heated surface ($t_1$) are given in degrees Centigrade, and the pressures in millimeters of mercury. The abscissas denote the deflections in scale divisions.

In accordance with equation 5.24 the plots on log-log scales should be linear with a slope of 45°. As shown in Fig. 5.34 the data for pressures below about 10 microns are in agreement with the theoretical deductions. Furthermore, it was observed that, in accordance with equation 5.25, $E_0$ varied linearly with the value of $(T_1/T_0)^{1/2}$, independently of the nature of the gas used.

The *Molvakumeter* devised by Gaede[105] is of interest because it incorporates features of both the viscosity type of gauge and the Knudsen type. Figure 5.35(*a*) gives a diagrammatic sketch of the construction used. A thin aluminum vane *a* (1 mg weight) is fastened by means of a fine glass fiber *b* to another glass fiber *c*, which carries a pointer *k*. This, in turn, is attached above to the fine quartz fiber *d*. Protection against too great a deflection and breakage in manipulation is provided, as shown in the figure, by supports attached to the wire *o*. The vane *a* is inside the tube *e*, which is connected with the vacuum system by means of the ground-glass opening *x*, and outside the region *e* are located two heating surfaces *f*, arranged as shown in Fig. 5.35(*b*).

The vane is set in oscillation by means of the very thin magnetized plate *h* and a current pulse in the coil *i*. Let $\tau_0$ denote the period of oscillation when the heating surfaces *f* are not operating, and $\tau$ the period when these are operated. Then, the pressure

$$P = A\left(\frac{1}{\tau^2} - \frac{1}{\tau_0^{\,2}}\right), \tag{5.26}$$

where *A* is an empirically determined constant. According to Gaede, the turning moment is proportional to the square of the radius of the foil *a* and inversely proportional to the radius of the tube *e*.

The maximum amplitude of the oscillations will decrease in the period *t* from $\alpha_0$ to $\alpha_n$. Since this is a pure viscosity effect,

$$P(M)^{1/2} = B\,\frac{\log\,(\alpha_0/\alpha_n)}{t}\,. \tag{5.27}$$

Hence by combining these two equations both *P* and *M* can be determined. These equations are applicable only for pressures lower than about $10^{-4}$ mm mercury, at which pressures *L*, the mean free path, is large compared to the distance from the glass wall *e* to the aluminum foil *a*.

Gaede has also described an optical arrangement for observing pressures higher than $10^{-4}$ mm. In the particular instrument used by him the values observed were as follows: $\tau_0 = 0.5$ min and $\tau = 0.25$ min at $P_{mm} = 10^{-5}$. From 20 successive oscillations the value of $\tau$ could thus be determined with an accuracy of 1 per cent, and *P* to within $10^{-7}$ mm.

## Theory of the Radiometer Type of Gauge

At the beginning of this section, a simple derivation was given for the relation between the pressure and the force exerted on a surface by a parallel plane surface at higher temperature. As stated above, equation 5.20 is valid only at pressures so low that the mean free path *L* of the molecules is greater than the distance between the cold and hot surfaces. It is only under these conditions that the molecules can transfer momentum directly from the hot to the cold surface. At higher pressures, the "hot" molecules must lose part of their energy by collisions with cold molecules before reaching the cold surface. Hence it would be expected that the force exerted on the cold surface, which is known as the radiometer pressure and which we shall designate by *F*, will be lower at higher pressures, in spite of the fact that the concentration of molecules is higher.

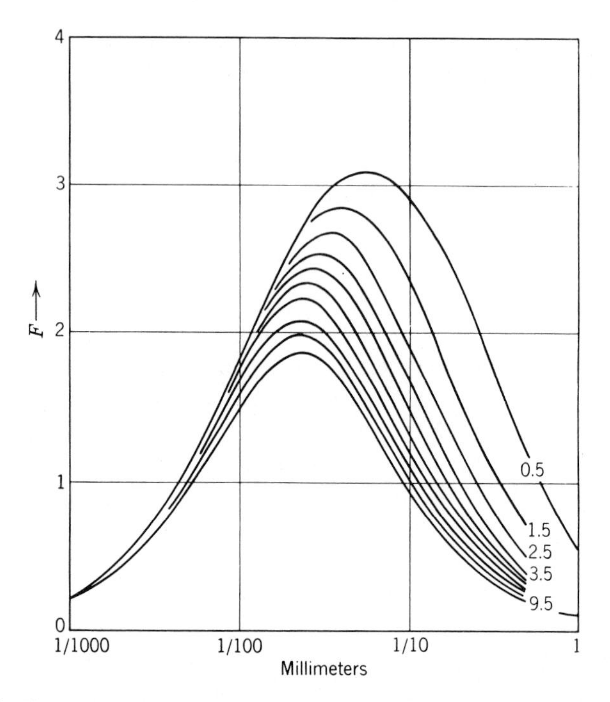


Fig. 5.36. The values of the force $F$ (in arbitrary units) on the movable vane in a Knudsen gauge, plotted against log $P_{mm}$, for a series of values of the distance between the heated surface and the vane. From observations by Brüche and Littwin.

A number of investigators have made measurements on the variation in the magnitude of $F$ with pressure. The results obtained by Brüche and Littwin[106] with a number of different gases are of particular interest. In their experiments they used a gauge in which the thin vane for measuring $F$ was located symmetrically between a hot surface and a cold surface. Figure 5.36 shows the variation in relative value of $F$ with pressures in millimeters (plotted on log scale) for different values of the distance $d$, in millimeters, between each surface and the vane. At any given pressure, the value of $F$ is decreased by increase in $d$, because of the decrease in the ratio $L/d$. Furthermore all the curves show that $F$ attains a maximum value, $F_m$, for a certain pressure, $P_m$.

From their observations Brüche and Littwin concluded that for any gas and given value of $d$ these could be best represented by a relation of the form

$$F = \frac{1}{a/P + P/b}, \tag{5.28}$$

where $a$ and $b$ are two empirically determined constants. At very low pressures, this assumes the form, similar to equation 5.20,

$$F = \frac{P}{a}, \tag{5.29}$$

and at high pressures the relation becomes

$$F = \frac{b}{P}. \tag{5.30}$$

Thus the variation in $F$ with $P$ is represented on linear scales by a curve such as $AMB$, which is shown in Fig. 5.37. At very low pressures, $F$ increases linearly with $P$ until the pressure attains a value corresponding approximately to the point $A$ (6–7 microns). With further increase in $P$, $F$ deviates more and more from the linear plot and reaches a maximum value at $M$ (about 50 microns). Beyond this point $F$ decreases slowly and approaches asymptotically the values given by the hyperbola $F = b/P$.

For constant value of $d$, and different gases, it was observed that the values of $F_m$ (the maximum value of $F$), and of the pressure at which this maximum occurs, vary in approximately the same manner as those

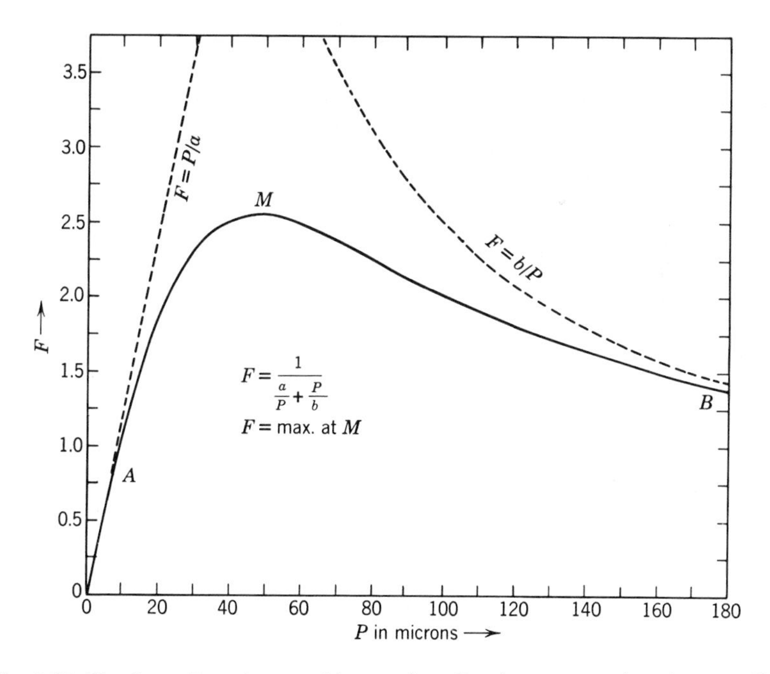


Fig. 5.37. The force $F$ on the movable vane in a Knudsen gauge, plotted versus $P_\mu$, illustrating the application of equation 5.28.

of $L$.[107] Thus, if $AMB$ represents the plot for helium, the plots for other gases would lie below it, and that for carbon dioxide would be below that for argon or oxygen.

Fredlund[108] found that, in general, the linear relation (equation 5.29) between $F$ and $P$ is valid for values of $L/d$ which are greater than about 13. For instance, $d = 0.485$ cm corresponds to the following maximum values of $P_\mu$ for different gases: $H_2$, 1.3; $N_2$, 0.73; $O_2$, 0.8; and Ar, 0.8. On the other hand, for $d = 0.095$ cm, the maximum values of $P_\mu$ for the applicability of the linear relations were as follows: $H_2$, 5.75; $N_2$, 3.25; $O_2$, 3.65.

It follows that the relation between $F$ and $P$ is a function of the ratio $L/d$, which approaches that given by the linear relation for $L/d$ very large and deviates from the linear relation more and more as $L/d$ becomes smaller (at increased pressures).

A factor that has not been taken into account in deriving equation 5.20 above is the *accommodation coefficient*, designated in section 1.10 by $\alpha$. If this coefficient has a value less than unity, then equation 5.20 has to be replaced by the relation

$$F = \frac{KP}{2}\left[\left(\frac{T_1}{T_0}\right)^{1/2} - 1\right], \tag{5.31}$$

where $K$ is a function of the values of $\alpha$ for each surface the value of which lies between 0 and 1. Hence equation 5.20 cannot be used to deduce a calibration for a Knudsen gauge from moments of inertia, etc., unless some definite value is assumed for the factor $K$. This point has been discussed at length by Fredlund, who reaches the conclusion that in most cases the factor $K$ is only slightly less than 1.

## 5.6. HEAT-CONDUCTIVITY MANOMETERS

Energy is dissipated from a heated surface in a gas in two ways: (1) by radiation:

$$E_R = 5.67 \cdot 10^{-12}(\epsilon_1 T_1^4 - \epsilon_0 T_0^4) \quad \text{watts} \cdot \text{cm}^{-2}, \tag{5.32}$$

where $\epsilon_1$ is the radiant emissivity and $T_1$ the temperature of the heated surface, and $\epsilon_0$ and $T_0$ refer to the cold surface; and (2) by conduction through the gas.

(*a*) At *normal pressure* the conductivity $\lambda$ is independent of the pressure, and the loss is given by the relation (see section 1.10)

$$E_c = 4.186\lambda \frac{dT}{dx} \quad \text{watts} \cdot \text{cm}^{-2}, \tag{5.33}$$

where $\lambda$ is expressed in calories per square centimeter per second per unit temperature gradient, at the surface, and $dT/dx$ is the gradient in degrees per centimeter.

(*b*) At *low pressures*, where the distance between hot and cold surfaces is less than the mean free path, the loss by conduction is given by the relation (see equation 1.138)

$$E_K = \frac{4}{3}\frac{\alpha}{2-\alpha}\Lambda_M P_\mu \left(\frac{273.2}{T_0}\right)^{1/2} (T_1 - T_0) \quad \text{watts} \cdot \text{cm}^{-2}, \quad (5.34)$$

where $\Lambda_M$ = molecular heat conductivity in watts per square centimeter at 0° C, and

$\alpha$ = accommodation coefficient.

Hence, for given values of $T_1$ and $T_0$, the energy loss by conduction in a gas at low pressures increases linearly with the pressure. Evidently the sensitivity of the conductivity method for the determination of pressure depends upon the relation between the two quantities $E_R$ and $E_K$.

Let us calculate the values of $E_R$ and $E_K$ for hydrogen ($\lambda_0 = 41.3 \cdot 10^{-5}$, $\Lambda_M = 45.54 \cdot 10^{-6}$), assuming $\epsilon_1 = \epsilon_2 = \alpha = 1$. Table 5.3 gives these values for a series of values of $T_1$ and $T_0$. The first two rows in the table give, for comparison, values of $E_c$. Thus for $T_1 = 373$, $T_0 = 273$, $E_K = E_R$ for $P_\mu = 12.92$.

**TABLE 5.3**

**Heat Loss (watts · cm$^{-2}$) from Plane Surface in Hydrogen**

| Pressure | $dT/dx$ | $T_1$ °K | $T_0$ °K | $E_R$ | $E_c$ | $E_K$ per Micron |
|---|---|---|---|---|---|---|
| 1 atm | 100 | 373 | 273 | 7.83 | 17.3 | ... |
| 1 atm | 50 | 323 | 273 | 3.02 | 8.7 | ... |
| 1 micron | ... | 373 | 273 | 7.83 | ... | 0.606 |
| 1 micron | ... | 323 | 273 | 3.02 | ... | 0.303 |
| 1 micron | ... | 323 | 90 | 5.25 | ... | 7.78 |

It follows that, at $T_1 = 373$, $T_0 = 273$, and $P_\mu = 10^{-2}$, the loss by temperature radiation from a "perfect" radiator is about 100 times that due to conduction by the gas. Hence, in order to reduce the effect of the term $E_R$ and to increase the relative effect of $E_K$, at low pressures, the surfaces used should have as low an emissivity as can be obtained, while it is advantageous to decrease $T_0$ to as low a value as convenient. For tungsten and platinum (outgassed at high temperature) the value of $\epsilon$ varies from about 0.03 at 0° C to about 0.10 at 500° C.

Gauges depending for their indication of pressure on the conductivity of the gas are, in general, operated under conditions in which the energy input is maintained practically constant. Hence the temperature of the heated surface decreases with increase in pressure, and the different types of conductivity gauges involve different methods for determining the change in temperature. In the thermocouple type, the temperature is determined by means of a thermocouple spot-welded at the center of a heated wire; in the resistance type, the change in temperature is determined, as the name implies, by measuring the change in resistance; and, finally, since the length of a wire varies with temperature, this leads to a third method of measuring the temperature.

### Thermocouple Gauges

The earliest application of this type of gauge was made by Voege,[109] who used a small thermocouple attached to a wire heated by constant alternating current. Rohn[110] adopted the novel idea of heating the thermocouple in the gas by an external source of radiation (an incandescent lamp maintained at constant voltage), while Rumpf[111] surrounded the thermocouple with a heated spiral.

A customary construction of gauge is that used by Haase, Klages, and Klumb.[112] Figure 5.38 taken from their paper, shows the arrangement, which consists of a ribbon or wire $H$, heated by current (alternating or direct), and a thermocouple $T$, the junction of which is spot-welded at the center of $H$. In this case $H$ was a Constantan ribbon, 0.72 mm wide, 0.05 mm thick, and 49 mm long, connected to the inner leads $a_1$ and $a_2$ sealed

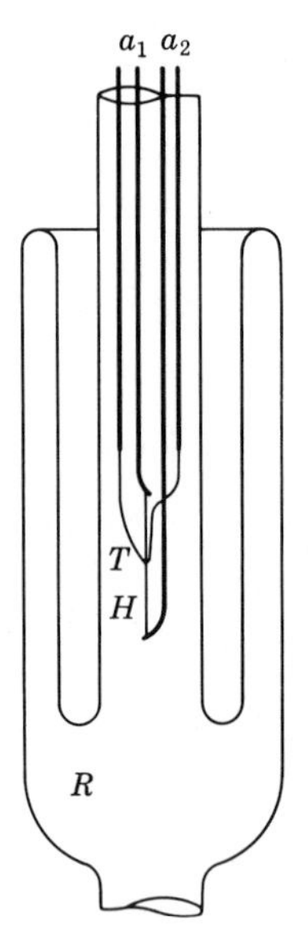


Fig. 5.38. Thermocouple gauge (Haase, Klages, and Klumb).

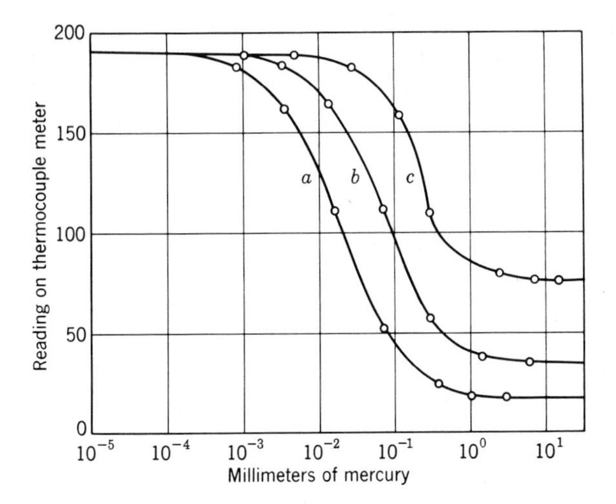


Fig. 5.39. Plots of readings on thermocouple meter versus log $P_{mm}$ (for air), at three values of the heating current, with gauge shown in Fig. 5.38.

in the glass tube $R$, which in turn was connected to the system in which the pressure was to be determined. An iron-Constantan couple was used, and an annular space was provided, as shown in the figure, for ice water or other liquid by means of which the glass walls could be maintained at constant temperature ($T_0$).

Figure 5.39 shows plots of the deflections of an indicating instrument connected to the leads of the thermocouples, as a function of the pressure of air. The three plots were obtained under the following conditions:

| Curve | Input (watts) | Temperature (° C) |
|---|---|---|
| *a* | 0.014 | 48 |
| *b* | 0.226 | 314 |
| *c* | 0.98 | 508 |

As shown by these plots, the range for air was from 1 mm down to a pressure between $10^{-3}$ and $10^{-4}$ mm. Observations with hydrogen and carbon dioxide showed that the change in temperature with pressure decreases with increase in molecular weight of the gas. This conclusion is in agreement with that obtained from an inspection of the relative values of $\Lambda_M$.

Moll and Burger,[113] with a similar design of gauge, in which a more sensitive thermocouple was used, obtained an operating range from about

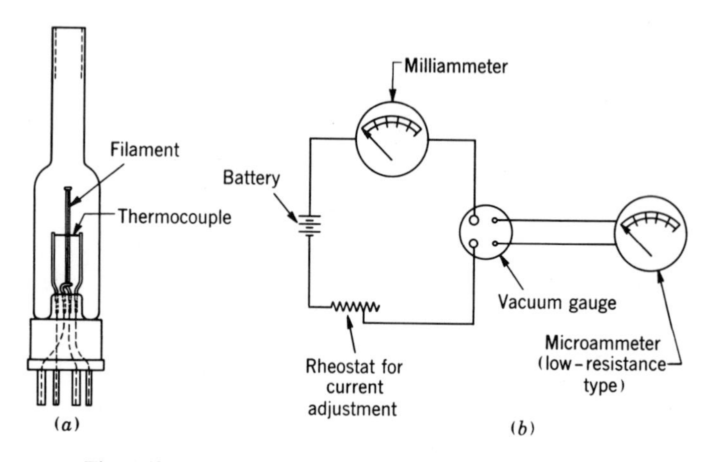


Fig. 5.40. Thermocouple gauge (*a*) and auxiliary circuit (*b*).

$10^{-1}$ to $10^{-4}$ mm. Weisz and Westmeyer[114] were able to extend the range down to about $10^{-5}$ mm, by using a very thin wire (0.04-mm diameter).

The construction employed by Bartholomeyczyk[115] is very simple. A four-lead glass stem is used. The heater, which is a 0.01-mm platinum wire, is fastened to the two outer leads, and at the center of the heater there is fastened the junction of the thermocouple, consisting of 0.01-mm iron (or copper) and Constantan wires. These are connected to the two inner leads of the stem. The range of operation is given as 10–0.1 mm.

The construction of a thermocouple gauge developed in the General Electric Research Laboratory by Gordon is illustrated in Fig. 5.40(*a*). It consists of a platinum-iridium ribbon 3.66 by 0.0234 by 0.0078 cm, to the center of which is welded a thermocouple of Nichrome-advance ribbon (0.01 by 0.00125 cm), all mounted on a four-lead stem in a tubular bulb. Figure 5.40(*b*) shows the circuit used. A constant current of approximately 0.030–0.050 amp, depending on the range of pressures to be measured, is passed through the platinum ribbon, and a sensitive microammeter is used to read the thermocouple emf. The heater filament operates at a temperature varying from 100 to 200° C.

Figure 5.41 shows plots of log $P_\mu$ versus microammeter readings, obtained by means of this gauge, for hydrogen, nitrogen, and xenon. In these observations the walls of the tube were at room temperature ($T_0 \approx 300$). As will be observed the gauge is most sensitive, under these conditions, for the range $10^{-2}$ to about $10^{-1}$ mm.[116]

For all-metal vacuum systems, Webber and Lane[117] have designed a thermocouple gauge which consists of Chromel and Alumel wires arranged on a support in the form of an X, welded at the center. Constant current is supplied to the lower arms of the X by a battery and resistance, and the temperature of the junction is measured by a galvanometer in series with a high resistance, connected to the upper arms of the X. The calibration curves resemble those shown in Fig. 5.41.

Figure 5.42 shows a glass and a metal thermocouple gauge together with a control box. The metal gauge is useful for measuring pressures between 1 and 1000 microns of mercury. The glass gauge measures up to 200 microns. The gauge operates from a small storage battery to give good voltage regulation and can be operated for as long as 60 hr without auxiliary power. Recharging of the battery occurs automatically when the circuit is plugged into the 115-volt a-c line. The control box contains a

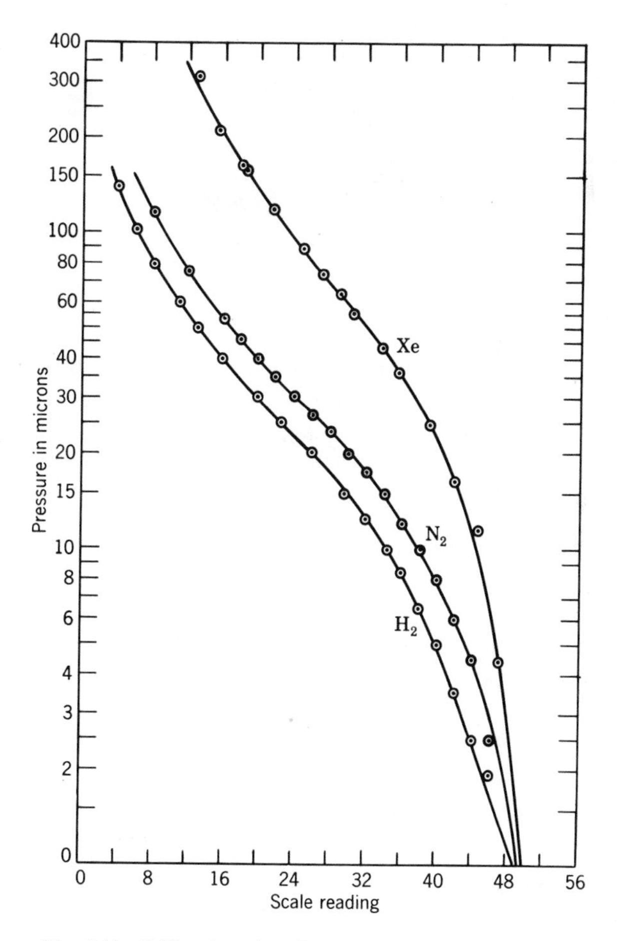


Fig. 5.41. Calibration plots for gauge shown in Fig. 5.40.

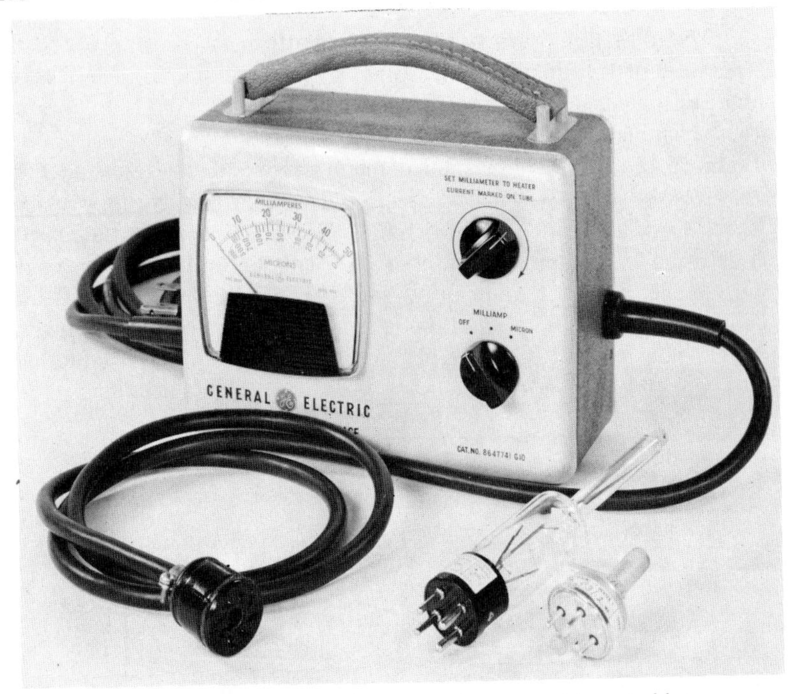


Fig. 5.42. Glass and metal thermocouple gauges with control box.

switch, filament rheostat, and a combination filament milliammeter and millivoltmeter calibrated in microns. The thermocouple gauge is often used to advantage on vacuum systems that require ionization gauges, since it indicates the pressure at which the hot-cathode ionization gauge may be safely turned on.

### Resistance Gauges (Pirani-Hale Type)

The fact that the heat conductivity of a gas at low pressures decreases linearly with the pressure was applied by Warburg, Leithauser, and Johansen[118] to the construction of a gauge in which the variation in temperature of a bolometer strip was determined from the change in resistance.

Pirani[119] pointed out that, in order to construct a gauge based on the relation between the heat conducted from a wire and the pressure, three different schemes could be used.

1. The voltage on the wire is maintained constant, and the change in current is observed as a function of the pressure.

2. The resistance (and consequently the temperature) of the wire is maintained constant, and the energy input required for this is observed as a function of the pressure.

3. The current is maintained constant, and the change in resistance is observed as a function of the pressure.

The first scheme was tried using an ordinary 110-volt tantalum lamp. Better results, however, were obtained when the tantalum wire was clamped tightly to the anchor wires in order to keep constant the heat loss through the supports. With the improved instrument the two other methods were tried, using a Wheatstone bridge arrangement to measure the resistance of the wire, and the third one was finally recommended as the most sensitive for making pressure measurements.

While the principle of the Pirani gauge is thus extremely simple, the sensitivity actually obtained by Pirani was not very great, the lower limit of accuracy being around 0.1 micron.

An improved form of this gauge, constructed by Hale,[120] is illustrated in Fig. 5.43. The following description is quoted from Hale's paper:

A piece of pure platinum wire, 0.028 mm in diameter and 450 mm long, is mounted upon a glass stem carrying two radial glass supports near the top and three at the bottom. The wire is anchored to these radial supports by means of short pieces of platinum wire 0.052 mm in diameter. The anchor is fused into the radial supports at one end, and the other end is made fast to the manometer wire either by an arc weld or by a tiny glass bead. The leading-in wires at $L$, to

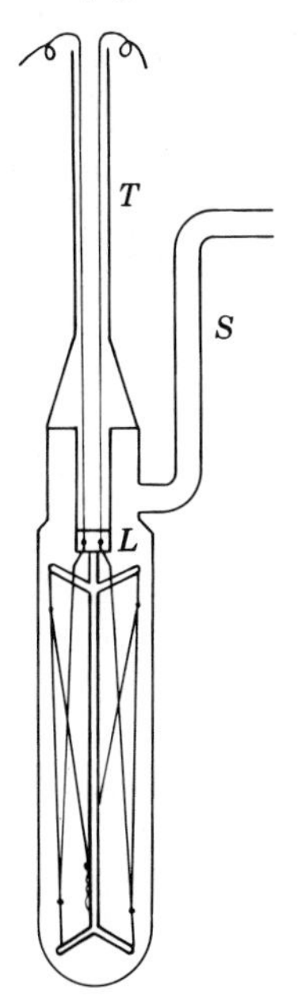


Fig. 5.43. Hale's improved form of Pirani gauge.

which the ends of the manometer wire are welded, are of platinum, 0.31 mm in diameter. All of the platinum wire employed in making the manometer was drawn from the same lot of larger wire and was assumed to be of uniform purity. The temperature coefficient of the manometer wire was found to be 0.00376 per cent per degree. The platinum leading-in wires are joined to heavy copper leads (1.1-mm diameter) by welded joints, and these joints are fused into the stem as in electric lamps. The stem is sealed into a tubular bulb 3.2 cm in diameter and 11.4 cm long. This size of bulb is easily obtained, since it is the size regularly used for 50-watt tubular lamps, such as are commonly employed for galvanometer illumination. At $S$ is a tube by which the manometer is connected with the system whose pressure is being studied. The upper end of the stem $T$ is considerably elongated to permit the complete immersion of the manometer in a constant-temperature bath, whose temperature was approximately 0° C. This stem tube is made of sufficient length to leave 15 cm of it above the level of the bath, a provision which we found to be necessary in order to avoid the condensation of atmospheric moisture upon the top of the tube and the leading-in wires during humid weather. For electrical insulation this tube is packed with purified dry asbestos wool.

A diagram of the electrical connections is shown in Fig. 5.44. A Wheatstone bridge arrangement was used for measuring the resistance changes; and, in order to increase the sensitivity of the gauge, an exact duplicate was exhausted as carefully as possible to an extremely low pressure, sealed off, and inserted in one arm of the bridge as a compensator. Both the compensator $C$ and manometer $I$ were kept immersed in the constant-temperature bath. $R_1$ was a Manganin wire resistance of 925.6 ohms, and $R_2$, a decade plug box containing 10,000 ohms. The strength of the current, as indicated by the milliammeter $A$, was maintained constant at $9.25 \cdot 10^{-3}$ amp by means of the battery and resistance $R_3$. This current was sufficient to raise the temperature of the wire in the manometer and compensator to about 125° at the lowest pressures.

In calibrating the gauge against a McLeod gauge care had to be taken to keep mercury vapor out by means of a liquid-air trap inserted between the manometer and the remainder of the system.

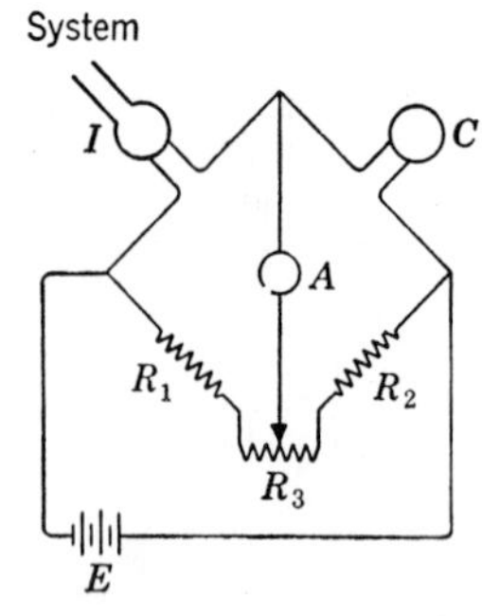


Fig. 5.44. Electrical circuit for Pirani gauge.

The indications of the manometer are dependent upon the thermal conductivity of the gas used. Hale's measurements showed that the lower limit of sensitivity for a gauge of this construction is about $10^{-5}$ mm.

Subsequently further measurements with a Pirani gauge were carried out by So[121] and Campbell.[122] The construction of the gauge used by Misamichi So differed in a few slight details from that of Hale. It was found that the sensitivity of the gauge is higher the lower the temperature of the surrounding bath. At 0° C, and with a heating current of 0.03 amp for a platinum wire 0.076 mm in diameter, the sensitivity, as measured by $(dR/dP)(1/R)$, was observed to be $1.38 \cdot 10^{-3}$ per $1 \cdot 10^{-4}$ mm of mercury. Furthermore, varying the heating current from 0.03 to 0.05 amp was found to produce no change in sensitivity.

Instead of measuring the change in resistance, Campbell measures the potential which must be supplied to the Wheatstone bridge to keep the resistance (and therefore the temperature) of the wire constant. Three Manganin resistances are chosen so that the bridge is balanced when the manometric wire is at a convenient temperature, say 100° C. A voltmeter is connected to the terminals of the bridge, and the potential across the whole bridge is varied by means of a rheostat in the battery circuit until a balance is obtained.

If $V_0$ is the potential required for a balance when $P = 0$, and $V$ is the potential at any pressure $P$, it is found that

$$\frac{V^2 - V_0^2}{V_0^2} = k \cdot f(P), \tag{5.35}$$

where $k$ is a constant for any given gauge and $f(P)$ is a function of the pressure. The function is actually found to be approximately proportional to $P$. For pressures ranging from 1 to 150 microns, as pointed out by Campbell, this method of using the Pirani gauge is specially suited. For such pressures the method used by Hale does not give a linear relation between change in resistance and pressure.

Of interest in this connection is the investigation by Murmann[123] on the variation in sensitivity with variation in the radius of wire $r$, current $I$, and temperature difference $T_1 - T_0$.

Let us consider the case of a metal filament at temperature $T_1$ in a tube the walls of which are at temperature $T_0$. The energy loss per unit area is given by the relation

$$\frac{I^2 \rho_0}{2\pi^2 r^3}[1 + \alpha(T_1 - T_0)] = \sigma_1 T_1^4 - \sigma_0 T_0^4 + KP(T_1 - T_0), \tag{5.36}$$

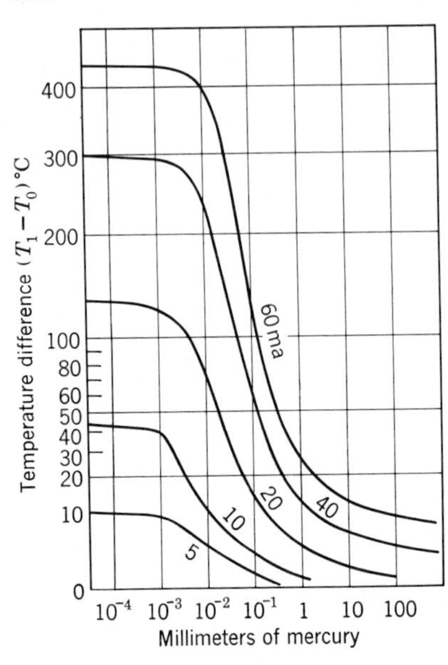


Fig. 5.45. Log-log plots of temperature difference (between wire and walls of Pirani gauge) versus pressure, for a series of values of the heating current.

where $\rho_0$ denotes the specific resistivity at $T_0$, $\alpha$ is the temperature coefficient of resistance, $K$ denotes a constant the value of which is determined from equation 5.34, and $\sigma_1$, $\sigma_0$ are radiation constants defined by means of equation 5.32.

For *constant watt input* and in good vacuum ($P = 0$), the value of $T_1 - T_0$ is constant (for a given filament material) for constant values of $I^2/r^3$; that is, in a vacuum, the temperature is a function of $I/r^{3/2}$. As the pressure is increased at *constant value of* $I^2/r^3$, $T_1 - T_0$ must decrease. This is illustrated by the plots in Fig. 5.45, taken on a wire (90% Pt + 10% Ir), having a diameter of $5.1 \cdot 10^{-3}$ cm (2 mil). The currents are given in milliamperes and the pressures in millimeters. Evidently the value of the ratio

$$S = \frac{\Delta(T_1 - T_0)}{(T_1 - T_0)} \cdot \frac{P}{\Delta P}$$

varies with $P$. From the plots, Murmann finds that, as $P_{mm} = 10^{-4}$, the maximum value of $S$ is obtained for $I = 22.3$ ma. On the other hand, as $P$ is increased, the value of $I$ for maximum value of $S$ must be increased. These conclusions are in agreement with the observations made by Knauer and Stern[124] in using a hot-wire manometer to determine the intensity of a molecular beam of a light gas.

An investigation on the design of a sensitive form of Pirani gauge was carried out by Stanley.[125] Two identical tubes were placed, as in Hale's arrangement, in the opposite arms of a Callendar-Griffith's bridge. The heated element in each tube was a loop of 0.001-in. platinum wire 10 cm in length, together with a compensating loop of the same wire, 2 cm in length. Constant current was supplied to the filaments. In using the gauge "the short loop of the manometer was placed in the arm of the bridge containing the longer loop of the compensator, and vice versa."

It was observed that, with the walls at constant temperature, the sensitivity (measured as $dR/dP$, where $R$ is the resistance, in bridge units) was greater for a filament temperature of 100° C than for a temperature of 30° C. Increasing the temperature beyond 100° C seemed to have little effect on the sensitivity. Variation of the temperature of the glass walls from 100° C to about −180° C showed that, the lower the temperature, the greater the sensitivity of the arrangement to variations of pressure. Pressures in the range $2 \cdot 10^{-3}$ mm to $4 \cdot 10^{-6}$ mm could be measured very satisfactorily.

In the General Electric Research Laboratory H. C. Thompson and E. F. Hennelly developed a simple form of direct-reading gauge which has proved very convenient in practice. The measuring tube and compensator each consists of a glass tube 1 in. in diameter and approximately 3 in. long in which four, or more, standard 25-watt, 115-volt tungsten-filament coils are welded to supports, the total resistance being about 15 ohms at room temperature. The resistances $R_1$ and $R_2$ (see Fig. 5.44) are about 14 ohms each, and the resistance $R_3$ about 2 ohms, which can be adjusted to obtain zero current in the ammeter $A$. A dry-cell battery supplies the constant source of potential.

Figure 5.46 shows plots of the current (in milliamperes) versus $P_\mu$ for a typical outfit. In order to prevent confusion three scales for pressure have been used. These scales are displaced along the axis of abscissas, so that the origin in each case is at $P_\mu = 0$. Curves $A$, $B$, $B'$, and $C$ refer to dry air at three different battery voltages; curves $D$ and $E$ refer to hydrogen and argon, respectively. It will be observed that, for pressures below about 40 microns, the plots are linear and practically independent of battery voltage.

With a sensitive milliammeter and care in construction of the tube and compensator, pressures as low as $10^{-3}$ micron may be measured by this arrangement. Ordinarily, however, where it is desired to use a relatively cheap type of indicating instrument, the lower limit of sensitivity is between $10^{-2}$ and $10^{-1}$ micron.

It should be added that Pirani gauges consisting of two tubes (one as compensator) and of a case containing the requisite electrical circuits are available from a number of manufacturers and have found wide application in vacuum technique.

The problem of designing a very sensitive gauge for measuring the intensity of a molecular beam of gas has been considered by Ellett and Zabel.[126]

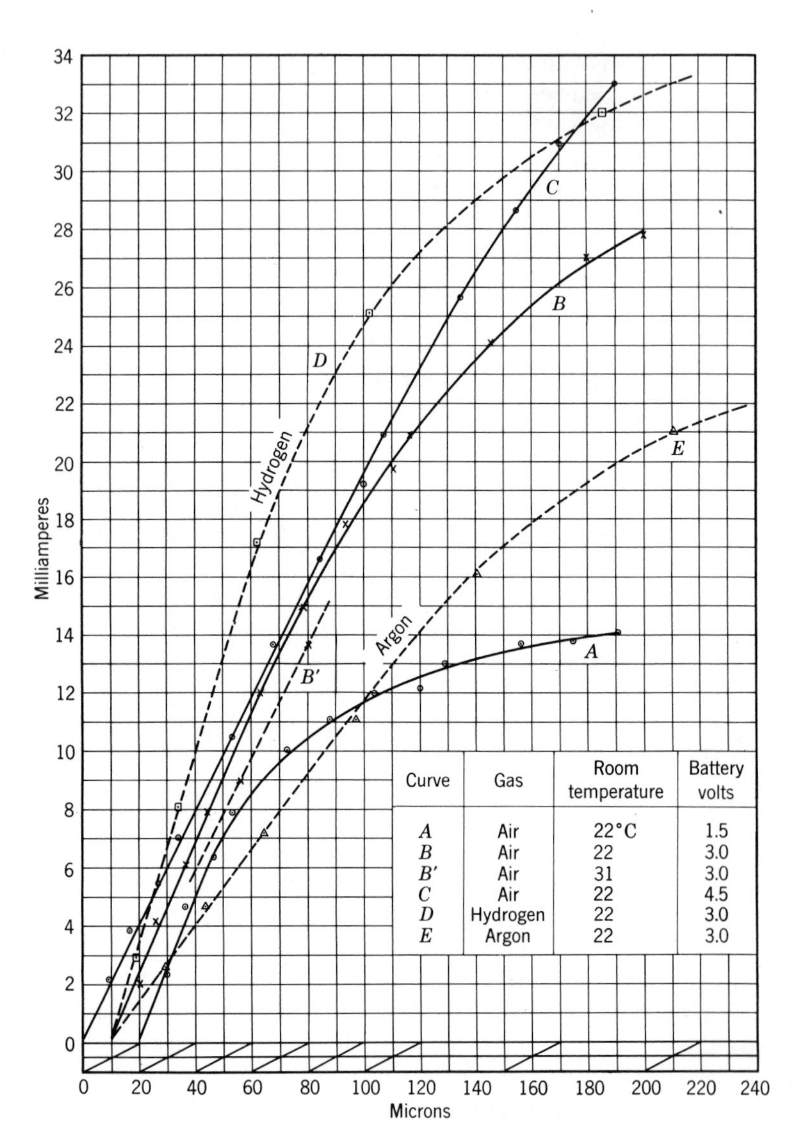


| Curve | Gas | Room temperature | Battery volts |
|---|---|---|---|
| A | Air | 22°C | 1.5 |
| B | Air | 22 | 3.0 |
| B′ | Air | 31 | 3.0 |
| C | Air | 22 | 4.5 |
| D | Hydrogen | 22 | 3.0 |
| E | Argon | 22 | 3.0 |

Fig. 5.46. Calibration plots for Pirani gauge.

If in equation 5.36 we assume that $\sigma_1 = \sigma_0$, and that the energy loss by conduction is negligible compared to that by radiation, then, at constant watt input,

$$\frac{dT}{dP} = \frac{C(T - T_0)}{T^3}, \tag{5.37}$$

where $C$ is a constant. The expression on the right-hand side of this equation has a maximum value for $T = \frac{3}{2}T_0$, and substituting this value in equation 5.37 we obtain the relation

$$\left(\frac{dT}{dP}\right)_{max} = \frac{C'}{T_0^2}, \tag{5.38}$$

where $C'$ is another constant.

Hence the value of the maximum temperature change for a given change in pressure varies inversely as the square of the absolute temperature of the walls of the tube in which the heater is sealed. Since the resistance $R$ varies linearly with $T$, it also follows that

$$\frac{dR}{R_0 dP} = \frac{C''(T - T_0)}{T^3}, \tag{5.39}$$

where $C''$ is a constant, and hence the change in resistance with pressure has the same maximum as $dT/dP$.

From these relations the following conclusions are drawn by Ellett and Zabel:

1. In a bridge circuit, of which the hot wire constitutes one arm, which is operated at constant watt input, the galvanometer deflection should be proportional to the pressure.
2. The sensitivity of the gauge is proportional to the square root of the area of the wire.
3. For maximum sensitivity the wire should be as long as convenient, and the diameter should be adjusted so that its resistance is of the same order of magnitude as that of the galvanometer.

The construction of a gauge which is described as sensitive to pressures of less than $10^{-5}$ micron is shown in Fig. 5.47. The wire used had a diameter of 0.001 in. The writers state,

> The sensitivity attained by the use of nickel wire is in all cases greater than that for tungsten. Nickel wire has the additional advantage that it may be flattened more easily than tungsten which greatly increases the sensitivity. . . . Immersing the walls of the gauge in liquid oxygen approximately multiplied the sensitivity of the gauge by 2 when the galvanometer resistance is high and by 7 when the galvanometer resistance is low.

The authors conclude that a compensating gauge is not as satisfactory as one gauge immersed in a constant-temperature bath.

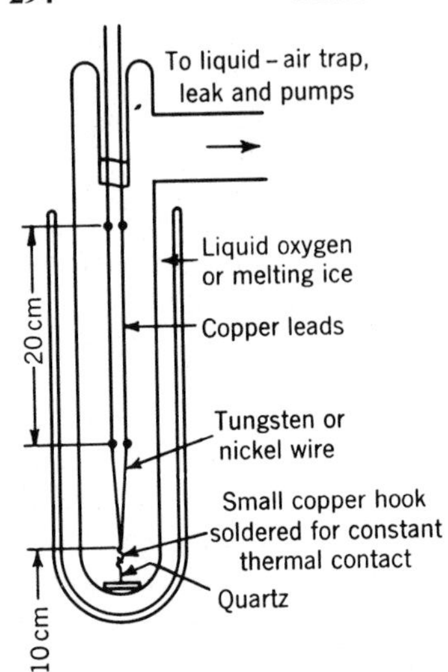


Fig. 5.47. Construction of ultrasensitive Pirani gauge (Ellett and Zabel).

For reading pressures as low as $2 \cdot 10^{-3}$ micron, Reimann[127] has used a manometer constructed as follows: To a four-lead stem in a tubular bulb were attached

. . . two independent stabilized tungsten spiral filaments, each of 0.047 mm diameter wire, of 28 cm length, wound into a spiral 0.9 mm in diameter and 6 cm in length. Each filament was flashed in a vacuum at 20 volts for 10 minutes prior to use. This tube was clamped to a similar evacuated and sealed-off tube and both were immersed in a water-bath at room temperature. The filaments in each bulb were connected up to form opposite arms of a Wheatstone bridge, across whose ends a P.D. of 2 volts was maintained. The bridge was balanced with a good vacuum in the manometer by adjusting a very small variable resistance inserted in one arm, and the "out-of-balance" current in a microammeter was then taken as a measure of the pressure. By inserting a resistance in series with the microammeter the manometer could be made less sensitive.

This gauge was used by Reimann to investigate rates and extent of clean-up of hydrogen by magnesium.[128]

A similar arrangement of two manometers has been described by Hunsmann.[129] Each manometer consists of a cylindrical bulb having a 0.2-mm nickel wire along the axis. In order to obtain reproducible results, even at higher pressures, the filament is attached at one end to the lead-in wire by a steel spring which is short-circuited by means of a copper wire.

Instead of using two tubes in the two arms of a Wheatstone bridge, Scott[130] has suggested measuring

. . . the resistance of the wire by feeding the voltage drop across it onto the grid of a triode tube. This allows the pressure to be read directly on the calibrated scale of a galvanometer contained in the plate circuit. Although the scale reading is a simple function of the pressure, the relation is by no means linear over the entire range.

Figure 5.48 shows the electrical circuit. The hot wire constitutes the resistance $R_1$,

> . . . which consists of the filaments of three 6-watt lamps connected in series to give about 700 ohms. The resistance $R_2$ is made up of a 20,000-ohm fixed resistor and a 2000-ohm variable resistor. The two resistances, $R_1$ and $R_2$, constitute a simple potentiometer, or bleeder. The resistance $R_3$ is a load resistance of 10,000 ohms. This serves to give the triode a more nearly linear characteristic for values of the grid voltage near cut-off. . . . The voltage $E_2 + E_3$ across the resistances $R_2 + R_3$ is the plate voltage of the tube. The voltage $E_1$ across the Pirani resistance $R_1$ is the grid voltage of the tube.

The resistance $R_2$ is adjusted to give a certain plate current for a known pressure, and as the pressure changes both $E_1$ and $E_2$ will vary in opposite directions while $E_1 + E_2$ remains constant. With decrease in pressure, $R_1$ and, consequently, $E_1$ increases, which makes the grid more negative and decreases the plate current. The author states that the arrangement "affords a simple, rapid, and reliable means of measuring pressures between $10^{-2}$ and 1 micron." Figure 5.49 shows the calibration curve obtained, presumably with air.

The observation that certain *semiconductors* possess a high (negative) temperature coefficient of resistance has been applied by Weise[131] to the construction of a Pirani gauge. In the earlier form the resistance consisted of a small thin-walled tube of magnesium-titanium spinel. A tube 1 mm in diameter and 30 mm long, with a wall thickness of 0.1 mm, has a resistance of 15,000 ohms. This resistance is in the form of a thin film, 20 microns thick and 10 cm² in area. The thin film is attached to two supporting spirals of tungsten wire [132] as shown in Fig. 5.50.

For high accuracy a bridge circuit is used, and water at constant temperature (city water) is circulated through the outer jacket. For ordinary tests, the resistor is connected to another series resistor, and the current indicated at constant voltage across the high resistance is a function of the pressure. The values chosen for the constant voltage and series

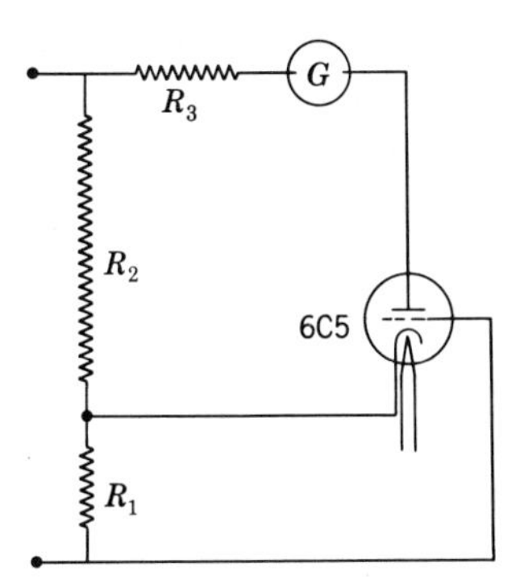


Fig. 5.48. Electrical circuit for automatic gauge (Scott).

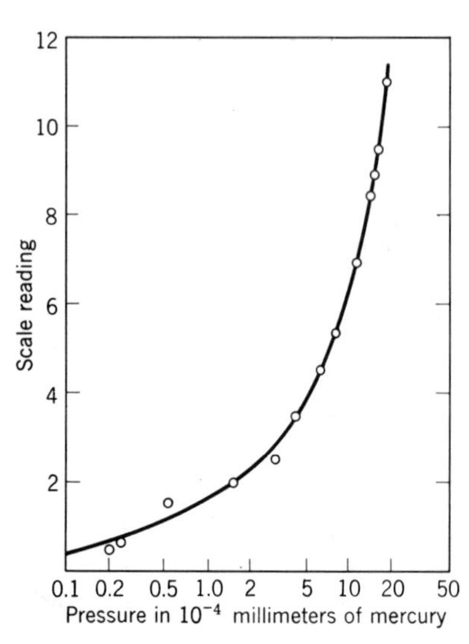


Fig. 5.49. Calibration curve for Pirani gauge circuit shown in Fig. 5.48.

resistor depend upon the desired pressure range. Figure 5.51 shows calibration curves for air and hydrogen for a resistor under conditions in which the pressure was varied from 1 mm to $10^{-3}$ mm. The author states that, by varying the voltage and series resistor, it is possible to vary the range of sensitivity from atmospheric pressure down to $10^{-6}$ mm. In the

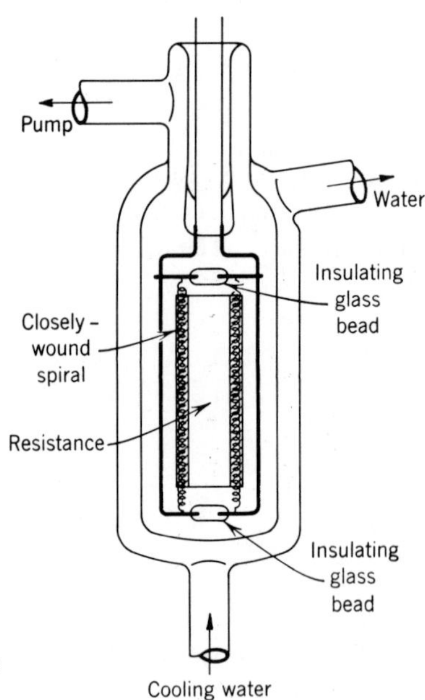


Fig. 5.50. Pirani gauge involving use of a semiconductor as heated element.

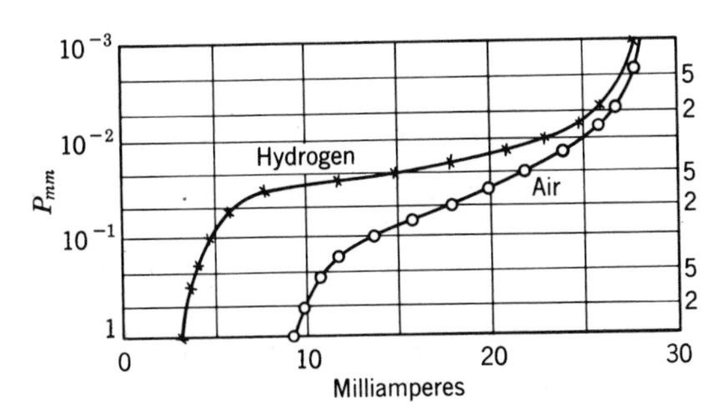


Fig. 5.51. Calibration curves for air and hydrogen with gauge shown in Fig. 5.50.

original paper there is also described a relay circuit for using the gauge for maintaining constant, low gas pressure in a system.

A semiconductor, having a high negative temperature coefficient of resistance, has also been utilized by Becker, Green, and Pearson in the development of the *thermistor manometer*.[133] The thermistor element consists of a small bead of a semiconducting material which is fastened to two leads and sealed into a small bulb. Two elements are used, with characteristics matched so that they can be operated in adjacent arms of a Wheatstone bridge, as shown in Fig. 5.52. One element (V-550A) is sealed off in good vacuum and gettered. This serves as a temperature compensator for the other element (V-550B), which is connected to the system in which the pressure is to be determined. Typical characteristics of an element at 25° C and $10^{-3}$ micron are as follows:

| | |
|---|---|
| Resistance at zero power | Approximately 50,000 ohms |
| Voltage at 0.5 ma | Approximately 2 volts |
| Voltage at 3.0 ma | Approximately 1 volt |

These thermistors are now manufactured by the Victory Engineering Corporation, Union, N.J. They are designated as A-58, and consist of a pair of glass bulbs containing thermistor beads with matched characteristics as described above.

The suggested circuit constants are as follows:

| | |
|---|---|
| Current, each thermistor in high vacuum | Approximately 3 ma |
| Voltage across bridge (ac or dc) | Approximately 9 volts |
| Bridge ratio arms | Approximately 2,000 ohms |
| Bridge balance potentiometer | Approximately 1,000 ohms |
| Meter resistance | Approximately 10,000 ohms |

The bridge is balanced under conditions of good vacuum in the manometer, and the off-balance current then becomes a measure of the pressure.

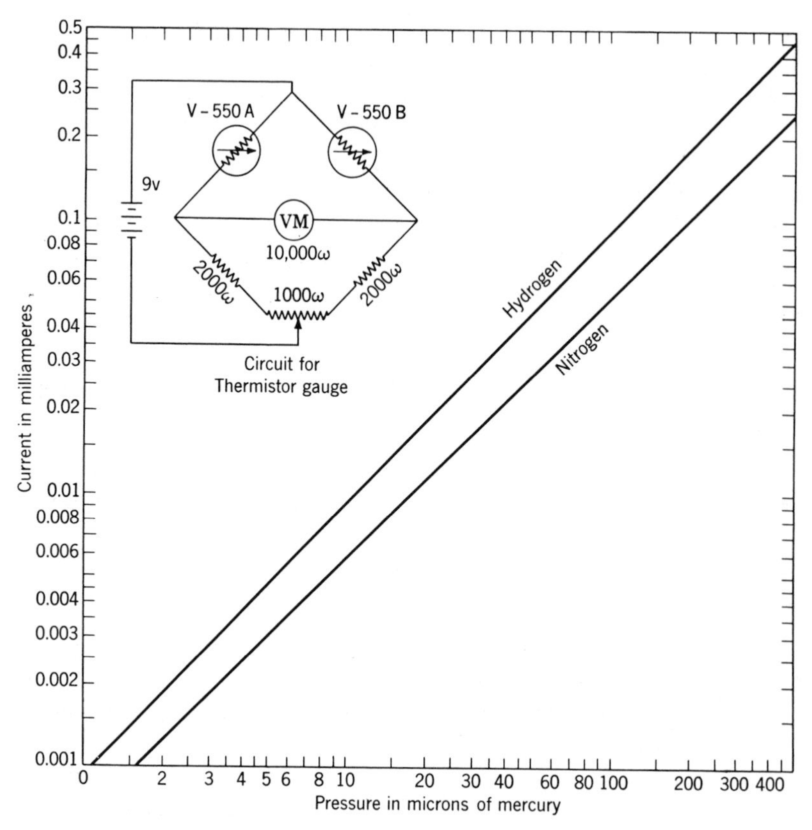


Fig. 5.52. Diagram of circuit and calibration plots for hydrogen and nitrogen with thermistor gauge.

The exact calibration varies with the nature of the gas, and Fig. 5.52 shows two such calibration curves, plotted on log-log scales, obtained on a thermistor manometer by A. H. Young in the General Electric Research Laboratory, by means of a McLeod gauge. The ordinates correspond to milliamperes through a 10,000-ohm resistance.

The fact that these plots are practically linear over the range of pressures from 1 to about 1000 microns gives this type of gauge an advantage over the thermocouple gauge and should make it extremely useful in vacuum technique.

## Gauges Which Involve Linear Expansion of Metal Wire or Strip

Since the linear expansion of a heated wire is a function of the temperature, it is evident that such observations may also be applied to measure the pressure of a gas in which the wire is located.

A manometer based on this principle has been described by Murmann.[134] Two very thin platinum-iridium wires are each fastened at both ends to a framework and connected in the center to a spiral which can be used to indicate the expansion of the wires when heated. The lower limit of sensitivity is apparently about $10^{-3}$ mm.

Much more practical are gauges in which bimetallic strips are used. Klumb and Haase[135] have described three different forms of such a gauge.

Let $a_1$ denote the thickness and $E_1$ the modulus of elasticity of one of the metals, and $a_2$ and $E_2$ denote the corresponding values for the second metal. Then, according to Michelson,[136]

$$\frac{a_1}{a_2} = \left(\frac{E_1}{E_2}\right)^{1/2}. \tag{5.40}$$

Furthermore, if $l$ denotes the total length, and $d$ the total thickness, of the strip, then the deflection $y$ for a temperature rise of $t - t_0$ degrees C is given by the relation

$$y = 10^{-4}k \cdot \frac{l^2}{d}(t - t_0), \tag{5.41}$$

where $k$ denotes the "specific" deflection for $l = 100$ mm, and $d = 1$ mm. The value of $k$ is proportional to the difference in coefficients of linear expansion and varies from 0.1 to 0.14 mm for industrially available bimetallic strips.

Figure 5.53 shows the construction of a gauge in which the deflection is observed by means of a microscope. The sketch at the right shows further details. The bimetallic strip has the dimensions 1 by 0.1 by 35 mm. To this is fastened a glass fiber about 70 mm long which carries a weight of about 100 mg. At 0.1 volt, the heating current in the strip is about 250 ma.

Figure 5.54 shows calibration curves for three gases in the range $10$–$10^{-5}$ mm. With the microscope used, 1-mm deflection was equal to 30 scale divisions.

In another design of the gauge the band is in the form of a spiral having 14 turns and 6-mm outside diameter. To the lower end of this spiral is attached a galvanometer mirror, and the deflections are observed optically. The zero position of the mirror is adjusted by means of the torsion head, at the top of the greased joint, and the heating current is applied at the upper and lower terminals. The plot of angular deflection of the mirror as a function of pressure exhibits a curve quite similar in its characteristics to that shown in Fig. 5.54.

In the third form, compensation is provided for variations in external temperature by means of a second heated spiral made of bimetallic strip.

Seeman[137] has pointed out that a similar design of low-pressure gauge

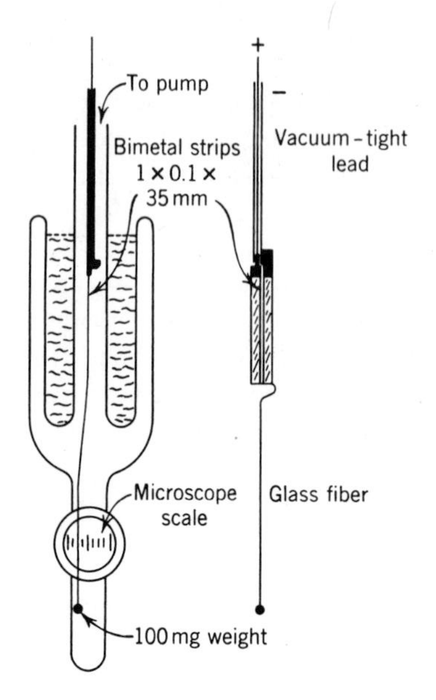


Fig. 5.53. Bimetallic strip manometer with microscopic scale for reading deflection (Klumb and Hasse).

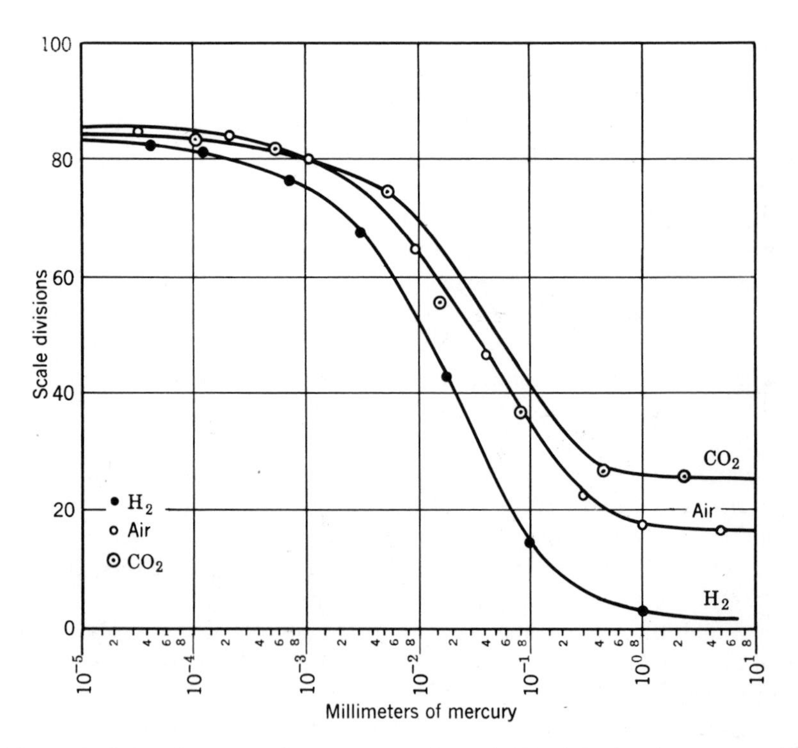


Fig. 5.54. Calibration curves for manometer shown in Fig. 5.53, for three gases, with cooling liquid at 0° C.

utilizing a compensation spiral has been developed by Brown, Boveri, and Company, Switzerland.

## 5.7. IONIZATION GAUGES

### Introduction

The kinetic energy acquired by an electron in passing through a potential difference of $V$ volts corresponds to $Ve$, where $e$ is the charge of the electron. When this energy exceeds a certain critical value, corresponding to the ionization potential $V_i$, there is a definite probability, which varies with $V$ and with the nature of the gas, that collisions between electrons and molecules will result in the formation of positive ions. The relatively high-velocity electron, on colliding with a gas molecule, drives an electron out of the molecule, leaving it positively charged. The values of $V_i$ for different monatomic vapors and gases vary from 3.88 volts for cesium to 24.58 volts for helium, which has the highest ionizing potential. For oxygen, nitrogen, hydrogen, and other diatomic gases, the values of $V_i$ range around 15 volts.

In a thermionic diode with a fixed anode voltage in excess of $V_i$, the number of positive ions produced by the electrons in traveling to the anode should be directly proportional to the gas pressure, provided that the pressure is so low that any one electron does not make more than one ionizing collision during the trip.[138] It would also be expected that for a constant pressure the number of positive ions produced would be proportional to the electron emission, if space-charge effects are negligible and do not alter the potential distribution in the diode. Consequently, if a determination of the rate of production of positive ions in the diode could be made, it would yield a measure of the pressure. A third electrode, operated at a fixed negative potential with respect to the cathode, may be added to the diode for this purpose. In practice, this ion-collector electrode will not collect all the ions formed, but only a constant fraction of them. A measure of the ion current to the ion collector then gives an indication of the gas pressure in the device. The shape and disposition of the three electrodes in this device, called an ionization gauge, have been the subject of much study and ingenious inventive effort.

Let $i_p$ denote the positive-ion current to the ion collector, $i_e$ the electron-emission current to the anode, and $P$ the pressure in the ionization gauge. Then, under the operating conditions discussed above,

$$i_p = Si_eP, \tag{5.42}$$

or

$$P = \frac{1}{S}\frac{i_p}{i_e}, \tag{5.43}$$

where the proportionality constant $S$ is called the *sensitivity* of the gauge.

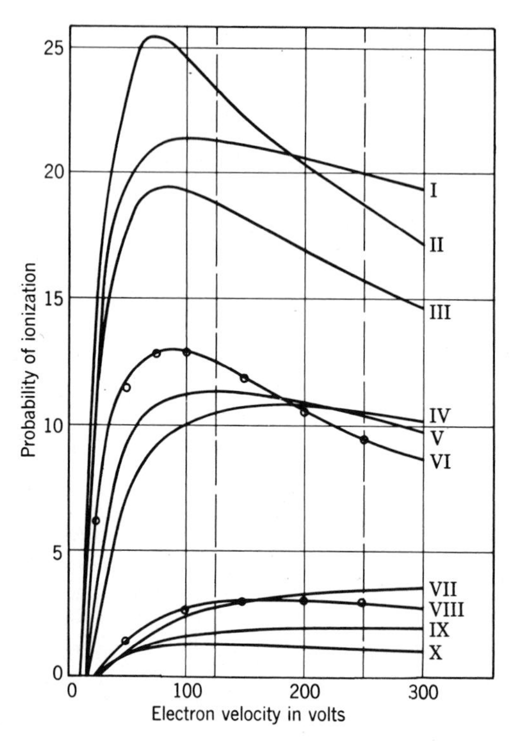


Fig. 5.55. Probability of ionization (ions per electron per centimeter path at 1-mm pressure) as function of electron velocity.

If $i_p$ and $i_e$ are measured in the same units, $S$ has the dimensions of reciprocal pressure.

Thus $S = m \cdot 10^{-3}$ per micron signifies that the sensitivity is $m$ microamperes per milliampere per micron.

It will be observed that

$$S \text{ per microbar} = 0.75S \text{ per micron.}$$

It is not uncommon for an ionization gauge, such as the VG-1A, to have a sensitivity of $S = 20$ per mm Hg for nitrogen. At a pressure of $10^{-6}$ mm Hg, substitution in equation 5.43 shows that only one ion is produced by every 50,000 electrons that travel from the cathode to the anode.

The probability of ionizing a gas by electron impact is shown in Fig. 5.55 for several different gases. Here $P_i$, the number of ion pairs (a singly charged positive ion and one electron) produced per centimeter of path length by an electron in traveling through a gas at 1 mm Hg pressure and 0° C, is plotted as a function of the initial kinetic energy of the electron expressed in electron volts.

From the definition of $P_i$ it follows that

$$N = P_i LP \quad \text{or} \quad L = \frac{N}{P_i P} \quad \text{for } N \leqslant 1, \tag{5.44}$$

where $N$ is the number of ion pairs produced, $L$ is the length of the path in centimeters traveled by an electron, and $P$ is the gas pressure in millimeters Hg. Although $P_i$ is commonly called the probability of ionization, it is in reality a differential ionization coefficient.[139] Thus, when a constant value of $P_i$, corresponding to the initial kinetic energy of an electron, is substituted in equation 5.44, this equation can be used only when $N \leqslant 1$. If $N = 1$, $L$ becomes the length of the mean free path traveled by an electron before making its first ionizing impact. After its first ionizing collision the electron will have lost kinetic energy, and a new value of $P_i$ will have to be used to compute the average distance traveled before the next ionizing collision occurs.

If it is assumed that in an ionization gauge the number of ions produced per electron is equal to the ratio of the ion current to the electron-emission current, then, since at normal operating pressures $i_p/i_e \ll 1$, it follows from equation 5.44 that

$$\frac{i_p}{i_e} = P_i LP. \tag{5.45}$$

This assumption is not strictly true. As explained above, there are more ions formed in the gauge than reach the ion collector. Also, in most ionization gauges $P_i$ is not a constant over the entire path length of the electrons because of changes in the electron velocity due to potential variations. However, one may use a value of $P_i$ for the gas involved corresponding to the kinetic energy the electrons would have if they were at anode potential. Under these conditions, $L$ in equation 5.45 may be considered as the effective path length traveled by the electrons in going from the cathode to the anode. Substitution of equation 5.45 in equation 5.42 gives

$$L = \frac{S}{P_i}. \tag{5.46}$$

Consider the VG-1A ionization gauge with a sensitivity of $S = 20$ per mm Hg for nitrogen with 150 volts on the anode. From Fig. 5.55, $P_i = 11$; hence on substitution of numerical values in equation 5.46 the effective path length of the electrons in this gauge is found to be 1.8 cm. Yet at a pressure of $10^{-6}$ mm Hg, for example, the mean free path of the electrons is 900 meters (computed from equation 5.44 for $N = 1$). Thus at this pressure most of the electrons make no ionizing collisions during their short flight from the cathode to the anode and only a few make even one collision.

## Design

It is of historical interest that in 1916 Buckley[140] published the following description of the construction of an ionization gauge:

> The manometer consists of three electrodes sealed in a glass bulb which serve as cathode, anode, and collector of positive ions. The cathode may be any source of pure electron discharge such as a Wehnelt cathode or a heated tungsten or other metallic filament. The exact forms of the electrodes are not of great importance. The collector is preferably situated between the other two electrodes and is of such form as not to entirely block the electron current to the anode. A milliammeter is used to measure the current to the anode and a sensitive galvanometer to measure the current from the collector which is maintained negative with respect to the cathode so as to pick up only the positive ions.
>
> Currents [Buckley continues] from 0.2 to 2.0 milliamperes were used with from 100 to 250 volts between cathode and anode. The collector was held at 10 volts negative with respect to the cathode. The resulting current to the collector at a pressure of $10^{-3}$ mm was about one-thousandth the current to the anode and at lower pressures was proportionately less.

Evidently, the lower limit of pressure that can be measured with such a gauge depends upon the sensitivity of the indicating instrument in the collector circuit, and this may be considered the most important feature of the ionization gauge from the point of view of laboratory vacuum technique. Certain disadvantageous features about the use of the gauge will be discussed in a subsequent section.

Any standard construction of three-electrode device, such as used in commercial radio tubes, may be utilized as an ionization gauge. Two different types of circuits have been suggested: (*a*) the "internal-control" type, in which the middle electrode (grid of a triode) functions as positive-ion collector; and (*b*) the "external-control" type, in which the outer electrode (plate of a triode) functions as positive-ion collector. These are illustrated in Figs. 5.56(*a*) and (*b*), respectively. The meters $m$ and $M$ indicate the positive-ion and electron currents, respectively.

The potentials may be varied from about 100 to 300 volts on the anode and from $-2$ to $-25$ volt on the ion collector. Also, instead of using an A battery to supply current to the filament and connecting the zero potential lead of the B battery to the midpoint of a high resistance across the filament, it is customary to employ an a-c step-down transformer and then connect the zero potential lead to the secondary winding, as shown in Fig. 5.57. It is also customary to use the gauge with the external-control type of circuit since it is more sensitive under these conditions.

A design of gauge utilized by Dushman and Found[141] in their investigation is shown in Fig. 5.58. The cathode and anode were made of two tungsten

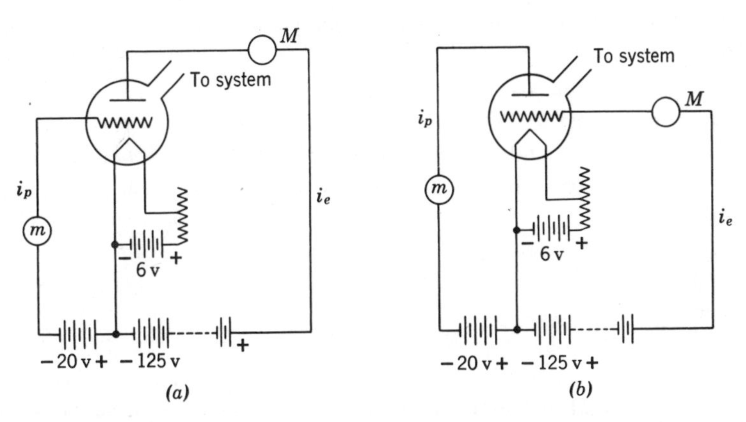


Fig. 5.56. Electrical circuits for ionization gauge. (*a*) Internal-control type. (*b*) External-control type.

filaments, each wound in the form of a double spiral and mounted coaxially on a four-lead stem sealed into the upper end of a glass tube about 4 cm in diameter and 12 cm long. The inner spiral (the cathode) was made of 5 turns of 0.125-mm wire wound on a 0.125-mm mandrel. The outer spiral was made of 3 turns of 0.125-mm wire wound on a 3.65-mm mandrel. The positive-ion collector surrounding the spirals was a molybdenum cylinder about 12 mm in diameter and 12 mm long, supported on a two-lead stem sealed in at the lower end of the tube. As will be observed from the figure, the external-control type of circuit was used, since this provides a longer electrical leakage path between the ion collector and the other electrodes.

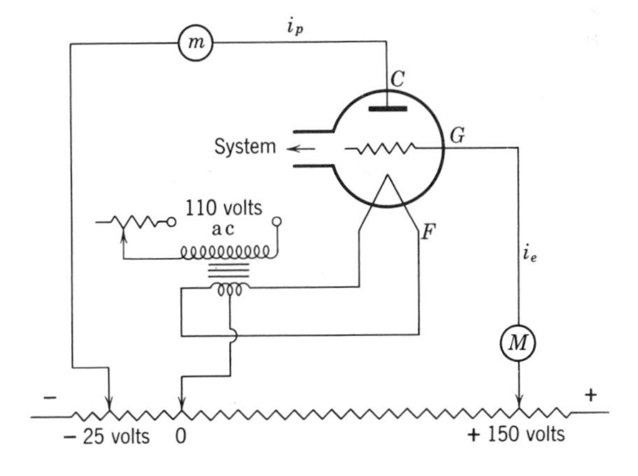


Fig. 5.57. Electrical circuit for ionization gauge involving use of filament transformer *C* denotes the positive-ion collector, *G* the electron collector, and *F* the filament.

The best conditions for operation of the gauge were found to be as follows:

*a.* For *very low pressures* (below about 1 micron): $V_g$ (voltage on electron collector) = 250, $V_c$ (voltage on positive-ion collector) = −20, and $i_e$ = 20 ma. Under these conditions $S = 5 \cdot 10^{-3}$ per micron for argon, which corresponds to 5 microamperes per milliampere per micron.

*b.* For *higher pressures* (1–10 microns): $V_g = 125$, $V_c = -20$, $i_e = 0.5$ ma. In this case, $S = 4.6 \cdot 10^{-3}$ per micron for argon.

Before using the gauge for any measurements it is absolutely necessary to bake out the tube to as high a temperature as practicable (450° C for Pyrex or Nonex), and to eliminate all occluded gases from the electrodes and leads. For the filaments this is accomplished by direct passage of current, and for the collector high-frequency heating is most convenient. To clean the leads supporting the spirals, one spiral is utilized as hot cathode to bombard the other.

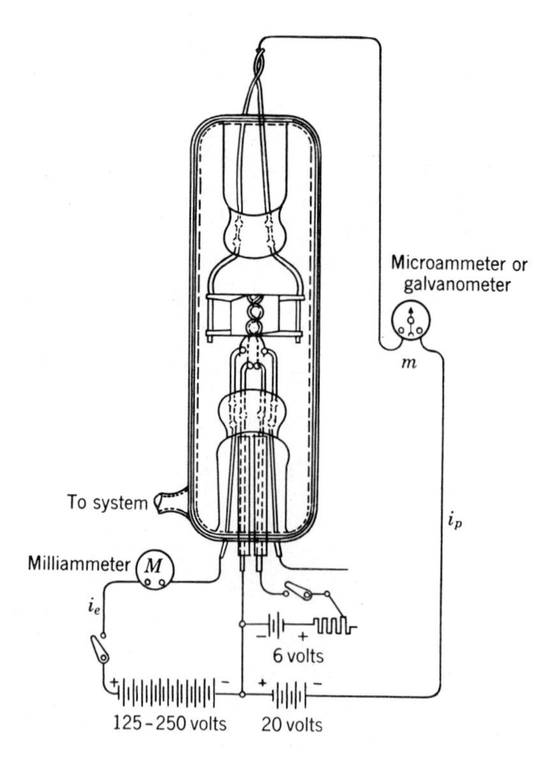


Fig. 5.58. Construction of ionization gauge and electrical circuits (Dushman and Found).

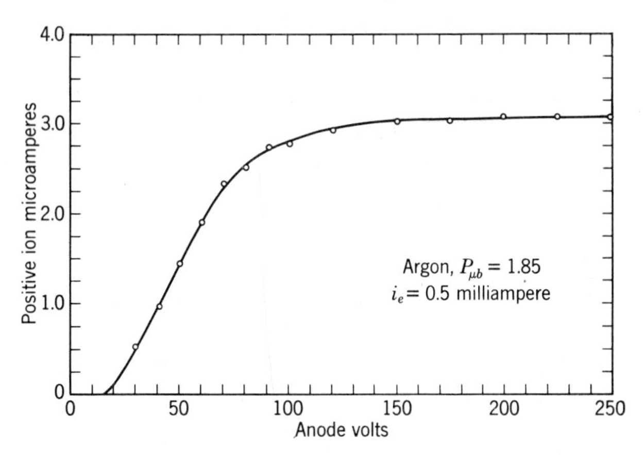


Fig. 5.59. Positive-ion current versus anode voltage at constant electron current and constant pressure, for gauge shown in Fig. 5.58. External-control type of circuit.

Where a grid instead of the outer spiral is the anode, electron bombardment should be used. During the series of measurements of pressure it is advisable to maintain the filament at a temperature approximately that required for the desired emission, and the anode voltage should be applied *only for the relatively short interval* necessary to adjust the cathode current for the emission at which it is desired to measure the ion current. This is necessary wherever there is likely to be voltage clean-up of the gas. (See the discussion in Chapter 9.)

Since the ionization gauge indicates the *total pressure* ("condensable" and "noncondensable" gases), extra care should be taken to eliminate completely, by means of a liquid-air trap, mercury, water-vapor, etc., from the gauge. The positive-ion current, at constant pressure, is a function of the nature of the gas used, of $V_g$, and of the electron current, and it also depends upon whether an external or internal type of control circuit is used. It is therefore of interest to discuss briefly some of the observations made by Found and Dushman on the characteristics of the gauge shown in Fig. 5.58.

With the external-control connection, it was observed that, other conditions remaining constant, a variation in $V_c$ from 0 to $-22$ volt resulted in only a small change in the ion current (about 10 per cent). However, it appeared advisable to use the higher negative voltage in order to prevent electrons from reaching the collector.

Figure 5.59 shows the relation between $i_p$ (positive-ion current in microamperes) and $V_g$ for a constant pressure of 1.85 microbars of argon and $i_e = 0.5$ ma.

The effect of varying $i_e$ at the same pressure and at the value $V_g = 250$ is

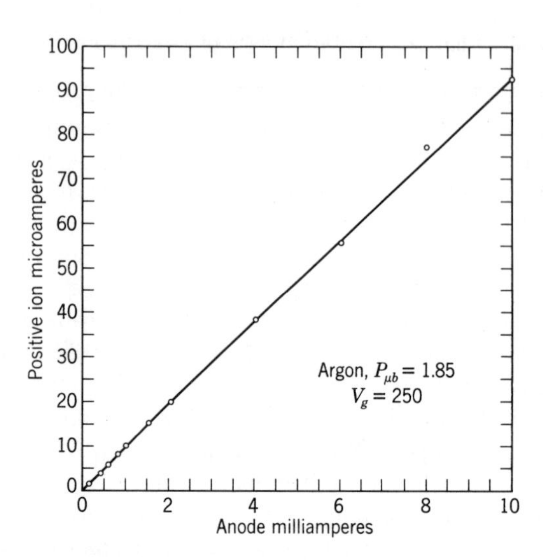


Fig. 5.60. Positive-ion current versus electron current at constant pressure and constant anode voltage, for gauge shown in Fig. 5.58. External-control type of circuit.

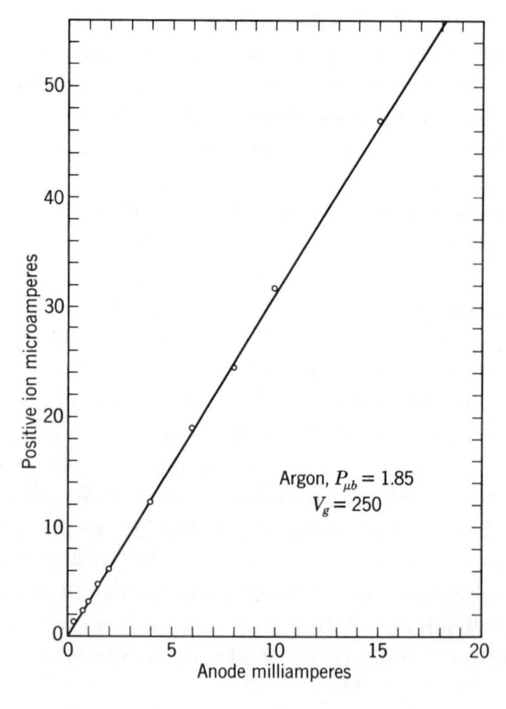


Fig. 5.61. Positive-ion current versus electron current at constant pressure and constant anode voltage, for gauge shown in Fig. 5.58. Internal-control type of circuit.

shown in Fig. 5.60. It will be observed that there was a slight deviation from proportionality at the higher values of $i_e$.

Using the internal-control connection at the same pressure and same value of $V_g$, the plot shown in Fig. 5.61 was obtained. The variation in $i_p$ with $V_g$, at this pressure and $i_e = 10$ ma, is shown in Fig. 5.62.

As will be observed, $S$ is greater for the external- than for the internal-control type of circuit. (This statement is valid for practically all the different types of construction of ionization gauge.) At 1.85 microbars (1.39 microns) argon, $S = 6.85 \cdot 10^{-3}$ per micron for the former, and is about one-third of this value for the latter. On the other hand, the proportionality between $i_p$ and $i_e$ is observed to be valid for higher pressures with the internal control than with the external control. Furthermore, at constant pressure, $i_p/i_e$ is practically constant over the range 125–250 volts on the anode for the internal control, whereas it is sensitive to variations in $V_g$ when the external control is used.

In order to obtain a calibration of the ionization gauge at pressures too low to measure with a McLeod gauge, a method based on Knudsen's laws of flow was used. Between a large bulb $A$ connected to a McLeod gauge and the ionization gauge which was connected to another large bulb $B$, there was inserted a "resistance" made of a long length of capillary tubing the conductance of which, $F$, could be calculated by means of Knudsen's equations. Let $C$ denote the total volume of bulb $B$ and the gauge. Then

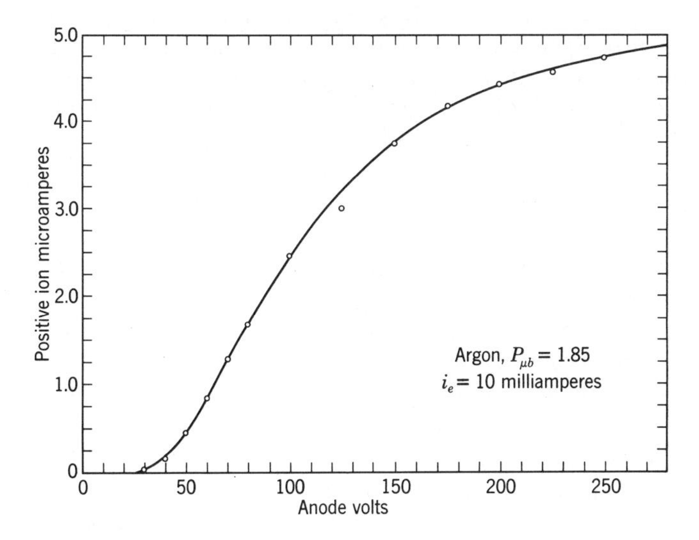


Fig. 5.62. Positive-ion current versus anode volts at constant pressure and constant electron current, for gauge shown in Fig. 5.58. Internal-control type of circuit.

the pressure $P$ in the gauge at any instant is determined by the relation

$$C\frac{dP}{dt} = F(P_A - P), \tag{5.47}$$

where $P_A$ = pressure in bulb $A$. Since, under the conditions chosen, $P$ was negligible compared to $P_A$, the rate of flow could be assumed to be practically constant, and therefore the rate of increase in $i_p$ (at constant value of $i_e$) as measured by the ionization gauge should be constant and proportional to the rate of increase in pressure calculated from the above relation. The observations confirmed this expected proportionality and showed that, for $i_e = 20$ ma and $V_g = 250$, $S = 5.07 \cdot 10^{-3}$ per micron of argon, down to the lowest pressure. (The variation in $S$ with the nature of the gas is discussed later in this section.)

The constancy of $S$ over the range $10^{-3}$ to $10^{-5}$ mm was also confirmed by So.[142]

In experiments with vacuum tubes it was discovered by Barkhausen and Kurz[143] that *high-frequency oscillations* could be maintained in a circuit of the Lecher wire type connected to the grid and plate. The mechanism of generation of these oscillations was investigated by Gill and Morrell.[144] When the grid is operated at a positive potential and the plate at a negative potential with respect to the cathode, electrons may oscillate coherently back and forth through the grid delivering high-frequency energy to the external circuit. Oscillations of this type may occur even in the absence of the Lecher wire system. Electrons oscillating through the grid out of phase with the main space-charge cloud may actually gain enough energy from the high-frequency field to reach the negative plate. When this occurs in an ionization gauge, the ion-collector (plate) current is reduced, and may actually reverse in sign if the number of electrons reaching the ion collector exceeds the number of ions. Trouble of this kind has plagued many workers in this field in the past.

Obviously the occurrence of this phenomenon depends upon the geometrical design of the ionization gauge, and Jaycox and Weinhart[145] have described a construction which is free of this highly undesirable feature. The oscillations are also strongly dependent on the power input to the ionization gauge and may often be avoided by reducing the filament emission and/or the grid voltage. In most of the commercially available ionization gauges today, oscillations seldom occur if the gauges are operated at recommended ratings.

An ionization gauge for the detection of molecular rays has been described by Copley, Phipps, and Glasser.[146] It consists of a 2-mil thoriated-tungsten filament, 1.5 cm long, along the axis of a double helix of 10-mil molybdenum or tungsten, which function as cathode and anode,

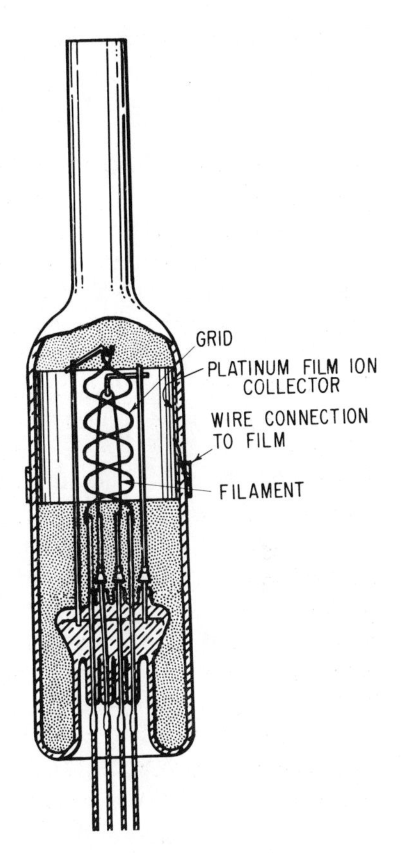


Fig. 5.63. The VG-1A triode ionization gauge. This gauge is outgassed by passing current through the grid and filament and by torching the envelope. The ion collector is a platinum film deposited on the wall of the envelope. (Morse and Bowie.)

respectively, while the collector is a thin layer of metal distilled on the walls of the cylindrical tube by evaporation from the anode (which is heated to a very high temperature for the purpose). A flush seal through the glass forms a connection to this deposit. The two inside electrodes can be degassed very readily by heating them to between 2000 and 2500° K, and the metallic deposit is degassed at the same time by the radiation from the helical electrode. The total volume of the gauge is 3 cm$^3$, and its sensitivity is $9.5 \cdot 10^{-3}$ per micron of argon at all electron current values up to 20 ma.

The construction of a gauge for atomic beam measurements has also been described by Huntoon and Ellett.[147] All electrodes had two leads sealed in, which permitted ready outgassing. According to the authors, with an amplifier and galvanometer in the collector circuit, beam pressure changes of the order of $10^{-8}$ micron could be detected.

A construction which was devised by Morse and Bowie[148] and which is widely used by workers in vacuum technique is illustrated in Fig. 5.63.

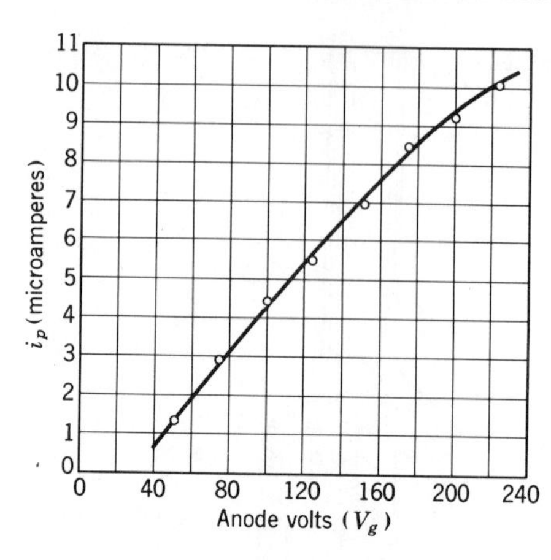


Fig. 5.64. Plot of positive-ion current versus anode voltage at a constant pressure of $7 \cdot 10^{-5}$ mm Hg and a constant electron current of 5 ma, for the gauge shown in Fig. 5.63.

Here, also, the electron collector is a filament, and the positive-ion collector consists of a platinum film deposited on a portion of the inner wall. "Electrical contact," as the authors state, "is made to this extremely thin, slightly translucent film by means of a fine wire entirely embedded in the glass. The external plate contact is in the form of a metallized ring about the center of the bulb." This construction thus introduces an extremely long leakage path between the inner electrodes and the positive-ion collector. Conditions for operation are as follows:

$V_g = 150$, $V_c = -25$, $i_e = 5 \cdot 10^{-3}$ amp, and the filament current (approximately 2.75 amp at 3.50 volts) is adjusted to yield the desired emission. The sensitivity is $S = 20 \cdot 10^{-3}$ per micron for nitrogen. The variation in $i_p$ with $V_g$, for constant values of $i_e$ and $P$, is shown in Fig. 5.64.

The upper pressure limit for most triode ionization gauges is about a micron. At higher pressures the ion current saturates, leveling off at a constant value independent of pressure. This is due to several effects. At high pressures the mean free path of the electrons between collisions with the gas molecules becomes comparable with the path length normally traveled by the electrons in going from the filament to the grid.[149] Under these conditions the energy lost by the electrons in nonionizing inelastic collisions becomes important. The velocity of the electrons is reduced, resulting in lower ionizing efficiencies during the latter part of their flight. The *effective* ionization potential,[150] expressed in electron volts or average energy lost by the electrons per ionizing collision, may well be equal to nearly half their initial kinetic energy of 150 ev.

The low-energy secondary electrons produced by ionizing collisions in a triode ionization gauge are, in most cases, not effective in producing further ionization. However, these electrons will be collected by the grid and measured with electrons emitted by the filament. If the total electron current to the grid is held constant during the operation of the gauge, as is often the case, the apparent gauge sensitivity will start to drop when these secondary electrons become comparable with the number of electrons emitted by the filament. It can be seen, by substituting numerical values in equation 5.42, that for an ionization gauge, such as the VG-1A, with a sensitivity of $S = 20$ per mm Hg, the ion current, and hence the secondary electron current, would just be equal to the electron-emission current at a pressure of $5 \cdot 10^{-2}$ mm Hg, if the gauge could retain its sensitivity at this pressure. Actually, when the pressure becomes greater than a few microns, positive-ion space charge builds up around the cathode and an arc-like discharge is established. Under these conditions a characteristic glow may be observed in the gauge. The ion collector becomes a probe immersed in a plasma and collects the random ion current arriving at the sheath surrounding it. This current is no longer proportional to the pressure and cathode emission current in the usual way.

Another factor that leads to nonlinearity in an ionization gauge at high pressures is a decrease in the ion-collection efficiency. At high pressures the ions may be scattered to other electrodes before reaching the collector. As a result a smaller fraction of the total ions produced are collected by the ion collector.

Schulz and Phelps[151] have designed a high-pressure ionization gauge that takes these factors into account. Their gauge, the Westinghouse WX4145, shown in Fig. 5.65, operates in the approximate pressure range of $10^{-5}$ to 1 mm Hg. It consists of a 5-mil thoriated iridium filament located halfway between two parallel plates of molybdenum $\frac{3}{8} \times \frac{1}{2}$ in. spaced $\frac{1}{8}$ in. apart. The electrons emitted by the filament pass across the 60-mil gap directly to the anode plate without oscillatory motion. This gives the electrons a short path length independent of pressure with little chance of making more than one collision in passing from the filament to the anode. The short path length and the low anode voltage (60 volts) give the gauge a sensitivity of only 0.6 per mm Hg for nitrogen. This limits the secondary electrons produced by ionizing collisions to less than 10 per cent of the electron-emission current, even at pressures of 0.15 mm Hg. To ensure efficient ion collection, the ion-collector electrode is made large compared to the filament, and its potential is adjusted to $-60$ volt to assure parallel plane equipotential surfaces between the electrodes.

This Westinghouse gauge has been used for measuring the pressure of chemically active gases as well as rare gases. Table 5.4, taken from Schulz and Phelps' paper, gives the sensitivity and maximum operating pressure of the gauge for several gases. The data on oxygen were furnished for this book by G. J. Schulz. The low-pressure deviation from linearity is due to

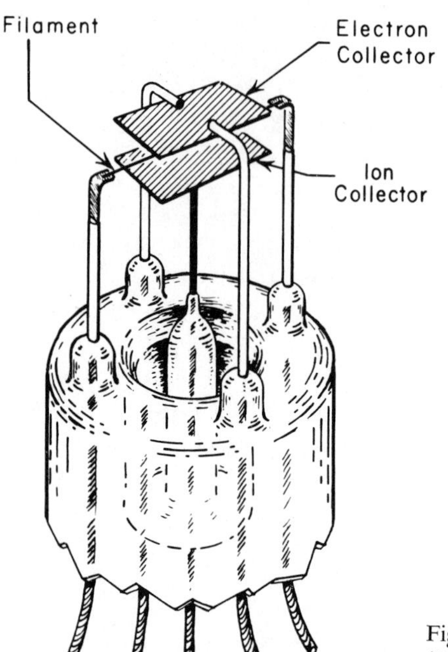


Fig. 5.65. High-pressure ionization gauge (Schulz and Phelps).

X-ray photoemission and occurs at $i_p/i_e = 4 \cdot 10^{-7}$ for a thoroughly outgassed ion-collector surface.

As has been mentioned previously, any standard type of triode radiotron may be used as a gauge, and in the literature there are a considerable

**TABLE 5.4**

**Sensitivity and Range of the WX4145 High-Pressure Ionization Gauge Shown in Fig. 5.65**

($S$ is the sensitivity, and $P_L$ is the pressure at which the gauge departs from linearity by 10 per cent.)

| Gas | $N_2$ | He | $H_2$ | $O_2$ | CO |
|---|---|---|---|---|---|
| $S$ (mm Hg)$^{-1}$ | 0.6 | 0.09 | 0.32 | 0.4 | 0.62 |
| $P_L$ (mm Hg) | 0.6 | 6.0 | 0.6 | 1.0 | 0.6 |

number of references to the applcation of other constructions of gauges than those described above. Many users of ionization gauges have preferred oxide-coated cathodes because of the fact that a sudden inrush of air, as the result of a leak, is not as likely to burn out this type of filament as one of tungsten; or, at least, it is possible to switch off the current supply before any permanent damage has been done to the cathode. One of the greatest disadvantages of the barium oxide-coated cathode type of ion gauge is that the cathode is likely to be poisoned or even destroyed by chemically active gases such as oxygen and hydrocarbon vapors. In many vacuum-technique operations it is necessary to introduce air from time to time in order to seal on a fresh tube. Then the ion gauge has to be re-exhausted, and the gases adsorbed by the electrodes in the presence of air have to be eliminated. It is often necessary under these conditions to reactivate the cathode, which may prove difficult.

A consideration of the filament burn-out problem has led Weinreich and Bleecker[152] to construct ionization gauges with thoria-coated iridium cathodes. Iridium is not readily oxidized, so that a filament made of this metal may be safely exposed to air at operating temperatures for short periods of time. The activation of the thoria apparently presents no problem. The RG-75 ionization gauge has a thoria-coated iridium filament and is made by the Veeco Vacuum Corporation, New Hyde Park, Long Island, N.Y.

Many of the difficulties associated with hot-cathode ionization gauges have been overcome by the *Philips ionization manometer* developed by Penning.[153] In this gauge, electrons are ejected from a *cold cathode* of zirconium, thorium, or other active surface by bombardment with positive ions which have been accelerated by passage through the cathode fall of potential. These secondary electrons are deflected by means of a magnetic field so that they travel in long helical paths before reaching the anode. The total length of path traveled by the electrons in going from the cathode to the anode is many hundreds of times the direct distance between the two electrodes. As a result the ionization produced per electron at any given pressure is considerably greater than would be obtained in the absence of a magnetic field. The magnitude of the total discharge current, which is the sum of the positive-ion current to the cathode and the electron current from the same electrode, is used as a measure of the pressure of gas present.

Figure 5.66 shows the construction of the gauge. The manometer tube $M$ contains a ring-shaped anode $R$ located between the cathode plates $P$. The magnetic field (about 370 oersteds) is applied by means of the permanent magnet $H$. A microammeter (to measure the current) and 1 megohm resistance are connected in series with the manometer across a d-c source of about 2000 volts.

For approximate measurement of the pressure, the microammeter may be replaced by a glow tube provided with a cathode having the form of a metal rod (Philips tube 4662). The length $l$ of the glow along this rod is proportional to the current.

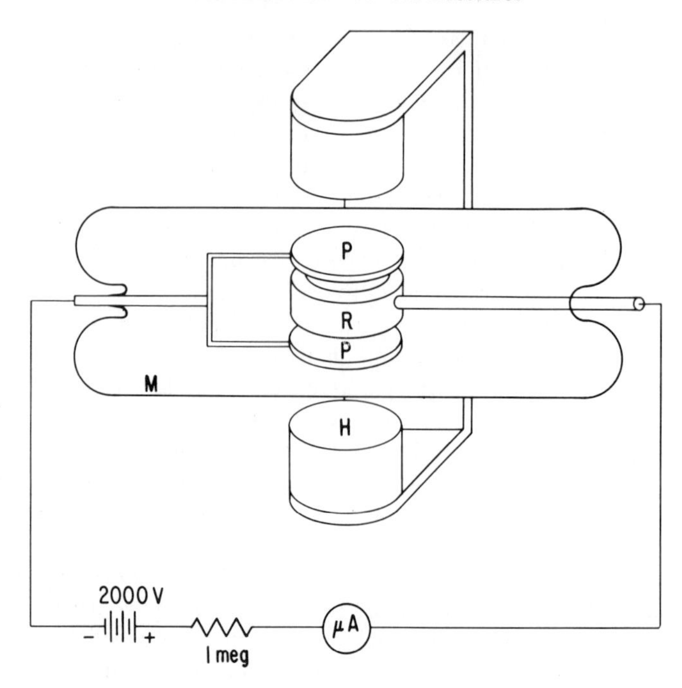


Fig. 5.66. Philips cold-cathode ionization gauge (Penning).

Figure 5.67, taken from Penning's paper, gives a calibration curve for the manometer, which is applicable to hydrogen, carbon monoxide, argon, and air. The Penning gauge does not have the accuracy of the hot-cathode triode gauge. Unstable oscillations generally occur in the glow discharge and cause unpredictable jumps in the calibration curve. These discontinuities, which appear to be present in all gauges of this type, may be as high as 10 per cent. In a later version of the gauge, Penning and Nienhauis[154] have improved the stability by replacing the ring anode with a cylinder.

Penning gauges operate over a pressure range of about $10^{-2}$ to $10^{-5}$ mm Hg. At lower pressures it becomes difficult to initiate the discharge, and in some cases it may fail completely to strike. The pumping speed of these cold-cathode discharge gauges is from 10 to 100 times greater than that of the hot-cathode triode gauge. For this reason they should be provided with a large tubulation with adequate conductance to prevent a pressure drop between the gauge and the vacuum system. Notwithstanding these disadvantages, the Penning gauge has found wide application because of its simplicity and ruggedness.

All-metal versions of the Philips gauge have been designed by the Consolidated Vacuum Corporation and the Veeco Vacuum Corporation.

They are provided with a control cabinet containing a rectifier and filter circuit for supplying the high-voltage direct current. A selector switch makes it possible to read either low or high pressures.[155]

A novel type of cold-cathode gauge was developed by Downing and Mellen[156] in the laboratories of the NRC Equipment Corporation of Newton, Mass., and designated the Alphatron, to indicate that the ionization is produced by alpha particles. Figure 5.68 shows a schematic diagram of the ionization chambers and electrical circuit. A d-c potential of 108 volts is applied between the ionization chambers and the ion collectors. The very small ionization current, produced by the alpha particles from the radium sources, is amplified so that it can be read on a standard type of microammeter. The d-c voltage as well as the power for operating the amplifier is obtained from the regular 60-cycle a-c, 110-volt supply. The radium sources are so constructed that a small quantity of radium in equilibrium with its disintegration products, radon, radium A, radium B, and so forth, radiates alpha particles at a constant rate. Since radon, the first disintegration product of radium, is a gas, the source must be

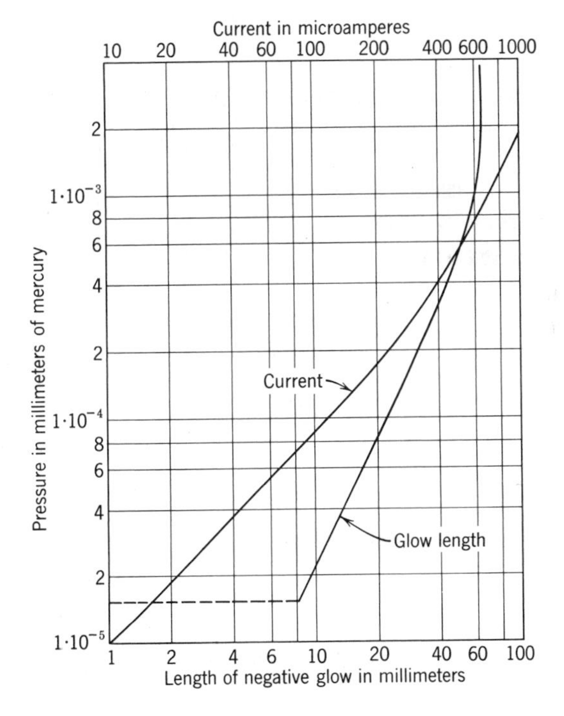


Fig. 5.67. Typical calibration curves of gauge shown in Fig. 5.66. Either the glow length or the discharge current may be used as a measure of the pressure.

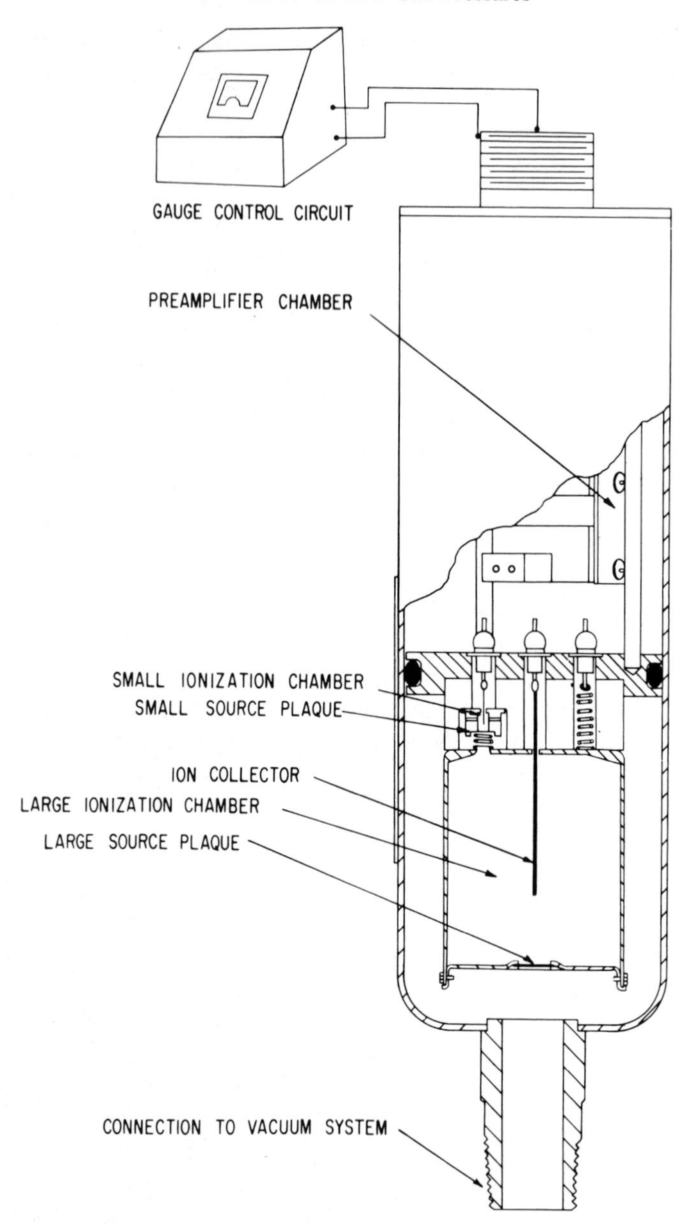


Fig. 5.68. Schematic sketch of the Alphatron ionization gauge, which uses a radioactive source (Dowing and Mellen).

sealed in a plaque to prevent the loss of radon gas so that the succeeding products may be held in the source and their alpha activity utilized.

By using two ionization chambers, the model 530 Alphatron gauge is made to operate with a linear response for air from $10^{-3}$ to $10^{3}$ mm Hg. The large chamber has a volume of 51 cc, and with a radium source of 100 microcuries has a sensitivity for air of about $10^{-10}$ amp per mm Hg. This chamber is used to cover the four low-pressure ranges from $10^{-3}$ to 1 mm Hg. In the three high-pressure ranges from 10 to 1000 mm Hg the small ionization chamber is used in order to keep down ionization current loss due to volume recombination. The small chamber has a volume of 0.2 cc and with a radium source of 1.5 microcuries has a sensitivity for air of about $1.5 \cdot 10^{-13}$ amp per mm Hg.

The sensitivity of the Alphatron depends on the kind of gas being measured. Conversion factors for the more common gases which are to be applied to the output-current meter are given in Fig. 5.69. Because of recombination, the readings for gases heavier than air are not linear in the pressure range of 100–1000 mm Hg.

An obvious advantage of this gauge is that there is no filament to burn out and no possibility of chemical reaction between the gas and the cathode. It will also measure higher pressure than the triode ionization gauges. On the other hand, there are certain precautions (discussed very fully in the operation instructions) which have to be carefully observed in using the gauge, in order to avoid any possible physiological effect arising from the radium emanation.

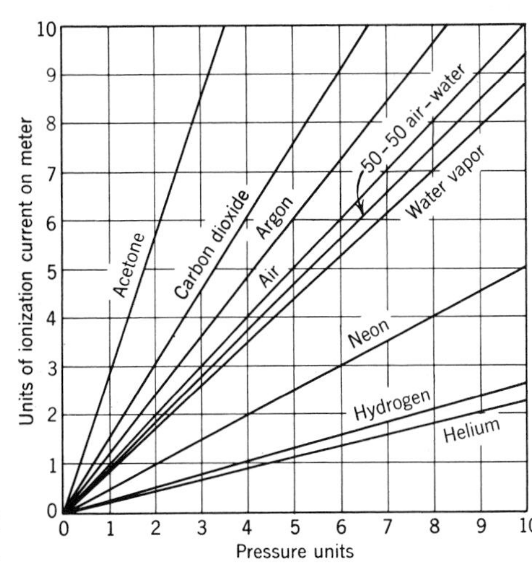


Fig. 5.69. Relative responses of Alphatron gauge to various gases.

Spencer, Boggess,[157] and their colleagues at the University of Michigan have developed a radioactive ionization gauge for making upper-atmosphere pressure measurements from a rocket. Their gauge uses tritium for the radioactive source. Tritium is a high-activity beta source emitting 17-kev particles and has a half-life of 12 years. It has several advantages over radium. The health hazard is reduced, since tritium has no high-intensity gamma rays. A higher sensitivity is obtained with tritium because a high-intensity source may be employed. At normal temperatures tritium is a gas. In order to get it in a form suitable for a source, it is reacted with a 1-micron-thick layer of titanium on stainless steel to form titanium tritide. A disk of this material is placed in one end of the ionization chamber as shown in Fig. 5.70. This source has an effective activity of about 1 curie per square inch.

By using two collector electrodes, the ionization gauge is made to operate with a linear response in air from $10^{-3}$ to 25 mm Hg. To keep down ionization-current losses at high pressures, due to volume recombination of the charged particles, a high-pressure collector electrode, in the form of a large ring, is placed directly adjacent to the source. At low pressures the background current is kept to a minimum by using a low-pressure collector electrode consisting of a small wire shielded from the source.

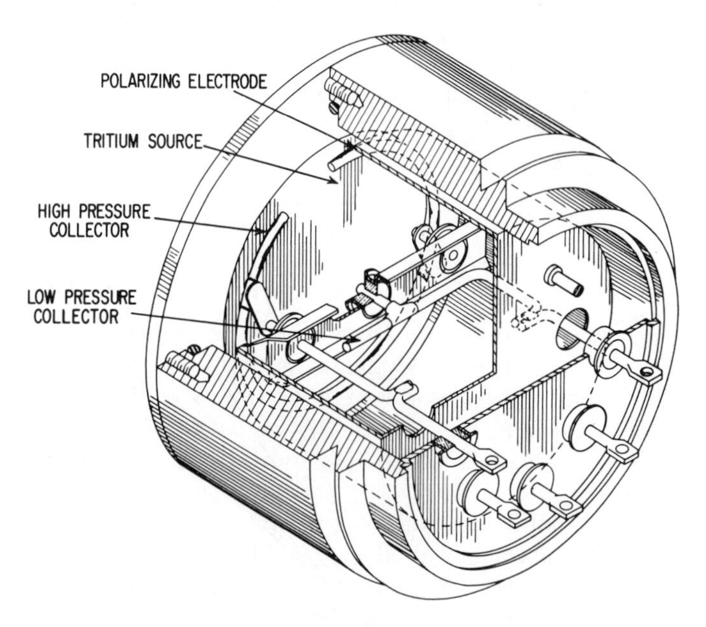


Fig. 5.70. Sketch of a wide-pressure-range ionization gauge using a tridium source (Spencer and Boggess).

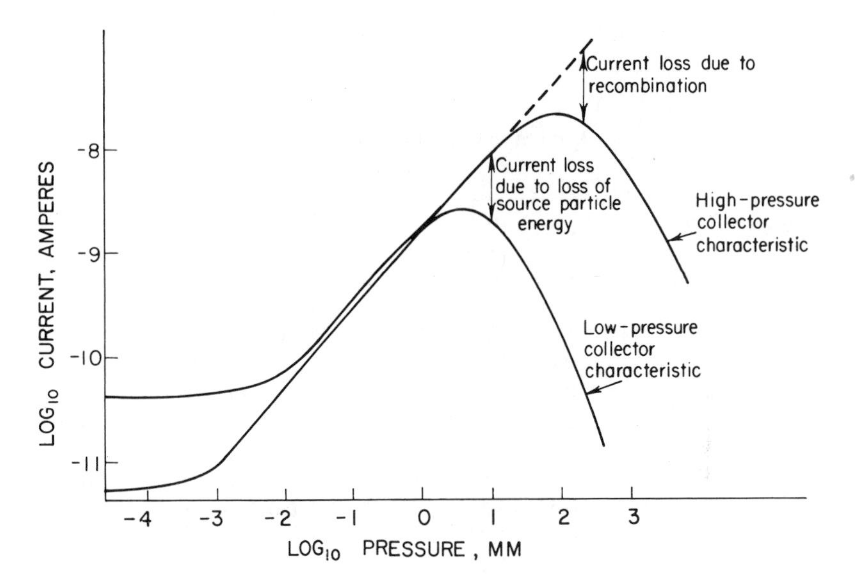


Fig. 5.71. High-pressure and low-pressure ionization-current characteristics for the radioactive ionization gauge shown in Fig. 5.70 (reprinted from *ARS Journal* by permission).

Shielding of the low-pressure collector is necessary to minimize secondary electron emission caused by source-electron bombardment. The collector wire is kept small to reduce photoelectric emission caused by X-rays that are produced by the impact of the source electrons on the polarizing electrode. Because of the relatively low energy of the beta particles, little ionization is produced in the vicinity of the low-pressure collector electrode at high pressures. These characteristics are shown for air in Fig. 5.71.

**Calibration of Hot-Cathode Gauge for Different Gases**

The sensitivity of any given design of gauge depends upon the nature of the gas used. In their early investigation on the calibration of an ionization gauge, Found and Dushman[158] concluded that the sensitivity $S$ varies directly with the number of electrons in the molecule. On the basis of the knowledge of the arrangement and number of electrons in the atoms of different gases such a conclusion seemed quite incomprehensible, and a subsequent investigation was carried out by Reynolds.[159]

Later Dushman and Young[160] repeated this work, with gauges similar in design to some of these used by Reynolds, also including the type VG-1, an earlier form of the VG-1A described above. A summary of the results obtained in both these investigations is given in Tables 5.5, 5.6, and 5.7. In these tables, $S$ signifies, as stated previously, the value of the ratio $i_p/i_e$ per micron.

**TABLE 5.5**

**Calibration of Ionization Gauge**

(Reynolds)

| Gauge or Tube | $10^3 S$ | | | | | $r = S/S_{Ar}$ | | | |
|---|---|---|---|---|---|---|---|---|---|
| | **Measurements at Higher Pressures ($i_e = 0.5$ ma)** | | | | | | | | |
| | He | Ne | Ar | Hg | $N_2$ | He | Ne | Hg | $N_2$ |
| Gauge 1 | | 0.88 | 6.93 | 12 | | | 0.127 | 1.73 | |
| 2 | | 0.75 | 5.87 | | | | 0.128 | | |
| 3 | | | 5.33 | | 3.73 | | | | 0.70 |
| 4 | | 0.67 | 4.80 | | 3.20 | | 0.140 | | 0.67 |
| 5 | | 0.56 | 4.53 | | 2.53 | | 0.124 | | 0.56 |
| 6 | 0.40 | 0.53 | 4.53 | | 2.93 | 0.09 | 0.117 | | 0.65 |
| 7 | 0.40 | 0.27 | 3.73 | | | 0.11 | 0.072 | | 0.61 |
| Average: | | | 5.10 | | | 0.10 | 0.118 | 1.73 | 0.64 |
| FP-62 Tube | 0.61 | 1.07 | 5.60 | | 4.53 | 0.11 | 0.191 | | 0.81 |
| Ditto | | 1.15 | | | 5.33 | | | | |
| | **Measurements at Lower Pressures ($i_e = 20$ ma)** | | | | | | | | |
| Gauge 1 | | 1.27 | 5.33 | 13.3 | | | 0.238 | 2.5 | |
| 3 | | | 5.86 | | 4.40 | | | | 0.75 |
| 4 | | 1.20 | 4.93 | | 4.00 | | 0.246 | | 0.81 |
| 5 | | 1.13 | 5.13 | | 3.33 | | 0.220 | | 0.65 |
| 6 | 0.65 | 1.40 | 5.00 | | 3.67 | 0.13 | 0.280 | | 0.73 |
| 7 | 0.61 | 1.00 | 4.67 | | 3.40 | 0.13 | 0.214 | | 0.73 |
| Average: | | | 5.15 | | | 0.13 | 0.24 | | 0.73 |

Table 5.5 gives the values of $S$ (per micron) obtained by Reynolds for four gases and for mercury vapor, with seven gauges of the design originally described by Found and Dushman (see Fig. 5.58). Data were also obtained for two FP-62 triode tubes. The oval flat plate was used as ion collector, the grid as anode, and the hairpin filament as cathode.

The measurements were carried out under two sets of conditions: (*a*) higher pressures, $i_e = 0.5$ ma, $V_g = 125$; and (*b*) lower pressures, $i_e = 20$ ma, $V_g = 250$.

In order to compare observations on the same gas with different gauges, the values of $r$ (= ratio of $S$ for the gas to $S$ for argon) are also given in the table. As will be observed the values of $r$ for any one gas, obtained with different gauges, do not differ greatly except in the data for neon at 0.5 ma.

The results obtained by Young and Dushman are given in Table 5.6. Comparing them with those given in Table 5.5, the agreement is found to be reasonably fair, except for mercury, for which the more recent data are

**TABLE 5.6**

**Calibration of Ionization Gauge**

(Dushman and Young)

($i_e$ = 1.0 ma)

| Gas | FP-62(1) | | FP-62(2) | | VG-1 | | Average Value | |
|---|---|---|---|---|---|---|---|---|
| | $10^3S$ | $r$ | $10^3S$ | $r$ | $10^3S$ | $r$ | $r$ | $r(P_i)$ |
| He | 0.77 | 0.140 | 0.70 | 0.133 | 2.88 | 0.127 | 0.133 | 0.14 |
| Ne | 1.08 | 0.196 | 1.08 | 0.205 | 4.65 | 0.206 | 0.202 | 0.21 |
| Ar | 5.50 | 1.0 | 5.27 | 1.0 | 22.6 | 1.0 | 1.0 | 1.00 |
| Kr | 8.9 | 1.62 | 8.4 | 1.59 | 32.9 | 1.46 | 1.56 | ... |
| Xe | 12.0 | 2.16 | 13.0 | 2.46 | 50.9 | 2.25 | 2.29 | ... |
| Hg | 15.0 | 2.73 | 14.5 | 2.75 | 72.0 | 3.18 | 2.89 | 1.90 |
| $H_2$ | 2.14 | 0.39 | 2.10 | 0.40 | 8.65 | 0.38 | 0.39 | 0.34 |
| $N_2$ | 4.90 | 0.89 | 4.40 | 0.83 | 18.2 | 0.81 | 0.84 | 0.90 |

to be considered more reliable. Of greatest interest is the fact that the values of $r$ for the different gases seem to be independent, to a large extent, of the design of gauge used. That is, for any one gauge it is only necessary to obtain a calibration with one gas (argon is best), and then, by means of the average values of $r$, the values of $S$ for the other gases may be deduced. Whether this conclusion is valid for all possible variations in design is questionable, since the observations have been made on only two or three designs of gauge which are fairly similar.

In his investigation Reynolds compared his results for the values of $S$ with the results derived by means of a graphical integration based on the probability of ionization $P_i$ of different gases as a function of electron velocity. The observations recorded in the literature on the values of $P_i$ as a function of electron velocity are shown in Fig. 5.55, taken from Reynolds' paper. The different curves were obtained from the following sources: I, mercury—Compton and Van Voorhis,[161] II, mercury—Bleakney;[162] III, mercury—Smith;[163] IV, nitrogen—Compton and Van Voorhis; V, argon—Compton and Van Voorhis; VI, argon—Smith; VII, neon—Compton and Van Voorhis; VIII, neon—Smith; IX, helium—Compton and Van Voorhis; X, helium—Smith. The circles represent results obtained by Bleakney for argon and neon.[164]

Assuming a linear potential distribution between the electrodes, Reynolds calculated from these curves, and the known distances between

**TABLE 5.7**

**Ionization-Gauge Sensitivity Ratios $r$ for Various Gases**

$$r = \frac{\text{Gauge sensitivity for gas}}{\text{Gauge sensitivity for argon}}$$

| Gas | Reynolds | Dushman and Young | Wagener and Johnson | Riddiford | Schulz |
|---|---|---|---|---|---|
| He | 0.10–0.13 | 0.13 | ... | 0.24 | 0.14 |
| Ne | 0.12–0.24 | 0.20 | ... | ... | 0.22 |
| Ar | 1.00 | 1.00 | ... | 1.00 | 1.00 |
| Kr | ... | 1.56 | ... | ... | ... |
| Xe | ... | 2.29 | ... | ... | ... |
| $H_2$ | ... | 0.39 | 0.44 | 0.36 | 0.28 |
| $N_2$ | 0.73–0.81 | 0.84 | 0.84 | 0.94 | 0.67 |
| $O_2$ | ... | ... | 0.71 | 1.07 | ... |
| Hg | 1.73–2.50 | 2.89 | ... | ... | ... |
| Dry air | ... | ... | ... | 0.76 | ... |
| CO | ... | ... | 0.90 | ... | ... |
| $CO_2$ | ... | ... | 1.15 | ... | ... |
| $H_2O$ | ... | ... | 0.75 | ... | ... |
| $SF_6$ | ... | ... | ... | ... | 1.7 |

the electrodes, the total probability of ionization per electron between the ionizing potential and the maximum anode voltage, $V_g$. Multiplying this value by $i_e$ gave the calculated value $i_p$ at 1 mm pressure. He found that $i_p$ (the observed value) was equal to 3.9 times that calculated, a factor which could readily be accounted for on the basis of theoretical considerations.

It will be observed that the values of $P_i$ shown in Fig. 5.55 are in the same order, for the different gases, as the values of $r$ given in Tables 5.5 and 5.6. Dushman found that these values are in rather remarkable agreement with the relative values for $P_i$ observed for 100-volt electrons by Langmuir and Jones.[165] These relative values are given in the last column of Table 5.6.

Table 5.7 gives a summary of the values of $r$ obtained by Reynolds, Dushman, and Young, Wagener and Johnson,[166] Riddiford,[167] and Schulz.[168] Wagener and Johnson's data are averages obtained on three different gauges. One of these gauges had an oxide cathode. The measurements were made in the $10^{-5}$ mm Hg pressure range. The anode voltage used was not given. With the exception of water vapor, the values of $r$ obtained for the three gauges for the same gas were in fair agreement. These original data were all divided by 1.19 to normalize them with Dushman and Young's value of $r$ for nitrogen. Riddiford's data were taken on a type 507 British-American Research Ltd. ionization gauge with 145 volts on the grid. His value of $r$ for helium would appear to be considerably higher than the values of other investigators. Schulz's data were taken on a Bayard-Alpert gauge (WL-5966) with 140 volts on the grid and with an electron-emission current of $10^{-4}$ amp at pressures above $10^{-5}$ mm Hg. On the whole, there appears to be a fair agreement between the values of $r$ obtained by these investigators for the rare gases. For the chemically active gases there is considerably more spread in the data. This is not surprising in view of the complicated reactions that may occur in a hot-cathode ionization gauge for these gases.

For the gases neon, argon, krypton, and xenon a plot of log $r$ (or log $S$ for the same gauge) versus $V_i$ is approximately linear. However, the values for helium, mercury, hydrogen, and nitrogen do not fall on this plot. In the case of hydrogen and nitrogen the values for $V_i$ obtained by Found[169] are 15.1 and 15.8 volts, respectively.

As is evident from Fig. 5.72, the values of $S$ per micron (for a standard gauge) versus $N_e$, the *total number of electrons per molecule*, for the different monatomic and diatomic vapors and gases lie on a fairly satisfactory straight line. Although these observations thus confirm the conclusion deduced some years ago by Found and Dushman, they are still just as difficult to explain.

A comprehensive discussion of the theory of the ionization gauge has been published by Morgulis.[170] He finds that the observations on the values of $P_i$ made by Compton and Van Voorhis[171] may be adequately represented by a relation of the form

$$P_i(V_g) = \alpha(V_g - V_i)\epsilon^{-(V_g - V_i)/\beta}, \tag{5.48}$$

where $P_i$ is the probability of ionization as a function of accelerating voltage $V_g$, and $\alpha$ and $\beta$ are constants, characteristic of each gas, for which the values are deduced from the experimental data. On the assumptions that the electrodes are in the form of concentric cylinders, that the internal-control type of circuit is used, and that the distribution of potential between the electrodes is linear, the relation derived for the gauge constant is of the form

$$S = \frac{BP}{V_g}\left[\beta - (V_g - V_i + \beta)\epsilon^{-(V_g - V_i)/\beta}\right], \tag{5.49}$$

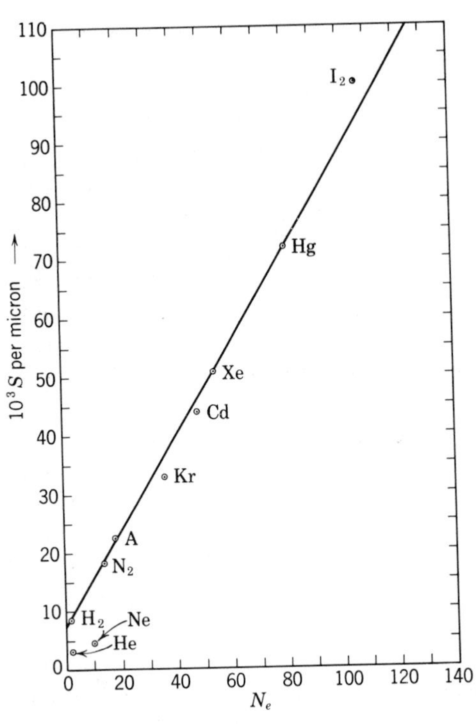


Fig. 5.72. Plot of sensitivity of ionization gauge versus number of electrons in molecule, for different gases and vapors. $10^3S$ per micron = microamperes per milliampere per micron.

where $B$ is a constant for the given design of tube and for any given gas.

For the external-control type of circuit, the relation for $S$ has the same form except that the constant $B$ is different.

A comparison of calculated with observed values of $S$ for a given construction of gauge shows good agreement in the case of the internal-control and less favorable accord for the external-control type of circuit. It is evident that the above equations do not lead to any simple relation between $r$ and $V_i$.

In this connection it is of interest to compare with the above data some observations on the values of $S$ made by Simon[172] with a Telefunken "R.E.11" triode. With the external-control type of circuit, the values of $S$ per micron (observed) and values of $r$ calculated from the values of $S$ were as follows:

| | Ar | $N_2$ | $H_2$ |
|---|---|---|---|
| $10^3S$ | 1.88 | 1.44 | 0.58 |
| $r$ | 1 | 0.77 | 0.31 |

Although obtained with quite a different design of triode, the values of $r$ for nitrogen and hydrogen are in fair agreement with those given in Tables

5.6 and 5.7. The actual values of $S$ obtained by Simon are obviously much lower than those recorded in Table 5.6.

Also deserving of comment are the plots for $P_i$ as a function of anode voltage which have been published by Boburieth[173] in a review of characteristics of ionization gauges. According to these plots, the maximum values of $P_i$ occur for hydrogen, nitrogen, and mercury at about 200 volts, whereas, according to the plots in Fig. 5.55, the maximum values are observed at anode potentials ranging between 75 and 150 volts. Since the author does not state the source of his plots, some question may be raised regarding their reliability. However, the actually observed value of $S$ for argon is stated to be about $62.5 \cdot 10^{-3}$ per micron with an anode potential of 250 volts, whereas the maximum value given in Table 5.6 is $22.6 \cdot 10^{-3}$ for 150 volts on the anode.

The gauge used in this case consisted of a V-shaped tungsten filament inside a cylindrical anode made of tungsten in the form of a spiral with only a few turns, while the collector was a molybdenum box of rectangular cross section outside the anode. Gauges of similar design, tried out in the General Electric Research Laboratory, have not proved to be as sensitive as the VG-1A construction described above.

When the ionization gauge is used to measure the pressure of gases or vapors which can react chemically or electrochemically (as ions) with the cathode or the other metal elements, the interpretation of the positive-ion currents observed becomes quite difficult. In the presence of water vapor, carbon monoxide, carbon dioxide, and halogen gases, a tungsten cathode is attacked very rapidly, even at pressures below $10^{-2}$ micron, and calibration of the gauge becomes almost impossible. In the presence of hydrocarbon vapors tungsten reacts to form carbides ($W_2C$ and WC), thus affecting both the resistance and the electron emissivity. Even at pressures of $10^{-2}$ micron, these reactions occur at a fairly rapid rate and ultimately lead to burn-out of the cathode.

Studies made by Young[174] of the General Electric Research Laboratory show that oxygen reacts with the carbon usually present in an incandescent tungsten filament to produce CO and $CO_2$. Hickmott,[175] also of the General Electric Research Laboratory, has found that atomic hydrogen, produced from molecular hydrogen by a hot tungsten filament, reacts with glass to produce CO, $H_2O$, and $CH_4$. Both these workers have found that these effects may be substantially reduced by substituting a lanthanum hexaboride-coated filament[176] for the tungsten filament normally used in spectrometers and ionization gauges. The reaction of oxygen and hydrogen with a tungsten filament, particularly with regard to filament life and gas clean-up, has been discussed by Leck.[177]

In his investigations on the ultimate pressures obtainable with oil-vapor

pumps, Hickman assumed that, for a given ratio of positive-ion to electro
current, the real vapor pressure relative to that of nitrogen, for the sam
value of $i_e$, is $28.0/M$, where $M$ denotes the molecular weight of the liqui
used. Thus, for the different esters, the positive-ion currents (for constan
values of $i_e$ and anode $V$) would be 11–13 times those observed at the sam
absolute pressure for nitrogen.[178]

Actually, the molecules of the vapor are decomposed by the hot cathod
into hydrogen, carbon monoxide, carbon dioxide, and hydrocarbon com
pounds of lower molecular weight. Obviously these molecules contribut
towards the observed positive-ion current, and the magnitude of thi
additional current must depend upon the rate at which the decompositio
products are removed from the immediate neighborhood of the electrodes
These conclusions are strikingly illustrated by observations made b
Blears,[179] which have been discussed in Chapter 4, in his investigation o
the ultimate pressures obtained in vapor pumps, and also by result
obtained in the General Electric Research Laboratory, which are discusse
below.

In Fig. 5.73, taken from Blear's paper, are given pressure-temperatur
plots for several oils. "For comparison purposes," as Blears states, "a
curve for mercury has been added (the 20° C value for this material wa
measured with the ionization gauge, but the slope is taken from th
standard tables)." All these pressures are expressed in terms of th
calibration of the gauge for nitrogen.

Assuming that these plots correspond to the vapor pressures of th
different liquids in the temperature range obtained by means of the coolin
water, it is possible to conclude that the equivalent nitrogen pressure i
anywhere from 20 to 200 times that of the actual vapor pressure as deduced
from the data in Table 3.6.[180]

In order to investigate this matter at further length a series of observations was made in the General Electric Research Laboratory by H.
Huthsteiner. The arrangement used in this investigation was the following.

A standard gauge, to the lower end of which was connected an appendix
containing a few cubic centimeters of the oil, was immersed in an oil bath
maintained at constant temperature. The upper end of the gauge was
connected through a short length of wide tubing to a large-diameter trap
and this, in turn, to a high-speed oil-vapor pump. In some of the experiments the trap was replaced by a U-tube containing activated charcoal.
Thus any decomposition products or volatile constituents in the oil under
investigation were removed continuously. The filament current and plate
voltage were applied only during the very short interval of time (less than a
minute) required to adjust the circuit for the desired value of $i_e$ and to read
the corresponding value of $i_p$.

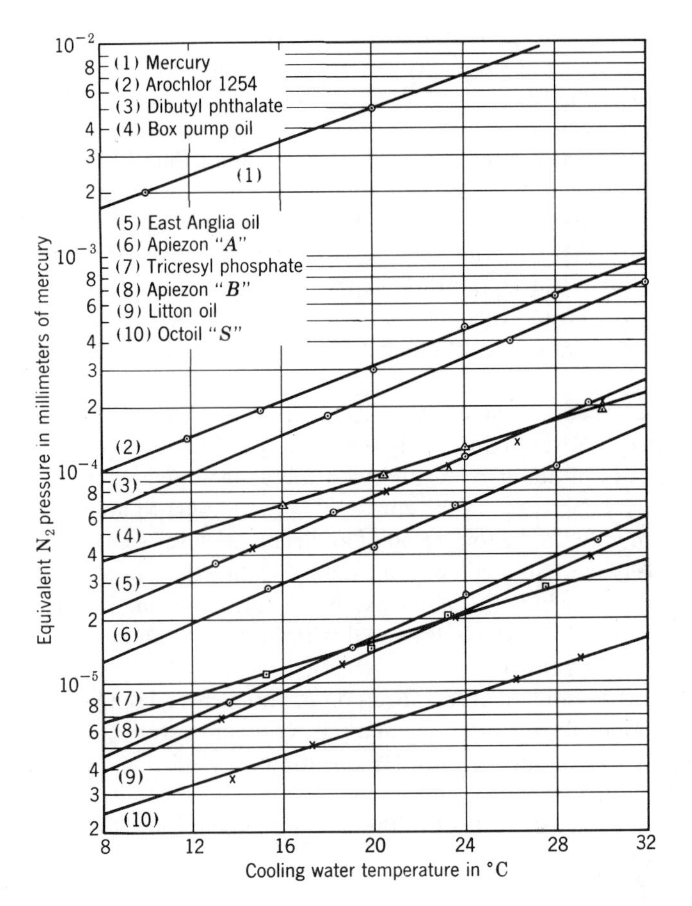


Fig. 5.73. Ultimate pressure of typical fluids as measured with "high-speed" ionization gauge (Blears).

In this manner a series of calibrations of the gauge was obtained for different temperatures of the bath, corresponding to vapor pressures of the oil ranging from about 0.1 to 1.0 micron. The values of the vapor pressures were obtained from the data in Table 3.6.

Before starting a series of measurements, the oil in the appendix was evacuated for an hour or longer at a temperature above the maximum used in the calibration. The voltages on the anode and collector were the same as those employed in the calibration with nitrogen, for which the sensitivity, under these conditions, was 18.3 microamperes per milliampere per micron.

In the case of Octoil-S a plot of log $i_p$ (at constant value of $i_e$) versus $1/T$ was observed to lie parallel to the plot of log $P_\mu$ versus $1/T$ (using the data obtained from Metropolitan-Vickers)[181] which corresponds to a value for $S$ of $91 \cdot 10^{-3}$ per micron. This value was obtained for $i_e = 0.1$ ma. For higher values of $i_e$, the value of $S$ tended to increase, especially at pressures of about 1 micron.

With Amoil-S the initial values of $S$ were around $130 \cdot 10^{-3}$ per micron. However, this value increased with time, even when the oil was maintained at constant temperature. Evidently this oil is adsorbed by the collector and grid, and hence the resulting positive-ion current is much greater than that corresponding to the equilibrium vapor pressure.

In the experiments with butyl sebacate a similar increase was observed from an initial value for $S$ of about $120 \cdot 10^{-3}$ to $300 \cdot 10^{-3}$ per micron and higher.

Some observations made with silicone oil were of special interest. The initial value of $S$ was about $50 \cdot 10^{-3}$ per micron. However, with the oil at a temperature of about 80° C and with the gauge operated continuously the sensitivity decreased in the course of several hours to about $10 \cdot 10^{-3}$, owing, evidently, to the formation of an insulating layer (presumably SiO) on the collector. This could be removed by heating the electrode with high frequency, after which the gauge exhibited normal sensitivity for quite a period (the length of which decreased with increase in the vapor pressure of the oil in the gauge).

On the other hand, in observations on naphthalene, a comparison of the values obtained for the ratio $i_p/i_e$ at a series of temperatures with values of the vapor pressure published by Mrs. Andrews[182] yielded an average value for $S$ of about $15.5 \cdot 10^{-3}$ per micron.

All these observations lead to the obvious conclusion that, in using an ionization gauge to measure vapor pressures of organic liquids and solids, considerable caution is necessary in the interpretation of the observed data.

### Gauges for Ultrahigh Vacua

The conventional ionization gauge used before 1948 was not capable of measuring pressures below about $10^{-8}$ mm Hg. However, Nottingham at M.I.T. and Apker in the General Electric Research Laboratory had observed, in attempting to evacuate electronic devices to extremely low pressures, that thermionic and photoelectric emission characteristics indicated pressures much lower than $10^{-8}$ mm Hg. In 1947 Nottingham[183] presented evidence of a residual current to the ion collector which is completely independent of the pressure. This current is caused by photoelectrons ejected from the ion collector by soft X-rays produced by 150-volt electrons striking the anode. Since the photoelectrons *emitted* from the ion collector cannot be distinguished on the current meter from the

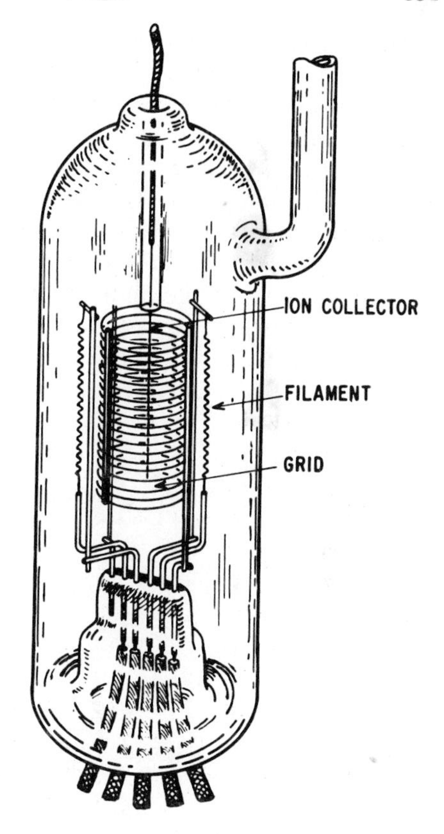


Fig. 5.74. The Bayard-Alpert inverted ionization gauge.

positive ions *incident* on the collector, there must exist a lower limit to the value of $i_p$, with decrease in pressure, which corresponds to pure photoelectric emission. For conventional gauges at normal operating voltages this X-ray photocurrent is about $2 \cdot 10^{-7}$ times the anode current. Thus, for a gauge with a sensitivity of $S = 20$ per mm Hg, the ion current to the collector would be just equal to the photoelectric current at a pressure of $10^{-8}$ mm Hg.

*Bayard-Alpert gauge.* A consideration of these problems led Bayard and Alpert[184] in 1950 to develop the inverted ionization gauge shown in Fig. 5.74. This design is of the external-control type but with the positions of the filament and the ion collector interchanged. The filament is placed outside the cylindrical grid, and the ion collector, consisting of a very fine wire, is suspended within the grid. The usual potentials are applied to the electrodes, +150 volts on the grid and −45 volt on the ion collector. Electrons from the filament are accelerated into the grid cylinder, where they make ionizing collisions. A large fraction of the ions thus formed inside the grid are collected by the center wire. With this arrangement the ion collector intercepts only a small fraction of the X-rays produced at the grid. The small surface area of the collector wire presents a solid angle to the X-rays from the grid that is several hundred times smaller than that for the conventional cylindrical collector.

Evidence for the reduction of the X-ray photoelectric current in the Bayard-Alpert gauge is shown in Fig. 5.75. In these curves the ion-collector current is plotted as a function of the grid voltage for (*a*) a conventional gauge and (*b*) a Bayard-Alpert gauge. The upper curve in Fig. 5.75(*a*) is similar to the curves for gas-ionization probability by electron impact shown in Fig. 5.55. The collector current rises rapidly with grid voltage up to 200 volts, and varies slowly with grid voltages above

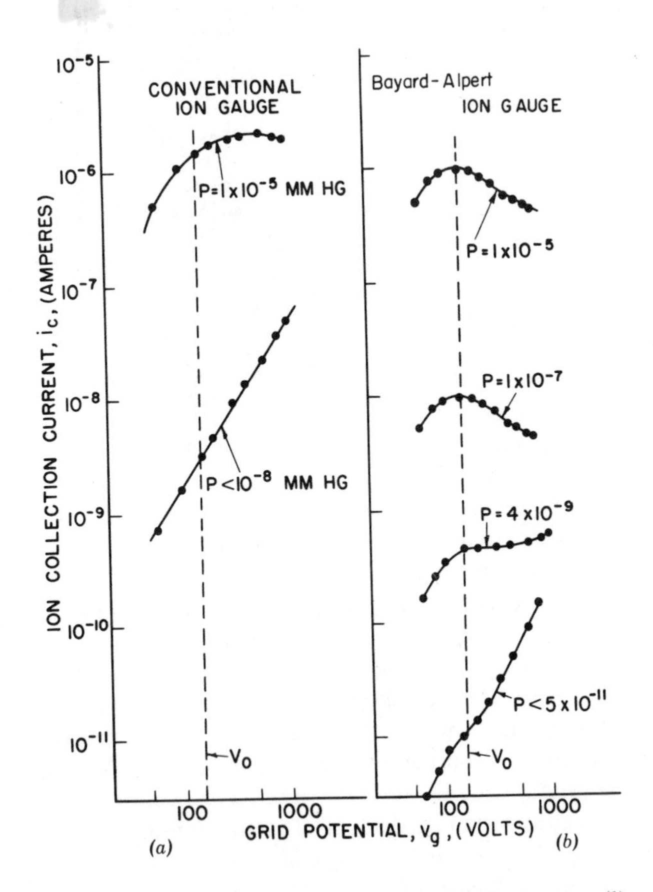


Fig. 5.75. Collector-current characteristics of the Bayard-Alpert gauge (*b*) compared with those of a conventional triode gauge (*a*).

this value. When the pressure is lowered below $10^{-8}$ mm Hg, the shape of the curve is changed radically, as shown by the lower curve in Fig. 5.75(*a*). This residual curve continues to rise rapidly with grid voltage, the slope on a log-log plot falling between 1.5 and 2. At intermediate pressure ranges the measured curve corresponds to a superposition of the "gas-ionization" curve and the "residual" curve. The same characteristics for a Bayard-Alpert gauge are shown in Fig. 5.75(*b*). At higher pressures the curve is similar to that for the conventional gauge, but it is evident that the Bayard-Alpert gauge continues to have typical gas-ionization characteristics at much lower pressures. At $4 \cdot 10^{-9}$ mm Hg the collector current is still predominantly due to gas ionization. Even at pressures below $5 \cdot 10^{-11}$ mm Hg, the lowest curve shows that the gauge has still not reached the point where the X-ray effect is predominant at normal operating voltages. At higher grid voltages the residual effect is clearly predominant.

A unique feature of the Bayard-Alpert gauge is the logarithmic potential distribution in the ionizing region between the cylindrical grid and coaxial ion collector. Since the potential inside the grid is nearly uniform, except in the immediate vicinity of the collector wire, the electrons travel within most of the grid volume with an efficient ionizing energy of nearly 150 electron volts. This is a distinct advantage over the conventional gauge, where the potential fall approximates a linear distribution between the grid and ion collector. In this case the electrons are decelerated to energies inefficient for ionization over a sizable portion of the volume between the grid and the ion collector.

The almost uniform energy of the electrons in the ionizing space of the Bayard-Alpert gauge has allowed Nottingham to identify residual gases by measuring the appearance potential for ionization.

The second filament in the Bayard-Alpert gauge is electrically separated from the first, and not only serves as a spare but also may be used in connection with the flash-filament technique of estimating low gas pressure, developed by Apker.[185]

The glass sleeve around the ion-collector lead prevents X-rays from striking the large-diameter lead-in wire and provides a long leakage path between the collector and the wall of the gauge, which may become charged or electrically conducting from metallic deposits.

Nottingham[186] and Alpert[187] have suggested modifications of the Bayard-Alpert gauge to improve its sensitivity. The cylindrical grid is closed at top and bottom, and a second grid, acting as a screen grid, is installed around all the electrodes. The purpose of closing the cylindrical grid is to prevent ions from escaping to the negatively charged glass wall. At low pressures the ions may oscillate about the collector wire many times before being collected. Since the ions are likely to have velocity components parallel to the collector wire, they may escape through the ends of the grid before being collected unless the grid is closed. The screen grid shields the gauge from the wall charges. Operation of the screen grid at a negative potential causes the electrons to oscillate several times through the positive grid before being collected. This increases their average path length and gives the gauge a higher sensitivity. In some cases the screen grid is made by coating the glass envelope with a conducting film that is connected to an external lead. These improvements have increased the sensitivity by factors of 2 or 3.

At the American Vacuum Society Meeting in 1960 Nottingham reported finding that at normal calibrating pressures the ion collector in the modified Bayard-Alpert gauge did not collect a constant fraction of the ions generated inside the grid. He found that for nitrogen at a pressure of $2.4 \cdot 10^{-3}$ mm Hg the gauge may be in error by as much as a factor of 2 for an emission current of 100 $\mu$a and a factor of 6 at 1 ma. Correct readings are obtained at 10 $\mu$a and less. Nottingham observed similar effects to a smaller degree in the WL-5966 type Bayard-Alpert gauge. As the gas pressure is reduced, the maximum permissible electron current increases inversely with the square root of the pressure. Nottingham has suggested that this effect may be due to positive-ion space charge in the vicinity of the ion-collector wire.

The Bayard-Alpert gauge, in the form of the Westinghouse WL-5966, has a sensitivity of $S = 12$ per mm Hg for nitrogen and has a linear calibration curve over the pressure range of $10^{-9}$ to $10^{-4}$ mm Hg. The helium sensitivity is a factor of 10 lower. Because of its simplicity of construction, ease of outgassing, and dual filaments, this gauge has found wide acceptance and use by workers in all phases of the high-vacuum field.

*Cold-cathode inverted magnetron gauge.* The Penning type of cold-cathode discharge gauge described on p. 315 would appear to have certain advantages over the hot-cathode triode ionization gauges for measurement of ultrahigh vacua. There is no limitation on pressure measurements due to X-ray photoemission, since the number of electrons that strike the anode and produce the X-rays decreases with the pressure. The vapor pressure of a hot tungsten filament also presents no limitations on the low-pressure limit.

However, there are certain inherent disadvantages in the operation of a Penning gauge at pressures below $10^{-5}$ mm Hg. In most gauges the discharge fails to strike below these pressures. The application of higher voltages to start and maintain the discharge leads to field emission from the ion-collector electrode which cannot be distinguished from ion current by the external measuring circuit. This field-emission current establishes a lower pressure limit for the operation of the Penning gauge, analogous to

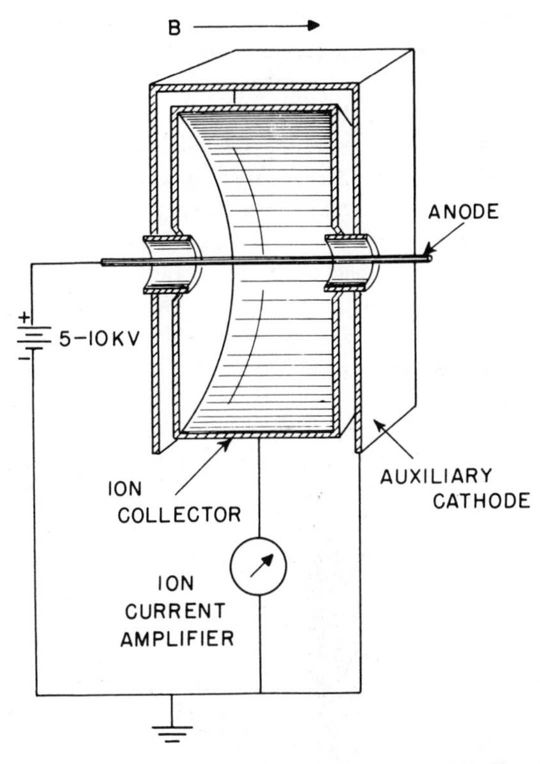


Fig. 5.76. Inverted magnetron gauge (Hobson and Redhead).

the X-ray limit for the Bayard-Alpert gauge. To circumvent these difficulties Hobson and Redhead[188] designed an ionization gauge employing a cold-cathode discharge in crossed electric and magnetic fields. Their gauge, shown in Fig. 5.76, has the structure of an inverted magnetron with an auxiliary cathode. This geometry provides efficient electron trapping in the discharge region, and the auxiliary cathode provides the initial field emission for starting and allows the positive-ion current to be measured independently of the field-emission current. The auxiliary cathode also acts as an electrostatic shield for the ion collector. The two short tubular shields, which project 2 mm into the ion collector from the auxiliary cathode, protect the end plates of the ion collector from the high electric fields and provide the field emission which initiates the discharge.

The gauge operates with an applied magnetic field of 2060 oersteds and a potential of 6 kv on the anode. Under these conditions the relationship between ion-collector current and pressure is given by

$$i_p = cP^n, \tag{5.50}$$

where $P$ is the pressure, $c$ is a constant, and $n$ lies between 1.10 and 1.4 for various gauges. According to Hobson and Redhead, the value of $n$ is essentially independent of the anode voltage and approaches unity as the magnetic field is increased. The dimensions of the gauge have little effect on the value of $n$. In the theory of the Townsend discharge in the inverted magnetron gauge, Redhead[189] has pointed out that this nonlinear relationship between the ion current and the pressure exists because the striking voltage is a slowly varying function of the pressure.

One of the gauges described by Hobson and Redhead had an anode made of a 0.25-mm-diameter tungsten wire. The ion collector was 30 mm in diameter and 15 mm long. The auxiliary shield cylinders were 6 mm in diameter. These parts were made of Nichrome V. Figure 5.77, taken from

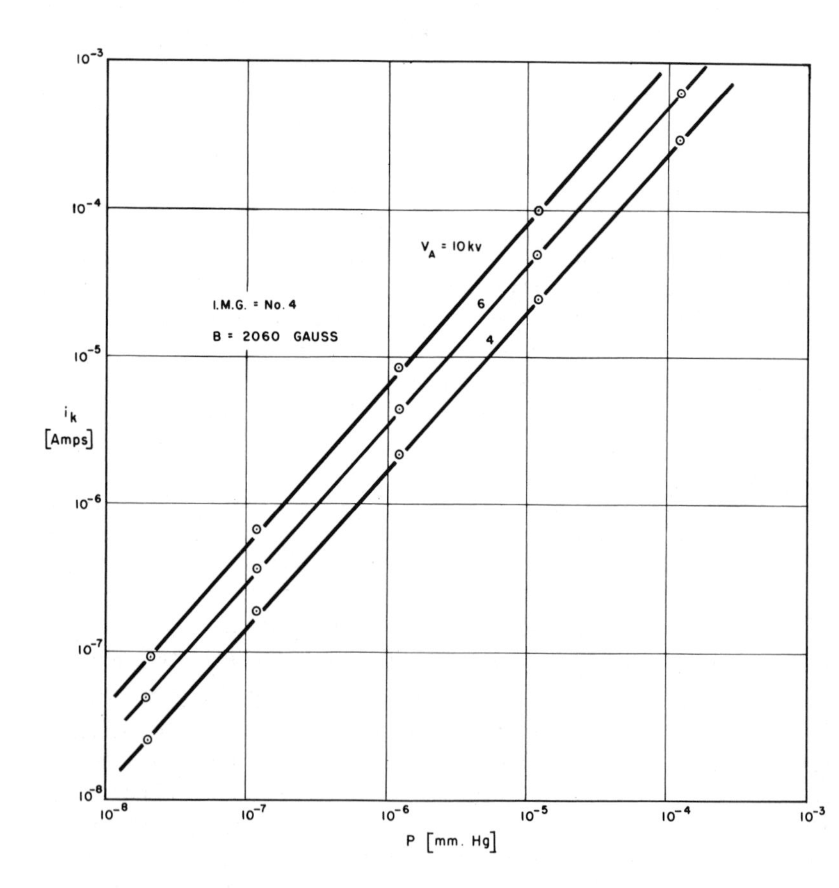


Fig. 5.77. Typical calibration curves for the inverted magnetron gauge. These curves have a slope of $n = 1.10$.

Hobson and Redhead's paper, shows the current-pressure characteristics of this gauge for dry air at several anode voltages.

Unlike the triode gauges, the sensitivity ratio *r* of the inverted magnetron gauge for various gases is a function of pressure. The sensitivity ratio for helium to air decreases with pressure and is also a function of the anode voltage.

Oscillations were observed in the inverted magnetron gauge under most operating conditions. Discontinuous jumps in the mode of oscillation from one frequency to another were reported to give only small discontinuities in the current-pressure characteristics.

The time lag between the application of anode voltage to the gauge and the initiation of the discharge becomes appreciable and variable at pressures below $10^{-8}$ mm Hg. At pressures near $10^{-12}$ mm Hg the time lag was as long as 10 min; however, it is reported that no gauges failed to strike.

Hobson and Redhead have shown the inverted magnetron gauge to be useful in the pressure range $10^{-3}$ to $10^{-12}$ mm Hg. The lower limit has not been determined with certainty and may extend below $10^{-12}$ mm Hg. The sensitivity of this gauge is about 1 amp per mm Hg.

A commercial version of this gauge, type 2205-02, is now manufactured by the NRC Equipment Corporation of Newton, Mass. The commercial gauge has a linear response over a pressure range from $10^{-4}$ to $10^{-12}$ mm Hg with a sensitivity of 4.5 amp per mm Hg for air at 6000 volts with a magnetic field strength of 1000 oersteds.

*Houston ionization gauge.* Houston[190] of the General Electric Research Laboratory has developed a new ultrahigh-vacuum ionization gauge which is useful in the pressure range below $10^{-8}$ mm Hg. The lower limit of this gauge has not yet been ascertained, but in preliminary experiments pressures as low as $2 \cdot 10^{-12}$ mm Hg have been indicated. The gauge itself was utilized as a pump in obtaining these low pressures.

A sketch of this gauge is shown in Fig. 5.78. It is similar to a Penning gauge with the addition of a hot cathode. The end plates *A* and *B* are maintained at negative potentials with respect to the hot filament *D*. The anode cylinder *C* is several hundred volts positive. The device is immersed in a uniform magnetic field *H*. Electrons emitted by the filament are prevented from going directly to the anode by the magnetic field. They travel instead in helical paths back and forth between the negative end plates millions of times before eventually drifting out to the anode. During their long trajectories the electrons have ionizing collisions with the gas molecules, producing positive ions which are collected by the end

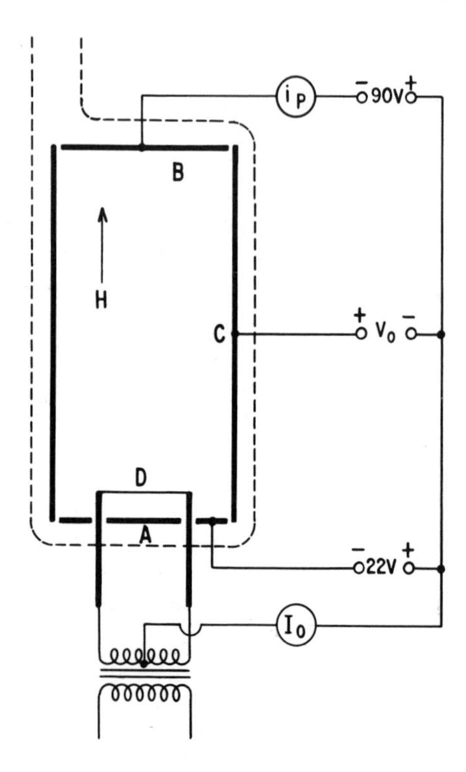


Fig. 5.78. Ultrahigh-vacuum ionization gauge (Houston).

plates. The pressure is determined by measuring the ion current to end plate *B*, which is maintained at −90 volt. End plate *A* is biased to only −22 volt, so that it will collect any high-energy electrons before they gain enough energy to reach end plate *B*. The tube is $3\frac{1}{2}$ in. in diameter and 7 in. long. The 3-mil-diameter tungsten filament is $\frac{11}{16}$ in. from the end plate.

The ion current in this guage was found to be proportional to the pressure for pressures below approximately $10^{-8}$ mm Hg. By varying the filament temperature, the ion-current output at a given pressure was found to be nearly independent of the electron-emission current $I_e$ over a range of $10^{-9}$ to $10^{-5}$ amp. While an increase in filament emission did not appear to increase the total number of electrons traveling back and forth in the central ionizing region of the gauge, it did increase the drift rate of the electrons across the magnetic field. If the emission current was kept below $10^{-7}$ amp, no instabilities were observed. With an applied magnetic field of 1700 oersteds and a potential of 1 kv on the anode the Houston gauge has a sensitivity of 10 amp per mm Hg for an emission current of $10^{-8}$ amp. The Bayard-Alpert gauge has a corresponding sensitivity of 0.1

amp per mm Hg for an electron-emission current of 10 ma. Thus at a given pressure the Houston gauge can generate an ion current that is 100 times larger than the Bayard-Alpert gauge with an electron emission that is 1,000,000 times smaller than that of the Bayard-Alpert gauge. To do this the electrons have to travel about $10^8$ times further in the Houston gauge in a volume that is approximately 100 times larger than the ionizing space in the Bayard-Alpert gauge.

Although X-ray photoelectric emission has not been measured for the Houston gauge, it is estimated that it will establish a lower pressure limit at about $10^{-15}$ mm Hg.

*Hot-cathode magnetron ionization gauge.* In order to extend the low-pressure limit of the conventional hot-cathode ionization gauge it is necessary, at a given emission current, to increase the ratio of the ion current to the X-ray photocurrent. As previously explained, this was accomplished in the Bayard-Alpert gauge by reducing the X-ray photocurrent without substantial loss of gauge sensitivity. This ratio may also be increased by improving the sensitivity of the ionization gauge. It is evident that, if the gauge is modified in such a way that the electrons travel in longer paths before they are collected by the positive grid or anode, the probability of their colliding with and ionizing a gas molecule will be greatly enhanced, and the sensitivity of the gauge will be improved with no increase in X-ray photoemission. Lafferty[191] has applied this principle in developing a hot-cathode magnetron ionization gauge for ultrahigh-vacuum use. In this arrangement, shown in Fig. 5.79, a cylindrical magnetron is operated with a magnetic field greater than cut-off. Two end plates maintained at a negative potential relative to the cathode prevent the escape of electrons. One or both of these plates may be used to collect the positive-ion current generated in the magnetron. Electrons emitted by the tungsten filament spiral around the axial magnetic field in the region between the negative end plates. If the magnetic field is sufficiently high, most of the electrons fail to reach the anode. Some of the electrons make many orbits around the cathode before being collected. The electron density is therefore increased by the presence of the magnetic field, and the probability of ionizing the gas is considerably increased. This is illustrated in Fig. 5.80, where the ion current to one end plate and the electron current to the anode are plotted as a function of magnetic field for a magnetron gauge with an anode $\frac{15}{16}$ in. in diameter and $1\frac{1}{8}$ in. long. The 8-mil hairpin tungsten filament is $\frac{3}{4}$ in. long, separated 40 mils at the base. The filament temperature is adjusted to give an emission current of $10^{-7}$ amp with zero magnetic field. Under these conditions the ion-collector current is nearly $10^{-13}$ amp at a pressure of $10^{-9}$ mm Hg. However, from measurements made in the $10^{-6}$ mm Hg pressure range, the sensitivity of the gauge is known to be 30 per mm Hg. Thus the true ion current at a pressure of $10^{-9}$ mm Hg would be only $3 \cdot 10^{-15}$ amp, as shown by the dotted line in Fig. 5.80. The actual current measured in the ion-collector circuit is 30 times this value and is essentially all photocurrent produced by X-rays from the 300-volt electrons striking the anode. From this it would appear that on the average one photoelectron is emitted by the ion collector for every million electrons striking the anode. As the magnetic field is increased and the electrons begin to miss the anode, their path length is increased, as indicted by a sharp rise in the ion current and a drop in the electron current collected by the anode. At a magnetic field strength of 250 oersteds, the ion current is enhanced 25,000 times over what it would be without the field, and the electron current collected drops to one-fiftieth of its former value. The ratio of ion current to X-ray photocurrent is thus increased $1.25 \cdot 10^6$ times by application of the magnetic field. Since the cut-off current to the anode remains constant

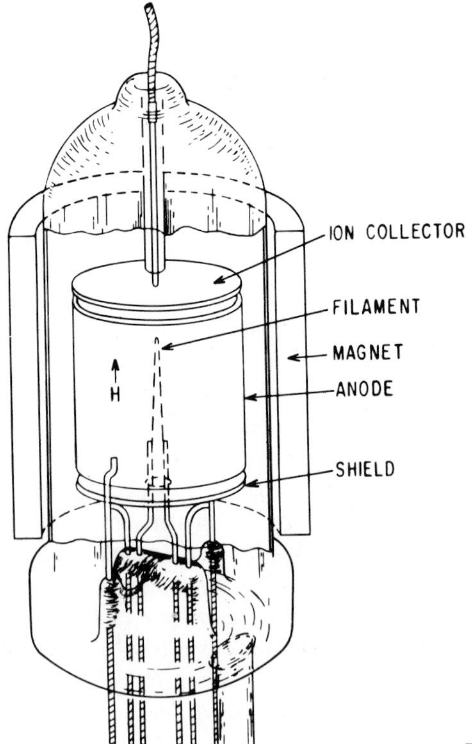


Fig. 5.79. Hot-cathode magnetron ionization gauge (Lafferty).

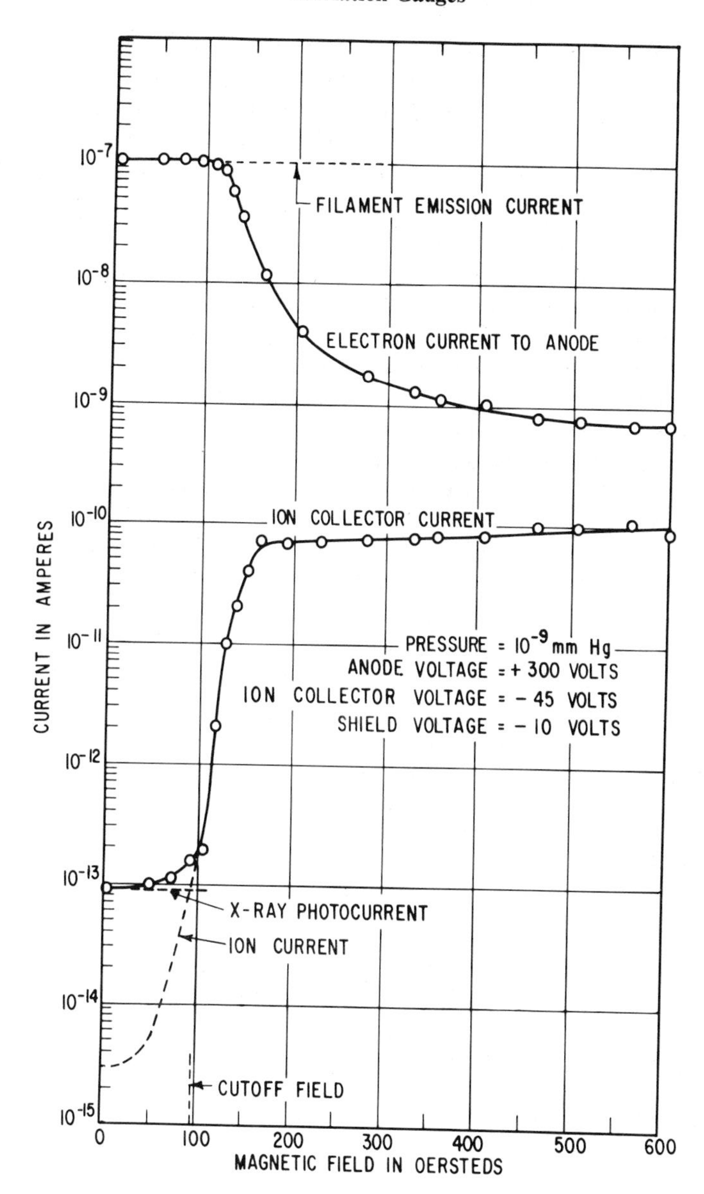


Fig. 5.80. Ion-collector current and anode electron current as a function of magnetic field for the hot-cathode magnetron gauge shown in Fig. 5.79.

independent of pressure at these low pressures, the ion current to the collector would just equal the X-ray photocurrent at a pressure of $2.4 \cdot 10^{-14}$ mm Hg. Thus, pressures at least this low could be measured. In practice the ability to read low pressure is limited by the sensitivity of the external circuit used to measure the ion current.

Hot-cathode magnetron gauges have been shown to have extremely high sensitivities when operated in the milliampere electron-emission range at pressures of the order of $10^{-5}$ mm Hg.[192] However, they have always been subject to excessive pumping action and unstable operation frequently associated with oscillations. Lafferty has found that these conditions may be avoided by operating the magnetron gauge at very low electron-emission levels.

*Omegatron*. While the ionization gauge gives a measure of the total residual pressure of the gases in a vacuum system, there are times when it would be highly desirable to know what these residual gases are. The omegatron, originally developed by Sommer, Thomas, and Hipple[193] for measuring atomic constants, has been used by Alpert and Buritz[194] as a mass spectrometer for measuring the partial pressures of the various gases in an ultrahigh-vacuum system. A simplified version of the omegatron capable of high-temperature outgassing, as developed by Alpert and Buritz, is shown in Fig. 5.81.

The operation of the omegatron is similar to that of the cyclotron. Electrons emitted from the hot filament are accelerated through a $\frac{1}{16}$-in.-diameter aperture to form a 4-$\mu$a beam at 90 volts. This beam, which produces ions in the ionization chamber, emerges through a second aperture and is collected by a collector plate biased positively with respect to the shield to suppress secondary emission. The parallel magnetic field $H$ has a strong collimating action on the electron beam. The shield box forms a 2-cm cube. The r-f plates, which form the top and bottom of the ionization chamber, are connected to a source of r-f voltage (about 1 or 2 volts). These plates produce an r-f field $E$ across the ionization chamber perpendicular to the magnetic field. Ions formed by electrons colliding with the gas molecules within the electron beam are caused to spiral around the magnetic lines of force by their own thermal and dissociation energies. The radii of these spirals are very small, and few ions would normally escape from the electron beam because of their own initial energy. However, the applied weak electric r-f field causes the ions to be accelerated in Archimedes-like spiral orbits of increasing size, provided that the frequency of the applied r-f voltage is the same as the cyclotron frequency of the ions. A trapping voltage that is positive with respect to the r-f plates is applied to the shield box to produce an electric field which retards the loss of ions in the axial direction of the magnetic field. This gives the r-f field an opportunity to act on the ions over a greater number of cycles. The spiral orbits terminate on a $\frac{1}{16}$-in.-square ion-collector plate. The ion current to this electrode, detected by a vibrating-reed electrometer, is a measure of the abundance of the gas with a cyclotron frequency equal to applied frequency.

Alpert and Buritz report that an omegatron of this design has a sensitivity of $S = 10$ per mm Hg, which is comparable to that of an ionization gauge. With a magnetic-field intensity of 2100 oersteds the r-f oscillator must be swept from 3.2 mc to 81 kc to cover a mass range from 1 to 40. There is some discrepancy between the observed and the calculated frequency for a given mass, presumably due to lack of uniformity of the r-f field. However, the instrument has adequate resolution up to at least mass 40.

An analysis of the ion motion in the omegatron has been made by Sommer, Thomas, and Hipple,[193] assuming a uniform sinusoidal varying

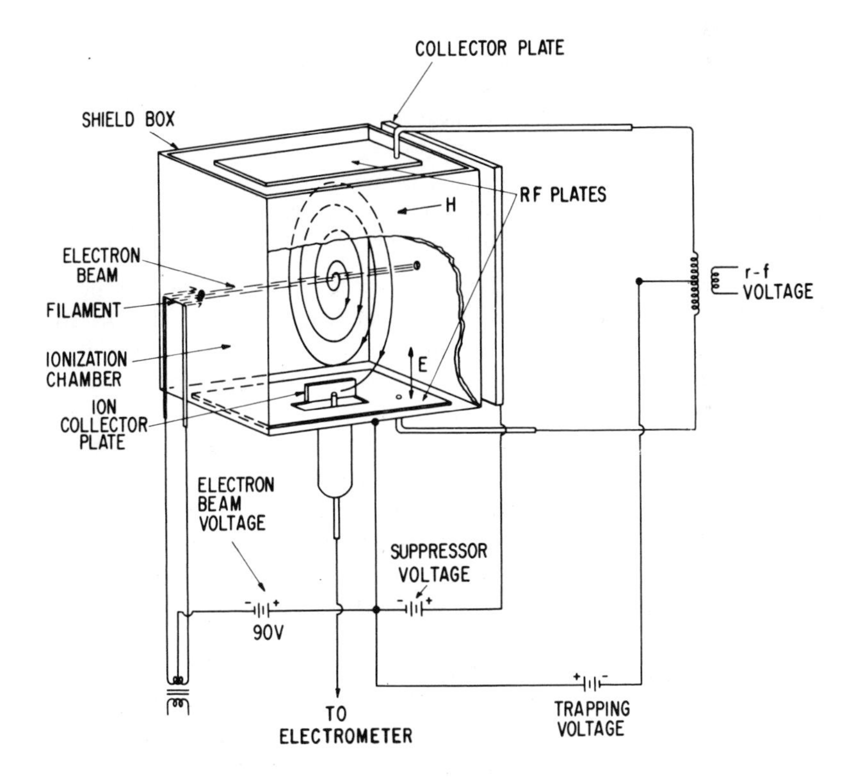


Fig. 5.81. Schematic diagram of a simplified version of the omegatron developed by Alpert and Buritz.

r-f field normal to a constant magnetic field. The results of their study may be summarized as follows:

Cyclotron or resonant frequency
$$f = 1.54 \frac{H}{M} \quad \text{kc} \cdot \text{sec}^{-1},$$

Number of revolutions made by ions before reaching collector at resonance
$$n = 3.06 \cdot 10^{-5} \frac{R_0 H^2}{E_0 M},$$

Time for ions to reach collector at resonance
$$t = 0.02 \frac{R_0 H}{E_0} \quad \mu\text{sec},$$

Length of spiral path for ions at resonance
$$L = 9.6 \cdot 10^{-5} \frac{R_0^2 H^2}{E_0 M} \quad \text{cm},$$

Final energy of resonant ions at collector
$$V = 4.8 \cdot 10^{-5} \frac{R_0^2 H^2}{M} \quad \text{ev},$$

Maximum radius attained by non-resonant ions differing in mass by the amount $\Delta M$ from the resonant ions of mass $M$
$$r_m = \frac{2}{\pi n} \frac{M}{\Delta M} R_0 \quad \text{cm},$$

Resolution (defined as: midpeak frequency ÷ width of resonant peak; theory assumes parallel-sided peaks)
$$\frac{M}{\Delta M} = 4.8 \cdot 10^{-5} \frac{R_0 H^2}{E_0 M},$$

where $E_0$ = peak value of sinusoidal r-f field in volts per centimeter,
$M$ = mass of ion in atomic mass units,
$R_0$ = distance from electron beam to ion collector in centimeters,
$H$ = magnetic-field intensity in oersteds.

If these equations are applied to the Alpert and Buritz omegatron, it is found that, if one assumes $E_0 = 1$ volt $\cdot$ cm$^{-1}$, $R_0 = 1$ cm, and $H = 2100$ oersteds, then for helium the required radiofrequency for resonance is 810 kc. At resonance the helium ions will spiral around 34 times, traveling about 1 meter in 42 $\mu$sec before being collected. The helium ions will gain 53 ev of energy from the r-f field. At the resonance frequency for helium, the molecular ions of hydrogen travel less than 0.4 mm away from the electron beam. The resolution is nearly 53.

The resolution of an omegatron may be improved by operating at low r-f voltages. However, under these conditions the ion path length becomes large, and the device must be operated at low pressures to prevent scattering. Further effects of varying the operating parameters on the performance of the omegatron are discussed by Edwards.[195]

Mass spectrometers of the magnetic deflection type have also been used in measuring the residual gases in ultrahigh-vacuum systems. Reynolds[196] has described a high-sensitivity mass spectrometer for noble-gas analysis. A nine-stage electron multiplier with magnesium-silver dynodes is used to give a gain in the range of $10^3$–$10^6$ electrons per ion. This increase

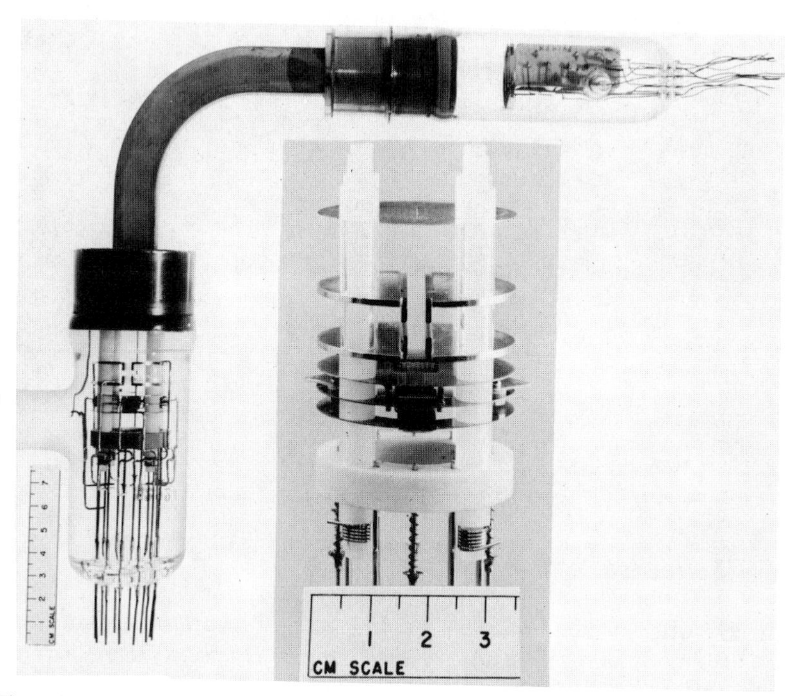


Fig. 5.82. A small, portable mass spectrometer for the study of transient pressure phenomena at ultrahigh vacua (Davis and Vanderslice).

in sensitivity permitted Reynolds to measure partial gas pressures of the order of $10^{-12}$ mm Hg. A total pressure of $5 \cdot 10^{-10}$ mm Hg was achieved during operation of the spectrometer after rigorous bake-outs.

A small portable magnetic-deflection type of mass spectrometer with a secondary emission-electron multiplier has also been developed by Davis and Vanderslice[197] of the General Electric Research Laboratory for studying residual gases at ultrahigh vacua and for making transient pressure studies. This device is shown in Fig. 5.82. Because of its small size, it has the advantage that it can be sealed directly to the tube or system being investigated and is easily transportable. The spectrometer may be baked out at 450° C and ultrahigh vacua obtained when properly processed.

The design consists of a 90° sector, 5-cm radius of curvature magnetic analyzer which will resolve adjacent mass peaks up to about mass 140. A Nier-type ion source is used, and the ion detector is a ten-stage DuMont 6467 electrostatically focused electron multiplier with a gain of one to ten million. The sensitivity of this spectrometer is in the range of 0.02–0.2 per

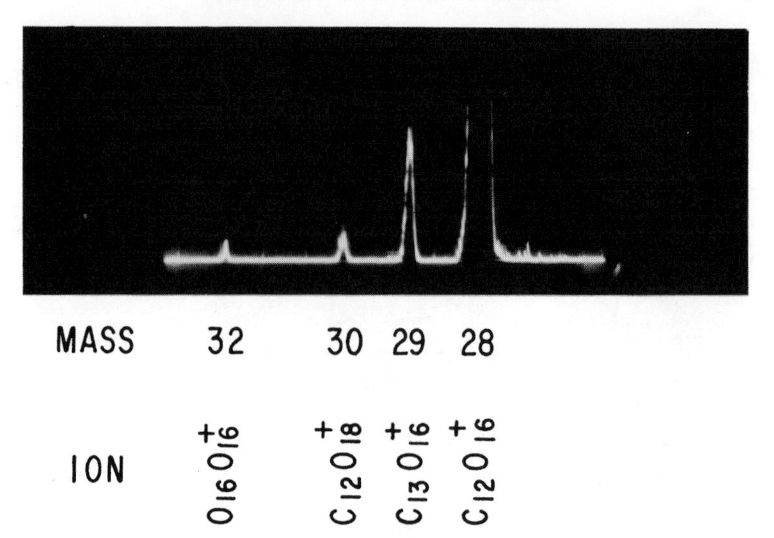


Fig. 5.83. Oscillogram of $CO^+$ isotopes taken at $6 \cdot 10^{-8}$ mm Hg with a sweep speed of 3 msec per unit mass with the spectrometer shown in Fig. 5.82.

mm Hg without the multiplier. Assuming that the lowest current which can be conveniently measured is $10^{-14}$ amp, this spectrometer, without the multiplier, could be used to detect partial pressures of about $10^{-10}$ mm Hg. With the multiplier, the output current is increased by a factor of $10^6$–$10^7$; but, because of the multiplier dark current, a gain in sensitivity of only $10^3$ can be realized at room temperature. Under normal conditions a dark current of $10^{-10}$ amp was obtained. With a multiplier gain of $10^7$ this corresponds to an equivalent ion current of $10^{-17}$ amp, which is equal to the signal current that would be produced by a partial pressure of $10^{-13}$ mm Hg. Cooling the multiplier in liquid nitrogen or counting individual ion pulses reduces the dark current by a factor of 100, making pressures of the order of $10^{-15}$ mm Hg measurable.

An additional advantage of the multiplier is that it raises the signal level to a point where a low output load resistance may be used to give a short response time. This makes it possible to scan the mass range and display the spectrum on an oscilloscope. Several ion species may thus be observed almost simultaneously during a transient pressure or composition phenomenon. The sawtooth voltage which is available on a Tektronix 545 scope and automatically synchronized with the display makes an excellent source of sweep voltage. Figure 5.83 shows an example of an oscilloscope display for the isotope spectrum of $CO^+$ obtained after bake-out at a pressure of $6 \cdot 10^{-8}$ mm Hg. The mass peaks 28, 29, and 30 correspond to ${C_{12}O_{16}}^+$, ${C_{13}O_{16}}^+$, and ${C_{12}O_{18}}^+$, respectively. These peaks occur in the ratio 100 : 1 : 0.2. This spectrum was taken at a sweep speed of 3 msec per unit mass. Sweep rates as high as 1.5 $\mu$sec per unit mass have been used successfully. At the higher sweep rates the familiar well-defined mass peaks degenerate into groupings of pulses because of the collection of individual ions.

## Control Circuits for Ionization Gauge

The ionization-gauge circuits in use today vary from the very simple circuits shown in Figs. 5.56, 5.57, and 5.58 to more elaborate circuits which operate entirely from the 115-volt a-c line. A circuit for a-c operation contains a full-wave rectifier power supply to furnish proper voltages for the ionization-gauge electrodes and the amplifier. A transformer is used to supply a-c heater current to the gauge filament. The ion current is amplified by means of a vacuum-tube amplifier to a level that may be readily measured on a rugged current meter. Bowie[198] has described such a circuit, in which high sensitivity and stability are achieved by means of a balanced amplifier, negative feedback, and a gas-tube voltage regulator.

A similar circuit that has proved very satisfactory in operation for many years in the General Electric Research Laboratory is shown in Fig. 5.84. It is very simple to construct and operate. A description follows.

The circuit shown in Fig. 5.84 may be divided into three parts, (1) the regulated power supply, (2) the balanced bridge d-c amplifier, and (3) the ionization gauge.

The d-c power supply voltage is held nearly constant by means of the voltage regulator tubes $V_2$ and $V_3$. The various voltages required are tapped off from the potentiometer network $R_2$ through $R_7$. The potentiometer $R_6$, $R_7$ supplies a 25-volt negative bias for the ionization gauge ion collector, $R_2$ and $R_3$ provide 150 volts for the grid. The potentiometer $R_4$, $R_5$ across $V_3$ gives 0.2 volt for calibrating the amplifier. A total potential of 255 volts across the two *VR* tubes ($V_2$ and $V_3$) is supplied to the amplifier tube $V_4$. $R_1$ is adjusted so that a total current of 60 ma flows through it.

The amplifier is a bridge-type vacuum-tube voltmeter. This type of circuit has the advantage of being quite stable against variations in plate and filament voltage. Both arms of the bridge are affected nearly the same by these variations and bridge balance is maintained. Further stabilization is obtained by means of negative feedback through the common self-biasing resistor $R_{12}$. Resistors $R_8$ and $R_{10}$ form two arms of the bridge circuit and the triodes form the other two arms. Positive-ion current from the gauge flowing through any one of the resistors $R_{14}$–$R_{17}$ produces a voltage drop which is applied to the grid of the left-hand triode. This produces an unbalancing of the bridge resulting in a current flow through the microammeter $M_3$. This current is indicative of the pressure in $V_5$.

The procedure for operating the circuit is as follows: $S_1$ is turned on and the circuit allowed to warm up for about 3 minutes. With the Variac turned down so that the ionization gauge filament is off, and the selector switch $S_2$ set at tap 1, the bridge is balanced by adjusting $R_9$ until the current through $M_3$ is zero. The amplifier is next calibrated by turning the selector switch to position 2. This

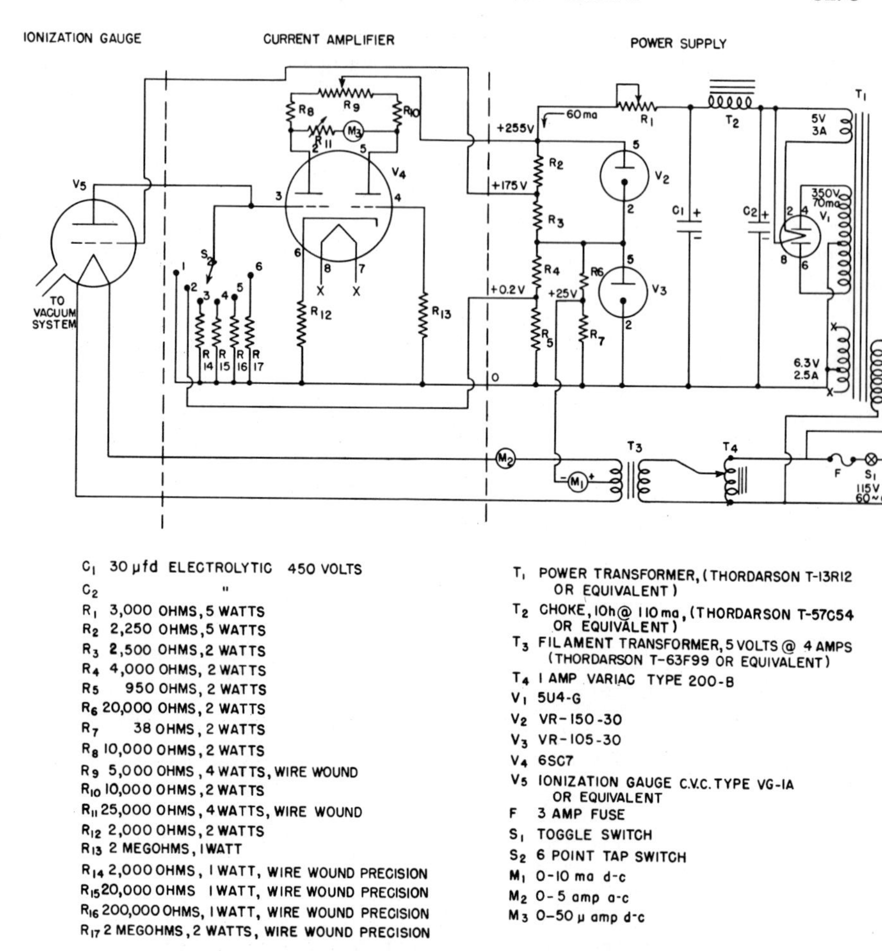


Fig. 5.84. Amplifier and power supply for ionization gauge (Lafferty).

applies 0.2 volt to the amplifier. By varying $R_{11}$, the current through $M_3$ is adjusted to give a full-scale deflection of 50 microamperes. The sensitivity of the amplifier is now such that a full-scale deflection of $M_3$ would be obtained on application of 100, 10, 1, or 0.1 microampere to the amplifier input with the selector switch on positions 3, 4, 5, and 6 respectively. The Variac is now turned up until the ionization gauge filament gives 5 milliamperes of electron emission as measured by $M_1$. The selector switch is turned to the proper tap and the pressure is read on $M_3$. If the ionization gauge has a sensitivity of 20 microamperes per micron, a full-scale deflection of $M_3$ on taps 3, 4, 5, and 6 corresponds to pressures of $5 \cdot 10^{-3}$, $5 \cdot 10^{-4}$, $5 \cdot 10^{-5}$, and $5 \cdot 10^{-6}$ mm of mercury respectively.

When the tap switch is set on the most sensitive position there may be a slight zero shift of $M_3$ due to the flow of positive-ion grid current from the amplifier tube $V_4$ through $R_{17}$. This may be corrected by turning the Variac down and readjusting the zero by means of $R_9$ with the tap switch set at position 6.

By placing a No. 3002 Sola constant-voltage transformer in the 115 volt line, the zero shift and changes in the sensitivity of the amplifier are negligible for a ±20 per cent change in line voltage.

The desirability of maintaining a constant value of electron emission $i_e$ in an ionization gauge for a series of pressure measurements has led to the development of a variety of electrical circuits which have emission-regulation features. These circuits automatically control the power supplied to the ionization-gauge filament and adjust its temperature to provide a constant electron emission independent of moderate variations in its thermionic activity.

An emission-regulating circuit has been described by Nelson and Wing.[199] Both Ridenour and Nelson[200] have introduced modifications in the control circuit which make it possible to use the arrangement either as a gauge or as a leak detector (see discussion in section 5.9).

Commercial ionization-gauge circuits are now available which not only have the emission-regulating feature but also provide power for outgassing the ionization-gauge electrodes, and during operation automatically turn off the filament when the pressure becomes too high. Such circuits are made by the Consolidated Vacuum Corporation and the Veeco Vacuum Corporation.

A radical type of control and indicating circuit has been described by Ridenour,[201] in which the grid control circuit is the same as that proposed previously by Ridenour and Lampson. The circuit is stated to have the following features:

(1) No meters whatever are required; the plate current of the gauge tube is indicated by a 6E5 tube,[202] both during normal operation and during the outgassing operation. (2) The grid current in the gauge is held constant at any desired value, regardless of changes in line voltage, gas pressure, or the nature of gas in the vacuum system. (3) Outgassing of the grid of the gauge tube may be done by throwing one switch. (4) The grid current is controlled automatically during outgassing. (5) Any type of tube may be used as gauge tube proper, without changes in the control and indicating circuit.

A modified construction of gauge, which is very interesting, is described by C. G. Montgomery and D. D. Montgomery.[203] They state,

Instead of using a single grid and accelerating the whole emission of the filament, it is possible to employ a gauge with two grids, and accelerate only those electrons which have passed the first grid. The number of such electrons can be regulated by the potential of the first grid, and is then largely independent of the emission of the filament and the filament current. It is then possible to control the potential of the first grid in such a way as to make this independence practically complete.

A standard type "47" vacuum tube is used for the gauge, and the sensitivity obtained has a value $S = 3.33 \cdot 10^{-2}$ per micron.

## 5.8 GENERAL REMARKS ON GAUGES

In the different problems encountered in high-vacuum technique the question usually arises as to which one or two of the assortment of gauges described in the previous sections would be most satisfactory. Even if some of the designs described are obviously either impractical or very limited in their application, there still remains a considerable variety from which a choice may be made. One of the most important factors that must be considered is the range of pressures to be measured. Another factor is the nature of the gas or gases present in the system. Is it necessary to eliminate condensable vapors from the gauge? How does the sensitivity of the gauge depend upon the nature of the gases or vapors?

Figure 5.85 shows the approximate range of pressures of the gauges described in this chapter. The full line indicates the usual range, and the dashed portions indicate sensitivities which have been, or may be, attained by specially designed constructions.

The two bottom rows give values of $n/(3.24)$, where $n$ denotes the number of molecules per cubic centimeter at 25° C, and values of $L$, the mean free

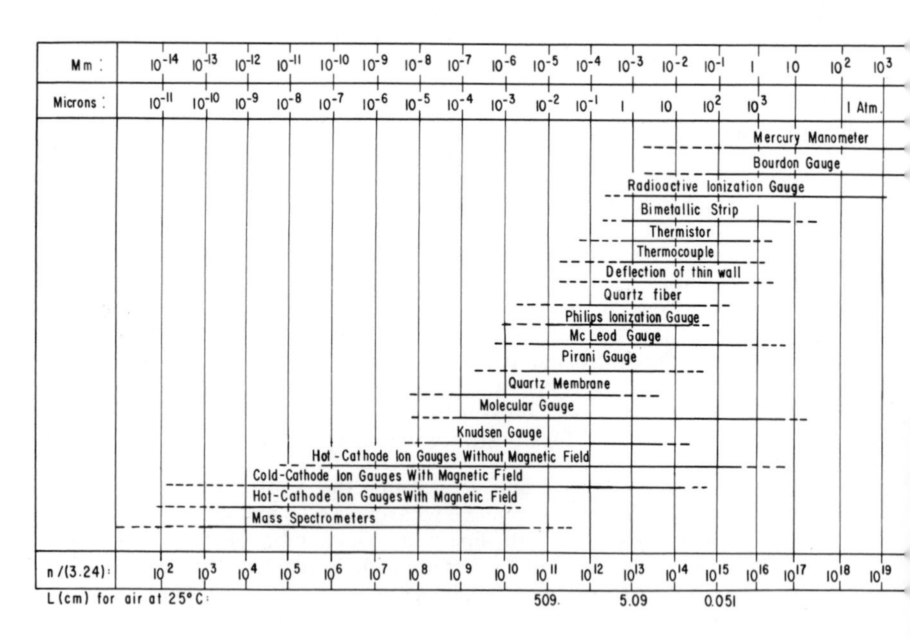


Fig. 5.85. Chart showing ranges of pressures for different types of gauges.

path (in centimeters) for air at 25° C. By means of this chart it should be possible to choose two or more types of gauges that meet the requirements of any special conditions in vacuum technical operations.

The advantages and disadvantages of the principal types of gauges for use with oil-vapor pumps have been stated in a paper by DuMond and Pickels.[204] Their discussion, which is limited to the McLeod, Pirani, ionization, and Knudsen types of gauges, is summarized in the following remarks.

The *McLeod gauge* has the following disadvantages. (1) It does not give continuous readings of the pressure. (2) It is insensitive to condensable vapors. (3) In order to eliminate mercury from the rest of the system it is necessary to insert a liquid-air trap. "Such traps," it is stated, "are never completely effective over long periods. (4) The readings are never very dependable below $10^{-4}$ or $10^{-5}$ mm Hg pressure."

The *ionization gauge* has a very wide range of pressures but possesses the following disadvantages. (1) It requires considerable auxiliary electrical equipment. (2) Its sensitivity varies for different gases and vapors. (3) The filament emission may be "poisoned" in the presence of certain gases. (4) The filament is susceptible to burn-out if exposed to air while hot. (5) "Ion bombardment destroys the filament more or less rapidly with continued use." (6) The electrodes must be outgassed very thoroughly before the gauge can be used to give reliable indications, especially at extremely low pressures. (7) The hot filament decomposes certain gases (especially hydrocarbons).

To these disadvantages may be added those due to chemical and electrical "clean-up" effects. As will be discussed more fully in Chapter 9, the gradual decrease in pressure in an electrical discharge tube is a well-known phenomenon, which is observed with cold as well as with hot cathodes. Such clean-up occurs with all gases and obviously must be taken into account in the interpretation of observations made by means of an ionization gauge. In fact, Schwarz[205] has made the following statement: "Everyone who measures pressures under $10^{-4}$ mm with the ionization gauge must regard his measurements with more or less skepticism."

However, in view of the experience of Dushman, the situation is not quite as bad as this statement would imply. If the plate voltage is applied only during the *very short interval* necessary to observe both the electron and positive-ion currents, then the amount of clean-up is reduced to a minimum. That is, *the gauge should not be operated continuously* when it is desired to determine the pressure in a sealed-off system.

There is little point in having an ionization gauge operate with great precision unless one knows accurately what gas is being measured. As shown in Table 5.7, the sensitivity of the ionization gauge varies by as much as 10 to 1 for the different gases. As low-pressure measurements have taken on more sophistication, there has been a trend towards the use of partial pressure gauges (mass spectrometers). In the ultrahigh-vacuum range the low pressure obtained may well be limited by gases generated or liberated by reactions involving the hot cathode of the gauge or mass spectrometer. Liberation of positive ions from electrode surfaces by electron bombardment is also a source of apparent residual pressure in ionization gauges.

As DuMond and Pickels point out, the *Pirani* or *hot-wire gauge* has two main disadvantages. (1) "It is very difficult with it to maintain a stable 'zero' point of pressure and it cannot therefore be used with confidence to measure pressures much below $10^{-4}$ mm Hg." (2) As the pressure is decreased below about $10^{-3}$ mm Hg, the loss of heat by conduction becomes negligibly small as compared with the radiation loss. It is stated:

> No modification in the geometrical design of the gauge can improve the ratio of energy loss by gas conduction to energy loss by radiation since at these low pressures the molecules of gas travel in straight lines between the wire and the walls just as the photons of radiation do.
>
> Now this is the chief reason for the superiority of the Knudsen gauge. At these pressures the energy transported by molecules of gas is small compared to the energy transported by photons of radiation, but the *momentum* transported by molecules is huge compared to the momentum transported by photons on account of the comparatively enormous mass of the molecules. The Pirani gauge measures the molecular *energy* transport while the Knudsen gauge measures the momentum transport, hence its superiority.

The main advantages of the *Knudsen* type are the following. (1) It requires no auxiliary electrical circuits, such as are used with the ionization gauge. (2) The deflections of the mirror are proportional to the pressure and correspond to absolute pressures, independently of the nature of the gas.

On the other hand, the use of a suspension involves difficulties similar to those met with in any sensitive galvanometer, especially when dealing with extremely low pressures. Such difficulties will arise whenever the gauge is subjected to external causes of vibration. Moreover, in a gauge such as that designed by DuMond and Pickels it is very difficult to outgas the parts, and any wax or grease joints are likely to be sources of constant gas evolution.

The fact is that each of the different types of gauges described in the previous sections possesses both advantages and disadvantages. The experimenter must decide upon the objective of any set of measurements and then choose that gauge or those gauges which will give him the desired information with the greatest convenience and to the desired degree of accuracy. Many times information of order of magnitude of the pressure is sufficient; at other times a continuous record of variation in pressure may be desired, without any reference to the nature of the gases present.

For instance, in the evacuation of electronic devices the factor of significance is the ratio $i_p/i_e$, irrespective of the nature of the ions. An ionization gauge gives this information directly, and, though the necessity for using auxiliary electrical circuits may appear a disadvantage, the actual fact is that such gauges have found extensive application in factory practice as well as in the laboratory.

The same statement applies to a Pirani gauge. In industrial applications of vacuum technique, as well as in purely scientific investigations, the hot-wire gauge is widely used, along with the thermocouple and McLeod gauges.

## 5.9. LEAK DETECTORS

To the experimenter interested in using vacuum systems, one of the greatest problems is leaks. Leaks may occur from a variety of causes, and the increasing complexity of tube design has multiplied the number of these causes to a great extent.

The following conversion factors will be found useful because of the wide variety of units employed in the literature to express leak rates.

| micron · liters · sec$^{-1}$ | micron · liters · hr$^{-1}$ | atm · cc · sec$^{-1}$ | atm · cc · hr$^{-1}$ | micron · cu ft · sec$^{-1}$ | micron · cu ft · hr$^{-1}$ |
|---|---|---|---|---|---|
| 1 | $3.60 \cdot 10^3$ | $1.32 \cdot 10^{-3}$ | 4.75 | $3.53 \cdot 10^{-2}$ | $1.27 \cdot 10^2$ |
| $2.78 \cdot 10^{-4}$ | 1 | $3.67 \cdot 10^{-7}$ | $1.32 \cdot 10^{-3}$ | $9.81 \cdot 10^{-6}$ | $3.53 \cdot 10^{-2}$ |
| $7.60 \cdot 10^2$ | $2.73 \cdot 10^6$ | 1 | $3.61 \cdot 10^3$ | 26.8 | $9.65 \cdot 10^4$ |
| $2.11 \cdot 10^{-1}$ | $7.60 \cdot 10^2$ | $2.78 \cdot 10^{-4}$ | 1 | $7.43 \cdot 10^{-3}$ | 26.8 |
| 28.3 | $1.02 \cdot 10^5$ | $3.74 \cdot 10^{-2}$ | $1.35 \cdot 10^2$ | 1 | $3.60 \cdot 10^3$ |
| $7.87 \cdot 10^{-3}$ | 28.3 | $1.04 \cdot 10^{-5}$ | $3.74 \cdot 10^{-2}$ | $2.78 \cdot 10^{-4}$ | 1 |

The time-honored method of detecting leaks in glass systems is by means of a Tesla spark-coil; a leak at any point is indicated by a characteristic pink color of the discharge, due to the presence of nitrogen in the system. (Occasionally, if the spark is too intense, the spark-coil itself will produce a fine hole!) If ether, carbon tetrachloride, or acetone is poured on the suspected spot, the presence of a leak will be indicated by the characteristic color of the glow due to the presence in the system of the vapors of these compounds.

The spark-coil test is, of course, useless for leaks in metal parts. Furthermore, this test is not very sensitive. In order to maintain a discharge the pressure in the system must be of the order of a few microns, and this pressure implies, even with a fairly moderate speed of exhaust, a high rate of leak of air into the system.

For any exhaust system operating under optimum conditions, there will be established a stationary pressure, $P_s$, which corresponds to equilibrium between rate of leak of gas into the system and rate of exhaust. This leakage of gas in the system may also arise, in "tight" systems, from evolution of gas at some part of the system which has not been baked out well, or it may be due to back diffusion from the pump.

The pressure is measured by some form of low-pressure gauge attached to the system, and any abnormal leak will be observed by the gauge as an increase in pressure above the value $P_s$. Under these conditions the location of the leak may be found by isolating parts of the system, if possible, or by coating suspected places with a suitable wax. Kuper[206] has observed that the best location for the gauge is immediately above the fore pump, and he has used for this purpose an ordinary 40-watt light bulb, as a Pirani gauge, with a duplicate bulb sealed off at 1 micron as compensator. The bridge circuit is fed from the a-c mains, and an audioamplifier and loud speaker are used to indicate gas evolution or leakage.

Since the Pirani gauge is most sensitive for hydrogen, a method used in many laboratories is to insert such a gauge in the position suggested by Kuper and pass a jet, through which a fine stream of hydrogen is flowing, over suspected points on the vacuum system. The fact that hydrogen diffuses through a hole more rapidly than air, combined with the much greater thermal conductivity of hydrogen, causes a marked change in the indication of the gauge when the jet strikes a leak. Obviously, a thermocouple or thermistor gauge may be used instead of the Pirani, although the Pirani is more sensitive.

The application of this method, using *a very sensitive Pirani gauge*, has been investigated by F. W. Reuter and C. Kenty in the Lamp Development Laboratory of the General Electric Company.[207] The method of measuring a leak, as described by these investigators, is to close off the main pump with a mercury cut-off and measure the rate of increase in pressure in the system as a function of the time, first with air flowing through the leak, and then with hydrogen. It is evident that with extremely small leaks the interval of time required to reach a decision regarding the actual occurrence of a leak must become quite large. Furthermore, it is always necessary to distinguish between rate of evolution of gas in the system itself and increase in pressure due to a leak: the rate of evolution will, in general, decrease with time and should ultimately become extremely small, whereas with a leak the pressure should increase linearly with time. Actually Reuter and Kenty were able to detect leaks of the order of magnitude of $1 \cdot 10^{-5}$ micron · liter · hr$^{-1}$.

However, especially in production, it is usually essential to detect a "leaker" in as short a time as possible; consequently other methods have been applied, which, though much less sensitive, are still both sensitive and speedy enough to detect a very high percentage of leaks.

The effect of gases like oxygen and halogen compounds on the electron emission from a tungsten or oxide-coated filament has been applied to detection of leaks. For instance, Ridenour[208] has observed that "a very small air leak" is sufficient to increase the current required by a "combined oxide" filament of a Western Electric (D79510) gauge tube for the standard emission value from 1.4–1.6 to 1.8–2.0 amp.

While testing for leaks by noting the response on an ion gauge caused by different liquids, Manley, Haworth, and Luebke[209] observed a decrease in emission resulting from entrance of the vapor at the leak. This observation led Lawton[210] of the General Electric Research Laboratory to develop a method of leak testing which depends upon *the effect of oxygen on the emission from a tungsten filament.*

As had been observed many years previously by I. Langmuir, an adsorbed layer of oxygen on tungsten decreases the emission to a tremendous extent. Thus, if a diode containing a heated tungsten filament as cathode is inserted in a vacuum system as near as possible to the device under test and a jet of oxygen is passed over a leak in some part of the device, the emission will suddenly decrease because of the effect of the increased concentration of oxygen in the diode.

Hydrogen can be used in the same manner as oxygen; it usually causes a decrease in emission because the atomic hydrogen formed at the incandescent filament (see discussion in Chapter 9 on chemical clean-up) reduces some of the oxides present, and the water vapor then affects the emission in the same manner as oxygen.

The following description of the leak detector based on this principle, and simple circuit used in the General Electric Research Laboratory by W. A. Ruggles, is given in a paper by Nelson.[211]

> The leak detector consists of a diode with a pure tungsten filament and a closely spaced plate. A potential of approximately 22 volts is used to draw saturated emission from the filament, which is heated by a storage battery. The emission current is measured by a microammeter supplied with a bucking voltage so that a sensitive meter may be used and small changes in the emission observed. Figure 5.86 shows the circuit diagram. With this apparatus, a leak of $10^{-3}$ cc of air per hour has been readily detected.

A leak of this magnitude corresponds to about 0.76 micron · liter · $hr^{-1}$. Actually L. R. Koller of the General Electric Research Laboratory has observed that it is possible to detect by this method rates of leak which are as low as 0.01 micron · liter · $hr^{-1}$.

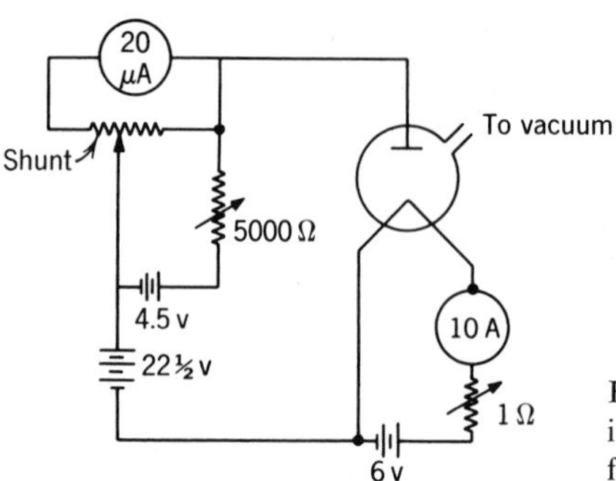


Fig. 5.86. Battery-operated leak detector involving effect of oxygen on emission from tungsten (Lawton).

Since the battery-operated detector presents certain practical disadvantages, Nelson developed an a-c operated detector which is shown in Fig. 5.87. The description of the circuit is as follows:

A control diode similar to the leak detector, but highly evacuated and sealed off, has its filament heated by the same source of a-c power as the leak detector filament. The saturated thermionic emission current of the "control diode" is fed back to regulate the a-c voltage which heats both filaments, thus compensating for line voltage changes in such a way as to keep constant emission in the control diode. Since we regulate the thermionic emission—the same rapidly changing function of filament power which we want to measure in the leak detector diode—the emission of the leak detector can be held quite constant as the line voltage varies.

The feedback circuit by which the emission current of the control diode regulates the a-c voltage is similar to the emission-regulating circuit for an

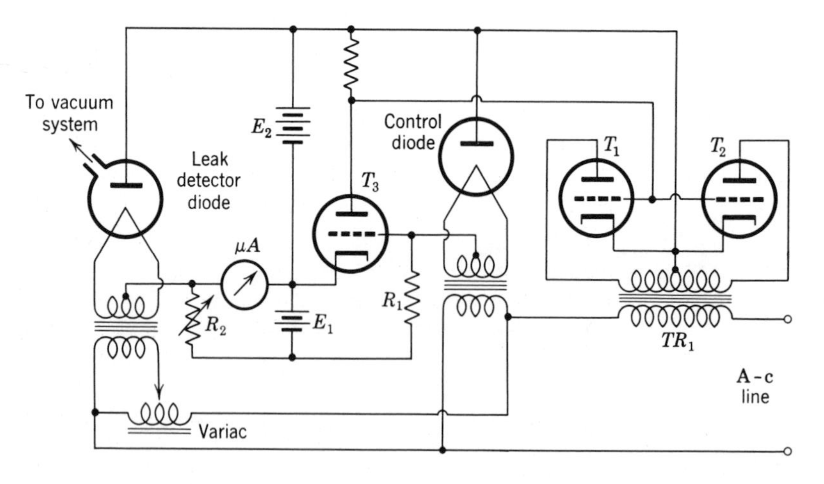


Fig. 5.87. Essential elements of a-c operated leak detector based on the same principle as the circuit in Fig. 5.86 (R. B. Nelson).

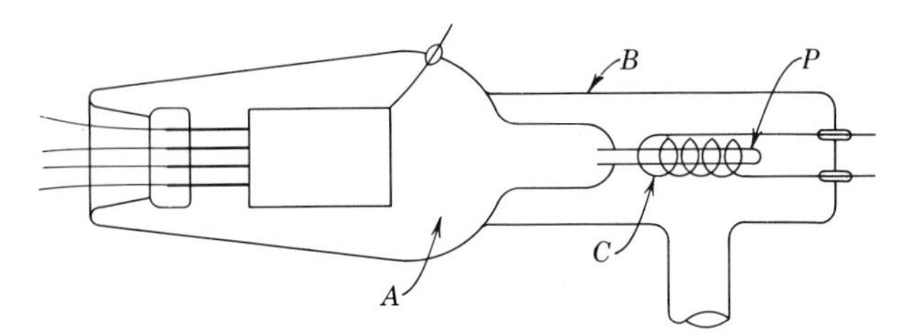


Fig. 5.88. Leak-detector tube involving use of palladium tube and ionization gauge (H. Nelson).

ionization gauge described by Ridenour and Lampson.[212] A transformer $TR_1$ has its primary in series with the a-c line which heats the filaments. Across the secondary in full-wave connection are two triodes $T_1$ and $T_2$. As the grid voltage of the triodes is raised, lowering their plate resistance, the impedance of the transformer primary is likewise lowered. This causes a decreased voltage drop across the primary and hence increased voltage on the filaments of the diodes.

The emission current of the control diode, passing through resistor $R_1$, produces the grid voltage of the d-c amplifier tube $T_3$, a high-$\mu$ triode. The cathode of this tube is held at a steady positive potential $E_1$, so that the tube is cut off until the voltage drop in $R_1$ is almost equal to $E_1$. Further increase in grid voltage causes $T_3$ to conduct, producing a negative bias on the grids of $T_1$ and $T_2$.

The control circuit thus regulates the emission current of the control diode to a value about equal to $E_1/R_1$. Since the leak-detector diode may not have an identical filament, its filament transformer runs from a Variac by which the emission may be adjusted. The steady emission is balanced out from the microammeter by current through $R_2$ so that a sensitive meter may be used and small changes in emission detected.

Experience has shown that the leak detector must be run for about 15 min before the emission becomes sufficiently stable to look for leaks.

As mentioned previously, Nelson has combined the circuit for the leak detector with that required to operate an ionization gauge, and a description of the combined circuit is given in the original publication. The whole system is arranged in a box provided on the panel with two indicating instruments and control switches, which can be plugged into the standard 115–127 volt 60-cycle a-c supply.

An interesting method for the detection of leaks described by Nelson[213] involves the use of a *hydrogen ionization gauge*, which is shown in Fig. 5.88. According to Nelson's description,

Part *A* is an evacuated and sealed-off ionization-gauge tube with a piece of palladium tubing *P* forming part of the envelope. Part *B* is made from Nonex glass tubing, ring-sealed onto the glass envelope of the ionization-gauge tube. *C* is a coil of platinum-clad molybdenum wire which serves to heat the palladium tubing.

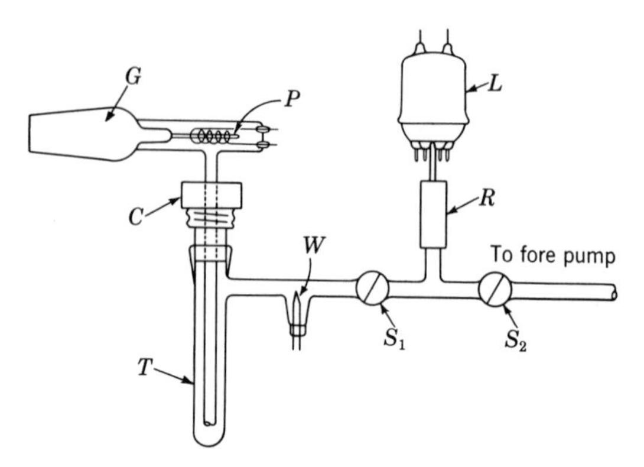


Fig. 5.89. Arrangement for use on exhaust system of leak detector shown in Fig. 5.88. $C$ = compression rubber port; $G$ = hydrogen gauge; $L$ = tube under test; $P$ = palladium; $R$ = rubber port; $S_1, S_2$ = stopcocks; $T$ = liquid-air trap; and $W$ = tungsten filament.

When this device is sealed to a vacuum system, it may be used for detecting and locating air leaks in a manner similar to that in which a conventional ionization gauge is used but with several important advantages. The palladium tubing is heated to a temperature of about 800° C at which it is exceedingly permeable to hydrogen but to no other gases or vapors. If a jet of hydrogen, illuminating, or forming gas is then used to probe the system for leaks, hydrogen will enter the vacuum system at the point of leakage, pass through the walls of the palladium tubing, and give an increased pressure reading in the ionization-gauge unit. After the leak has been located and the hydrogen jet removed, the hydrogen is pumped out from the ionization-gauge tube through the walls of the hot palladium tubing. Since a far higher vacuum may be maintained in the sealed-off ionization-gauge tube with aid of getter than normally exists in a vacuum system, it follows that far smaller leaks can be detected by the technique described above than by the conventional technique in which the ionization-gauge tube is directly connected to the vacuum system.

For detecting and locating air leaks in vacuum-tube envelopes the arrangement shown diagrammatically in Fig. 5.89 is used.

With $T$ immersed in liquid air and with all stopcocks open, the system is pumped to a pressure of about $10^{-4}$ mm of mercury. (The use of a mercury- or oil-diffusion pump is not required.) The palladium tubing of the hydrogen gauge is then heated to about 800° C, a hydrogen hood is lowered over the tube under test, and the ion current of the sealed-off ionization gauge is watched for indication of a leak. If an increase in the ion current indicates the presence of a leak, the hydrogen hood is raised and suspected points on the tube are probed by a fine hydrogen jet until a further increase in the ion current indicates the exact location of the leak. When a leak is too small to be located by the above method, increased sensitivity may be obtained by closing the stopcock $S_2$ before applying hydrogen to the tube under test. To obtain maximum sensitivity, hydrogen may be allowed to diffuse through the leak in the tube under test with stopcocks $S_1$ and $S_2$ both closed for a period of one or more minutes, after which stopcock $S_1$ is opened to allow the accumulated hydrogen to give indication of leakage.

Nelson has drawn attention to the fact that lack of sensitivity may be observed because of adsorption of oxygen on the palladium or because of the presence of oil vapors in the space surrounding the tube. To avoid the latter difficulty it is recommended that the hydrogen gauge should be removed before the trap is allowed to reach room temperature.

With regard to the sensitivity of the detector, the following statement is made.

During the period that the leak detector has been in use, leaks of widely different magnitude have been located. Leaks large enough to cause a vacuum tube to lose completely its vacuum in less than 24 hours after seal-off can usually be located in less than 2 or 3 minutes after the tube is put on the detector. Leaks of smaller magnitude require more time for their location but they can usually be found if they are large enough to cause a tube to lose its vacuum completely in less than two weeks after seal-off or if the rate of flow of hydrogen through the leak is greater than about $1 \cdot 10^{-4}$ micron · liter per second. [This corresponds to about 0.36 micron · liter per hour.]

The Radio Corporation of America has built a hydrogen-sensitive leak detector based on Nelson's findings. This device, model 722-SS, is portable, weighing only 25 pounds. It uses an RCA-1945 hydrogen ionization gauge and has a sensitivity of $1 \cdot 10^{-5}$ micron · liter · sec$^{-1}$ or $1.3 \cdot 10^{-8}$ cc · sec$^{-1}$.

*An ionization gauge with helium* has been found by S. Dushman to be a very useful and convenient leak detector in factory practice. As pointed out in section 5.7, the sensitivity of the ionization gauge for helium is about one-sixth that for nitrogen (or air). If a gauge is connected to the exhaust system at some point between the origin of the leak and the vapor pump, it will indicate a pressure (in microns) of air, which is determined by the relation

$$Q = SP_0, \tag{5.51}$$

where $Q$ = rate of leak of air in, for instance, micron · liters per second, and $S$ = speed of exhaust in liters per second. The value of $P_0$ will obviously be higher than the ultimate pressure obtainable for a "tight" system. If, now, helium at atmospheric pressure is directed by means of a fine jet at the point where the leak is present, then the values of both $Q$ and $S$ are increased in the ratio $(29/4)^{1/2}$ and therefore the absolute value of $P_0$ is not affected. However, since this pressure is due to helium instead of air, the ion current is considerably smaller.

If $Q$ is very small, so that the stationary value of $P_0$ is only slightly higher than it would be if no leak were present, the magnitude of $P_0$ may be increased, at constant value of $Q$, by closing the main stopcock to the pump and letting the evacuation take place through a smaller-diameter tube which is connected in parallel with the stopcock between the gauge and pump.

Actual tests have shown that it is possible by means of this procedure to demonstrate the presence of leaks which correspond to about 0.2 micron · liter · $hr^{-1}$. One advantage of this method is that a number of tubes may be connected to the same pumping system at the same time, and it is then possible by means of the ionization gauge and a helium jet to determine which particular tube or tubes leak. Another advantage is the fact that no time lag is involved. The change in the value of the positive-ion current on replacing the air by helium occurs practically instantaneously.

It should also be mentioned that replacing the helium by vapor of a volatile organic liquid such as ether or acetone causes an *increase* in the ion current over that observed with air. This observation is evidently in agreement with the results, mentioned in section 5.7, which have been obtained by Blears, and also in the General Electric Research Laboratory, on the magnitudes of the positive-ion currents in the presence of organic vapors.

A summary of vacuum testing methods in use before 1943, quoted from the paper by Jacobs and Zuhr,[214] is given in Table 5.8.

In the first three methods the probe fluids were chosen to ensure maximum sensitivity of the gauge indications. With increase in "tightness" of vacuum systems it became possible to replace the hot-wire gauge by one of the ionization type (method 4). The authors state,

> To obtain the ultimate sensitivity of these methods the dynamic method was discarded and one of the following methods adopted: (1) the "backing space" technique, or (2) the apparent change in rate of pressure build-up when probe gas displaces air flowing through the leak. Both methods achieve the improved sensitivity at the expense of an increase in testing time.
>
> Backing space is the name given to the process of multiplying the effect of the probe gas by permitting the diffusion pump to compress the gas into a dead volume. This is achieved by shutting the valve in the foreline leading to the mechanical pump. The method further requires judgment in the analysis of the signal indicated on the gauge controls.

Furthermore the operation of this method requires highly trained personnel. In methods 7 and 8 the system has to be outgassed for a prolonged period.

A very *sensitive method for the detection of leaks*, and one which also gives an indication in a minimum of time, was developed during World War II.[215] It involves the *use of a mass spectrometer with helium*,[216] and the principle by which it operates may be described briefly as follows.

When a beam of positive ions, accelerated by a potential $V$, is passed through a magnetic field of strength $H$, the ions are sorted out, according to their values of $e/m$, where $m$ is the mass of the ion, and $e$ the charge.

**TABLE 5.8**

**Résumé of Vacuum Testing Methods in Use before 1943**

(Jacobs and Zuhr)

| Leak Detector | Operating Range (microns) | Size of Leak Discernible (micron · liter · min$^{-1}$) | Commonly Used Probes | Remarks |
|---|---|---|---|---|
| 1. Tesla coil | 50–1000 | ... | ... | Useful only on glass systems |
| 2. Discharge tube | 100–1000 | ... | Acetone, methanol, $CO_2$, $H_2$ | Residual gases confusing |
| 3. Hot-wire gauge | $<100$ | 4.7–470 | Acetone, methanol, $H_2$ | Affected by pressure changes and residual vapor |
| 4. Ionization gauge | $<0.5$ | 0.47–47 | Gaseous hydrocarbons, $H_2$, $O_2$ | Affected by pressure changes and residual vapor |
| 5. Hot wire with backing space | $<100$ | 24–47 | Gaseous hydrocarbons, $H_2$, $O_2$ | Time-consuming |
| 6. Ionization gauge with backing space | $<0.5$ | 0.5–47 | Gaseous hydrocarbons, $H_2$, $O_2$ | Time-consuming |
| 7. Hot wire with pressure build-up | $<50$ | 0.5–47 | Gaseous hydrocarbons, $H_2$, $O_2$; masking with vacuum putty | Extensive outgassing required |
| 8. Ionization gauge with pressure build-up | $<0.5$ | 0.5–47 | Gaseous hydrocarbons, $H_2$, $O_2$; masking with vacuum putty | Extensive outgassing required |

The radius of curvature $R$ of the path of any given type of *singly charged* ion is determined by the relation

$$R = 143.9\,\frac{(MV)^{1/2}}{H}, \tag{5.52}$$

where $R$ is expressed in centimeters, $V$ in volts, $H$ in oersteds, and $M$ is the molar mass of the ion in grams.

For example, for $R = 4.0$ cm and $H = 1500$ oersteds, the values of $V$ for helium ($M = 4.003$) and nitrogen ($M = 14.008$, since $N^+$ is formed), are found to be 434 volts and 124 volts, respectively. Thus, for a given value of the radius of curvature, the ions reaching a collector may be differentiated by varying the value of $V$, and the magnitude of the ion current at any given voltage setting will depend on the concentration of the ions in the beam, that is, on the rate at which the corresponding molecules leak into the spectrometer.

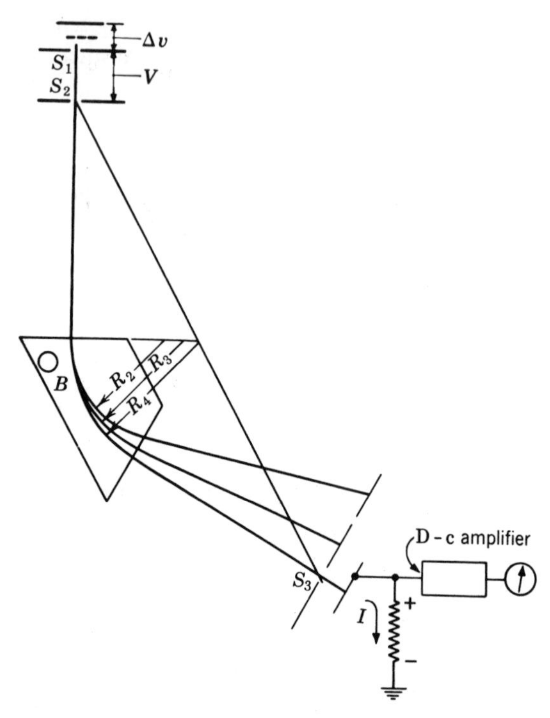


Fig. 5.90. Sixty-degree mass spectrometer. The ions are repelled through the slit $S_1$ by a small positive voltage $\Delta v$, and are accelerated between $S_1$ and $S_2$ by the potential $V$. Helium ions sorted according to the value of $R$, the radius of curvature, enter the slit $S_3$ and are discharged on the collector plate, causing a positive current, $I$, to flow through the high resistance to ground.

For the detection of leaks helium has been chosen as testing gas for the following reasons:

1. Because of the low value of $M$, the rate of diffusion through a leak is greater than that of any other gas except hydrogen.
2. Helium occurs in the atmosphere to the extent of only 1 part in 200,000 parts of air.
3. There is no possibility that an ion due to any other gas will give an indication that can be mistaken for helium.

The application of the mass spectrometer to the detection of leaks has been described in two papers, one by Worcester and Doughty,[217] and the other by Thomas, Williams, and Hipple.[218]

The type of spectrometer developed by Worcester and Doughty is best illustrated by Fig. 5.90. The positive ions, produced by ionization of gas molecules by means of electrons emitted from a hot tungsten filament, are accelerated by means of a potential of about 270 volts applied to the slits $S_1$ and $S_2$ and deflected through an angle of 60° by the field of about 900 oersteds due to a permanent magnet with trapezoidal pole pieces, which is inside the spectrometer tube.[219] For a given type of ion, the radii of curvature corresponding to different values of the accelerating voltage are represented by $R_2$, $R_3$, and $R_4$. With the values of $V$ and $H$ given above, the value of $R$ for helium ions is 5.25 cm, and, if the slit $S_3$ is set for this value of $R$, only helium ions will reach the collector, whereas all other types of ions will be deflected either to the right or to the left of the slit.

A modern version of the helium mass-spectrometer leak detector, made by the Instrument Department of the General Electric Company, is shown in Fig. 5.91. This instrument is capable of detecting leaks as small as $10^{-10}$ cc of air per second in evacuated enclosures. Such a leak is so small that more than 5000 years would be required for 1 cu in. of air at atmospheric pressure to pass through the opening. For pressurized enclosures leaks down to $10^{-6}$ cc per second can be located. In this

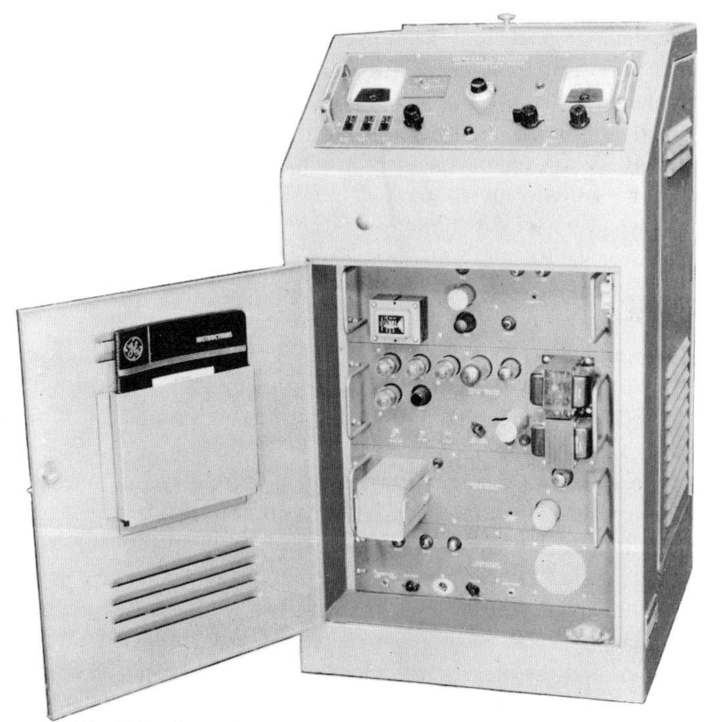

Fig. 5.91. General Electric mass-spectrometer leak detector.

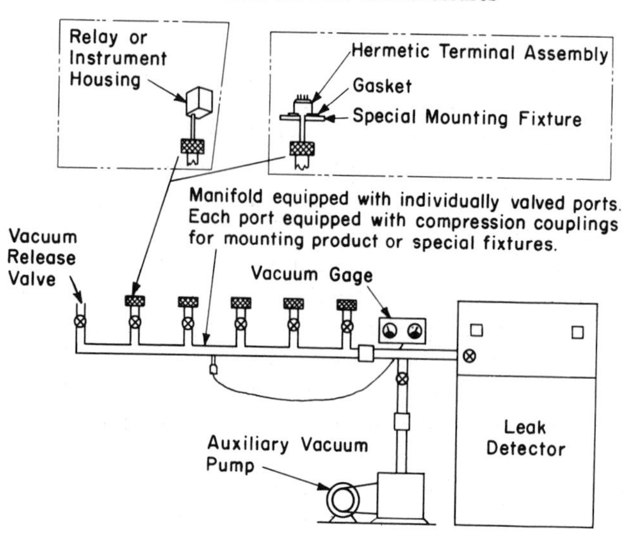


Fig. 5.92. Simple leak-test manifold.

spectrometer the helium ions are deflected through 90 degrees by the magnetic field. The current generated by collection of the helium ions is amplified and used to indicate a leak on the leak ratemeter and on an audible alarm.

Fast leak testing is made possible with this spectrometer by the incorporation of an automatic balance circuit. This circuit allows the leak detector to respond only to a rapid increase in helium signal such as is obtained when a jet of helium passes near a leak. When testing large systems, large leaks are sometimes encountered with the jet of helium which cause the mass-spectrometer vacuum system to become partially saturated with helium. Normally the operator would have to wait a considerable time before the helium could be pumped out of the system to a point where leak testing could be resumed. However, with the automatic balance circuit the helium background is balanced out and a leak signal is given only when the operator passes the jet of helium near a leak.

Figure 5.92 shows the arrangement for connecting components to the mass spectrometer for leak checking. The components are first exhausted by the mechanical roughing pump (shown at the left), and when the pressure, as indicated, for instance, by a thermocouple gauge connected to the test manifold is sufficiently low the throttle valve to the spectrometer tube is opened. A fine jet of helium is passed over suspected parts in each of the tubes, and the presence of a leak is indicated on the leak ratemeter.

A new method of leak-testing hermetically sealed components nondestructively by utilizing a radioactive gas has been described by Cassen and Burnham.[220] Leakage rates in the order of $10^{-12}$ cc · $sec^{-1}$ can be measured under favorable conditions by this process, called "Radiflo." Equipment for this leak-detection system is manufactured by the Analytical and Control Division of Consolidated Electrodynamics Corporation, Pasadena, Calif. The basic principle involves the detection of radioactive krypton 85 which has been allowed to diffuse into the leaky components. Krypton 85 has a half-life of 10.3 years. Over 99 per cent of the disintegrations involve the emission of beta rays with a maximum energy of 0.67 Mev, while only 0.7 per cent of the disintegrations give 0.54-Mev gamma rays. The gamma-emitting disintegrations are used to detect the leaks, since the gamma rays will pass readily through most materials. Detection of the beta rays may be used under some circumstances to determine the presence of occluded krypton on the surface of the components under test. The testing procedure is as follows.

The components to be tested are placed in an activating tank which is then sealed. Air in the tank is evacuated down to approximately 2 mm Hg, and diluted krypton 85 is pumped into the tank under pressure up to 7 atm. The radioactive gas diffuses into existing leaks in the components. After a prescribed "soaking" period, which may be from several minutes to a few hundred hours, the krypton is pumped out of the activating tank and returned to storage for reuse. If the leaks follow Poiseuille's law (Chapter 2), the quantity of krypton diffusing into leaky components will increase with the square of the pressure and linearly with time. Thus a suitable combination of krypton pressure and "soaking" time may be selected to give the desired sensitivity. Next, an air wash is circulated over the components to remove any residual krypton from the external surfaces. The components are then removed from the activating tank. Those with leaks will retain some radioactive atoms which emit gamma radiation. This radiation is detected by a scintillation counter and the intensity determined by a ratemeter. When this radiation intensity is related to the conditions of the activation process, the leak rate may be determined. A photograph of the activating equipment is shown in Fig. 5.93.

Some components may have absorptive surfaces such as organic coatings, gaskets, or insulation which will retain krypton 85 for various lengths of time after the pressure is released. The gamma radiation emitted by this absorbed gas could make the components appear as leakers when, in fact, they are not. A routine check of rejected parts with a thin-window Geiger-Müller counter tube for beta-ray activity reveals surface contaminations.

Quite different in principle from all the other methods of leak detection described above is that involved in the *positive-ion detector for halogen compounds* which has been described by White and Hickey[221] of the General Electric Research Laboratory.

Some of the earliest investigators in the field of thermionic emission, such as O. W. Richardson, had observed that platinum, even in air, at a temperature of a red heat, emits positive ions, and that the rate of ion emission increases with temperatures according to a relation similar to that observed for electron emission from incandescent cathodes.[222] This positive-ion emission is most probably due to the presence in the anode of

Fig. 5.93. Activating equipment for the "Radiflo" leak-detection system.

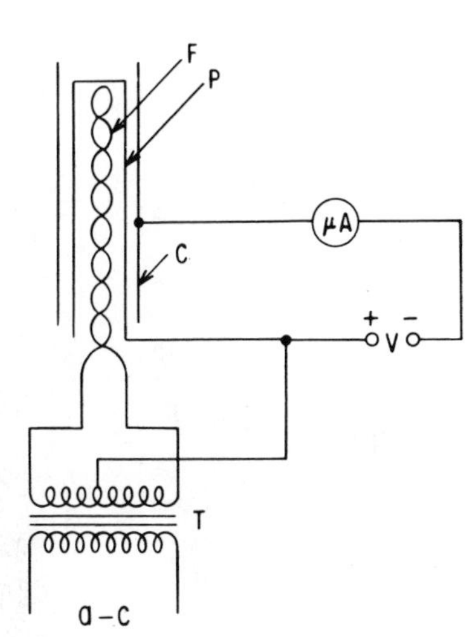


Fig. 5.94. Schematic diagram of the thermionic detector and circuit used for detection of halogen compounds.

salts of the alkali metals, although this may not be the correct interpretation of the mechanism of operation of the device described below.

It was observed by C. W. Rice of the General Electric Company that this emission, at any given anode temperature, of positive ions in air is increased very markedly when vapors of compounds containing a halogen strike the electrode surface. This observation forms the basis of the detector developed and described by White and Hickey.

A diagrammatic sketch of the device and the simple circuit used for its operation is shown in Fig. 5.94. The detector consists of a platinum cylinder $P$, which is heated by an insulated platinum filament $F$, the low voltage required for this purpose being supplied by the transformer $T$. A metal cylinder $C$, concentric with $P$, is connected through a microammeter ($\mu A$) to the negative end of a d-c source of voltage (50–500 volts), while $P$ is connected through the midpoint of the secondary of $T$ to the positive end of the voltage source.

In using the device for detection of leaks, air or any other suitable gas containing a halogen vapor is introduced into the system at a positive pressure. This positive pressure forces the air containing the halogen out through the leak, where it may be picked up by the device and is indicated by the meter in the detector circuit. Or the detector may be sealed on the vacuum system in series with the pump (preferably between the rough and the fine pumps). Air containing a halogen vapor is forced under pressure through a very small jet and directed at any suspected spot in the system; it then is picked up by the detector.

In using the detector, the temperature of the anode should be between 850 and 950° C. At lower temperatures, the positive-ion emission is too

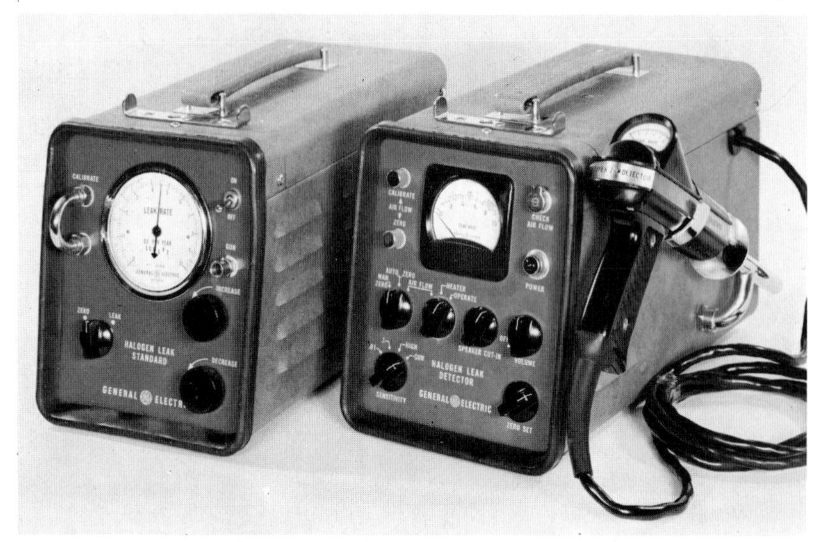


Fig. 5.95. Gun-type halogen leak detector and leak standard used for calibration.

small, and at higher temperatures the emission becomes unstable. Typical halogen compounds, which are fairly volatile at room temperatures and therefore applicable in using the positive-ion detector, are "Freon," carbon tetrachloride, and chloroform.

A convenient form of the halogen leak detector, made by the Instrument Department of the General Electric Company, is shown in Fig. 5.95. The unit contains a power supply, amplifier, sensing element, and air pump. The detector probe is connected to the control unit through a flexible cable. Air drawn in at the probe passes through the sensing element and is exhausted through the pump. When the probe is moved so as to pass near a leak, the sudden increase of halogen gas mixing with the air going into the probe will cause an increase in the detector ion emission. This current is amplified and measured on the leak ratemeter. The detector can be calibrated by use of the leak standard also shown in Fig. 5.95. The leak standard consists of a capillary through which $CCl_2F_2$ leaks to the atmosphere. This reference leak has an adjustable leakage rate from zero to $3 \cdot 10^{-4}$ cc per second. The detector sensitivity is adjusted so that, when the tip of the probe sniffs the leak standard, the detector leak ratemeter reads the same as the standard leak rate dial.

The leak detector contains an automatic balance circuit which permits operation in an atmosphere with a constant or slowly changing contamination. Thus the detector may be used to locate leaks down to $5 \cdot 10^{-5}$ cc per second in an atmosphere with contamination up to 1000 parts of halogen gas per million parts of air.

The halogen leak detector has a special advantage in that no helium is required. This feature has given this method preference in instances where scarcity of helium has been a problem.[223]

The halogen leak detector may also be used for locating leaks in vacuum systems by employing a special vacuum-head detector unit. The detector unit is usually installed in the pipe line between the system to be tested and the vacuum pumps. After the system has pumped down, leaks are located by spraying the system with a fine jet of halogen gas. When the jet passes over a leak, the gas will enter the system and be detected. By probing, the operator can accurately locate the leak. Leaks of the order of $10^{-6}$ cc per second may be detected.

## REFERENCES AND NOTES

1. In the first edition of this book Dushman stated that in the preparation of this chapter he had made considerable use of the article entitled "High-Vacuum Gauges" by M. Pirani and R. Neumann, *Electronic Engineering* (England), December 1944, January, February, and March 1945, which contains a very comprehensive list of references. A condensation of the contents of this chapter as it appeared in the first edition was published by Dushman in *Instruments*, **20**, 234 (1947).

In the revision of this chapter only minor changes have been made in the sections on liquid manometers, mechanical manometers, viscosity manometers, radiometers, and heat-conductivity manometers. Major revisions have been made in the sections on ionization gauges and leak detectors. A considerable amount of new technology has been developed since 1948 in ultrahigh-vacuum ionization gauges, mass spectrometers, and leak-detection methods. The operating pressure range for ionization gauges has been extended in both directions from about 1 to $10^{-13}$ mm Hg. Partial pressure gauges have been developed for analyzing the residual gases in vacuum systems. Radioactive gas has been used to detect leaks in sealed vacuum devices, making nondestructive tests possible. Mass-spectrometer leak detectors have been made more reliable and simple to operate. The inclusion of this new material has required the elimination and condensation of some of the earlier work.

The reader may also be interested in the following general references on the subject of low-pressure measurements:

Daniel Alpert, "Production and Measurement of Ultra-high Vacuum," *Handbuch der Physik*, Vol. XII, Springer, Berlin, 1958.

W. E. Barr and V. J. Anhorn, *Instruments*, **19**, 666, 734 (1946).

A. Farkas and H. W. Melville, *Experimental Methods in Gas Reactions*, Cambridge University Press, 1939, pp. 71–102.

A. Guthrie and R. K. Wakerling, *Vacuum Equipment and Techniques*, McGraw-Hill Book Company, New York, 1949.

H. Klumb and H. Schwarz, *Z. Physik*, **122**, 418 (1944).

J. H. Leck, *Pressure Measurement in Vacuum Systems*, Chapman and Hall, London, 1957.

G. Mönch, *Vakuum Technik im Laboratorium*, Weimer, 1937, published by J. W. Edwards, Ann Arbor, Mich., by authority of Alien Property Custodian.

A. Venema and M. Bandringa, "The Production and Measurement of Ultra-high Vacua," *Philips Tech. Rev.*, **20**, 145 (1958).

J. Yarwood, *High Vacuum Techniques*, John Wiley & Sons, New York, 3rd edition revised, 1955.

J. Yarwood, "Ultra-high Vacua," *J. Sci. Instr.*, **34,** 297 (1957).

2. H. Stintzing, *Z. physik. Chem.*, **108,** 70 (1914); Pirani and Neumann (Ref. 1); Mönch, *op. cit.*, p. 54.
3. J. Yarwood, *High Vacuum Techniques*, John Wiley & Sons, New York, 3rd edition revised, 1955, pp. 73–77.
4. G. Burrows, *J. Sci. Instr.*, **20,** 21 (1943).
5. Developed by the Skaneateles Manufacturing Company, Syracuse, N.Y.
6. *Phil. Trans.*, **A, 196,** 208 (1901); *Z. physik. Chem.*, **37,** 713 (1901).
7. K. Scheel and W. Heuse, *Z. Instrumentenk.*, **29,** 344–349 (1909). K. Jellinek, *Lehrbuch d. physik. Chem.*, I, 1, p. 321. *Ann. Physik*, **29,** 723 (1909).
8. M. Thiesen, *Z. Instrumentenk.*, **6,** 89 (1886), and **24,** 276 (1904).
9. E. Hering, *Ann. Physik*, (4), **21,** 319 (1906).
10. C. F. Mündel, *Z. Physik. Chem.*, **85,** 435 (1913).
11. K. Prytz, *Ann. Physik*, (4), **16,** 735 (1905).
12. C. F. Mündel (Ref. 10) and M. Knudsen, *Ann. Physik*, (4), **33,** 1435 (1910).
13. J. E. Shrader and H. M. Ryder, *Phys. Rev.*, **13,** 321 (1919).
14. E. K. Carver, *J. Am. Chem. Soc.*, **45,** 59 (1923).
15. M. L. Hamlin, *J. Am. Chem. Soc.*, **47,** 709 (1925).
16. K. Newbury and C. L. Utterback, *Rev. Sci. Instr.*, **3,** 593 (1932).
17. M. C. Johnson and G. O. Harrison, *J. Sci. Instr.*, **6,** 305 (1929).
18. T. G. Pearson, *Z. physik. Chem.*, **A, 156,** 86 (1931); A. Farkas and H. W. Melville, *Experimental Methods in Gas Reactions*, Cambridge University Press, 1939, p. 84.
19. K. C. D. Hickman, *J. Opt. Soc. Am.*, **18,** 305 (1929). Mention should be made, in this connection, of the Dubrovin gauge, manufactured by W. M. Welch Scientific Company. This consists of a glass tube, about half-filled with mercury, connected to the system. Floating on the mercury is a metal tube closed at the top by a flat disk, used as an indicator, and loosely closed at the other end, which is immersed in the mercury, by a glass ball. The interior of the metal tube is evacuated, and therefore the tube will float high on the mercury when the pressure is reduced and low when the pressure is high. The gauge may be used to measure pressures ranging from 0.1 to 20 mm of mercury.
20. G. Burrows, *J. Sci. Instr.*, **20,** 21 (1943).
21. H. Klumb and Th. Haase, *Z. tech. Physik*, **13,** 372 (1932).
22. A. Simon and F. Fehér, *Z. Elektrochem.*, **35,** 162 (1929). The original paper contains a comprehensive list of references to other methods available for telemetering of pressure differences.
23. J. J. Manley, *Proc. Phys. Soc.*, **40,** 57 (1927–1928).
24. H. McLeod, *Phil. Mag.*, **48,** 110 (1874).
25. Detailed drawings of a number of designs of McLeod gauges are shown by W. E. Barr and V. J. Anhorn, in *Instruments*, **19,** 666 (1946).
26. C. Hagen, *Physik. Z.*, **27,** 47 (1926), *Z. tech. Physik*, **8,** 599 (1927).
27. P. Rosenberg, *Rev. Sci. Instr.*, **9,** 258 (1938).
28. G. Haase, *Z. tech. Physik*, **24,** 27, 53 (1943).
29. M. H. Armbruster, *J. Am. Chem. Soc.*, **68,** 1342 (1946).
30. Similar observations have been recorded, as Miss Armbruster notes, by J. C. W. Frazer, W. A. Patrick, and H. E. Smith, *J. Phys. Chem.*, **31,** 897 (1927), and A. S. Coolidge, *J. Am. Chem. Soc.*, **49,** 708 (1927).
31. For a discussion of the procedure for measuring the pressure of water vapor see E. W. Flosdorf, *Ind. Eng. Chem., Anal. Ed.*, **17,** 198 (1945).

32. K. Scheel and W. Heuse, *Ber. deut. physik. Ges.*, **6,** 785 (1908); **7,** 1 (1909).
33. Rayleigh, *Phil. Trans.*, **A, 196,** 208 (1901).
34. K. Scheel and W. Heuse, *Z. Instrumentenk.*, **29,** 344–349 (1909); K. Jellinek, *Lehrbuch d. physik. Chem.*, **I,** 1, p. 321. *Ann. Physik*, **29,** 723 (1909).
35. W. Gaede, *Ann. Physik*, **41,** 289 (1913).
36. M. Francis, *Trans. Faraday Soc.*, **31,** 1325 (1935).
37. See *Vacuum Practice*, by Louis Dunoyer, translated by J. H. Smith, D. Van Nostrand Company, New York, 1926, p. 71, for illustration of this behavior of a fresh gauge.
38. Specially selected capillary tubing of uniform bore is now available commercially.
39. See Dunoyer (Ref. 37), p. 68, for method of correction required for the form of the meniscus.
40. R. J. Clark, *J. Sci. Instr.*, **5,** 126 (1928).
41. Personal communication.
42. See especially the paper by M. Pirani and R. Neumann cited in Ref. 1.
43. A. H. Pfund, *Phys. Rev.*, **18,** 78 (1921).
44. E. W. Flosdorf, *Ind. Eng. Chem., Anal. Ed.*, **17,** 198 (1945).
45. M. Bruner, *Helv. Chim. Acta*, **13,** 915 (1930). An earlier form similar to this in many respects was described by H. J. Reiff, *Z. Instrumentenk.*, **34,** 97 (1914).
46. Made by W. Edwards and Company, England.
47. J. Yarwood (Ref. 3), p. 78.
48. Described by A. Farkas and H. W. Melville (Ref. 18), p. 93.
49. H. Moser, *Physik. Z.*, **36,** 1 (1935).
50. A very comprehensive list of references on this type of manometer has been given by C. Kenty, *Rev. Sci. Instr.*, **11,** 377 (1940), in connection with the description of a manometer with thin quartz wall designed for pressures ranging from a few centimeters of mercury to 50 atm.
51. V. H. Dibeler and F. Cordero, *J. Research Natl. Bur. Standards*, **46,** 1 (1951).
52. Wallace and Tiernan Incorporated, Belleville, N.J.
53. D. Alpert, C. G. Matland, and A. O. McCoubrey, *Rev. Sci. Instr.*, **22,** 370 (1951).
54. E. Ladenburg and E. Lehmann, *Verhandl. deut. physik. Ges.*, **8,** 20 (1906).
55. F. M. G. Johnson and D. McIntosh, *J. Am. Chem. Soc.*, **31,** 1138 (1909); F. M. G. Johnson, *Z. physik. Chem.*, **61,** 457 (1908).
56. For further details regarding this type of manometer, refer to K. Jellinek, *Lehrb. physik. Chem.*, I, 1, p. 638, also to the references given in footnotes 54 and 55.
57. M. Bodenstein and W. Dux, *Z. physik. Chem.*, **85,** 297 (1913).
58. A. Farkas and H. W. Melville (Ref. 18), p. 86.
59. An extensive account of similar manometers made of glass or quartz has been given by B. P. Harteck, *Wien-Harms Handbuch der experimental Physik*, Vol. III, 2, p. 254, 1929.
60. S. G. Foord, *J. Sci. Instr.*, **11,** 126 (1934).
61. R. Grigorovici, *Z. tech. Physik*, **20,** 102 (1939).
62. K. Scheel and W. Heuse, *Z. Instrumentenk.*, **29,** 14 (1909); *Ber. deut. physik. Ges.*, **11,** 1 (1909).
63. E. A. Stewardson, *J. Sci. Instr.*, **7,** 217 (1930).
64. See R. Seeliger, *Z. tech. Physik*, **1,** 20 (1920), and G. Kornfeld and E. Klingler, *Z. physik. Chem.*, **B, 4,** 37, (1929).
65. K. Sommermeyer, *Z. physik. Chem.*, **A, 155,** 208 (1931).
66. W. H. Rodebush and C. E. Coons, *J. Am. Chem. Soc.*, **49,** 1953 (1927).
67. V. Deitz, *J. Chem. Phys.*, **4,** 575 (1936).
68. A. Farkas and H. W. Melville (Ref. 18), p. 93.

69. K. C. D. Hickman, J. C. Hecker, and N. D. Embree, *Ind. Eng. Chem., Anal. Ed.*, **9**, 264 (1937).

70. F. H. Verhoek and A. L. Marshall, *J. Am. Chem. Soc.*, **61**, 2737 (1939). The results obtained in this investigation as well as the preceding one have been discussed in Chapter 3, especially Table 3.6.

71. M. Volmer, S. Heller, and K. Neumann, *Z. physik. Chem., Bodenstein-Festschrift*, p. 863, 1931.

72. K. Neumann and E. Volker, *Z. physik. Chem.*, **A, 161**, 33 (1932).

73. W. Sutherland, *Phil. Mag.*, **43**, 83 (1897).

74. J. L. Hogg, *Proc. Am. Acad. Arts Sci.*, **42**, 115 (1906), and **45**, 3 (1909). *Contributions from the Jefferson Physical Lab.*, 1906, No. 4, and 1909, No. 4.

75. P. E. Shaw, *Proc. Phys. Soc. London*, **29**, 171 (1917).

76. I. Langmuir, *J. Am. Chem. Soc.*, **35**, 105 (1913).

77. F. Haber and F. Kerschbaum, *Z. Elektrochem.*, **20**, 296 (1914).

78. A. Henglein, *Z. anorg. allgem. Chem.*, **123**, 145 (1922).

79. D. H. Scott, *Phil. Mag.*, **47**, 32 (1924).

80. A. S. Coolidge, *J. Am. Chem. Soc.*, **45**, 1637 (1923).

81. Mrs. M. R. Andrews, *J. Phys. Chem.*, **30**, 1497 (1926).

82. E. Bolton King, *Proc. Phys. Soc. London*, **38**, 80 (1925).

83. E. Brüche, *Physik. Z.*, **26**, 717 (1925); *Ann. Physik*, **79**, 695 (1926).

84. G. Wetterer, *Z. tech. Physik*, **20**, 281 (1939); *Wiss. Veröffentl. Siemens-Werken*, **19**, 68 (1940).

85. J. W. Beams, J. L. Young III, and J. W. Moore, *J. Appl. Phys.*, **17**, 886 (1946).

86. I. Langmuir, *Phys. Rev.*, **1**, 337 (1913).

87. S. Dushman, *Phys. Rev.*, **5**, 212 (1915).

88. A. Timiriazeff, *Ann. Physik*, **40**, 971 (1913).

89. H. Riegger, *Z. tech. Physik*, **1**, 16 (1920); *Sci. Abst.*, **24**, 153 (1921).

90. Sir James Dewar, *Proc. Roy. Soc. London*, **A, 79**, 529 (1907).

91. The theory of the radiometer, especially at medium pressures, has been dealt with rather fully by G. D. West, *Proc. Phys. Soc.* (*London*), **25**, 324 (1912–1913); **28**, 259 (1915–1916); **31**, 278 (1918–1919); **32**, 166, 222 (1919–1920). More recent discussion of the same topic is given at the end of this section.

92. M. Knudsen, *Ann. Physik*, **32**, 809 (1910).

93. G. W. Todd, *Phil. Mag.*, **38**, 381 (1919). See also derivation by A. E. Lockenvitz, *Rev. Sci. Instr.*, **9**, 417 (1938).

94. Martin Knudsen, *Ann. Physik*, **44**, 525 (1914).

95. E. v. Angerer, *Ann. Physik*, **41**, 1 (1913).

96. J. W. Woodrow, *Phys. Rev.*, **4**, 491 (1914).

97. J. E. Shrader and R. G. Sherwood, *Phys. Rev.*, **12**, 70 (1918).

98. L. F. Richardson, *Proc. Phys. Soc. London*, **31**, 270 (1918–1919).

99. H. Riegger, *Z. tech. Physik*, **1**, 66 (1920).

100. E. Fredlund, *Ann. Physik*, **13**, 802 (1932); **14**, 617 (1932); **30**, 99 (1937).

101. Jesse W. M. DuMond and W. M. Pickels, Jr., *Rev. Sci. Instr.*, **6**, 362 (1935).

102. A. E. Lockenvitz, *Rev. Sci. Instr.*, **9**, 417 (1938).

103. A. L. Hughes, *Rev. Sci. Instr.*, **8**, 409 (1937).

104. H. Klumb and H. Schwarz, *Z. Physik*, **122**, 418 (1944).

105. W. Gaede, *Z. tech. Physik*, **15**, 664 (1934).

106. E. Brüche and W. Littwin, *Z. Physik*, **52**, 318 (1928).

107. See Tables 1.6 and 1.7 for values of $L$ at 1 micron and 25° C.

108. E. Fredlund, *Ann. Physik*, **13**, 802 (1932); **14**, 617 (1932); **30**, 99 (1937).

109. W. Voege, *Physik. Z.*, **7**, 498 (1906).

110. W. Rohn, *Z. Elektrochem.*, **20,** 539 (1914).
111. E. Rumpf, *Z. tech. Physik*, **7,** 224 (1926).
112. Th. Haase, G. Klages, and H. Klumb, *Physik. Z.*, **37,** 440 (1936).
113. W. J. H. Moll and H. C. Burger, *Z. tech. Physik*, **21,** 199 (1940).
114. C. Weisz and H. Westmeyer, *Z. Instrumentenk.*, **60,** 53 (1940).
115. W. Bartholomeyczyk, *Z. tech. Physik*, **22,** 25 (1941).
116. A similar type of gauge has been described by G. C. Dunlap and J. G. Trump, *Rev. Sci. Instr.*, **8,** 37 (1937).
117. R. T. Webber and C. T. Lane, *Rev. Sci. Instr.*, **17,** 308 (1946).
118. E. Warburg, G. Leithauser, and E. Johansen, *Ann. Physik*, **24,** 25 (1907).
119. M. Pirani, *Verhandl. deut. physik. Ges.*, **8,** 686 (1906).
120. C. F. Hale, *Trans. Am. Electrochem. Soc.*, **20,** 243 (1911).
121. Misamichi So, *Proc. Phys. Math. Soc. Japan*, 3rd Ser., **1,** 152 (1919).
122. N. R. Campbell, *Proc. Phys. Soc. London*, **33,** 287 (1921).
123. H. Murmann, *Physik*, **86,** 14 (1933).
124. F. Knauer and O. Stern, *Z. Physik*. **53,** 766 (1929).
125. L. F. Stanley, *Proc. Phys. Soc., London*, **41,** 194 (1928–1929).
126. A. Ellett and R. M. Zabel, *Phys. Rev.*, **37,** 1102, 1112 (1931).
127. A. L. Reimann, *Phil. Mag.*, **16,** 673 (1933).
128. A. L. Reimann, *Phil. Mag.*, **18,** 1117 (1934).
129. W. Hunsmann, *Z. Elektrochem.*, **44,** 540 (1938).
130. E. J. Scott, *Rev. Sci. Instr.*, **10,** 349 (1939).
131. E. Weise, *Z. tech. Physik*, **18,** 467 (1937); **24,** 66 (1943).
132. No description is given in the original paper of the method used for producing such thin films of the semiconducting material.
133. A description and samples of the thermistors described here were furnished to Dushman by Becker in advance of the publication of the following papers on this subject: J. A. Becker, C. B. Green, and G. L. Pearson, *Trans. A.I.E.E.*, **65,** 711 (1946), and *Bell System Tech. J.*, **26,** 170 (1947).
134. H. Murmann, *Z. tech. Physik*, **14,** 538 (1933).
135. H. Klumb and Th. Haase, *Physik. Z.*, **37,** 27 (1936).
136. W. A. Michelson, *Physik. Z.*, **9,** 18 (1908).
137. W. Seeman, *Physik. Z.*, **37,** 446 (1936).
138. It should be observed that, for low-velocity electrons such as are obtained in the ionization gauge, the mean free path for electrons is $4(2)^{1/2}$ times the mean free path for the molecules, while for ions the mean free path is $2^{1/2}$ times that for the molecules. See J. D. Cobine, *Gaseous Conductors*, Dover Publications, 1958.
139. See J. D. Cobine (Ref. 138), p. 78.
140. O. E. Buckley, *Proc. Natl. Acad. Sci. U.S.*, **2,** 683 (1916).
141. S. Dushman and C. G. Found, *Phys. Rev.*, **17,** 7 (1921).
142. Misamichi So, *Proc. Phys. Math. Soc. Japan*, **1,** 76 (1919).
143. H. Barkhausen and K. Kurz, *Physik. Z.*, **21,** 1 (1920).
144. E. W. B. Gill and J. H. Morrell, *Phil. Mag.*, **44,** 161 (1922).
145. E. K. Jaycox and H. Weinhart, *Rev. Sci. Instr.*, **2,** 401 (1931).
146. M. J. Copley, T. E. Phipps, and J. Glasser, *Rev. Sci. Instr.*, **6,** 371 (1935).
147. R. D. Huntoon and A. Ellett, *Phys. Rev.*, **49,** 381 (1936).
148. R. S. Morse and R. M. Bowie, *Rev. Sci. Instr.*, **11,** 91 (1940). This gauge is made by Consolidated Vacuum Corporation, Rochester, N.Y., and is designated Type VG-1A.
149. The mean free path between ionizing collisions may be computed from equation 5.44. If $N = 1$, then $L$ becomes equal to the mean free path and is equal to $1/P_iP$. For 150-volt electrons in nitrogen, $P_i = 11$. Hence at a pressure of $5 \cdot 10^{-2}$ mm Hg, the

mean free path for the first ionizing collision is 1.8 cm. This is equal to the effective path length for electrons in the VG-1A.

150. J. D. Cobine (Ref. 138), 81–82.

151. G. J. Schulz and A. V. Phelps, *Rev. Sci. Instr.*, **28**, 1051 (1957).

152. O. A. Weinreich, *Phys. Rev.*, **82**, 573 (1951); H. Bleecker, *Rev. Sci. Instr.*, **23**, 56 (1952).

153. F. M. Penning, *Physica*, **4**, 71 (1937); J. Yarwood (Ref. 3), p. 96.

154. F. M. Penning and K. Nienhauis, *Philips Tech. Rev.*, **11**, 116 (1949).

155. A circuit diagram and calibration curve for a Philips gauge are also shown in the paper by H. A. Thomas, T. W. Williams, and J. A. Hipple, *Rev. Sci. Instr.*, **17**, 368 (1946).

156. J. R. Downing and G. Mellen, *Rev. Sci. Instr.*, **17**, 218 (1946); also *Electronics*, April 1946, p. 142.

157. N. W. Spencer and R. L. Boggess, *ARS Journal*, **29**, 68 (1959).

158. C. G. Found and S. Dushman, *Phys. Rev.*, **23**, 734 (1924).

159. N. B. Reynolds, *Physics*, **1**, 182 (1931).

160. S. Dushman and A. H. Young, *Phys. Rev.*, **68**, 278 (1945).

161. K. T. Compton and C. C. Van Voorhis, *Phys. Rev.*, **26**, 436 (1925); **27**, 724 (1926).

162. W. Bleakney, *Phys. Rev.*, **34**, 157 (1929); **35**, 139, 1180 (1930); **36**, 1303 (1930).

163. P. T. Smith, *Phys. Rev.*, **36**, 293 (1930); **37**, 809 (1931). Also a curve for helium is given by P. T. Smith, *Phys. Rev.*, **39**, 270 (1932).

164. Similar plots, including a few more recent observations, are given in the review by R. B. Brode, *Revs. Modern Phys.*, **5**, 257 (1933).

165. I. Langmuir and H. A. Jones, *Phys. Rev.*, **31**, 357 (1928), Table VIII.

166. S. Wagener and C. B. Johnson, *J. Sci. Instr.*, **28**, 278 (1951).

167. L. Riddiford, *J. Sci. Instr.*, **28**, 375 (1951).

168. G. J. Schulz, *J. Appl. Phys.*, **28**, 1149 (1957).

169. C. G. Found, *Phys. Rev.*, **16**, 41 (1920).

170. N. Morgulis, *Physik. Z. Sowjetunion*, **5**, 407 (1934).

171. K. T. Compton and C. C. Van Voorhis, *Phys. Rev.*, **26**, 436 (1925), and **27**, 724 (1925).

172. H. Simon, *Z. tech. Physik*, **5**, 221 (1924); also A. Goetz, *Physik und Technik des Hochvakuums*, Braunschweig, 1926, p. 197.

173. A. Boburieth, *Le Vide*, **1**, 61 (1946).

174. J. R. Young, *J. Appl. Phys.*, **30**, 1671 (1959).

175. T. W. Hickmott, *J. Appl. Phys.*, **31**, 128 (1960).

176. J. M. Lafferty, *J. Appl. Phys.*, **22**, 299 (1951).

177. J. H. Leck, *Pressure Measurement in Vacuum Systems*, Chapman and Hall, London, 1957, 81 and 93.

178. This assumption was deduced from the statement made in 1922 by Found and Dushman regarding the calibration of an ionization gauge for different gases.

179. J. Blears, *Proc. Roy. Soc. London*, **A, 188**, 62 (1947).

180. In correspondence with Dushman, Blears has advanced valid reasons against drawing such a conclusion.

181. See reference 3, Table 5.2.

182. Mrs. M. R. Andrews, *J. Phys. Chem.*, **27**, 270 (1933).

183. W. B. Nottingham, Conference on Physical Electronics, 1947.

184. R. T. Bayard and D. Alpert, *Rev. Sci. Instr.*, **21**, 571 (1950).

185. L. Apker, *Ind. Eng. Chem.*, **40**, 846 (1948).

186. W. B. Nottingham, *Vacuum Symposium Transactions*, Committee on Vacuum Techniques, Boston, 1954, p. 76.

187. D. Alpert, *J. Appl. Phys.*, **24,** 860 (1953).
188. J. P. Hobson and P. A. Redhead, *Can. J. Phys.*, **36,** 271 (1958).
189. J. P. Hobson and P. A. Redhead, *Can. J. Phys.*, **36,** 255 (1958).
190. J. M. Houston, *Bull. Am. Phys. Soc.*, **II, 1,** 301 (1956).
191. J. M. Lafferty, *J. Appl. Phys.*, **32,** 424 (1961). U.S. Patent 2,884,550, filed Oct. 17, 1957.
192. G. K. T. Conn and H. N. Daglish, *J. Sci. Instr.*, **31,** 412 (1954).
193. H. Sommer, H. A. Thomas, and J. A. Hipple, *Phys. Rev.*, **82,** 697 (1951).
194. D. Alpert and R. S. Buritz, *J. Appl. Phys.*, **25,** 202 (1954).
195. A. G. Edwards, *Brit. J. Appl. Phys.*, **6,** 44 (1955).
196. J. H. Reynolds, *Rev. Sci. Instr.*, **27,** 928 (1956).
197. W. D. Davis and T. A. Vanderslice, *Proceedings of the Seventh National Symposium of the American Vacuum Society*, October, 1960, p. 417.
198. R. M. Bowie, *Rev. Sci. Instr.*, **11,** 265 (1940).
199. R. B. Nelson and A. K. Wing, Jr., *Rev. Sci. Instr.*, **13,** 215 (1942).
200. R. B. Nelson, *Rev. Sci. Instr.*, **16,** 55 (1945).
201. L. N. Ridenour, *Rev. Sci. Instr.*, **12,** 134 (1941).
202. This is also known as a "magic eye" tube, since it contains a fluorescent screen on which a pattern is produced.
203. C. G. Montgomery and D. D. Montgomery, *Rev. Sci. Instr.*, **9,** 58 (1938).
204. J. W. DuMond and W. M. Pickels, *Rev. Sci. Instr.*, **6,** 362 (1935).
205. H. Schwarz, *Z. Physik*, **122,** 437 (1943).
206. J. B. H. Kuper, *Rev. Sci. Instr.*, **8,** 131 (1937).
207. Located at Nela Park, Cleveland. The remarks in this paragraph are based on a personal communication.
208. L. N. Ridenour, *Rev. Sci. Instr.*, **12,** 134 (1944).
209. J. H. Manley, L. J. Haworth, and E. A. Luebke, *Rev. Sci. Instr.*, **10,** 389 (1939).
210. E. J. Lawton, *Rev. Sci. Instr.*, **11,** 134 (1940).
211. R. B. Nelson, *Rev. Sci. Instr.*, **16,** 55 (1945).
212. L. N. Ridenour and C. W. Lampson, *Rev. Sci. Instr.*, **8,** 162 (1937).
213. H. Nelson, *Rev. Sci. Instr.*, **16,** 273 (1945).
214. R. B. Jacobs and H. F. Zuhr, *J. Appl. Phys.*, **18,** 34 (1947). The original rates of leak, expressed in micron · cubic feet per hour, have been converted into micron · liters per minute by means of the conversion factor, 1 micron · cu ft · $hr^{-1}$ = 0.472 micron · liter · $min^{-1}$.
215. R. B. Jacobs and H. F. Zuhr, *J. Appl. Phys.*, **18,** 34 (1947).
216. A. O. Nier, C. M. Stevens, A. Hustrulid, and T. A. Abbott, *J. Appl. Phys.*, **18,** 30 (1947).
217. W. G. Worcester and E. G. Doughty, *Trans. Am. Inst. Elect. Engrs.*, **65,** 946 (1946).
218. H. A. Thomas, T. W. Williams, and J. A. Hipple, *Rev. Sci. Instr.*, **17,** 368 (1946); *Westinghouse Engr.*, **6,** 108 (1946).
219. This design of spectrometer was based on that described by A. O. C. Nier, *Rev. Sci. Instr.*, **11,** 212 (1940).
220. B. Cassen and D. Burnham, *Intern. J. Appl. Radiation and Isotopes*, **9,** 54 (1960).
221. W. C. White and J. S. Hickey, *Electronics*, **21,** 100 (1948).
222. See discussion in section 9.9, as well as references to literature given there.
223. A. Weber, *Glas- u. Hochvakuum-Tech.*, **2,** 259 (1953).

# 6 Sorption of gases and vapors by solids

Revised by George L. Gaines, Jr.

## 6.1. ADSORPTION, ABSORPTION, OCCLUSION, AND SORPTION[1]

One of the most important problems in high-vacuum technique is the removing of gases and vapors which are present both on the surface and in the interior of glass walls and metal parts. The fact that charcoal and other finely divided solids, after thorough evacuation, can take up considerable volumes of different gases and vapors has found applications in a great number of industrial processes.

There are a number of mechanisms by which a gas or vapor may be taken up by a solid. The solid may actually react with the gas or vapor chemically. For example, phosphorus pentoxide ($P_2O_5$) removes water vapor by combining with it chemically to form phosphoric acid ($H_3PO_4$). Other commonly used laboratory desiccants undergo hydration reactions in which the water is bound in the crystal lattice of the solids; calcium chloride and magnesium perchlorate are in this category. Some types of such "clean-up" by chemical reaction will be discussed in Chapter 9.

There are, however, other mechanisms by which gas or vapor disappears in the presence of an evacuated solid. The gas may enter into the *interior of the solid* in much the same manner as gas dissolving in a liquid; this is known as *absorption* or (less commonly) *occlusion*. Metals in particular exhibit this kind of behavior with many gases, and examples of such interactions will be given in Chapter 8. Alternatively, the gas may interact only with the surface of the solid; the deposition of a layer (having a thickness of one or more molecules) *on the surface* is known as *adsorption*. In many cases the gas taken up by the solid may be present both on the surface and in the bulk; sometimes it is difficult to determine the exact nature of the mechanism involved. Consequently the term *sorption*, introduced by J. W. McBain in 1909, is often used as a generic term which includes both adsorption and absorption.

The solid which takes up the gas is known as the *sorbent* (*adsorbent*, *absorbent*); the gas or vapor removed from the gas phase is known as the *sorbate* (*adsorbate*, *absorbate*).

The process of removing gas from a sorbent is ordinarily designated *desorption*.

With respect to the interaction of a gas with the surface of a solid, it is of course apparent that the sorptive capacity of any adsorbent depends largely upon the extent of surface per unit mass. The amount actually adsorbed per unit mass for any adsorbent is much greater for porous substances, such as charcoal, silica gel, and finely divided oxides and metals, than for plane surfaces, such as glass, mica, and smooth metals.

The intermolecular forces which cause adsorption differ for different gas-solid systems. In some cases interactions very similar to those involved in chemical reactions may occur. Sufficient disturbance of the electronic configuration of the gas molecules adsorbed and of the molecules of the solid surface takes place to lead to the formation of a valence bond between the adsorbate and the absorbent. Such *chemisorption* is associated with a high degree of adsorbent-adsorbate specificity, with large heats of adsorption, and tends to be irreversible or reversible with great difficulty.[2]

*Physical adsorption*, on the other hand, is a much more general phenomenon. Even gases which can have no chemical interaction with a solid will be adsorbed by that solid as a result of purely physical forces. Such relatively weaker interactions as polarization forces or van der Waals forces are involved. These forces, of course, are those associated with liquefaction of gases, and there are many correlations between physical adsorption behavior and liquefaction behavior. Physical adsorption increases with decrease in temperature, and in general the adsorption is greater at any given temperature for those gases that are more readily condensable or have higher boiling points. This fact is illustrated by the data in Table 6.1,[3] taken from a paper by Hene.[4]

A number of criteria, both phenomenological and mechanistic, have been offered to distinguish between physical adsorption and chemisorption. Although in most cases these allow an adequate differentiation, in some instances it is difficult to decide whether a given phenomenon involves chemisorption or not. Under proper conditions, too, sorption of additional gas molecules caused by physical forces will be superimposed on a primary chemisorption process. The investigation by Ehrlich[5] of the adsorption of nitrogen on tungsten provides an interesting example of the latter behavior.

TABLE 6.1

**Adsorption of Gases by Charcoal**

(Volume per gram adsorbent, temperature 15° C)

| Gas | Volume Adsorbed ($cm^3$) | Boiling Point* (° C) | Critical Temperature* (° C) |
|---|---|---|---|
| $COCl_2$ | 440 | +8.3 | 182 |
| $SO_2$ | 380 | −10.0 | 157.2 |
| $CH_3Cl$ | 277 | −24.2 | 143.1 |
| $NH_3$ | 181 | −33.4 | 132.4 |
| $H_2S$ | 99 | −61.8 | 100.4 |
| HCl | 72 | −83.7 | 51.4 |
| $N_2O$ | 54 | −88.5 | 36.5 |
| $C_2H_2$ | 49 | −83.6 (subl) | 36 |
| $CO_2$ | 48 | −78.5 (subl) | 31.1 |
| $CH_4$ | 16 | −161.5 | −82.5 |
| CO | 9 | −192 | −139 |
| $O_2$ | 8 | −183 | −118.8 |
| $N_2$ | 8 | −195.8 | −147.1 |
| $H_2$ | 5 | −252.8 | −239.9 |

* Values given in *Handbook of Physics and Chemistry*, 40th edition.

For these reasons, physical adsorption and chemisorption frequently cannot be clearly differentiated, and any separate consideration of these phenomena must be subject to the qualification that they may overlap in real systems. De Boer[6] has discussed this point in some detail.

We are concerned with sorption from the point of view of its importance in high-vacuum technique. The present chapter will treat the general phenomena of sorption; physical adsorption will be discussed first and then chemisorption. Succeeding chapters will contain specific information about the sorptive behavior of materials commonly used in vacuum systems.

## 6.2. SOME TYPICAL ADSORPTION ISOTHERMS

In order to investigate the sorption of a gas or vapor by a given adsorbent, a definite weight or area of the adsorbent is first of all desorbed at as high a temperature as practicable, in a good vacuum, in order to remove all adsorbed or absorbed gases. Then, while the adsorbent is maintained at a constant known temperature, the sorbate gas is introduced into the system, and a decrease in pressure is noted when equilibrium is obtained.[7]

It is observed that there exists in most cases a definite relation between the amount of gas adsorbed per unit mass or unit area of adsorbent and the pressure, which can be expressed in the form

$$V = f(P, T),$$

where $V$ is the amount of gas (commonly expressed as the volume) and $f$ is an algebraic function of the variables pressure ($P$) and temperature ($T$). A plot of $V$ versus $P$, for example, at *constant temperature* is known as an *adsorption isotherm*. From the isotherms at a series of temperatures, it is possible to plot *adsorption isosteres* which give the variation of the equilibrium pressure with temperature for a *constant amount of gas adsorbed.* The isosteres are thus analogous to vapor-pressure curves of liquids, and, as in the case of the latter, a form of the *heat of adsorption* ($Q_S$) is given by the Clausius-Clapeyron relation:

$$Q_S = -R_0 \frac{d \ln P}{d(1/T)}, \tag{6.1}$$

where $R_0$ is the gas constant per mole.

From the isotherms it is also possible to plot *isobars*, which indicate the amounts adsorbed at a series of temperatures for a given *constant pressure.* Illustrations of both isosteres and isobars are given in the subsequent discussion.

In the following sections the form of the adsorption isotherm is discussed for those cases in which the adsorbed gas forms a monolayer or a layer only a few molecules thick. Adsorption isosteres and heat of adsorption are discussed in subsequent sections.

The data on the amount of adsorption are usually expressed in terms of $V_0$, the volume in cubic centimeters at 0° C and 760 mm (STP) per gram of adsorbent, or per square centimeter of smooth surface (glass, mica, metals). Some observers record their data in terms of $V_{20}$, the volume per gram (or per unit area) at 20° C and 760 mm. Other investigators have given the amounts adsorbed in terms of $x$ (sometimes written as $x/m$), the mass in *milligrams per gram* of adsorbent, or in terms of $N_M$, the number of *moles per gram.* Tables 6.2 and 6.3 give the values of conversion factors which will be found useful in comparing data from different sources. The fifth column in Table 6.3 gives the number of molecules required to form one square centimeter of a layer one molecule thick (unimolecular layer or monolayer), while $V_{20}$ and $V_0$ give the values of the corresponding volumes in cubic centimeters at 760 mm and 20° C and 0° C, respectively.

Figure 6.1 shows adsorption isotherms for nitrogen on activated coconut charcoal, as observed by Homfray.[8] Values of $V_0$ are plotted against $P$, the pressure in millimeters, for the range 83–373° K (−190 to +100° C).

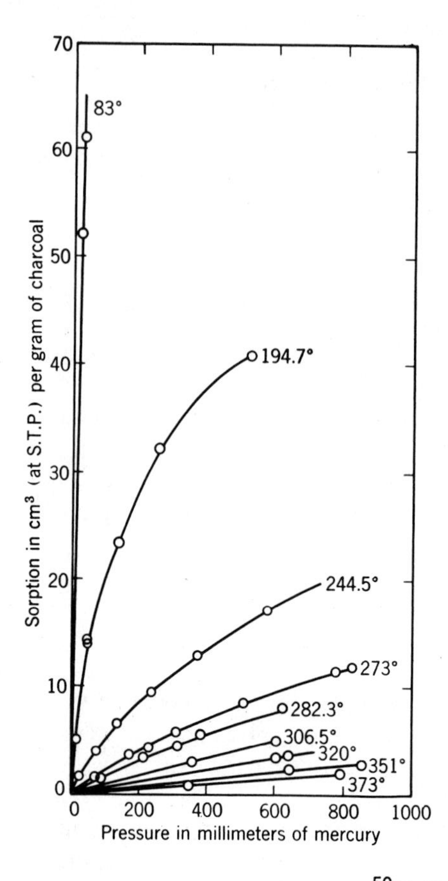


Fig. 6.1. Isotherms for the sorption of nitrogen by coconut charcoal (Homfray).

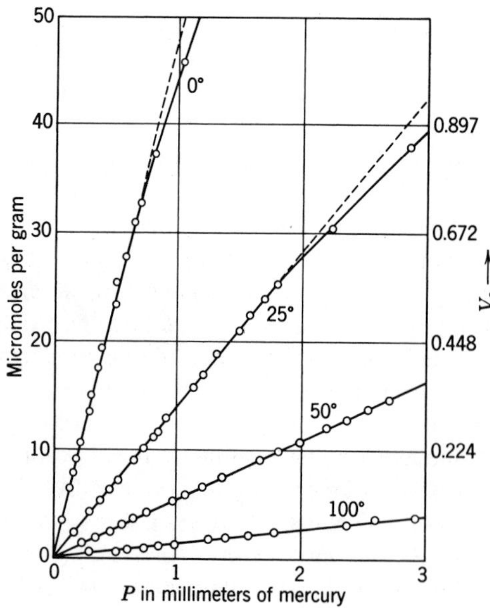


Fig. 6.2. Isotherms for the sorption at low pressures of carbon dioxide by wood charcoal (Magnus and Kratz).

TABLE 6.2

**Conversion Factors for Amount Adsorbed**

$V_{20} = V_0 \times 293.2/273.2 = 1.074V_0.$
$V_0 = 0.9317V_{20}.$
$Q_{\mu l}$ = number of micron · liters at $t°$ C = $760V$,
where $V$ = volume in cubic centimeters at 1 atm and $t°$ C.
$N_M = V_0/22{,}415$ = number of moles per gram,
$= V_{20}/24{,}050.$
$10^6 N_M$ = number of micromoles per gram,
$x = 10^3 N_M M$ = milligrams per gram,
$= V_0 M/(22.415) = V_{20}M/(24.050).$
$n_S$ = number of molecules adsorbed per gram,
$= 6.023 \cdot 10^{23} N_M,$
$= 2.687 \cdot 10^{19} V_0,$
$= 2.504 \cdot 10^{19} V_{20}.$
$V_t = 1.3158 \cdot 10^{-3} Q_{\mu l}$ (at $t°$ C).

TABLE 6.3

**Number of Molecules per Monolayer and Equivalent Volumes**

| Gas | $M$ | $M/(22.415)$ | $M/(24.050)$ | $10^{-14}N_S$* | $10^5 V_{20}$ | $10^5 V_0$† |
|---|---|---|---|---|---|---|
| $H_2$ | 2.016 | 0.0900 | 0.08381 | 15.22 | 6.08 | 5.67 |
| He | 4.003 | 0.1790 | 0.1664 | 24.16 | 9.65 | 8.99 |
| Ar | 39.94 | 1.7820 | 1.661 | 8.54 | 3.41 | 3.18 |
| $N_2$ | 28.02 | 1.250 | 1.165 | 8.10 | 3.24 | 3.02 |
| $O_2$ | 32.00 | 1.428 | 1.330 | 8.71 | 3.48 | 3.24 |
| CO | 28.01 | 1.250 | 1.165 | 8.07 | 3.23 | 3.00 |
| $CO_2$ | 44.01 | 1.963 | 1.830 | 5.34 | 2.13 | 1.99 |
| $CH_4$ | 16.03 | 0.7152 | 0.6664 | 5.23 | 2.09 | 1.95 |
| $NH_3$ | 17.03 | 0.7597 | 0.7078 | 4.56 | 1.82 | 1.70 |
| $H_2O$ | 18.02 | 0.8041 | 0.7492 | 5.27 | 2.11 | 1.96 |

* Values of $N_S$ (estimated for viscosity) taken from Table 1.6. Values for CO, $CH_4$, and $NH_3$ were calculated from the values of $\eta_0$, $1.659 \cdot 10^{-4}$, $1.022 \cdot 10^{-4}$, and $0.92 \cdot 10^{-4}$, respectively. See also Table 1.9.

† To deduce the values at 25° C, multiply the values given in this column by 1.091.

Figure 6.2 shows isotherms at low pressures for carbon dioxide on wood charcoal, as observed by Magnus and Kratz.[9] Figure 6.3 gives the isotherms over a higher range of pressures. The ordinates give micromoles ($10^6 N_M$) per gram of adsorbent, and the corresponding values of $V_0$ are indicated on the right-hand scale. ($V_0 = 0.02242 \times$ micromoles.)

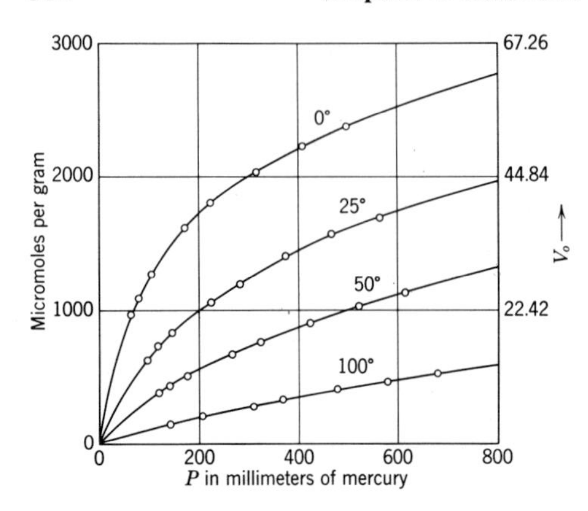


Fig. 6.3. Isotherms for the sorption at higher pressures of carbon dioxide by wood charcoal (Magnus and Kratz).

Figure 6.4, taken from a paper by Armbruster[10] gives low-temperature adsorption isotherms for carbon monoxide, nitrogen, and argon on smooth silver (area = 8002 cm²), while Fig. 6.5 shows plots of observations made by Frankenburg[11] on the adsorption of hydrogen on tungsten powder. From the latter set of measurements, it was concluded that the "true saturation value" for the particular sample of powder used was $6.60 \cdot 10^{-6}$ mole hydrogen per gram (instead of $3.60 \cdot 10^{-6}$, as would follow from the isotherm for −195° C). Table 6.4 gives the adsorbed amounts in percentage of this value as a function of the equilibrium pressures at various temperatures.

**TABLE 6.4**

**Adsorption of Hydrogen by Tungsten Powder, Expressed as Percentage of Saturation Value**

| Equilibrium $P_{mm}$ | Temperatures (° C) | | | | | | | |
|---|---|---|---|---|---|---|---|---|
| | 0 | 100 | 200 | 300 | 400 | 500 | 600 | 700 |
| $1 \cdot 10^{-6}$ | ... | ... | 4.4 | 0.8 | ... | ... | ... | ... |
| $1 \cdot 10^{-4}$ | 21.0 | 15.5 | 7.5 | 3.3 | 0.9 | ... | ... | ... |
| $1 \cdot 10^{-3}$ | 23.5 | 17.5 | 10.0 | 4.6 | 1.8 | 0.6 | ... | ... |
| $1 \cdot 10^{-2}$ | 25.5 | 19.0 | 12.0 | 6.5 | 3.1 | 1.2 | 0.4 | 0.15 |
| $1 \cdot 10^{-1}$ | 29.5 | 23.5 | 16.0 | 10.0 | 5.6 | 2.4 | 0.9 | 0.5 |
| 1 | 33.0 | 27.5 | 20.5 | 14.0 | 9.5 | 5.0 | 2.4 | 1.5 |
| 10 | 38.0 | 32.5 | 26.5 | 20.0 | 16.0 | 10.5 | 7.0 | 4.5 |

It will be observed that in all cases the amount adsorbed at any given temperature increases more or less rapidly with increase in pressure at very low pressures, increases less rapidly at higher pressures, and finally tends to reach a limiting value at still higher pressures. Furthermore, for the

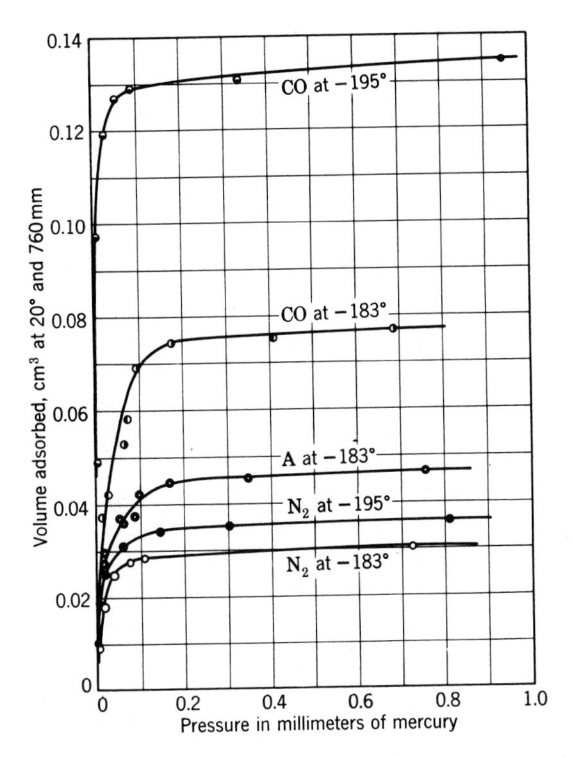


Fig. 6.4. Isotherms for the adsorption of carbon monoxide, argon, and nitrogen on smooth silver at −183° C and −195° C (Armbruster).

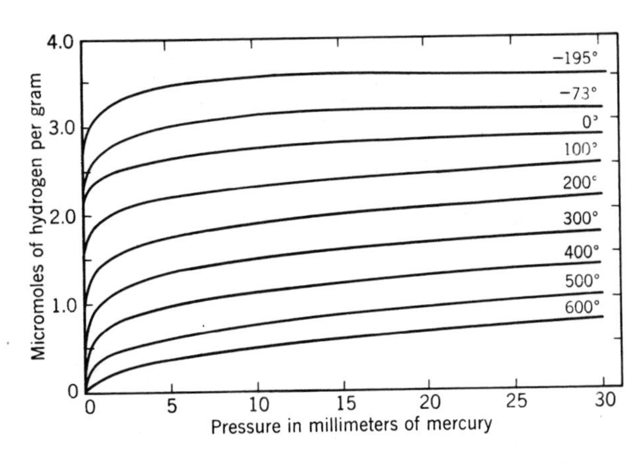


Fig. 6.5. Isotherms for the adsorption of hydrogen on tungsten powder (Frankenburg).

same equilibrium pressure, the amount adsorbed decreases with increase in temperature. These isotherms are qualitatively characteristic of almost all observations on adsorption over a range of pressures and temperatures at which the amount adsorbed does not exceed that corresponding to approximately a unimolecular layer. For cases of multimolecular adsorption the isotherms exhibit a more or less continuous increase with pressure up to a value $P_S$, which corresponds to the saturation vapor pressure of the adsorbate at the given temperature. In some instances there may be a plateau region at moderate pressures wherein the adsorption increases only slowly with increasing pressure, before beginning a steeper rise as saturation is approached. This type of behavior will be discussed in further detail below.

Although in the above discussion it has been implied that the amount adsorbed is a continuous function of the equilibrium pressure, some experimentally determined isotherms have exhibited discontinuities. While the significance of some of these observations is not altogether clear, they may be associated with phase transitions in the adsorbed layer and/or mark the completion of one adsorbed layer before another is begun.

Such discontinuities were first observed by Allmand and Chaplin[12] in measurements on the sorption of carbon tetrachloride by charcoal. Subsequent observations by Allmand and Burrage[13] on the sorption of both water and carbon tetrachloride by charcoal confirmed the earlier results.[14] Indications of discontinuities have been found in systems such as heptane-ferric oxide[15] and krypton-sodium bromide.[16] Other observers, however, have failed to confirm some of the observed discontinuities, and Corrin and Rutkowski[17] have described one system in which either a stepwise isotherm or a continuous one could be obtained, depending on the detailed experimental technique used.

## 6.3. PARABOLIC AND HYPERBOLIC EQUATIONS FOR ADSORPTION ISOTHERM

A number of algebraic expressions have been deduced by different investigators to describe the variation (at constant temperature) in amount adsorbed with the pressure. A review of these different relations is given in the following remarks.

### Parabolic Adsorption Isotherm

In 1909 Freundlich[18] proposed the semiempirical relation

$$V = kP^{1/n}, \tag{6.2}$$

where $k$ and $n$ are constants that depend upon the nature of both adsorbent

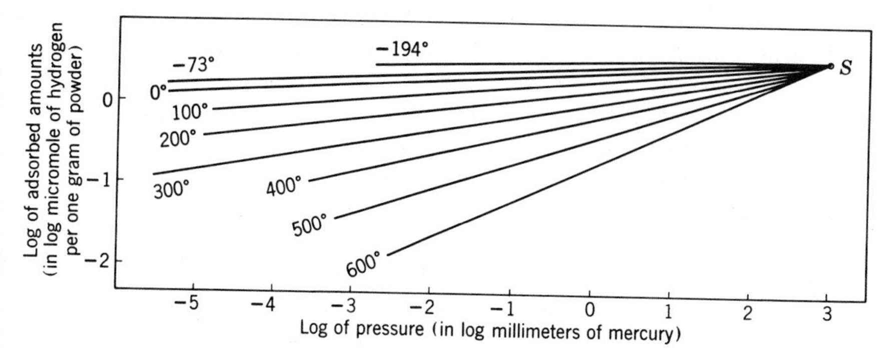


Fig. 6.6. Log-log plots of isotherms shown in Fig. 6.5.

and adsorbate and also upon the temperature. This relation is usually expressed in the form

$$\log V = \log k + \frac{1}{n} \log P, \tag{6.3}$$

so that a plot of log $V$ versus log $P$ should yield a straight line, the slope of which gives the value of $n$ in accordance with the relation

$$\frac{d(\log V)}{d(\log P)} = \frac{1}{n}. \tag{6.4}$$

In all cases $n$ is greater than or equal to unity. Figure 6.6 shows plots of log $N_M$ versus log $P$ for the isotherms plotted in Fig. 6.5. Table 6.5,

**TABLE 6.5**

**Values of $n$ for Adsorption of Hydrogen by Tungsten Powder**

| Temperature (° C) | Powder 9798 | Powder 9799 | |
|---|---|---|---|
| −194 | (67.4) | Not measured | ... |
| −73 | 24.1 | Not measured | ... |
| 0 | 16.85 | 16.41 | ... |
| 100 | 10.88 | 10.74 | ... |
| 200 | 7.80 | 7.36 | ... |
| 300 | 5.80 | 5.40 | ... |
| 400 | 4.04 | 4.30 | 2.02 |
| 500 | 3.08 | 3.04 | 2.04 |
| 600 | 2.30 | 2.32 | 1.98 |
| 650 | 2.09 | 2.23 | 2.00 |
| 700 | 1.85 | 1.85 | 2.06 |
| 750 | 1.71 | 1.73 | 2.03 |

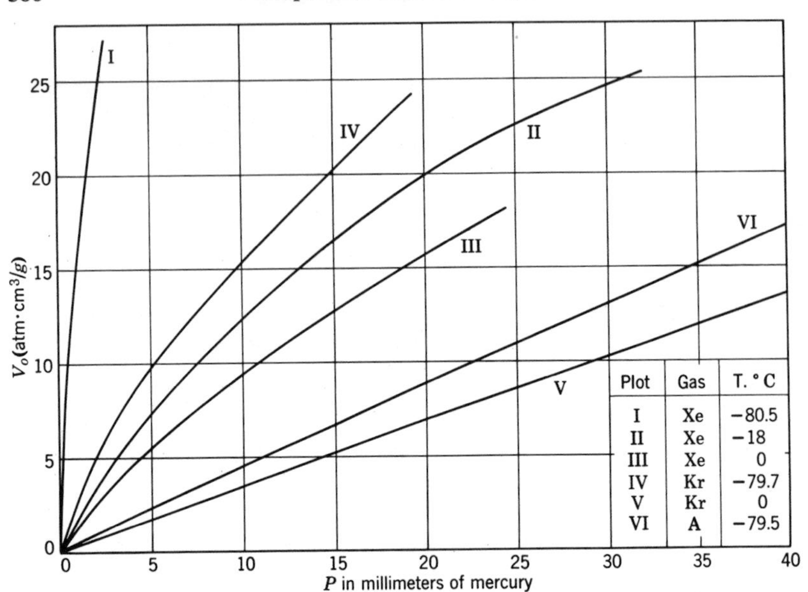


Fig. 6.7. Isotherms for adsorption of argon, krypton, and xenon on charcoal (Peters and Weil).

taken from the paper by Frankenburg (see Table 6.4), gives values, for a series of temperatures, of the slope ($d \log P/d \log N_M$) for two tungsten powders. The values given in the second and third columns were obtained for total amounts adsorbed greater than 0.8 per cent of the saturation value; the values in the last column were obtained for adsorptions less than 0.8 per cent of saturation. (The values plotted in Figs. 6.5 and 6.6 are for powder 9798.)

As Brunauer points out,[19]

> In general a large number of the experimental results in the field of van der Waals adsorption[20] (and even in chemisorption) can be expressed by means of the Freundlich equation in the middle pressure range. A still larger number cannot be expressed satisfactorily. Where the equation is obeyed, it can be used as an interpolation formula.

As an example Brunauer cites the observations of Peters and Weil[21] on the adsorption of argon, krypton, and xenon on charcoal. The adsorption isotherms for different temperatures are shown in Fig. 6.7, while Fig. 6.8 shows the plots of log $V_0$ (cubic centimeters STP) versus log $P_{mm}$. The values of $k$ and $1/n$ for the three gases are given in Table 6.6, where $k = \text{cm}^3/(\text{mm})^{1/n}$.

As will be observed, the value of $k$ for any one gas decreases with increase in temperature, while the value of $1/n$ increases and ultimately reaches the

TABLE 6.6

Adsorption Constants of Rare Gases (per Gram Charcoal)

| Gas | $T = 193°$ K | | $T = 255.2°$ K | | $T = 273.2°$ K | |
|---|---|---|---|---|---|---|
| | $k$ | $1/n$ | $k$ | $1/n$ | $k$ | $1/n$ |
| Argon | 0.500 | 0.950 | (0.0764) | (1.0) | (0.0581) | 1.0 |
| Krypton | 2.927 | 0.711 | (0.497) | (0.885) | 0.340 | 1.0 |
| Xenon | 15.99 | 0.574 | 2.458 | 0.692 | 1.583 | 0.77 |

value unity. For different gases, the values of $k$ increase in general, and those of $1/n$ decrease, with rise in boiling point, as shown by the data in Table 6.7,[22] where $k = \text{cm}^3/(\text{mm})^{1/n}$.

A similar conclusion has been drawn by Langmuir[23] from his observations on the adsorption of a number of gases on mica and glass surfaces at low temperatures. If the gases are arranged in order, according to the

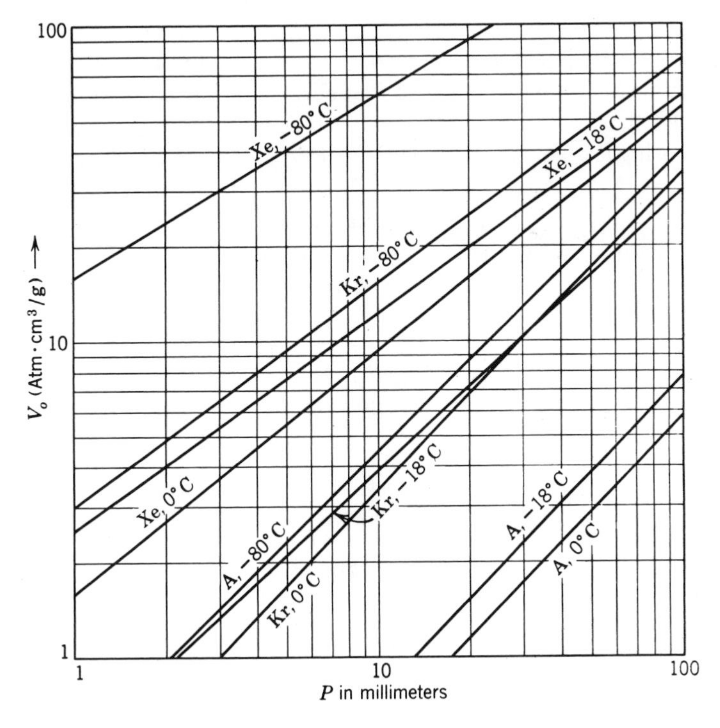


Fig. 6.8. Log-log plots of isotherms shown in Fig. 6.7.

TABLE 6.7

**Adsorption Constants of Gases on Coconut Charcoal**

(1 g of adsorbent at 0° C)

| Gas | $k$ | $1/n$ | Boiling Point (° C) |
|---|---|---|---|
| $N_2$ | 0.26 | 0.87 | −195.8 |
| CO | 0.56 | 0.76 | −191.6 |
| Ar | 0.22 | 0.88 | −185.7 |
| $CH_4$ | 2.69 | 0.56 | −161.5 |
| $C_2H_4$ | 23.7 | 0.23 | −103.9 |
| $CO_2$ | 8.25 | 0.53 | −78.5 |

amounts required to saturate the surface, it is found that this order is the same as that of the boiling points.

For $n = 1$, the amount adsorbed varies directly as the pressure. A similar relation has been observed to apply to the solubility of gases in liquids; it is known as *Henry's law*.[24] As will be shown in the subsequent discussion, there is considerable evidence that for many cases of adsorption the linear relation between $V$ and $P$ applies at very low pressures. Figure 6.2 shows that, in the sorption of carbon dioxide by coconut charcoal at 0° C and higher temperatures, Henry's law is valid in the range of a few

TABLE 6.8

**Application of Henry's Law to Adsorption of Carbon Dioxide on Charcoal**

| $t$° C | $10^6 N_M/P_{mm}$* | Approximate Upper Limit of $P_{mm}$ |
|---|---|---|
| 0 | 47.11 | 0.7 |
| 25 | 14.13 | 1.7 |
| 50 | 5.445 | 3.0 |
| 100 | 1.389 | 11.8 |

* Values given by Magnus and Kratz (Ref. 9).

millimeters pressure. Table 6.8 gives average values of the ratio $10^6 N_M/P_{mm}$ (micromoles per gram per millimeter) and the upper values of $P_{mm}$ at which the ratio begins to decrease below the value in the table (as shown in Fig. 6.3).

Similar illustrations of the applicability of Henry's law have been observed by Lowry and Morgan[25] in the adsorption of nitrogen on graphite at 56.7 and 100° C. In most cases (as will be apparent in our discussion of surface heterogeneity below), however, the true range of applicability of Henry's law is very short; in fact it may be below the range of experimental accessibility.

## Hyperbolic Adsorption Isotherm

While the parabolic equation is to a large extent empirical in origin, Langmuir developed a conception of adsorption phenomena which led not only to the formulation of an extremely important relation for the adsorption isotherm but also to a logical interpretation of the nature of the forces existing at the surfaces of solids and liquids.

Because of the fundamental importance of Langmuir's views, we quote at length from the original paper,[26] in which these new ideas were set forth in detail:

During the year 1914, in connection with studies of electron emission and chemical reactions at low pressures, I became much interested in the phenomena of adsorption, and developed a theory which has been strikingly verified by a large number of experiments carried out since that time. According to this theory there is an abrupt change in properties in passing through the surface of any solid or liquid. The atoms forming the surface of a solid are held to the underlying atoms by forces similar to those acting between the atoms inside the solid. From Bragg's work on crystal structure and from many other considerations we know that these forces are of the type that have usually been classed as chemical. In the surface layer, because of the asymmetry of the conditions, the arrangement of the atoms must always be slightly different from that in the interior. These atoms will be unsaturated chemically and thus they are surrounded by an intense field of force.

From other considerations, I was led to believe that when gas molecules impinge against any solid or liquid surface they do not in general rebound elastically, but condense on the surface, being held by the field of force of the surface atoms. These molecules may subsequently evaporate from the surface. The length of time that elapses between the condensation of a molecule and its subsequent evaporation depends on the intensity of the surface forces. Adsorption is the direct result of this time lag. If the surface forces are relatively intense, evaporation will take place at only a negligible rate, so that the surface of the solid becomes completely covered with a layer of molecules. In cases of true adsorption this layer will usually be not more than one molecule deep, for as soon as the surface becomes covered by a single layer the surface forces are chemically saturated. Where, on the other hand, the surface forces are weak the evaporation may occur so soon after condensation that only a small fraction of the surface becomes covered by a single layer of adsorbed molecules. In agreement with the chemical nature of the surface forces, the range of these forces has been found to be extremely small, of the order of $10^{-8}$ cm. That is, the effective range of the forces is usually much less than the diameter of the molecules. The molecules thus usually orient themselves in definite ways in the surface layer since they are held to the surface by forces acting between the surface and particular atoms or groups of atoms in the adsorbed molecule.

On the basis of this conception of the nature of the forces acting in adsorption, it follows that the material adsorbed on the surface will rarely exceed that contained in one layer of molecules (a *monolayer*), and the relation developed by Langmuir for the adsorption isotherm forms a logical consequence.

At any given pressure $P_{\mu b}$, the number of molecules striking unit area per second is given by the relation

$$\nu = \frac{P_{\mu b}}{(2\pi mkT)^{1/2}} = \frac{3.513 \cdot 10^{22} P_{mm}}{(MT)^{1/2}} = cP, \tag{6.5}$$

where the value of $c$ depends on the unit of pressure.

Let $\alpha$ = fraction of molecules incident on surface that condense, and let $\mu$ = number of *adatoms*[27] evaporating per square centimeter per second. At equilibrium,

$$\alpha\nu = \mu. \tag{6.6}$$

Let $\sigma_1$ = total number of "elementary spaces," or *sites*, available for adsorption, and let $\sigma$ = number of adatoms per square centimeter at pressure $P$.

Assuming that a single adatom can occupy one site, then the fraction of surface covered at equilibrium is given by

$$\theta = \frac{\sigma}{\sigma_1}. \tag{6.7}$$

Let us assume also that (1) "the adsorption sites are all identical and (2) the potential energy of an adatom in a site is independent of the presence of adatoms in other sites; in other words, the adatoms in separate sites exert no forces on one another."[28] Then

$$\mu = \mu_1\theta, \tag{6.8}$$

where $\mu_1$ is the rate of evaporation for a completely covered surface. Also,

$$\alpha\nu = \alpha_0\nu(1 - \theta), \tag{6.9}$$

where $\alpha_0$ is a coefficient of the order unity. Hence, at equilibrium,

$$\mu_1\theta = \alpha_0\nu(1 - \theta) \tag{6.10}$$

or

$$\frac{\theta}{1 - \theta} = \frac{\alpha_0\nu}{\mu_1}$$

while

$$\frac{1}{1-\theta} = \frac{\mu_1 + \alpha_0 \nu}{\mu_0}.$$

It follows that

$$\theta = \frac{\alpha_0 \nu}{\mu_1 + \alpha_0 \nu} \tag{6.11}$$

is the equation for the adsorption isotherm.

Let

$$b = \frac{\alpha_0 c}{\mu_1}, \tag{6.12}$$

where $c$ is defined by equation 6.5.

Since $\mu_1$ is a function of the temperature, it follows that $b$ is a constant for any given temperature.

The equation 6.11 takes the form

$$\theta = \frac{bP}{1 + bP} \tag{6.13a}$$

$$= b\left(\frac{1}{P} + b\right)^{-1}. \tag{6.13b}$$

This is the equation of a rectangular hyperbola, hence the designation "hyperbolic" isotherm.

From 6.13*a* it follows that

$$\theta = bP(1 - \theta). \tag{6.14}$$

It is more convenient to write the equation for the isotherm in terms of $V$, the volume adsorbed, or $x$, the amount adsorbed. Let $V_s$ and $x_s$ denote the maximum values, when the surface is completely covered. Then

$$\theta = \frac{V}{V_s}\left(= \frac{x}{x_s}\right), \tag{6.15}$$

and equation 6.14 becomes

$$V = \frac{V_s bP}{1 + bP}, \tag{6.16a}$$

which may also be written in the form

$$P = \frac{V}{b(V_s - V)}. \tag{6.16b}$$

In order to test the validity of the hyperbolic equations for any series of adsorption data, it is convenient to write equation 6.16 in the form

$$\frac{P}{V} = \frac{1}{bV_s} + \frac{P}{V_s} \tag{6.17a}$$

or

$$\frac{1}{V} = \frac{1}{bV_s} \cdot \frac{1}{P} + \frac{1}{V_s}. \tag{6.17b}$$

That is, a plot of $P/V$ versus $P$ should yield a straight line, the slope of which corresponds to $1/V_s$, while the intercept on the ordinate axis for $P = 0$ is equal to $1/(bV_s)$.

It also follows from equation 6.17*b* that a plot of $1/V$ versus $1/P$ should yield a straight line, which is another method of expressing the equation for a rectangular hyperbola.

Equation 6.16*b* shows that a straight line should also be obtained by plotting $V/P$ versus $V$. In that case, the intercept on the ordinate axis for $V = 0$ is equal to $bV_s$, while the slope is equal to $-b$.

It is evident that for very low pressures, where $bP \ll 1$ and $\theta$ *is extremely small* ($V$ very much less than $V_s$),

$$V = V_s bP. \tag{6.18}$$

That is, the amount adsorbed is proportional to $P$, and the adsorption therefore obeys Henry's law.

It will be noted that, in the above equations, $b$ has the reciprocal dimension of the pressure, while $V$ and $V_s$ must be expressed in the same units of volume or mass.

The hyperbolic equation has been derived on the basis of thermodynamical considerations by Volmer[29] and by statistical mechanics methods by Fowler[30] and a number of other investigators.

It is of interest to quote the summary of Fowler's derivation given by Brunauer.[31]

Fowler emphasizes the fact that the original kinetic derivation is apt to obscure the essentially thermodynamic character of the [Langmuir] equation and may lead to the erroneous belief that the form of the equation depends on the mechanism of condensation and evaporation of the adsorbed molecules. As a matter of fact, it depends only on the entire set of states, adsorbed and gaseous, that are accessible to the molecules at equilibrium. The statistical derivation is based on three assumptions: (1) the molecules of the gas are adsorbed without dissociation to definite points of attachment on the surface of the adsorbent; (2) each point of attachment can accommodate one and only one adsorbed molecule; and (3) the energies of the states of any adsorbed molecules are independent of the presence or absence of any other adsorbed molecules on neighboring points of attachment. These assumptions are similar to those of

Langmuir. The first assumption, although not stated, is implied in the kinetic derivation. When dissociation occurs, the form of the Langmuir equation is different from equation 6.13.[32] The second assumption corresponds to Langmuir's assumption of unimolecularity, embodied in equation 6.9. The third corresponds to the assumption of no interaction between adsorbed molecules.

This, as mentioned already, leads to equation 6.16.

Langmuir also showed that the same considerations as those used in deriving the hyperbolic equation lead to modifications of this equation in special cases.

One such case is that in which the surface contains several kinds of elementary spaces or sites. Let $V_1$, $V_2$, etc., denote the maximum values of the adsorptions for each of these varieties of sites. Then the expression for the isotherm is

$$\frac{V}{P} = \frac{V_1 b_1}{1 + b_1 P} + \frac{V_2 b_2}{1 + b_2 P} + \cdots, \tag{6.19}$$

where $b_1$, $b_2$, etc., refer to the different types of sites, and

$$V_1 + V_2 + \cdots = V_s. \tag{6.20}$$

Another very important case is that in which the molecules are adsorbed as atoms. Langmuir writes,

In the cases considered above, molecules of adsorbed substance have acted as indivisible units; that is, when a molecule evaporated from the surface it contained the same atoms as when it was first adsorbed. There is good evidence that the forces which hold adsorbed substances act primarily on the individual atoms rather than on the molecules. When these forces are sufficiently strong, it may happen that the atoms leaving the surface become paired in a different manner from that in the original molecules.

For example, consider the adsorption of a diatomic gas such as oxygen on a metallic surface. Let us assume that the atoms are individually held to the metal, each atom occupying one elementary space. The rate of evaporation of the atoms is negligibly small, but occasionally adjacent atoms combine together and thus nearly saturate each other chemically, so that their rate of evaporation becomes much greater. The atoms thus leave the surface only in pairs as molecules. Starting with a bare surface, if a small amount of gas is adsorbed, the adjacent atoms will nearly always be the atoms which condensed together when a molecule was adsorbed. But from time to time two molecules will happen to be adsorbed in adjacent spaces. One atom of one molecule and one of the other may then evaporate from the surface as a new molecule, leaving two isolated atoms which cannot combine together as a molecule and are therefore compelled to remain on the surface. When a stationary state has been reached there will be a haphazard distribution of atoms over the surface. The problem may then be treated as follows:

Let $\theta$ be the fraction of the surface covered by adsorbed atoms, while $(1 - \theta)$ is the fraction which is bare. In order that a given molecule approaching the surface may condense (and be retained for an appreciable time) on the surface, two particular elementary spaces must be vacant. The chance of one of these spaces being vacant is $(1-\theta)$; that both shall be vacant is $(1-\theta)^2$. The rate of condensation is thus equal to $\alpha_0\nu(1-\theta)^2$. Evaporation only occurs when adsorbed atoms are in adjacent spaces. The chance that an atom shall be in a given space is $\theta$. Therefore, the chance that atoms shall be in adjacent spaces is proportional to $\theta^2$. The rate of evaporation of molecules from the surface is therefore equal to $\mu_1\theta^2$, where, as before, $\mu_1$ is the rate of evaporation from a completely covered surface. For equilibrium we then have the condition

$$\alpha_0\nu(1-\theta)^2 = \mu_1\theta^2. \tag{6.21}$$

This leads to the relations

$$\theta = \frac{(bP)^{1/2}}{1+(bP)^{1/2}} \tag{6.22}$$

and

$$V = \frac{V_s(bP)^{1/2}}{1+(bP)^{1/2}}. \tag{6.23}$$

Thus [as Langmuir states], in this case, even at relatively low pressures, the total amount of adsorbed gas varies in proportion to the square root of the pressure. If three instead of two elementary spaces are occupied by the atoms of each molecule, then the square root relation expressed by equation 6.23 becomes a cube root relation.

Hence equation 6.23 may be expressed in the more general form

$$V = \frac{V_s(bP)^{1/n}}{1+(bP)^{1/n}}, \tag{6.24}$$

where $n \geqslant 1$, and this can also be written

$$\theta^n = bP(1-\theta)^n. \tag{6.25}$$

For small values of $\theta$, this evidently becomes identical in form with the Freundlich equation 6.2.

By analogy with the hyperbolic equation (for which $n = 1$), it is evident that, for the purpose of testing the validity of equation 6.24, it may be expressed in the two alternative forms:

$$\frac{P^{1/n}}{V} = \frac{1}{c_1} + \frac{P^{1/n}}{c_2} \tag{6.26}$$

or

$$\frac{V}{P^{1/n}} = k_1 - k_2V, \tag{6.27}$$

where $n$, and the constants $c_1$, $c_2$, $k_1$, and $k_2$, are determined from the slopes and intercepts on the axis of ordinates, for the linear plots.

The hyperbolic equation has been found to give good agreement with experimental data in a number of adsorption studies. However, it was soon found that difficulties arose in reconciling the actually derived values of the constants $V_s$ and $b$ with the physical significance attached to them by Langmuir. For example, a number of investigators working with materials whose surface areas were known or could be reasonably estimated have found that the area covered by adsorbed gas (as estimated from $V_s$ and the gas molecular diameter evaluated from the viscosity of the gas) is much smaller than the true area of the material. Langmuir himself, in fact, observed this type of behavior in the adsorption of a number of gases on mica and glass at low pressures.[33] Another observation frequently made is that the values of $V_s$ obtained with different gases on the same adsorbent differ widely. For example, the data of Markham and Benton[34] for the adsorption of oxygen, carbon monoxide, and carbon dioxide on silica show such differences; the observed values ranged from 60 $cm^3$ to 426 $cm^3$.

A number of attempts have been made to explain these low values of saturation adsorption and the variations in the value for a given substrate. Langmuir suggested that "the adsorption of most of these gases does not occur over the whole surface." In more recent years, however, it has appeared that the simple postulates of the Langmuir theory are insufficient to cover the extreme complexity of the phenomena involved in physical adsorption on even relatively simple surfaces. A number of modifications have been made to allow for variations in the forces between adsorbate molecules and the substrate. Some of these will be discussed in a subsequent section.

Like the Freundlich equation, the Langmuir isotherm is useful in many cases for the fitting of experimental data. In some systems, furthermore, the value of $V_s$ obtained may give a realistic evaluation of the surface area of solids studied. Moreover, in chemisorption, the Langmuir postulates may be more nearly fulfilled, and the equation may have both greater practical and greater theoretical significance.

The great importance of the Langmuir theory, however, rests on a much broader basis. Its publication marked one of the earliest attempts to treat adsorption phenomena theoretically. The hypothesis of a unimolecular adsorbed layer, which Langmuir invoked on the basis of the short range of molecular forces, is of far-reaching significance. The method of treating the model, furthermore, has proven to be extremely useful, and many of the other theories of adsorption phenomena have started from the techniques of Langmuir.

## 6.4. MULTIMOLECULAR ADSORPTION: THE BRUNAUER-EMMETT-TELLER EQUATION

Before we return to a consideration of some of the other isotherm equations which have been derived for unimolecular layer adsorption, it is

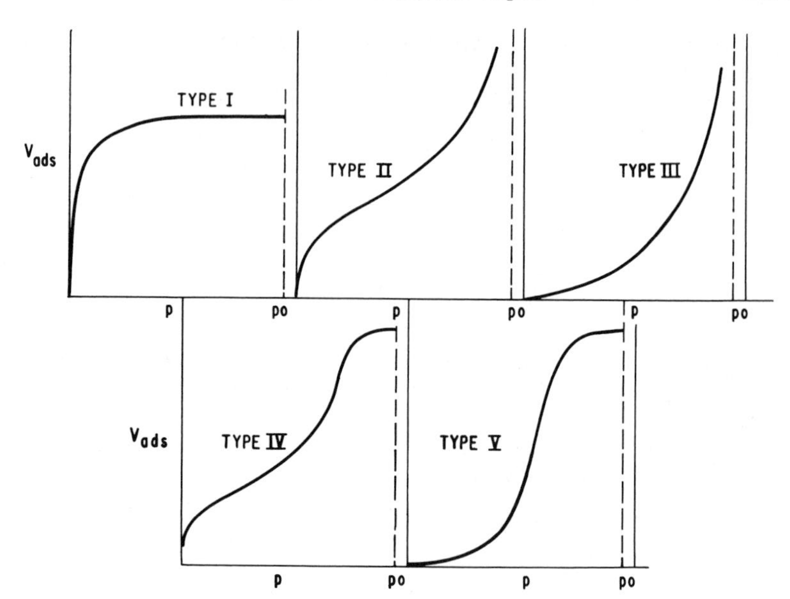


Fig. 6.9. The five isotherm types observed in physical adsorption (Brunauer).

pertinent to consider the equations for multimolecular-adsorption isotherms derived by Brunauer, Emmett, and Teller.[35] These workers concluded that many of the experimentally observed adsorption isotherms reported in the literature for physical adsorption showed evidence of multimolecular layer adsorption. (Brunauer[36] has discussed some of the evidence for the adsorption of multimolecular films.) These isotherms could be characterized as belonging to the five types illustrated in Fig. 6.9. The isotherms of type 1 are, of course, Langmuir isotherms and are characteristic of adsorption of only a single molecular layer. The other four types are now generally considered to represent multilayer adsorption.

The derivation of the simple form of the Brunauer-Emmett-Teller (BET) equation is based on the same assumptions as those involved in the derivation of the Langmuir equation, with the added hypothesis that the formation of successive layers can occur by the adsorption of molecules on top of an already formed previous layer. Ries[37] has described the derivation simply as follows:

The BET multilayer theory is essentially an extension of the Langmuir interpretation of monomolecular adsorption. The derivation is based on the same kinetic picture and the assumption that condensation forces are the principal forces in physical adsorption. As in the Langmuir theory, the rate of evaporation of the molecules in the first layer is considered to be equal to the rate of condensation on the uncovered surface. It is then similarly argued that the rate of evaporation of each succeeding layer is equal to the rate of condensation on the preceding layer. A summation is then performed. The heat of adsorption is involved exponentially in each of the equilibrium-rate expressions. It is assumed that the heat of adsorption in each layer other than the first is equal to the heat of liquefaction of the bulk adsorbate material. In other words, the van der Waals forces of the adsorbent are transmitted to the first layer only. Expressions are obtained for adsorption on a free surface and on a restricted surface.

The equation obtained for a free surface is:[38]

$$\frac{P}{V(P_0 - P)} = \frac{1}{V_m c} + \frac{(c - 1)}{V_m c} \cdot \frac{P}{P_0}, \tag{6.28}$$

where $V_m$ and $c$ are constants at any temperature, and $P_0$ is the saturation pressure of the gas at the given temperature.

Thus, for an adsorption isotherm which obeys this relation, a plot of the expression $PV^{-1}(P_0 - P)^{-1}$ versus $P/P_0$ should yield a straight line for which the intercept is $1/(V_m c)$ and the slope is $(c - 1)/(V_m c)$.

If the adsorption is limited to $n$ layers, equation 6.28 must be replaced by the relation

$$V = \frac{V_m c x}{1 - x} \cdot \frac{1 - (n + 1)x^n + nx^{n+1}}{1 + (c - 1)x - cx^{n+1}}, \tag{6.29}$$

where $x = P/P_0$.

It should be noted that this equation has two limiting cases: when $n = \infty$, it reduces to equation 6.28 (since $x$ must be $<1$); whereas, if $n = 1$ (that is, if only one layer can be adsorbed), it assumes the form

$$V = \frac{V_m(c/P_0)P}{1 + (c/P_0)P}, \tag{6.30}$$

that is, the Langmuir equation with Langmuir's constant $b$ having the value $c/P_0$.

If it is assumed that this limitation of adsorption occurs because of the presence of capillaries, and that bulk liquid condenses in the capillaries as the pressure of the adsorbate gas approaches saturation, a rather more complex equation can be derived. This was done by Brunauer, Deming, Deming, and Teller.[39] Their extended equation, although not commonly used, is of interest because it represented the first equation to be derived which could describe isotherms of all of the five types shown in Fig. 6.9.

Brunauer[40] has given many examples of the applicability of the equations derived from the BET theory. He has also made the following comments on the advantages and disadvantages of the theory:[41]

The multimolecular adsorption theory is the first attempt to arrive at a unified theory of physical adsorption. It describes the entire course of an adsorption isotherm, including unimolecular adsorption, multimolecular adsorption, and capillary condensation, whereas previous theories have dealt only with one or another of these adsorption regions. At the same time it supplies an isotherm equation that can describe all the five different isotherm types, whereas previous equations have dealt only with one type of isotherm at a time, and no equations have been offered at all for three of the five isotherm types.

The theory has several limitations. At low pressures it reduces to the Langmuir equation, consequently all the criticism that has been levelled against the Langmuir theory can be directed equally well against the multimolecular adsorption theory in the low-pressure region of the adsorption isotherm. The most active parts of the surfaces of most adsorbents are strongly heterogeneous, with strongly varying heats of adsorption, therefore the Langmuir equation is not obeyed. For most adsorbents the theory breaks down in the region from zero pressure to $P/P_0 = 0.05$, and for some adsorbents to as high as 0.10. It may be restated that in this region the only theory that can deal successfully with physical adsorption is the potential theory.

The multimolecular adsorption theory is at its best in the middle adsorption region, i.e., in the region preceding and following the building up of a unimolecular layer. There are several reasons for this. In the first place the less active parts of the surfaces of most adsorbents are roughly homogeneous, therefore an average value of $E_1$ can be used without introducing a serious error. In the second place the heterogeneity of the surface does not show up any more in the second adsorbed layer, and so $E_L$ becomes a very good approximation to the heat of adsorption in that layer. Finally, since most adsorbents have capillaries that are at least several molecular diameters wide, the difficulties due to capillary condensation do not yet begin to appear in this region. Thus the two-constant equation (6.28) is obeyed very closely for many adsorbents to at least $P/P_0 = 0.35$ and sometimes to 0.50. This fact enables one to evaluate accurately the surface area of the adsorbent.

In recent years the BET equation, particularly the simple equation 6.28, has found widespread use for the determination of surface areas. It has been found that the adsorption isotherms for nitrogen and the rare gases argon and krypton at low temperatures (commonly liquid-nitrogen temperature, 77.4° K, is used) on a large variety of adsorbents fit the BET equation over a considerable range. A large number of investigations have yielded reasonable values for $V_m$ on various adsorbents when these isotherms are determined and plotted according to the BET equation. The measurement of "BET surface areas" has become a standard technique in many laboratories. Emmett[42] has given a comprehensive survey of the methods used and has discussed the evidence supporting the significance of BET surface areas.

The determination of a surface-area value involves the measurement of adsorption, commonly in a volumetric system, over a range of pressures up to $P/P_0$ about 0.3. These data are then plotted in accord with equation

6.28 as $P/V(P_0 - P)$ versus $P/P_0$. The slope and intercept of the resulting linear plot then give the values of $V_m$ and $c$. On the model used to derive the equation, $V_m$ is the volume adsorbed in the first monolayer; $c$ is a constant which is approximately equal to exp. $(E_1 - E_L)/RT$, where $E_1$ is the average heat of adsorption in the first layer and $E_L$ is the heat of liquefaction of the bulk adsorbate.

Since $V_m$ represents a known amount of adsorbate in the monolayer and hence a known number of molecules, the evaluation of the sample surface area then depends only on assigning an area to a single adsorbed molecule. Emmett and Brunauer[43] first suggested that such areas could be calculated from the density of the liquefied adsorbate. As has been pointed out in Chapter 1, a number of other molecular cross sections might be derived from various physical properties of the adsorbate gas. Further work (some of which will be discussed in section 6.8) has shown that such values are not always satisfactory. It is found, for example, that different gases may frequently give different areas for the same adsorbent if molecular areas calculated from the adsorbate density are used. One possible explanation for this is that different molecules will assume different packing on the same adsorbent surface. Walker and Zettlemoyer[44] have discussed this possibility in some detail and have given experimental evidence supporting it for the adsorption of various gases on magnesium oxide. Gaines and Cannon have proposed an alternative explanation based on the characteristics of the adsorbed film.[45] It appears, however, that average values can be assigned for most of the simpler adsorbate gases which will give values accurate to 10 per cent or so, as long as linear BET plots are obtained.

The constant $c$ in the BET equation has less direct practical significance than does $V_m$. It can, however, provide criteria for the applicability of the BET equation.[45] As indicated above, it can also give an indication of the heat of adsorption in the first monolayer. The heat value obtained in this way, however, is a sort of average value which can be identified only with the more homogeneous and less energetic part of the solid surface. Furthermore, $E_1$ can be evaluated only roughly from the experimental data, both because the constant $c$ is exponential in nature and because small errors in experimental results lead to large deviations in its value. The $c$ value, however, may be quite useful in comparing different adsorbents. A high value of $c$ corresponds to a considerable sorbent-sorbate interaction, while low $c$ values indicate weak adsorption forces.

Some typical BET plots obtained for the adsorption of krypton on several adsorbents are shown in Fig. 6.10. The nature of the adsorbents and the values of surface area and the BET parameters obtained from the plots are given in Table 6.9.[45]

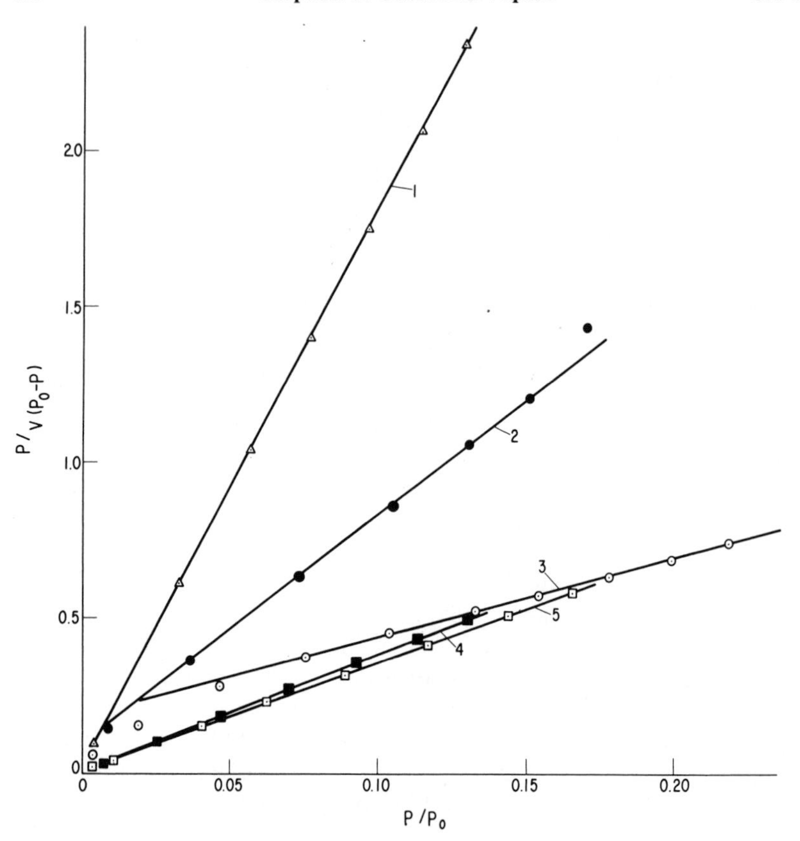


Fig. 6.10. Adsorption of krypton on several adsorbents at −196°, plotted according to the BET equation (Gaines and Cannon).

**TABLE 6.9**

**Nature of Adsorbents and Parameters Obtained from Fig. 6.10**

(Gaines and Cannon)

| Curve | Sample | Weight of Sample (g) | Surface Area ($m^2/g$)* | $c$ |
|---|---|---|---|---|
| 1 | Sintered Ag | 34.30 | 0.0085 | 52 |
| 2 | Glass powder | 6.24 | 0.11 | 80 |
| 3 | Mica paper | 1.00 | 1.92 | 15 |
| 4 | $MoS_2$ | 0.170 | 8.31 | 64 |
| 5 | W powder | 1.70 | 0.90 | 215 |

* With krypton atomic area = 19.5 $A^2$.

## 6.5. OTHER EXPRESSIONS FOR THE ADSORPTION ISOTHERM

It is of course apparent from our previous discussion that the Freundlich, Langmuir, and BET equations have many drawbacks in defining an adsorption isotherm over its entire course from $P = 0$ to saturation pressure. The empirical nature of the parabolic equation and the very limited generality of the models on which the Langmuir and BET equations are based permit these treatments to contribute little to the detailed understanding of adsorption processes. A large number of other equations, both empirical and theoretical, have been proposed to fit adsorption data. In the present section we shall discuss a number of these. Our treatment will be necessarily brief, not because many of these equations do not have considerable importance, but rather because, in the present state of scientific knowledge, their utility for practical purposes is invariably limited. The reader who wishes further details of these isotherm equations should consult the original literature.

### Modifications of the Langmuir and BET Theories

A number of attempts have been made to improve the models on which the Langmuir and BET equations are based by introducing additional assumptions. If adsorbate molecules occupy more than one site of the solid surface, or if there are interactions, either repulsive or attractive, between these adsorbed molecules, appropriate corrections may be made in the equations. Langmuir examined a number of such corrections in a comprehensive paper published in 1940.[46] His discussion included the Williams-Henry equation,[47] the Magnus equation,[48] the Tonks treatment,[49] and Langmuir's own modifications of the hyperbolic equation (equations 6.19, 6.23, and 6.24).

Hill[50] has summarized a number of the modifications of the BET equation. He has emphasized particularly the attempts of Hill[51] and Halsey[52] to introduce the contribution of lateral interactions in the adsorbed film.

We should also mention here the isotherm equations which have been derived on the basis of assumptions about the variation of the heat of adsorption with coverage. Such equations are of particular interest with regard to chemisorption. Zeldowitch[53] found that the Freundlich isotherm is obtained if the sites on the adsorbent surface are distributed exponentially with respect to the energy of the adsorption.

If a linear variation in heat is observed with increasing coverage, the Temkin isotherm

$$V = k \log P \tag{6.31}$$

is obtained.[54]

Another approach which has shed light on this point is the attempt to deduce the distribution of site energies from the experimentally determined isotherms. Although no completely general method is available, the treatments of Sips[55] and Graham[56] have given information about some systems.

## The Potential Theory

In 1914, Eucken[57] and Polanyi[58] developed a theory according to which adsorption is due to strong attractive forces exerted by the adsorbent upon the gas in its neighborhood. As Brunauer describes the assumptions involved,[59]

> The forces of attraction reaching out from the surface are so great that many adsorbed layers can form on the surface. These layers are under compression, partly because of the attractive force of the surface, partly because each layer is compressed by all the layers adsorbed on top of it. The compression is the greatest on the first adsorbed layer, less on the second, and so on until the density decreases to that of the surrounding gas.
>
> The adsorption potential at a point near the adsorbent is defined (by Polanyi) as the work done by adsorption forces in bringing in molecules from the gas phase to that point.

For gas pressures much below the critical value, this potential is given by the relation

$$\epsilon_i = R_0 T \ln \left(\frac{P_0}{P_x}\right), \tag{6.32}$$

where $P_0$ is saturation pressure, and $P_x$ the pressure in the gas phase at the point at which the *weight* of adsorbed vapor per unit area is given by $x$. At this point, the *volume*, $\phi_i$, of the adsorbed vapor per unit area is given by the relation

$$\phi_i = \frac{x}{\delta_T}, \tag{6.33}$$

where $\delta_T$ is the density of the liquid at temperature $T$.

At the surface of the adsorbent, $\phi = 0$, and $\epsilon_i$ has its maximum value, $\epsilon_0$. As we pass outwards, towards the gas, $\epsilon_i$ decreases and $\phi_i$ increases until, at the boundary of the adsorbed layers, the density drops abruptly to that of the gas, $\epsilon_i = 0$, and $\phi$ reaches a maximum value. Thus the different layers parallel to the surface of the adsorbent form a series of equipotential surfaces at each of which $\epsilon_i$ and $\phi_i$ have definite values.

The potential theory assumes that the adsorption potential does not change with temperature,

$$\frac{d\epsilon_i}{dT} = 0,$$

from which it follows that

$$\frac{d\phi_i}{dT} = 0.$$

This means that the curve representing the potential distribution in the adsorption space, i.e., $\epsilon = f(\phi)$, is the same for all temperatures. For this reason the curve is called the *characteristic curve*.

Since $\epsilon$ and $\phi$ can be expressed in principle as functions of the pressure, the temperature, and the amount of gas adsorbed, the equation $\epsilon = f(\phi)$ can be regarded as equivalent to an isotherm equation.[59]

In practice such a plot of $\epsilon$ versus $\phi$ is usually derived from one experimentally determined isotherm, and all other isotherms are then calculated from this curve.

In summarizing this topic, Brunauer makes the following remarks:

> The potential theory applies to both unimolecular and multimolecular adsorption. It is the only theory of physical adsorption that can handle quantitatively adsorption on a strongly heterogeneous surface. Since the theory does not attempt to formulate an isotherm equation, the scope of information obtainable from it is limited. One cannot determine with its help the extent of the surface of the adsorbent or the pore size distribution; indeed, one cannot even tell whether the adsorption is unimolecular or multimolecular.

### The Polarization Theory and the Capillary Condensation Theory

These two treatments are of considerable interest, since each emphasizes one aspect of the adsorption process. The polarization theory, primarily due to de Boer and Zwikker,[60] explains adsorption on ionic or polar adsorbents by assuming that dipoles are induced in the adsorbate molecules by the forces of the surface. These induced dipoles, in turn, induce dipoles in the next layer of adsorbed gas molecules and so on.

The capillary condensation theory, originally proposed by Zsigmondy,[61] and extended by Patrick and his coworkers,[62] assumes that adsorption occurs primarily because of the condensation of bulk liquid adsorbate in very fine pores of the adsorbent. The data are then treated with the aid of the Kelvin equation (section 7.6). Both these theories have been discussed in some detail by Brunauer.[63]

### The Frenkel-Halsey-Hill Treatment

A quite different approach, which concerns itself with the problem of adsorption at very high coverages (that is, $\theta > 2$), is due to Frenkel,[64] Halsey,[52] Hill,[65] and McMillan and Teller.[66] These authors all considered the variation in the potential energy of a molecule located in an adsorbed layer separated from the solid surface by several adsorbed layers. For such a molecule, the environment is not very different from that of corresponding molecules in the bulk liquid. The isotherm equation which is derived from this treatment is

$$\ln \frac{P}{P_0} = -\frac{k}{\theta^n}. \tag{6.34}$$

When fitted to experimental data, this equation usually gives a value of $n$ between 2 and 4 and the coefficient $k$ is generally about 4, as has been shown by an approximate argument by McMillan and Teller. A similar equation with $n = 2$ was used by Harkins and Jura[67] in their "relative" surface-area-measurement method.[68]

### The Liquid-State Treatment

Attempts have been begun to derive a completely rigorous and completely general treatment of the adsorption of gas molecules which interact with each other as well as with the adsorbent surface. Wheeler[69] and Ono[70] have both attempted this derivation for a mathematically uniform surface.

Two problems are still not resolved in this treatment. One is the lack of a suitable theory of the liquid state; the best currently available treatments are used. Second, the complexity introduced by trying to treat an actual surface, rather than a mathematically uniform surface, is completely insurmountable at the present time. Hill[50] and Ries[71] have both discussed the Wheeler-Ono approach.

## 6.6. HEAT OF SORPTION; TEMPERATURE VARIATION OF SORPTION AND SORPTION THERMODYNAMICS

In a physical adsorption process, for a given equilibrium pressure, the amount adsorbed always decreases with an increase in temperature. This is shown, for example, by the adsorption isotherms at a series of temperatures; cf. Fig. 6.3. As mentioned previously, the *isostere* represents the variation in pressure with temperature for a constant amount sorbed; the *isobar* is a plot of the amount adsorbed as a function of temperature, for constant pressure. In Figs. 6.11 and 6.12 the data of Magnus and Kratz for the adsorption of carbon dioxide on wood charcoal, shown in Fig. 6.3, are plotted as isobars and isosteres.

This decrease of sorption with increased temperature, of course, is a reflection of the fact that heat is evolved when the gas is adsorbed and taken up when the sorbed gas is removed. The Clausius-Clapeyron equation may therefore be applied to the isostere to determine the heat of adsorption, just as the heat of vaporization may be determined from a

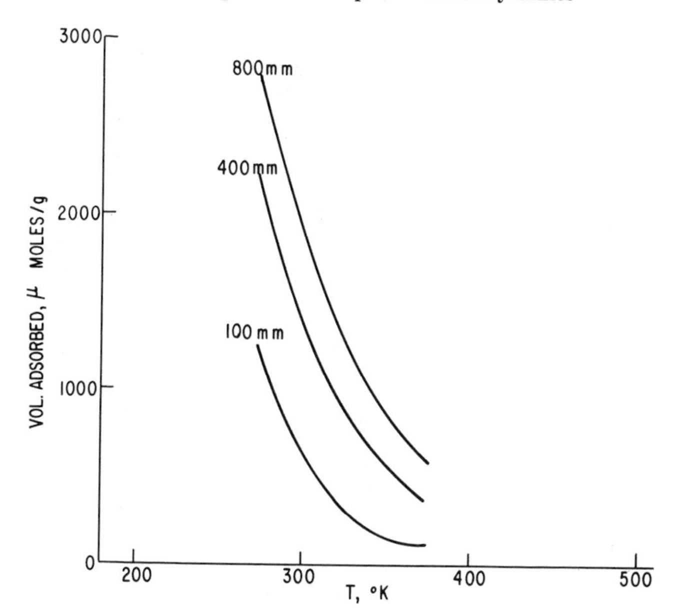


Fig. 6.11. Isobars for the sorption of carbon dioxide by wood charcoal (Magnus and Kratz).

vapor-pressure curve. The *isosteric* heat of adsorption is thus determined by means of the relation

$$Q_S = -R_0 \frac{d \ln P}{d(1/T)} \tag{6.35}$$

$$= \frac{4.574(\log P_2 - \log P_1)}{1/T_1 - 1/T_2}, \tag{6.36}$$

where $Q_S$ is expressed in calories per mole, and $P_1$ and $P_2$ are equilibrium pressures at temperatures $T_1$ and $T_2$, respectively, for a given amount adsorbed.

The heats of sorption (in calories per mole) may be calculated from the isosteres by means of equation 6.36, or by plotting $\log P$, for constant amount adsorbed, against $1/T$, but they have also been determined by calorimetric methods.[72] Table 6.10, taken from a review of this topic by Kruyt and Modderman,[73] illustrates the kind of agreement obtained by the two methods. The data were obtained by Titoff for carbon dioxide on charcoal at 0 and 30° C.

The fact that heat must be supplied to a surface to desorb it "has found application," as McBain observes,[74] "in the production of low temperatures, as, for example, reaching 6° abs by pumping off hydrogen sorbed at the temperature of liquid air or liquid hydrogen."[75]

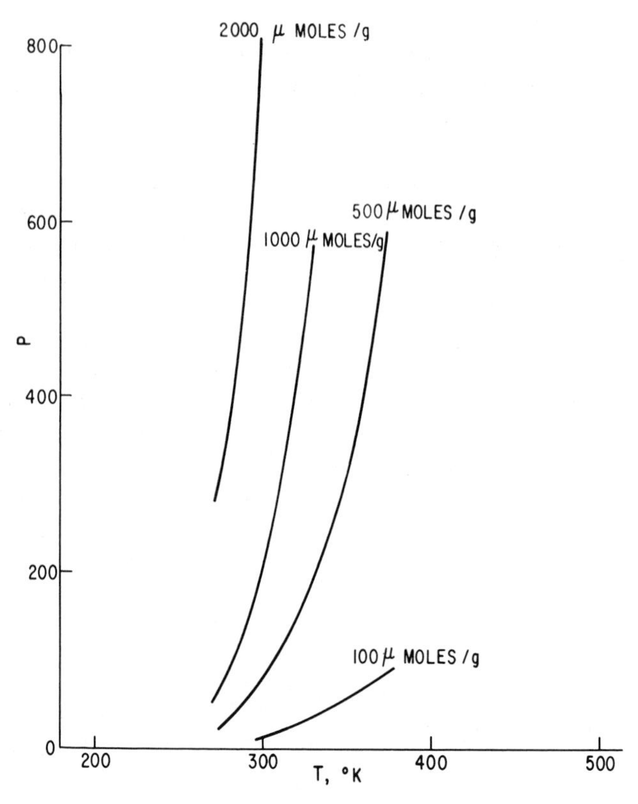


Fig. 6.12. Isosteres for the sorption of carbon dioxide by wood charcoal (Magnus and Kratz).

**TABLE 6.10**

**Heat of Adsorption for Carbon Dioxide on Charcoal**

| $V_0$ | $Q_S$ (cal/mole) | $Q$ (calorimetric) (cal/mole) |
|---|---|---|
| 1.82 | 8130 | 7727 |
| 4.02 | 7705 | 7280 (4.02–10.3) |
| 10.3 | 6966 | |
| 19.72 | 6548 | 6831 |
| 28.87 | 6292 | 6720 |
| 34.07 | 6452 | 6585 (34.07–41.4) |
| 38.14 | 6383 | |
| 41.4 | 6495 | |
| 44.9 | 6428 | 6540 |

As mentioned in section 6.4, it is possible to derive a sort of average heat of adsorption from the value of the constant $c$ in the BET equation.

In addition, by rigorous thermodynamic arguments, it is possible to define other heat functions. These include both *differential* heats (for example, the isosteric heat), which measure the increment in heat content for the adsorption of an infinitesimal quantity of gas, at a given surface coverage, and *integral* heats, which represent the total amount of heat evolved by adsorption up to a given coverage. Corresponding energy and entropy functions can also be defined. Hill[76] has discussed this thermodynamic system in detail.

These heat values are of considerable importance because they provide a measure of the energy of interaction between the adsorbed molecules and the surface atoms. It is, of course, apparent that, if we consider the interaction of molecules from the gas phase with a solid surface, there will be variations in the magnitude of this energy as the molecules are adsorbed. Even on a perfectly smooth, perfectly uniform solid surface, over which there is no variation of the attractive force for gas molecules, the heat of adsorption will change as the surface becomes covered with sorbate because of lateral interactions between the sorbed molecules themselves. Since, in addition, it is apparent that no real solid exhibits a perfectly homogeneous smooth surface, there will also be variations in the interaction energy due to *surface heterogeneity*.

In recent years the interpretation of adsorption measurements has yielded much information about the nature of gas-solid interactions. Hill, Emmett, and Joyner[77] reported such an examination of the thermodynamic functions (heat and entropy) for the adsorption of nitrogen on a graphitized carbon black. Such an adsorbent is of considerable interest because its surface exhibits a high degree of energetic homogeneity. The data of Hill, Emmett, and Joyner indicate this clearly. Another example is given by the results of Gaines[78] for the adsorption of rare gases (argon and krypton) on mica. The variations of adsorption heat and entropy in the first monolayer could be correlated with the structure of the mica surface.

In order for adsorption to occur at all, instead of bulk condensation of the adsorbate, it is of course obvious that the free energy of the adsorbed phase must be less than the free energy of the bulk sorbate. Since there may be differences in entropy between these two states, it is possible that the *heat* of adsorption may be smaller than the heat of condensation. Graham[79] has discussed this point. In some systems, however, the forces between molecules in the liquid state may be stronger than the forces which would otherwise hold those molecules at the surface of an adsorbent. In some experiments by Wood[80] and Knudsen[81] this phenomenon has been observed.

In Wood's experiments a stream of cadmium vapor in a well-exhausted bulb was allowed to strike a glass surface at different temperatures. No visible deposit was observed unless the glass was held at a temperature below about −90° C. On the other hand, once a deposit was started by cooling the glass at that spot with liquid air, the deposition of cadmium continued even after the deposit was warmed to room temperature.

From these and similar observations Wood concluded that, whereas cadmium atoms condense on a cadmium surface at any temperature, they condense on glass only if the temperature is below about −90° C. At higher temperatures all the atoms are reflected.

These observations were interpreted by these investigators as indicating that at some temperature the condensation coefficient changes very rapidly from zero to unity. Langmuir, however, repeated Wood's experiments[82] and concluded from his observations that this deduction was not justifiable.

In his experiments, Langmuir used cadmium vapor in a well-exhausted bulb and investigated the effect of cooling a portion of the glass surface to different temperatures. He found that "traces of residual gas may prevent the growth of the deposit, particularly in those places which have been most effectively cooled. This is probably due to the adsorption of the gas by the cooled metal deposit." By using a side tube containing charcoal immersed in liquid air this effect was eliminated.

If all the cadmium [he states] is distilled to the lower half of the bulb and this is then heated to 220° in an oil bath while the upper half is at room temperature, a fog-like deposit is formed on the upper part of the bulb in about 15 seconds. This deposit is very different from that obtained by cooling the bulb in liquid air. Microscopic examination shows that it consists of myriads of small crystals. According to the condensation-evaporation theory, the formation of this fog is readily understood. Each atom of cadmium, striking the glass at room temperature, remains on the surface for a certain length of time before evaporating off. If the pressure is very low, the chance is small that another atom will be deposited, adjacent to the first, before this has had time to evaporate. But at higher pressures this frequently happens. Now if two atoms are placed side by side on a surface of glass, a larger amount of work must be done to evaporate one of these atoms than if the atoms were not in contact. Not only does the attractive force between the cadmium atom and the glass have to be overcome, but also that between the two cadmium atoms. Therefore, the rate of evaporation of atoms from pairs will be much less than that of single atoms. Groups of three and four atoms will be still more stable. Groups of two, three, four atoms, etc., will thus serve as nuclei on which crystals can grow. The tendency to form groups of two atoms increases with the square of the pressure, while groups of three form at a rate proportional to the cube of the pressure. Therefore the tendency for a foggy deposit to be formed increases rapidly as the pressure is raised or the temperature of the condensing surface is lowered.

On the other hand, according to the reflection theory, there seems to be no satisfactory way of explaining why the foggy deposit should form under these conditions.

Experiments show clearly that, when a beam of cadmium vapor at very low pressure strikes a given glass surface at room temperature, no foggy deposit is formed, although when the same quantity of cadmium is made to impinge against the surface in a shorter time (and therefore at higher pressure) a foggy deposit remains. This fact constitutes strong proof of the condensation-evaporation theory.

A deposit of cadmium of extraordinarily small thickness will serve as a nucleus for the condensation of more cadmium at room temperature. Let all the cadmium be distilled to the lower half of the bulb. Now heat the lower half to 60° C. Apply a wad of cotton, wet with liquid air, to a portion of the upper half for 1 minute, and then allow the bulb to warm up to room temperature. Now heat the lower half of the bulb to 170° C. In about 30 seconds a deposit of cadmium appears which rapidly grows to a silver-like mirror. This deposit only occurs where the bulb was previously cooled by liquid air.

Langmuir calculated from the vapor-pressure data for cadmium as follows:

A deposit which forms in 1 minute with the vapor from cadmium at 60° contains only enough cadmium atoms to cover 3/1000 of the surface of the glass. Yet this deposit serves as an effective nucleus for the formation of a visible deposit.

At lower temperatures, where the vapor pressure is much smaller, the probability that the atoms striking the glass will fall into positions adjacent to atoms already on the surface becomes very much less and the atoms evaporate before new ones arrive; consequently there is no apparent condensation. Langmuir stated the difference between his point of view and that of Wood and the others quite clearly:

When an atom strikes a surface and rebounds elastically from it, we are justified in speaking of this process as a reflection. Even if the collision is only partially elastic, we may still use this term. The idea that should be expressed in the word "reflection" is that the atom leaves the surface by a process which is the direct result of the collision of the atom against the surface.

On the other hand, according to the condensation-evaporation theory, there is no direct connection between the condensation and subsequent evaporation. The chance that a given atom on a surface will evaporate in a given time is not dependent on the length of time that has elapsed since the condensation of that atom. Atoms striking a surface have a certain average "life" on the surface, depending on the temperature of the surface and the intensity of the forces holding the atom. According to the "reflection" theory, the life of an atom on the surface is simply the duration of a collision, a time practically independent of temperature and of the magnitude of the surface forces.

The above experiments prove [as stated by Langmuir] that the range of atomic forces is very small and that they act only between atoms practically in contact with each other. Thus a surface covered by a single layer of cadmium atoms behaves, as far as condensation and evaporation are concerned, like a surface of massive cadmium.

In this manner it is possible therefore to account very well for the apparent reflection of cadmium and mercury from glass surfaces, and it is seen that these observations are not at all in contradiction to Langmuir's theory.

From their experiments, Volmer and Estermann[83] concluded that, for liquid mercury, the condensation coefficient (that is, the ratio designated $\alpha_0$ in the derivation of the hyperbolic equation for the sorption isotherm) is unity, while for the solid it is slightly smaller. They also observed that for other substances, such as sulfur, phosphorus, and benzophenone, the value of the condensation coefficient varied from 0.2 to 0.5. On the whole they expressed views in accord with Langmuir in concluding that the first stage in the condensation of molecules on a crystalline surface is adsorption.

Much effort has been devoted to the study of the processes occurring when gaseous molecules collide with solid surfaces. The advent of modern atomic- and molecular-beam techniques[84] has greatly aided these studies. Such methods have also led to experimental examination of the interactions between particles in the gas phase.[85] As a result of this work, Sears and Cahn[86] have pointed out that the finite rate of energy exchange between a gas molecule and a surface must be considered in order to understand the condensation-evaporation process in detail.

In *physical adsorption* the heats evolved are ordinarily of the order of 1 or 2 times the heat of vaporization of the bulk adsorbate. If *chemisorption* occurs (that is, if the bond between the adsorbed gas molecule and the surface is of the same sort as that involved in a chemical reaction), the observed heat values are very much higher. Heats appropriate to the heats of chemical reaction (that is, some tens of kilocalories per mole adsorbed) are then observed.

In the system oxygen-charcoal both van der Waals sorption and chemisorption occur, as illustrated by Fig. 6.13, taken from a paper by Keyes and Marshall.[87] From an initial value of 72,000 cal per mole, the heat of sorption decreases rapidly with increase in amount sorbed to about 4000 cal per mole for very large amounts sorbed. We shall make some further comments on heats of chemisorption in section 6.9.

In general, the heats of physical adsorption for the *same gas* on different adsorbents are roughly the same. Also the heats of sorption for different gases increase, in general, with increase in boiling point. For instance, as Brunauer points out,

> Helium, boiling at 4.2° K, has a heat of adsorption of about 140 cal per mole; hydrogen, boiling at 20.4° K, has a heat of adsorption of about 1500 cal per mole, and the four gases, argon, nitrogen, oxygen, and carbon monoxide, boiling in the temperature range 77.3–90.1° K, have heats of adsorption of about 3000–4000 cal per mole.

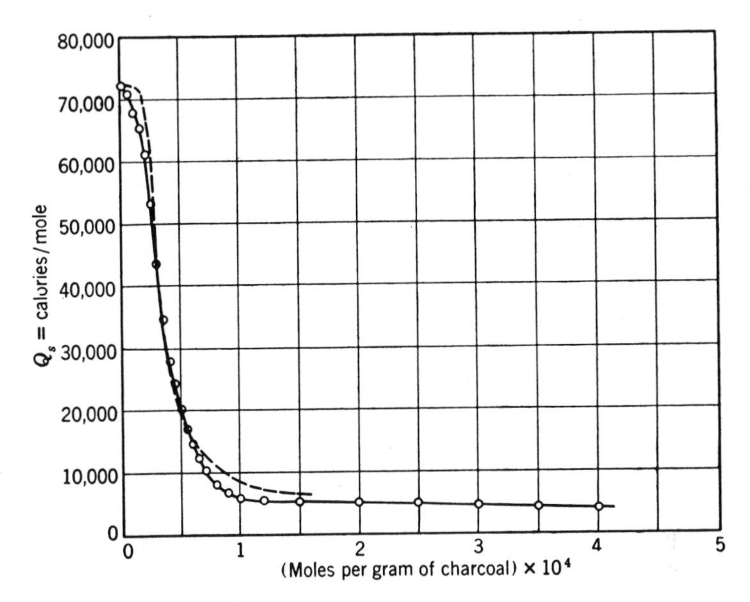


Fig. 6.13. Heat of adsorption for oxygen on charcoal, as a function of amount adsorbed (Keyes and Marshall).

For ammonia, carbon dioxide, and nitrous oxide Favre obtained heats of adsorption of 7200, 7300, and 7400 cal per mole, respectively. For ethylene, acetylene, carbon dioxide, and nitrous oxide Gregg obtained values of 6600, 8100, 6900, and 7200 cal per mole, respectively. These heats of adsorption were obtained on charcoal. The boiling points of the five gases lie between 169° and 240° K.[88]

Similarly, for a number of organic vapors which have their boiling points in the range 300–350° K, the observed heats of adsorption on charcoal fall between 12,000 and 15,500 cal per mole.[89]

For the adsorption of nitrogen, argon, and carbon monoxide on smooth surfaces of steel and silver, Armbruster and Austin[90] deduced, from the observed adsorption isotherms at −183 and −195° C, the values of adsorption heats shown in Table 6.11. The last column gives values which these investigators deduced from data of Wilkins.[91]

It will be observed that the range of values is the same as that given above by Brunauer for the same gases.

Comparison of the values for heats of sorption obtained by different investigators with those for the corresponding latent heats of vaporization and sublimation leads to the conclusion that the forces between adsorbed molecules and the surface atoms of the adsorbent are of the same nature as the van der Waals forces of cohesion. Actually, it has been found

TABLE 6.11

Values of Heat of Adsorption (Calories per Mole) on Metal Surfaces

| | Fe | | | |
|---|---|---|---|---|
| | Unreduced | Reduced | Ag | Pt |
| Ar | 3280 | 3200 | 3500 | 3280 |
| $N_2$ | 3280 | 3120 | 3600 | 3385 |
| CO | | 3400 | 3170 | 3600 |

possible to calculate the magnitudes of these forces from theoretical considerations and to deduce from these conclusions approximate values of heats of sorption.

For example, for the adsorption of argon and carbon dioxide on ionic crystals (such as those of the alkali halides) the experimental and theoretical heats of sorption are found to be in very fair agreement. Similar results have been obtained for the heats of sorption of hydrogen, argon, and nitrogen on graphite.[92] The most elaborate calculation of this nature has been carried out by Orr[93] on the adsorption of argon, nitrogen, and oxygen on crystals of potassium chloride and cesium iodide. For each of these gases the maximum values of the heats of sorption (with range from 2000 to 3000 cal per mole) are observed for the completed monolayers ($\theta = 1$), with more or less rapid decreases for $\theta > 1$.

Let us now consider the course of a typical adsorption process on a finely divided adsorbent material. We shall examine the heats of adsorption which are observed as the process proceeds, starting from an adsorbent which has been thoroughly outgassed and therefore has a clean surface.[94] The adsorbate will be a relatively inert gas such as nitrogen, which has no permanent dipole moment.

When the first portion of the adsorbate is admitted to the system, chemisorption may occur if there are parts of the solid surface which are sufficiently reactive. An example of such behavior has already been given; cf. Fig. 6.13. Whether or not chemisorption occurs, the first physically adsorbed gas will be taken up with a relatively high heat. The first gas molecules adsorbed will occupy the sites of highest energy; even if the surface is chemically homogeneous, geometrical heterogeneity will lead to higher force fields near cracks or steps in the surface crystal lattice. de Boer and Custers[95] showed that the van der Waals force field in a pore may be several times that above a plane surface. Later, Barrer[96] discussed this point.

As the amount of gas adsorbed increases, and the less energetic part of the surface begins to be covered, the heat of adsorption will decrease. In some cases a minimum in the heat versus coverage curve is observed at about $\theta = 0.5$. Commonly, heats of the order of the heat of vaporization of the bulk adsorbate, or a few hundred calories more, are observed here. As the first monolayer becomes more nearly filled, some of the adsorbed molecules may be grouped in patches large enough to produce *cooperative interactions* with newly arriving adsorbate molecules. This may lead to a rise in the heat of adsorption. At coverages nearly equal to a full monolayer, each gas molecule which arrives at the surface will find itself surrounded by its already bound fellows. There may, therefore, be a maximum in the heat of adsorption in this range.

After the completion of the first monolayer, the force field of the solid is screened. The heat of adsorption, therefore, may decline rather quickly after this point. Ordinarily, it approaches the value of the heat of liquefaction of the adsorbate fairly soon after.

It is apparent, of course, from what we have said, that this "typical" adsorption process is subject to great variation. If a major fraction of the solid surface exhibits appreciable energetic heterogeneity, the decline in the heat of adsorption may continue from low coverages up to the monolayer point. The relative magnitudes of sorbent-sorbate interactions and sorbate-sorbate interactions will determine the detailed course of the heat curve. They will also affect the behavior of the adsorbed film. As can be inferred from our previous discussion of adsorption isotherms, the adsorbed molecules either may be fixed at sites on the solid surface, or may have appreciable translational freedom (that is, may exist in a two-dimensional liquid-like state on the surface). Recently, a number of studies have been made of a variation of the thermodynamic functions for adsorption, in an effort to learn about gas-solid interactions. Everett[97] has discussed these studies in some detail.

## 6.7. RATE OF SORPTION

Some of the earliest experiments on rate of sorption were made by Blythswood and Allen.[98] They observed that the rate of sorption of air by charcoal at the temperature of liquid air could be represented quite accurately by the first-order reaction equation

$$\frac{dx}{dt} = k(A - x), \tag{6.37}$$

where $x$ = amount adsorbed at time $t$,

$A$ = total amount adsorbed when equilibrium was attained,

$k$ = constant.

Integrating equation 6.37, we obtain the relation

$$\log (A - x) = \log A - 0.4343kt. \tag{6.38}$$

That is, a plot of $\log (A - x)$ versus $t$ should yield a straight line, for which the intercept at $t = 0$ is $\log A$, and the slope

$$-\frac{\Delta \log (A - x)}{\Delta t} = 0.4343k.$$

An equation similar to equation 6.37 was deduced by Langmuir[99] for the rate of adsorption of molecules to form a monolayer.

Rates of sorption at constant pressure for ammonia, carbon dioxide, sulfur dioxide, and nitrous oxide by glass surfaces have been investigated by Bangham and Burt[100]. They derived the empirical expression

$$\frac{d \log s}{d \log t} = \frac{1}{n}, \tag{6.39}$$

where $s$ = amount adsorbed, and $n$ is a constant.

Since this relation would indicate an infinite amount of sorption at infinite time, the expression used for longer periods is

$$\log \frac{\sigma}{\sigma - s} = kt^{1/n}, \tag{6.40}$$

where $\sigma$ is the estimated saturation value. For $n = 1$, this evidently becomes identical with equation 6.38, while for small values of $s$ and $t$ equation 6.40 becomes identical with equation 6.39.

The rate of sorption of hydrogen by charcoal at the temperature of liquid air has been specially investigated by Firth[101] and McBain.[102] Both investigators observed that equilibrium is attained only after a lapse of many hours. Most of the gas is apparently condensed practically instantaneously on the surface. This is followed by a gradual diffusion of the hydrogen into the charcoal, the rate of which, as shown by McBain, is in agreement with the laws of diffusion.

The rate of occlusion of gases by chabasite and other members of the zeolite group of minerals has been investigated by Barrer and Ibbitson.[103] In these minerals, as will be discussed in greater detail in Chapter 7, the rate of diffusion of molecules along pores, or very fine channels, is the controlling factor, as shown by the fact that the rate of occlusion depends upon the relative diameters of the molecules and the pores, as well as on the external pressure and temperature.

These and similar observations will be discussed more fully in connection with the remarks on specific adsorbents in Chapter 7. As Brunauer[104] observes,

In general, it seems safe to conclude that in van der Waals adsorption the gas molecules are adsorbed as rapidly as they can reach the surface. The slow effects are either due to chemisorption, chemical reaction, solution, or to the inability of the molecules to get in contact with the surface of the adsorbent.

A problem of interest in connection with the slow rate of sorption which follows the practically instantaneous adsorption as a unimolecular layer is the time required for molecules to cover the surface of a pore or very narrow channel. If the rate of re-evaporation of adsorbed molecules is very small, then, as shown by Clausing[105] the time $t$ necessary to cover the surface of the pore with molecules is

$$t = \frac{3l^2}{8a^2 S\nu}, \tag{6.41}$$

where $l$ = length, $a$ = radius of pore, $S$ = cross-sectional area of a molecule, and $\nu$ = rate at which molecules strike a surface at the pressure $P$.

For instance, let us assume that $l = 0.1$ cm, $a = 10^{-5}$ cm, and $P_\mu = 1$, and that the gas is nitrogen at $T = 80$. Then

$$\nu = 7.41 \cdot 10^{17} \text{ sec}^{-1} \text{ cm}^{-2},$$

$$S = 16.2 \cdot 10^{-16} \text{ cm}^2,$$

and

$$t = 8.68 \text{ hr}.$$

Since $\nu$ varies linearly with $P$, $t$ for 10 microns would be 52.1 min. Thus, for long pores of very small diameter, the time required to saturate the walls may be quite long.

## 6.8. SURFACE OF ADSORBENTS

As mentioned in a previous section, the heat of physical adsorption for any given gas is approximately the same for different adsorbents. Thus the only other important factor by which the adsorption on different adsorbents can be influenced is the extent of surface per unit mass of adsorbent.

We have already noted (section 6.4) that surface area can be estimated by the application of the Brunauer-Emmett-Teller (BET) equation to low-temperature gas adsorption data. A number of other methods have been devised to determine this important quantity; reviews have been given by Brunauer[106], Emmett[107], and Orr and Dallavalle[108]. In this section we shall discuss briefly three of these techniques in addition to the BET method.

The method developed by Bowden and Rideal[109] and by Bowden and O'Connor[110] is applicable only to metals and is described by Brunauer as follows:

When a metal is made the cathode in a dilute acid and current is passed through the solution, the potential changes due to accumulation of hydrogen on the electrode. This phenomenon is called polarization. Bowden and Rideal noted that the change in potential was proportional to the amount of hydrogen deposited on the electrode,

$$-\frac{\Delta E}{\Delta V} = K, \tag{6.42}$$

where $E$ is the potential, and $V$ the amount of hydrogen on the surface of the cathode. They also made the important discovery that $K$ was not dependent on the nature of the metal but only on its surface area. All liquid metals investigated (mercury, amalgamated silver, platinized mercury, liquid gallium, liquid Wood's metal) required about $6 \cdot 10^{-7}$ coulomb of electricity per square centimeter surface to produce a change of 100 millivolts in the electrode potential. It seemed therefore safe to assume for all metals that

$$-E = \frac{KV}{A} + \text{constant}, \tag{6.43}$$

where $A$ is the surface area accessible to hydrogen ions and $K/A$ is a constant for all metals. $A$ will be referred to as the *true* surface of the metal.

In the electrolytic method of surface determination one measures the quantity of electricity necessary to produce a change of 100 millivolts in the electrode potential. When this value is divided by $6 \cdot 10^{-7}$ coulomb one obtains the ratio between the true surface of the metal and its apparent surface.

Table 6.12[111] gives values of this ratio for a number of cases, as observed by Bowden, Rideal, and O'Connor.

Harkins and Jura[112] devised a calorimetric method of surface-area measurement which is generally referred to as the "Harkins-Jura absolute method." Orr and Dallavalle[113] have described the technique as follows:

A small crystal without cracks or crevices when suspended in the saturated vapor of a liquid will become coated with an adsorbed film. The adsorbed film will thicken until its surface energy becomes the same as that of the liquid in bulk. Now consider a system composed of the crystal and a liquid with its saturated vapor at equilibrium in a sensitive calorimeter. When the crystal coated with its adsorbed film is dropped into the liquid, the only energy change involved is that due to the disappearance of the adsorbed film. If the liquid were water with a surface energy of 118.5 ergs/cm$^2$, the energy released in the calorimeter, expressed in ergs and divided by 118.5, would give the area of the adsorbed film of water. To obtain the area of the uncoated particle, a very small correction must be made for the thickness of the adsorbed film. This is essentially the procedure developed by Harkins and Jura.

The term "absolute" applied to this method refers, of course, to the fact that it, unlike most measurements involving adsorbed films, requires

TABLE 6.12

**The Ratio between True Surface and the Apparent Surface or That of a Liquid as Found by the Method of Electrolytic Polarization**

| Surface | Ratio |
|---|---|
| Platinum, bright foil | 2.2 |
| Platinum, bright foil, cleaned in acid and heated in a flame | 3.3 |
| Platinum, platinized | 1830 |
| Gallium, solidifying undisturbed | 1.7 |
| Nickel, polished, new | 75 |
| Nickel, polished, old | 9.7 |
| Nickel, activated by alternate oxidation and reduction, new | 46 |
| Nickel, activated by alternate oxidation and reduction, old | 29 |
| Nickel, activated, then annealed, new | 10.8 |
| Nickel, activated, then annealed, old | 7.7 |
| Nickel, repolished and electroplated, new | 12 |
| Nickel, repolished and electroplated, old | 9.5 |
| Nickel, rolled, new | 5.8 |
| Nickel, rolled, old | 3.5 |
| Wood's metal, solidifying undisturbed | 1.4 |
| Wood's metal, sandpapered | 6.3 |
| Wood's metal, corroded by etching with nitric acid | 800–1000 |
| Silver, freshly etched with dilute nitric acid | 51 |
| Silver, etched with dilute nitric acid, after 20 hr | 37 |
| Silver, finely sandpapered | 16 |
| Silver, amalgamated, after 1 hr | 1.2 |
| Silver, amalgamated, after 20 hr | 1.3 |
| Silver, amalgamated, after 150 hr | 1.8 |
| Arc carbon rod | 328–366 |

no assumptions about molecular area or the nature of an adsorbed monolayer. Unfortunately, the method has limited applicability, both because of the experimental difficulty attendant on the precise calorimetric measurements, and because of the fact that only a few solids exhibit the inertness, high surface area, and freedom from cracks and pores required for accurate determinations. Since it was first proposed, the method has been applied to only a few materials. These are, however, of considerable interest because they provide "standards" for other methods.

In the Harkins-Jura "relative method,"[114] an equation developed by these same workers is applied to gas-adsorption data in much the same way as in the BET method. The equation

$$\ln \frac{P}{P_0} = B - \frac{A}{V^2}, \tag{6.44}$$

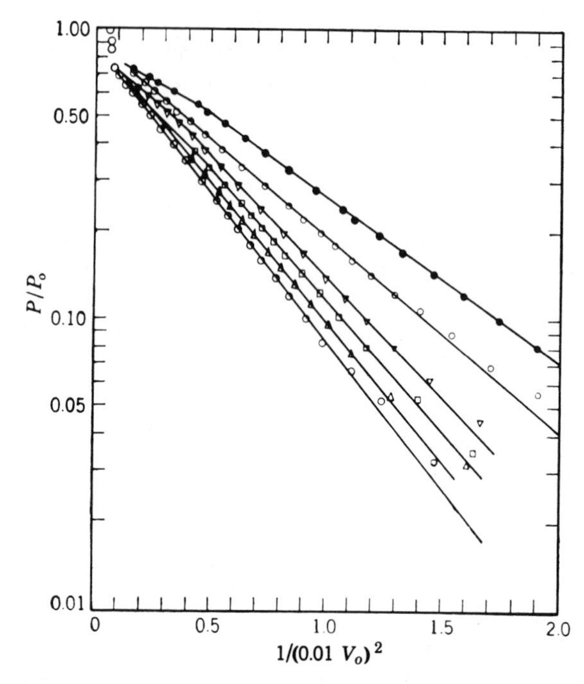


Fig. 6.14. Nitrogen isotherms on porous solids. The area $A_m$, square meters per gram, is given by the relation $A_m = 4.06(S)^{1/2}$, where $S$ is defined by equation 6.40 as the slope of the linear portion of the plot, and $V_0$ denotes the volume adsorbed (cm³, STP) per gram of adsorbent (Harkins and Jura).

where the constant $A$ is a function of the surface area of the sorbent, was derived from a consideration of the analogy between adsorbed films on solids and the behavior of thin films on liquid surfaces.[115]

If this equation is obeyed, a plot of log $P$ versus $1/V^2$ should be a straight line, and the surface area is given by

$$\text{Area} = k(S)^{1/2}, \tag{6.45}$$

where $S$ is the slope of the line, and $k$ is a constant which depends on the gas used and the temperature. The values obtained are: for water vapor at 25° C, $k = 3.83$; for nitrogen at $-195.8°$, $k = 4.06$; for $n$-butane at 0° C, $k = 13.6$; and for $n$-heptane at 25° C, $k = 16.9$.

Figure 6.14, from the publication by Harkins and Jura, shows adsorption isotherms of nitrogen at $-195.8°$ C on some porous solids. Using the value $k = 4.06$, the areas calculated from the observed values of $S$, beginning at the uppermost plot, are 321, 365, 395, 409, 438, and 455 square meters per gram.

Harkins and Jura have shown that the areas calculated for several adsorbents by the BET method are in very satisfactory agreement (within 20 per cent) with the values obtained by equation 6.45.

Corrin[116] has shown, however, that in some cases more than one straight line is obtained for the same system over different pressure ranges.

As noted in section 6.4, the BET method is currently widely used for surface-area measurement. Instead of attempting to determine $V_m$, Brunauer and his collaborators have also used isotherms directly for a determination of surface area. This is illustrated by the adsorption isotherms for six different gases on an iron catalyst, shown in Fig. 6.15.[117] As Brunauer points out:

> The isotherms are S-shaped with three distinct regions: the low-pressure region is concave to the pressure axis, the high-pressure region is convex, and between the two there is a linear portion. . . . It will be noted that the five smaller molecules of approximately equal size ($N_2$, $O_2$, Ar, CO, and $CO_2$) show approximately equal adsorptions around 50 mm pressure, but at higher pressures the isotherms diverge widely. At first it was believed that the extrapolation of the intermediate straight-line portion to zero pressure (point *A*) corresponded to the volume of gas necessary to cover the surface with a unimolecular adsorbed

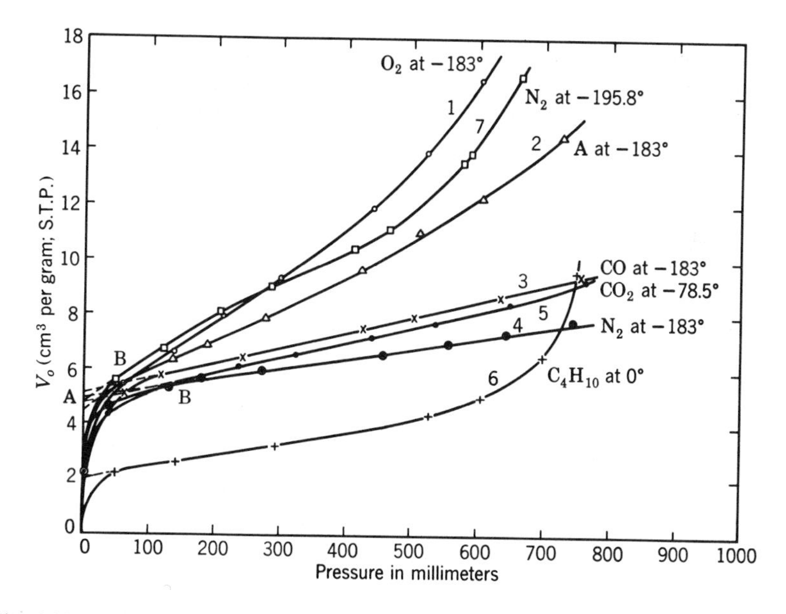


Fig. 6.15. Determination of the surface area of an iron catalyst by the gas adsorption method. The value of $V_m$ at the point *B* is regarded as that corresponding to the volume ($cm^3$, STP) of a monolayer (Brunauer).

layer,[118] but later the beginning of the straight portion (point $B$) was chosen as the most likely point to correspond to the monolayer.[119]

Granted that point $B$ corresponds to the number of molecules in a complete monolayer, the surface area of the adsorbent can then be calculated as follows:

Assuming that the adsorbed molecules have the same packing on the surface as the molecules of the solidified gas have in their plane of closest packing, we obtain for the area covered by a molecule,

$$\text{Area } (S) = \frac{4(3)^{1/2}}{2}\left(\frac{M}{4\sqrt{2}\, N_A \rho_s}\right)^{2/3} \text{cm}^2, \tag{6.46}$$

where $N_A$ is the Avogadro constant, and $\rho_s$ is the density of the solidified gas.

Hence,

$$S = 1.091\left(\frac{M}{N_A \rho_s}\right)^{2/3} \text{cm}^2 \tag{6.47}$$

$$= 1.530 \cdot 10^{-16}\left(\frac{M}{\rho_s}\right)^{2/3} \text{cm}^2. \tag{6.48}$$

The coefficient 1.091 has been designated the "packing factor," and, as shown by Livingston,[120] its value varies with the type of packing. For closest packing,

$$S = \frac{3^{1/2}}{2}\,\delta^2 = 0.866\delta^2, \tag{6.49}$$

where $\delta$ = molecular diameter. Hence, we obtain from equation 6.48 the relations

$$\delta = 1.074(S)^{1/2}$$
$$= 1.329 \cdot 10^{-8}\left(\frac{M}{\rho_s}\right)^{1/2}, \tag{6.50}$$

and

$$S = 1.103\,\frac{\pi\delta^2}{4}. \tag{6.51}$$

That is, the cross-sectional area is 1.103 times the projected area of the molecule if it is regarded as a sphere of diameter $\delta$.

The *number of molecules which can be adsorbed per square centimeter* to form a monolayer is given by $N_S = 1/S$, and the *volume* of gas at STP is given by

$$V_0 = \frac{10^{-19}N}{2.687} = 3.722 \cdot 10^{-20} N_S$$
$$= 3.722 \cdot \frac{10^{-20}}{S}, \tag{6.52}$$

where $S$ is the area per molecule calculated by means of equation 6.48

either from the density of the solid (giving $S_s$) or from that of the liquid (giving $S_l$).

Table 6.13 gives values of $S$ and $1/S$, as deduced from the values of $\rho_s$ and $\rho_l$ in rows 1 and 4.[121] The temperatures corresponding to these values of the density are given in rows 2 and 5. Row 3 gives the values of the lattice constant $a$, in centimeters.

Corresponding values of $S_s$ and $S_l$, and their reciprocals, are given in rows 6–9 inclusive. For comparison there are also given in row 10 values of $N_S{}^g$ derived on the basis of kinetic-theory relations from $\eta_0$, the coefficient of viscosity at 0° C.[122]

Rows 11 and 12 give the values of $V_0$, the volume adsorbed per square centimeter corresponding to a monolayer as calculated from values of $1/S_s$ and of $N_S{}^g$, respectively. The values in row 11 marked with an asterisk were derived from the values of $1/S_l$.[123]

Row 13 gives values of $1/S_b$, as derived from values of the van der Waals constant $b_0$, which are given in the standard tables. The relation used is

$$\frac{1}{S_b} = \frac{1.695 \cdot 10^{13}}{b_0}. \tag{6.53}$$

For example, for argon, $b_0 = 1.437 \cdot 10^{-3}$. Hence $S_b = 7.51 \cdot 10^{-16}$, $1/S_b = 13.31 \cdot 10^{14}$, and $\delta = 2.95 \cdot 10^{-8}$.

The last row in Table 6.13 gives the values of $1/S_{exp}$, which are calculated from the molecular areas given by Livingston.[124] These values represent the "best" estimates based on both values calculated from density of the appropriate condensed phase and those obtained by measurements of $V_m$ with the various gases on solids of known area, as determined by the absolute method.

From the values of $V_m$, the volume (STP) adsorbed per gram, as determined from the point $B$ in Fig. 6.15, it is possible to calculate the effective surface area, $A_m$, in *square meters per gram*, of the adsorbent, by means of the relation

$$A_m = 2.687 \cdot 10^{15} V_m S. \tag{6.54}$$

Thus, for nitrogen at $-195°$ C, $S = 16.2 \cdot 10^{-16}$ cm². Hence,

$$A_m = 4.352 V_m.$$

Table 6.14 gives values of $A_m$, thus deduced from the observed values of $V_m$, for the iron catalyst used in determining the plots in Fig. 6.15.

The adsorption method has been used by Brunauer and Emmett as well as other investigators to determine surface areas of a variety of adsorbents. Table 6.15, taken from Brunauer,[125] shows that "the method is applicable through a 100,000 fold variation in specific surface, from

**TABLE 6.13**
**Molecular Areas and Number of Molecules per Square Centimeter for Various Gases**

| Data | Unit | $N_2$ | $O_2$ | Ar | CO | $CO_2$ | $CH_4$ | $NH_3$ | $H_2O$ | $N_2O$ | NO | $SO_2$ |
|---|---|---|---|---|---|---|---|---|---|---|---|---|
| 1. $\rho_s$ | $g \cdot cm^{-3}$ | 1.026 | 1.426 | 1.65 | ... | 1.565 | ... | ... | ... | ... | ... | ... |
| 2. $t$ | °C | −252.5 | −252.5 | −253 | −252.5 | −80 | −253 | −80 | ... | ... | ... | ... |
| 3. $10^8 a$ | cm | ... | ... | ... | 5.63 | ... | 5.89 | 5.19 | ... | 5.77 | ... | ... |
| 4. $\rho_l$ | $g \cdot cm^{-3}$ | 0.808 | 1.14 | 1.374 | 0.763 | 1.179 | 0.392 | 0.688 | 1.00 | 1.199 | 1.269 | 0.642 |
| 5. $t$ | °C | −195.8 | −183 | −183 | −183 | −56.6 | −140 | −36 | 3.98 | −80 | −150 | −50 |
| 6. $10^{16} S_s$ | $cm^2$ | 13.8 | 12.1 | 12.8 | 13.7 | 14.1 | 15.0 | 11.7 | ... | 14.4 | ... | ... |
| 7. $10^{16} S_l$ | $cm^2$ | 16.2 | 14.1 | 14.4 | 16.8 | 17.0 | 18.1 | 12.9 | 10.51 | 16.8 | 12.5 | 3.29 |
| 8. $10^{-14}/S_s$ | $cm^{-2}$ | 7.25 | 8.26 | 7.81 | 7.30 | 7.09 | 6.67 | 8.55 | ... | 6.94 | ... | ... |
| 9. $10^{-14}/S_l$ | $cm^{-2}$ | 6.17 | 7.09 | 6.94 | 5.95 | 5.88 | 5.53 | 7.75 | 9.52 | 5.95 | 8.00 | 3.04 |
| 10. $10^{-14} N_S^g$ | $cm^{-2}$ | 8.10 | 8.71 | 8.54 | 8.07 | 5.35 | 5.23 | 4.56 | 5.27 | 5.36 | 10.23 | 3.73 |
| 11. $10^5 (V_0)_s$ | $cm^3$ | 2.70 | 3.07 | 2.91 | 2.72 | 2.64 | 2.48 | 3.18 | 3.54* | 2.58 | 2.98* | 1.13* |
| 12. $10^5 (V_0)_N$ | $cm^3$ | 3.02 | 3.24 | 3.18 | 3.00 | 1.99 | 1.95 | 1.70 | 1.96 | 2.00 | 3.81 | 1.39 |
| 13. $10^{-14}/S_b$ | $cm^{-2}$ | 11.69 | 13.42 | 13.31 | 11.54 | 11.04 | 11.01 | 12.12 | 13.80 | 10.78 | 14.65 | 9.16 |
| 14. $10^{-14}/S_{exp}$ | $cm^{-2}$ | 6.33 | 6.85 | 6.45 | 6.13 | ... | ... | 6.85 | ... | 4.90 | ... | ... |

TABLE 6.14

Surface Area of Iron Catalyst

| Gas | Temperature of Isotherm (° C) | Boiling Point of Gas (° C) | Point $B$ (cm³ at 0° C and 760 mm) | Calculated Area (m²/g) | |
|---|---|---|---|---|---|
| | | | | Solid Packing | Liquid Packing |
| $N_2$ | −183 | −195.8 | 5.5 | 0.444 | 0.548 |
| $N_2$ | −195.8 | ... | 5.5 | 0.444 | 0.522 |
| CO | −183 | −191.6 | 5.7 | 0.457 | 0.561 |
| Ar | −183 | −185.7 | 5.8 | 0.435 | 0.489 |
| $O_2$ | −183 | −183.0 | 5.5 | 0.390 | 0.454 |
| $CO_2$ | −78.5 | −78.5 | 5.95 | 0.491 | 0.593 |
| $C_4H_{10}$ | 0 | −0.3 | 2.2 | 0.411 | 0.411 |

unreduced $Fe_3O_4$ to Darco G." (It should be noted, however, that the area for Darco G is probably somewhat too high, a value of 1295 m²/g having more recently been obtained.[126])

Assuming that the adsorbent is constituted of microscopic (or ultramicroscopic) spheres of diameter $d$ (centimeters), a relation for $A$ (square centimers per gram) in terms of $d$ and the density $\rho$ (grams per cubic centimeter) is obtained as follows:

Let $n$ = number of particles per gram.

$$A = n\pi d^2,$$

$$1 = n\rho \frac{\pi}{6} d^3.$$

Hence

$$A = \frac{6}{d\rho}. \tag{6.55}$$

For example, the density of activated charcoal[127] varies over a considerable range. Assuming that $\rho = 2$ and $A = 4 \cdot 10^6$ cm²/g, it follows from equation 6.55 that $d = 7.5 \cdot 10^{-7}$ cm.

For the measurement of relatively small surfaces, such as those of oxide-coated cathodes, Wooten and Brown[128] have developed a method which represents an extension of that of Emmett and Brunauer. Instead of using nitrogen at temperatures such that $P_0$ is of the order of 1 atm, these investigators have used butane at −116° C ($P_0$ = 0.17 mm), and ethylene at −183° C ($P_0$ = 0.0305 mm) and at −195.8° C ($P_0 = 7.3 \cdot 10^{-4}$ mm). For ethylene, $S = 17.55 \cdot 10^{-16}$ cm², and for butane, $S = 32.0 \cdot 10^{-16}$ cm². At these low pressures, it is possible to measure accurately the much

**TABLE 6.15**

**Specific Surface Areas of Various Adsorbents**

| Adsorbent | Specific Surface ($m^2/g$) |
|---|---|
| 1. $Fe_3O_4$ catalyst (unreduced) | 0.02 |
| 2. Fe catalyst 973, sample I, unpromoted | 0.55 |
| 3. Fe catalyst 973, sample II, unpromoted | 1.24 |
| 4. Fe catalyst 954, 10.2% $Al_2O_3$ | 11.03 |
| 5. Fe catalyst 424, 1.03% $Al_2O_3$, 0.19% $ZrO_2$ | 9.44 |
| 6. Fe catalyst 931, 1.3% $Al_2O_3$, 1.59% $K_2O$ | 4.78 |
| 7. Fe catalyst 958, 0.35% $Al_2O_3$, 0.08% $K_2O$ | 2.50 |
| 8. Fe catalyst 930, 1.07% $K_2O$ | 0.56 |
| 9. Fused Cu catalyst | 0.23 |
| 10. Commercial Cu catalyst | 0.42 |
| 11. Pumice | 0.38 |
| 12. Ni on pumice, 91.8% pumice | 1.27 |
| 13. NiO on pumice, 89.8% pumice | 4.28 |
| 14. $Cr_2O_3$ gel | 228 |
| 15. $Cr_2O_3$ "glowed" | 28.3 |
| 16. KCl (finer than 200 mesh) | 0.24 |
| 17. $CuSO_4 \cdot 5H_2O$ (40–100 mesh) | 0.16 |
| 18. $CuSO_4$ anhydrous | 6.23 |
| 19. Cecil soil, 9418 | 32.3 |
| 20. Cecil soil colloid, 9418 | 58.6 |
| 21. Barnes soil, 10,308 | 44.2 |
| 22. Barnes soil colloid, 10,308 | 101.2 |
| 23. Glaucosil | 82 |
| 24. Silica gel I (nonelectrodialyzed) | 584 |
| 25. Silica gel II (electrodialyzed) | 614 |
| 26. Dried bacteria | 0.17 |
| 27. Dried bacteria (pulverized) | 3.41 |
| 28. Granular Darco B | 576 |
| 29. Granular Darco G | 2123 |
| 30. Activated charcoal | 775 |
| 31. Lampblack | 28 |
| 32. Acetylene black | 64 |
| 33. Grade 3 rubber black | 135 |
| 34. Carbolac 1 color black | 947 |
| 35. Graphite | 30.47 |
| 36. Cuprene | 20.7 |
| 37. Paper | 1.59 |
| 38. Cement | 1.08 |
| 39. $TiO_2$ | 7.88 |
| 40. $BaSO_4$ | 4.30 |
| 41. $ZrSiO_4$ | 2.76 |
| 42. Lithopone | 34.8 |
| 43. Lithopone, calcined | 1.37 |
| 44. Lithopone, calcined and ground | 3.43 |
| 45. Porous glass | 125.2 |

smaller quantities of gas adsorbed by small areas. The adsorption isotherms obtained by plotting $V$ adsorbed versus $P/P_0$ resemble those obtained, for large areas, with nitrogen at $-195.8°$ C, and it is possible by application of equation 6.28 to calculate $V_m$. In this manner, Wooten and Brown found that, for oxide-coated cathodes, the surface areas are approximately 10 $cm^2/mg$ of coating. Areas as low as 10 $cm^2$ could be measured by this method.

Krypton has been used in a similar manner by Beebe, Beckwith, and Honig.[129] The saturation pressure of this gas at the temperature of liquid nitrogen is about 2 mm. By comparing adsorption isotherms for nitrogen and krypton on the same sample of $TiO_2$, it was concluded that the cross-sectional area for krypton is $(19.5 \pm 0.4) \cdot 10^{-16}$ $cm^2$. From adsorption isotherms for krypton with other materials in the range of values of $P/P_0$ below about 0.3, surface areas as low as $1.6 \cdot 10^{-2}$ $m^2/g$ were obtained, as shown in the second column of Table 6.16. The third column gives, for comparison, values obtained with nitrogen.

**TABLE 6.16**

**Surface Areas of Adsorbents**

| | Square Meters per Gram | |
|---|---|---|
| Adsorbent | Krypton | Nitrogen |
| Marble | $1.69 . 10^{-2}$ | |
| Quartz | $3.70 . 10^{-2}$ | |
| Glass wool | $5.94 . 10^{-1}$ | |
| Cork | $7.10 . 10^{-1}$ | |
| Cotton | $8.01 . 10^{-1}$ | |
| Anatase | 13.92 | 13.92 |
| Porous glass | 110 | 121 |

The utility of krypton has been further extended by Rosenberg[130]. Gaines and Cannon[45] have given some comments on the significance of surface areas measured with this and other gases. It is also of interest to mention that Beebe and his associates found, from their adsorption data, that for krypton at liquid-nitrogen temperature the value of the Harkins-Jura constant $k$ in equation 6.45 is 4.20.

## 6.9. CHEMISORPTION

We have so far considered primarily sorption due to physical or van der Waals forces, with only an occasional comment on sorption processes involving the formation of chemical bonds.

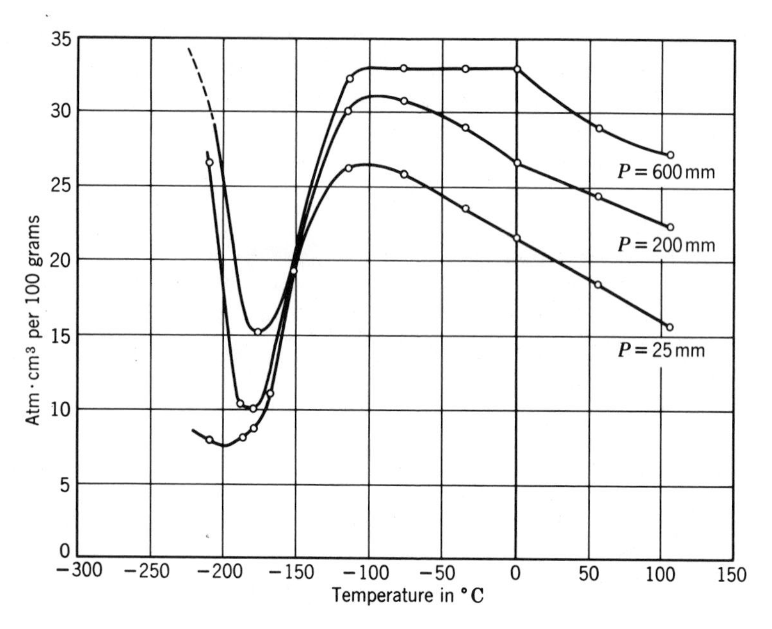


Fig. 6.16. Isobars for the adsorption of hydrogen on nickel powder at three different pressures (Benton and White).

In true physical adsorption the amount sorbed increases with decrease in temperature, as stated in the previous sections. However, it has been observed that frequently the sorption, after decreasing with increase in temperature to a minimum value, exhibits a reversal in this respect with still further increase in temperature until a maximum value is attained, after which further increase in temperature results in decreased adsorption once more.

Figure 6.16 shows isobars for hydrogen on nickel powder, which were observed by Benton and White.[131] Whereas equilibrium was reached almost instantaneously at −180° C, about 1–2 hr was required at −100° C; and, as will be observed, the amount adsorbed at −100° C was over 3 times that absorbed at −180° C.

Similar observations have been made by Benton and White[132] on the sorption of hydrogen by nickel and copper (see Fig. 6.17) and by Taylor and McKinney[133] on the sorption of carbon monoxide by palladium. As indicated in Fig. 6.17, the sorption of hydrogen by nickel is a maximum at 0° C. There are many examples of sorption in which the physical adsorption at low temperatures is succeeded by a much greater amount of sorption at higher temperatures, and the rate of this sorption is usually very slow. This second stage has been designated *activated adsorption*[134] and also *chemisorption*, since the corresponding energy evolved is of the same order of magnitude as that evolved in chemical reactions.[135]

As would be expected, on the basis of this interpretation, activated adsorption has not been observed for the rare gases, for nitrogen on copper, or for hydrogen on gold and silver. Smithells[136] states,

> It would appear therefore that this type of adsorption depends on a chemical attraction between the metal surface and the gas, and is most strongly marked in such systems as oxygen-silver, carbon monoxide-nickel, and hydrogen with copper, nickel, or iron. . . . Although it cannot actually be chemical combination, since no new phase is formed, it is probably a necessary preliminary to chemical action which may occur at still higher temperatures. In the systems just mentioned, for example, definite compounds $Ag_2O$, $Ni(CO)_4$, $NiH_2$, etc., between gas and metal are known.

Smithells further writes:[137]

> Whilst the process of physical adsorption is perfectly reversible with respect to changes in temperature and pressure, this is not the case with activated adsorption. So far as pressure changes are concerned, the isotherms usually show reversibility, although this is not always so. . . .
>
> Activated adsorption is never reversible with respect to changes in temperature. If the equilibrium value is found by admitting gas to the adsorbent at a temperature $T_1$, and it is then heated, still in contact with the gas, to a higher temperature

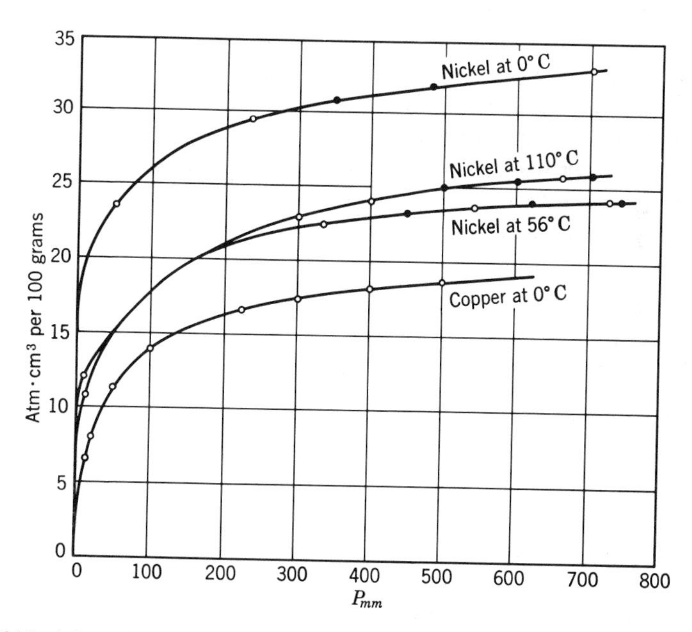


Fig. 6.17. Adsorption isotherms for hydrogen on nickel and copper, illustrating chemisorption (Benton and White).

$T_2$, on again cooling to $T_1$ the equilibrium value is higher than before. . . . It is significant that the gas adsorbed as the result of thermal activation cannot be removed by evacuation at $T_1$, but only at higher temperatures. This is, of course, in accordance with the usual practice of baking during evacuation to remove gas adsorbed on glass and metal parts of vacuum apparatus.

An illustration of activated adsorption which is of importance in the exhaust of vacuum tubes is found in the investigations of Kingdon and Langmuir[138] on the effect of adsorbed oxygen on the thermionic emission from tungsten filaments. At low temperatures and low pressures oxygen is adsorbed and decreases the emission to values that range from $10^{-2}$ to $10^{-5}$ of that from pure tungsten. It is only on heating the filament in vacuum to about 1700° C or higher that the adsorbed layer is removed with a resulting restoration of the emission to its original value. According to Langmuir,

> An oxygen film on tungsten at 1500° K does not evaporate in less than a year. . . . The fact that a monatomic film of oxygen on tungsten does not react with hydrogen at 1500° K is a striking indication that the oxygen is in a condition far different from that of gaseous oxygen.

Roberts[139] has also obtained evidence for the activated adsorption of hydrogen on a clean tungsten surface. Desorption is observed only at temperatures above 400° C.

Since chemisorption, even more than physical adsorption, will be strongly influenced by the chemical and physical state of the solid surface, much effort has recently been devoted to attempts to prepare and define solid surfaces. The cleanliness, that is, lack of any foreign contaminants, of a solid is very difficult to assure. Ultrahigh-vacuum techniques (cf. Chapter 9) are now extensively used with adsorbent systems prepared by evaporation or very high-temperature outgassing. Only if the pressure in the system is maintained in the ultrahigh-vacuum range (below $10^{-8}$ mm), is it possible to assume that a significant amount of contamination does not arrive from the gas phase before or during the experiment.

Cleaning the surface before an experiment may involve outgassing, reduction of metal oxides, ion bombardment, or other methods.[140] Metal filaments which can be electrically heated to incandescence to remove contaminants, and metal films prepared by evaporation *in vacuo*, have been extensively used. Some interesting analytical data for the gases found after evaporation of metal films in ultrahigh vacuum have been given by Gentsch.[141]

It must be remembered, however, that both chemical and geometrical factors will affect the interaction of a gas molecule with a surface, so that crystal perfection and surface roughness are important. The development of the field emission microscope[142] has led to much work on the different adsorptive properties of different crystal faces of a metal. Such variations are doubtless due, at least in part, to the difference in packing of the metal atoms on the different faces.[143]

More than one chemisorbed state may be involved in the adsorption process. One of the simplest examples is provided by diatomic gases, which may be adsorbed as molecules or be dissociated into atoms. More than one type of bonding has been suggested, for example, by Ehrlich[144] to account for his observations on the desorption of nitrogen from a tungsten filament. Ehrlich has also given a useful summary of dissociative chemisorption.[145]

Studies in which more than one adsorbed state has been suggested for hydrogen adsorbed on platinum are those of Mignolet[146] and Suhrmann *et al.*[147] Jones and Pethica[148] have suggested that the different chemisorbed states which they find for nitrogen on tungsten can be correlated with the electronic state of the surface. This concept, the so-called "*d*-orbital hypothesis," was first proposed by Dowden[149] and by Eley and coworkers.[150] A discussion is given by Trapnell.[151]

The chemisorption of more complex molecules is intimately concerned with the processes of heterogeneous catalysis, and much work has been reported in the literature of that field.[152]

In vacuum technology, systems wherein chemisorption may occur are common, and specific data on some of these are given in succeeding chapters.

## REFERENCES AND NOTES

1. The following monographs are cited in this chapter:

*a.* S. Brunauer, *The Adsorption of Gases and Vapors*, Vol. I, *Physical Adsorption*, Princeton University Press, Princeton, N.J., 1943.

*b.* J. W. McBain, *The Sorption of Gases and Vapours by Solids*, George Routledge and Sons, London, England, 1932.

Reference *b* contains a great deal of detailed information on charcoal and silica gel, as well as other adsorbents. Reference *a* deals more with the theoretical aspects and with specific adsorbents only as they illustrate the theoretical deductions.

*c.* B. M. W. Trapnell, *Chemisorption*, Butterworths Scientific Publications, London, England, 1955.

Data and references on the adsorption of gases are given in *International Critical Tables*, McGraw-Hill Book Company, New York, 1928, Vol. 3, pp. 250–251. In addition, the reader seeking information on specific systems will find an extensive collection of abstracts from the literature in "The Bibliography of Solid Adsorbents," V. R. Dietz, Editor, Vol. 1, 1900–1942, Bone Char Research Project, Inc., 1944; Vol. 2, 1943–1953, *Natl. Bur. Standards Circ. No.* 566, 1956.

2. In this connection, see also Ref. 1*c*, Chapter 1.
3. Reference 1*a*, p. 12; Ref. 1*b*, p. 102.
4. W. Hene, dissertation, Hamburg, 1927.
5. G. Ehrlich, *J. Phys. Chem.*, **60,** 1388 (1956).
6. J. H. de Boer, *Advances in Catalysis*, **VIII,** 18 (1956).

7. For description of the methods used in obtaining sorption data see references in note 1.

8. I. Homfray, *Z. physik. Chem.*, **74,** 129 (1910); Ref. 1*b*, p. 101, Fig. 28.

9. A. Magnus and H. Kratz, *Z. anorg. Chem.*, **184,** 241 (1929).

10. M. H. Armbruster, *J. Am. Chem. Soc.*, **64,** 2545 (1942).

11. W. G. Frankenburg, *J. Am. Chem. Soc.*, **66,** 1827 (1944).

12. A. J. Allmand and R. Chaplin, *Proc. Roy. Soc. London*, **A, 129,** 235 (1930).

13. A. J. Allmand and L. J. Burrage, *J. Phys. Chem.*, **35,** 1692 (1931).

14. See also A. J. Allmand, L. J. Burrage, and R. Chaplin, *Trans. Faraday Soc.*, **28,** 218 (1932).

15. G. Jura, E. H. Loeser, P. R. Basford, and W. D. Harkins, *J. Chem. Phys.*, **14,** 117 (1946).

16. B. A. Fisher and W. G. McMillan, *J. Chem. Phys.*, **28,** 549 (1958).

17. M. L. Corrin and C. P. Rutkowski, *J. Phys. Chem.*, **58,** 1089 (1954).

18. H. Freundlich, *Kapillarchemie*, Leipzig, 1930, Vol. I, pp. 153–172.

19. Reference 1*a*, p. 57.

20. The terminology indicates that the forces between gas molecules and atoms on the surface of the adsorbent are of the same nature as those leading to deviations from the ideal-gas laws for gases at high pressures. Heats of evaporation and of adsorption constitute a measure of the magnitudes of these forces. (See section 6.6.)

21. K. Peters and K. Weil, *Z. physik. Chem.*, **A, 148,** 1 (1930).

22. H. Freundlich (Ref. 18), p. 170.

23. I. Langmuir, *J. Am. Chem. Soc.*, **40,** 1361 (1918).

24. Proposed by William Henry (1803–1805).

25. H. H. Lowry and S. O. Morgan, *J. Phys. Chem.*, **29,** 1105 (1925).

26. I. Langmuir, *J. Am. Chem. Soc.*, **40,** 1361 (1918). A review of the theory and further experimental evidence for the adsorption of gases on metals as monatomic films are also given in the extremely important paper, "Surface Chemistry," *Chem. Revs.*, **13,** 147 (1933).

27. A designation introduced subsequently by J. A. Becker and Langmuir for *adsorbed atoms* or *molecules*.

28. I. Langmuir, *J. Chem. Soc.*, **1940,** 511; the Seventeenth Faraday Lecture, entitled "Monolayers on Solids."

29. M. Volmer, *Z. physik. Chem.*, **115.** 253 (1925).

30. R. H. Fowler, *Proc. Cambridge Phil. Soc.*, **31,** 260 (1935); **32,** 144 (1936).

31. Reference 1*a*, pp. 66–67.

32. See subsequent discussion.

33. I. Langmuir, *J. Am. Chem. Soc.*, **40,** 1361 (1918).

34. E. C. Markham and A. F. Benton, *J. Am. Chem. Soc.*, **53,** 497 (1931).

35. S. Brunauer, P. H. Emmett, and E. Teller, *J. Am. Chem. Soc.*, **60,** 309 (1938).

36. Reference 1*a*, pp. 317–329.

37. H. E. Ries in *Catalysis*, P. H. Emmett, Editor, New York, Reinhold Publishing Corporation, 1954, Vol. 1, p. 14.

38. See Ref. 1*a*, pp. 151–155, for derivation; also *J. Am. Chem. Soc.*, **60,** 309 (1938).

39. S. Brunauer, L. S. Deming, W. E. Deming, and E. Teller, *J. Am. Chem. Soc.*, **62,** 1723 (1940).

40. Reference 1*a*, pp. 155–177.

41. Reference 1*a*, pp. 177–178.

42. P. H. Emmett (Ref. 37), Chapter 2.

43. P. H. Emmett and S. Brunauer, *J. Am. Chem. Soc.*, **59,** 1553 (1937).

44. W. C. Walker and A. C. Zettlemoyer, *J. Phys. Chem.*, **57,** 1826 (1953).

45. G. L. Gaines, Jr., and P. Cannon, *J. Phys. Chem.*, **64,** 997 (1960).
46. I. Langmuir, *J. Chem. Soc.*, **1940,** 511.
47. A. M. Williams, *Proc. Roy. Soc. London*, **A, 96,** 287, 298 (1919), and D. C. Henry, *Phil. Mag.*, **(6), 44,** 689 (1922).
48. A. Magnus, *Z. physik. Chem.*, **A142,** 401 (1929).
49. L. Tonks, *Phys. Rev.*, **50,** 955 (1936), and *J. Chem. Phys.*, **8,** 477 (1940).
50. T. L. Hill, *Advances in Catalysis*, **4,** 228 (1952).
51. T. L. Hill, *J. Chem. Phys.*, **15,** 767 (1947).
52. G. D. Halsey, *J. Chem. Phys.*, **16,** 931 (1948).
53. J. Zeldowitch, *Acta Physicochim. U.R.S.S.*, **1,** 961 (1935).
54. M. I. Temkin, *J. Phys. Chem. U.S.S.R.*, **15,** 296 (1941).
55. R. Sips, *J. Chem. Phys.*, **18,** 1024 (1950).
56. D. Graham, *J. Phys. Chem.*, **57,** 665 (1953); **58,** 869 (1954).
57. A. Eucken, *Verhandl. deut. physik. Ges.*, **16,** 345 (1914).
58. M. Polanyi, *Verdandl. deut. physik. Ges.*, **16,** 1012 (1914).
59. Reference 1*a*, p. 96. This reference, as well as Chapter XVII in Ref. 1*b*, should be consulted for detailed discussion of the topic.
60. J. H. de Boer and C. Zwikker, *Z. phys. Chem.*, **B3,** 407 (1929).
61. R. Zsigmondy, *Z. anorg. Chem.*, **71,** 356 (1911).
62. J. McGavack, Jr., and W. A. Patrick, *J. Am. Chem. Soc.*, **42,** 946 (1920).
63. Reference 1*a*, pp. 120–149.
64. J. Frenkel, *Kinetic Theory of Liquids*, Oxford University Press, Oxford, 1946.
65. T. L. Hill, *J. Chem. Phys.*, **17,** 590, 668 (1949).
66. W. G. McMillan and E. Teller, *J. Chem. Phys.*, **19,** 25 (1951); *J. Phys. Colloid Chem.*, **55,** 17 (1951).
67. W. D. Harkins and G. Jura, *J. Am. Chem. Soc.*, **66,** 1362, 1366 (1944).
68. Further discussion of this topic will be given in section 6.8.
69. A. Wheeler, paper presented before the Division of Physical and Inorganic Chemistry, American Chemical Society, Atlantic City, N.J., September, 1949, and paper presented before the Division of Colloid Chemistry, American Chemical Society, New York, September, 1951.
70. S. Ono, *J. Chem. Phys.*, **18,** 397 (1950); *J. Phys. Soc. Japan*, **5,** 232 (1950), **6,** 10 (1951).
71. H. E. Ries (Ref. 37), p. 21.
72. These are described in Ref. 1*a*, Chapter III, and Ref. 1*b*, Chapter XIV.
73. H. R. Kruyt and J. G. Modderman, *Chem. Revs.*, **7,** 259 (1930); Ref. 1*a*, p. 223.
74. Reference 1*b*, p. 400.
75. F. Simon, *Physik. Z.*, **27,** 790 (1926). See also D. H. Andrews, R. K. Witt, and E. Crigler, *Refrig. Eng.*, **19,** 177 (1930).
76. T. L. Hill, *J. Chem. Phys.*, **17,** 520 (1949), and **18,** 246 (1950); *Trans. Faraday Soc.*, **47,** 376 (1951).
77. T. L. Hill, P. H. Emmett, and L. G. Joyner, *J. Am. Chem. Soc.*, **73,** 5102 (1951).
78. G. L. Gaines, Jr., *J. Phys. Chem.*, **62,** 1521, 1526 (1958).
79. D. Graham, *J. Phys. Chem.*, **60,** 1022 (1956).
80. R. W. Wood, *Phil. Mag.*, **30,** 300 (1915); **32,** 364 (1916).
81. M. Knudsen, *Ann. Physik*, **50,** 472 (1916).
82. I. Langmuir, *Proc. Natl. Acad. Sci. U.S.*, **3,** 141 (1917).
83. M. Volmer and I. Estermann, *Z. Physik*, **7,** 1, 13 (1921).
84. See, for example, the review by S. Wexler, *Revs. Modern Phys.*, **30,** 402 (1958).
85. See, for example, E. A. Mason and J. T. Vanderslice, *J. Chem. Phys.*, **28,** 253 (1958).

86. G. W. Sears and J. W. Cahn, *J. Chem. Phys.*, **33**, 494 (1960).
87. F. G. Keyes and M. J. Marshall, *J. Am. Chem. Soc.*, **49**, 156 (1927); Ref. 1*a*, p. 228, Fig. 83.
88. Reference 1*a*, pp. 240–241.
89. Reference 1*a*, p. 241. A very comprehensive survey of data on heats of adsorption has been given by H. R. Kruyt and J. G. Modderman in *International Critical Tables*, Vol. V, pp. 139–141. The values of $Q$, given in joules, can be converted into calories by multiplying by the factor 0.2392. Also calories per mole = 22,400 × (calories per cubic centimeter).
90. M. H. Armbruster and J. B. Austin, *J. Am. Chem. Soc.*, **66**, 159 (1944).
91. F. J. Wilkins, *Proc. Roy. Soc., London*, **A, 164**, 496 (1938).
92. See Ref. 1*a*, p. 215.
93. W. J. C. Orr, *Trans. Faraday Soc.*, **35**, 1247 (1939).
94. The significance of the word "clean" will be considered further in the subsequent discussion.
95. J. H. de Boer and J. F. H. Custers, *Z. phys. Chem.*, **B25**, 225 (1934).
96. R. M. Barrer, *Nature*, **181**, 176 (1958).
97. D. H. Everett, *Proc. Chem. Soc.* (London), **1957**, 38.
98. Lord Blythswood and H. S. Allen, *Phil. Mag.*, **10**, 497 (1905).
99. I. Langmuir, *J. Am. Chem. Soc.*, **40**, 1361 (1918).
100. D. H. Bangham and F. P. Burt, *Proc. Roy. Soc. London*, **A, 105**, 481 (1924). See also Ref. 1*b*, p. 48.
101. J. B. Firth, *Z. physik. Chem.*, **86**, 294 (1914).
102. J. W. McBain, *Z. physik. Chem.*, **68**, 471 (1909).
103. R. M. Barrer and D. A. Ibbitson, *Trans. Faraday Soc.*, **40**, 195 (1944).
104. Reference 1*a*, p. 8.
105. P. Clausing, *Ann. Physik*, **7**, 489, 521 (1930). This reference is mentioned in the publication by O. Beeck, A. E. Smith, and A. Wheeler, on catalytic activity, crystal structure, and adsorptive properties of evaporated metal films, *Proc. Roy. Soc. London*, **A, 177**, 62 (1940–1941).
106. Reference 1*a*, Chapter IX.
107. P. H. Emmett (Ref. 37), Chapter 2.
108. C. Orr, Jr., and J. M. Dallavalle, *Fine Particle Measurement*, New York, The MacMillan Company, 1959.
109. F. P. Bowden and E. K. Rideal, *Proc. Roy. Soc. London*, **A, 120**, 59, 80 (1928).
110. F. P. Bowden and E. A. O'Connor, *Proc. Roy. Soc. London*, **A, 128**, 317 (1930).
111. Reference 1*b*, p. 341.
112. W. D. Harkins and G. Jura, *J. Am. Chem. Soc.*, **66**, 1362 (1944).
113. C. Orr, Jr., and J. M. Dallavalle (Ref. 108), p. 236.
114. W. D. Harkins and G. Jura, *J. Am. Chem. Soc.*, **66**, 1366 (1944).
115. See, for example, N. K. Adam, *The Physics and Chemistry of Surfaces*, Oxford University Press, 3rd edition, 1941, Chapter 2, for a discussion of the properties of such films.
116. M. L. Corrin, *J. Am. Chem. Soc.*, **73**, 4061 (1951).
117. Reference 1*a*, p. 286, Fig. 100; *J. Am. Chem. Soc.*, **59**, 1553 (1937).
118. S. Brunauer and P. H. Emmett, *J. Am. Chem. Soc.*, **57**, 1754 (1935).
119. P. H. Emmett and S. Brunauer, *J. Am. Chem. Soc.*, **59**, 1553 (1937).
120. H. K. Livingston, *J. Am. Chem. Soc.*, **66**, 569 (1944).
121. All the values of the density, with the exception of that for sulfur dioxide, are taken from the paper by Emmett and Brunauer, *J. Am. Chem. Soc.*, **59**, 1553 (1937). The value of $\rho_l$ for sulfur dioxide is taken from standard tables.

122. Some of the values are taken from Table 1.6. For the following gases the values of $\eta_0$ used for the calculation of $N_S{}^g$ are taken from the summary by A. E. Schuil, *Phil. Mag.*, **28,** 679 (1939).

| Gas: | $N_2O$ | NO | $SO_2$ | $CH_4$ | $NH_3$ | CO |
|---|---|---|---|---|---|---|
| $10^5\eta_0$: | 13.88 | 17.77 | 11.58 | 10.22 | 9.16 | 16.59 |

123. It is of interest to observe that Langmuir, *J. Am. Chem. Soc.*, **40,** 1361 (1918), used the following values of $1/S$:

| Gas: | $N_2$ | $O_2$ | Ar | CO | $CO_2$ | $NH_3$ |
|---|---|---|---|---|---|---|
| $10^{-14}/S$: | 6.6 | 7.7 | 7.7 | 6.6 | 6.1 | 6.3 |

Also Brunauer (Ref. 117) gives for $H_2O$ the value $S = 11.5 \cdot 10^{-16}$ cm$^2$; that is, $1/S = 8.70 \cdot 10^{14}$.

124. H. K. Livingston, *J. Colloid Sci.*, **4,** 450 (1949).

125. S. Brunauer, *J. Am. Chem. Soc.*, **59,** 1553 (1937); Ref. 1*a*, p. 298. See also determinations (by means of this method) of surface area for "pigments, carbon blacks, cements and miscellaneous finely divided or porous materials," by P. H. Emmett and T. DeWitt, *Ind. Eng. Chem., Anal. Ed.*, **13,** 28 (1941). Using the same technique, W. R. Smith, F. S. Thornhill, and R. I. Bray, *Ind. Eng. Chem.*, **33,** 1303 (1941), obtained values for the surface area of carbon black ranging from 15 to 64 m$^2$/g. It is of interest to note that, although the calculated diameters (see equation 6.50) show good agreement in many cases with electron microscope observations, there is poor agreement in other cases.

126. W. V. Loebenstein and V. R. Deitz, *J. Research Natl. Bur. Standards*, **46,** 51 (1951).

127. The value of $\rho$ for charcoal is discussed in Ref. 1*b*, pp. 79–92.

128. L. A. Wooten and J. R. C. Brown, *J. Am. Chem. Soc.*, **65,** 113 (1943).

129. R. A. Beebe, J. B. Beckwith and J. M. Honig, *J. Am. Chem. Soc.*, **67,** 1554 (1945).

130. A. J. Rosenberg, *J. Am. Chem. Soc.*, **78,** 2929 (1956).

131. A. F. Benton and T. A. White, *J. Am. Chem. Soc.*, **52,** 2325 (1930); C. J. Smithells, *Gases and Metals*, Chapman and Hall, London, 1937, p. 32, Fig. 22.

132. A. F. Benton and T. A. White, *J. Am. Chem. Soc.*, **54,** 1373 (1932).

133. H. S. Taylor and P. V. McKinney, *J. Am. Chem. Soc.*, **53,** 3604 (1931).

134. H. S. Taylor, *J. Am. Chem. Soc.*, **53,** 578 (1931).

135. See C. J. Smithells (Ref. 131), p. 48, for a summary of values of heats of activated adsorption, which range from 4000 to over 100,000 cal/g-mole of gas.

136. C. J. Smithells (Ref. 131), p. 30.

137. This quotation and that in the next paragraph are from C. J. Smithells (Ref. 131) pp. 38–39.

138. K. H. Kingdon and I. Langmuir, *Phys. Rev.*, **24,** 510 (1924). See also Langmuir and D. S. Villars, *J. Am. Chem. Soc.*, **53,** 486 (1931) and Langmuir, *Chem. Revs.*, **13,** 147 (1933), for discussion of this topic. The remarks quoted are taken from this extremely important review.

139. J. K. Roberts, *Proc. Roy. Soc. London*, **A, 152,** 445 (1935).

140. See, for example, Ref. 1*c*, pp. 16–20.

141. H. Gentsch, *Z. physik. Chem.*, N.F., **24,** 55 (1960).

142. See reviews by E. W. Müller, *Ergeb. exakt. Naturwiss.*, **27,** 290 (1953), and J. A. Becker and C. D. Hartman, *J. Phys. Chem.*, **57,** 157 (1953).

143. Some of this work has been reviewed by J. A. Becker, *Advances in Catalysis*, **7,** 135 (1955).

144. G. Ehrlich, *J. Phys. Chem.*, **60,** 1388 (1956).
145. G. Ehrlich, *J. Chem. Phys.*, **31,** 1111 (1959).
146. J. C. P. Mignolet, *J. chim. phys.*, **54,** 19 (1957).
147. R. Suhrmann, G. Wedler, and H. Gentsch, *Z. physik. Chem.*, N.F., **17,** 350 (1958).
148. P. L. Jones and B. A. Pethica, *Proc. Roy. Soc. London*, **A, 256,** 454 (1960).
149. D. A. Dowden, *Research*, **1,** 239 (1948).
150. M. H. Dilke, E. D. Maxted, and D. D. Eley, *Nature*, **161,** 804 (1948).
151. Reference 1*c*, Chapter 7.
152. See, for example, K. J. Laidler, "Chemisorption," Chapter 3 in Ref. 37; and Ref. 1*c*, Chapters 9 and 10.

# 7 Sorption of gases by "active" charcoal, silicates (including glasses), and cellulose

Revised by George L. Gaines, Jr.

## 7.1. DATA ON SORPTION OF GASES

In Chapter 6 we examined the nature of sorption phenomena and the molecular processes which are involved in them. In this chapter, typical data are given on the sorption of various gases on some nonmetallic materials which are of importance in vacuum technology. As might be expected on the basis of the comments in Chapter 6, the sorptive properties of chemically complex materials like cellulose, charcoal, and glasses will depend greatly on the method of their preparation. This will also be apparent from an examination of the data in this chapter.

There has been a very marked increase in the number of studies in this area reported in the literature. "The Bibliography of Solid Adsorbents,"[1] for example, contains 1647 references to publications on the "adsorption of gases and vapors on solid adsorbents" in the years 1900–1942; for the period 1943–1953, the number of such abstracts is 2385.

For these reasons, the present discussion is not intended to be comprehensive, but rather to present typical data on these materials, together with a few interpretive comments. The reader who wishes detailed information should consult the original literature.

## 7.2. PREPARATION AND STRUCTURE OF ACTIVATED CHARCOAL

The capacity of charcoal for sorption of gases depends both on the nature of the source used in the preparation and on the type of heat treatment. Baerwald[2] showed that by heating charcoal above 500° C its adsorptive power is increased considerably; also that charcoal from coconut shell is a much better adsorbent than that from the soft part of the nut or from wood. Bergter[3] measured the rate of adsorption of nitrogen, oxygen, and air on coconut charcoal at about 18° C, and before World War I a number of other investigators, including Dewar, Travers,[4] McBain,[5] Homfray,[6] Titoff,[7] and Claude,[8] carried out important investigations on the sorption of different gases by charcoal, the results of which will be mentioned subsequently. However, the incidence of the war in 1914 and the necessity of developing a highly efficient adsorbent for use in gas masks provided a tremendous stimulus to further investigations in this field. As a result there was evolved a technique for the production of a highly activated form of charcoal, and, furthermore, comprehensive studies were carried out on the relation between the structure of the adsorbent and its adsorptive capacity.

The procedure used by McBain and others in preparing coconut charcoal was as follows. The soft part was heated in a muffle furnace for several hours at just below a red heat, until no more evolution of vapor could be observed. Then the temperature was raised to a dull red heat for 30 sec. Before use as an adsorbent the charcoal was heated to 440° C for several hours, *in vacuo*.

Lemon[9] observed in 1915 that different samples of charcoal made from the same source (coconut shell) showed very wide variations in sorptive power. It was found that this was due to variations in heat treatment, and that increased activation could be produced by repeated evacuations at 650° C, each evacuation being followed by sorption of air at the temperature of liquid air. On the other hand, a decrease in sorptive capacity resulted from heating the charcoal to as high as 800–900° C during evacuation.

The theory was advanced that the successive sorptions of air oxidize nonvolatile hydrocarbons present in the charcoal. As a result, an air process of activation was evolved which, as described by Dorsey,[10] consisted essentially of the following operations: (1) initial distillation of cracked coconut hulls to a temperature of 850–900° C, and (2) "air treating" this carbonized material, screened 6 to 14 mesh, at 350–400° C, for a certain length of time.

The essential characteristics of an active charcoal, according to Lamb, Wilson, and Chaney,[10] are (1) high and fine-grained porosity, (2) the presence of amorphous base carbon, (3) freedom from adsorbed hydrocarbons.

To secure these objects it is necessary to use dense woods, carry out the distillation at relatively low temperatures, and then oxidize the hydrocarbons without injuring the carbon base to any measurable extent. "The permissible range of temperatures for the latter operation is a relatively narrow one, only about 50–75°." For air oxidation this lies between 350 and 450° C. Subsequently a steam process of activation was adopted, and for this reaction the optimum temperature is between 800 and 1000° C. Other methods of activation have been used in Europe. All these processes yield charcoal which is much more active than that obtained by the simple distillation process used at one time.

From a study of the slope of the vapor pressure curves of liquids adsorbed upon such charcoal, the indications are that the pores have, if a cylindrical form be assumed, an average diameter of about $5 \cdot 10^{-7}$ cm. On this basis, 1 $cm^3$ of active charcoal would contain about 1000 $m^2$ of surface.[10]

A review of methods of preparation of adsorbent carbons and their properties has been given by Kipling.[11]

The density of activated charcoal from coconut shells is about 0.4. Hence 1 g would contain about 2500 $m^2$ of surface. Assuming that the clean-up by charcoal is due to a condensation of gas molecules on the surface, and that the diameter of a hydrogen molecule is about $2 \cdot 10^{-8}$ cm, it would require approximately 2000 $cm^3$ (measured at 0° C and 760 mm) to cover the surface of 1 g of charcoal. By comparison, the adsorptions obtained by Homfray, Titoff, and Firth, even at atmospheric pressure, are very low, which may be accounted for partly by the smaller porosity of the charcoal used by them.

In view of his theory, discussed in Chapter 6, that true adsorption consists in the formation of a monolayer, Langmuir has pointed out:[12]

Truly porous bodies, such as charcoal, probably consist of atoms combined together in branching chains of great complexity. The fibers of cellulose from which charcoal is usually formed consist of practically endless groups of atoms

```
 H  H  H  H
—C—C—C—C—
 O  O  O  O
 H  H  H  H
```

held together by primary valences in the direction of their length and by secondary valence in the transverse directions. When the hydrogen and oxygen atoms are driven out by heat, the carbon atoms for the most part remain in their chains, but a certain number of cross linkages occur between these chains. *The porosity of the charcoal thus undoubtedly extends down to atomic dimensions.* The unsaturated state of the remaining carbon atoms explains the practical impossibility of removing the last traces of oxygen and hydrogen from any form of amorphous carbon.

Hence it is evident that, with a structure of this kind, it is meaningless to talk about the surface on which the adsorption can take place.

On plane surfaces the maximum adsorption would correspond to a layer one molecule deep; but in the case of charcoal, there is no definite surface that can be covered by a unimolecular layer. The atoms of carbon would be separated by spaces which might hold one or more molecules of the gas, or, on the other hand, might be too small to hold even one molecule.

Briggs,[13] as a result of his investigations on the adsorption of gas on charcoal and silica gel, concludes that, while the chemical characteristics of a material affect its properties as a gas adsorbent, there are, with any given material, two factors which also influence the adsorptive power: "(*a*) the degree of canalization of the substance, i.e., its porosity on the microscopic or ultramicroscopic scale, and (*b*) the degree of porosity on the molar scale." From the measurements on the size of the pores made by Lamb, Wilson, and Chaney,[10] which have already been mentioned, Briggs infers that "the greater part of the internal gaseous space of an efficient adsorbent consists of passages which are not greatly larger than the gas molecules."

The nature of the pore-size distribution in activated charcoals has received much attention, especially by Russian workers.[14] Dubinin has suggested that the pores can be divided into micropores (those of molecular dimensions), macropores (those with diameters greater than about 1 micron), and transitional pores.[14] Macropores contribute relatively little to the adsorbing surface of the charcoal; Juhola and Wiig[15] estimated that, in charcoals having areas of $\sim$1500 $m^2/g$, macropores contribute only $\sim$3 $m^2/g$.

Micropores, on the other hand, may be extremely important, since pores of molecular diameter may admit certain molecules while rejecting others. The Saran charcoals[16] have an extremely narrow pore-size distribution, and as a result show "molecular sieve" properties.[17]

Adsorption in transitional pores is less selective and weaker than that in micropores; for this reason, it has been suggested that charcoals with well-developed microporosity should be more useful in vapor sorption than those with larger pores.[11] The latter type are to be preferred, on the other hand, for purification of solutions, for example.

Sheldon[18] has shown that, although ordinarily nitrogen is more readily adsorbed on charcoal than hydrogen, it is possible by suitable treatment to produce a charcoal in which the relative adsorptive capacities for the two gases are reversed. He has also observed that evacuation of charcoal at 600° C, followed by adsorption of oxygen at low temperatures, and this again by re-evacuation at the higher temperature, leads to considerable improvement in the adsorptive capacity of charcoal.

Philip, Dunnill, and Workman[19] have made similar observations with wood charcoal. The greater the facilities for the access of air to the material during heating, the greater the resulting increase that was obtained in the sorptive power. They also observed that the increased activity was accompanied by a decrease in bulk density, indicating, according to Briggs, an increasing degree of "molar" porosity.

These observations are in accord with results obtained in the General Electric Research Laboratory. A given sample of charcoal shows marked improvement as a clean-up reagent in high-vacuum work after it has been evacuated and saturated with gas several times in succession.

It should be mentioned in this connection that Magnus and Kratz[20] have developed a method for the purification of charcoal by which a material free from oxides may be obtained. Bruns and Frumkin[21] have also described a method for obtaining ash-free charcoal in which very pure sucrose is used as starting material. The charcoal obtained by charring the sucrose in a silica dish is activated by heating for 4–5 hr at 1100° C, in a current of carbon dioxide.

In general, a preliminary sorption of the gas followed by desorption at relatively high temperatures, ranging from 500 to 800° C, improves the sorptive capacity for the particular gas. This is probably due to elimination of other gases that may be present in a sorbed condition in the adsorbent.

The effect of activation on a number of charcoals is shown by the data in Table 7.1, which were obtained by Barker.[22] The most important result is that shown in the second column, viz., the increase in surface. Furthermore, true density and pore space are increased by activation.

**TABLE 7.1**

**Effect of the Activation of Charcoal on Pore Space**

| Charcoal | Adsorption of $CCl_4$ (mg/g) | Granule Density (g/cm³) | True Density (g/cm³) | Volume (cm³/g) | Charcoal Substance (cm³/g) | Pore Space (cm³/g) | Pore Space Filled by Adsorbed $CCl_4$ | |
|---|---|---|---|---|---|---|---|---|
| | | | | | | | cm³/g | % |
| Primary coconut | 47 | 0.96 | 1.46 | 1.04 | 0.685 | 0.355 | 0.032 | 9.1 |
| Activated coconut | 630 | 0.84 | 2.15 | 1.19 | 0.465 | 0.725 | 0.420 | 58 |
| Primary ironwood | 30 | 0.89 | 1.46 | 1.123 | 0.685 | 0.438 | 0.020 | 4.6 |
| Activated ironwood | 1160 | 0.72 | 2.15 | 1.39 | 0.465 | 0.925 | 0.774 | 83.6 |
| Primary lignite | 30 | 1.09 | 1.43 | 0.92 | 0.70 | 0.220 | 0.020 | 9.1 |
| Activated lignite | 640 | 0.89 | 2.15 | 1.12 | 0.465 | 0.655 | 0.426 | 65 |
| Extremely activated lignite | 2715 | 0.31 | 2.15 | 3.23 | 0.465 | 2.765 | 1.81 | 65 |

Barker[23] has also discussed at length the relation between the effect of nature of source and type of treatment on the sorptive capacity. Specific mention is made of observations by Briggs[24] on the sorption by various charcoals of nitrogen and hydrogen, which are shown in Table 7.2. The table gives the volumes (cubic centimeters, STP) per gram charcoal adsorbed at 1 atm pressure.

**TABLE 7.2**

**Values of $V_0$ for Adsorption of Nitrogen and Hydrogen by Charcoal at Liquid-Air Temperature and Atmospheric Pressure**

| Substance | $N_2$ | $H_2$ |
|---|---|---|
| Coconut charcoal (activated by steaming) | 247 | 127 |
| Birch charcoal (activated) | 202 | 123 |
| German impregnated charcoal 1918 | 303 | 63.8 |
| Briquetted coal, charcoal, and dusts | 206 | 96.7 |
| Cleveland (U.S.) activated anthracite | 89.7 | 56.3 |
| Colloidal silica (Briggs) | 376 | 51.6 |

## 7.3. SORPTION DATA AT HIGHER PRESSURES

Most of the investigations on the sorption of gases by charcoal have been concerned with pressures above 1 mm, and there are relatively few published data on sorption at lower pressures. Though such data are discussed in the next section, it is of importance to review those obtained at higher pressures.

The adsorption data obtained by Peters and Weil[25] were mentioned at the beginning of Chapter 6 in connection with the application of the Freundlich or parabolic equation for the isotherms.

From the values given in Table 6.6 for the constants $k$ and $1/n$, we deduce the values given in Table 7.3 for the sorptions at the temperatures

**TABLE 7.3**

**Sorption of Rare Gases by Charcoal at Low Temperatures**

| | | $V_0$ at $P_{mm}$ = | |
|---|---|---|---|
| Gas | $t$° C | 1 | 10 |
| Argon | −80.2 | 0.5 | 4.37 |
| | −18.0 | 0.076 | 0.764 |
| | 0 | 0.058 | 0.581 |
| Krypton | −79.7 | 2.93 | 15.03 |
| | −18.0 | 0.497 | 3.81 |
| | 0.0 | 0.340 | 3.40 |
| Xenon | −80.5 | 15.99 | 59.98 |
| | −18.0 | 2.458 | 12.10 |
| | 0 | 1.583 | 9.32 |

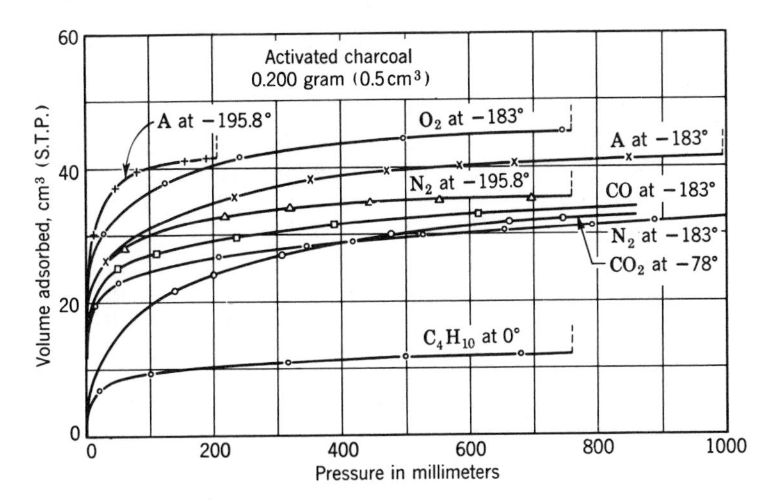


Fig. 7.1. Low-temperature adsorption isotherms on charcoal (Brunauer and Emmett).

for which the isotherms were observed. In this and the following tables, the values under $V_0$ correspond to the volume (cubic centimeters per gram charcoal) measured at 0° C and 760 mm (STP). As will be observed, the sorption at any given temperature increases with increase in boiling point of the liquefied gas.

Data on the sorption of the more common gases by charcoal have been published by Miss Homfray,[26] Titoff,[27] and Richardson.[28] Tables 7.4 and 7.5 give a survey of the observations made by the first two of these investigators. These results indicate the variations to be expected from different samples of charcoal.

Data on the sorption of the more common gases by charcoal, at very low temperatures, have been obtained by Brunauer and Emmett.[29] Figure 7.1, taken from their paper, shows the actual isotherms. Table 7.6 gives values of $V_m$ derived by Brunauer and Emmett using the BET equation, and the corresponding values of the surface area per gram for both solid and liquid packing.

It will be observed from the data given in Table 7.6 that the values of the surface area deduced from the different isotherms are in very good agreement. In fact, as Brunauer and Emmett point out, "If the surface area of the charcoal used in these experiments is taken as 845 m²/g, the *maximum* deviation from this value is less than 6 per cent."

In the range 1–10 mm, oxygen is adsorbed by charcoal at −183° C in about three or four times as great an amount as nitrogen, and the adsorbing power for air is about three times that for nitrogen.

**TABLE 7.4 Sorption Data for Activated Charcoal**
(Homfray)

| $P_{mm}$ | $V_0$ | $P_{mm}/V_0$ |
|---|---|---|
| | *Helium* | |
| | −190 °C | |
| 120 | 0.337 | 356 |
| 171 | 0.466 | 367 |
| 235 | 0.809 | 290 |
| 427.6 | 1.167 | 366 |
| 705.0 | 1.839 | 383 |
| | *Argon* | |
| | −190° C | |
| 4.6 | 15.9 | 0.289 |
| 17.0 | 66.4 | 0.256 |
| 379 | 77.7 | 0.488 |
| 410 | 93.5 | 0.439 |
| | −128° C | |
| 4.0 | 3.82 | 1.047 |
| 6.0 | 5.12 | 1.170 |
| 12.8 | 9.92 | 1.283 |
| 13.4 | 13.05 | 1.028 |
| 19.4 | 15.91 | 1.220 |
| 27.0 | 19.23 | 1.404 |
| 36 | 25.08 | 1.435 |
| 97 | 48.8 | 1.99 |
| 228 | 57.8 | 3.95 |
| 360 | 64.3 | 5.59 |
| 790 | 76.9 | 10.27 |
| | −78.5° C | |
| 8 | 1.28 | 6.27 |
| 19 | 3.77 | 5.04 |
| 24 | 5.09 | 4.71 |
| 54.2 | 10.02 | 5.41 |
| 98.4 | 15.56 | 6.46 |
| 129 | 18.81 | 6.85 |
| 218 | 24.84 | 8.78 |
| 295 | 29.14 | 10.13 |
| 564 | 39.88 | 14.15 |
| 758 | 47.47 | 15.96 |
| | −37° C | |
| 309.6 | 11.98 | 25.85 |
| 417 | 14.41 | 28.94 |
| 568.4 | 17.23 | 32.98 |
| 815 | 22.15 | 36.81 |

| $P_{mm}$ | $V_0$ | $P_{mm}/V_0$ |
|---|---|---|
| | *Nitrogen* | |
| | −190 °C | |
| 13 | 52.4 | 0.248 |
| 22 | 61.8 | 0.348 |
| 33 | 80.7 | 0.409 |
| 343 | 90.4 | 3.796 |
| | −78.5° C | |
| 14 | 5.06 | 2.77 |
| 46 | 14.27 | 3.15 |
| 135 | 23.61 | 5.72 |
| 253 | 32.56 | 7.77 |
| 518 | 40.83 | 12.69 |
| | 0° C | |
| 72 | 1.53 | 47.10 |
| 173 | 3.61 | 47.92 |
| 229 | 4.39 | 52.22 |
| 312 | 5.94 | 52.55 |
| 510 | 8.57 | 59.53 |
| | *Carbon Monoxide* | |
| | −78.5° C | |
| 6 | 6.2 | 0.96 |
| 30 | 17.3 | 1.74 |
| 40 | 20.9 | 1.91 |
| 72 | 27.7 | 2.60 |
| 117 | 34.1 | 3.43 |
| 148 | 34.3 | 4.32 |
| 187 | 40.5 | 4.62 |
| 228 | 46.7 | 4.88 |
| 442 | 53.1 | 8.32 |
| | −33.7° C | |
| 101 | 9.2 | 11.0 |
| 188 | 14.5 | 13.0 |
| 320 | 19.6 | 16.4 |
| 430 | 22.6 | 19.0 |
| 540 | 25.6 | 21.1 |
| 670 | 28.3 | 23.6 |
| | 0° C | |
| 73 | 2.53 | 28.9 |
| 180 | 5.57 | 32.3 |
| 309 | 8.47 | 36.5 |

| $P_{mm}$ | $V_0$ | $P_{mm}/V_0$ |
|---|---|---|
| | *Carbon Monoxide* | |
| | 0° C | |
| 540 | 12.85 | 42.0 |
| 713 | 17.43 | 40.9 |
| 882 | 17.64 | 50.0 |
| | 20° C | |
| 123 | 2.36 | 52.1 |
| 300 | 5.20 | 57.8 |
| 495 | 7.83 | 63.2 |
| 856 | 11.81 | 72.5 |
| | *Carbon Dioxide* | |
| | 0° C | |
| 60 | 21.2 | 2.83 |
| 87 | 26.4 | 3.30 |
| 123 | 32.2 | 3.82 |
| 179 | 38.6 | 4.65 |
| 242 | 38.1 | 6.35 |
| | 20° C | |
| 20 | 4.86 | 4.12 |
| 29 | 6.58 | 4.41 |
| 63 | 12.8 | 4.91 |
| 85 | 15.7 | 5.43 |
| 128 | 21.0 | 6.09 |
| 186 | 26.1 | 7.11 |
| 288 | 31.9 | 9.04 |
| 374 | 38.2 | 9.79 |
| 500 | 44.3 | 11.30 |
| | *Ethylene* | |
| | 0° C | |
| 70 | 39.5 | 1.77 |
| 93 | 42.8 | 2.17 |
| 168 | 49.9 | 3.37 |
| 319 | 56.5 | 5.65 |
| 616 | 64.5 | 9.56 |
| | 20° C | |
| 15 | 10.4 | 1.45 |
| 32 | 20.8 | 1.54 |
| 52 | 25.6 | 2.03 |
| 79 | 30.5 | 2.59 |
| 171 | 39.2 | 4.37 |

TABLE 7.4 (*Continued*)

| $P_{mm}$ | $V_0$ | $P_{mm}/V_0$ | $P_{mm}$ | $V_0$ | $P_{mm}/V_0$ | $P_{mm}$ | $V_0$ | $P_{mm}/V_0$ |
|---|---|---|---|---|---|---|---|---|
| | *Ethylene* | | | *Methane* | | | *Methane* | |
| | 20° C | | | −18° C | | | 0° C | |
| 220 | 42.4 | 5.19 | 70 | 12.5 | 5.60 | 358 | 22.3 | 16.1 |
| 388 | 49.1 | 7.90 | 103 | 15.8 | 6.51 | 451 | 25.3 | 17.8 |
| 685 | 55.2 | 12.41 | 151 | 19.6 | 7.71 | 578 | 28.4 | 20.4 |
| | | | 190 | 22.9 | 8.31 | 702 | 31.0 | 22.5 |
| | *Methane* | | 242 | 26.1 | 9.29 | | | |
| | −33° C | | 311 | 29.4 | 10.6 | | 20° C | |
| 36 | 12.6 | 2.85 | 393 | 32.4 | 12.1 | 40 | 2.7 | 14.1 |
| 53 | 16.0 | 3.31 | 484 | 35.4 | 13.7 | 69 | 4.3 | 16.0 |
| 76 | 19.9 | 3.82 | | | | 127 | 7.2 | 17.7 |
| 102 | 23.2 | 4.40 | | 0° C | | 248 | 11.9 | 20.9 |
| 132 | 26.5 | 4.98 | 67 | 7.4 | 9.1 | 355 | 15.0 | 23.7 |
| 171 | 29.9 | 5.72 | 134 | 12.3 | 10.8 | 497 | 18.4 | 26.9 |
| 215 | 33.1 | 6.50 | 191 | 15.2 | 12.6 | 640 | 21.3 | 30.0 |
| 270 | 36.2 | 7.43 | 274 | 19.1 | 14.3 | 795 | 24.2 | 32.9 |

According to Lemon and Blodgett,[30]

Oxygen, which boils at −182.95° C, is adsorbed and expelled so much more readily than nitrogen, which boils at −195.8° C, that Dewar, after finding that the gas adsorbed by charcoal from air was 57 per cent oxygen, recommended "that one of the most rapid means of extracting a high percentage of oxygen from atmospheric air is to adsorb it in charcoal at low temperatures and then to expel it either rapidly or slowly by heating the mass of charcoal to the ordinary temperature."[31]

Lemon and Blodgett found that in a mixture the two gases are not adsorbed independently, and that "there is a nearly linear relation between the logarithm of the equilibrium pressure and the per cent of oxygen in the mixture." Table 7.7 shows the values of the final pressure obtained under the same conditions from mixtures of the two gases.

The sorption of oxygen by pure sugar charcoal has been investigated by Lendle.[32] In agreement with a number of previous investigators he observed that the attainment of equilibrium requires considerable periods of time. Taylor[33] interpreted this slow sorption as due to an activation process.

Table 7.8 gives sorption data at 0 and 25° C obtained by Lendle, using charcoal which had been degassed for 150 min at 780° C. The values given were obtained at the end of 90 min. Simultaneously with the values of $V_0$, observations were made in a calorimeter on the integral molar heat of sorption, $Q_S$. Comparing these values with those observed by Titoff and Homfray for nitrogen at 0° C (see Tables 7.4 and 7.5), it is obvious that the sorption of oxygen by charcoal is considerably higher, especially at very low pressures.

**TABLE 7.5 Sorption Data for Activated Charcoal (Titoff)***

| $P_{mm}$ | $V_0$ | $P_{mm}/V_0$ |
|---|---|---|
| | *Hydrogen* −78.5° C | |
| 7.9 | 0.059 | 134.8 |
| 19.0 | 0.148 | 128.4 |
| 67.5 | 0.531 | 127.1 |
| 141.9 | 1.121 | 126.3 |
| 236.9 | 1.892 | 125.2 |
| 347.9 | 2.787 | 124.4 |
| 471.8 | 3.607 | 130.7 |
| 561.9 | 4.276 | 131.6 |
| 721.6 | 5.414 | 133.3 |
| | 0° C | |
| 17.4 | 0.039 | 452.5 |
| 39.3 | 0.098 | 400.0 |
| 66.9 | 0.149 | 446.4 |
| 119.4 | 0.270 | 442.5 |
| 206.9 | 0.451 | 456.6 |
| 322.8 | 0.698 | 462.9 |
| 427.5 | 0.914 | 467.3 |
| 537.3 | 1.139 | 471.7 |
| 642.1 | 1.343 | 478.5 |
| 744.2 | 1.554 | 478.5 |
| | *Nitrogen* −78.5° C | |
| 1.5 | 0.145 | 10.35 |
| 4.6 | 0.894 | 5.15 |
| 12.5 | 3.468 | 3.60 |
| 66.4 | 12.04 | 5.52 |
| 149.5 | 20.03 | 7.47 |
| 271.4 | 27.94 | 9.27 |
| 388.4 | 33.43 | 11.62 |
| 542.9 | 38.39 | 14.15 |
| 740.6 | 43.51 | 17.02 |
| | 0° C | |
| 4.3 | 0.111 | 38.75 |
| 12.1 | 0.298 | 40.61 |
| 39.3 | 0.987 | 39.82 |
| 129.8 | 3.043 | 42.67 |
| 229.4 | 5.082 | 45.09 |
| 340.1 | 7.047 | 48.26 |
| 562.3 | 10.31 | 54.54 |
| 774.6 | 13.05 | 59.36 |

| $P_{mm}$ | $V_0$ | $P_{mm}/V_0$ |
|---|---|---|
| | *Nitrogen* 30° C | |
| 8.6 | 0.082 | 104.9 |
| 20.3 | 0.227 | 89.4 |
| 64.5 | 0.766 | 84.2 |
| 144.6 | 1.718 | 81.2 |
| 239.6 | 2.764 | 86.7 |
| 347.6 | 3.875 | 89.7 |
| 455.4 | 4.959 | 91.8 |
| 650.0 | 6.646 | 97.8 |
| 752.3 | 7.490 | 100.4 |
| | *Ammonia* −23.5° C | |
| 0.3 | 5.496 | 0.055 |
| 2.9 | 17.17 | 0.169 |
| 16.3 | 44.97 | 0.362 |
| 112.2 | 119.3 | 0.941 |
| 770.2 | 151.3 | 5.09 |
| | −23.5° C | |
| 2.5 | 8.34 | 0.30 |
| 26.6 | 62.99 | 0.42 |
| 389.1 | 147.3 | 2.64 |
| 536.0 | 151.8 | 3.53 |
| 790.6 | 157.5 | 5.02 |
| | 0° C | |
| 2.9 | 5.41 | 0.536 |
| 28.8 | 30.16 | 0.955 |
| 78.7 | 60.39 | 1.30 |
| 161.0 | 90.29 | 1.78 |
| 319.2 | 115.7 | 2.76 |
| 490.0 | 127.1 | 3.86 |
| 636.4 | 132.4 | 4.81 |
| 746.7 | 135.9 | 5.49 |
| | 0° C | |
| 2.6 | 5.54 | 0.47 |
| 29.1 | 30.58 | 0.95 |
| 77.5 | 60.98 | 1.27 |
| 245.6 | 107.8 | 2.28 |
| 466.0 | 126.3 | 3.69 |
| 764.5 | 135.4 | 5.65 |

| $P_{mm}$ | $V_0$ | $P_{mm}/V_0$ |
|---|---|---|
| | *Ammonia* 30° C | |
| 1.2 | 2.62 | 0.46 |
| 20.5 | 8.64 | 2.37 |
| 62.0 | 19.81 | 3.13 |
| 151.3 | 40.43 | 3.74 |
| 269.1 | 60.02 | 4.49 |
| 408.2 | 79.97 | 5.11 |
| 679.0 | 97.63 | 6.96 |
| 785.4 | 103.5 | 7.59 |
| | *Carbon Dioxide* −76.5° C | |
| 0.3 | 15.26 | 0.020 |
| 1.8 | 41.51 | 0.043 |
| 41.3 | 86.74 | 0.476 |
| 168.6 | 104.1 | 1.615 |
| 483.1 | 111.5 | 4.343 |
| 691.0 | 114.1 | 6.056 |
| | 0° C | |
| 0.5 | 0.849 | 0.589 |
| 3.2 | 3.460 | 0.925 |
| 10.9 | 8.506 | 1.28 |
| 25.4 | 15.15 | 1.68 |
| 83.0 | 27.78 | 2.99 |
| 173.5 | 39.90 | 4.35 |
| 315.9 | 50.24 | 6.29 |
| 457.2 | 56.82 | 8.05 |
| 589.1 | 61.37 | 9.60 |
| 703.2 | 64.53 | 10.90 |
| 755.1 | 65.85 | 11.29 |
| | 30° C | |
| 1.3 | 0.507 | 2.56 |
| 5.3 | 1.823 | 2.91 |
| 14.7 | 4.016 | 3.56 |
| 50.7 | 10.29 | 4.93 |
| 138.5 | 19.72 | 7.03 |
| 283.3 | 28.87 | 9.81 |
| 398.9 | 34.07 | 11.70 |
| 498.1 | 38.14 | 13.06 |
| 621.8 | 41.45 | 15.01 |
| 758.6 | 44.93 | 16.89 |

* For $NH_3$ two sets of data are given for each of the two temperatures, −23.5 and 0° C.

TABLE 7.6

**Surface Area of Activated Charcoal Derived by BET Method**

| Gas | Temperature (°C) | $V_m$ | Square Meters per Gram: Solid Packing | Square Meters per Gram: Liquid Packing |
|---|---|---|---|---|
| $N_2$ | −195.8 | 181.5 | 677 | 795 |
| $N_2$ | −183 | 173.0 | 646 | 795 |
| Ar | −195.8 | 215.5 | 746 | 804 |
| Ar | −183 | 215.5 | 746 | 839 |
| $O_2$ | −183 | 234.6 | 767 | 894 |
| CO | −183 | 179.5 | 665 | 820 |
| CO | −78 | 185.5 | 707 | 853 |

TABLE 7.7

**Sorption by Charcoal of Mixtures of Oxygen and Nitrogen at Liquid-Air Temperature**

| % Oxygen | % Nitrogen | Final Pressure (mm) |
|---|---|---|
| 100 | . . . | 0.0797 |
| 75 | 25 | 0.240 |
| 50 | 50 | 0.924 |
| 35 | 65 | 2.14 |
| 20 | 80 | 4.89 |
| 10 | 90 | 8.86 |
| . . . | 100 | 18.6 |

TABLE 7.8

**Sorption of Oxygen by Sugar Charcoal**

| 0° C: $P_{mm}$ | 0° C: $V_0$ | 0° C: $Q_S$ | 25° C: $P_{mm}$ | 25° C: $V_0$ | 25° C: $Q_S$ |
|---|---|---|---|---|---|
| $7.71 \cdot 10^{-3}$ | 0.303 | 67,760 | $6.52 \cdot 10^{-3}$ | 0.248 | 79,400 |
| $4.3 \cdot 10^{-2}$ | 0.474 | 51,856 | 0.1554 | 0.858 | 63,060 |
| 0.540 | 1.215 | 39,000 | 0.5664 | 1.195 | 54,860 |
| 3.020 | 2.121 | 32,640 | 4.658 | 2.608 | 43,200 |
| 7.732 | 3.725 | 25,660 | 6.116 | 2.760 | 39,340 |
| 10.29 | 4.764 | 19,300 | 9.40 | 4.015 | 36,900 |
| 20.8 | 8.460 | 16,010 | 18.20 | 6.019 | 31,430 |
| 39.7 | 13.93 | 13,170 | 70.90 | 10.46 | 20,770 |
| 62.2 | 20.18 | 10,727 | 94.1 | 17.99 | 13,510 |
| 82.0 | 25.45 | 10,230 | 113.8 | 18.97 | 13,200 |

At $-80°$ C, equilibrium is established very rapidly, but as the temperature is increased the rate at which equilibrium is reached decreases.

As will be observed, the heat of adsorption decreases from about 80,000 cal/mole initially to about 10,000 cal/mole, corresponding to the fact that the type of sorption changes from chemisorption to purely physical adsorption.

An important feature of the sorption of oxygen by charcoal, for which the change in value of $Q_S$ is an explanation, is that, unlike most other cases of sorption, it is not reversible. When charcoal which has taken up oxygen is heated, only a portion of the gas is recovered as oxygen, the remainder being re-evolved as carbon monoxide and dioxide. It has been shown by Lowry and Hulett[34] that, while some of the oxygen is adsorbed on the surface and may be recovered by heating, the remainder is attached to carbon atoms by valence forces. It is these chemically bound oxygen molecules which can be removed from the surface only in the form of

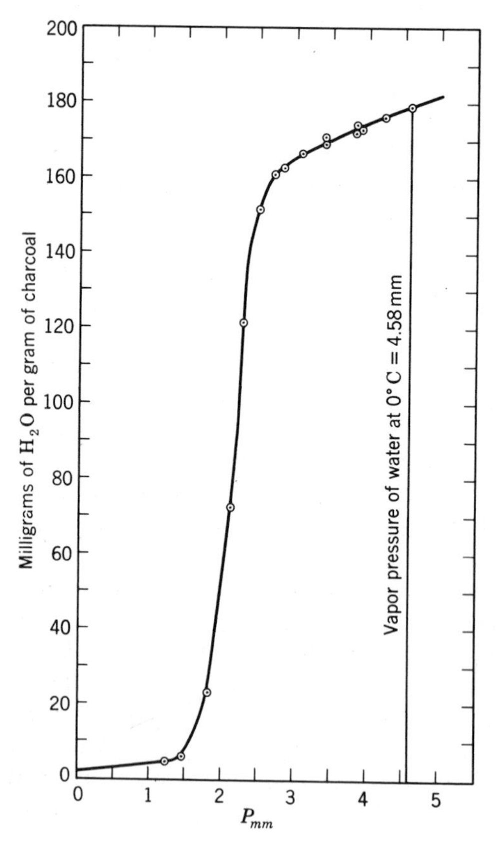


Fig. 7.2. Adsorption isotherm for water vapor on sugar charcoal at 0° C (Coolidge).

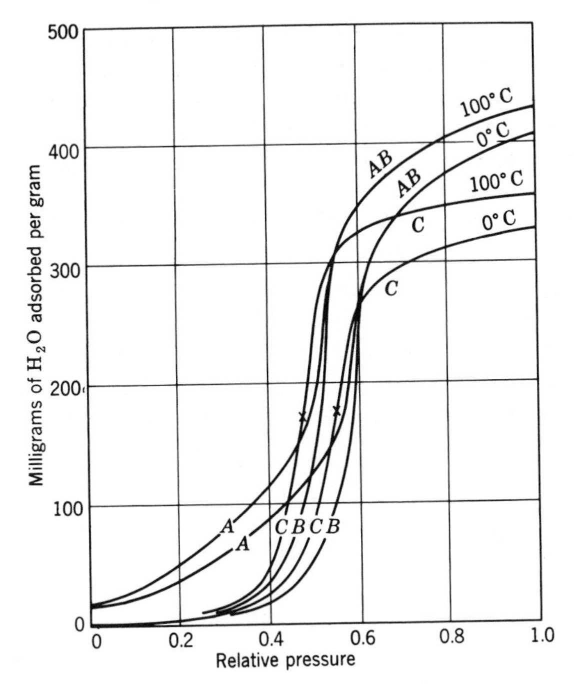


Fig. 7.3. Isotherms at 0° and 100° C for the sorption of water vapor by sugar charcoal and coconut charcoal (Coolidge).

carbon monoxide and carbon dioxide by heating the charcoal above 200° C.

The sorption of *water vapor* by charcoal exhibits a behavior quite different from that observed for the less readily condensable gases. Figure 7.2[35] shows an isotherm for sugar charcoal at 0° C as determined by Coolidge.[36] At pressures below about 1.5 mm, the amount adsorbed is very small and obeys Henry's law. Between 1.5 and 2.5 mm, the sorption increases very rapidly; this is followed by very slowly increasing sorption until the saturation vapor pressure of water at 0° C (4.58 mm) is reached. The isotherm is typical of type V in Brunauer's classification (cf. Fig. 6.9).

Figure 7.3 shows a series of isotherms at 0° C and 100° C for charcoal prepared by different methods. The amounts adsorbed, in milligrams, are plotted against relative pressures, $P/P_0$, where $P_0$ denotes the vapor pressure for water at the given temperature. The lower of the two isotherms designated $C$ represents the same data as shown in Fig. 7.2, while the upper isotherm $C$ represents the sorption data at 100° C. The charcoal used for determining these two isotherms was prepared from "pure cane sugar recrystallized from conductivity water with the aid of alcohol, and charred in a platinum basin at about 1000° C, in a vacuum." The isotherms

*AA* were obtained with untreated coconut charcoal; those designated *BB* were obtained with the same charcoal after treatment with hydrofluoric acid to remove mineral ash. In each case, the upper curve *A* (or *B*) corresponds to determinations at 100° C, and the lower curve *A* (or *B*) corresponds to determinations at 0° C.

As is evident from these isotherms, the adsorption on purified charcoal is extremely low for values of $P/P_0$ less than about 0.3. Coolidge explains this by the hydrophobic nature of carbon; that is, carbon is difficultly wet by water. On the other hand, he points out that silica, silicates (including glass), and inorganic compounds in general, are hydrophilic. He states,

> Therefore it is not surprising that small amounts of hydrophilic impurities, probably siliceous, should exercise a profound influence upon the adsorption of water by carbon at low concentrations, and should be able to cause a serious irreversible retention. At higher concentrations, where the determining factor is the volume of capillary space available, the effect would not be noticed.

The presence of ash can modify the surface properties of carbon greatly.[37] Even in the absence of ash, the adsorption of molecules such as water seems to depend primarily on the presence of polar sites (for example, residual oxygenated complexes) on the otherwise nonpolar and hydrophobic surface. Kipling has discussed the evidence for this conclusion,[11] and Smith has reviewed the types of carbon-oxygen complexes which may be found on graphite and charcoal surfaces.[38]

A series of isotherms for different temperatures, obtained with the sugar charcoal, is plotted in Fig. 7.4. In this figure, the amounts adsorbed, in milligrams per gram, are plotted against relative pressure. From the plots of the isosteres the following values of the molal heat of adsorption were derived:

| Temperature (°C) | −15 | 10 | 40 | 80 | 128 | 187 |
|---|---|---|---|---|---|---|
| Molal heat | 11,100 | 10,000 | 9300 | 8300 | 7200 | 5200 |

At the lower temperatures the molal heat of adsorption is about the same as the latent heat of evaporation of liquid water.

From the data given by Coolidge it would seem that about 10 mg $H_2O$ per gram of charcoal corresponded to a complete monolayer, while the maximum amount adsorbed was about 180 mg/g. According to the data in Table 6.13, a monolayer of $H_2O$ corresponds to $2.8 \cdot 10^{-5}$ mg/cm$^2$. It would therefore follow that the adsorbing area of the charcoal used by Coolidge was about 350 m$^2$/g, and that the maximum amount adsorbed corresponded to about 18 monolayers. This is in qualitative accord with the observation by Lowry and Hulett[39] that, in the sorption of water by charcoal, the adsorbed water corresponded to a layer about 13 molecules thick.

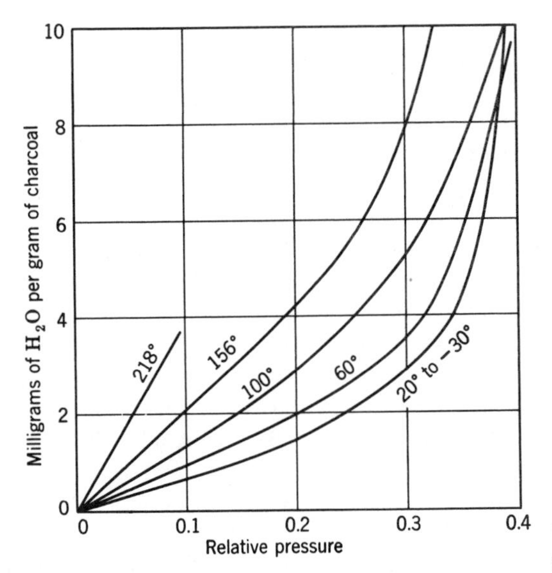


Fig. 7.4. Isotherms at a series of temperatures for sorption of water vapor by sugar charcoal (Coolidge).

The sorption of water vapor by activated charcoals has also been investigated by Allmand and his associates,[40] both in the presence and in the absence of other gases. They observed that, after absorption of water, the charcoal retains a certain amount (2.3–10 mg/g) on desorption. This hysteresis effect is illustrated by the data in Table 7.9 for sorption and

**TABLE 7.9**

**Sorption and Desorption of Water for Charcoal at 25° C**

| Sorption | | Desorption | |
|---|---|---|---|
| $P_{mm}$ | $H_2O$ (mg/g) | $P_{mm}$ | $H_2O$ (mg/g) |
| 3.64 | 10.6 | 18.68 | 731.4 |
| 9.24 | 23.8 | 17.35 | 659.4 |
| 12.79 | 44.9 | 16.75 | 571.6 |
| 15.41 | 84.0 | 15.97 | 438.8 |
| 18.15 | 222.2 | | |
| 18.40 | 345.9 | | |
| 19.42 | 593.0 | | |
| 23.70 | 780.2 | | |

desorption at 25° C, in which the charcoal had been previously outgassed at 800° C. At 25° C the saturation vapor pressure of water is 23.756 mm.

For different temperatures of outgassing, the amount adsorbed, at the saturation pressure for 25° C, varied from 285 to 851 mg $H_2O$ per gram, but in all cases a similar difference was observed between the sorption and desorption data.

The sorption of hydrogen by coconut charcoal at the *temperature of liquid air* has been investigated by McBain[41] and Firth.[42] In both cases it was observed that equilibrium is attained only after a lapse of many hours. According to McBain, most of the gas is condensed instantaneously on the surface (true adsorption); this is followed by a gradual diffusion of the hydrogen into the charcoal at a rate which obeys the regular law of diffusion. This last stage corresponds to (interstitial) solution of the gas between the carbon atoms, and McBain concludes that this solubility is proportional to the square root of the pressure, indicating that the dissolved gas is in the atomic condition. Furthermore, McBain concludes that the *solubility* of hydrogen at liquid-air temperature is at least 100 times that observed at room temperature. Maggs and coworkers[43] have estimated the amount of helium adsorbed on several carbons and concluded that some activated charcoals adsorb significant amounts of helium even at room temperature. As they point out, this observation is of some importance because of the common use of helium for measurements of free volume in vacuum systems.

A number of investigators have compared the sorptive capacity of graphite with that of coconut charcoal. In the case of carbon dioxide, Magnus and Kratz[44] have observed that the amount adsorbed on graphite at any given temperature and pressure is about 1/700 of that adsorbed on charcoal.

Similar observations have been made by Lowry and Morgan[45] on the sorption of carbon dioxide, nitrogen, and hydrogen by graphitic carbon. Even at the temperature of liquid air, the hydrogen adsorption is only about 1/10 of that observed for active charcoal. This doubtless reflects the much smaller extent of surface available for adsorption in graphite.

Carbon blacks prepared by incomplete combustion or decomposition of gaseous hydrocarbons ("soot" is a good example) have received much attention. As was pointed out in Chapter 6, certain carbon blacks which have been "graphitized" by high-temperature heat treatment seem to have very homogeneous surfaces.[46] The surface areas of such materials are commonly in the range 10–200 $m^2/g$.

The adsorption of water vapor on carbon black has been investigated by Emmett and Anderson.[47] Two commercial varieties of carbon black were used. Figure 7.5 shows results obtained for one of these. The

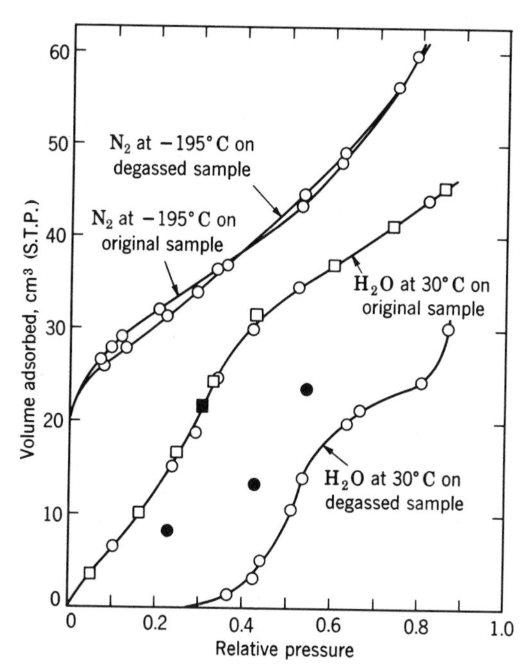


Fig. 7.5. Adsorption of nitrogen at −195° C and water vapor at 30° C on a sample of carbon black (Grade 6 Spheron) before and after degassing at 1000° C (Emmett and Anderson).

specific areas for the original sample and that for the sample degassed at 1000° C, as determined by the adsorption isotherm with nitrogen at −195° C, exhibited hardly any change, but the adsorption of water vapor was decreased considerably by the degassing treatment. In the figure, data indicated by circles were obtained by volumetric measurements; those indicated by squares were obtained gravimetrically. Solid circles and squares are desorption points.

Similar results were obtained with the other sample of carbon black. From these observations it is evident that in these cases the heat treatment, though producing little or no change in specific area of the adsorbent, did affect its adsorption capacity for water vapor.

## 7.4. SORPTION DATA AT LOW PRESSURES

Observations on the sorption of gases by charcoal at low pressures have been published by Dewar,[48] Claude,[49] and Rowe.[50] Some of the data obtained by Rowe for sorption at room temperature are shown in Table 7.10.

**TABLE 7.10**

**Sorption by Charcoal at Room Temperature**

(Rowe)

| $CO_2$ at 15° C | | $N_2$ at 15° C | | $O_2$ at 22° C | |
|---|---|---|---|---|---|
| $P_\mu$ | $q$* | $P_\mu$ | $q$* | $P_\mu$ | $q$* |
| 0.326 | 5.165 | 3.321 | 0.1313 | 0.136 | 2.299 |
| 1.164 | 25.21 | 5.727 | 0.2212 | 0.298 | 9.204 |
| 4.250 | 56.10 | 10.88 | 0.3145 | 0.608 | 22.48 |
| 9.989 | 75.48 | 20.76 | 0.7782 | 1.011 | 39.50 |
| 17.85 | 92.70 | 45.10 | 1.734 | 1.830 | 76.35 |
| 42.78 | 131.2 | 99.64 | 5.066 | 5.921 | 203.2 |
| 73.02 | 150.5 | 117.0 | 6.642 | 74.49 | 670.1 |

* $q$ = cubic centimeters at 1 mm of mercury and 15° C, per gram.
$V_0 = 1.206 \cdot 10^{-3} q$ = cubic centimeters at 760 mm and 0° C.

From the data published by Rowe, it follows that, for $P_{mm} = 10^{-2}$ and $P_{mm} = 10^{-1}$, the volumes adsorbed at *room temperature* (measured in terms of *cubic centimeters at STP*), per gram, are as follows:

| | $P_{mm} = 10^{-2}$ mm | $P_{mm} = 10^{-1}$ mm |
|---|---|---|
| $CO_2$ | 0.090 | 0.199 |
| $N_2$ | 0.00036 | 0.006 |
| $O_2$ | 0.360 | 0.900 |

Table 7.11 gives the sorption data obtained by Rowe at −183° C (liquid oxygen).

**TABLE 7.11**

**Sorption by Activated Charcoal at −183° C**

(H. Rowe)

| Nitrogen | | | Carbon Monoxide | | | Hydrogen | | |
|---|---|---|---|---|---|---|---|---|
| $P_{mm}$ | $V_0$* | $10^4 P_{mm}/V_0$ | $P_{mm}$ | $V_0$* | $10^5 P_{mm}/V_0$ | $P_{mm}$ | $V_0$* | $P_{mm}/V_0$ |
| $3.0 \cdot 10^{-5}$ | 0.223 | 1.35 | $2.4 \cdot 10^{-5}$ | 0.339 | 7.08 | $7.00 \cdot 10^{-4}$ | 0.0046 | 0.152 |
| 6.4 | 0.336 | 1.91 | 9.8 | 2.475 | 3.96 | 9.44 | 0.0073 | 0.129 |
| $1.16 \cdot 10^{-4}$ | 1.095 | 1.06 | $1.87 \cdot 10^{-4}$ | 4.007 | 4.67 | $1.51 \cdot 10^{-3}$ | 0.0122 | 0.124 |
| 2.67 | 2.539 | 1.05 | 6.77 | 17.96 | 3.77 | 2.57 | 0.0231 | 0.111 |
| 4.69 | 4.318 | 1.09 | $1.20 \cdot 10^{-3}$ | 30.01 | 4.00 | 3.40 | 0.0304 | 0.112 |
| 8.15 | 8.832 | 0.92 | 1.46 | 35.33 | 4.13 | 4.25 | 0.0403 | 0.105 |
| $1.32 \cdot 10^{-3}$ | 13.54 | 0.97 | 2.08 | 47.78 | 4.35 | 5.83 | 0.0563 | 0.104 |
| 1.99 | 22.14 | 0.90 | 3.08 | 57.17 | 5.39 | 8.07 | 0.0781 | 0.103 |
| 4.38 | 39.11 | 1.12 | 5.17 | 67.53 | 7.66 | 12.21 | 0.126 | 0.097 |
| 6.66 | 44.16 | 1.51 | 11.80 | 81.88 | 14.41 | 15.72 | 0.170 | 0.093 |
| 9.72 | 49.21 | 1.98 | 20.75 | 91.82 | 22.60 | 19.36 | 0.212 | 0.091 |
| 13.11 | 54.48 | 2.40 | 40.24 | 100.96 | 39.87 | $22.81 \cdot 10^{-3}$ | 0.255 | 0.090 |
| 18.60 | 60.73 | 3.06 | 55.68 | 104.75 | 53.16 | | | |
| 22.58 | 65.78 | 3.44 | $83.01 \cdot 10^{-3}$ | 111.23 | 74.64 | | | |
| 31.09 | 70.82 | 4.39 | | | | | | |
| 46.06 | 75.13 | 6.14 | | | | | | |
| 55.99 | 77.33 | 7.25 | | | | | | |
| 70.30 | 79.77 | 8.82 | | | | | | |
| $79.75 \cdot 10^{-3}$ | 81.31 | 9.80 | | | | | | |

* $V_0$ = cubic centimeters (STP) per gram.

Table 7.12 gives sorption data for nitrogen, hydrogen, and neon obtained by Claude. In the case of helium, the volume sorbed at 27 mm and $-195.5°$ C was 0.21 $cm^3/g$.

TABLE 7.12

**Sorption by Activated Charcoal at Very Low Temperatures**
(Claude)

| Nitrogen (−182.5° C) | | | Hydrogen (−195.5° C) | | | Neon (−195.5° C) | | |
|---|---|---|---|---|---|---|---|---|
| $P_{mm}$ | $V_0$ ($cm^3$) | $P_{mm}/V_0$ | $P_{mm}$ | $V_0$ ($cm^3$) | $P_{mm}/V_0$ | $P_{mm}$ | $V_0$ ($cm^3$) | $P_{mm}/V_0$ |
| 0.004 | 9.35 | $4.28 \cdot 10^{-4}$ | 0.006 | 0.105 | 0.057 | 0.45 | 0.105 | 4.29 |
| 0.010 | 18.70 | 5.35 | 0·0115 | 0.21 | 0.055 | 0.88 | 0.21 | 4.19 |
| 0.032 | 37.4 | 8.56 | 0.0205 | 0.42 | 0.049 | 1.30 | 0.32 | 4.06 |
| 0.088 | 46.6 | $1.89 \cdot 10^{-3}$ | 0.036 | 0.84 | 0.043 | 1.74 | 0.42 | 4.14 |
| 0.385 | 56 | 6.88 | 0.083 | 2.05 | 0.040 | 3.5 | 0.84 | 4.17 |
| 1.107 | 65.3 | $1.35 \cdot 10^{-2}$ | 0.176 | 3.71 | 0.047 | 5.3 | 1.22 | 4.34 |
| 11.50 | 93 | 0.124 | 0.475 | 8.40 | 0.054 | 7.2 | 1.63 | 4.42 |
| 33.2 | 103 | 0.322 | 1.06 | 14 | 0.076 | 11.3 | 2.44 | 4.63 |
| 90 | 112 | 0.804 | 3.50 | 28 | 0.125 | 15.5 | 3.25 | 4.77 |
| 247 | 121 | 2.04 | 8.7 | 42 | 0.207 | 19.4 | 4.06 | 4.78 |
| | | | 20.6 | 56 | 0.368 | 30.5 | 6.18 | 4.94 |
| | | | 43.7 | 63 | 0.694 | 40.5 | 8.01 | 5.06 |

All these observations agree in the conclusion that nitrogen (or air) is sorbed by charcoal at liquid-air temperature to a considerably larger extent than hydrogen.

In the General Electric Research Laboratory, H. Huthsteiner has measured the adsorption of nitrogen and hydrogen by activated charcoal at $-195.8°$ C. The charcoal was heated during exhaust to about 500° C, and after the McLeod gauge indicated an extremely low residual pressure the charcoal tube was immersed in liquid nitrogen and successive known amounts of the gas were admitted. In less than 5 min equilibrium was attained and the pressure noted.

For hydrogen, over the range 0–250 microns, the volume adsorbed per gram of charcoal (STP) was found to be in agreement with the relation

$$V_0 = 1.6 \cdot 10^{-2} P_\mu.$$

For nitrogen, the following data were obtained.

| $P_\mu$: | 1 | 2 | 3 | 4 | 5 |
|---|---|---|---|---|---|
| $V_0$: | 4 | 16 | 30 | 45 | 60 |

A comparison with the data in Table 7.11 shows the effect of the lower temperature in increased amounts adsorbed.

The adsorption of nitrogen at $-195$ to $-205°$ C has been studied by Deitz and his coworkers.[51] In their studies, it was found necessary to adsorb and desorb several times before reproducible values were obtained, and equilibrium was attained slowly, more than 6 hr being required at pressures below 0.5 micron.

## 7.5. USE OF CHARCOAL AT LOW TEMPERATURES IN HIGH-VACUUM SYSTEMS

From the sorption data given in the previous section it is possible to calculate the reduction in pressure that should be obtained by means of charcoal at very low temperature connected to the system.

Let $V$ = volume (cubic centimeters) of system,
$w$ = weight of charcoal (grams),
$P_{mm1}$ = initial pressure,
$P_{mm2}$ = equilibrium pressure after immersing charcoal tube in low-temperature bath,
$V_a$ = volume adsorbed (cubic centimeters at STP).

Evidently,

$$\frac{VP_{mm1}}{760} = \frac{VP_{mm2}}{760} + V_a. \tag{7.1}$$

If the adsorption data may be expressed either by a hyperbolic isotherm (I) or by Henry's law (II), we may then write

I. For $V_a = \dfrac{wbV_sP_{mm2}}{1 + bP_{mm2}}$ and $P_{mm1} \gg P_{mm2}$.

$$\frac{1}{P_{mm2}} = \frac{760wbV_s}{VP_{mm1}} - b. \tag{7.2}$$

II. For $V_a = kwP_{mm}$,

$$P_{mm2} = \frac{VP_{mm1}}{760}\left(\frac{1}{V/760 + kw}\right). \tag{7.3}$$

Woodrow[52] measured, with a Knudsen gauge (see Chapter 5), the amount of clean-up of different gases by charcoal at liquid-air temperatures. Neither the volume of the apparatus nor the weight of the charcoal is given. The charcoal was heated under simultaneous evacuation until the pressure fell to $1.5 \cdot 10^{-6}$ mm of mercury ($2 \cdot 10^{-3}$ microbar), and hydrogen, oxygen, or nitrogen was then introduced at an initial pressure of about 0.65 microbar. The rate of clean-up in each case as followed with the gauge is shown in Table 7.13.

Miss Daly and Dushman also carried out some experiments, in the General Electric Research Laboratory, on the adsorption of gases by

TABLE 7.13

**Rate of Clean-up by Charcoal at about −190° C**
(Pressure in Microbars)

| Time | Hydrogen | Oxygen | Nitrogen |
|---|---|---|---|
| 0 | 0.647 | 0.667 | 0.600 |
| 5 sec | | 0.387 | |
| 10 sec | | 0.227 | |
| 1 min | 0.613 | 0.0027 | 0.020 |
| 5 min | 0.547 | 0.002 | |
| 20 min | 0.387 | | |
| 1 hr | 0.253 | | 0.007 |
| 3 hr | 0.180 | 0.002 | |
| 10 hr | 0.180 | 0.002 | 0.007 |

charcoal at low temperatures. Specially activated products, supplied by F. M. Dorsey, were used in this work. Pressures were determined by means of the ionization gauge, and the rate of clean-up of residual gases present after sealing off the pump was observed. The gauge and tube containing 5 g charcoal with a large bulb (total volume = 300 cm³) were well exhausted on the condensation pump, with simultaneous heating of the charcoal for over an hour to 360° C. After this system was sealed off the pump, the pressure was measured, and then liquid air put on the charcoal. The pressures (in microbars) before cooling the charcoal, and after, were as follows:

| $V$ (cm³) | Initial Pressure | Final Pressure |
|---|---|---|
| 3000 | 0.022 | 0.0004 |
| 3000 | 0.036 | 0.0004 |
| 3000 | 0.92 | 0.0006 |

It may be noted that the sensitivity of the galvanometer used with the ionization gauge was such that 0.0004 microbar was about the lowest pressure that could actually be measured.

In carrying out experiments with hydrogen, the same apparatus was used, except that another side tube was sealed on in which was contained a small thin-walled glass pellet (volume 3 cm³) filled with hydrogen at a known pressure. After the gauge, charcoal tube, and bulb were exhausted, this system was sealed off, the hydrogen pellet broken by shaking, and the residual pressure observed after immersing the charcoal tube in liquid air. Five grams of activated charcoal were used, as in the previous experiments. The residual gas pressure was also measured before breaking the pellet. Table 7.14 shows some of the results obtained.

TABLE 7.14

**Clean-up of Hydrogen by Activated Charcoal at −190° C**

(Pressures in Microbars)

| $V$ (cm$^3$) | Pressure after Sealing Off | Initial Pressure of Hydrogen | Final Pressure at Room Temperature | Pressure at Liquid-Air Temperature |
|---|---|---|---|---|
| 3025 | 0.0180 | 0.31 | 0.014 | 0.0004 |
| 100 | 0.104 | 8.64 | 0.02 | 0.0004 |
| 3025 | 0.022 | 8.33 | 2.0 | 0.15 |
| 100 | 0.28 | 17.7 | 0.24 | 0.0016 |

In general it required about an hour to attain equilibrium. In accord with McBain's observations it was found that the initial condensation was followed by a slow diffusion of gas into the charcoal.

A certain fraction of this clean-up was no doubt due to the action of the gauge itself. As mentioned in Chapter 5 and also in Chapter 9, an electrical clean-up occurs in all hot-cathode devices. Furthermore, in the presence of a heated filament atomic hydrogen is formed, which is cleaned up at the temperature of liquid air. A blank experiment carried out without the use of charcoal gave the following results:

$$V = 340; \text{ pressure after seal-off} = 1.60 \text{ microbars.}$$

On breaking the hydrogen pellet, the pressure rose to 7 microbars, then fell, owing to clean-up by the gauge, to 2 microbars. Liquid air was put on the side tube, and the pressure fell in the course of 6 hr to 0.16 microbar. On removal of the liquid air, the pressure came back to 1.2 microbars.

In the measurements with charcoal the gauge filaments were lighted only during the time necessary to take a pressure reading, so that any error due to clean-up in the gauge was certainly not very large.

It is evident from these observations that with a charcoal immersed in liquid air it is possible to adsorb appreciable volumes of hydrogen and obtain residual gas pressures of less than 0.0001 micron. Winkler[53] has investigated the clean-up of nitrogen and hydrogen at a temperature of −183° C, by charcoal and silica gel. He observed that sorption and desorption curves did not coincide, but that the volume sorbed for a given value of $P$ increased in successive runs. The limits actually observed for *nitrogen* on charcoal are shown as curves 1 and 2 in Fig. 7.6, while Rowe's data are plotted as curve 3. In the case of *hydrogen*, the sorption and desorption were found to be completely identical for a given sample of

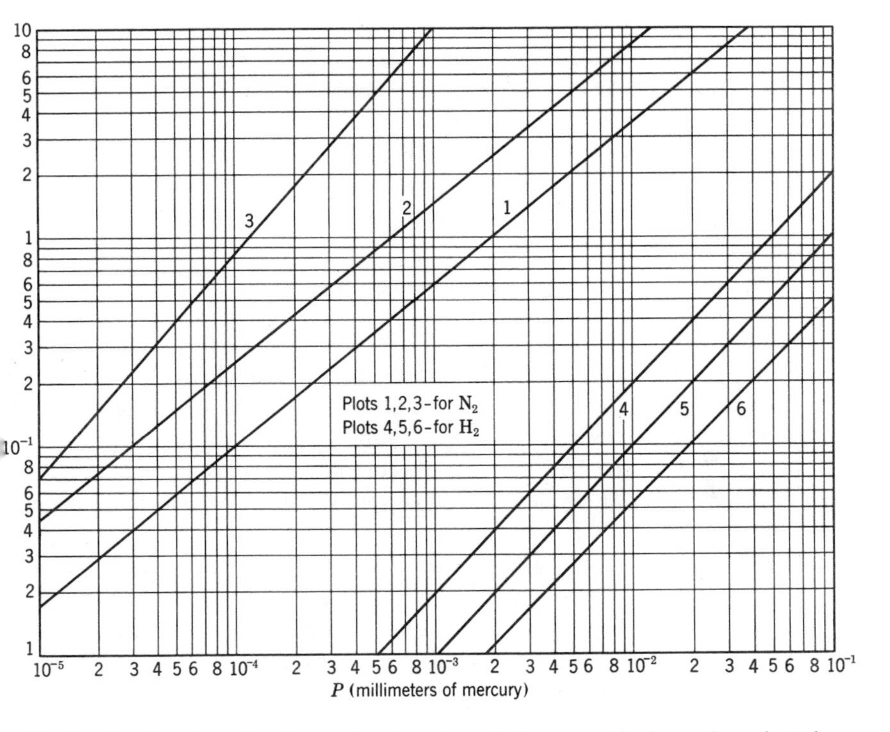


Fig. 7.6. Log-log plots of isotherms for adsorption of nitrogen and hydrogen by activated charcoal at −183° C and low pressures. Plots 1 and 2 for nitrogen and 4 and 5 for hydrogen represent limits observed by Winkler. Plot 3 for nitrogen and 4 and 6 for hydrogen were obtained by Rowe.

charcoal. For different samples the isotherms were observed to lie between the two limiting plots 4 and 5, whereas Rowe's observations lie on plots 6 and 4. Winkler also observed, in agreement with Firth and McBain, that for the attainment of equilibrium in both sorption and desorption of hydrogen a period of as much as 1–3 hr was required.

Some extremely interesting observations on the sorption by carbon dust have been made by Savage[54] in the General Electric Research Laboratory.

> The dust, formed by the wear of graphite rods rubbing against a rotating base in vacuum, has been found to adsorb hydrogen, nitrogen, oxygen, carbon monoxide, carbon dioxide, and methane *irreversibly* at room temperature. The hydrogen clean-up, 2 cc/g of dust, is $10^5$ times the adsorption shown by representative activated charcoals at room temperature, and pressures of the order of 10 microns.

The total surface of the dust, as determined, in accordance with the BET method, by adsorption of nitrogen at $-195°$ C, was found to be about 400 $m^2/g$, whereas the *chemically active* specific surface of the dust, as indicated by the hydrogen clean-up, was about 5 $m^2/g$.

The authors have interpreted these observations on the assumptions that the dust consists of ultramicroscopic plates of graphite which are about 20 angstroms thick and about 3500 angstroms in diameter and that the hydrogen is adsorbed on "unsaturated carbon valences at points of cleavage at right angles to the main (001) cleavage plane."

## 7.6. SORPTION BY SILICA GEL AND SILICATES

Extensive use is now being made of various porous silicas and complex silicates as sorbents. *Silica gels* are prepared by the dehydration of silicic acid, which is obtained as a hydrous gel when solutions of soluble silicates are acidified. Dehydration may be accomplished either by simple evaporation, leading to *xerogels*, or by more complex methods of replacing the liquid phase with vapor, designed to prevent the shrinkage which occurs in ordinary drying, to yield *aerogels*.

Finely divided silica powders, useful as adsorbents, are also prepared by the burning or hydrolysis of a volatile silicon compound, for example, silicon tetrachloride or ethyl silicate, in the vapor phase. The distinction between these *fumed silicas* and silica gels is analogous to the difference between carbon blacks and charcoals. Iler[55] has discussed the preparation and properties of these forms of silica and others.

Just as in the case of the carbon adsorbents, the sorptive properties of silicas vary greatly, depending on the details of preparation of the materials. Recent work indicates that such variations can be caused by differences in the pore-size distribution and the chemical nature of the surface. In charcoals, the presence of carbon-oxygen complexes can alter the sorptive properties; in the case of silica, the primary inherent chemical variable is the degree of hydration of the surface. Isolated surface hydroxyl groups remain on silicas even after they have been ignited at 940° C;[56] materials which have not been subjected to such treatment may have almost completely hydroxylated surfaces. Such substrates can adsorb polar gases such as water very strongly.

According to S. J. Gregg,[57]

The characteristics of adsorption on silica gel differ somewhat from those of charcoal, particularly with respect to the Freundlich isotherm, for in the case of many gases, e.g., sulfur dioxide, nitrogen peroxide, butane—the Freundlich equation holds fairly accurately. Moreover, deviations from Henry's law even seem to be much less than for charcoal.

Figure 7.7[58] shows isotherms for the sorption of carbon dioxide by silica gel, as determined by Patrick, Preston, and Owens.[59] Similar observations have been made by Magnus and his coworkers.[60]

For adsorption of *argon* on silica gel, Henry's law is obeyed to several hundred millimeters pressure. At 0° C, the actually observed amount, according to the results obtained by Kälberer and Schuster,[61] is approximately $0.1 \cdot 10^{-6}$ mole/g/mm ($= 2.24 \cdot 10^{-3}$ atm · cm³ g/mm). Similarly, in respect to other common permanent gases, the sorption by most silica gels is less than that by typical charcoals.[62]

Figure 7.8, taken from the publication by Coolidge,[63] shows the relative adsorption of water vapor by a number of different adsorbents, including charcoal and silica gel. The remarkably high sorption by chabasite and the sorption by glass are discussed below.

One especially interesting characteristic of silica gel and of similar gels prepared from $TiO_2$, $Al_2O_3$, $SnO_2$, and other oxides, as well as with some charcoals, is also illustrated in Fig. 7.8. This is the phenomenon of hysteresis in sorption and desorption isotherms.

The hysteresis in the sorption of water vapor by silica gel has been investigated very fully by Pidgeon[64] and Rao.[65] A typical plot of the sorption data at 20° C, obtained by Pidgeon, is shown in Fig. 7.9. The arrows indicate the order in which the observations were made. Pidgeon also concluded from his observations "that the hysteresis loop which appears during water sorption by silica gel has a real existence and is not due to the failure to eliminate air and other gases from the system"—an explanation which has been suggested by previous investigators. Rao also

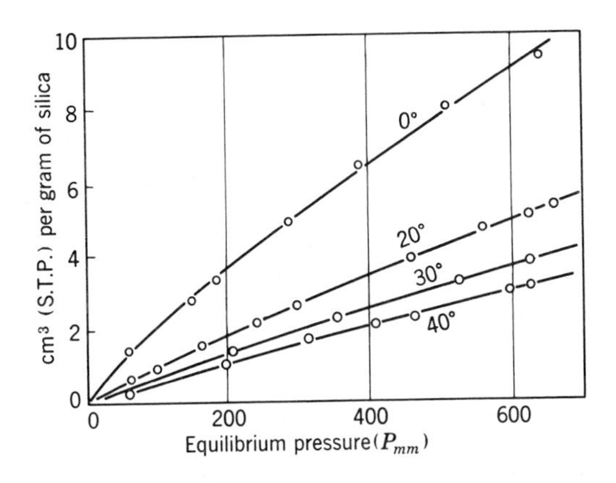


Fig. 7.7. Isotherms for the sorption of carbon dioxide by silica gel (Patrick, Preston, and Owens).

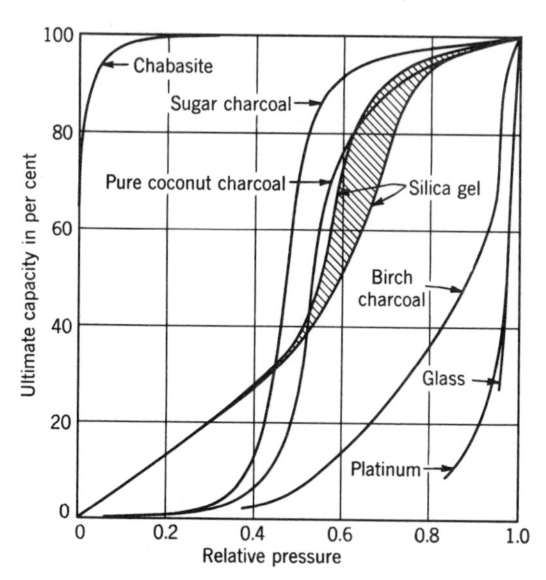


Fig. 7.8. Isotherms for the relative sorption of water vapor by different adsorbents, illustrating the hysteresis phenomenon observed with silica gel (Coolidge).

confirmed the existence of such a hysteresis loop at values of $x/m$ above 15 per cent, where $x/m$ = weight of $H_2O$ per unit weight of gel.

Hysteresis may be due to condensation in pores, in which case the Kelvin equation, which evaluates the lowering of the sorbate vapor pressure due to the concavity of the liquid meniscus in the pore, can be applied. In its simple form,

$$\ln \frac{P}{P_0} = - \frac{2\sigma V \cos \theta}{rR_0T}, \qquad (7.4)$$

where $\sigma$ is the surface tension of the liquid sorbate, $V$ is its molal volume, $\theta$ is its contact angle against the sorbent, and $r$ is the pore radius, the Kelvin equation assumes no adsorption except that due to capillary condensation. This is the basis of the capillary-condensation theory of adsorption cited in Chapter 6. By applying a correction for noncapillary adsorption, such as has been done by Barrett, Joyner, and Halenda[66] and others, it is possible to estimate the pore-size distribution from the isotherm.

Occasionally some hysteresis may also be found with apparently nonporous sorbents. An example is shown in Fig. 7.10, where the data obtained by Young[67] for the adsorption of water vapor on a fumed silica are shown. He found that in the initial adsorption measurement appreciable hysteresis occurred, but that this decreased in successive runs. This change he attributed to changes in the packing (and hence number and size of interstitial pores *between* the particles) of the sample as the experiment continued. Some of the final hysteresis loop may be due to the nature of the adsorption process, which seems to involve cooperative interactions between sorbate molecules, Young suggested.

Certain aluminosilicate minerals are important as adsorbents, Chief among these are the *zeolites*, a group including a number of naturally occurring minerals as well as some synthetic varieties.

As shown by a number of investigators, by use of X-ray methods, three types of zeolite framework have been recognized: (1) three-dimensional networks (chabasite, analcite); (2) fibrous structures (natrolite, etc.); (3) plate-like structures (heulandite). "These frameworks," as Barrer[68] states, "are anionic Si-O-Al networks, containing interstitial and exchangeable cations, and also neutral molecules of water, which are replaceable in some instances by molecules of ammonia, iodine, mercury, or gases."

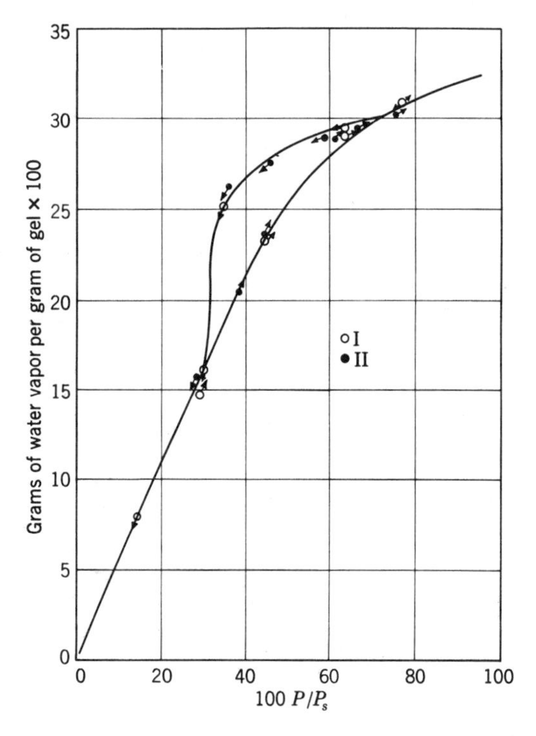


Fig. 7.9. Sorption and desorption curves for water vapor and silica gel at 20° C. Sorption plotted versus relative pressure, $P/P_s$, where $P_s$ = saturation pressure of water vapor (Pidgeon).

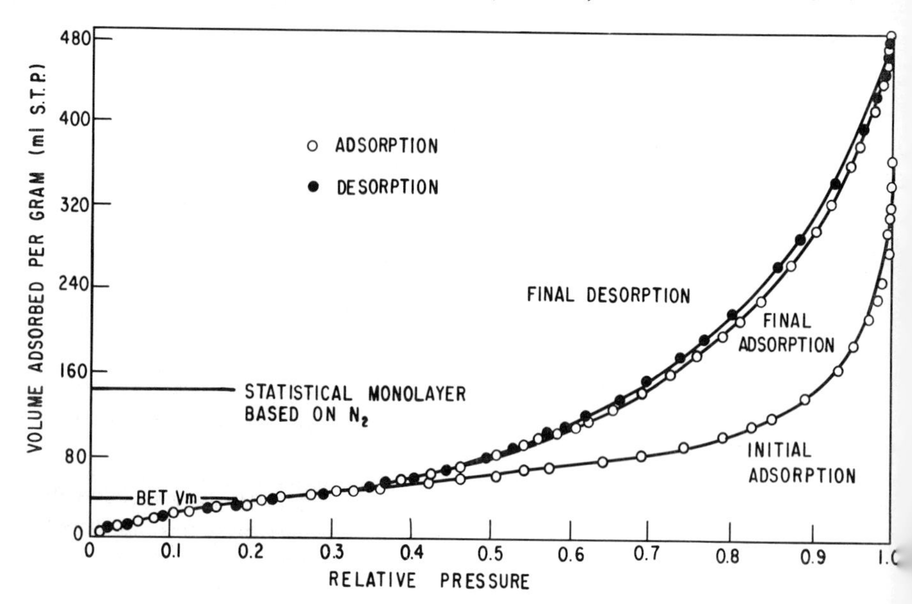


Fig. 7.10. Adsorption of water vapor by fumed silica (prepared by flame hydrolysis of $SiCl_4$) at 25° C (Young).

These minerals are of special interest because of two significant characteristics. The first of these, as described by Gregg.[69] is that "unlike ordinary crystals containing water of crystallization they can be dehydrated without any change in the form of their crystal lattice." As a result, molecules of different gases can occupy the spaces left vacant by the removal of water, and the zeolites are therefore good adsorbents. This is, however, true only for certain gases and involves the second characteristic of the zeolites: that these minerals exhibit a phenomenon known as *persorption.* According to Brunauer,[70]

> Persorption may be defined as adsorption in pores that are only slightly wider than the diameter of the adsorbate molecules. By using molecules of different diameters and noting the extent to which adsorption under given experimental conditions decreases as the size of the molecule increases one can study the structure of the finest pores of the adsorbent.

The amount of gas taken up by chabasite (or any of the other zeolites) varies, at constant temperature and pressure, with the degree of dehydration. Table 7.15 presents a summary of observations made by Lamb and Woodhouse[71] on the sorption by chabasite, at 0° C, and 1 atm pressure, of three gases. The first column for each gas gives the temperature (° C) at which the mineral was heated with simultaneous evacuation;

the second column, the percentage dehydration ($D$); and the third column, the amount occluded $V_0$ (cubic centimeters, STP) per gram of anhydrous mineral. From these observations it is concluded that maximum occlusion of hydrogen, oxygen, and carbon dioxide at 0° C occurs when 97.8, 96.9, and 92.9 per cent, respectively, of the original water has been removed by suitable heat treatment. A similar maximum sorption was observed for carbon dioxide at the same percentage dehydration, at 34.5, 61.2, and 100°C. (The maximum values of amount adsorbed are indicated in Table 7.15 by asterisks.)

TABLE 7.15

**Sorption of Hydrogen, Oxygen, and Carbon Dioxide by Dehydrated Chabasite at 0° C and 760 mm**

(Lamb and Woodhouse)

| Hydrogen | | | Oxygen | | | Carbon Dioxide | | |
|---|---|---|---|---|---|---|---|---|
| $t$° C | Per Cent $D$ | $V_0$ | $t$° C | Per Cent $D$ | $V_0$ | $t$° C | Per Cent $D$ | $V_0$ |
| 180 | 61.4 | 0.1 | 250 | 40.0 | 1 | 120 | 33.2 | 33 |
| 255 | 78.4 | 0.5 | 306 | 74.0 | 5.8 | 180 | 61.2 | 77 |
| 380 | 91.3 | 1.0 | 390 | 90.7 | 9.7 | 250 | 77.0 | 103 |
| 630 | 96.3 | 2.6 | 460 | 94.5 | 15.2 | 390 | 90.7 | 128 |
| 634 | 96.7 | 2.75* | 575 | 95.5 | 21.3 | 380 | 91.3 | 130* |
| 640 | 97.5 | 2.7 | 635 | 96.3 | 27.0* | 512 | 93.8 | 118 |
| | | | 635 | 96.7 | 23.2 | 608 | 95.7 | 119 |
| | | | 730 | 97.8 | 18.0 | 635 | 96.3 | 115 |
| Sorption by charcoal at 0° C and 760 mm: | | 1.5 | | | 20 | | | 87 |

This variation in amount of adsorption with extent of dehydration is explained on the assumption that the pore sizes and extent of corresponding inner surface change with percentage of dehydration. In fact, as shown in the bottom row in Table 7.15, the specific sorption of dehydrated chabasite approaches closely that of charcoal.

Figure 7.11 shows typical isotherms at 0° C for hydrogen, oxygen, and carbon dioxide, as observed in this investigation, for different percentages of dehydration ($D$).

Barrer[72] has given data on the sorption isotherms for hydrogen, nitrogen, and argon by dehydrated chabasite and analcite[73] at temperatures ranging

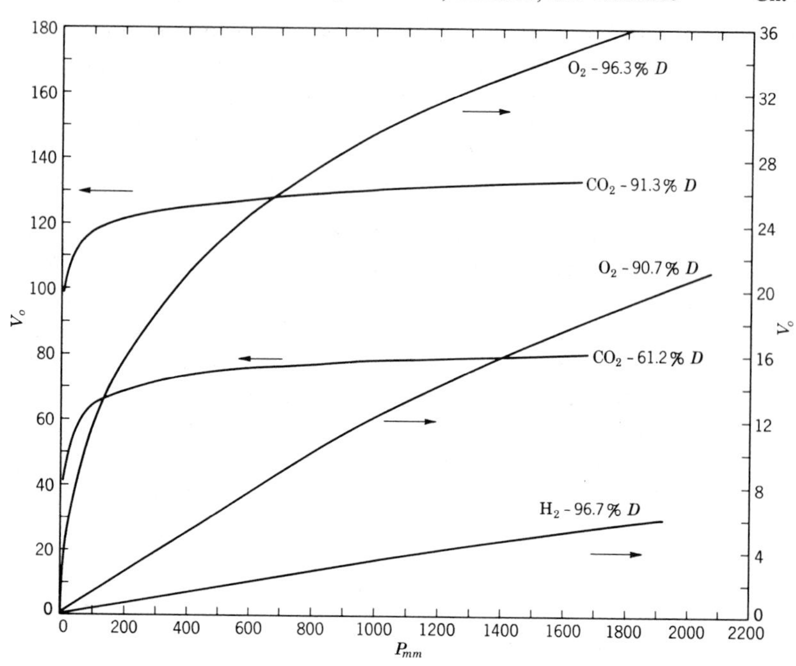

Fig. 7.11. Typical isotherms at 0° C for the sorption of hydrogen, carbon dioxide, and oxygen by dehydrated chabasite. Per cent dehydration ($D$) is given for each isotherm. Scale for $V_0$ is indicated by an arrow. (Lamb and Woodhouse.)

from −184 to 110° C. The saturation amounts observed at the lowest temperature, per gram of *original* chabasite, were as follows:

| | |
|---|---|
| Hydrogen approximately | 280 cm³ (STP) |
| Argon approximately | 250 cm³ (STP) |
| Nitrogen approximately | 164 cm³ (STP) |

Since the original chabasite contains 21.34 per cent, by weight, of $H_2O$, this would correspond to 266 cm³ water vapor (STP) per gram of original chabasite, or 338 cm³ per gram of *completely dehydrated* chabasite.

For analcite the volume of water vapor is 0.078 cm³ (STP) per gram of original material; according to Barrer's observations, the saturation values are considerably less than those for chabasite. In both cases, the sorption isotherms for the above gases, at temperatures higher than that of liquid air, show characteristics similar to those obtained with other adsorbents.

However, as mentioned previously, while chabasite is an excellent adsorbent for vapors or gases constituted of smaller molecules, such as water, oxygen, nitrogen, and hydrogen, vapors or gases constituted of larger molecules are not adsorbed at all, or only with difficulty. Thus Schmidt[74] observed that, at 20° C and 760 mm pressure, the volumes of gas (STP) adsorbed per gram of chabasite are as follows:

| $H_2$ | Ar | $O_2$ | $N_2$ | CO | $CH_4$ | $CO_2$ | $NH_3$ | $H_2O$ |
|---|---|---|---|---|---|---|---|---|
| 3.3 | 28.1 | 34.3 | 52.2 | 74.5 | 80.4 | 282.0 | 567.0 | 702.0 $cm^3$ |

From these and similar observations for the larger molecules of organic compounds he concluded that the pore diameter for chabasite is about $3.5 \cdot 10^{-8}$ cm.

This characteristic behavior of chabasite, which has been designated above by the term "persorption," has been investigated more extensively by Barrer and Ibbitson.[75] They have determined both occlusion isotherms and rates of occlusion of hydrocarbons by active chabasite and analcite over a range of temperatures. The isotherms resemble those obtained by Lamb and Woodhouse, which have been described. Figure 7.12 shows

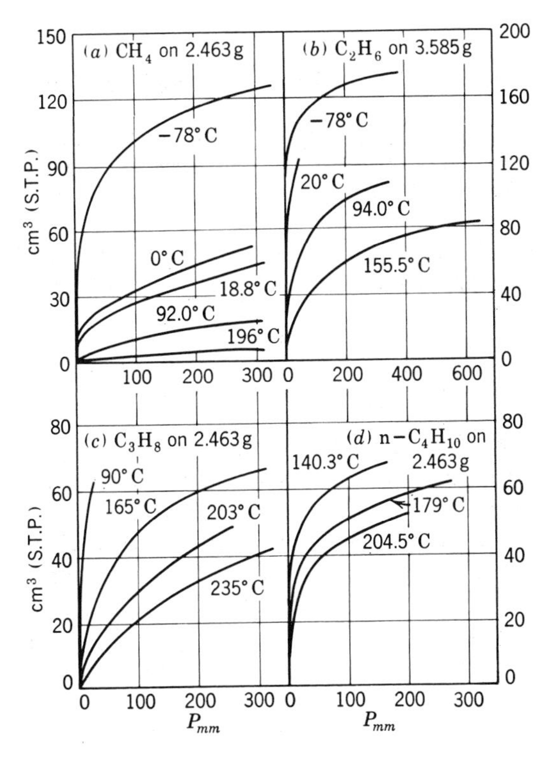


Fig. 7.12. Isotherms for sorption by chabasite of four normal-chain hydrocarbons (Barrer and Ibbitson).

typical isotherms obtained for four normal-chain hydrocarbons. Table 7.16 gives a summary of saturation values, as observed by the above investigators, for a number of gases in both chabasite and analcite, and, for comparison, the diameter or length in angstrom units (1 A = $10^{-8}$ cm).

As will be observed the saturation values in this table for hydrogen and argon are less than those given by Barrer in the previous publication. The decrease in the saturation value with increase in the length of the molecule is the most important conclusion to be deduced from the data in Table 7.16.

TABLE 7.16

**Saturation Values for Sorption by Chabasite and Analcite**

| Gas | Diameter or Length (A) | Cubic Centimeters per Gram (STP) | |
|---|---|---|---|
| | | Chabasite | Analcite |
| $H_2O$ | 2.76 | 266 | 97 |
| $NH_3$ | 3.60 | 193 | 72 |
| $H_2$ | 3.74 | 186 | 69 |
| Ar | 3.84 | 181 | 67 |
| $O_2$ | 3.83 | 181 | 67 |
| $N_2$ | 4.08 | 170 | 63 |
| $CH_4$ | 4.00 | 173 | 64 |
| $C_2H_6$ | 5.54 | 125 | 46.4 |
| $C_3H_8$ | 6.52 | 106 | 39.5 |
| $n$-$C_4H_{10}$ | 7.78 | 89 | 33.1 |
| $n$-$C_5H_{12}$ | 9.04 | 77 | 28.5 |
| $n$-$C_6H_{14}$ | 10.34 | 67 | 25.0 |
| $n$-$C_7H_{16}$ | 11.56 | 59 | 22.2 |

Kington and Laing[76] have discussed the relation between the structure and sorptive properties of chabasite in detail.

Of special interest are the results obtained by Barrer and Ibbitson[77] for rates of occlusion of hydrocarbons. If $Q_0$, $Q_t$, and $Q_\infty$ denote the amounts occluded at $t = 0$, $t$, and $t = \infty$, respectively, then it was observed that, for values of $y$ less than about 0.2,

$$y = \frac{Q_t - Q_0}{Q_\infty - Q_0} = k(t)^{1/2}, \tag{7.5}$$

where $k$ is a constant the value of which depends upon the particle size and the coefficient of diffusion $D$. It was also observed that $k$ decreased with increase in $y$. This is interpreted as follows.

Let $\theta$ denote the fraction of interstitial sites on which occlusion takes place. Then $D_\theta$, the diffusion constant corresponding to a given value of $\theta$, is related to $D_0$, the value for $\theta = 0$, by the equation

$$D_\theta = D_0(1 - \theta), \tag{7.6}$$

and, since the constant $k$ varies linearly with $(D_\theta)^{1/2}$, the value of this constant must decrease with increase in the value of $\theta$, that is, with increase in the value of $y$.

It was also observed that $k$ increases with increase in temperature, and, from a plot of $\log k$ versus $1/T$, a value for the energy of activation could be deduced.

For the effect of pressure, it was observed that the rate of sorption follows a "hyperbolic" relation of the form

$$\frac{dQ}{dt} = \frac{k_1 P}{1 + k_2 P}. \tag{7.7}$$

That is, with increase in pressure, the rate increases, at first practically linearly with $P$, then, as $P$ is increased still more, the increase in rate becomes less and less, until finally the rate attains a constant value.

A comparison of rates of occlusion for different hydrocarbons under given conditions of temperature and pressure shows that the rate decreases with increase in the number of carbon atoms in the hydrocarbon chain and also becomes extremely small for branched-chain hydrocarbons. This result is interpreted by Barrer and Ibbitson as follows:

The rate of sorption depends upon the cross section of the solute molecules. In Fig. 7.13 the cross sections are shown to scale for $CH_4$ (or $C_2H_6$); for a *n*-hydrocarbon (propane); and for a branched-chain hydrocarbon (isobutane). These represent classes of solute which enter the zeolitic lattice respectively with great rapidity, slowly and with an energy of activation, or not at all. Since the cross-sectional diameters[78] are, in order, 4.00, 4.89, and 5.58 A, one sees that the narrowest cross section of the interstitial channel in both chabasite and active analcite must lie between 4.89 and 5.58 A. . . . The three types of molecule given in Fig. 7.13 may then be used as yardsticks to measure the narrowest cross section of any zeolite channel, or in a zeolite thus "calibrated" one may fix within narrow limits the cross sections of other molecules.

Confirming observations obtained by other investigators, Barrer and Ibbitson also observed that the rate of sorption was greatly affected by the conditions of dehydration of the original mineral and the fineness of subdivision of the mineral.

As will be noted, Barrer and Ibbitson refer to "solute molecules" rather than occluded or adsorbed molecules. The reasons for this designation are discussed by them in the following remarks:

McBain [they state] used the term persorption to describe the uptake of gases by porous solids such as charcoal and silica gel, in which the pore systems are

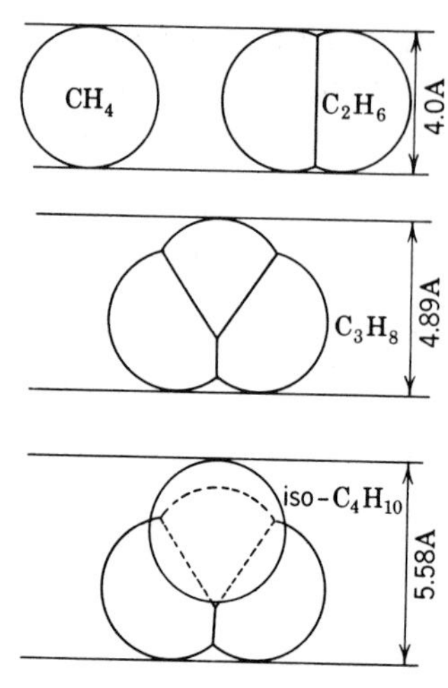


Fig. 7.13. The dimensions (in angstrom units) of representative molecules, respectively very rapidly occluded ($CH_4$ and $C_2H_6$), occluded by activated diffusion ($C_3H_8$), and excluded (*iso*-$C_4H_{10}$) (Barrer and Ibbitson).

quite irregular. In zeolites on the other hand the occluded molecules form a regularly disposed, frequently mobile, interstitial component of the dehydrated crystal lattice and the sorption process is better described as solid solution.

The term "interstitial" is used to indicate that the solute molecules occupy interstices left by the removal of $H_2O$ from the original lattice structure, and Barrer and Ibbitson conclude that "zeolitic solid solution is very similar to the solutions of $H_2$ in Pd. It differs from the latter in that there is in zeolitic solution no *specific* interaction between solute and solid such as characterizes hydrogen-metal systems."[79] Thus sorption of gases by dehydrated zeolitic minerals is a purely physical interaction. A number of other zeolites, in which the lattices contain voids of different sizes, have been studied by Barrer and his coworkers.[80]

Synthetic zeolites are now commercially available[81] and have found use in vacuum technology.[82] Cannon has studied the adsorption of fluorinated methanes[83] on these materials, as well as the adsorption of argon and water on a synthetic zeolite "modified" by the preadsorption of $CCl_2F_2$.[84] The zeolite was found to adsorb a significant amount (0.19 g/g) of $CCl_2F_2$ very strongly; this led, however, to only slight differences in the amount of subsequent adsorption of argon and water, although the energetics of these adsorption processes was altered. The most striking change in the behavior of the sorbent after "modification" was the loss of its ability to irreversibly adsorb[85] and decompose $CHClF_2$.

The sorption of water vapor by other silicates and quartz is a topic that has received considerable attention. For instance, R. W. Bunsen observed in 1885 that, even at 500° C, silicates retain appreciable amounts of water vapor. Briggs[86] measured the sorption of water vapor by quartz powder.

Fifty grams were used, having a superficial area of 20 m². The amounts adsorbed at different pressures of water vapor at 30° C are indicated in Table 7.17.

TABLE 7.17

**Sorption of Water Vapor by Quartz Powder**

| $P_{mm}$ | Adsorbed $H_2O$ (mg) | Number of Monolayers |
|---|---|---|
| 0.2 | 0.5 | 0.9 |
| 10.7 | 2.9* | 5.1 |
| 19.6 | 4.6 | 8.1 |
| 26.1 | 9.0 | 15.9 |
| 31.4 | 26.7 | 47.0 |

* Value obtained with samples dried to constant weight at 110° C.

The saturation vapor pressure at 30° C is 31.82 mm. Assuming, as deduced in Table 6.13, that a monolayer of water corresponds to $3.5 \cdot 10^{-5}$ cm³ per cm², that is, $2.84 \cdot 10^{-5}$ mg $H_2O$ per cm², it follows that the number of monolayers adsorbed at each pressure was that given in the last column of Table 7.17. It should be observed that the value marked with an asterisk was obtained with samples which had previously been dried to constant weight at 110° C.

Similar results have been obtained by Katz[87] for the sorption of water vapor by pulverized synthetic quartz and anorthite ($CaO \cdot Al_2O_3 \cdot 2SiO_2$). "The amount of water taken up reaches a fairly definite limit when the vapor pressure of the water is about 0.7 of the saturated vapor. The quantities of water adsorbed per square centimeter of surface under these conditions were $1.3 \cdot 10^{-6}$ g for quartz and $6.2 \cdot 10^{-6}$ g for anorthite." These correspond to 46 and 177 monolayers, respectively.

Palmer and Clark[88] have concluded that in the sorption of certain organic vapors by vitreous silica ($A_m = 0.469$ m²/g) the films consist of two complete monolayers when the relative pressure is 0.5–0.6. (Relative pressure $= P/P_0$, where $P_0 =$ saturation vapor pressure.) In a subsequent investigation, Palmer[89] has shown that at saturation pressure the films are probably as many as four molecules thick.

## 7.7. EVOLUTION AND SORPTION OF GASES BY GLASS[90]

### Observations on Evolution of Gases

The problem of completely removing adsorbed or absorbed water vapor from the walls of glass vessels is one of the most important in high-vacuum technique. Since glass is related to the silicates, it is not surprising that its behavior with respect to water vapor is similar to that of quartz and the silicates, which were discussed in the previous section.

Thus Bunsen (1885) observed that a glass surface that had previously been dried thoroughly at 20° C evolved 22.3 mg from an area of 2.11 $m^2$, which corresponds to a layer of adsorbed water about 37 molecules thick.

According to Langmuir, the sorption by glass is to be regarded as a process of solution of the water in the glass, in much the same manner as the sorption of moisture by sodium silicate and the gels. It is also quite possible that in very fine powders the moisture may be actually condensed as a liquid in fine capillary spaces between the grains. The presence of such relatively large amounts of water vapor on glass surfaces (and even metal surfaces, as will be mentioned in Chapter 8) means that, in experimenting at very low pressures, special care must be taken to remove water vapor by heating all parts to as high temperatures as possible with simultaneous absorption of the vapor in a liquid-air trap or phosphorus pentoxide.

The evolution of water vapor and other gases from glass was investigated very extensively by Langmuir. His earlier work on the evolution of gas from the walls of bulbs, such as were used for incandescent lamps about that time, has been described by him in the following remarks:[91]

> On heating bulbs of 40-watt lamps for 3 hours to a temperature of 200° C, after having dried out the bulbs at room temperature for 24 hours by exposing in a good vacuum to a tube immersed in liquid air, the following average quantities of gas were given off.:
>
> 200 cu mm water vapor
> 5 cu mm carbon dioxide
> 2 cu mm nitrogen
>
> These are the quantities of gas, liberated by the heating, expressed in cubic millimeters at room temperature and atmospheric pressure.
>
> By raising the temperature of the bulbs from 200° C to 350° C an additional quantity of water vapor was obtained, so that the total now became
>
> 300 cu mm water vapor
> 20 cu mm carbon dioxide
> 4 cu mm nitrogen
>
> A subsequent heating of the bulbs to 500° C caused the total amount of gas evolved to increase to
>
> 450 cu mm water vapor
> 30 cu mm carbon dioxide
> 5 cu mm nitrogen
>
> At each temperature the gas stopped coming off the glass after a half hour of heating, only to begin again whenever the temperature was raised to a higher value than that to which the bulb had been previously heated.

It therefore seems that, even by heating the bulb to 500° C, not all of the water vapor can be removed, but it does seem probable that after this treatment the amount of water vapor that can come off a bulb at ordinary temperatures must be extremely small.

The internal surface of this bulb was about 200 sq cm. The number of molecules of gas given off per sq cm was thus $56 \cdot 10^{15}$ molecules of $H_2O$; $37 \cdot 10^{15}$ molecules of $CO_2$, and $0.6 \cdot 10^{15}$ molecules of $N_2$. If we calculate the number of molecules of each of the gases necessary to cover a *square centimeter one molecule deep* (taking the molecules to be cubical in shape) we find $1.0 \cdot 10^{15}$ for $H_2O$; $0.77 \cdot 10^{15}$ for $CO_2$, and $0.67 \cdot 10^{15}$ for $N_2$. Thus the quantities of gas obtained from this bulb correspond to: a layer of water 55 molecules deep, a layer of carbon dioxide 4.8 molecules deep, and a layer of nitrogen 0.9 molecule deep.[92]

On the other hand Langmuir has observed[93] that microscope cover glasses (2.5 cm in diameter and 0.14 mm thick), previously heated to the softening point and then heated *in vacuo*, gave off amounts corresponding to the following numbers of layers of molecules: 4.5 for water vapor, 1.05 for carbon dioxide, and 0.9 for nitrogen.[94] It will be observed that, for the last two gases, the amounts adsorbed corresponded to monolayers.

Langmuir also investigated the optimum conditions for the evolution of water vapor from glass. It was observed that certain lamps made of sodium magnesium borosilicate glass (G-702-P), and consisting of high-wattage filaments in very small bulbs, blackened very rapidly if they were baked out at 550–600° C during exhaust, while lamps baked out at 400–500° C did not blacken so rapidly. The effect was ascribed to water vapor evolved from the glass during the life of the lamp, and experiments were therefore undertaken to try to remedy this condition.

The following description of the experiments is taken from Langmuir's patent specifications:[95]

> Three lots of lamps were made with the same structural details and operating characteristics; the first lot was exhausted at approximately 450° C, the second lot at 550° C, and the third lot at 550° C, at first, and then at 400° C. The average life of the first lot was approximately 575 hours, of the second lot 300 hours, and of the third lot over 900 hours, the conditions of operation with all three lots being the same.

The explanation given of this result is as follows:

> Apparently the treatment at 400° C to 500° C liberates the water vapor only from a comparatively thin surface layer of the glass. If, however, the exhaust is continued at 400° C to 500° C, no more water vapor will be drawn out of the deeper layers, and that which remains in the surface layer will be liberated.

The main conclusion arrived at by Langmuir is that, in order to remove water vapor efficiently from the walls of glass vessels, the *heating during exhaust should be carried out in two or more stages of gradually decreasing temperatures.* He found that ½-hr treatment at each of the above temperature ranges was sufficient, and he made the interesting observation, which is in accord with that made by Sherwood (see below), that, whereas the gas evolution at temperatures below 500° C practically ceases at the end of ½ hr, the evolution of water vapor at higher temperatures continues indefinitely

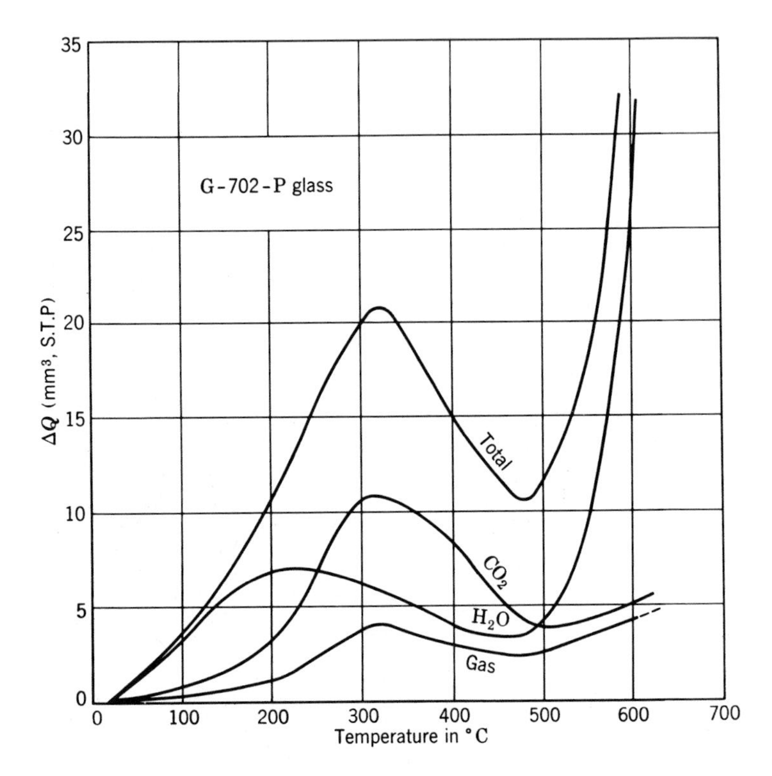


Fig. 7.14. Evolution of gas from Corning G-702-P glass, showing amount, $\Delta Q$ ($mm^3$ STP), liberated at a series of increasing temperatures (Sherwood). In this and the two subsequent figures, "gas" refers to the noncondensable volume (consisting of hydrogen, nitrogen, oxygen, and carbon monoxide).

no matter how long the heating period. Apparently the glass actually suffers a chemical decomposition at higher temperatures.

An extensive series of investigations on the gases and vapors evolved from glass was carried out by Sherwood[96] and Shrader.[97] Sherwood measured the amounts of water vapor, carbon dioxide, and gases noncondensable in liquid air liberated from different kinds of glass at various temperatures. Figure 7.14 shows the results obtained with Corning

G-702-P glass—a high-melting-point glass used extensively in the manufacture of the gas-filled type of incandescent lamp. The samples of glass used in these measurements had a total area of about 350 cm², and the curves show the amounts of gas liberated at different temperatures. The period of heating at each temperature was 3 hr. Figures 7.15 and 7.16

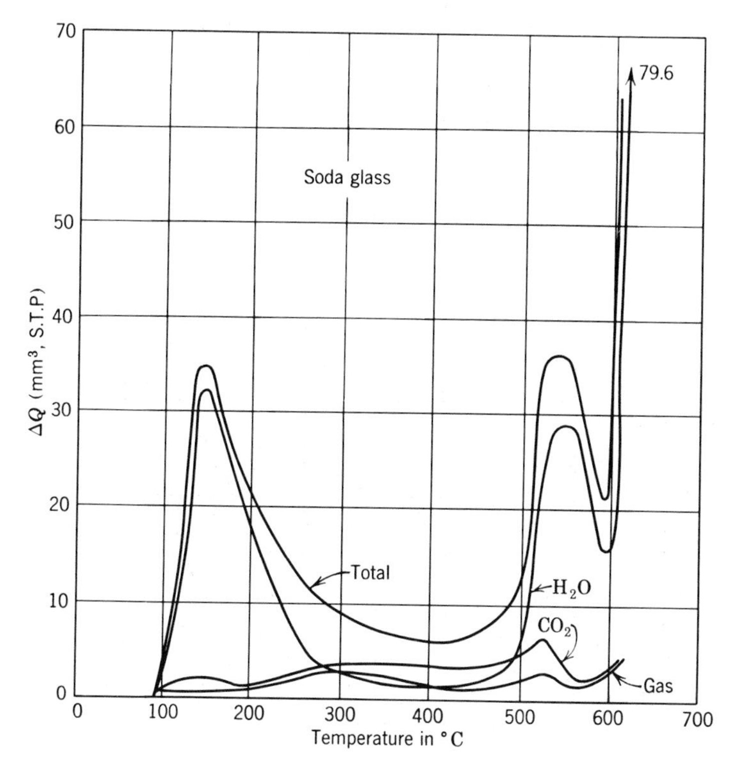


Fig. 7.15. Evolution of gas from soda glass (Sherwood).

show similar data with samples of soda glass and lead glass, respectively. It will be observed that in all samples the gas evolution first reaches a maximum which is at about 300° C for G-702-P, 150° C for soda glass, and 175° C for lead glass, then decreases, and again rises rapidly at a temperature which is above the softening point of the glass. Sherwood concluded that the products removed below 300° C are adsorbed gases, while at higher temperatures there is an actual decomposition of the glass itself. In other experiments, it was observed that at the higher temperatures the gas evolution continued even after the samples were heated for 24 hr and

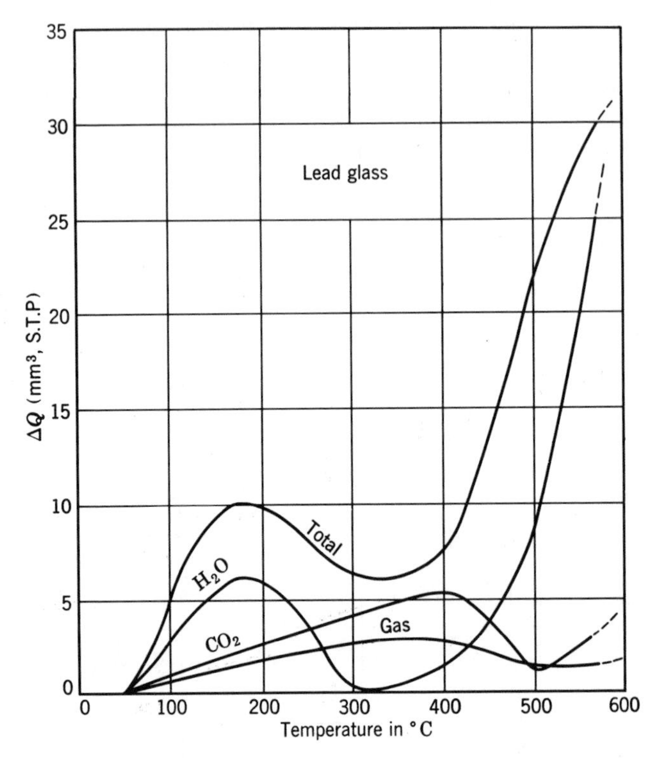


Fig. 7.16. Evolution of gas from ordinary lead glass (Sherwood).

longer. By previously annealing the glass at very high temperatures, the subsequent gas evolution in vacuum was decreased considerably, a result in accord with certain observations made by Langmuir and mentioned before. Similar results have been obtained in the General Electric Research Laboratory by Mrs. M. Andrews and J. Pangburn in investigating the gases evolved from lamp bulbs.

Analyzing Sherwood's data, we find that, for instance, in the case of soda glass, the total gas evolved up to 200° C was about 50 mm³, or about 0.15 mm³/cm², most of which was $H_2O$; this would correspond to a layer of gas about four molecules deep. Sherwood concluded that the gases which are removed fairly rapidly at lower temperature are genuine adsorption products, as they correspond to quantities represented by a layer of gas that does not exceed one or two molecules in thickness. As the temperature of the glass is raised to the softening point, the gas evolved consists practically wholly of water vapor; undoubtedly this arises, as mentioned above, from the chemical decomposition of the glass.

Investigations by C. A. Kidner and H. A. Huthsteiner, in the General Electric Research Laboratory, on the rate of evolution of water vapor and noncondensable gases from glass bulbs have shown that, at the highest temperatures at which the exhausted bulb can be heated, most of the adsorbed gas is evolved in the first 2 or 3 min. Preheating in dry air is also beneficial in removing most of this adsorbed gas. The preheating at ordinary pressure possesses the advantage that the glass can be heated to much higher temperatures, as there is no danger of the bulb's sucking inward.

Some interesting measurements were carried out by Sherwood on the adsorption of water vapor and other gases by dry surfaces of glass. Dry air could be removed very rapidly at ordinary temperature, whereas, for either moist air or air mixed with carbon dioxide, the rate of leakage at ordinary temperature was very slow. However, on heating to a high temperature practically all this adsorbed gas could be removed in a few minutes. It is interesting to observe that, in one experiment, after a pressure of about $10^{-4}$ mm $H_2O$ had been reached by evacuation, the bulb (of about 9000 $cm^3$ capacity) was sealed off; after standing 10 hr the pressure rose to 0.0095 mm, owing to the gas leakage from the walls, but did not materially increase subsequently. This gas most probably was adsorbed air, since it could not be condensed in liquid air. Dushman's experience has shown that invariably there is a slight increase in pressure after the bulb is sealed off. Part of this increase is due to gases adsorbed on the glass near the constriction, which is heated to a high temperature during sealing off, and a portion is due to gradual leakage from the walls. Even with the utmost precaution in baking out at high temperature and low exhaust pressure, there is always a slight increase in pressure in the sealed-off device.

In an investigation on the minimum pressure attainable with a Gaede molecular pump, Dushman[98] observed that, unless care was taken to torch[99] the tubing connecting the gauge to the pump, it was impossible to get below about 0.025 micron, because of the slow evolution of water vapor at ordinary temperature. When, however, the tubing was baked out at 330° C, the pressure could be reduced to $5 \cdot 10^{-4}$ micron. Similar results have been reported by Shrader. The volume of the system exhausted in his experiments was about 2 liters. The effect of heat treatment on the vacuum obtainable after pumping until equilibrium was reached at that temperature is shown by the following results:

| Temperature, °C: | 20 | 100 | 200 | 300 | 500 |
|---|---|---|---|---|---|
| $P_{mm}$: | $1 \cdot 10^{-5}$ | $1.9 \cdot 10^{-6}$ | $1.7 \cdot 10^{-7}$ | $1.2 \cdot 10^{-7}$ | $2.4 \cdot 10^{-8}$ |

Shrader also observed that not only does the vacuum in sealed vessels gradually deteriorate with time, at first rapidly and then more slowly, but

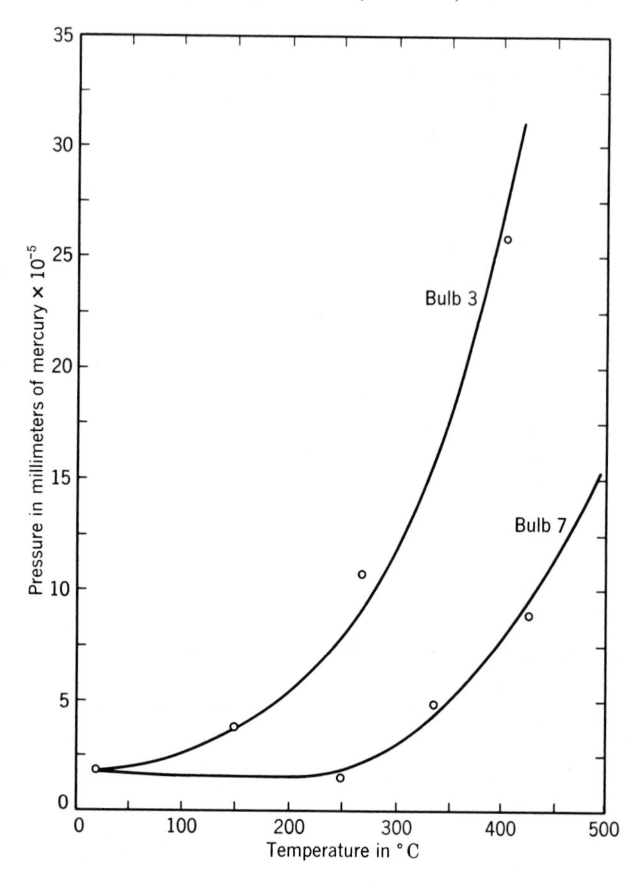


Fig. 7.17. Increase in pressure in sealed-off, well-evacuated bulb after reheating at a series of increasing temperatures (Shrader).

also that "subsequent heating even at temperatures lower than the heat-treating temperature (on the pump) results in increase of pressure due to further liberation of gases and vapors from the glass." Figure 7.17 shows the effect of heating a sealed-off system consisting of a 1500 cm$^3$ bulb and gauge of 500 cm$^3$ capacity for 1 hr at increasing successive temperatures. In each case the bulbs had previously been heated at 500° C on the pump.

A comprehensive investigation has been carried out by Harris and Schumacher[100] on the relation between chemical composition of glasses and the gas evolution on heating to 400–500° C. Table 7.18 gives a list of the types of glass used and their chemical compositions in per cent by weight of each constituent. The $B_2O_3$ was determined by difference. Glasses 3 and 5 contained traces of fluorine, and glass 6 contained 0.013

TABLE 7.18

**Chemical Analyses of the Glasses Used for Determination of Amounts and Compositions of Gases Evolved**

(Harris and Schumacher)

| | 1 | 2 | 3 | 4 | 5 | 6 |
|---|---|---|---|---|---|---|
| $SiO_2$ | 69.93 | 69.40 | 64.64 | 61.50 | 72.05 | 65.47 |
| $Al_2O_3$ | 1.54 | 0.78 | 0.20 | 0.57 | 2.21 | 2.99 |
| $Fe_2O_3$ | 0.19 | 0.14 | 0.04 | 0.11 | 0.05 | 0.51 |
| PbO | 1.44 | Trace | 21.66 | 22.55 | 6.11 | 20.20 |
| CaO | 3.17 | 5.15 | 0.02 | 0.21 | 0.06 | 0.22 |
| MgO | 0.03 | 4.09 | 0.02 | 0.36 | 0.09 | 0.13 |
| $Na_2O$ | 21.02 | 16.67 | 9.10 | 8.14 | 4.23 | 6.40 |
| $K_2O$ | 0.10 | 0.20 | 3.20 | 3.76 | 1.12 | 3.59 |
| $P_2O_5$ | 0.08 | 0.16 | 0.75 | 0.34 | Trace | Trace |
| $Sb_2O_3$ | 0.05 | 0.10 | ... | Trace | ... | ... |
| $MnO_2$ | 0.09 | 0.19 | ... | 0.19 | 0.01 | 0.073 |
| $B_2O_3$ | 2.36 | 3.12 | 0.37 | 2.27 | 14.07 | ... |

per cent sulfur trioxide as well as a trace of barium oxide. Glasses 1 and 2 were soda-lime glasses; 3, 4, and 6 were soda-lead glasses; and glass 5 was a borosilicate of lead and soda.

The procedure for determining the gas evolution is described by the authors as follows:

For the purpose of determining the amount of gases given up by the various types of glass, the glass was first cleaned with chromic acid, and washed thoroughly with water. It was then placed in the sample container, made of the same glass as that being tested—and the pump operated for several hours in order to dry the sample thoroughly. The pump—a mercury-vapor pump in conjunction with two oil pumps—was operated until the pressure within the apparatus after being trapped off from the pump remained constant at about $1 \cdot 10^{-6}$ mm for a period of at least 2 hr. The glass was then heated to a temperature as close to the softening range as it was possible to go without causing the container to collapse; it was kept at this temperature until the gas pressure became constant, when the volumes of the gases were determined in the manner outlined above. The period required for completely driving off the gases ranged from 65 to 80 hr.

The temperature to which the glasses were heated was 500° C for sample 5, and 400° C for all the other samples. The amount and composition of the gas evolved in each case are shown in Table 7.19.

For both carbon dioxide and water vapor, the volume of gas (STP) required for a monolayer is about $2 \cdot 10^{-5}$ cm³/cm², while for oxygen, nitrogen, and carbon monoxide it is about $3 \cdot 10^{-5}$ cm³/cm² (as shown in Table 6.13). The data in Table 7.19 indicate that the noncondensable gas

TABLE 7.19

**Amounts* and Compositions of Gases Evolved during Heat Treatment in Vacuum**

(Harris and Schumacher)

| Sample | % Total Alkali | Composition of Gas, % | | Atm · $cm^3$ · $10^4$ per $cm^2$ Area | Atm · $cm^3$ · $10^4$ per $cm^3$ Glass |
|---|---|---|---|---|---|
| 1 | 21.12 | $H_2O$ | 88.5 | 46.6 | 726 |
| | | $CO_2$ | 10.5 | 5.5 | 86 |
| | | P.G.† | 1.0 | 0.5 | 8 |
| | | | Total | 52.6 | 820 |
| 2 | 16.87 | $H_2O$ | 92.6 | 30.0 | 508 |
| | | $CO_2$ | 6.3 | 2.0 | 34 |
| | | P.G. | 1.1 | 0.4 | 6 |
| | | | Total | 32.4 | 548 |
| 3 | 12.30 | $H_2O$ | 96.4 | 23.7 | 506 |
| | | $CO_2$ | 2.2 | 0.5 | 11 |
| | | P.G. | 1.4 | 0.4 | 8 |
| | | | Total | 24.6 | 525 |
| 4 | 11.90 | $H_2O$ | 97.2 | 25.4 | 568 |
| | | $CO_2$ | 1.4 | 0.4 | 8 |
| | | P.G. | 1.4 | 0.4 | 8 |
| | | | Total | 26.2 | 584 |
| 5 | 5.35 | $H_2O$ | 33.3 | 0.6 | 12 |
| | | $CO_2$ | 44.5 | 0.9 | 16 |
| | | P.G. | 22.2 | 0.4 | 8 |
| | | | Total | 1.9 | 36 |
| 6 | 9.99 | $H_2O$ | 49.6 | 1.0 | 19 |
| | | $CO_2$ | 49.6 | 1.0 | 19 |
| | | P.G. | 0.8 | 0.02 | 0.3 |
| | | | Total | 2.02 | 38.3 |

* $10^{-3}$ atm · $cm^3$ = 0.76 micron · liter.

† P.G. = Noncondensable at −190° C.

must have been present as an adsorbed layer approximately one molecule thick.

Although it would appear from the table that the gas content decreases with decreasing alkali content, there are other factors that affect the gas content. For instance, glass 6 had been subjected to a heat treatment of between 1500 and 1600° C for a period of 1 hr.

Arranged in order of increasing temperature at which the glass begins to soften, glasses 3, 4, and 6 would be lowest, 5 highest, and 1 and 2 intermediate.[101] Thus the fact that glass 5 "undoubtedly received a higher heat treatment in the melting process, because of its higher melting point and greater viscosity, than did the other glasses that were tested," would account for its low gas content.

A further set of experiments was carried out in which the pressures of the gases evolved were measured at intervals of 100° C, from 100° C to the softening point of the glass. The results of these experiments are shown in Fig. 7.18. The numbers on the curves refer to the corresponding numbers of the glass samples (see Table 7.18). A run made with Pyrex glass is shown as curve 7. This is a borosilicate glass that is "practically free from alkali and heavy metals."

It is seen [the investigators state] that the *adsorbed*[102] gases for the lime and lead glasses are practically all given up at a temperature of 200° C, while 300° C is required in the case of the borosilicate glasses. The *absorbed* gases begin to come off at the softening points of the various glasses, 400° C for the lead and lime glasses, and 600° C for the borosilicate glasses. . . . In this connection it should be stated that the amount of absorbed gases found in the above experiments represents only that portion of the dissolved gases which lies nearest the surface of the glass. Owing to the great viscosity of the glass at the temperatures used, the rate of diffusion of the gas would be altogether too slow to permit any considerable portion to reach the surface.

To obtain further data on the volumes *adsorbed*, "the glasses were heated to a temperature high enough to drive off all the adsorbed gases, as indicated by the curves in Fig. 7.18 (200° C for curves 1, 2, 3, and 4, and 300° for 5) and the amount of carbon dioxide and permanent gases determined." The results are shown in Table 7.20.

The authors summarized the results of their investigation as follows:

1. Glasses whose compositions run high in alkali give off more gas during their heat treatment than do those of lower alkali content.
2. A definite relation appears to exist between the amount of water vapor held by a glass and its alkali content.
3. A relation, although not as pronounced as that mentioned above, appears to exist between the amount of carbon dioxide held by a glass and its alkali content.
4. Adsorbed carbon dioxide seems to be held to glass primarily by primary valence forces.

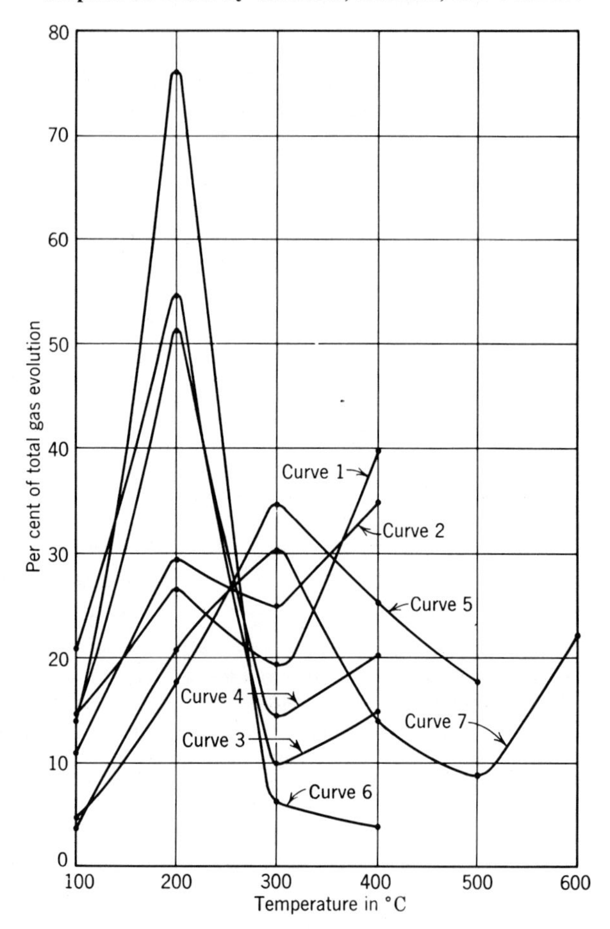


Fig. 7.18. Percentage (by volume) of total gas evolved from different samples of glass (analysis of samples 1–6 given in Table 7.18) on heating at a series of temperatures. Curve 7 is for a sample of Pyrex. (Harris and Schumacher.)

5. Adsorbed permanent gases seem to be held to glass primarily by secondary valence forces.

6. Glass relatively free from adsorbed gas can be produced by means of heating the glass during its melting process to a sufficiently high temperature.

Results obtained by C. H. Cartwright on three types of glass are shown in Figs. 7.19, 7.20, and 7.21.[103] The samples used are described as follows:

1. Glass 008, Corning brand, clear lime bulb glass.
2. Glass 774, Pyrex brand, heat-resistant clear chemical glass.
3. Glass 791, Vycor brand, 96 per cent silica, ultraviolet-transmitting glass.[104]

TABLE 7.20

**Amounts* of Carbon Dioxide and Noncondensable Gases Adsorbed by Different Samples of Glass**
(Harris and Schumacher)

| Glass No. | Volume in $10^{-5}$ $cm^3/cm^2$ of Adsorbed $CO_2$ | Volume in $10^{-5}$ $cm^3/cm^2$ of Adsorbed Noncondensable Gas |
|---|---|---|
| 1 | 19.3 | 1.300 |
| 2 | 10.4 | 1.800 |
| 3 | 7.5 | 2.650 |
| 4 | 7.5 | 2.350 |
| 5 | 4.5 | 0.480 |
| 6 | 6.5 | 0.009 |

* $10^{-4}$ atm · $cm^3$ = 0.076 micron · liter.

Figure 7.19 shows the amounts of gas[105] (micron · liters/100 $cm^2$ of surface at 25° C) evolved from glasses 008 and 774 as a function of the temperature.

Only one glass sample was used in obtaining each curve and that sample was heated for 1 hour at each of the temperatures indicated. The curves show a maximum at 150° C and increased evolution at the strain point. This is in agreement with the work of Sherwood and of Harris and Schumacher.

The rate curves shown in Fig. 7.20 were calculated from the results obtained in 008—run 1. As will be observed, these curves show that on baking out at constant temperature most of the gas is evolved very quickly.

Figure 7.21 compares the gas evolutions from glass 008 and glass 791. The comparatively small gas evolution from 791 is no doubt due to the fact that this glass is fined at high temperatures.

One further important observation made in this investigation was that "heating the lime glass to 500° C at atmospheric pressure removes 80–90 per cent of the gas that would otherwise be evolved in the bake-out." This observation has been confirmed repeatedly by similar observations in the General Electric Research Laboratory.

Morey[106] states,

> The volatile substances found in solution in glass are all components of the original batch: water, in combination as hydroxides or hydrated salts, or as "wet" batch; carbon dioxide in combination as carbonates; oxygen, in combination as nitrate or introduced mechanically as entrapped air; sulfur dioxide introduced as sodium sulfate; and nitrogen, from entrapped air. During the melting of the glass, these substances are given off in large quantities; that which remains in the glass is only a small proportion of that introduced into the batch.

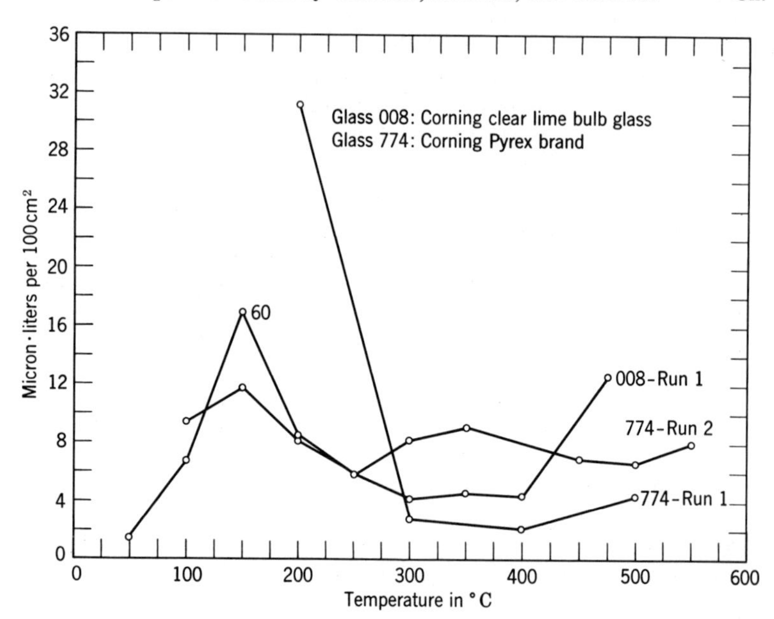


Fig. 7.19. Amounts of gas evolved from two samples of glass as a function of the temperature (Cartwright).

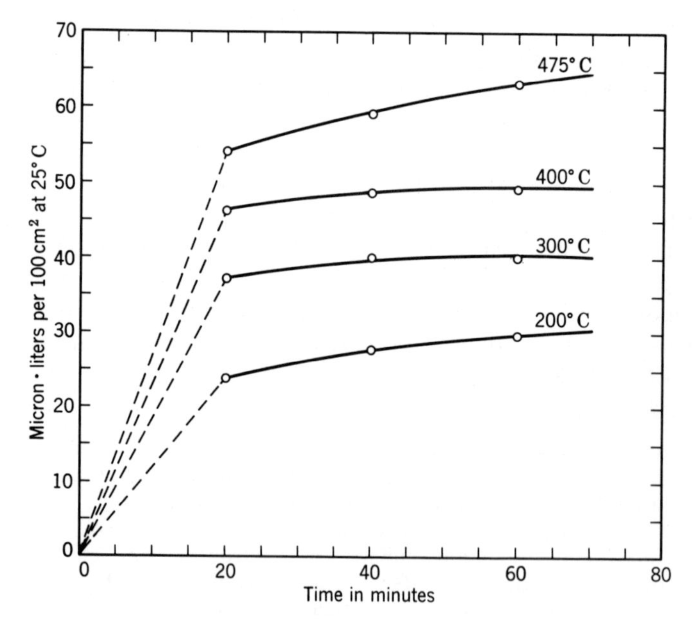


Fig. 7.20. Rate curves derived from plot in Fig. 7.19 for glass 008 (Cartwright).

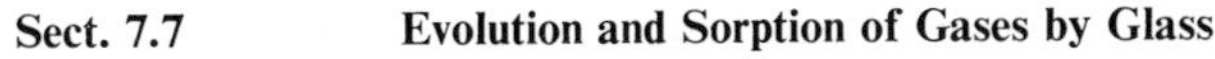

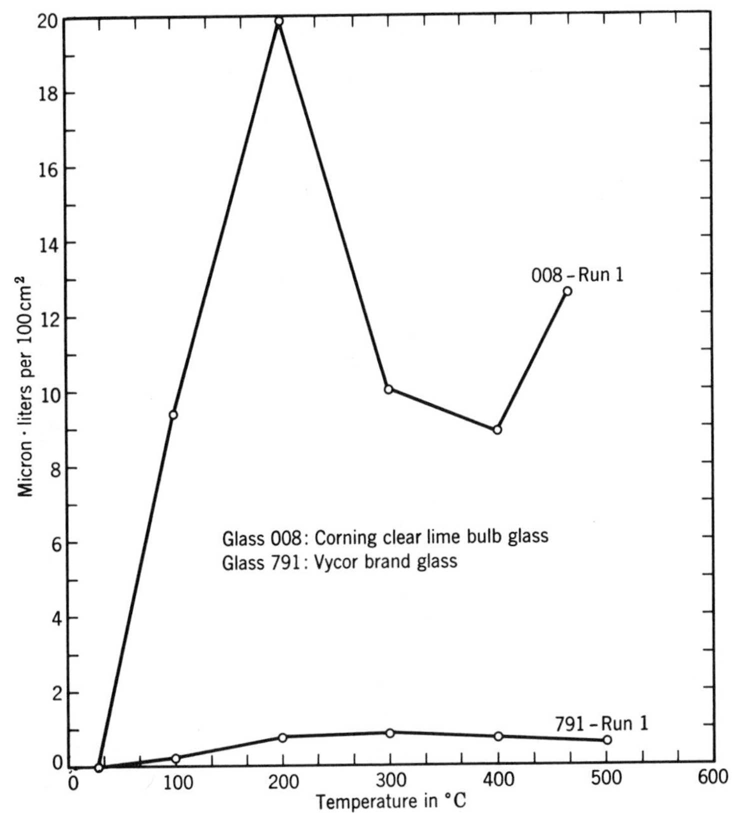


Fig. 7.21. Comparison of gas evolution as a function of the temperature for a Corning brand lime glass (008) and Vycor brand glass (791) (Cartwright).

As shown by Niggli,[107] the amount of gas in solution in the glass depends on the temperature of treatment, the pressure of each gas in the atmosphere above the melt, and the composition of the glass.

Washburn, Footitt, and Bunting[108] found that glass molten in vacuum gave off amounts of different gases which are shown in Table 7.21. The last four columns give the volumes in cubic centimeters at 0° C and 760 mm pressure, per 100 g of glass. The solubilities in water at 0° C are given for comparison at the bottom of the table. It is evident that the amounts of gases dissolved in molten glass are considerable. Compared with the volumes of gases actually evolved from glass vessels in high-vacuum exhaust, even at the highest temperatures practicable, the volumes observed for molten glass are much larger. This probably accounts to some extent for the observed increases in pressure in sealed-off glass vessels.

Dalton[109] devised an apparatus for analyzing the gases evolved from

TABLE 7.21

**Amounts and Compositions of Dissolved Gases in Finished Glass**

| Glass | Weight Per Cent | | | | Volume ($cm^3$, STP, per 100 g) | | | |
|---|---|---|---|---|---|---|---|---|
| | $O_2$ | $CO_2$ | $N_2$ | Total | $O_2$ | $CO_2$ | $N_2$ | Total |
| Barium flint 1 | 0.035 | 0.011 | | 0.046 | 24.5 | 8.8 | | 33.3 |
| Barium flint 2 | 0.015 | 0.0045 | | 0.020 | 10.5 | 3.6 | | 14.1 |
| Light flint | 0.0045 | 0.014 | 0.0025 | 0.21 | 3.2 | 11.2 | 2.0 | 16.4 |
| Borosilicate | 0.0036 | 0.0035 | 0.0031 | 0.010 | 2.5 | 2.8 | 2.5 | 7.8 |
| Water at 0° C | | | | | 5.15 | 179.2 | 2.24 | |

glass at 1400° C. Table 7.22 gives the amount and composition of the gas from several commercial glasses, and from one experimental glass, No. 8, containing an unusually large amount of arsenic. The third column gives the volume in cubic centimeters (STP) per gram of glass.

TABLE 7.22

**Volumes and Compositions of Gases Evolved from Glass at 1400° C**

(Dalton)

| No. | Glass Type | $V_0$ | Percentage Composition of Gases* | | | | |
|---|---|---|---|---|---|---|---|
| | | | $H_2O$ | $SO_2$ | $CO_2$ | $O_2$ | R |
| 1 | Barium (optical) | 0.71 | 23 | | 38.5 | 38.5 | 0.1 |
| 2 | Soda-lime (milk bottle) | 0.93 | 51 | 33.5 | 3 | 11.5 | 0.1 |
| 3 | Borosilicate (heat-resistant) | 0.40 | 91.5 | | 5 | 3.5 | |
| 4 | Soda-lime (bulb) | 0.90 | 44 | 35.5 | 5.75 | 14.5 | |
| 5 | Borosilicate (heat-resistant) | 0.36 | 90.5 | 0.3 | 3.5 | 6 | |
| 6 | Borosilicate (bulb) | 0.74 | 94 | 2.5 | 3 | 3.25 | |
| 7 | Lead (sign tubing) | 0.70 | 36 | 8 | 0.7 | 56.5 | |
| 8 | Soda-lime (experimental) | 1.41 | 28 | 10 | 8 | 63 | |
| 9 | Borosilicate (heat-resistant) | 0.44 | 93 | 2 | 4 | 4 | |

* The values given are, in each case, the average values of two sets of determinations given in the original paper. The symbol R refers to nitrogen and other noncondensable gases.

Table 7.23 summarizes observations made by Hahner, Voigt, and Finn[110] on different glasses at the temperatures shown in the second column. The first group includes optical glasses; the second group, commercial glasses.

TABLE 7.23

**Volumes* and Compositions of Gases from Glasses and Compositions of the Glasses**

| | $t°$ C | Volume per 100 g of Glass | | | Percentage Composition of Glass | | | | | | | |
|---|---|---|---|---|---|---|---|---|---|---|---|---|
| | | $H_2O$ | $O_2$ + R† | $CO_2 + SO_2$ | $SiO_2$ | $Na_2O$ | $K_2O$ | CaO | ZnO | BaO | $B_2O_3$ | As |
| Borosilicate crown A | 1325 | 49 | 0.2 | <0.1 | 65.6 | 7.91 | 12.8 | ... | 1.8 | ... | 11.4 | 0.43 |
| Borosilicate crown B | 1345 | 29 | 0 | 0 | 65.6 | 7.9 | 12.8 | ... | 1.8 | ... | 11.4 | 0.5 |
| Light barium crown A | 1350 | 20 | 1.0 | 0 | 47.1 | 1.9 | 7.7 | 2.1 | 8.5 | 27.9 | 4.3 | 0.54 |
| | 1325 | 31 | 0.8 | 0 | | | | | | | | |
| Light barium crown B | 1345 | 21 | 1.2 | 0.7 | 46.6 | 0.5 | 7.1 | ... | 7.8 | 33.7 | 3.7 | 0.6 |
| Light barium crown C | 1345 | 17 | 1.1 | <0.1 | 46.9 | 0.5 | 7.1 | ... | 7.8 | 33.2 | 3.7 | 0.2 |
| Light crown A | 1300 | 33 | 1.7 | 4.0 | 71.0 | 15.4 | ... | 13.1 | ... | ... | ... | 0.32 |
| Light crown B | 1350 | 38 | 5.4 | 4.1 | 71.0 | 15.4 | ... | 13.1 | ... | ... | ... | 0.5 |
| | | | | | $SiO_2$ | $Na_2O$ | $K_2O$ | CaO | MgO | $SO_3$ | $R_2O_3$ | |
| Window glass A | 1345 | 16 | 10.2 | 43.1 | 72.1 | 13.5 | ... | 10.0 | 3.3 | 0.35 | 0.7 | |
| Window glass B | 1345 | 28 | 17.1 | 17.0 | 72.6 | 15.0 | ... | 8.0 | 3.6 | 0.47 | 0.7 | |
| Window glass C | 1345 | 22 | 32.5 | 26.6 | 73.1 | 12.4 | 0.1 | 13.3 | 0.3 | 0.44 | 1.0 | |
| Window glass D | 1345 | 26 | 22.2 | 51.2 | 70.9 | 12.2 | 0.1 | 11.2 | 4.4 | 0.41 | 0.9 | |
| Window glass E | 1345 | 32 | 7.9 | 40.2 | 71.5 | 15.0 | 0.2 | 10.6 | 1.8 | 0.45 | 0.9 | |
| Window glass F | 1345 | 26 | 3.3 | 40.4 | 72.3 | 15.1 | 0.3 | 10.0 | 1.0 | 0.50 | 0.6 | |
| Soda-lime tubing | 1345 | 23 | 3.0 | 14.0 | ... | ... | ... | ... | ... | 0.19 | ... | |

* Volume in cubic centimeters (STP).

† The symbol R refers to other noncondensable gases.

The last column, headed As, refers to the total percentages of $As_2O_3$ and $As_2O_5$. (The original table gives the separate values for each of these oxides.) The term $R_2O_3$ refers to oxides of aluminum, iron, etc. The light barium crown C also contained 0.5 per cent $Sb_2O_3$. The last sample was analyzed only for $SO_3$.

In Tables 7.18, 7.22, and 7.23 the compositions have been given for a number of types of glass, in connection with the amount and composition of gases evolved on heating to high temperatures. It is therefore of interest to note the data in Table 7.24, for the chemical composition (per cent by

**TABLE 7.24**

**Chemical Compositions of Glasses Used in High-Vacuum Devices**

| Type | $SiO_2$ | $Al_2O_3$ | $B_2O_3$ | PbO | CaO | MgO | $Na_2O$ | $K_2O$ | BaO |
|---|---|---|---|---|---|---|---|---|---|
| A | 70.5 | 1.8 | ... | ... | 6.7 | 3.4 | 16.7 | 0.8 | ... |
| B | 69.0 | 4.0 | ... | ... | 5.8 | 1.6 | 17.5 | 1.9 | ... |
| C | 69.3 | 3.1 | 1.2 | ... | 5.6 | 3.4 | 16.8 | 0.6 | ... |
| D | 80.1 | 3.0 | 12.0 | ... | 0.2 | ... | 3.9 | 0.3 | ... |
| E | 80.6 | 2.7 | 12.2 | ... | ... | ... | 4.2 | ... | ... |
| F | 71.0 | 7.4 | 13.7 | ... | 0.3 | ... | 5.3 | 2.4 | ... |
| G | 71.6 | 5.7 | 11.0 | ... | 3.6 | 0.6 | 3.5 | 3.9 | ... |
| H | 56.5 | 1.5 | ... | 29.0 | 0.2 | 0.6 | 5.6 | 6.6 | ... |
| I | 57.0 | 1.5 | ... | 29.4 | 0.2 | 0.4 | 4.1 | 7.3 | ... |
| J | 54.5 | 21.1 | 7.4 | ... | 13.5 | ... | ... | ... | 3.5 |
| K | 58.7 | 22.4 | 3.0 | ... | 5.9 | 8.4 | 1.1 | 0.2 | ... |
| L | 22.6 | 23.7 | 37.0 | ... | 10.0 | ... | 6.5 | 0.2 | ... |

weight) of glasses used in high-vacuum devices, taken from a paper by Douglas[111] dealing with the thermal, mechanical, and electrical properties of these glasses.

The customary classification and applications are described as follows:

A, B, and C are *soft soda glasses*, used in lamp and radio tubes, also in tubing for neon sign lights.

D, E, F, and G are *hard borosilicate glasses*, used in high-wattage lamps and vacuum tubes, also in vapor-stream pumps and in chemical apparatus.

H and I are *lead glasses*, used in vacuum seals of lamps and tubes.

J and K are *extra-hard* glasses, used in mercury-vapor discharge lamps.

L refers to a special glass used in sodium-vapor lamps because of its resistance to chemical attack by sodium.

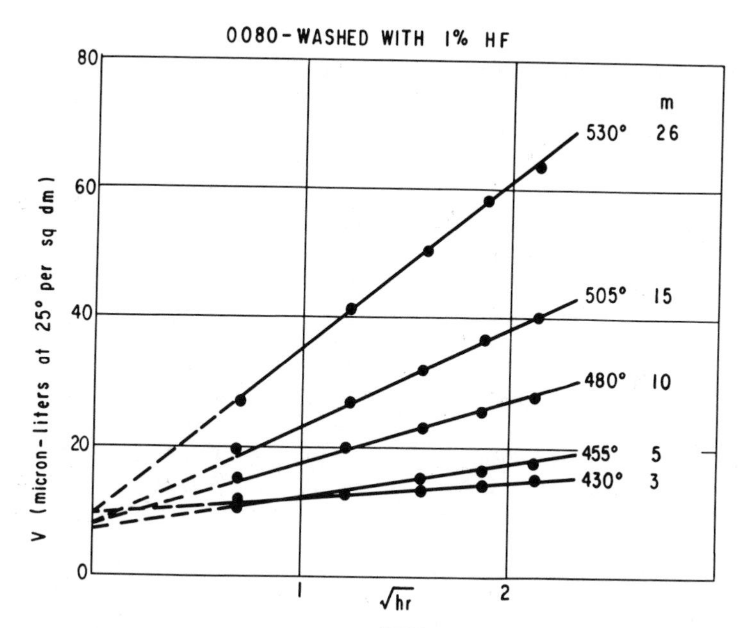


Fig. 7.22. Plots of gas evolution against √time, for the evaluation of $m$ for No. 0080 glass (Todd).

In two papers[112,113] Todd has reported similar results for the evolution of gases from a number of glasses. In every case studied by Todd, water constituted by far the major component of the evolved gases. He found that his data agreed with earlier suggestions that the gas evolved represented a combination of adsorbed and dissolved material, and advanced the further important possibility that these effects could be separated. When $V$, the volume of water evolved by unit area of glass surface at constant temperature, is plotted against $(\text{time})^{1/2}$, a curve such as that shown in Fig. 7.22 is obtained. Todd proposed[112] that the *adsorbed* gas, at least at higher temperatures, was the portion given off quickly, while the linear part of the curve represented water evolved after diffusion out of the glass. Therefore, the data at constant temperature could be represented by the equation

$$V = mt^{1/2} + s, \tag{7.8}$$

where $s$ represents the volume of adsorbed gas, and $m$ is related to the diffusion coefficient in the glass. Furthermore, $m$ was measured as a function of temperature and found to obey a typical Arrhenius equation

$$\log m = -\frac{A}{T} + B, \tag{7.9}$$

where $T$ is absolute temperature, and $A$ and $B$ are constants. Table 7.25

TABLE 7.25

**Constants for the Calculation of $m$ [micron (Hg) · liters at 25° C per square decimeter per (hour)$^{1/2}$] at an Arbitrary Temperature from the Equation, $\log m = (-A/T) + B$**
(Todd)

| Glass Code No. | Glass Type | $A$ (° K) | $B$ | $\Delta H_a$ (kcal/mole) |
|---|---|---|---|---|
| 7911 | Vycor brand 96% silica | 6230 | 5.397 | 57 |
| 7910 | Vycor brand 96% silica | 8240 | 9.772 | 75 |
| 1720 | Lime-aluminum | 7000 | 7.952 | 64 |
| 7740 | Borosilicate | 4510 | 6.310 | 41 |
| 7720 | Lead-borosilicate | 4150 | 5.983 | 38 |
| 0080 | Soda-lime | 5420 | 8.153 | 50 |
| 0120 | Potash-soda-lead | 3910 | 6.208 | 36 |
| 9014 | Potash-soda-barium | 4840 | 7.799 | 44 |

shows the values of the constants obtained by Todd for the glasses which he studied.

Todd also pointed out that this relationship could be used to estimate the amount of water which would be evolved in a system after any given bake-out. Suppose that the system had been evacuated at temperature $T_1$ (for which the corresponding outgassing constant $m$ was known or could be calculated from Table 7.25) for a time $t_1$. The volume of gas given off per unit area of glass surface in a subsequent interval $t_2$, when the temperature was $T_2$, would then be

$$V_2 = (m_1^2 t_1 + m_2^2 t_2)^{1/2} - m_1 t_1^{1/2}. \tag{7.10}$$

Todd has also shown[113] that the process of water evolution is reversible, and that after the water content of a glass sample has been decreased by outgassing, it may be increased again by exposure to a humid atmosphere at elevated temperature. The amount of adsorbed water is also increased by such treatment.

With the increasing use of well-baked ultrahigh-vacuum systems (see Chapter 9), there has been renewed interest in the composition of residual gases in such systems. For example, Gentsch[114] has studied the nature of the gases in an ultrahigh-vacuum system before and after the evaporation of metal films in the system.

## Adsorption on Glass

A number of workers have reported data for the adsorption of gases on glass surfaces. Langmuir[115] and Zeise[116] both studied the uptake of methane, oxygen, and nitrogen by glass at low temperature (90° K) and pressures. Langmuir also examined the adsorption of carbon monoxide and argon, while Zeise gave a few data for hydrogen.

Durau[117] investigated the adsorption on carefully dried and evacuated glass powder. Even after treatment in vacuum at 570° C, the powder continued to give off gas at higher temperatures (up to 1060° C). At each temperature an equilibrium was reached at which the gas evolution was practically zero. After the high-temperature treatment the powder did not adsorb nitrogen, hydrogen, or dry air at 18° C. It was observed that carbon dioxide was adsorbed, but only to the extent of 16.4 per cent of the surface at 1 atm pressure. The surface area of the powder was determined from measurements of the adsorption of dyes.[118]

The sorption of water vapor by glass has been investigated by McHaffie and Lehner[119] by a method which, as Brunauer describes it,

> . . . consists simply in measuring at different temperatures the pressure of a known amount of vapor enclosed in a vessel of known volume and known surface area. At the lower temperatures part of the vapor must be in the liquid form; at the higher temperature all must be in the gaseous state.

At these higher temperatures the pressure decreases linearly with decrease in temperature; at the lower temperatures the pressure will correspond to that of the saturation vapor pressure of the liquid. When adsorption occurs, the transition in the plot of $P$ versus $T$ from the high-temperature to the low-temperature range will be not sharp, but gradual.

From their observations, McHaffie and Lehner arrived at the following conclusion: For values of $P/P_0$ less than about 0.7, the number of molecular layers adsorbed is one or less; but as $P$ is increased and approaches $P_0$ the number of layers increases rapidly and approaches a value of the order of 100.

Frazer, Patrick, and Smith[120] have, however, criticized this conclusion because of the fact that McHaffie and Lehner used glass washed with chromic acid. This undoubtedly increased the effective surface area manyfold. For instance, Emmett[121] found by the method of nitrogen adsorption that washing small glass spheres with chromic acid increased their surface by 40 per cent. Similar observations have been made by Frank[122] in the case of Pyrex glass treated with chromic acid cleaning solution. But, even if the actual quantitative data obtained by McHaffie and Lehner are in error, it may be concluded from their observations and from those made by Frazer[123] that, for relatively large values of $P/P_0$, the adsorbed layers of water are many molecular diameters thick.

The low-pressure adsorption of water vapor on glass, at 25° C, has been measured by Frank.[122] As mentioned above, the Pyrex surface was treated with chromic acid and then washed with distilled water and evacuated at ordinary temperature. Water vapor was then let in at low pressure, and the equilibrium pressure was determined as well as the amount adsorbed. The observations are summarized in Table 7.26. The volume in cubic centimeters (STP) per square centimeter is denoted by $V_0$.

**TABLE 7.26**

**Adsorption of Water Vapor by Glass at 25° C**

| $P^{\mu}$ | $10^5 V_0$ | $n$ |
|---|---|---|
| 5.2 | 1.157 | 0.327 |
| 13.4 | 2.201 | 0.622 |
| 46.2 | 4.340 | 1.226 |
| 64.8 | 5.231 | 1.478 |
| 87.0 | 5.969 | 1.686 |

The last column gives the thickness of the adsorbed layer in molecular diameters. That is,[124]

$$n = 10^5 \frac{V_0}{3.54}. \tag{7.11}$$

Assuming that the effective surface was greater than that calculated from the geometrical dimensions, these results are entirely compatible with the conclusion that the adsorbed water formed a layer that did not exceed about one molecule in thickness, at these pressures.

A number of studies have shown that both the surface area and sorptive properties of glass can be markedly altered by the pretreatment of the surface. For example, Thompson, Washburn, and Guildner[125] found that the surface area of glass beads (as measured by the BET method, using carbon dioxide) increased by a factor of 10 or more when the beads were exposed to liquid water for 33 days. Razouk and Salem[126] found three- to tenfold differences in surface area (measured by water-vapor adsorption) between acid-treated and untreated glass.

Finally, observations made by Hamaker, Bruining, and Aten[127] on a very probable cause of slumping of oxide cathodes in vacuum tubes should be mentioned. According to these investigators there is a slow evolution of hydrogen chloride from glass heated to 400° C, due to the reaction

$$2NaCl + SiO_2 + H_2O \rightarrow 2HCl + Na_2SiO_3.$$

The hydrogen chloride reacts with the carbonates or oxides on the cathode to form the corresponding chlorides.

When the cathode is subsequently heated these chlorides evaporate and condense on the grid and the anode. Under electron bombardment these chlorides decompose, thereby producing chlorine atoms or positive chlorine ions which poison the cathode emission in a very pronounced way.

The sodium chloride is introduced in the manufacture of glass as a slight impurity in the sodium carbonate. The reaction between the salt and water vapor occurs "on a production scale" at red to dull-red temperatures. But at 400° C hydrogen chloride is evolved at a sufficient rate to produce the poisoning effect mentioned above. However, if the temperature of the glass during bake-out on the pump is not raised to much above 200° C, the rate of reaction is decreased enormously. Hamaker and his coworkers have observed that, in order to prevent the deleterious effects of hydrogen chloride on the cathode, "a large patch of barium carbonate somewhere in the bulb proved effective" as well as "evaporation of some of the barium getter before baking-out or rinsing the interior of the bulb with caustic potash solution before sealing-in."

## 7.8. DIFFUSION OF GASES THROUGH QUARTZ AND GLASSES[128]

It has been observed that thin-walled vessels of quartz or glass are permeable to helium, hydrogen, and a few other gases, especially at higher temperatures.

Let $q$ denote the *volume in cubic centimeters* (STP) flowing through the wall, *per square centimeter per second.* Since $q$ varies linearly with the pressure of the gas,[129] and inversely as the wall thickness $d$, we can express these relations in the form

$$q = \frac{KP}{d}, \tag{7.12}$$

where $K$ is designated the *permeability.* In the literature on this topic $P$ is usually expressed in centimeters of mercury (as $P_{cm}$), and $d$ in millimeters. Hence $K$ denotes *the volume in cubic centimeters (STP) per second per square centimeter, per millimeter thickness, per centimeter pressure.*

The value of $K$ increases rapidly with increase in $T$, and it has been observed that it is possible to express $K$ as a function of $T$ by means of the relation

$$K = K_0 \epsilon^{-E/R_0 T}, \tag{7.13}$$

where $E$ is the activation energy for the reaction, expressed in calories per mole. Equation 7.13 can be written in the form

$$\log K = A - \frac{B}{T}, \tag{7.14}$$

where

$$A = \log K_0$$

and

$$B = \frac{E}{4.574}. \tag{7.15}$$

Instead of expressing the rate of flow in atmosphere · centimeters³ per second, which is a magnitude of the order of $10^{-8}$ to $10^{-10}$ or less, it is more convenient from a practical point of view to express the rate in terms of $Q_{\mu l}$ *micron · liters* (at 0° C) *per hour per square centimeter, per millimeter thickness*, for $P_{cm} = 76$. Hence

$$Q_{\mu l} = 3600 \cdot 760 \cdot 76K = 2.08 \cdot 10^8 K. \tag{7.16}$$

For the rate in micron · liters at 25° C, the value obtained by means of equation 7.16 should be multiplied by the factor 298/273 = 1.091. From equations 7.15 and 7.16 it follows that

$$\log Q_{\mu l} = 8.318 + A - \frac{B}{T}, \tag{7.17}$$

by means of which it is possible to derive the value of $T$ corresponding to any given value of $Q_{\mu l}$.

As an illustration of the method used in deriving values of $A$, $B$, $E$, and $Q_{\mu l}$ from observed values of $K$ we shall use the data given in Table 7.27 for

**TABLE 7.27**

**Values of the Permeability ($10^9 K$) for He-$SiO_2$**

| $t$° C | $10^3/T$ | Ref. 1* | Ref. 2* | Ref. 3* |
|---|---|---|---|---|
| 150 | 2.364 | 0.73 | 0.78 | ... |
| 200 | 2.114 | 1.39 | 1.52 | ... |
| 300 | 1.745 | 3.15 | 4.13 | 0.48 |
| 400 | 1.486 | 6.15 | 8.25 | 0.99 |
| 500 | 1.294 | 10.4 | 13.8 | 1.72 |
| 600 | 1.145 | 16.4 | 19.3 | 3.00 |
| 700 | 1.028 | 21.9 | ... | 4.25 |
| 800 | 0.922 | 28.5 | ... | 5.50 |
| 900 | 0.853 | 36.2 | ... | 6.72 |
| 1000 | 0.786 | 45.4 | ... | 8.42 |

* See references below Table 7.28.

the permeability of helium in quartz, which are taken from Barrer's treatise.

Figure 7.23 shows corresponding plots of log $K$ versus $10^3/T$. From these plots are derived the values of the constants $A$, $B$, and $E$, from which

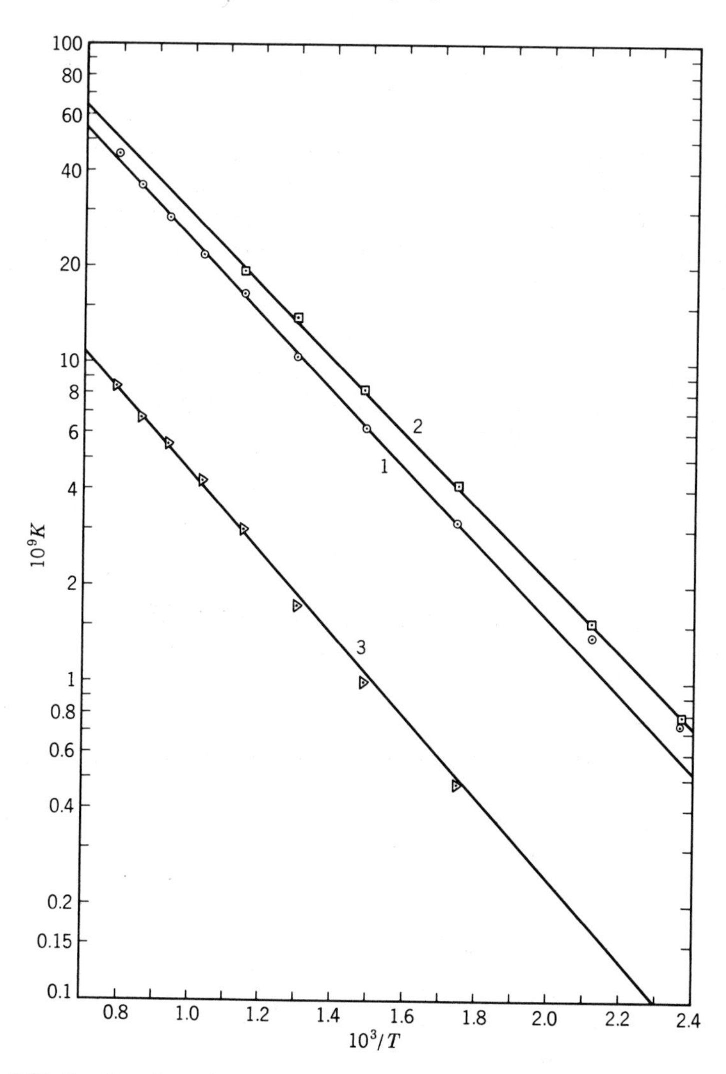


Fig. 7.23. Semilog plots of permeability versus $1/T$ for helium in quartz, according to three different publications. (Numbers on plots correspond to references listed below Table 7.28).

in turn are derived the values of the temperature in degrees Centigrade for $Q_{\mu l} = 0.1$ and 1.0 (micron · liter/*per hour*). Values thus calculated are shown in Table 7.28.

Barrer has deduced values for $E$ which are somewhat higher than those given in Table 7.28. This is probably due to the fact, which is

TABLE 7.28

Permeabilities of Quartz and Pyrex

| System | Ref. | $9 + A$ | $B$ | $E$ | $8.318 + A$ | $t°$ C for $Q_{\mu l}$ = | |
|---|---|---|---|---|---|---|---|
| | | | | | | 1.0 | 0.1 |
| $He\text{-}SiO_2$ (150–1000° C) | 1 | 2.570 | 1186 | 5,429 | 1.888 | 355 | 134 |
| | 2 | 2.609 | 1147 | 5,247 | 1.927 | 322 | 119 |
| | 3 | 1.912 | 1266 | 5,793 | 1.229 | 757 | 295 |
| $He\text{-}SiO_2$ (−200–150° C) | 2 | 2.363 | 1042 | 4,771 | 1.681 | 347 | 116 |
| | 2 | 2.349 | 1059 | 4,846 | 1.667 | 362 | 124 |
| $Ne\text{-}SiO_2$ | 1 | 1.864 | 2094 | 9,583 | 1.182 | ... | 686 |
| $H_2\text{-}SiO_2$ | 3 | 2.417 | 1888 | 8,640 | 1.735 | 815 | 417 |
| | 4 | 2.373 | 1694 | 7,756 | 1.691 | 729 | 356 |
| | 5 | 2.125 | 1778 | 8,141 | 1.443 | 959 | 455 |
| $N_2\text{-}SiO_2$ | 3 | 4.086 | 4820 | 22,060 | 3,404 | 1143 | 821 |
| | 5 | 4.778 | 5508 | 25,210 | 4.096 | 1072 | 808 |
| He-Pyrex | 6 | 1.580 | 1170 | 5,356 | 0.898 | ... | 314 |
| Air-porcelain | 7 | 4.280 | 7800 | 35,700 | 3.598 | ... | 1440 |

*References for Tables 7.27 and 7.28*

1. L. S. T'sai and T. Hogness, *J. Phys. Chem.*, **36,** 2595 (1932).
2. E. O. Braaten and G. Clark, *J. Am. Chem. Soc.*, **57,** 2714 (1935).
3. R. M. Barrer, *J. Chem. Soc.*, **1934,** 378.
4. G. A. Williams and J. B. Ferguson, *J. Am. Chem. Soc.*, **44,** 2160 (1922); **46,** 635 (1924).
5. J. Johnston and R. Burt, *J. Optical Soc. Am.*, **6,** 734 (1922).
6. W. D. Urry, *J. Am. Chem. Soc.*, **54,** 3887 (1932).
7. W. F. Roeser, *Bur. Standards J. Research*, **7,** 485 (1931) (RP 354).

evident from an inspection of the plots in Fig. 7.23, that a certain amount of personal judgement is involved in drawing the "best" straight line through the set of points for a given temperature.

Norton[130] has reported additional permeation measurements in which not only the amount of gas but also its composition was studied with the aid of a mass spectrometer. His observations on the helium-silica system covered the range −78 to 700° C, and agree fairly well with those of T'sai and Hogness. In addition, he studied helium permeation through a number of glasses, whose compositions are detailed in Table 7.29, as well

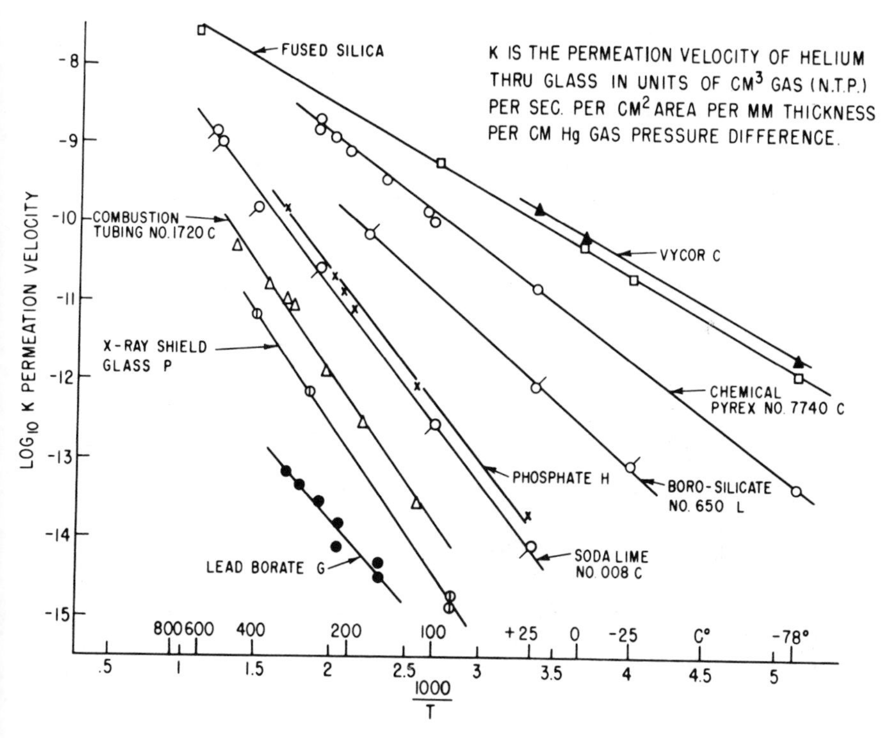


Fig. 7.24. Permeability of various glasses to helium (Norton).

as the transport of several other gases through silica. The results of his helium-permeability measurements are shown in Fig. 7.24, and the data reported for other gases are given in Tables 7.30 and 7.31. Data for the permeation of helium, neon, hydrogen, and nitrogen through Vycor glass have been given by Leiby and Chen.[131]

It should be noted that Alpert has shown that permeation of atmospheric helium is one of the limiting factors in the attainment of ultrahigh vacua.[132] (See also section 9.11.)

Urry reported that Pyrex is permeable to hydrogen, and this result was confirmed by Taylor and Rast,[133] who reported a permeability of $1.86 \cdot 10^{-13}$ g/cm²/sec/mm/atm. They also emphasized the importance of the history of the glass sample in affecting the permeability observed. In this connection, it has been found, for example, that the permeation rates are altered if the glass is under stress.[134]

Barrer has observed permeation of air, oxygen, and argon in quartz. For argon the rate was extremely slow and was accelerated by treatment of the surface with hydrofluoric acid.

**TABLE 7.29**

**Glass Compositions**

(Norton)

| Constituent | 1 Fused $SiO_2$ (G)* | 2 Vycor Brand (C)* | 3 No. 7740 (C)* | 4 No. 650 (L)* | 5 Phosphate (H)* | 6 No. 0080 (C)* | 7 No. 1720 (C)* | 8 X-ray (P)* | 9 Pb Borate (G*) |
|---|---|---|---|---|---|---|---|---|---|
| $SiO_2$ | 100 | 96 | 81 | 90 | 0 | 72 | 62 | 31 | 0 |
| $B_2O_3$ | | 3 | 13 | | 5 | | 5 | | 22 |
| $P_2O_5$ | | | | | 77 | | | | |
| $Al_2O_3$ | | 1 | 2 | 3 | 11 | 1 | 18 | | |
| CaO | | | | | | 10 | 15 | | |
| MgO | | | | | | | | | |
| BaO | | | | | | | | 8 | |
| PbO | | | | | | | | 61 | 78 |
| ZnO | | | | | 7 | | | | |
| $Na_2O$ | | | 4 | 7 | | 17 | | | |
| $K_2O$ | | | | | | | | | |
| Sum of % of glass-formers $SiO_2$ + $B_2O_3$ + $P_2O_5$ | 100 | 99 | 94 | 90 | 82 | 72 | 67 | 31 | 22 |

* Letters refer to the source of the sample:

G—General Electric Co.
C—Corning Glass Works.
L—Owens-Illinois Glass Co.
H—Haverford Glass Co.
P—Pittsburgh Plate Glass Co.

TABLE 7.30

**Permeability Constant $K$ through Fused Silica**
(Norton)

| Gas | 700° C | 600° C |
|---|---|---|
| Helium | $2.1 \cdot 10^{-8}$ | ... |
| Hydrogen | $2.1 \cdot 10^{-9}$ | $1.25 \cdot 10^{-9}$ |
| Deuterium | $1.7 \cdot 10^{-9}$ | ... |
| Neon | $4.2 \cdot 10^{-10}$ | $2.8 \cdot 10^{-10}$ |
| Argon | *Under* $10^{-15}$ | ... |
| Oxygen* | *Under* $10^{-15}$ | ... |
| Nitrogen | *Under* $10^{-15}$ | ... |

* See also Ref. 157.

**Table 7.31**

**Hydrogen Permeation Constant $K$**
(Norton)

| Glass | 665° C |
|---|---|
| No. 1720 (Corning) | $4.5 \cdot 10^{-12}$ |
| Mullite | $1.8 \cdot 10^{-11}$ |

In order to show the selectivity of the permeability of silica towards a number of gases, Barrer has used the observations upon which the constants given in Table 7.28 are based to calculate the permeabilities of a number of gases at 900° C.[135] Values of $K$ and derived values of $Q_{\mu l}$ are given in Table 7.32. (It should be noted that the values for nitrogen and argon are probably too high; cf. Table 7.30.)

TABLE 7.32

**Values of the Permeabilities of Different Gases in Fused Quartz at 900° C**

| Gas | $10^9 K$ | $Q_{\mu l}$ |
|---|---|---|
| Helium | 36.2 | 7.52 |
| Hydrogen | 6.4 | 1.33 |
| Neon | 1.18 | 0.245 |
| Nitrogen | 0.95 | 0.198 |
| Argon | 0.58 | 0.120 |

These observations, of course, suggest the use of quartz for purifying helium, and Young and Whetten have described[136] a quartz-tube assembly for admitting helium to a vacuum system. With this device, they found that the only impurity remaining above one part per million was hydrogen.

Barrer has also compared the permeability of different materials for the same gas (helium) at around 300° C. These values and derived values of $Q_{\mu l}$ are given in Table 7.33.

**TABLE 7.33**

**Permeabilities of Different Glasses and Quartz for Helium at about 300° C**

| Material | Temperature (°C) | $10^9 K$ | $Q_{\mu l}$ |
|---|---|---|---|
| $SiO_2$ | 300 | 3.15 | 0.655 |
| Pyrex | 300 | 0.38 | 0.079 |
| Soda-glass | 283 | 0.0098 | 0.00204 |
| Lead-glass | 283 | 0.0037 | 0.00077 |

Van Voorhis[137] investigated the permeability of a number of glasses to helium. He observed that acidic oxides, such as boric oxide and silica, increased the permeability, whereas basic oxides decreased it approximately in proportion to their amount.

Norton[130] also found that the permeability increased with increased amounts of the "glass-forming" oxides, $SiO_2$, $B_2O_3$, and $P_2O_5$. He pointed out that this behavior could be accounted for by the filling of the vacancies of the amorphous glassy lattice by the small cations of the basic oxides added to the "glass-formers," with a resultant hindrance to diffusion when the composition includes large amounts of the basic oxides. Altemose[138] has reported measurements of helium permeation in several glasses and has shown that, if log (permeation rate) is plotted against mole per cent ($SiO_2 + B_2O_3 + P_2O_5$) (at a given temperature), a reasonably good straight line is obtained.

The results obtained by Roeser on the passage of air through porcelain varied for different samples, and his values of log $K$ versus $1/T$ do not give a satisfactory linear plot. Hence, the values for the permeability constants given in the last row of Table 7.28 should be considered as only approximate.

According to Russell and Stokes,[139] the diffusion of hydrogen through Vycor glass, in the range 700–975° C, is about the same as that of hydrogen through quartz, in the same range of temperatures. The rate of diffusion of oxygen was observed to be about one-twelfth that of hydrogen.

The mechanism of diffusion in glasses and silica has been discussed comprehensively by Barrer.[140] He states,

The diffusing gases can migrate according to two mechanisms. Helium, hydrogen, and neon can pass through the "lattice" of fused silica at high temperatures. The heavy gases, oxygen, nitrogen, and argon, migrate mainly through slip planes. At low temperatures, helium, hydrogen, and neon also show slip-plane diffusion. There is evidence that migration proceeds from the adsorbed layer, not directly from the gas phase.

Barrer has shown that the values observed for the energy of activation can be accounted for on the basis of the known laws of interaction of the rare gases, nitrogen, and hydrogen with glass and quartz.

A similar opinion regarding the mechanism of diffusion in glass is favored by Weyl.[141] He states,

As in chemical reactions the molecule jumps from place to place and the kinetics of this process can be treated from the point of view of reaction rates. The activation energy of the chemical process corresponds to the energy required to jump over an energy hill; that is, to squeeze through the electric fields of its neighbors. If the constituents of the solid exert strong forces upon the diffusing units, the activation energy is high and the process is slow. The chemical concept accounts for the fact that hydrogen diffuses much more slowly through the oxygen network of a glass than does helium because of the smaller "chemical affinity" for the oxygen ions.

Furthermore, it is possible on this basis to interpret the observations made by Smith and Taylor[142] on the diffusion of a gas through glasses of the same composition but different heat treatment. According to Weyl, "A glass which has been chilled from a high temperature can be considered an unstable modification having a higher energy content. The higher chemical affinity of such a chilled glass increases the activation energy of the diffusing units and slows down the diffusion rate." It has also been observed that, "with the exception of lead-containing glass, all showed a marked decrease in the diffusion after chilling."

The experiments indicate that chilling exerts two antagonistic effects upon the diffusion of gases through a glass:

*a.* The more spacious network of the rapidly chilled glass (lower density) enables gas atoms or molecules to flow faster than through the more "compacted" well-annealed glass.

*b.* The chemical forces of a chilled glass are stronger, or the valence forces of the ions are less "saturated," than those of the annealed glass. The stronger forces delay diffusion by increasing the activation energy, or raise the energy barriers over which the diffusing atoms have to pass.

For soda-lime silicate and borosilicate glasses the second factor apparently dominates the diffusion process and as a result the diffusion is faster in the annealed glass than in the chilled one.

Finally, it is important to note that the permeability is actually proportional to the diffusion constant of the gas in the material and to the solubility of the gas, in accordance with the relation

$$K = Dc \frac{\Delta P}{d}, \tag{7.18}$$

where $D$ = diffusion constant,

$c$ = solubility,

$\Delta P/d$ = pressure gradient across the thickness.

The solubility of hydrogen and helium in fused silica is approximately 0.01 cm³ (STP) per cm³ at 1 atm and temperatures above about 700° C. Substituting this value in equation 7.18, it is possible, from the observed values of $K$, to derive values of $D$. Thus, for helium in fused silica, the value of $D$ at 500° C lies between $10^{-7}$ and $10^{-8}$ cm² · sec⁻¹. Since the value of $c$ may be assumed to be constant over a considerable range of temperature, it follows that in the relation

$$D = D_0 \epsilon^{-E/R_0 T} \tag{7.19}$$

the values of $E$ are approximately the same as those derived from the temperature variation in $K$.

## 7.9. SORPTION AND DESORPTION PHENOMENA IN CELLULOSE[143]

Occasionally the problem arises in vacuum technique of evacuating a device in which a cellulosic material such as cotton, paper, or wood is present. The principal difficulty encountered in that connection is due to the fact that water vapor, as well as carbon dioxide and many organic compounds (especially those containing a hydroxyl or —OH group), is strongly adsorbed by cellulose.

Cellulose[144] has the chemical composition $C_6H_{10}O_5$. Its avidity for water is shown in the formation of a hydrate, $(C_6H_{10}O_5)_2H_2O$. When heated to a temperature above about 130° C in vacuum, cellulose begins to decompose at an appreciable rate with consequent loss of mechanical and electrical strength. It is therefore impractical, in general, to evacuate cellulose materials at temperatures in excess of the above value.

Like the sorption and desorption isotherms for silica gel and other adsorbents mentioned in the previous sections, the isotherms for water and cellulose present a "hysteresis" loop. Figure 7.25 shows plots of observations made by Filby and Maass,[145] on standard cellulose at 20° C, in which percentage moisture is plotted against percentage relative humidity.

Table 7.34, taken from a publication by Sheppard and Newsome,[146] gives typical data for the sorption by cotton cellulose at 23° C. Before

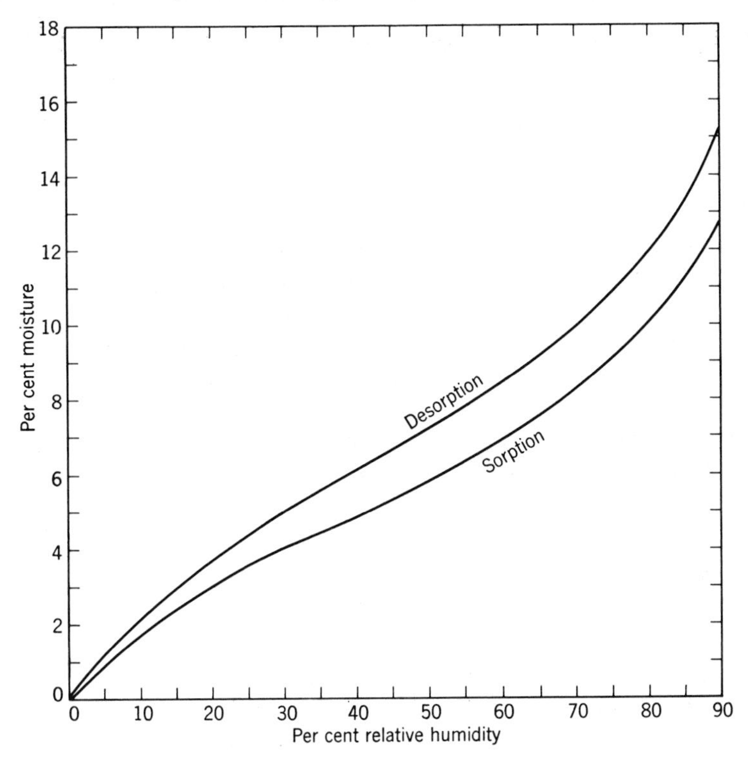


Fig. 7.25. Sorption and desorption isotherm for water vapor and standard cellulose at 20° C (Filby and Maass).

these observations, the samples had been evacuated and dried at 105° C. The three sets of data refer to results published by different investigators.

Of special significance in connection with vacuum technique are the observations made by Neale and Stringfellow.[147] They determined the amounts of water taken up by cotton *at low pressures*, for a series of temperatures. Table 7.35 summarizes their results, which were obtained with samples of yarn that had been evacuated at 90° C for several hours. A definite amount of water was then admitted to the cotton, the temperature raised to the desired value, and the pressure read periodically until a steady value was obtained. "Steady readings were obtained within an hour or two after a change of temperature." The system was re-evacuated with the cotton at 90° C before the next admission of water.

The last line gives the vapor pressure of water, at each temperature, in millimeters of mercury, from which the equilibrium pressure may be derived in terms of relative humidity. For instance, an equilibrium

TABLE 7.34

Sorption of Water Vapor by Cellulose Cotton at 23° C

| Relative Humidity | $P_{mm}$ of $H_2O$ Vapor | Sorption: Milligrams $H_2O$ per Gram Cotton | | |
|---|---|---|---|---|
| 0.05 | 1.054 | 14.4 | 10.4 | 12.5 |
| 0.10 | 2.107 | 20.4 | 17.1 | 19.0 |
| 0.15 | 3.161 | 24.4 | 21.6 | 23.2 |
| 0.20 | 4.214 | 28.4 | 25.8 | 26.4 |
| 0.25 | 5.268 | 33.4 | 29.8 | 29.6 |
| 0.30 | 6.321 | 37.4 | 33.6 | 32.8 |
| 0.40 | 8.428 | 45.8 | 40.8 | 39.4 |
| 0.50 | 10.54 | 54.4 | 48.1 | 46.0 |
| 0.60 | 12.64 | 63.8 | 56.9 | 53.3 |
| 0.70 | 14.75 | 75.8 | 66.5 | 62.0 |
| 0.80 | 16.86 | 93.8 | 81.5 | 71.6 |
| 0.90 | 18.96 | 125.6 | 115.4 | 84.0 |
| 1.00 | 21.07 | | | |

TABLE 7.35

Equilibrium Pressures for Different Amounts of Water Adsorbed by Cotton at a Series of Temperatures

| Water Adsorbed (mg/g) | Pressure (microns) 20° C | 30 | 40 | 50 | 60 | 70 | 80 |
|---|---|---|---|---|---|---|---|
| 0.120 | (1.08) | 2.00 | 4.33 | 9.00 | 19.9 | 39.0 | 77.0 |
| 0.234 | 1.65 | 3.60 | 8.5 | 19.2 | 40.8 | 81 | 157 |
| 0.517 | 4.06 | 10.3 | 23.4 | 51 | 107.5 | 219 | 440 |
| 0.795 | 7.65 | 17.8 | 40.8 | 88 | 185 | 380 | 725 |
| 1.060 | 11.9 | 27.8 | 62.5 | 139 | 306 | 586 | ... |
| 1.595 | 20.15 | 47.0 | 111.4 | 250 | 510 | (880) | ... |
| Vapor pressure of water (mm) | 17.54 | 31.82 | 55.32 | 92.5 | 149.4 | 233.7 | 355.1 |

pressure of 250 microns at 50° C corresponds to $250 \cdot 10^{-3} \cdot 100/92.5 = 0.27$ per cent relative humidity.

Figure 7.26 shows plots of these data, and from them we conclude that, if a device is sealed off under such conditions that the pressure of water vapor at 30° C is 17.8 microns, the amount of water retained in the cotton is 0.795 mg/g.

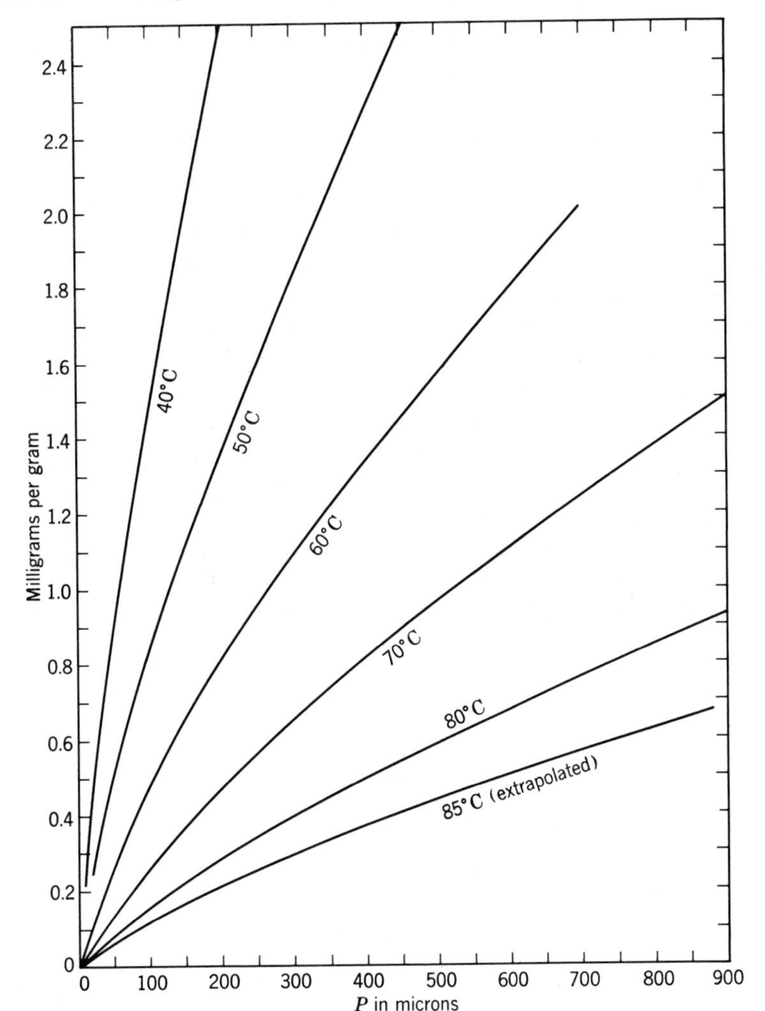


Fig. 7.26. Sorption isotherms for water vapor by cotton at a series of temperatures (Neale and Stringfellow).

Neale and Stringfellow also showed that, for the isosteres (curves for constant amount adsorbed), values of log $P$ plotted against $1/T$ yield straight lines with the same slope, independent of the amount sorbed, corresponding to a heat of adsorption, $Q = 15{,}700$ cal/mole.[148] That is, for *constant amount adsorbed*,

$$\log P_t = \log P_{70} + \frac{15{,}700}{4.574}\left(\frac{1}{343} - \frac{1}{T}\right),$$

where the value of $P$ at 70° C, for the given amount of sorption, is obtained from the plot of Fig. 7.26, and $P_t$ is the equilibrium pressure at $t$° C.

Extrapolating, by means of the last equation, to higher temperatures, we obtain the sorption isotherm for 85° C, plotted in Fig. 7.26, and the isotherms for higher temperatures shown in Fig. 7.27. Assuming that for small amounts of water adsorbed there is a linear variation with pressure, it follows that, for a vapor pressure of 1 mm at, for instance, 140° C, the amount of water retained by the cotton would be 0.05 mg/g.

It is interesting to compare these sorption data for cotton, extrapolating from observations at temperatures below 100° C, with observations made by Houtz and McLean[149] on the sorption of water vapor by paper, at temperatures ranging from 100° C to 150° C. The samples were weighed in the dry condition after being heated at 100° C, in vacuum, for at least 8 hr. Figure 7.28 shows the isotherms, obtained with linen rag paper, in which the amount sorbed (milligrams per 100 g paper) is plotted against the pressure of water vapor.

Observations, made under similar conditions, with wood pulp gave somewhat higher values for the amount sorbed. Table 7.36 gives data obtained at relatively low vapor pressures for the two samples of paper

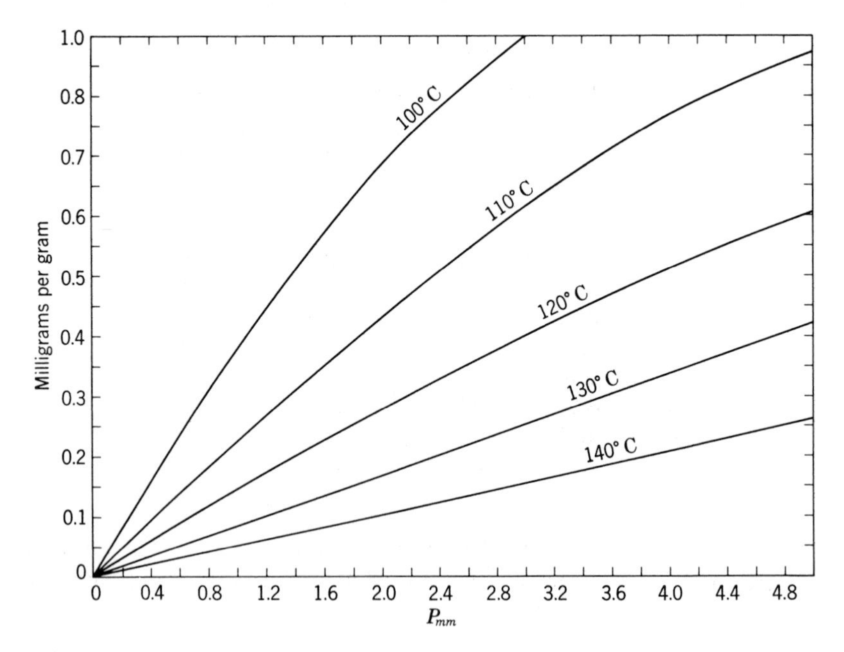


Fig. 7.27. Sorption isotherms for water vapor by cotton, extrapolated to higher temperatures from the isotherms in Fig. 7.26.

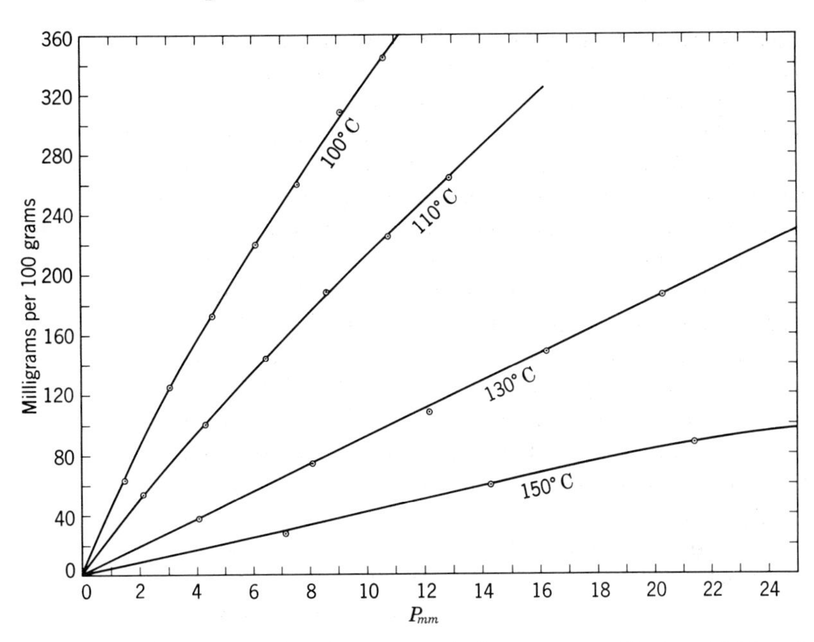


Fig. 7.28. Isotherms for sorption of water vapor by linen rag paper (Houtz and McLean).

**TABLE 7.36**

**Sorption of Water by Paper and Cotton in the Range 100–150° C**

| $t°$ C | $P_{mm}$ | Milligrams per Gram | | |
|---|---|---|---|---|
| | | Linen Rag | Wood Pulp | Cotton |
| 100 | 1.52 | 0.63 | 0.92 | 0.55 |
| | 3.04 | 1.25 | 1.60 | 0.91 |
| 110 | 2.15 | 0.54 | 0.77 | 0.46 |
| | 4.30 | 1.00 | 1.32 | 0.80 |
| 120 | 2.98 | ... | 0.72 | 0.40 |
| | 5.96 | ... | 1.20 | ... |
| 130 | 4.05 | 0.38 | 0.56 | 0.34 |
| 140 | 4.05 | ... | ... | 0.21 |
| 150 | 7.14 | 0.28 | 0.40 | ... |

and, for comparison, values taken from the plots in Fig. 7.27 for cotton.

Houtz and McLean found that their isotherms could be represented by the Freundlich relation

$$\log x = \log A + \frac{1}{n} \log P, \tag{7.20}$$

where $x$ = milligrams of $H_2O$ per gram of paper adsorbed at pressure $P$, and $A$ and $1/n$ are constants. The values of $1/n$ varied from 0.99 to 0.767, depending on the temperature and on whether linen or wood paper was used.

In a subsequent publication[150] the same investigators observed that by heating the paper at 150° C during evacuation the sorption was reduced to a reproducible minimum, which varied from 85 to 70 per cent of the amount sorbed under similar conditions of temperature and pressure by the same material heated at 100° C. Variations obtained in the values of $1/n$ were similar to those in the earlier series of measurements.

Also a relation was observed for the isosteres similar to that observed by Neale and Stringfellow. From these linear plots of log $P$ versus $1/T$ for constant values of $x$, it was concluded that the heat of adsorption $Q$ (in calories per mole $H_2O$) decreased with $x$ as follows:

| $x$ | $Q$ |
|---|---|
| 0.25 | 15,800 |
| 0.50 | 15,300 |
| 1.00 | 14,800 |
| 2.00 | 14,300 |
| 3.50 | 14,000 |

It will be observed that the highest value observed for $Q$ in this investigation is in remarkable agreement with the value 15,700 deduced by Neale and Stringfellow. The results in these investigations are typical of those obtained by other observers.[151]

In an investigation carried out in the General Electric Research Laboratory by A. H. Young and Dushman, on the rate and extent of desorption of samples of fiber and wood pulp, the following results were obtained. For samples evacuated at both room temperature and 140° C, the rate of decrease in weight $w$ was observed to follow the first-order equation

$$-\frac{dw}{dt} = k(w - w_0), \tag{7.21}$$

where $w_0$ = weight at $t = \infty$, and $k$ denotes the "rate constant." That is, a plot of log $(w - w_0)$ versus $t$ yields a straight line, the slope of which is proportional to $k$. At room temperature, the total time required to remove 90 per cent of the adsorbed water from an approximately 1 g sample of fiber was of the order of 8 hr, thus indicating that the rate of desorption was governed by the rate of diffusion of water vapor through the pores or capillaries.

The percentage loss by weight during evacuation for a typical sample was as follows:

| | |
|---|---|
| At 27° C | About 2.5 per cent |
| Between 27 and 140° C | About 2.5 per cent |
| At 140° C | About 1.2 per cent |
| Total = | 6.2 per cent |

This percentage corresponds to that sorbed by cotton, at room temperature, in an atmosphere of 60 per cent relative humidity.

In the case of wood pulp, the observed rate of desorption, for the same size sample as that of fiber or paper, was much faster; and of a total loss in weight, on evacuation, of approximately 5 per cent, 4.5 per cent was obtained at room temperature.

When desorbed samples of fiber and wood pulp were exposed to an atmosphere containing 30 per cent relative humidity at room temperature, it was observed that the rates of sorption were so slow that even after several hours the gain in weight was only about 2 per cent. Even after 24 hr, the amount of water adsorbed was less than that given off during desorption.

Of special importance in the evacuation of cellulose is the fact that the material undergoes thermal decomposition at a noticeable rate at 100° C and at a much higher rate as the temperature is increased above 100° C. The following discussion is based on the data published by Murphy.[152]

The gases evolved during thermal decomposition of cellulose and their relative proportions are given by the approximate molar ratios: $10H_2O : 2CO_2 : 1CO$. As Murphy has stated,

It is difficult to determine when paper is dry. As the drying process progresses, the rate at which water vapor comes off steadily decreases, but water vapor continues to come off, however long the drying is continued. Raising the temperature increases the rate of drying, but it also increases the rate of thermal decomposition. Water vapor coming from the thermal decomposition of cellulose is not distinguishable directly from water vapor coming from the adsorbed state. Thus the fact that water vapor continues to come off after protracted drying at any temperature above about 100° C does not mean that the paper can be dried further. It is not feasible to reach a condition where no more water vapor is evolved by paper.

Dryness, or the lowest possible concentration of adsorbed water, can be recognized when water vapor comes off in association with $CO_2$ and CO in proportions appropriate to the temperature of the paper. At temperatures of the order of 130° C, paper which is yielding 2 or 3 molecules of water in association with about 1 molecule of $CO_2$ and 1 of CO (or some proportions of this order) would be accounted dry. The degree of dryness could not be increased at this temperature without changing the composition of the paper. Thus the composition of the gas evolved can be used to determine an "end point" in the drying operation.

A plot of the rate of gas evolution versus $1/T$ on semilog paper yields a straight line. For *kraft paper*, the relation between rate of gas evolution in *cubic millimeters* (at normal temperature and pressure) *per hour per gram* (denoted by $q$) is given by the relation, deduced from the plot,

$$\log q = 18.24 - \frac{7424}{T},$$

corresponding to the activation energy, $Q = 33{,}950$ cal/mole.

Assuming that 70 per cent by volume is $H_2O$, this rate corresponds in mass to

$$m = 8.050 \cdot 10^{-4} \cdot 0.7q \text{ mg } H_2O/\text{hr/g}$$
$$= 5.635 \cdot 10^{-4} q \text{ mg } H_2O/\text{hr/g}.$$

Hence

$$\log m = 14.751 - \frac{7424}{T}.$$

From the last equation we obtain the values for $m$, in *milligrams water per hour per gram of paper*, given in Table 7.37. For linen paper, the rate

**TABLE 7.37**
**Rate of Evolution of Water from Paper, for a Series of Temperatures**
(Murphy)

| $t°$ C | $10^6 m$ | $t°$ C | $10^4 m$ |
|---|---|---|---|
| 80 | 0.537 | 120 | 0.741 |
| 90 | 2.00 | 130 | 2.14 |
| 100 | 7.08 | 140 | 6.03 |
| 110 | 24.0 | | |

of decomposition, according to Murphy, is about 55 per cent of that for kraft paper.

It has also been observed by Murphy that paper reacts slowly with oxygen even at as low a temperature as 80° C. After exposing paper at 135° C to air at atmospheric pressure for 1 hr the subsequent rate of evolution of gas in vacuum was increased considerably.

The original report also contains a discussion of the effect of increased temperatures of desorption on the mechanical properties of paper.

## 7.10. POLYMERS; SORPTION AND PERMEABILITY

With the increasing technological importance of polymers, particularly synthetic materials, it is appropriate to make a few comments about the sorption and permeability behavior of such materials.

Relatively few studies have been made on *adsorption* by polymers. In fact, because of the comparatively high solubility and diffusion rates for many gases in most such materials (see below), it is a matter of some experimental difficulty to determine what part of the gas taken up by a sample of a polymer is really retained on the surface.

Zettlemoyer, Chand, and Gamble[153] have reported low-temperature nitrogen and krypton adsorption measurements on polyethylene, nylon, and collagen. Their results show typical type II isotherms with no unusual or specific effects.

On the other hand, a large amount of work has been done on rates of permeation of polymers by various gases. Reports of such studies are widely scattered in the literature. As Norton has stated.[154]

There seems to be one generalization in this field, and that is that all gases, including the rare gases, will permeate all polymers to some degree. There are wide variations in the rates and some strange specificities. An added complication is that as temperature is raised the polymers may evolve solvents, or pyrolytic break-up of the polymer itself will occur, greatly complicating the permeation picture.

It has been mentioned that permeability is proportional to a diffusion factor and a solubility factor. This solubility factor is probably involved in the high permeability of cellulose polymers to water and their low hydrogen permeability. So also may be the case for polythene and its rate for benzene, which is 10,000 times greater than the permeation rate for nitrogen.

Figure 7.29 illustrates some typical data; this figure, together with many other useful facts on this topic, is to be found in Barrer's treatise.[155]

In Figure 7.30, taken from a paper by Norton,[154] some permeability constants for polymers are compared with values for glass and metal systems.

**TABLE 7.38**

**Water-Vapor Permeability of Polymers at 25° C**

(Doty, Aiken, and Mark)

| Polymer | Permeability ($cm^3/cm^2/sec/mm/cm\ Hg \cdot 10^8$) |
|---|---|
| Cellophane | 4.7 |
| Koroseal | 5.6 |
| Nylon | 3.7–260 |
| Pliofilm | 1.3 |
| Polyethylene | 2.1 |
| Saran | 0.1 |
| Ethyl cellulose | 510 |

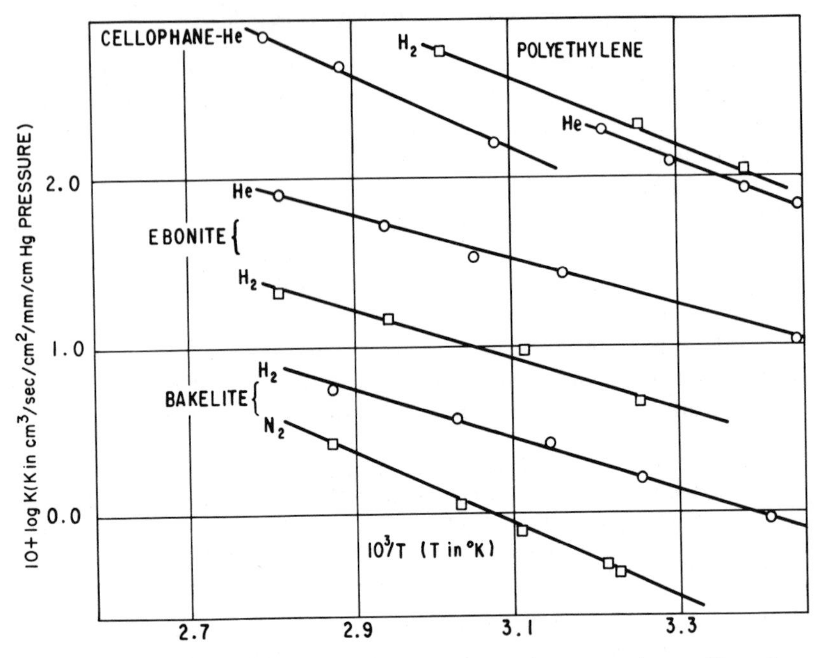


Fig. 7.29. Temperature dependence of permeability for some polymers (Barrer).

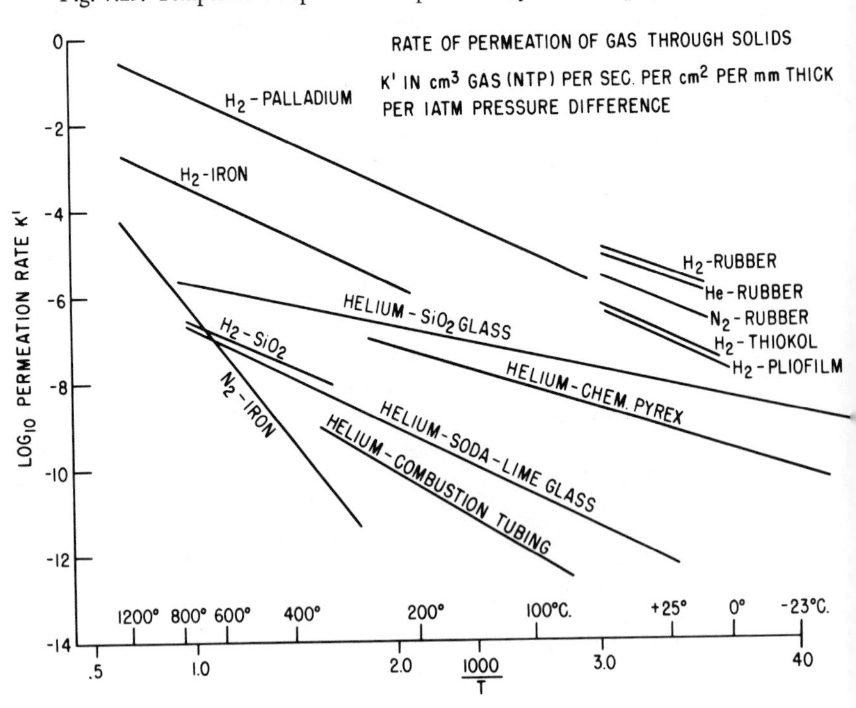


Fig. 7.30. Permeation rates for several systems (Norton).

Some values for water-vapor permeability of typical polymers are given in Table 7.38. These data are from the study of water permeation by Doty, Aiken, and Mark,[156] who have pointed out that wide variations are observed even between different samples of the same polymer.

## REFERENCES AND NOTES


1. V. R. Deitz, Editor, Vol. 1, 1900–1942, Bone Char Research Project, Inc., 1944; Vol. 2, 1943–1953, *Nat. Bur. Standards Cir. No.* 566, 1956.
2. H. Baerwald, *Ann. Physik*, **23,** 84 (1907).
3. F. Bergter, *Ann. Physik*, **37,** 472 (1912).
4. M. W. Travers, *Proc. Roy. Soc. London*, **A, 78,** 9 (1907).
5. J. W. McBain, *Phil. Mag.*, **18,** 916 (1909).
6. I. F. Homfray, *Z. physik. Chem.*, **74,** 129 (1910).
7. A. Titoff, *Z. physik. Chem.*, **74,** 641 (1910).
8. G. Claude, *Compt. rend.*, **158,** 861 (1914).
9. H. B. Lemon, *Phys. Rev.*, **14,** 281 (1919).
10. The results of the investigations on this subject carried out by the Chemical Warfare Service, U.S. Army, have been published mainly in the following papers:
A. B. Lamb, R. E. Wilson, and N. K. Chaney, *J. Ind. Eng. Chem.*, **11,** 420 (1919), on gas-mask absorbents.
F. M. Dorsey, *J. Ind. Eng. Chem.*, **11,** 281 (1919), on the development of activated charcoal.
N. K. Chaney, *Trans. Am. Electrochem. Soc.*, **36,** 91 (1919).
11. J. J. Kipling, *Quart. Revs.* (*London*), **10,** 1 (1956).
12. I. Langmuir, *J. Am. Chem. Soc.*, **38,** 2221 (1916).
13. H. Briggs, *Proc. Roy. Soc. London*, **A, 100,** 88 (1921–1922).
14. Cf. the review by M. M. Dubinin, *Quart. Revs.* (*London*), **9,** 101 (1955).
15. A. J. Juhola and E. O. Wiig, *J. Am. Chem. Soc.*, **71,** 2078 (1949).
16. C. Pierce, R. N. Smith, J. W. Wiley, and H. Cordes, *J. Am. Chem. Soc.*, **73,** 4551 (1951).
17. J. R. Dacey and D. G. Thomas, *Trans. Faraday Soc.*, **50,** 740 (1954).
18. H. H. Sheldon, *Phys. Rev.*, **16,** 165 (1920).
19. J. C. Philip, S. Dunnill, and O. Workman, *Trans. Chem. Soc. London*, **117,** 362 (1920).
20. A. Magnus and H. Kratz, *Z. anorg. Chem.*, **184,** 241 (1929).
21. B. Bruns and A. Frumkin, *Z. physik. Chem.*, **A, 141,** 141 (1929).
22. M. E. Barker, *J. Ind. Eng. Chem.*, **22,** 926 (1930).
23. See remarks, J. W. McBain, *The Sorption of Gases and Vapours by Solids*, G. Routledge and Sons, London, 1932, Chapter 4, p. 58.
24. H. Briggs, *Proc. Roy. Soc. London*, **A, 100,** 58 (1921); McBain (Ref. 23), p. 79.
25. K. Peters and K. Weil, *Z. physik. Chem.*, **A, 148,** 1 (1930).
26. I. F. Homfray, *Z. physik. Chem.*, **74,** 129 (1910).
27. A. Titoff, *Z. physik. Chem.*, **74,** 641 (1910).
28. L. B. Richardson, *J. Am. Chem. Soc.*, **39,** 1829 (1917).
29. S. Brunauer and P. H. Emmett, *J. Am. Chem. Soc.*, **59,** 2682 (1937).
30. H. B. Lemon and K. Blodgett, *Phys. Rev.*, **14,** 394 (1919).
31. It is of interest to observe in this connection that the selective sorption of different gases by charcoal may be used for the purification of gases. Thus, where it is desirable to obtain pure helium, charcoal in liquid air may be used to remove

other gases such as nitrogen, oxygen, and even hydrogen. Similarly hydrogen may be freed of traces of nitrogen and oxygen. See, for instance, R. E. Wilson, *Phys. Rev.*, **16**, 8 (1920).

32. A. Lendle, *Z. physik. Chem.*, **A, 172,** 77 (1935).
33. H. S. Taylor, *J. Am. Chem. Soc.*, **53,** 578 (1931).
34. H. H. Lowry and G. A. Hulett, *J. Am. Chem. Soc.*, **42,** 1408 (1920). This paper contains also a large number of references to previous literature on this topic.
35. 1 mg $H_2O = 3.37 \cdot 10^{19}$ molecules.
36. A. S. Coolidge, *J. Am. Chem. Soc.*, **49,** 708 (1927).
37. See, for example, A. Blackburn and J. J. Kipling, *J. Chem. Soc.* (*London*), **1955,** 4103.
38. R. N. Smith, *Quart. Revs.* (*London*), **13,** 287 (1959).
39. H. H. Lowry and G. A. Hulett, *J. Am. Chem. Soc.*, **42,** 1393 (1920).
40. P. G. T. Hand and D. O. Shiels, *J. Phys. Chem.*, **32,** 441 (1929); A. J. Allmand, R. Chaplin, and D. O. Shiels, *J. Phys. Chem.*, **33,** 1151 (1929); A. J. Allmand and P. G. T. Hand, *J. Phys. Chem.*, **33,** 1161 (1929); A. J. Allmand, P. G. T. Hand, J. E. Manning, and D. O. Shiels, *J. Phys. Chem.*, **33,** 1682 (1929); A. J. Allmand, P. G. T. Hand, and J. E. Manning, *J. Phys. Chem.*, **33,** 1694 (1929).
41. J. W. McBain, *Z. physik. Chem.*, **68,** 471 (1909); *Phil. Mag.*, **18,** 916 (1909).
42. J. B. Firth, *Z. physik. Chem.*, **86,** 294 (1914).
43. F. A. P. Maggs, P. H. Schwabe, and J. H. Williams, *Nature*, **186,** 956 (1960).
44. A. Magnus and H. Kratz, *Z. anorg. Chem.*, **184,** 241 (1929).
45. H. H. Lowry and S. O. Morgan, *J. Phys. Chem.*, **29,** 1105 (1925).
46. M. H. Polley, W. D. Schaeffer, and W. R. Smith, *J. Phys. Chem.*, **57,** 469 (1953).
47. P. H. Emmett and R. B. Anderson, *J. Am. Chem. Soc.*, **67,** 1492 (1945).
48. J. Dewar, *Proc. Roy. Inst. Gt. Brit.*, **18,** 751 (1907).
49. G. Claude, *Compt. rend.*, **158,** 861 (1914).
50. H. Rowe, *Phil. Mag.*, **1,** 109, 1042 (1926).
51. J. deD. Lopez-Gonzalez, F. G. Carpenter, and V. R. Deitz, *J. Research Natl. Bur. Standards*, **55,** 11 (1955).
52. J. W. Woodrow, *Phys. Rev.*, **4,** 491 (1914).
53. O. Winkler, *Z. tech. Physik*, **14,** 319 (1933).
54. R. H. Savage, *J. Appl. Phys.*, **19,** 1 (1948). The quotation is from a paper by Savage and J. R. C. Brown, *J. Am. Chem. Soc.*, **70,** 2362 (1948).
55. R. K. Iler, *The Colloid Chemistry of Silica and Silicates*, Cornell University Press, Ithaca, N.Y., 1955.
56. R. S. McDonald, *J. Phys. Chem.*, **62,** 1168 (1958).
57. S. J. Gregg, *The Adsorption of Gases by Solids*, D. Van Nostrand Company, Princeton, N.J., 1934, p. 11.
58. McBain (Ref. 23), p. 196, Fig. 69.
59. W. A. Patrick, W. C. Preston, and A. E. Owens, *J. Phys. Chem.*, **29,** 421 (1925).
60. A. Magnus and R. Kieffer, *Z. anorg. allgem. Chem.*, **179,** 215 (1929); A. Magnus and H. Kratz, *Z. anorg. allgem. Chem.*, **184,** 241 (1929).
61. W. Kälberer and C. Schuster, *Z. physik. Chem.*, **A, 141,** 270 (1929).
62. Sorption data for oxygen on silica gel at −49 to −69° C are given by A. Magnus and K. Grähling, *Z. physik. Chem.*, **A, 145,** 27 (1929).
63. A. S. Coolidge, *J. Am. Chem. Soc.*, **49,** 708 (1927).
64. L. M. Pidgeon, *Can. J. Research*, **10,** 713 (1934).
65. K. S. Rao, *J. Phys. Chem.*, **45,** 513 (1941).
66. E. P. Barrett, L. G. Joyner, and P. P. Halenda, *J. Am. Chem. Soc.*, **73,** 373 (1951).
67. G. J. Young, *J. Colloid Sci.*, **13,** 67 (1958).

68. R. M. Barrer, *Proc. Roy. Soc. London*, **A, 167,** 392 (1938).

69. Gregg (Ref. 57), p. 12.

70. S. Brunauer, *The Adsorption of Gases and Vapors*, Vol. I, *Physical Adsorption*, Princeton University Press, 1945, p. 366.

71. A. B. Lamb and J. C. Woodhouse, *J. Am. Chem. Soc.*, **58,** 2637 (1936).

72. R. M. Barrer, *Proc. Roy. Soc.* (*London*), **A, 167,** 392 (1938).

73. "The dehydration of chabasite was carried out *in vacuo* at progressively increasing temperatures for several days. The temperature was finally raised to 480° C for 1 day." In the case of analcite ($Na_8Al_8Si_{12}O_{48} \cdot 12H_2O$) dehydration *in vacuo* was carried out at 330° C for 3 days.

74 O. Schmidt, *Z. physik. Chem.*, **133,** 263 (1928); also Mc Bain (Ref. 23), pp. 166–176, which contains a comprehensive discussion of this topic.

75. R. M. Barrer and D. A. Ibbitson, *Trans. Faraday Soc.*, **40,** 195 (1944).

76. G. L. Kington and W. Laing, *Trans. Faraday Soc.*, **51,** 287 (1955).

77. R. M. Barrer and D. A. Ibbitson, *Trans. Faraday Soc.*, **40,** 206 (1944).

78. According to L. Pauling, *The Nature of the Chemical Bond*, Cornell University Press, Ithaca, N.Y., 1940, p. 189.

79. These systems are discussed in Chapter 8.

80. See, for example, R. M. Barrer, *Trans. Faraday Soc.*, **40,** 555 (1944) (gmelinite and mordenite); and R. M. Barrer, F. W. Bultitude, and J. W. Sutherland, *Trans. Faraday Soc.*, **53,** 1111 (1957) (faujasite).

81. See D. W. Breck, W. G. Eversole, R. M. Milton, T. B. Reed, and T. L. Thomas, *J. Am. Chem. Soc.*, **78,** 5963 (1956).

82. See, for example, M. A. Biondi, *Rev. Sci. Instr.*, **30,** 831 (1959), and section 4.4.

83. P. Cannon, *J. Phys. Chem.*, **63,** 160 (1959).

84. P. Cannon and C. P. Rutkowski, *J. Phys. Chem.*, **63,** 1292 (1959).

85. P. Cannon, *J. Am. Chem. Soc.*, **80,** 1766 (1958).

86. L. J. Briggs, *J. Phys. Chem.*, **9,** 617 (1905).

87. J. R. Katz, *Proc. Amsterdam Acad.*, **15,** 445 (1912). The quotation is from Langmuir's paper, *J. Am. Chem. Soc.*, **38,** 2283 (1916).

88. W. G. Palmer and R. E. D. Clark, *Proc. Roy. Soc. London*, **A, 149,** 360 (1935).

89. W. G. Palmer, *Proc. Roy. Soc. London*, **A, 160,** 254 (1937).

90. A comprehensive review of this topic is given in the treatise entitled *The Properties of Glass*, by G. W. Morey, Reinhold Publishing Corporation, New York, 1938, pp. 89–101.

91. I. Langmuir, *Trans. Am. Inst. Elec. Engrs.*, **32,** 1921 (1913), and *J. Am. Chem. Soc.*, **38,** 2221 (1916).

92. These values are to be compared with those given in Table 6.13 for the number of molecules ($1/S_l$) required to form a monolayer. These are $0.95 \cdot 10^{15}$ for water vapor, $0.71 \cdot 10^{15}$ for carbon dioxide, and $0.73 \cdot 10^{15}$ for nitrogen.

93. I. Langmuir, *J. Am. Chem. Soc.*, **40,** 1361 (1918).

94. These values are, of course, based on Langmuir's values for the number of molecules required to form a monolayer.

95. U.S. Patent 1,273,629, July 23, 1918.

96. R. G. Sherwood, *J. Am. Chem. Soc.*, **40,** 1645 (1918); *Phys. Rev.*, **12,** 448 (1918).

97. J. E. Shrader, *Phys. Rev.*, **13,** 434 (1919). See also abstract of a paper by D. Ulrey, *Phys. Rev.*, **14,** 160 (1919), which discusses the same subject.

98 S. Dushman, *Phys. Rev.*, **5,** 212 (1915).

99. In *Rev. Sci. Instr.*, **18,** 856 (1947), J. Rothstein has drawn attention to the use of silver chloride as a temperature indicator for hand torching in high-vacuum work. Some silver chloride is fused to the glass to form a thin layer. "When solid, the silver

chloride has a dull, opaque, whitish matte appearance, which becomes successively dark, shiny, and transparent when fusion occurs" at 455° C.

It is of interest to mention in this connection that in the production of vacuum incandescent tungsten lamps certain organic compounds were frequently coated on the button of the glass stem supporting the filament to indicate by the change in color the maximum temperature attained inside the bulb during bake-out on the pump. The composition of the coating used varied with the range of temperatures to be indicated.

100. J. E. Harris and E. E. Schumacher, *J. Ind. Eng. Chem.*, **15,** 174 (1923).

101. Personal communication from Dr. L. Navias of the General Electric Research Laboratory.

102. Italicized by Dushman.

103. We are indebted for this information to Dr. J. T. Littleton, Director of the Research Laboratory of the Corning Glass Works, and Dr. B. J. Todd, who prepared the summary and the plots shown in Figs. 7.19, 7.20, and 7.21.

104. According to L. Navias and R. L. Green, *J. Am. Ceram. Soc.*, **29,** 267 (1946), this glass contains 3 per cent $B_2O_3$ and a "very low" content of $Na_2O$.

105. 1 mm · cm³ = 1 micron · liter.

$1 \text{ mm} \cdot \text{cm}^3/10^2 \text{ cm}^2 = 0.01 \text{ micron} \cdot \text{liter/cm}^2$.

106. G. W. Morey (Ref. 90).

107. P. Niggli, *J. Am. Chem. Soc.*, **35,** 1693 (1913).

108. E. W. Washburn, F. F. Footitt, and E. N. Bunting, *Univ. Illinois Eng. Expt. Sta. Bull.* 118 (1920).

109. R. H. Dalton, *J. Am. Ceram. Soc.*, **16,** 425 (1933); *J. Am. Chem. Soc.*, **57,** 2150 (1935). G. W. Morey (Ref. 90), p. 100.

110. C. Hahner, G. Q. Voigt, and A. N. Finn, *J. Research Natl. Bur. Standards*, **19,** 95 (1937).

111. R. W. Douglas, *J. Sci. Instr.*, **22,** 81 (1945).

112. B. J. Todd, *J. Appl. Phys.*, **26,** 1238 (1955).

113. B. J. Todd, *J. Appl. Phys.*, **27,** 1209 (1956).

114. H. Gentsch, *Z. physik. Chem.*, N.F., **24,** 55 (1960).

115. I. Langmuir, *J. Am. Chem. Soc.*, **40,** 1361 (1918).

116. H. Zeise, *Z. physik. Chem.*, **136,** 385 (1928).

117. F. Durau, *Z. Physik*, **37,** 419 (1926).

118. See description of this method in Brunauer (Ref. 70), p. 277, and McBain (Ref. 23), p. 334.

119. I. R. McHaffie and S. Lehner, *J. Chem. Soc.*, **127,** 1559 (1925); *J. Phys. Chem.*, **31,** 719 (1927), see Brunauer, (Ref. 70), p. 320, for detailed discussion.

120. J. C. W. Frazer, W. A. Patrick, and H. E. Smith, *J. Phys. Chem.*, **31,** 897 (1927).

121. See E. O. Kraemer, *Advances in Colloid Science*, Interscience Publishers, New York, 1942, Chapter 1.

122. H. S. Frank, *J. Phys. Chem.*, **33,** 970 (1929).

123. J. H. Frazer, *Phys. Rev.*, **33,** 97 (1929).

124. The values of $n$ thus deduced are less than those given by Frank because of the different value of molecular cross section used by Dushman.

125. J. B. Thomson, E. R. Washburn, and L. A. Guildner, *J. Phys. Chem.*, **56,** 979 (1952).

126 R. I. Razouk and A. S. Salem, *J. Phys. & Colloid Chem.*, **52,** 1208 (1948).

127. H. C. Hamaker, H. Bruining, and A. H. W. Aten, Jr., *Philips Research Repts.*, **2,** 171 (1947).

128. The discussion in this section is based largely on the comprehensive review of the topic by R. M. Barrer, *Diffusion In and Through Solids*, The Macmillan Company,

New York, 1941, 117–143. See also discussion of diffusion of gases in metals in section 8.10.

129. In this discussion it is assumed that $P = 0$ on one side of the wall through which diffusion occurs. Otherwise $P$ in the following equations should be replaced by $\triangle P$, the pressure difference.

130. F. J. Norton, *J. Am. Ceram. Soc.*, **36,** 90 (1953).

131. C. C. Leiby, Jr., and C. L. Chen, *J. Appl. Phys.*, **31,** 268 (1960).

132. D. Alpert and R. S. Buritz, *J. Appl. Phys.*, **25,** 202 (1954). W. A. Rogers, R. S. Buritz, and D. Alpert, *J. Appl. Phys.*, **25,** 868 (1954), have also given data for the diffusion coefficient, solubility, and permeation rate for helium in glass.

133. N. W. Taylor and W. Rast, *J. Chem. Phys.*, **6,** 612 (1938).

134. K. B. McAfee, Jr., *J. Chem. Phys.*, **28,** 218, 226 (1958).

135. Values for helium and neon are also given by E. L. Jossem, *Rev. Sci. Instr.*, **11,** 164 (1940).

136. J. R. Young and N. R. Whetten, *Rev. Sci. Instr.*, **32,** 453 (1961).

137. C. C. Van Voorhis, *Phys. Rev.*, **23,** 557 (1927); R. M. Barrer (Ref. 128), p. 129.

138. V. O. Altemose, *J. Appl. Phys.* **32,** 1309 (1961).

139. A. S. Russell and J. J. Stokes, Jr., *J. Am. Chem. Soc.*, **69,** 1316 (1947).

140. R. M. Barrer, *J. Chem. Soc.*, **1934,** 378; also J. H. Simons and W. R. Ham, *J. Chem. Phys.*, **7,** 899 (1939).

141. W. A. Weyl, *Research*, **1,** 50 (1947).

142. R. L. Smith and N. W. Taylor, *J. Am. Ceram. Soc.*, **23,** 139 (1940).

143. For discussion of earlier work on this topic see McBain (Ref. 23), Chapter XII, p. 358.

144. For a discussion of the chemical and physical characteristics of cellulose and its derivatives, the reader should consult the treatise *Cellulose and Cellulose Derivatives* by E. Ott, *High Polymers*, Vol. 5, Interscience Publishers, 1943. This contains a chapter, "Sorption of Water and Other Vapors by Cellulose," by E. I. Valko. *The Chemistry of Cellulose* by E. Heuser, John Wiley & Sons, New York, 1944, contains a chapter on the reactions of cellulose with water.

145. E. Filby and O. Maass, *Can. J. Research* **B13,** 1 (1935).

146. S. E. Sheppard and P. T. Newsome, *J. Ind. Eng. Chem.*, **26,** 285 (1934).

147. S. M. Neale and W. A. Stringfellow, *Trans. Faraday Soc.*, **37,** 525 (1941).

148. The values of the heat of evaporation of water at 40 and 100° C are 10,332 and 9712 cal/mole, respectively.

149. C. C. Houtz and D. A. McLean, *J. Phys. Chem.*, **43,** 309 (1939).

150. C. C. Houtz and D. A. McLean, *J. Phys. Chem.*, **45,** 111 (1941).

151. See, for instance, the discussion and references given by E. I. Valko (Ref. 144) and references in the "Bibliography of Solid Adsorbents" (Ref. 1).

152. E. J. Murphy, *Trans. Am. Electrochem. Soc.*, **83,** 161 (1943).

153. A. C. Zettlemoyer, A. Chand, and E. Gamble, *J. Am. Chem. Soc.*, **72,** 2752 (1950).

154. F. J. Norton, *J. Appl. Phys.*, **28,** 34 (1957).

155. R. M. Barrer (Ref. 128), p. 405.

156. P. M. Doty, W. H. Aiken, and H. Mark, *Ind. Eng. Chem., Anal. Ed.*, **16,** 686 (1944).

157. F. J. Norton has reported that the permeation rate of oxygen through fused silica, K, is $9 \times 10^{-13}$ at 900° C [*Nature*, **191,** 701 (1961)].

# 8 Gases and metals

Revised by Francis J. Norton

## 8.1. INTRODUCTORY REMARKS

In his early work on the reactions of tungsten and molybdenum with different gases at low pressures and his further investigations on thermionic emission, Langmuir recognized the importance of eliminating from these metals all traces of residual gases. Subsequent developments in the production of electronic devices have served to emphasize this conclusion. In addition, later investigations have also demonstrated that the mechanical and other physical characteristics of metals may be seriously affected by the presence in them of occluded or dissolved gases.

Hence, a knowledge of the manner in which metals may take up gases and of the conditions under which these gases may be removed is of extreme importance in vacuum technique. In this chapter the subject will be discussed under the following headings:

I. Solubility of gases in metals.
II. Diffusion of gases through metals.
III. Degassing of metals.

Surface sorption of gases on metals has been discussed in Chapters 6 and 7.

There are several treatises which are valuable for further details on these subjects, among them works by Smithells,[1] Barrer,[2] Jost,[3,4] Smith, Eastwood, Carney, and Sims,[5] Nikuradse and Ulbrich,[6] Seith,[7] Cupp,[8] and Ziliani.[9]

Data on gas content, permeation, and degassing temperatures for materials important in vacuum technology, especially ultrahigh vacuum, are summarized in *Tabellen der Electronphysik*,[10] as well as by Jaekel, and Alpert,[11] Schweitzer,[12] and Yarwood.[13]

Another useful book is *Glossary of Terms Used in Vacuum Technology*, Pergamon Press, 1958.

## 8.2. SOLUBILITY OF GASES IN METALS

Except for the rare gases (He, Ne, Ar, Kr, Xe) many gases may react with metals to form compounds or effect solution of the gas in solid or liquid metal. Under purely thermal conditions, the rare gases do not dissolve in any metal, either liquid or solid. However, ionization of the gas and acceleration of these ions, or nuclear disintegration processes, may put the rare gases into solution in the metal. This subject, of great interest in the field of nuclear fuels, is discussed in section 8.17.

A good summary of the reactivity of various gases to metals is given by Barrer,[2] p. 146. The relationships between solubility, permeation, and diffusion are discussed by Van Amerongen,[14] Norton,[15] and Smithells.[16]

The relationships between the fundamental equations for gas diffusion in solids and heat flow are discussed at the end of section 8.10.

There are four different types of interaction between *hydrogen* and metals.

1. With the alkali and alkaline-earth metals (Groups IA and IIA) hydrides are formed, such as NaH and $CaH_2$, which are ionic in behavior, with hydrogen as the negative ion.
2. With elements such as C, Si, S, Se, As, and other metals of Groups IVB, VB, and VIB, covalent hydrides are formed, such as $H_2S$, $AsH_3$, and $SiH_4$.
3. With a number of metals such as Cu, Ag, Cr, Mo, W, Fe, Co, Ni, Al, and Pt, hydrogen forms true solutions, as indicated by (*a*) the observation that the solubility varies as $P^{1/2}$, and (*b*) the increase in solubility with increase in temperature. We shall designate these metals as *Group A*.
4. With elements of Group IIIA (Ce, La, etc.), Group IVA (Ti, Zr, Th, Hf), and Group VA (V, Cb, Ta) hydrogen forms pseudo hydrides. While the solubility varies as $P^{1/2}$, for certain ranges of values of $P$, it decreases with increase in temperature. Under certain conditions palladium behaves with respect to hydrogen in the same manner. These metals will be designated in the following discussion as *Group B*.

Hydrogen is not absorbed by Au, Zn, Cd, In, or Tl.

In the case of *oxygen* it is often difficult to distinguish between solution of oxygen and solution of oxides. However, the formation of true solutions has been determined in Ag, Cu, Co, and a few other metals.

*Nitrogen* dissolves only in those metals which form nitrides at higher temperatures, for example, Al (molten), Zr, Ta, Mn, and Fe.

*Carbon monoxide* behaves in a similar manner. Thus it is absorbed by Ni and Fe, each of which is capable of forming a carbonyl.

The observations on the solubility of sulfur dioxide in molten copper are discussed in a subsequent section.

In discussing the sorption of gases by metals, the solubility (or gas content) may be expressed in terms of a variety of units. These units and conversion factors in terms of other units are as follows:

$v_0$ = cubic centimeters (STP) per 1 g metal.

$s$ = cubic centimeters (STP) per 100 g metal $= 100\,v_0$.

$$V_g = \text{volume of gas (STP) per 1 volume of metal} = v_0\rho = 10^{-2}s\rho, \text{ where } \rho = \text{density of metal.} \tag{8.1}$$

$$x = \text{milligrams per 100 g} = 4.461 \cdot v_0 M = 4.461 \cdot 10^{-2} sM, \tag{8.2}$$

where $M$ = molecular weight of gas. Hence

$$s = 22.42x/M \tag{8.3}$$

$$p_w = \text{per cent by weight} = 10^{-3}x = 4.461 \cdot 10^{-5} sM. \tag{8.4}$$

| Gas: | $H_2$ | $O_2$ | $N_2$ or CO | $H_2O$ | $CO_2$ |
|---|---|---|---|---|---|
| $c_M = 4.461 \cdot 10^{-2}M$: | $8.995 \cdot 10^{-2}$ | 1.428 | 1.250 | 0.8041 | 1.963 |
| $1/c_M = 22.42/M$: | 11.12 | 0.700 | 0.800 | 1.244 | 0.5093 |

$$s = \frac{10^3}{c_M} p_w. \tag{8.5}$$

$$\mu m = \text{number of } \textit{micromoles} \text{ per 100 g} = 44.61s. \tag{8.6}$$

$$s = 2.242 \cdot 10^{-2}\,\mu m. \tag{8.7}$$

For a *diatomic* gas in a metal of atomic mass $A$,

$$r = \text{number of atoms per atom of metal} = 2 \cdot 10^{-5}\frac{Ax}{M} = 8.922 \cdot 10^{-7}As \tag{8.8}$$

$$= 8.922 \cdot 10^{-5} V_g \frac{A}{\rho}. \tag{8.9}$$

## 8.3. SOLUBILITY OF HYDROGEN IN METALS OF GROUP A

Typical metals of this group are iron and nickel. The values of $s$ are quite small, as compared with those for metals of Group B. Consequently, the magnitude of $r$ is of the order of $10^{-4}$ (1 atom of hydrogen for $10^4$ atoms of metal). As mentioned above, the value of $s$ increases at constant temperature as $P^{1/2}$, from which it follows that the hydrogen is *absorbed as atoms*. At constant pressure, the solubility increases with increase in temperature, indicating that *energy is absorbed* in the process of sorption. The solubility can thus be expressed, as a function of pressure and temperature, by a relation of the form

$$s = s_0 P^{1/2} \epsilon^{-Q_S/(2R_0T)}, \tag{8.10}$$

where $s_0$ is a constant.

The factor 2 in the denominator of the exponential takes into account the fact that hydrogen enters the metal lattice as atoms, and $Q_S$ therefore denotes the heat absorbed in calories *per gram-mole* hydrogen.

Equation 8.10 may be expressed in terms of ordinary logs in the form

$$\log s = \log s_0 + 0.5 \log P - \frac{Q_S}{9.148T}, \tag{8.11}$$

and evidently the value of $Q_S$ may be derived from the slope of the linear plot of log $s$ (at constant pressure) versus $1/T$.

A large number of investigators have worked on this topic, and a list of these, together with reference numbers, is given in List I. In the following text, numbers in parentheses refer to items in this list.

*List I. References on Solubility of Hydrogen in Metals of Group A*

IRON

1. A. Sieverts, *Z. Metallkunde*, **21,** 37 (1929).
2. E. Martin, *Arch. Eisenhüttenw.*, **3,** 407 (1929/30).
3. A. Sieverts and H. Hagen, *Z. physik. Chem.*, **A, 155,** 314 (1931).
4. L. Luckemeyer-Hasse and H. Schenck, *Arch. Eisenhüttenw.*, **6,** 209 (1932/33).
5. F. Pihlstrand, *C.A.*, **31,** 8471 (1937).
6. W. Baukloh and R. Müller, *Arch. Eisenhüttenw.*, **11,** 509 (1937/38).
7. K. Iwasé and M. Fukusima, *Science Repts., Tôhoku Imp. Univ.*, **27,** 162 (1938–1939).
8. A. Sieverts, G. Zapf, and H. Moritz, *Z. physik. Chem.*, **A, 183,** 19 (1938/39).
9. J. H. Andrew, H. Lee, and A. G. Quarrell, *J. Iron Steel Inst. London*, **146,** No. II, 181 (1942).
10. M. H. Armbruster, *J. Am. Chem. Soc.*, **65,** 1043 (1943).
11. R. M. Barrer, *Discussions Faraday Soc.*, No. 4, 68 (1948).
12. Bibliography, 233 refs. Atomic Energy Research Estab. (Gt. Britain) 1956, IGRL-IB/R17.
13. M. Smialowski, *Neue Hütte*, **2,** 621 (1957).

A very comphrehensive bibliography on gas solubility in metals is given by O. Kubaschewski, *Z. Elektrochem.*, **44,** 152 (1938).

NICKEL

See Refs. 1, 4, 10, 14.

14. J. Smittenberg, *Rec. trav. chim.*, **53,** 1065 (1934).

COBALT

See Refs. 1 and 15.

15. A. Sieverts and H. Hagen, *Z. physik. Chem.*, **A, 169,** 237 (1934).

COPPER

See Refs. 1 and 16.

16. P. Roentgen and F. Moeller, *Metallwirtschaft*, **13,** 81, 97 (1934).

ALUMINUM

See Refs. 1, 16, and 17–21.

17. J. Czochralski, *Z. Metallkunde*, **14,** 277 (1922).
18. K. Iwasé, *Science Repts., Tôhoku Imp. Univ.*, **15,** 531 (1926).
19. P. Roentgen and H. Braun, *Metallwirtschaft*, **11,** 459, 471 (1932).
20. L. L. Bircumshaw, *Trans. Faraday Soc.*, **31,** 1439 (1935).
21. H. Kostron, *Z. Metallkunde*, **43,** 269 (1952).

MANGANESE

See Refs. 4, 6, and 22.

22. A. Sieverts and H. Moritz, *Z. physik. Chem.*, **A, 180,** 249 (1937).

SILVER

23. E. W. R. Steacie and F. M. G. Johnson, *Proc. Roy. Soc. London*, **A, 117,** 662 (1928).

MOLYBDENUM

See Refs. 2 and 24.

24. A. Sieverts and K. Bruning, *Arch. Eisenhüttenw.*, **7,** 641 (1933–1934).

CHROMIUM

See Refs. 2 and 4.

MISCELLANEOUS ALLOYS

25. L. Kirschfeld and A. Sieverts, *Z. Electrochem.*, **36,** 123 (1930).
26. A. Sieverts and K. Brüning, *Z. Physik. Chem.*, **A, 168,** 411 (1934).
27. A. Sieverts and H. Hagen, *Z. physik. Chem.*, **A, 174,** 247 (1935).
28. K. H. Lieser and H. Witte, *Z. physik. Chem.*, **202,** 321 (1954).
29. K. H. Liesen and G. Rinck, *Z. Elecktrochem.*, **61,** 357 (1957).
30. W. Hofman and J. Maatsch, *Neue Hütte*, **2,** 648 (1957).

## Solubility of Hydrogen in Iron ($A = 55.85$, $\rho = 7.88$)

Table 8.1 gives values of $s$, $x$, $\mu m$, $V_g$, and $r$ obtained by Sieverts (1), for the range 400–1650° C, at 1 atm. The last column gives later values obtained by Sieverts, Zapf, and Moritz (8).

Figure 8.1, taken from reference 8, shows a plot (*a*) of the solubility data at 1 atm over the range of 400–1200° C. The insert (*b*) shows data obtained with both increasing and decreasing temperatures (above 1250° C). The plots also show data obtained by other investigators.

As is well known, iron exhibits three critical temperatures for the solid, corresponding to the phase transformations shown in Table 8.2[17] In this table are indicated the associated lattice structure and range of stability for each phase. The last column in Table 8.1 and the plot in Fig. 8.1 show that at each critical temperature and at the melting point (1535° C) there is a change in solubility. At the $\alpha - \gamma$ transition and at the melting point the solubility increases considerably. Even at the melting point, the

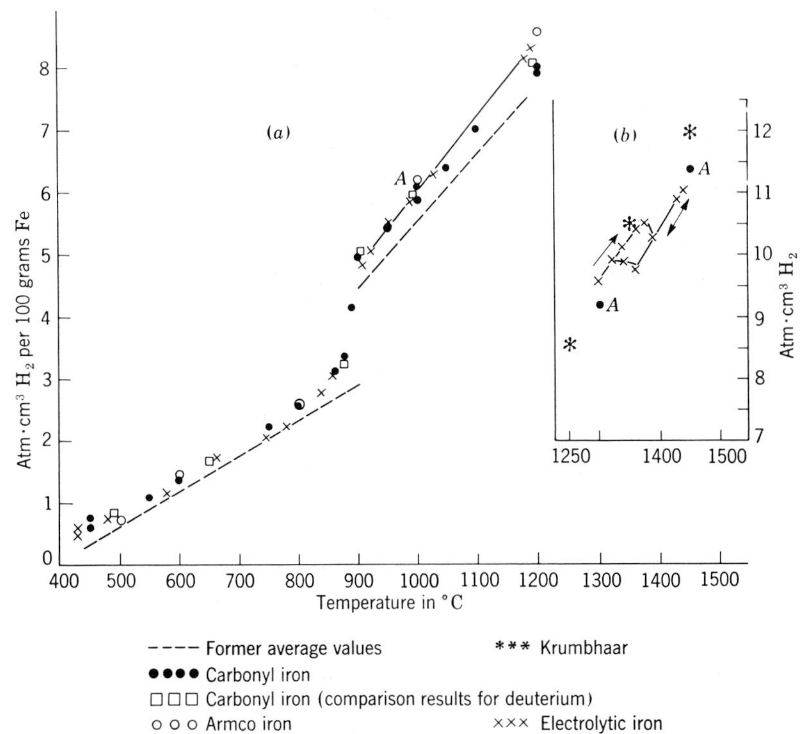


Fig. 8.1. Solubility of hydrogen (at 1 atm) in iron, as a function of the temperature (Sieverts *et al.*).

value of $V_g$ is approximately 1, while that of $r$ for the $\delta$ phase, just below the melting point, is $6.64 \cdot 10^{-4}$, or about 1 hydrogen atom per 1500 iron atoms.

The greater value of $s$ for the $\gamma$ than for the $\alpha$ phase is accounted for by Luckemeyer-Hasse and Schenck (4) as due to the fact that the hydrogen atom (which, as mentioned subsequently, is actually present as a proton) can be more readily located in an interstitial site in the looser face-centered structure than in the more closely packed body-centered structure.

This topic has also been investigated by Miss Armbruster (10). She determined the values of $s$ for the three temperatures, 400, 500, and 600° C, and for the range 18.9–629 mm at 600° C. Figure 8.2, taken from her publication, shows that the $P^{1/2}$ law is valid down to very low pressures. It was found that all the observations at the three temperatures are in agreement with the relation

$$\log \mu m = 1.946 + 0.5 \log P_{mm} - \frac{1454}{T}, \tag{8.12}$$

TABLE 8.1

**Solubility of Hydrogen in Iron at 1 atm***

(Sieverts *et al.*)

| $t°$ C | $s$ | $100x$ | $\mu m$ | $V_g$ | $10^4 r$ | $s(8)$ | |
|---|---|---|---|---|---|---|---|
| 400 | 0.35 | 3.15 | 15.6 | 0.0276 | 0.175 | | |
| 500 | 0.75 | 6.75 | 33.5 | 0.0590 | 0.374 | 0.6 | |
| 600 | 1.20 | 10.80 | 53.5 | 0.0945 | 0.597 | | |
| 700 | 1.85 | 16.65 | 82.5 | 0.146 | 0.922 | 1.8 | |
| 800 | 2.45 | 22.05 | 109.2 | 0.193 | 1.22 | | |
| 900 | 4.05 | 36.45 | 182 | 0.319 | 2.02 | 3.0 ($\alpha$) | 4.7 ($\gamma$) |
| 1000 | 5.50 | 49.50 | 245 | 0.433 | 2.74 | | |
| 1100 | 6.60 | 59.40 | 294 | 0.519 | 3.29 | 7.0 | |
| 1200 | 7.95 | 71.55 | 355 | 0.626 | 3.96 | 8.2 | |
| 1300 | 9.50 | 85.50 | 424 | 0.748 | 4.73 | 10.1 | |
| 1400 | 11.20 | 100.8 | 500 | 0.881 | 5.58 | 10.5 ($\gamma$) | 10.1 ($\delta$) |
| 1500 | 12.75 | 114.8 | 569 | 1.00 | 6.36 | | |
| 1535 ($s$) | 13.32 | 119.9 | 594 | 1.05 | 6.64 | 14 ($\delta$) | |
| ($l$) | 26.64 | 239.8 | 1188 | 2.10 | 13.28 | 25 ($l$) | |
| 1550 | 27.75 | 249.8 | 1238 | 2.18 | 13.82 | 26 | |
| 1650 | 30.97 | 278.7 | 1381 | 2.44 | 15.43 | | |

* Values of $\mu m$, $V_g$, and $r$ calculated by Dushman.

TABLE 8.2

**Allotropic Forms and Transformation Points for Iron**

| Phase Designation | Lattice Structure | Range of Stability (° C) |
|---|---|---|
| Alpha ($\alpha$) | Body-centered cubic | To 906 |
| Gamma ($\gamma$) | Face-centered cubic | 906–1403 |
| Delta ($\delta$) | Body-centered cubic | 1403–1537 |
| Liquid | | Above 1537 |

where the solubility is measured in *micromoles per 100 g* iron. That this equation is valid for all temperatures up to 900° C is shown by the data in Table 8.3, taken from Miss Armbruster's publication, in which a comparison has been made of values calculated by means of equation 8.12 (data for carbonyl iron in the first row) with those observed by previous investigators.

The second-last column gives values of the heat of solution, in *calories per gram-mole hydrogen*, as derived from each set of observations. From equation 8.12 it follows that

$$Q_S = 9.148 \times 1454 = 13{,}300.$$

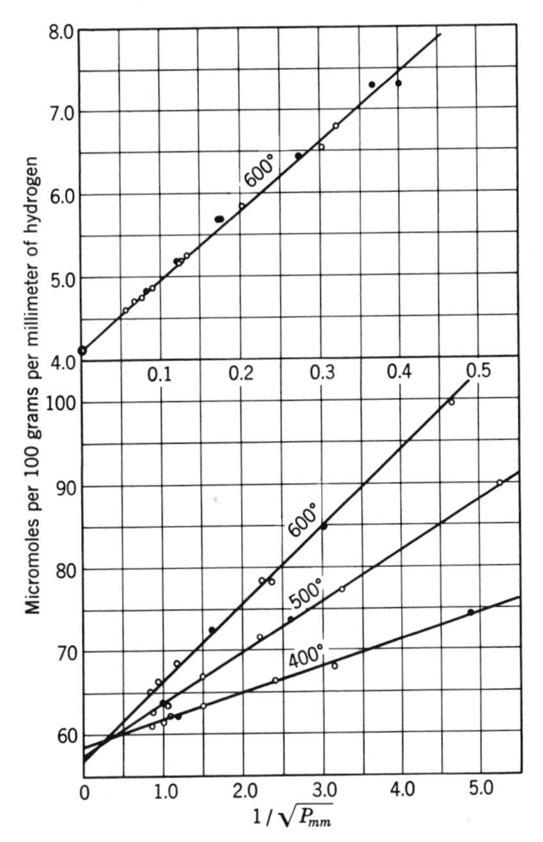


Fig. 8.2. Plot of solubility data for hydrogen in iron. Ordinates denote values of micromoles per 100 g iron per mm pressure. Abscissas give values of $1/(P_{mm})^{1/2}$. Upper section shows plot at 600° C and higher pressure range; lower section shows plots at 600, 500, and 400° C for lower pressure range. Open and closed circles designate measurements with increasing and decreasing pressure, respectively; double circle is value of intercept as determined by a blank measurement with argon. (Armbruster.)

According to Sieverts, Zapf, and Moritz (8), the solubility of deuterium ($D_2$) in both Armco and carbonyl iron is *lower* than that of ordinary hydrogen. The difference in the value of $s$ was observed to vary from 0.1 to 0.9 over the range 600–1450° C.

### Solubility of Hydrogen in Nickel ($A = 58.69$, $\rho = 8.9$)

The earliest work on the solubility of hydrogen in nickel was carried out by Sieverts and his associates (1907–1911). The measurements were repeated by Smittenberg (14), using a much more refined technique. To

TABLE 8.3

**Comparison of Solubility Determinations of Hydrogen at 1 atm in Different Samples of Iron**

(Armbruster)

(Solubility in micromoles per 100 g)

| Form of Sample | Temperatures (° C) | | | | | | | $Q_S$ | Ref. |
|---|---|---|---|---|---|---|---|---|---|
| | 300 | 400 | 500 | 600 | 700 | 800 | 900 | | |
| Carbonyl sheet | 7.06 | 16.8 | 32.0 | 52.6 | 78.0 | 107 | 140 | 13,300 | 10 |
| Wire 2 mm | ... | 15.2 | 29.4 | 50.4 | 73.4 | 96.8 | 138 | 13,900 | 5 |
| Wire and powder | ... | 15.6 | 26.8 | 53.5 | 80.3 | 106 | 134 | 13,700 | 8 |
| Chips | 4.4 | 8.9 | 29.0 | 51.3 | 78.0 | 105 | 129 | 13,800 | 4 |
| Sheet 0.3 mm | ... | 8.0 | 26.3 | 53.1 | 78.1 | 103 | 125 | 12,300 | 2 |
| Tubing | 18.5 | 29.9 | 44.6 | 59.8 | 80.3 | 110 | 138 | 11,300 | 9 |
| Electrolytic sheet | ... | 27.6 | 51.3 | 81.2 | 113 | 169 | 223 | 13,500 | 6 |
| Electrolytic block | ... | 11.2 | 33.4 | 57.1 | 91 | 125 | 163 | 14,500 | 6 |
| Reduced iron | ... | 13.4 | 44.6 | 78.0 | 110 | 145 | 195 | 13,000 | 6 |
| Electrolytic | ... | — | 22.3 | 44.6 | 71.4 | 96.8 | 125 | 14,400 | 7 |
| Reduced | ... | 8.9 | 31.2 | 62.4 | 84.7 | 114 | 143 | 11,400 | 7 |

secure a large surface, 12,000 m of wire, about 0.02 mm diameter (total weight = 42 g, and total surface = 8430 cm$^2$) was used. This was made of vacuum-fused metal. Before exposing to hydrogen, the wire was heated in vacuum in steps of increasing temperature up to 900° C. The gas evolved during this treatment consisted almost entirely of carbon monoxide.

After a 24-hour evacuation at 900°, the temperature was gradually lowered to 600° and the flushing-out (with hydrogen) repeated several times at this temperature. Only after this treatment was it possible to obtain accurate and reproducible results for the amount of hydrogen sorbed by the nickel at different pressures and temperatures.

It was shown that, for 300 and 600° C, the isotherms obeyed the $P^{1/2}$ law. Furthermore, the isobar at 0.1 mm was found to be in good agreement with Sievert's data when they were recalculated for the much lower pressure used by Smittenberg. Table 8.4 gives the values of $s$ and $\mu m$ at 1 atm based on the data obtained by Sieverts and Smittenberg.

The noteworthy features about these data are (1) the fact that the solubility of hydrogen in nickel is greater than that in iron, especially at the lower temperatures, and (2) the considerable increase in solubility at the melting point.

Later, solubility determinations were carried out by Miss Armbruster (10) on a sample of 99.5 per cent nickel in the form of strip 0.14 mm thick

TABLE 8.4

**Solubility of Hydrogen at 1 atm in Nickel***

(Sieverts and Smittenberg)

| $t°$ C: | 200 | 300 | 400 | 500 | 600 | 700 | 800 | 900 | 1000 |
|---|---|---|---|---|---|---|---|---|---|
| $s$: | 1.70 | 2.35 | 3.15 | 4.10 | 5.25 | 6.50 | 7.75 | 9.10 | 9.80 |
| $\mu m$: | 76 | 105 | 141 | 183 | 234 | 290 | 346 | 406 | 437 |
| $V_g$: | 0.151 | 0.209 | 0.280 | 0.365 | 0.467 | 0.578 | 0.689 | 0.809 | 0.873 |
| $10^4 r$: | 0.89 | 1.23 | 1.65 | 2.15 | 2.75 | 3.40 | 4.06 | 4.76 | 5.13 |

| $t°$ C: | 1100 | 1200 | 1300 | 1400 | 1452 | 1500 | 1600 |
|---|---|---|---|---|---|---|---|
| $s$: | 12.15 | 14.25 | 14.7 | 16.2 | (mp) | 41.6 | 43.1 |
| $\mu m$: | 542 | 636 | 656 | 723 | ... | 1860 | 1920 |
| $V_g$: | 1.08 | 1.27 | 1.31 | 1.44 | ... | 3.70 | 3.84 |
| $10^4 r$: | 6.36 | 7.46 | 7.70 | 8.48 | ... | 21.8 | 22.6 |

* Values of $\mu m$, $V_g$, and $r$ calculated by Dushman.

(weight = 337.7 g). The observations were made at 400, 500, and 600° C, over the pressure range 0.001–1.8 mm. It was found that the results obtained could be represented by the relation, similar to that obtained for iron,

$$\log \mu m = 1.732 + 0.5 \log P_{mm} - \frac{645}{T}. \tag{8.13}$$

Table 8.5 (10) presents a comparison of the values of $\mu m$, calculated by means of this equation for 1 atm pressure, with those of previous investigators.

The value derived by Miss Armbruster is $Q_S = 5900$ cal/g-mole $H_2$.

TABLE 8.5

**Comparison of the Determinations of the Solubility of Hydrogen at 1 atm in Different Samples of Nickel**

(Armbruster)

(Solubility in micromoles per 100 g)

| Form of Sample | 300 | 400 | 500 | 600 | 700 | 800 | 900 | $Q_S$ | Ref. |
|---|---|---|---|---|---|---|---|---|---|
| | Temperatures (° C) | | | | | | | | |
| Sheet | 111 | 164 | 218 | 271 | 323 | 373 | 419 | 5900 | 10 |
| Wire | 135 | 183 | 244 | 297 | 349 | 393 | 436 | 5300 | (14) |
| Wire and chips | 105 | 161 | 186 | 245 | 301 | 356 | 411 | 6100 | (1) |
| | 89 | 112 | 145 | 192 | 251 | 312 | 381 | 4300 (300–500°) 9000 (500–900°) | (4) |

## Solubility of Hydrogen in Alloys of Iron and Nickel

Figure 8.3 (10) shows plots (according to Miss Armbruster) of log $\mu m$ versus $1/T$, at $P_{mm} = 760$, for nickel, iron, and a number of iron-nickel alloys.

According to the data in Tables 8.3 and 8.4, the values of $\mu m$ at 400° C and 1 atm are 17 for iron and 164 for nickel. (Values of $s$ are 0.38 and 3.68, respectively.) The plots for the iron-nickel alloys lie between those for the two metals, and for each alloy the solubility can be expressed by an equation similar to equations 8.12 and 8.13. From the slopes of these plots it is seen that the value of $Q_S$ decreases with increase in nickel content, as would be expected from a comparison of the values for the pure metals.

The alloys tested by Miss Armbruster could be divided into three groups: (I), the ferritic type, for which the values of $s$ observed are substantially the same as for pure iron: (II), which comprised a 28 per cent steel and a 13 per cent Mn steel; and (III), the austenitic type, which comprised two stainless steels, Cr 18, Ni 8; and Cr 13, C 0.3; as well as nickel itself.

Similar observations on the effects of alloying elements have been made by other investigators. Figure 8.4 (9) shows an isobar at 1 atm for a 5 per cent Ni alloy compared with that for vacuum-melted iron.

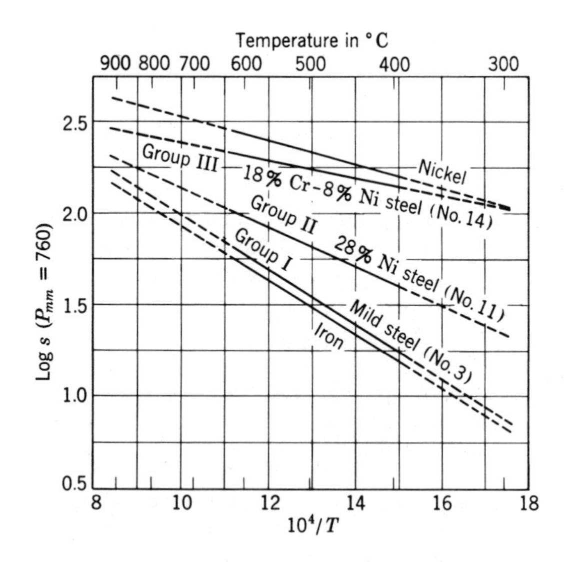


Fig. 8.3. Log of solubility of hydrogen at 1 atm in iron, nickel, and a typical steel from each of the three groups, plotted versus $1/T$. From the slopes of the straight lines, values of the heat of solution ($Q_S$) are derived. (Armbruster.)

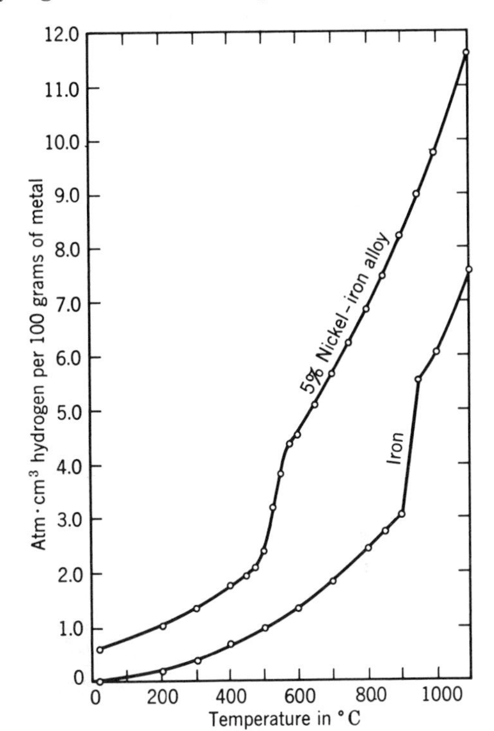


Fig. 8.4. Plots of solubility data for hydrogen at 1 atm in a 5 per cent nickel-iron alloy and in vacuum-melted iron, as a function of the temperature (Andrew, Lee, and Quarrell).

Luckemeyer and Schenck (4) have determined the solubility of hydrogen in Fe-Cr-Ni alloys of different compositions. The values of $s$ range from slightly above that for iron (most cases) to slightly lower. The same investigators also observed (in agreement with Miss Armbruster's observations) that, in Fe-Ni alloys ranging from 3.3 per cent Ni to 8.48 per cent Ni, $s$ increases with increase in nickel content.

### Solubility of Hydrogen in Cobalt ($A = 58.94$, $\rho = 8.8$)

The values of the solubility at 1 atm as observed by Sieverts and Hagen (15) are given in Table 8.6. As is evident from a comparison of the data in Tables 8.6 and 8.1, the solubility of hydrogen in cobalt is less than that in iron. From a plot of log $s$ versus $1/T$, the value derived for the heat of absorption is $Q_S = 14{,}550$ cal/g-mole $H_2$.

### Solubility of Hydrogen in Copper ($A = 63.57$, $\rho = 8.95$)

The values obtained by Sieverts (1) are given in Table 8.7. These are slightly higher than the values obtained subsequently by Roentgen and Moeller (16).

TABLE 8.6

**Solubility of Hydrogen at 1 atm in Cobalt***

(Sieverts and Hagen)

| $t°$ C: | 600 | 700 | 800 | 900 | 1000 | 1100 | 1150 | 1200 |
|---|---|---|---|---|---|---|---|---|
| $x$: | 0.08 | 0.110 | 0.167 | 0.227 | 0.298 | 0.390 | 0.437 | 0.490 |
| $s$: | 0.89 | 1.22 | 1.85 | 2.52 | 3.21 | 4.33 | 4.85 | 5.44 |
| $10^4 r$: | 0.47 | 0.65 | 0.98 | 1.34 | 1.76 | 2.30 | 2.58 | 2.89 |
| $V_g$: | 0.08 | 0.108 | 0.164 | 0.223 | 0.293 | 0.383 | 0.430 | 0.481 |

* Values of $s$, $r$, and $V_g$ calculated by Dushman.

TABLE 8.7

**Solubility of Hydrogen at 1 atm in Copper***

(Sieverts)

| $t°$ C: | 400 | 500 | 600 | 700 | 800 | 900 | 1000 |
|---|---|---|---|---|---|---|---|
| $s$: | 0.06 | 0.16 | 0.30 | 0.49 | 0.72 | 1.08 | 1.58 |
| $V_g$: | 0.0054 | 0.0143 | 0.0268 | 0.0439 | 0.0644 | 0.0967 | 0.141 |
| $10^4 r$: | 0.053 | 0.0908 | 0.170 | 0.278 | 0.409 | 0.613 | 0.896 |
| $t°$ C: | 1083 (mp) | | 1100 | 1200 | 1300 | 1400 | 1500 |
| $s$: | 2.10 ($s$) | 6.00 ($l$) | 6.3 | 8.1 | 10.0 | 11.8 | 13.6 |
| $V_g$: | 0.188 | 0.537 | 0.564 | 0.725 | 0.893 | 1.05 | 1.22 |
| $10^4 r$: | 1.19 | 3.40 | 3.58 | 4.60 | 5.67 | 6.70 | 7.71 |

* Values of $V_g$ and $r$ calculated by Dushman.

Below the melting point, the solubility of hydrogen is less in copper than in iron, but at the melting point the solubility increases to almost three times that for the solid; and, as the temperature is raised, $s$ approaches closely that for hydrogen in iron at the same temperature.

The heat of absorption, deduced from the plot of log $s$ versus $1/T$, is $Q_S = 18{,}300$ cal/g-mole $H_2$.

Roentgen and Moeller state that the addition of small percentages of aluminum to copper decreases the solubility of hydrogen proportionally. Figure 8.5 (1) shows the effect of alloying with other metals on the solubility of hydrogen in *molten* copper. (Note that $s = 11.12x$.)

Allen[18] has observed that the solubility of hydrogen in molten copper is decreased considerably when oxygen is also present in solution. Thus, for 0.05 per cent by weight of oxygen (0.35 cm$^3$ per 100 g), the value of $s$ for hydrogen is decreased to 0.45 cm$^3$ (STP) per 100 g.

The solubility of hydrogen in *fused* copper and copper-tin alloys (bronzes) has been investigated by Bever and Floe.[19] Their results are summarized in Table 8.8.

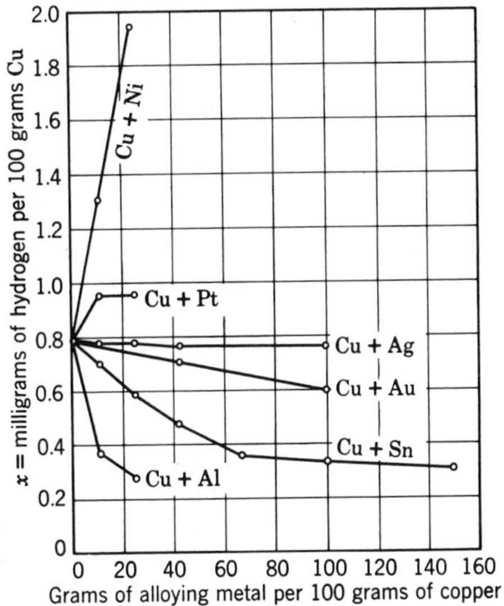


Fig. 8.5. Effect of alloying metals on the solubility of hydrogen in molten copper (Sieverts).

The values of $s$ for pure copper are in good agreement with those observed by Roentgen and Moeller, which, as stated previously, are slightly lower than those given in Table 8.7. The last column in Table 8.8 gives values of $s$ for fused tin; as will be observed, the solubility values for the bronzes decrease with increase in percentage of tin.

**TABLE 8.8**

**Solubility of Hydrogen at 1 atm in Copper, Copper-Tin Alloys, and Tin**

(Bever and Floe)

$s$ = cubic centimeters (STP) per 100 g metal

| | | Bronzes Containing $p$ Atomic Per Cent Sn | | | | | |
|---|---|---|---|---|---|---|---|
| $t°$ C | Pure Cu | $p = 3.3$ | 6.5 | 12.9 | 26.4 | 39.3 | 100 |
| 1000 | ... | ... | 3.09 | 2.11 | 0.53 | 0.50 | 0.04 |
| 1100 | 5.73 | 4.80 | 4.11 | 2.97 | 0.94 | 0.76 | 0.09 |
| 1200 | 7.33 | 6.28 | 5.35 | 3.94 | 1.50 | 1.15 | 0.205 |
| 1300 | 9.37 | 7.81 | 6.85 | 5.10 | ... | 1.61 | 0.355 |
| % Sn by wt: | | 5.9 | 11.5 | 21.7 | 40.2 | 54.8 | 100 |

According to Bever and Floe, "the investigation points definitely to hydrogen as a major source of porosity in bronze castings." These results also indicate that caution must be exercised in the use of bronze solders in a hydrogen atmosphere for devices that are to be evacuated.

For tin, $A = 118.70$ and $\rho = 7.28$. Hence, at 1300° C, $r = 3.76 \cdot 10^{-5}$ and $V_g = 0.0258$.

**Solubility of Hydrogen in Manganese ($A = 54.93$, $\rho = 7.2$)**

Figure 8.6 (22) shows the isobar at 760 mm as obtained by Sieverts and Moritz (22) for very pure metal. As for the other metals discussed above,

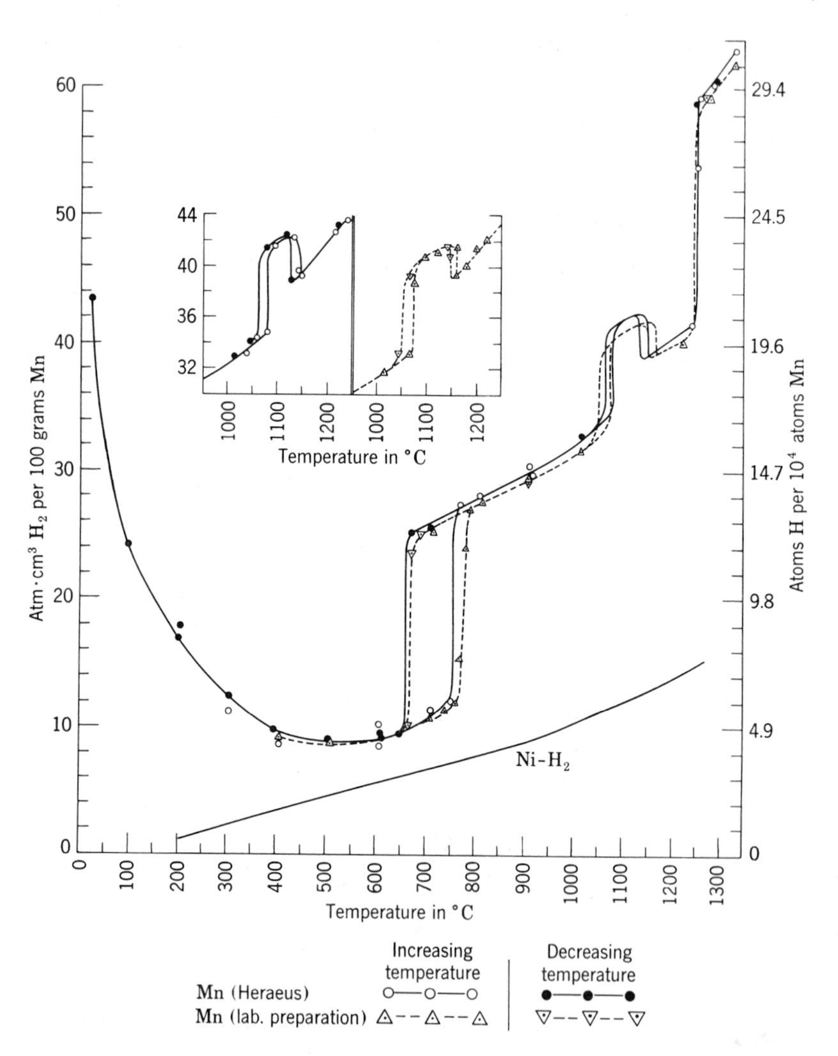


Fig. 8.6. Isobars (at 1 atm) for solubility of hydrogen in two samples of pure manganese at various temperatures. Inset shows plots for the range 1000–1250° C. (Sieverts and Moritz.)

the $P^{1/2}$ law was found to apply to the Mn-$H_2$ system, as shown by isotherms at 192, 608, and 912° C.

Comparing the isobar at 1 atm for this system with that for Ni-$H_2$ (see the lowest curve in Fig. 8.6), it is evident that the solubility of hydrogen in pure manganese is much greater than that observed for nickel.

An interesting and new feature, observed only in this system, is the occurrence of a *minimum* in the solubility at about 550° C. The character of the isobar is accounted for by the fact that manganese has four crystal modifications as follows:

$\alpha$-manganese, cubic, 58 atoms per elementary cube.
$\beta$-manganese, cubic, 20 atoms per elementary cube.
$\beta'$-manganese, cubic face-centered.
$\gamma$-manganese, tetragonal face-centered.

Table 8.9 gives the solubility data at temperatures below and above each of the transition temperatures and at the melting point (1242° C). Data are given for two samples of "pure" metal, one obtained from Heraeus and the other prepared by the investigators (Sieverts and Moritz), and also for two samples of "technical" manganese.

**TABLE 8.9**

**Solubility of Hydrogen at 1 atm in Manganese**

(Sieverts and Moritz)

$s$ = cubic centimeters (STP) per 100 g metal

| Transition Point | ° C | Pure Mn: Heraeus | Pure Mn: S.M. | Technical Mn: 97.8% Mn | Technical Mn: 96.4% Mn |
|---|---|---|---|---|---|
| $A_1$ | 600 | 9 | 9 | 7.5 | 0.5 |
| | 800 | 27 | 27 | 23 | 13.0 |
| $A_2$ | 1050 | 34 | 33 | 30 | 19.5 |
| | 1100 | 42 | 41 | 35.5 | (23) |
| $A_3$ | 1125 | 42.5 | 41.5 | ... | ... |
| | 1165 | 40 | 39.5 | ... | ... |
| Mp | 1242 | cryst. 44 | 44 | 37 | ... |
| | | liquid 59 | 59 | 52 | ... |
| | 1320 | 62 | 63 | (58) | ... |

The solubility in the $\alpha$ form *decreases* with increase in temperature; that in the $\beta$ and $\beta'$ forms *increases* with increase in temperature.

Evidently the presence of impurities in the metal decreases the solubility of hydrogen, in agreement with the fact that the solubility data obtained

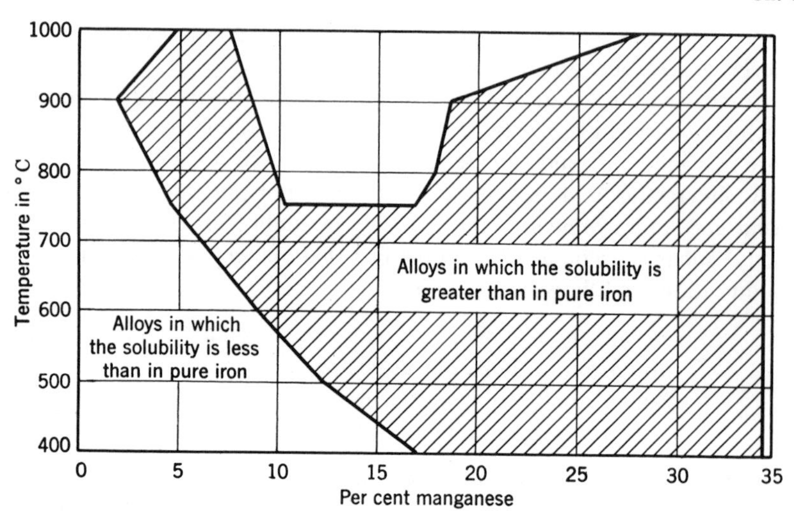


Fig. 8.7. Composition of iron-manganese alloys in which the solubility of hydrogen is *less* than that in pure iron (clear region) and alloys in which the solubility is *greater* than that in pure iron (shaded region) (Baukloh and Müller).

by Luckemeyer-Hasse and Schenck are still lower than those given in Table 8.9.

The value $s = 40$ corresponds to the values $V_g = 2.88$ and $r = 19.6 \cdot 10^{-4}$.

The solubility of hydrogen in iron-manganese alloys has been determined by Baukloh and Müller (6). Figure 8.7 taken from their publication, presents a comparison of the solubility data for different percentages of manganese in iron over the ranges of temperatures 400–1000° C. The shaded area corresponds to compositions of alloys in which hydrogen is more soluble than in pure iron; the unshaded area corresponds to alloy compositions in which hydrogen is less soluble.

As will be observed, there is a range of compositions (between 10 and 20 per cent manganese) in which the solubility decreases at temperatures above 750° C, owing to the transformation from an $\alpha$ to an $\epsilon$ phase, in which hydrogen is much less soluble than in $\alpha$-iron.

**Solubility of Hydrogen in Chromium and Molybdenum (For Cr, $A = 52.01$, $\rho = 7.1$; for Mo, $A = 95.95$, $\rho = 10.2$)**

An isobar at 1 atm for the solubility of hydrogen in *chromium* is shown by E. Martin (2). Since the original data are not given, approximate values of $s$ were taken from the plot. These appear along with values of $V_g$ and $r$ in Table 8.10. Thus it is seen that the solubility data for chromium lie between those for cobalt and copper.

TABLE 8.10

**Solubility of Hydrogen at 1 atm in Chromium**

(Martin)

$s$ = cubic centimeters (STP) per 100 g

| $t°$ C: | 600 | 800 | 900 | 1000 | 1100 |
|---|---|---|---|---|---|
| $s$: | 0.5 | 1.0 | 2.0 | 3.0 | 4.2 |
| $V_g$: | 0.035 | 0.07 | 0.14 | 0.42 | 0.3 |
| $10^4 r$: | 0.23 | 0.46 | 0.92 | 1.4 | 2.0 |

The solubility of hydrogen in *molybdenum* has been investigated by Martin (2) and by Sieverts and Brüning (24). The values obtained by the latter two, which are somewhat lower than those quoted by Martin, are given in Table 8.11. The solubility values are considerably lower than

TABLE 8.11

**Solubility of Hydrogen at $P_{mm}$ = 753 in Molybdenum**

(Sieverts and Brüning)

| $t°$ C: | 420 | 617 | 763* | 791 | 983* | 995 | 998 | 1095* |
|---|---|---|---|---|---|---|---|---|
| $x$: | 0.015 | 0.017 | *0.023* | 0.022 | 0.029 | 0.051 | *0.044* | *0.056* |
| $s$: | 0.167 | 0.189 | 0.255 | 0.244 | 0.322 | 0.566 | 0.488 | 0.622 |
| $10^4 r$: | 0.14 | 0.16 | 0.22 | 0.21 | 0.28 | 0.49 | 0.42 | 0.53 |

* Values obtained with molybdenum powder; the other values were obtained with wire.

those for hydrogen in cobalt. A plot of log $x$ versus $1/T$ yields values that are too scattered for an accurate determination of $Q_S$. From the data in italics in the table (which agree with a smooth curve in the original paper), the value deduced is $Q_S$ = 14,000 cal/mole $H_2$.

For alloys of molybdenum and iron, the solubility of hydrogen increases almost linearly with the iron content up to 52 per cent iron.[20] For smaller molybdenum content, the solubility is only slightly less than that in pure iron. The plots of log $s$ versus $1/T$ for the series up to 52 per cent iron are practically parallel. Actually these linear plots, which are much better than those for molybdenum, give the same value for $Q_S$ as that deduced above.

**Solubility of Hydrogen in Aluminum ($A$ = 26.97, $\rho$ = 2.70)**

The gas is practically insoluble in the solid metal, but is soluble to a small extent in the molten metal (mp, 659° C). Determinations of $s$ have been made by Bircumshaw (20), Roentgen and Moeller (16), and Roentgen and Braun (19). These three sets of data are in approximate agreement. The solubilities measured by Iwasé (18) are so much greater than those obtained by the other investigators that they should be disregarded.

The subject has also been investigated by Robertson[21] and by Azarian and Epremian.[22] Epremian confirmed the validity of the $P^{1/2}$ law over the range $P = 40$ to $P = 760$ mm of mercury. Azarian and Epremian's values of $s$ (cm³/100 g at 1 atm) are given in the second and third columns of Table 8.12, along with the values obtained by the other investigators, including Czochralski (17).

**TABLE 8.12**

**Solubility of Hydrogen at 1 atm in Molten Aluminum**

| $t°$ C | Azarian and Epremian | | Ref. 20 | | Ref. 16 Ref. 19 | | Ref. 17 | |
|---|---|---|---|---|---|---|---|---|
| | $s$ | $V_g$ | $s$ | $V_g$ | $t°$ C | $s$ | $t°$ C | $s$ |
| 700 | 0.15* | ... | 0.23 | 0.006 | 690 | 0.27 | 1200 | ~2.5 |
| 800 | 0.493 | 0.013 | 0.89 | 0.024 | 866 | 1.45 | 1500 | ~5.5 |
| 900 | 1.480 | 0.040 | 1.87 | 0.051 | 1096 | 5.00 | Ref. 18 | |
| 940 | 1.867 | 0.050 | ... | ... | 1233 | 8.02 | 800 | ~10 |
| 1000 | 3.0* | ... | 3.86 | 0.104 | ... | ... | 900 | ~17 |

* Extrapolated from a plot of log $s$ versus $1/T$.

From the plot of log $s$ versus $1/T$, Bircumshaw deduced the value $Q_S = 43{,}400$ cal/g-mole $H_2$, whereas Azarian and Epremian deduced the value $Q_S = 50{,}800$.

According to Roentgen and Moeller, copper alloyed with the aluminum reduces the value of $s$ by an amount that varies linearly with the percentage of copper, so that for about 30 atomic per cent of aluminum $s$ is negligibly small.[23]

It has also been reported[24] that silicon, manganese, and nickel reduce the solubility, while the addition of magnesium increases the solubility of hydrogen in aluminum.

Smithells[25] has stated, "Hydrogen accounts for about 80 per cent of the gas which can be extracted from commercial aluminum by vacuum treatment, although it cannot all be present in solution." Rapid cooling of the fused metal probably leaves some of the excess gas occluded in small pores.

**Solubility of Hydrogen in Platinum and Silver (For Pt, $A = 195.23$, $\rho = 21.45$; for Ag, $A = 107.88$, $\rho = 10.50$.)**

The data on platinum, given in Table 8.13, were obtained by Sieverts (1). For the range 1000–1350° C, $Q_S = 34{,}500$.

TABLE 8.13

**Solubility of Hydrogen at 1 atm in Platinum**

(Sieverts)

| $t°$ C: | 409 | 827 | 1033 | 1136 | 1239 | 1342 |
|---|---|---|---|---|---|---|
| $s$: | 0.067 | 0.10 | 0.233 | 0.40 | 0.61 | 0.93 |
| $V_g$: | 0.014 | 0.022 | 0.050 | 0.086 | 0.131 | 0.201 |
| $10^4r$: | 0.12 | 0.18 | 0.41 | 0.70 | 1.07 | 1.64 |

Table 8.14 gives data on the solubility of hydrogen in silver obtained by Steacie and Johnson (23). From the values of $V_g$, at 80 cm pressure, given in their publication, values of $s$ and $10^4r$ were calculated for atmospheric pressure.

TABLE 8.14

**Solubility of Hydrogen in Silver**

(Steacie and Johnson)

| | | | Solubility at 1 atm | | |
|---|---|---|---|---|---|
| $t°$ C | $V_g(P_{cm} = 80)$ | $(P_{cm})^{1/2}/V_g$ | $s$ | $V_g$ | $10^4r$ |
| 400 | 0.006 | 1500 | 0.055 | 0.0058 | 0.054 |
| 500 | 0.012 | 750 | 0.110 | 0.0116 | 0.107 |
| 600 | 0.019 | 470 | 0.176 | 0.0185 | 0.170 |
| 700 | 0.025 | 358 | 0.232 | 0.0244 | 0.224 |
| 800 | 0.036 | 248 | 0.334 | 0.0351 | 0.322 |
| 900 | 0.046 | 195 | 0.427 | 0.0448 | 0.411 |

The values in the third column in Table 8.14 are average values. From the plot of log $s$ versus $1/T$, the value derived for the heat of absorption is $Q_S = 11{,}900$ cal/g-mole $H_2$. A comparison with the data for the other metals shows that the solubility of hydrogen at 1 atm in silver is about one-tenth that (at the same pressure) in cobalt.

Table 8.15 presents a summary of the solubility data for the metals of Group A (with the exception of Mn), and Fig. 8.8 shows the isobars at 1 atm for Ni, Fe, Co, and Cu. The plot for Cr would be located between those for Co and Cu, and the plot for Al below that for Cu.

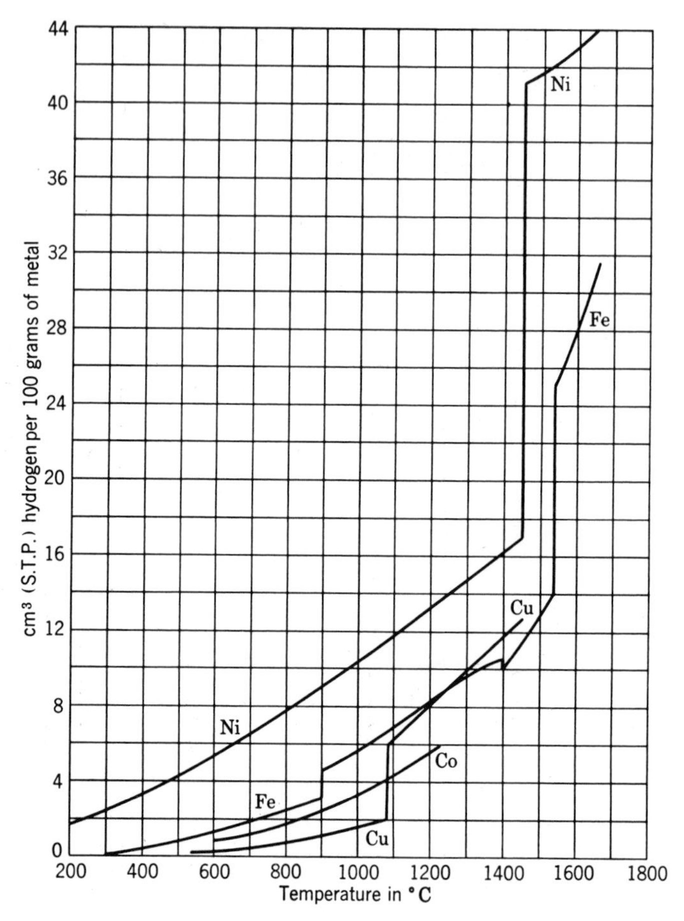


Fig. 8.8. Comparison of isobars for the solubility of hydrogen at 1 atm in nickel, iron, cobalt, and copper.

## 8.4. SOLUBILITY OF HYDROGEN IN METALS OF GROUP B

Solubility values, in terms of cubic centimeters (STP) *per gram* ($= 10^{-2}s$), are shown in Table 8.16.[26] Figure 8.9 shows plots of the isobars at 1 atm pressure. As with metals of Group A, $s$ varies as $P^{1/2}$, leading to the conclusion that, in both cases, hydrogen enters the lattice structure as atoms, or rather as protons.

There are, however, the following striking differences between the values of $s$ for the two groups:

*a.* The solubility of hydrogen in metals of Group B is $10^3$–$10^4$ times that of the solubility in metals of Group A. This is indicated by a comparison

## TABLE 8.15

### Solubility ($cm^3$/100 g) of Hydrogen at 1 atm, in Metals of Group A

| $t°$ C | Ni | Fe | Co | Cr | Cu | Al | Ag | Mo | Pt |
|---|---|---|---|---|---|---|---|---|---|
| 200 | 1.70 | | | | | | | | |
| 300 | 2.35 | 0.16 | | | | | | | |
| 400 | 3.15 | 0.35 | | | 0.06 | | 0.055 | 0.165 | 0.065 |
| 500 | 4.10 | 0.75 | | | 0.16 | | 0.110 | 0.175 | |
| 600 | 5.25 | 1.20 | 0.89 | 0.5 | 0.30 | | 0.176 | 0.186 | 0.075 |
| 700 | 6.50 | 1.85 | 1.22 | | 0.49 | 0.15 | 0.232 | 0.21 | |
| 800 | 7.75 | 2.45 | 1.85 | 1.0 | 0.72 | 0.49 | 0.334 | 0.25 | 0.095 |
| 900 | 9.10 | 3.0 (α) | 2.52 | 2.0 | 1.08 | 1.48 | 0.427 | 0.29 | 0.135 |
| | | 4.7 (γ) | | | | | | | |
| 1000 | 9.80 | 5.50 | 3.21 | 3.0 | 1.58 | 3.0 | | 0.50 | 0.20 |
| 1100 | 12.15 | 7.00 | 4.33 | 4.2 | 6.3 (*l*) | | | 0.62 | 0.35 |
| 1200 | 14.25 | 8.25 | 5.44 | | 8.1 | | | | 0.52 |
| 1300 | 14.7 | 10.1 | | | 10.0 | | | | 0.77 |
| 1400 | 16.2 | 10.5 (γ) | | | 11.8 | | | | 1.15 |
| | | 10.1 (δ) | | | | | | | |
| 1500 | 41.6 (*l*) | | | | 13.6 | | | | |
| 1535 | | 14 (δ) | | | | | | | |
| | | 25 (*l*) | | | | | | | |
| 1600 | 43.1 | 28 | | | | | | | |
| 1650 | | 31 | | | | | | | |
| $10^{-3}Q_S$ | 5.9 | 13.3 | 15.55 | | 18.30 | 50.80 | 11.90 | 14.00 | 34.50 |
| $V_g(800)$ | 0.69 | 0.19 | 0.164 | 0.071 | 0.066 | 0.024 | 0.025 | 0.036 | 0.020 |
| $10^4 r(800)$ | 4.06 | 1.22 | 0.98 | 0.46 | 0.41 | 0.214 | 0.33 | 0.214 | 0.165 |
| Mp (°C) | 1452 | 1535 | 1480 | 1615 | 1083 | 659 | 961 | 2620 | 1755 |

## TABLE 8.16

### Sorption of Hydrogen by Metals of Group B

$10^{-2}s$ = cubic centimeters (STP) per gram, at 1 atm pressure

| Temperature (°C) | Ti | V | Zr | Cb | La | Ce | Ta | Th | Pd |
|---|---|---|---|---|---|---|---|---|---|
| 20 | 407.4 | 150 | 235.5 | 55.0 | 223 | 215 | 46 | 148 | (60) |
| 300 | | 65 | ... | 44.4 | 192 | 184 | 33 | ... | (3.3) |
| 400 | 387.7 | 38 | ... | 36.8 | 182 | 176 | 25 | ... | 2.3 |
| 500 | 366.0 | 19 | ... | 22.7 | 172 | 168 | 14 | ... | 1.9 |
| 600 | 334.7 | 10 | 184 | 9.88 | 163 | 160 | 7 | 91 | 1.8 |
| 700 | 183.9 | 6.4 | 176 | 5.11 | 153 | 152 | 4.2 | 88 | 1.70 |
| 800 | 140.9 | 4.4 | 165 | 3.30 | 143 | 145 | 2.5 | 81 | 1.62 |
| 900 | 98.2 | 3.2 | 138 | 2.17 | 134 | 138 | 1.8 | 77 | 1.57 |
| 1000 | 66.1 | 2.5 | 78 | 1.63 | 123 | 130 | 1.4 | 26 | 1.55 |
| 1100 | 45.9 | 2.1 | 47 | ... | 111 | 113 | 1.1 | 19 | 1.54 |
| 1200 | ... | ... | 32 | ... | 41 | 53 | 1.0 | 17.5 | — |
| $A$ | 47.90 | 50.95 | 91.22 | 92.91 | 138.92 | 140.13 | 180.88 | 232.12 | 106.7 |
| $\rho$ | 4.52 | 5.87 | 6.4 | 8.4 | 6.15 | 6.90 | 16.62 | 11.2 | 11.40 |
| $V_g$ | 1800 | 900 | 1525 | 460 | 1490 | 1450 | 765 | 1660 | 700 |
| $r$ | 1.75 | 0.72 | 1.92 | 0.47 | 2.76 | 2.69 | 0.78 | 3.07 | 0.59 |
| $\rho_{HM}$ | 3.91 | 5.30 | 5.67 | ... | 5.83 | 5.7 | 15.10 | ... | 10.76 |

of values of $V_g$ and $r$. In Table 8.16, the values of these quantities are given for room temperature. These correspond to "hydrides" of the composition $MH_r$, where M denotes a metal of this group, and *r is not an integer.*

*b.* The solubility *decreases* with increase in temperature. That is, *heat is evolved* when hydrogen enters into solution, and hence the reaction involved in the formation of any one of the hydrides is exothermic, as in the formation of the chemically well-defined hydrides of the alkali and alkaline-earth metals. The magnitude of $Q$, obtained from a plot of log $s$ versus $1/T$, corresponds to the heat of formation of the "hydride."

*c.* The density of the saturated solution, indicated by $\rho_{HM}$ in the lowest row of the table, is uniformly *less* than that of the metal itself (designated by $\rho$). The values of $\rho$ given in the second row at the bottom of Table 8.16 are taken from the standard handbook and are not the same, in general, as those given by Sieverts and Gotta.[27]

A bibliography of references to the literature on the solubility data for

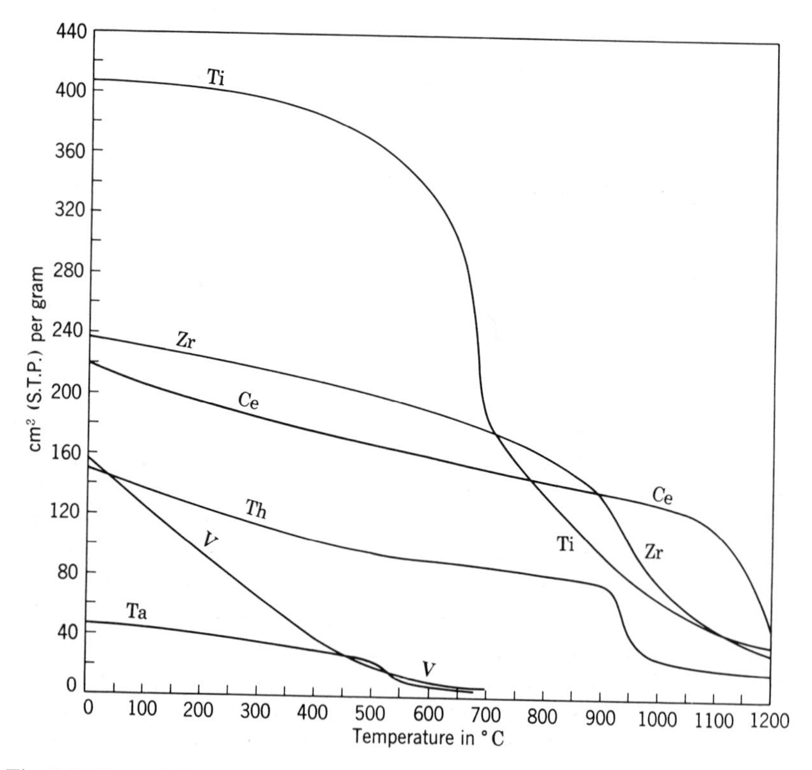


Fig. 8.9. Plots of the data for the solubility of hydrogen at 1 atm (cm³, STP, per g metal) in metals of Group B at various temperatures.

hydrogen in metals of Group B is given in List II. In the following text, numbers in parentheses refer to items in this list.

1. (Ta, W) A. Sieverts and E. Bergner, *Ber. deut. chem. Ges.*, **44**, 2394 (1911); also see reference 17 below.
2. (Ce) A. Sieverts and G. Müller-Goldegg, *Z. anorg. Chem.*, **131**, 65 (1923).
3. (Ce, La, Ce-Mg) A. Sieverts and E. Roell, *Z. anorg. Chem.*, **146**, 149 (1925).
4. (Pr, Nd) A. Sieverts and E. Roell, *Z. anorg. Chem.*, **150**, 261 (1926).
5. (Zr, Th) A. Sieverts and E. Roell, *Z. anorg. Chem.*, **153**, 289 (1926). Also see reference 18 below.
6. (V, Ti) H. Huber, L. Kirschfeld, and A. Sieverts, *Ber. deut. chem. Ges.*, **59B**, 2891 (1926).
7. (La, Ce, Pr) A Sieverts and A. Gotta, *Z. anorg. Chem.*, **172**, 1 (1928).
8. (Ti) L. Kirschfeld and A. Sieverts, *Z. physik. Chem.*, **A**, **145**, 227 (1929).
9. A. Sieverts, *Z. Metallkunde*, **21**, 37 (1929)—summary of previous results.
10. (Cb) H. Hagen and A. Sieverts, *Z. anorg. Chem.*, **185**, 225 (1930).
11. (Zr, Ta) A. Sieverts and A. Gotta, *Z. anorg. Chem.*, **187**, 155 (1930).
12. (V) L. Kirschfeld and A. Sieverts, *Z. Elektrochem.*, **36**, 123 (1930).
13. (Ti) A. Sieverts and A. Gotta, *Z. anorg. Chem.*, **199**, 384 (1931).
14. (Pd) H. Brüning and A. Sieverts, *Z. physik. Chem.*, **A**, **163**, 409 (1932).
15. (Pd) L. J. Gillespie and F. P. Hall, *J. Am. Chem. Soc.*, **48**, 1207 (1926).
16. (Pd) L. J. Gillespie and L. S. Galstaun, *J. Am. Chem. Soc.*, **58**, 2565 (1936).
17. (Ta) A. Sieverts and H. Brüning, *Z. physik. Chem.*, **A**, **174**, 365 (1935).
18. (Zr) M. N. A. Hall, S. L. H. Martin, and A. L. G. Rees, *Trans. Faraday Soc.*, **41**, 306 (1945). This paper contains an extensive list of references to previous publications.

For a comprehensive bibliography on the solubility of gases in these metals see O. Kubaschewski, *Z. Electrochem.*, **44**, 152 (1938).

We shall now consider, in more detail, the solubility data for each of the metals mentioned in Table 8.16 and the interpretation of the phenomena involved.

## Palladium

Of all the metals for which data are given in Table 8.16, *palladium* has received the most attention.[28] Whether in the form of foil, sponge, or palladium "black," the metal takes up as much as 900 volumes of hydrogen at normal pressure and temperature. The actual amount varies with the method for preparation as well as with the nature of the previous heat and exhaust treatment. The sorption at the temperature of liquid air is about ten times that observed at room temperature and is much greater than that observed for charcoal at the same low temperature.

Figure 8.10 shows plots of the isotherms as calculated by J. R. Lacher (see subsequent remarks) compared with the actual data (indicated by circles) obtained by Gillespie and Galstaun (16). The abscissas give values of the variable $\theta = r/0.59$ (used by Lacher in his interpretation of the observations). As will be observed, each isotherm exhibits a horizontal

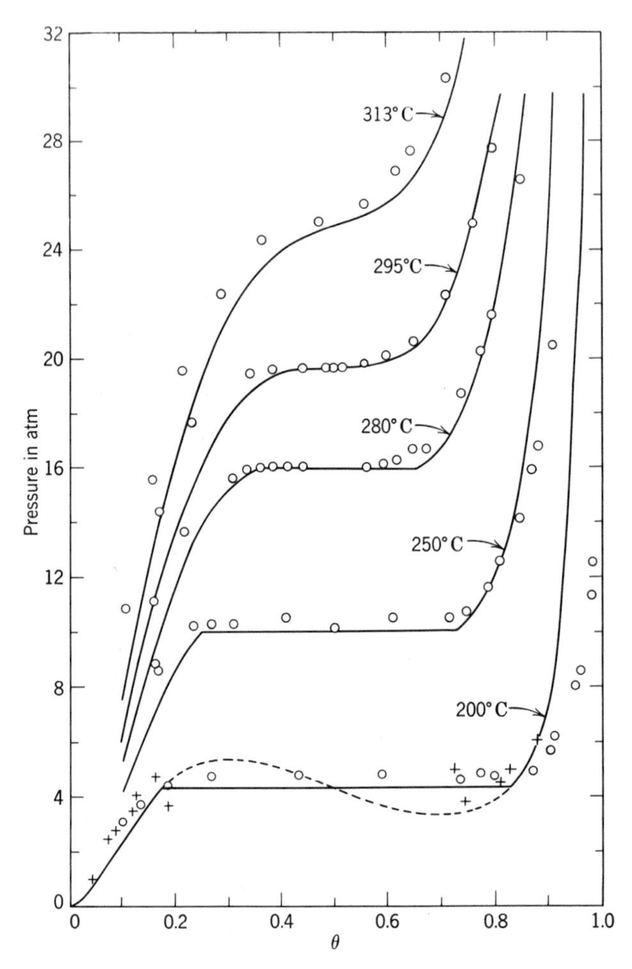


Fig. 8.10. Pressure-solubility isotherms for system hydrogen-palladium at a series of temperatures. $\theta = r/0.59$, where $r$ = number of hydrogen atoms per atom of palladium. The observed values are indicated by circles; the plots are those calculated according to theory. (Lacher).

portion for which the pressure of hydrogen is constant over a range of values of $r$. At room temperature the maximum value of $r$ for the horizontal portion is about 0.6. With increase in temperature the range of values of $r$ for which $P$ is constant decreases, and finally, at 295° C and 19.87 atm, there remains only a point of inflection at $r = 0.27$. Table 8.17 gives the values of $P$ associated with the horizontal portions at a series of temperatures.

TABLE 8.17

Dissociation Pressures of $PdH_r$ Alloys

| Temperature (° C) | $P_{atm}$ | $P_{mm}$ |
|---|---|---|
| 295 | 19.77 | |
| 250 | 10.28 | |
| 200 | 4.295 | |
| 150 | 1.455 | |
| 135 | 1.0 | 760 |
| 100 | | 223.4 |
| 80 | | 145.9 |
| 50 | | 46.8 |
| 30 | | 19.3 |
| 0 | | 4.0 |

According to Gillespie and Galstaun, these dissociation-pressure data are represented very satisfactorily by the relation

$$\log P_{atm} = 4.6018 - \frac{1877.82}{T}, \tag{8.14}$$

where $T =$ absolute temperature.

Above 300° C, the solubility of hydrogen in palladium decreases with increase in temperature, as shown in Table 8.16. It is of interest to note that the solubility of deuterium ($D_2$) is somewhat less than that of ordinary hydrogen.

Solubility values for hydrogen and deuterium in palladium at 760 mm and for the temperature range 300–1200° C have been determined by Sieverts and Zapf[29] and Sieverts and Danz.[30] These data are given in Table 8.18.

TABLE 8.18*

Values of (Solubility) $s$ for $H_2$ and $D_2$ at 1 atm, in Palladium

| $t$° C | $10^{-2}s_H$ | $10^{-2}s_D$ | $s_D/s_H$ |
|---|---|---|---|
| 300 | 1.64 | 1.10 | 0.67 |
| 400 | 1.26 | 0.938 | 0.74 |
| 500 | ... | 0.831 | ... |
| 600 | 0.927 | 0.758 | 0.82 |
| 700 | ... | 0.742 | ... |
| 800 | 0.84 | 0.729 | 0.87 |
| 900 | ... | 0.719 | ... |
| 1000 | 0.785 | 0.715 | 0.91 |
| 1100 | ... | 0.695 | ... |
| 1200 | 0.712 | ... | ... |

For the temperature range 600–1200° C, and pressures up to 1 atm, both hydrogen and deuterium obey the $P^{1/2}$ law.

A number of investigators have suggested explanations for these observations on the palladium-hydrogen system, but the most satisfactory interpretation is that given by Lacher.[31] As will be observed from the shape of the isotherms in Fig. 8.10, there is a region of values of $r$ for which the pressure remains constant. According to Lacher, "When $r$ is less than $r_\alpha$ or greater than $r_\beta$,[32] the pressure increases with increasing concentration." While equilibrium is established very rapidly in the regions of $r < r_\alpha$ and $r > r_\beta$, hysteresis effects have been observed for the region $r_\alpha < r < r_\beta$, which indicate that equilibrium is established much more slowly in this range of compositions. Also there is X-ray evidence for the conclusion that, in this region of constant pressure, the hydrogen-palladium system consists of "two coexistent expanded palladium lattices."

Furthermore, "at the critical temperature $r$ is 0.270. The value of $r_\alpha$, at which the second phase appears, decreases as the temperature decreases; the value of $r_\beta$, at which the first phase disappears, increases."

The fact that the observed maximum value of $r$ is about 0.6 is explained by Lacher on the basis of the electron configuration of the palladium atom as follows:

> Palladium is a transition metal and has two bands of allowed electron states, namely the $s$ and $d$ bands. These bands overlap. The $s$ band can accommodate two electrons per atom and the $d$ band ten. Palladium furnishes only ten, with the result that the two bands are not completely filled. Gold has one more electron than palladium, and when Pd is alloyed with Au, measurements of the susceptibility and electrical resistance can be explained by assuming that the extra electron goes into the $d$ band which is filled up when between 0.55 and 0.60 electrons per atom have been added.[33] Similarly, when palladium is alloyed with hydrogen, its paramagnetic susceptibility decreases and becomes zero when between 0.53 and 0.66 hydrogen atom per palladium atom have been added. This suggests that the dissolved hydrogen is almost completely ionized and that the electrons go into the $d$ shell. It suggests further that a definite process of hydrogen absorption will reach completion when $r = 0.6$ approximately.

In order to account for the observed isotherms, Lacher assumes the following mechanism for the process of solution of hydrogen in palladium. Let $M_s$ denote the number of hydrogen atoms per unit volume which enter into solution. These will be present in the metal as protons (nuclei of hydrogen atoms, that is, ionized atoms) and electrons. Let $N_s$ denote the number of available energy "holes" into which these protons may enter. "In order to explain the occurrence of critical phenomena, we must assume further that the energy of absorption depends on the number of holes filled, increasing as the number of holes filled increases." This assumption is similar to the one introduced to account for critical temperatures (or Curie points) in order-disorder phenomena.

The problem thus becomes similar to problems in that group of phenomena. Assuming that "the rate of increase in absorption energy is directly proportional to the fraction of holes filled" (that is, to $\theta = M_s/N_s$), and solving the problem by the methods of statistical mechanics, Lacher deduces a relation between $\theta$, $P$, and $T$ of the form

$$\log P_{mm} = 7.4826 + \frac{2 \log \theta}{1 - \theta} - \frac{891.2 + 1973.4\theta}{T}. \qquad (8.15)$$

From the value of the coefficient of $1/T$, it follows that the exothermic heat of solution, in calories per gram-mole of hydrogen, is

$$\begin{aligned} Q &= 4080 + 9028\theta \\ &= 8594 \quad \text{for} \quad \theta = 0.5, \end{aligned}$$

which is in good agreement with the value $Q = 8588$ derived from the measurements made by Gillespie and Galstaun. It is also in agreement with the observation that the energy evolution varies linearly with the concentration of hydrogen at lower temperatures.

The isotherms calculated by means of equation 8.15 are practically identical with those observed by Gillespie and Galstaun (16) and also by Brüning and Sieverts (14).

We shall quote from Lacher's paper once more, with regard to the physical interpretation of the equations which he has deduced.

When we add hydrogen to palladium, the pressure will increase continuously (according to the $\sqrt{P}$ law at first) until $M_s/N_s = \theta_\alpha$. In the region $0 \leqslant \theta \leqslant \theta_\alpha$, the hydrogen is distributed uniformly among the available holes, and there is only one solid phase—the α-phase. As more hydrogen is added, the pressure remains constant until $\theta = \theta_\beta$. In the region $\theta_\alpha < \theta < \theta_\beta$ the hydrogen cannot be uniformly distributed among the available holes. Instead, there are regions where the density is $\theta_\alpha$ and others where it is $\theta_\beta$—the α- and β-phases co-exist. It is in this concentration range that X-ray analysis shows the co-existence of two expanded palladium lattices. When one adds hydrogen, the β-phase increases at the expense of the α-phase until the system becomes homogeneous again at $\theta = \theta_\beta$. This, again, is what is found by X-ray analysis. When $\theta$ is slightly greater than $\theta_\alpha$, the intensity of the lines due to the α-phase is greater than those due to the β-phase. When hydrogen is added, the intensity of the lines due to the α-phase falls off while simultaneously the intensity due to the β-phase lines increases. When still more hydrogen is added, the equilibrium pressure increases again, and in the region $\theta_\beta \leqslant \theta \leqslant 1$ only the pure β-phase exists. As the temperature is raised, both $\theta_\alpha$ and $\theta_\beta$ approach each other, until at the critical temperature $\theta_\alpha = \theta_\beta = \frac{1}{2}$. Above this temperature only one phase can exist.

On the basis of this theory of the coexistence of two phases, it is possible, as Lacher shows, to interpret the hysteresis effects mentioned above.

For temperatures above 600° C, the solubility (cubic centimeters per 100 g Pd) of hydrogen ($s_{\mathrm{H}}$) and deuterium ($s_{\mathrm{D}}$) can be represented by the following equations:

$$s_{\mathrm{H}} = 48\epsilon^{1150/R_0T}\,(P_{atm})^{1/2}$$

$$s_{\mathrm{D}} = 60\epsilon^{410/R_0T}\,(P_{atm})^{1/2},$$

that is,

$$\log s_{\mathrm{H}} = 0.241 + 0.5 \log P_{mm} + \frac{251.4}{T}, \qquad (8.16)$$

$$\log s_{\mathrm{D}} = 0.338 + 0.5 \log P_{mm} + \frac{89.6}{T}. \qquad (8.17)$$

The solubility of hydrogen in a number of alloys of palladium has been investigated. Since a comprehensive discussion of the results obtained has been given by Smithells,[34] only some of the more salient features need be mentioned about these observations.

In alloys with platinum, the solubility of hydrogen is lower, and the decrease in solubility increases with increase in percentage of platinum. In alloys with silver, the solubility at constant temperature increases with increase in content of silver to a maximum value at about 40 per cent silver, and then decreases to a very low value for 70 per cent silver. The maximum values of the solubilities decrease with increase in temperature. Alloys of palladium with gold show similar characteristics with regard to solubility at lower temperatures (below about 400° C), but at higher temperatures the value of the solubility decreases continuously, with increase in percentage of gold, to a very low value for 80 per cent gold.

In boron-palladium alloys, at temperatures above 160° C, the solubility of hydrogen reaches a maximum value of about 1300 cm³/100 g for 6.9 atomic per cent of boron. For both smaller and larger contents of boron, the value of $s$ is lower, decreasing at all temperatures to a very low value for 20 atomic per cent of boron. With increase in temperature, the solubility decreases in the same manner as for pure palladium.

The literature on the palladium-hydrogen system has remarkably increased. There is a review by Darling.[35] Work by Norberg,[36] using nuclear magnetic resonance methods, gave information on the diffusion of the protons in the lattice.

Another review is by Lewis.[37] Commercial development of a palladium-alloy diffusion cell has been described.[38]

### Titanium (Refs. 6, 8, 9, 13)

Only the hydrides of lithium, sodium, and calcium have higher hydrogen content. Above 800° C, the solubility follows the $P^{1/2}$ law. Sorption

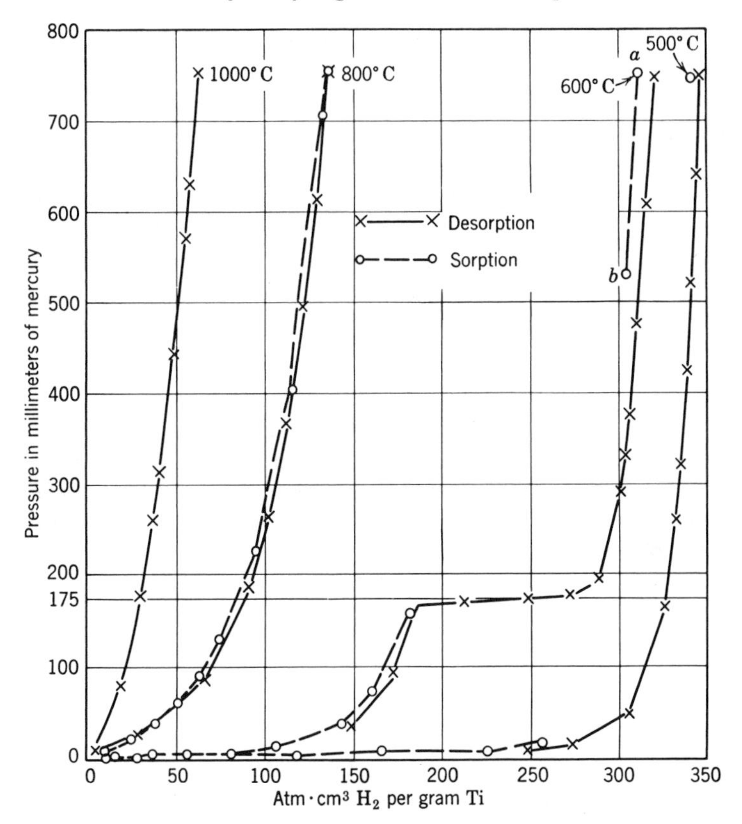


Fig. 8.11. Isotherms at four different temperatures for the system hydrogen-titanium (Sieverts).

begins at 375° C and occurs at a rapid rate above 400° C. Figure 8.11 shows isotherms at four different temperatures. The shape of the 600° C isotherm is accounted for, as in palladium, by the presence of two phases. Hägg[39] has demonstrated by X-ray investigation that the solid solutions consist of an $\alpha$ phase up to 35 atomic per cent, and a $\beta$ phase for 50 atomic per cent of hydrogen and higher. The $\alpha$ phase is a closely packed hexagonal lattice; the $\beta$ phase, a face-centered cubic lattice.

The maximum content of hydrogen in the solid solution corresponds to the formula $TiH_{1.75}$; and, as mentioned in the last row in Table 8.16, the density of the "hydride" is 3.91 as compared with $\rho = 4.52$ for titanium. The decrease in density varies linearly with the hydrogen content, and this fact indicates that the entrance of hydrogen atoms into interstitial sites in the metal lattice effects an expansion of the lattice. According to Hägg, the average value for the radius of the hydrogen atom is $0.44 \cdot 10^{-8}$ cm.

It is of interest to note that this value is *less* than the Bohr radius of the orbit of the electron ($0.5292 \cdot 10^{-8}$ cm) in the normal state of the hydrogen atom.

It is important to observe that in titanium, as in the other metals of Group B, maximum sorption is obtained only after the metal has been heated in vacuum for a prolonged period at a very high temperature. The metal thus degassed absorbs hydrogen very rapidly at a moderately high temperature. The sorption under these conditions is completely reversible. Sorption at lower temperatures is usually effected by cooling the metal, which has been saturated at the higher temperatures, in hydrogen.

### Vanadium (Refs. 6, 12)

The maximum hydrogen content corresponds to the chemical composition $VH_{0.72}$, and the density of this solid solution is 5.30 as compared with 5.87 for the metal. Figure 8.12 shows isotherms for solutions of hydrogen in vanadium. As indicated on the curves, the isotherms are reversible, and all of them follow the $(P)^{1/2}$ law.

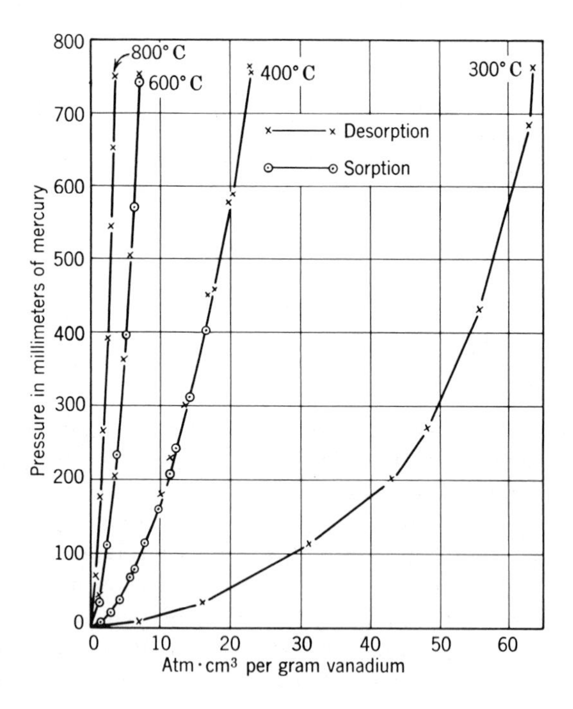


Fig. 8.12. Isotherms at four different temperatures for the system hydrogen-vanadium (Kirschfeld and Sieverts).

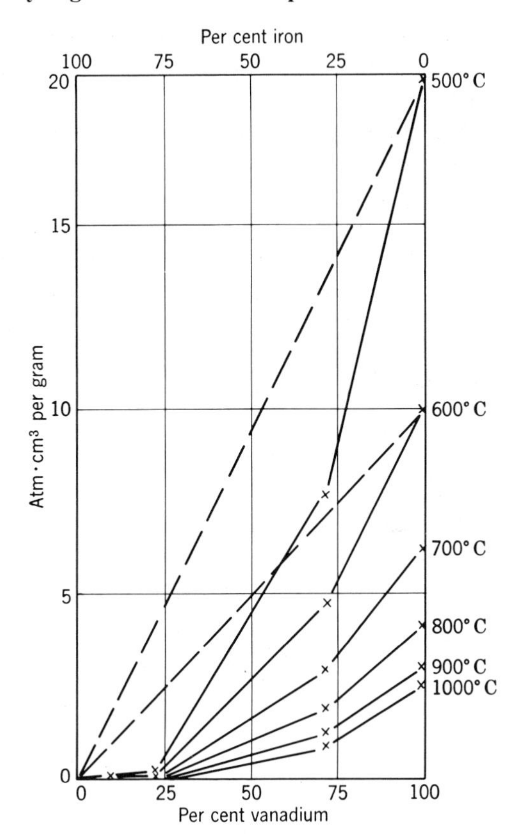


Fig. 8.13. Isobars at 1 atm and a series of temperatures for the sorption of hydrogen in iron-vanadium alloys (Kirschfeld and Sieverts).

Kirschfeld and Sieverts also investigated the solubility of hydrogen in three ferrovanadium alloys. Isobars for these alloys at a series of temperatures are shown in Fig. 8.13.

## Zirconium (Refs. 5, 11, 18)

Sieverts and Roell (5) gave the metal a preliminary treatment at 800–1100° C in vacuum, and rapid absorption of hydrogen at atmospheric pressure was observed above 700° C. Above 1000° C, the isotherms follow the $P^{1/2}$ law. The maximum hydrogen content corresponds to the composition $ZrH_{1.92}$, and the density of this composition is 5.67 as compared with 6.4 for pure zirconium.

Hägg[39] has concluded, from X-ray investigations, that, beginning with an $\alpha$ phase (close-packed hexagonal lattice) at 5 atomic per cent hydrogen, a $\gamma$ phase of similar structure but more expanded lattice dimensions is observed for a solution containing 33 atomic per cent hydrogen. At higher temperatures a $\beta$ phase (face-centered cubic lattice) is observed at about 20 atomic per cent hydrogen. At 50 atomic per cent, a $\delta$ phase (face-centered cubic lattice) is observed, and at the highest concentration an $\epsilon$ phase is formed in which the metal atoms form a face-centered tetragonal lattice.

Observations on the solubility of hydrogen in zirconium have also been

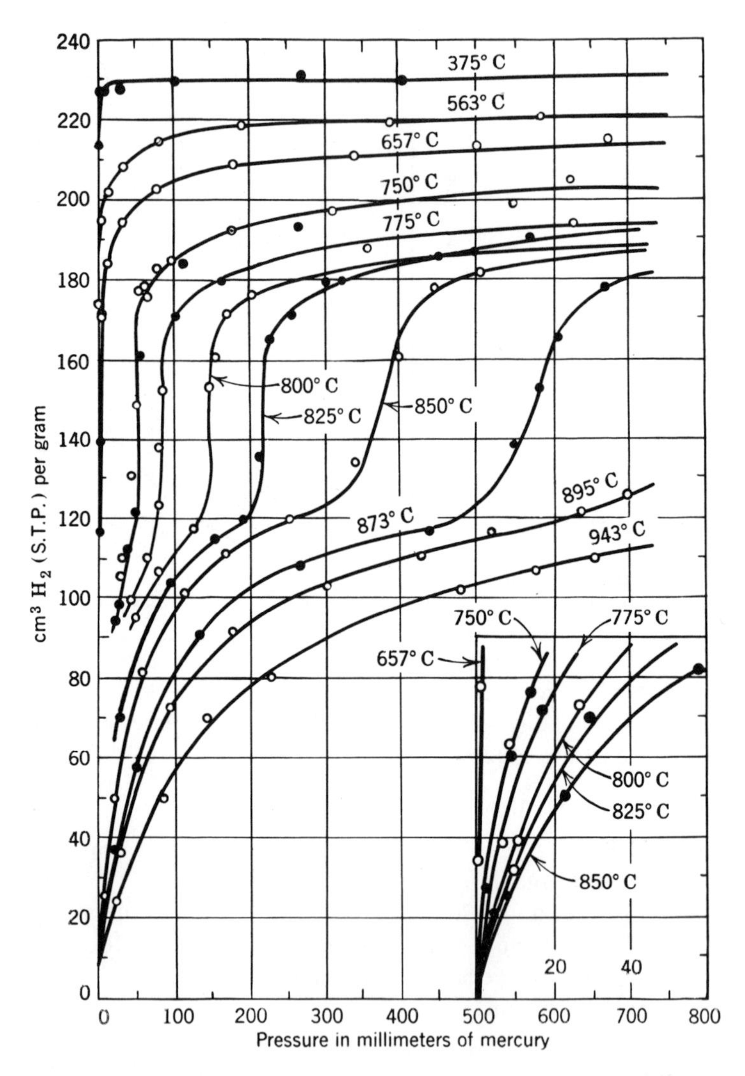


Fig. 8.14. Isotherms at various temperatures for the sorption of hydrogen by compact sample of zirconium containing 0.023 atom oxygen per atom zirconium. Whole series on same sample of 0.04093 g, with duplicate sets of points for 750° C. *Inset:* initial portion of isotherms at lower pressures. (Hall, Martin, and Rees.)

published by Hall, Martin, and Rees (18). Figure 8.14 shows isotherms for a compact sample of the metal containing 0.023 atom oxygen per atom zirconium. The inset shows the initial portions of the isotherms on an expanded pressure scale. The ordinates give the solubility in cubic centimeters $H_2$ (STP) per gram of metal.

Isotherms obtained for ductile zirconium wire and zirconium powder exhibit similar characteristics, although the values for the solubility at the same temperature and pressure may differ by a few per cent. The authors ascribe these variations to contamination by oxygen and nitrogen. From observations on metal containing different amounts of oxygen up to about 1 atom oxygen per atom zirconium, the tentative conclusion is drawn that "with zirconium-oxygen solid solutions the volume of hydrogen sorbed at saturation is decreased by a volume equivalent to that of oxygen present."

The isobar at 760 mm for zirconium containing 0.058 atom oxygen or less is substantially the same as that observed by Sieverts and Roell (5), and the volume of hydrogen dissolved at 20° C (on cooling from 400 or 825° C) was observed to be about 240 cm³ (STP) per gram.

**Columbium (Ref. 10)**

The isotherms for 300 and 400° C do not follow the $P^{1/2}$ law, which applies approximately to isotherms at 500° C and higher. On heating the metal to 1100° C in a good vacuum and letting it cool to room temperature, sorption of hydrogen occurs very rapidly. On repeated sorption and desorption, the amount going into solution decreases (aging phenomenon). The maximum solubility corresponds to $r = 0.47$, which is the lowest value observed for any of the metals of Group B.

**Cerium (Refs. 2, 3, 7, 9), Lanthanum (3, 7), Praseodymium (4), and Neodymium (4)**

The isobars for the first three metals are almost identical; that for neodymium is somewhat lower. All these metals form hydrides for which the values of $r$ range around 2.7. As will be observed from the data in Table 8.16 and Fig. 8.9, the solubility of hydrogen in cerium and lanthanum, as well as in zirconium, decreases only slowly with increase in temperature. Even at 1000° C, the solubility in lanthanum and cerium is only about half of that observed at room temperature.

The isotherms resemble to a certain extent those observed for the hydrogen-palladium system. At higher pressures, they exhibit a steep increase in pressure with increase in solubility of hydrogen, of the same nature as that observed with the hydrogen-palladium systems for the $\beta$ phase.

The following data are given to illustrate the relatively large sorptions observed at low pressures:

| Metal | $P_{mm}$ | $t^\circ$ C | $cm^3/g$ |
|---|---|---|---|
| LaCe[40] | 1.0 | 600 | 140 |
| | 0.2 | 800 | 3.2 |
| Ce | 0.2 | 800 | 3.24 |
| | 0.5 | 600 | 135.6 |

**Tantalum (Refs. 1, 11, 17)**

Pirani[41] observed that this metal when heated in hydrogen could occlude about 740 times its volume of the gas. On subsequent heating in vacuum about 550 volumes were given off, and the rest of the gas could be removed only at the melting point of the metal. Pirani also observed that the sorbed hydrogen makes a tantalum filament quite brittle and increases its electrical resistivity—observations that will be referred to in a subsequent section.

A more careful investigation by Sieverts and Bergner (1) showed that a wire heated in a vacuum to 1200° C absorbs hydrogen slowly at temperatures above 500° C. However, once the wire is saturated with gas at higher temperatures, it absorbs very much more readily at lower temperatures. The same investigators also observed that the sorption follows the $P^{1/2}$ law. The values of $v_0$ obtained by these investigators at a series of temperatures are shown in Table 8.19.

TABLE 8.19

**Sorption of Hydrogen at 1 atm by Tantalum**

(Sieverts and Bergner)

| ° C | $v_0$ ($cm^3/g$) | ° C | $v_0$ ($cm^3/g$) |
|---|---|---|---|
| 100 | 44.5 | 730 | 3.71 |
| 183 | 41.9 | 830 | 2.25 |
| 263 | 36.4 | 930 | 1.66 |
| 314 | 33.0 | 1030 | 1.32 |
| 474 | 17.5 | 1130 | 1.07 |
| 530 | 11.9 | 1230 | 0.89 |
| 630 | 5.69 | | |

As observed with titanium and zirconium, the alloys with hydrogen form three phases. Up to about 12 atomic per cent hydrogen the structure is body-centered cubic ($\alpha$ phase); between 31 and 35 atomic per cent hydrogen, the structure is that of a closely packed hexagonal lattice

($\beta$ phase); and between 48 and 52 atomic per cent a $\gamma$ phase is formed, which has the structure of a slightly deformed body-centered cubic lattice.

**Thorium (Ref. 5)**

The metal powder used by Sieverts and Roell was 96 per cent pure. After a preliminary heating at 800–1100° C in vacuum, sorption of hydrogen was observed to occur at 400° C. At 475° C the rate of sorption was quite rapid.

The sorption and desorption isotherms did not coincide at any temperature. This is illustrated by the observations recorded in Table 8.20. The

TABLE 8.20

**Sorption and Desorption Data for Hydrogen in Thorium**

| $t$° C | $P_{mm}$ | $v_0$ (cm³/g) | $t$° C | $P_{mm}$ | $v_0$ (cm³/g) |
|---|---|---|---|---|---|
| 530 | 760 | 95.6 | 800 | 754 | 82.4 |
| | Cooled in $H_2$ | | | Cooled in $H_2$ | |
| 25 | 754 | 136.6 | 25 | 752 | 141.0 |
| 800 | 754 | 86.4 | 1100 | 756 | 27.5 |
| | 298 | 84.0 | | 288 | 23.6 |
| | 167 | 79.8 | | 0 | 19.6 |
| | 69.7 | 42.6 | | 467 | (29.5) |
| | 27.8 | 25.9 | | 757 | (40.3) |
| | 2.1 | 17.6 | | | |

metal was first heated in vacuum to 1100° C, and cooled. Hydrogen was then admitted and the metal heated to 530° C.

As for the other metals of Group B, the maximum solubility of hydrogen in thorium does not correspond to a stoichiometric composition, since $r$ has the value 3.07.

Summarizing the data given in Table 8.16 and the isobars shown in Fig. 8.9, it is of interest to compare the relation between these isobars and the positions of the metals in the periodic arrangement. According to the periodic arrangement the metals of Group B fall in the following three groups: (*a*) Ti, Zr, Th; (*b*) La, Ce, Pr, Nd; (*c*) V, Cb, Ta.

In each group, the solubility decreases with increase in atomic number of the metal; while the solubility is highest for titanium, it is lowest for tantalum.

As has been mentioned previously, *heat is evolved* in the formation of these solid solutions. The values of these heats of formation, $Q$, in calories per gram-mole $H_2$ are given in Table 8.21; for comparison, there are also

TABLE 8.21

**Heats of Formation of Hydrogen Alloys and Hydrides**

| Hydrogen Alloys | $Q$ (cal/mole $H_2$) | Hydrides | $Q$ (cal/mole $H_2$) |
|---|---|---|---|
| $PdH_{0.59}$ | 8,588 | LiH | 43,000 |
| $TiH_{1.75}$ | 36,000 (13) | NaH | 26,280 |
| $ZrH_{1.92}$ | 40,500 (11) | KH | 28,200 |
| $LaH_{2.8}$ | 40,090 (7) | $CaH_2$ | 48,880 |
| $CeH_{2.7}$ | 42,260 (7) | $SrH_2$ | 42,060 |
| $PrH_{2.8}$ | 39,520 (7) | $BaH_2$ | 40,860 |

given values for the hydrides of the alkali and alkaline-earth metals. The numbers in parentheses give the references in List II.

As will be observed, the values of $Q$ for the "hydrides" of the metals of Group B are comparable with those for the hydrides of the alkali and alkaline-earth metals. (However, the similarity between the two classes of hydrides does not extend beyond the fact that the heats of solution are of the same order of magnitude as the heats of formation of the chemically defined hydrides. On heating, the latter dissociate, and for each temperature there exists a definite dissociation pressure of hydrogen which can be expressed as a function of the temperature by a relation similar to that used for expressing the vapor pressure of a solid or liquid. Consequently these observations are discussed in Chapter 11 along with the observations on the dissociation of nitrides and oxides.)

Because of the high values of the solubility of hydrogen, as well as other gases, in titanium, zirconium, and thorium, these metals have been utilized in the production of electronic tubes as "getters," since they improve the vacuum in a sealed-off device. (This topic is discussed at greater length in Chapter 9.)

One other characteristic of the metals of Group B must also be mentioned in this connection. All of them, including alloys which contain fairly appreciable percentages of other metals, exhibit an *increase in the electrical resistivity* when hydrogen is taken up.

Denoting the ratio of the resistivity of the hydride to that of the pure metal by $Z$, Sieverts and Brüning[42] showed that in the case of *tantalum*

$$Z - 1 = Kr, \tag{8.18}$$

where $r$ has the same significance as in Table 8.16, and $K = 1.6$ for 500° C and 600° C for all values of $r$. (For 400° C, $K = 1.8$.) That is, the resistivity increases linearly with the concentration of hydrogen.

From observations on palladium wires, the same investigators[43] found that equation 8.18 is valid only for lower concentrations of hydrogen, because of the phase changes which occur at higher concentrations. The maximum value of $Z$, observed for $r = 0.5$, was about 1.5, independently of the temperature in the range 160–310° C.

Hagen and Sieverts[44] obtained similar results with palladium at higher temperatures and higher pressures of hydrogen, as well as with Pd-Ag and Pd-Au alloys.[45]

The fact that drawn tantalum wire becomes brittle when heated in hydrogen is well known. This is no doubt the result of the straining of the lattice structure due to hydrogen absorption. Other metals, for example, copper, are also embrittled by hydrogen.

In general, dissolved gases affect the mechanical properties of metals,[46] and a large number of investigations published in metallurgical journals have dealt with this topic.

## 8.5. DERIVATION OF RELATIONS FOR THE SOLUBILITY OF HYDROGEN IN METALS

By applying the methods of statistical mechanics, Fowler and Smithells[47] have been able to deduce relations for the solubility of hydrogen which apply to each of the two groups A and B.

The plots shown in Fig. 8.15, taken from their paper, serve to emphasize the radical differences between the two groups. The ordinates give the values of $\log_e s$, where $s = 10^2 v_0 = \text{cm}^3$ (STP) per 100 g; and the abscissas give values of $10^4/T$. Thus the slopes of the lines are proportional to the heats of solution. As will be observed, these slopes for metals of Group A are in opposite direction to those for metals of Group B, which corresponds, as mentioned previously, to the fact that heat is absorbed in metals of Group A when hydrogen goes into solution, and heat is evolved in metals of Group B.

Fowler and Smithells assume that in metals of Group A the protons move quite freely throughout the lattice, because of their low concentration, whereas in metals of Group B the protons become "localized" as a result of their high concentration, which brings into play binding forces between protons and metal atoms.

For *metals of Group A*, the relation deduced for $s$ is

$$\log s = 2.774 + 0.5 \log P_{mm} - 0.25 \log T - \log \rho - \frac{Q_S}{9.148T}, \quad (8.19)$$

where ordinary logs are used, $\rho$ = density, and $Q_S$ = heat of solution in calories per gram-mole of hydrogen. (Values of $Q_S$ are given in one of the lower rows of Table 8.15.)

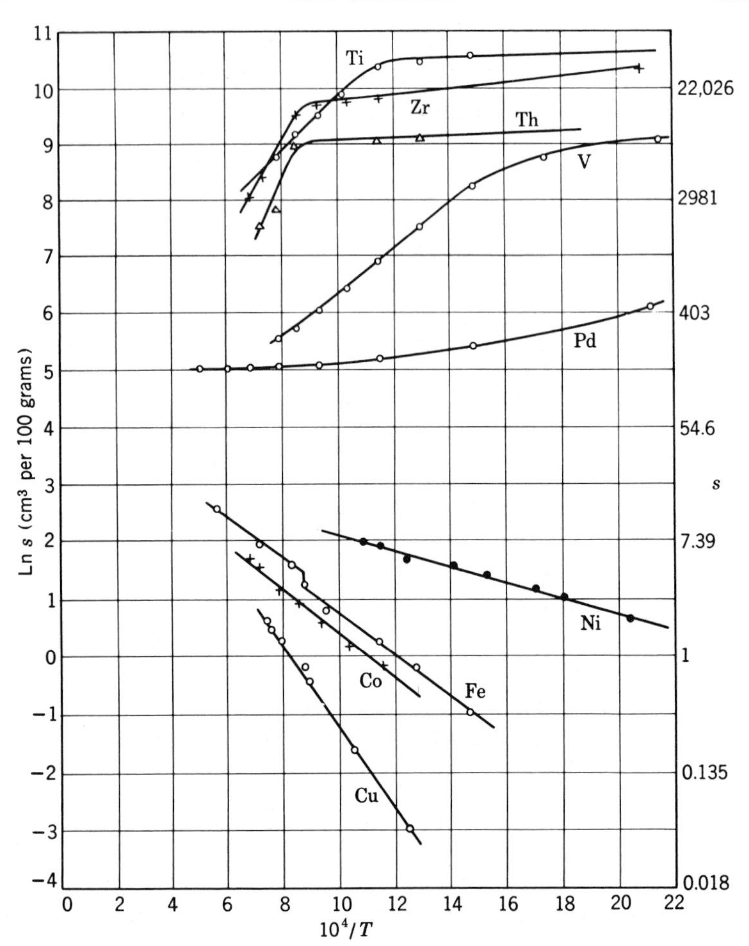


Fig. 8.15. Plots of ln $s$ versus $1/T$ for metals of Groups A and B. The values of $s$ (cm³, STP, per 100 g) are indicated for corresponding values of ln $s$. (Fowler and Smithells.)

For *metals of Group B*, the relation deduced is

$$\log\left(\frac{s}{s_0 - s}\right) = 0.242 + 0.5 \log P_{mm} - 1.75 \log T + \frac{Q}{9.148T}, \quad (8.20)$$

where $s_0$ = solubility at saturation, at room temperature, and $Q$ = heat evolved in calories per gram-mole of hydrogen.

Many of the values of $Q_S$ and $Q$ derived by Fowler and Smithells (designated by FS) are quite different from those given in Table 8.15 and 8.21, as shown by the data in Table 8.22.

TABLE 8.22

Heats of Solution of Hydrogen in Metals

| | $Q_S$ (cal/mole $H_2$) | | | $Q$ (cal/mole $H_2$) | |
|---|---|---|---|---|---|
| Metal | Table 8.15 | FS | Metal | Table 8.20 | FS |
| Cu | 18,300 | 14,100 | Ti | 36,000 | 10,000 |
| Co | 15,500 | 7,300 | Zr | 40,500 | 17,500 |
| Fe | 13,300 | 7,000 | Th | ... | 22,500 |
| Ni | 5,900 | 5,600 | V | ... | 7,700 |
| Al | 43,400 | 45,500 | Pd | 8,588 | 2,040 |
| Pt | 34,500 | 85,400 | | | |
| Mo | 14,000 | 3,500 | | | |
| Ag | 11,900 | 11,600 | | | |

Applying equation 8.19 to the solubility data for nickel and iron and using the values of $Q_S$ given in Table 8.15, the calculated values for $s$ are found to be about 10 times those observed.

Similar results are obtained by application of equation 8.20 to vanadium, for instance, by using the value for $Q$ given in Table 8.21 and the value of $s_0$ given in Table 8.16.

## 8.6. SOLUBILITY OF HYDROGEN IN PALLADIUM-LIKE METALS AND PALLADIUM BLACK

Although none of the other metals of the palladium and platinum group (the "noble" metals) behave like palladium with respect to hydrogen, they do absorb hydrogen to a very noticeable extent even at room temperature, when in a *finely divided state*. Table 8.23 gives absorption data at 1 atm

TABLE 8.23

Sorption Data for Hydrogen by "Noble" Metals at 18° C and 1 atm

| Metal: | Ru | Rh | Pd | Os | Ir | Pt |
|---|---|---|---|---|---|---|
| $v_0$: | 124 | 101 | 79 | 74 | 49 | 35 |
| $V_g$: | 1520 | 1240 | 910 | 1660 | 1100 | 760 |
| $A$: | 101.7 | 102.9 | 106.7 | 190.2 | 193.1 | 195.2 |
| $\rho$: | 12.2 | 12.5 | 11.4 | 22.48 | 22.42 | 21.45 |
| $r$: | 1.12 | 0.93 | 0.75 | 1.26 | 1.0 | 0.61 |

observed by Müller and Schwabe.[48] The metals were obtained in powder form by reduction of the oxides, and were then saturated with nitrogen and evacuated at 18° C.

It will be observed that the hydrogen absorbed corresponded to a value of the order of 1 atom hydrogen per atom of metal. The values given in the table were the highest sorptions observed. Temperature of reduction and duration of treatment in hydrogen affected the sorption to a considerable extent.

With the exception of palladium (for which the value of the amount absorbed is even higher than that recorded in Table 8.16), it was observed that the sorption capacity depended upon the extent of the surface. This indicates that the sorption in these metals is due either to physical adsorption or, probably, to chemisorption. Müller and Schwabe have used as an argument in favor of this conclusion the data obtained by Mond, Ramsay, and Shields[49] on the sorption of hydrogen at room temperature by platinum foil which had been "activated" by heat treatment in hydrogen. These data are given in Table 8.24.

**TABLE 8.24**

**Sorption ($v_0$) at Room Temperature for Hydrogen by Activated Pt Foil**

| Series I | | | Series II | | |
|---|---|---|---|---|---|
| $P_{mm}$ | $v_0$ | $P_{mm}/v_0^2$ | $P_{mm}$ | $v_0$ | $P_{mm}/v_0^2$ |
| 11.1 | 1.877 | 3.150 | 22.4 | 1.888 | 6.284 |
| 30.9 | 1.941 | 8.204 | 59.5 | 1.937 | 15.85 |
| 176.5 | 2.000 | 44.11 | 125.7 | 1.963 | 32.60 |
| 371.4 | 2.015 | 91.47 | 264.5 | 2.012 | 65.31 |
| 767.0 | 2.030 | 186.1 | 416.9 | 2.085 | 95.85 |
| | | | 767.0 | 2.085 | 176.4 |

Müller and Schwabe found that the two series of observations recorded in Table 8.24 could be represented by the Zeise equation (see Chapter 7),

$$v_0 = \left(\frac{k_1 k_2 P}{1 + k_2 P}\right)^{1/2} = \text{cm}^3/\text{g}. \tag{8.21}$$

In accordance with this relation, a plot of $P/v_0^2$ against $P$ is found to be linear.

Isobars for some of the metals under discussion have also been determined by Gutbier and Schieferdecker.[50] From the plots of these observations at 1 atm the interpolated values have been obtained which are given in Table 8.25.

The fourth column gives the results for a Heraeus gray powder, and the next column gives results for the same powder after being heated to 460° C. As will be observed, the values of $v_0$ decrease approximately linearly with the temperature.

TABLE 8.25

**Sorption ($v_0$) at 1 atm for Hydrogen by Pd-like Metals**

| $t$° C | Pd | Ir (black) | Ir (powder) | Ir (460° C) | Ru (black) | Os (black) |
|---|---|---|---|---|---|---|
| 20 | 70 | 5.61 | 1.10 | 0.58 | 1.20 | 3.15 |
| 60 | 64 | 4.90 | 0.85 | 0.47 | 0.90 | 2.57 |
| 100 | 59 | 4.45 | 0.80 | 0.43 | 0.78 | 2.24 |
| 140 | 52 | 4.05 | 0.74 | 0.40 | 0.66 | 2.06 |
| 150 | 32.8 | 3.95 | 0.73 | 0.38 | 0.62 | 2.02 |
| 160 | 11.4 | 3.80 | 0.71 | 0.37 | 0.60 | 2.00 |
| 180 | 5.7 | 3.60 | 0.69 | 0.35 | 0.55 | 1.94 |
| 190 | 4.6 | 3.55 | 0.67 | 0.33 | 0.52 | 1.92 |

By heating the powders to temperatures not exceeding 350° C, the sorption at lower temperatures was increased. It was also observed that the sorption of carbon monoxide by iridium was about the same as that of hydrogen.

Benton[51] observed that "platinum black" at 25° C and 1 atm absorbed 36.7 volumes of hydrogen, 37.8 volumes of carbon monoxide, and 20.4 volumes of oxygen. The sorption of hydrogen per gram was about 1.68 cm³ at 760 mm and about 1.38 cm³ at 1 mm.

The sorption by *palladium black* at liquid-air temperature is about ten times that observed at room temperature and exceeds that observed for charcoal. The data obtained by Valentiner[52] for the sorption of hydrogen by palladium black at three temperatures are shown in Table 8.26.

TABLE 8.26

**Sorption of Hydrogen by Palladium Black**

(Valentiner)

| $t = 20$° C | | $t = -20$° C | | $t = -190$° C | |
|---|---|---|---|---|---|
| $P_{mm}$ | $v_0$ | $P_{mm}$ | $v_0$ | $P_{mm}$ | $v_0$ |
| 0.001 | 0.10 | 0.014 | 0.27 | 0.0005 | 2.05 |
| 0.005 | 0.26 | 0.031 | 0.33 | 0.0015 | 2.11 |
| 0.037 | 0.40 | 0.056 | 0.37 | 0.001 | 3.06 |
| 0.110 | 0.52 | 0.087 | 0.41 | 0.001 | 9.1 |
| 0.190 | 0.59 | 0.184 | 0.49 | 0.002 | 33.0 |
| 0.315 | 0.70 | 0.30 | 0.55 | 0.005 | 40.0 |
| 0.52 | 0.82 | 0.52 | 0.61 | 0.012 | 47.2 |
| 0.76 | 0.92 | 0.88 | 0.67 | 0.025 | 63.0 |

Since 1 $cm^3$ (STP) is equal to 760 micron · liters (at 0° C) these sorptions, especially at room temperature and lower temperatures, and at pressures as low as 0.5 micron, are remarkably high. Consequently experiments were carried out by Dushman and his associates, during 1918, on the possibility of utilizing palladium as an adsorbent for the evacuation of small radio vacuum tubes.

The method of preparation has been described by C. Hoitsema.[53] With slight variations this method was used by Dushman as follows.

The palladium in the form of sheet or wire is dissolved in aqua regia and evaporated on a water bath till acid vapors have disappeared; the solution is then diluted and warmed, and concentrated solution of sodium carbonate is added to neutralize free acid. A slight amount of acetic acid is then added, the solution is warmed, and a warm concentrated solution of sodium formate is added. The palladium comes down as a black flocculent precipitate which settles rapidly at the bottom of the beaker. The supernatant liquid is decanted, and the precipitate washed with distilled water till the wash water shows no traces of chlorides. The palladium "black" is then washed with alcohol and transferred to a U-tube, where it is dried by blowing air over it and evacuated on a rough pump. The U-tube ought to have side tubes through which gas can be passed and constrictions at which it can be sealed off later. After the rough evacuation (with slight warming of the U-tube), hydrogen is passed over the palladium black for some time, and while the gas is still passing through the tube the tube is sealed off at the constrictions. This leaves the palladium black in equilibrium with hydrogen, and it can be kept active for a long time.

For exhaust work, the U-tube is opened and a sample transferred to a tube such as is used in sorption by charcoal. It is well to cover the top of the palladium black with glass wool in order to prevent it from being drawn into the rest of the apparatus when vacuum is applied.

A number of experiments were carried out in the General Electric Research Laboratory by A. G. Huntley, Miss M. Daly, and Dushman. While the behavior of palladium black was found to be extremely erratic, the results, as illustrated by the following experiment, showed that it is possible to obtain samples possessing very high absorbing power.

An ionization gauge with an appendix containing about 1 g of palladium black was well exhausted on a condensation pump and sealed off at a residual gas pressure of about 0.2 micron (the gas consisting probably of nitrogen and hydrogen). On immersion of the appendix in liquid air, the pressure decreased to 0.003 micron with the gauge filament lighted. On turning off the filament for some time, and then lighting it for an instant, the pressure was observed to have decreased still further to 0.0004 micron. Apparently, there is a continual slight evolution of gas from the walls of the gauge and filament leads even after the metal parts have been bombarded for a long time. In other experiments pressures as low as 0.0001 micron were obtained in a sealed-off gauge with a palladium tube immersed in liquid air.

A number of experiments were carried out using palladium black for absorbing the residual gases in a small hot-cathode diode exhausted on an oil pump only. An ordinary lamp exhaust system was used, giving an exhaust of about 1 micron. A few *milligrams* of palladium black were placed in a diode (about 100 $cm^3$ volume) which contained a 6-volt 2.5-amp tungsten filament and a cylindrical molybdenum anode about $\frac{1}{4}$ in. in diameter by $\frac{7}{8}$ in. in length. The tube was exhausted on the oil pump, with simultaneous heating in an oven for 30 min at 360° C, and sealed off. The metal cylinder was then bombarded to a white heat by making the filament cathode. The gases evolved were absorbed rapidly by the palladium black, in spite of its being above room temperature, and finally the vacuum became so good that excellent space-charge characteristics were obtained. Special experiments showed that in order to obtain this condition the pressure must be at least as low as 0.05 micron. Thus even with a few milligrams of palladium black at room temperature it was possible to clean up appreciable quantities of gas. Similar results were obtained repeatedly. In fact a large number of small diodes and triodes were exhausted in this manner, with the regular exhaust system used in lamp factories, and without a mercury-vapor pump or liquid air.

As subsequent investigation showed that the same results could be obtained with the very much cheaper activated charcoal, and, furthermore, as some samples of palladium black absolutely failed for some undetermined reason to act as absorbents, this method was used for only a short time. The results, however, suggest interesting possibilities in the production of high vacua by means of palladium black, and further investigation ought to be carried out to determine definitely the conditions under which it can be made active. It has been shown by Maxted[54] that hydrogen sulfide inhibits the absorbing efficiency of palladium black. Similar facts have been known for a long time about various metallic catalysts, and probably the same causes influence the behavior of palladium black.

## 8.7. SOLUBILITY OF NITROGEN IN METALS

According to Smithells,[55]

> Nitrogen is only soluble in those metals which are capable of forming nitrides. . . . Nitrogen has been shown to be insoluble, within the limits of the experimental method, in cobalt, copper, silver, and gold. Those metals which, like iron and molybdenum, absorb nitrogen show the same kind of changes in

lattice structure, mechanical and electrical properties as are shown when the hydride-forming elements absorb hydrogen.

When iron and other nitride-forming metals are heated in nitrogen, the amount of nitrogen which actually enters into solution is of the order of 1 per cent or less by weight, and there is no tendency to form a nitride.

The most reliable data on the solubility of *nitrogen in iron* are those obtained by Sieverts[56] and by Sieverts, Zapf, and Mortiz.[57] Table 8.27

TABLE 8.27

**Solubility (cm³/100 g) of Nitrogen at 1 atm in Iron**

| | | | | | | | | | | |
|---|---|---|---|---|---|---|---|---|---|---|
| °C: | 750 | 890 | 900 | | | | | | | |
| $s$: | 0.32 | 1.60 | 20. | | | | | | | |
| °C: | 1200 | 1300 | 1400 | 1450 | 1390 | | | | | |
| $(P_{mm} = 754)s$: | 19.2 | 18.2 | 13.9 | 9.3 | 17.2 | | | | | |
| °C: | 1300 | 1390 | 1400 | 1420 | 1460 | 1390 | 1070 | 920 | 840 | 805 |
| $(P_{mm} = 758)s$: | 16.8 | 16.6 | 10.6 | 8.1 | 8.5 | 16.6 | 19.0 | 9.3 | 4.6 | 1.1 |
| °C: | 1310 | 1380 | 1440 | 1500 | 1540 | 1460 | | | | |
| $(P_{mm} = 744s)$: | 17.4 | 17.1 | 8.8 | 9.7 | 24.6 | 10.5 | | | | |
| °C: | 1540 | 1560 | 1560 | | | | | | | |
| $(P_{mm} = 760)s$: | 24.6 | 26.7 | 27.9 | | | | | | | |

For $s = 25$, $V_g = 1.57$ and $10^4 r = 10$; also $x = 1.25s$.

gives the values of $s$ at approximately 1 atm pressure, as observed on a sample of Armco iron. (The actual pressures in millimeters are indicated in parentheses.) The data for the temperatures 750–900° C are those published in 1931; the other data are those given in the second paper. Figure 8.16, taken from this paper, shows the isobar at 1 atm. As shown by this plot, the value of $s$ exhibits a considerable increase at the $\alpha$-$\gamma$ transition, a decrease at the $\gamma$-$\delta$ transition, and, again, quite an increase at the melting point.

The solubility obeys the $P^{1/2}$ law, which signifies that the nitrogen dissolves in iron as *atoms*.

Comparing the solubility of nitrogen with that of hydrogen in iron, it is seen that hydrogen is more soluble in $\alpha$-iron, but above 900° C the solubilities are of the same order of magnitude.

The solubility of nitrogen in *molten iron* has been investigated by Chipman and Murphy.[58] At about 740 mm pressure, the solubility is 0.039 per cent (by weight) just above the melting point, and 0.04 per cent at 1760° C. This corresponds to a value $s = 32$ cm³/100 g iron. It was observed that in this case also $s$ varies as $P^{1/2}$.

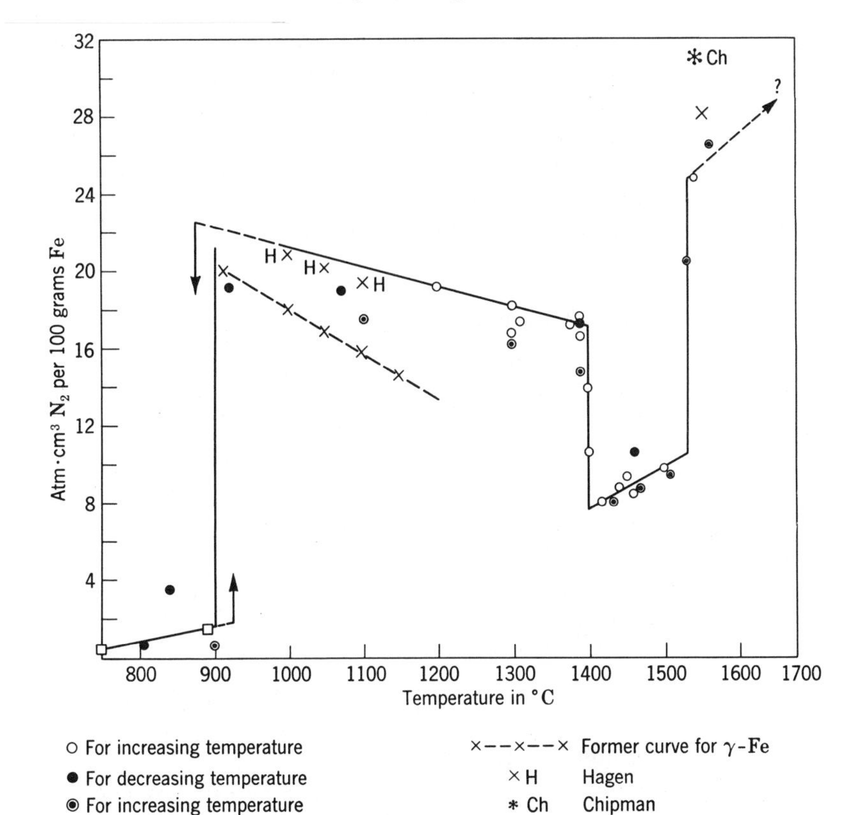


Fig. 8.16. Isobar for the solubility of nitrogen at 1 atm in Armco iron. Data obtained by other investigators are also shown. (Sieverts, Zapf, and Moritz.)

These investigators also observed that the rate of solution of nitrogen in molten iron is given by the simple relation

$$\frac{ds}{dt} = k(s_0 - s), \tag{8.22}$$

where $s_0$ is the saturation solubility. Small concentrations of aluminum or silicon in the iron caused the rate of solution to increase ten- to twentyfold.

The solubility data are in agreement with the observation that, on degassing iron, the volume of gas evolved rarely exceeds about 20 cm³/100 g. The rate of gas evolution is very low at 800° C on account of the low rate of diffusion of the dissolved gas but increases rapidly with the temperature.

Although carbon monoxide is probably soluble in iron to some extent, no solubility data are available. The observed evolution of carbon monoxide and dioxide on degassing iron may be accounted for (as in the case of nickel) by diffusion of carbon and reduction of oxides in the metal.

The solubility of *nitrogen in molybdenum* has been investigated in the range 936–1168° C by Sieverts and Brüning,[59] and in the range 1200–2400° C by Norton and Marshall.[60] Table 8.28 gives the solubility data observed by the first-mentioned investigators for molybdenum wire. The solubility follows the $P^{1/2}$ law at all temperatures.

**TABLE 8.28**

**Solubility of Nitrogen at 753 mm in Molybdenum Wire**

| $t$° C | $s$ = cm³/100 g | $x$ = mg/100 g* | $10^3 r$* |
|---|---|---|---|
| 936 | 58.8 | 73.51 | 50.4 |
| 1020 | 29.1 | 36.34 | 24.9 |
| 1045 | 24.7 | 30.90 | 21.2 |
| 1081 | 20.1 | 25.11 | 17.2 |
| 1118 | 16.5 | 20.65 | 14.2 |
| 1142 | 15.1 | 18.88 | 12.9 |
| 1168 | 13.0 | 16.27 | 11.2 |

* $x = 1.25s$; $10^4 r = 0.686x$.

On the other hand, molybdenum sheet absorbed much smaller amounts, 1.59 mg/100 g at 1127° C and 3.69 mg/100 g at 829° C. These results are in agreement with those obtained by Martin.[61]

Table 8.29 gives the solubility data obtained by Norton and Marshall for nitrogen in *molybdenum* and *tungsten*.

**TABLE 8.29**

**Solubility of Nitrogen at 1 atm in Molybdenum and Tungsten**

| | Molybdenum | | | Tungsten | | |
|---|---|---|---|---|---|---|
| ° C | mg/100 g | cm³/100 g | $V_g$ | mg/100 g | cm³/100 g | $V_g$ |
| 2400 | 20. | 16. | 1.65 | 0.38 | 0.304 | 0.059 |
| 2000 | 10.5 | 8.4 | 0.87 | 0.11 | 0.088 | 0.017 |
| 1600 | 4.3 | 3.44 | 0.35 | 0.019 | 0.015 | 0.003 |
| 1200 | 1.05 | 0.84 | 0.09 | 0.0013 | 0.001 | 0.0002 |
| Heat of solution (cal/mole $N_2$) | | 38,500 | | | 74,700 | |

It is difficult to understand the reason for the radical differences between the two sets of observations, both with respect to order of magnitude of $s$ and also with respect to the variation with temperature. It should be stated that Sieverts and Brüning also investigated the solubility of nitrogen in molybdenum-iron alloys. At 1100° C and 1150° C the solubility increases with addition of molybdenum, from that in pure iron (about 4 cm³/100 g) to about 16 cm³/100 g at 10 atomic per cent molybdenum. Further increase in molybdenum content causes the solubility to decrease to a minimum at 40 atomic per cent molybdenum, after which the solubility increases until it reaches that for pure molybdenum.

Nitrogen is soluble in *molten* aluminum, and the solubility increases with the temperature. At the melting point, $s = 1$ cm³/100 g.

According to the observations of de Boer and Fast,[62] *zirconium* heated in nitrogen takes up 1 atom of nitrogen to 11 of zirconium, forming a solid solution of the nitride ZrN. The electrical resistivity of this solution is greater than that of the metal; also, instead of a sharp transition from the $\alpha$ to the $\beta$ phase at 865° C, the transition is extended over several hundred degrees.

Oxygen is absorbed in a similar manner by zirconium, to the extent of 1 atom of oxygen to 9 atoms of zirconium, and the effect on the electrical resistance is the same as that of nitrogen.

De Boer and Fast state that "oxygen as well as nitrogen dissolves homogeneously in metallic zirconium. When a zirconium rod which is covered with a thick white oxide layer is heated *in vacuo*, the metallic luster reappears. The oxide has dissolved in the metal, and a homogeneous phase has apparently once more been formed." Along with the increase in electric resistivity there is a rise in the melting point and the lattice becomes expanded.

With many of the metals, nitrogen reacts to form nitrides. The thermodynamic properties of these compounds are discussed in Chapter 11.

## 8.8. SOLUBILITY OF OXYGEN AND OTHER GASES

The difficulty in the experimental determination of the solubility of oxygen in metals has been stated by Smithells[63] as follows:

> Oxygen is soluble to some extent in most metals, but except in the case of the noble metals, an oxide phase also appears when the limit of solid solubility is exceeded. The solid solubility of oxygen in the common metals is usually considered as small in comparison with the total amount of oxygen which may be present in the metal. Many molten metals are capable of dissolving large quantities of oxygen (or of their own oxides, which amounts to the same thing), but on freezing the excess oxygen is precipitated as oxide. The solubility of oxygen in the solid metals is of the same order as that of hydrogen. It is, however, much more difficult to determine accurately.

The system oxygen-copper has been investigated by Rhines and Mathewson.[64] The melting points of the metal and $Cu_2O$ are 1083° C and

1235° C, respectively. The eutectic melts at 1065° C. Table 8.30 gives solubility data observed below this temperature, according to Rhines and Mathewson, and also the data observed by Phillips and Skinner,[65] who used samples of copper from different sources; the table shows the range of values observed for the solubilities. The main reason for the difference between the two sets of data seems to be actual experimental difficulties in the analytical determination of such small percentages of dissolved gas.

**TABLE 8.30**

**Solid Solubility of Oxygen in Copper**

| Rhines and Mathewson | | | Phillips and Skinner | |
|---|---|---|---|---|
| $t$° C | cm³/100 g | Weight Per Cent | $t$° C | Weight Per Cent |
| 600 | 5.0 | 0.0071 | 700 | 0.0022 |
| 800 | 6.6 | 0.0094 | 850 | 0.0025–0.0029 |
| 950 | 7.0 | 0.0100 | 900 | 0.0027–0.0035 |
| 1050 | 10.9 | 0.0156 | 950 | 0.0034–0.0046 |
| | | | 1000 | 0.0044–0.0055 |
| | | | 1050 | 0.0072–0.0077 |

The values of the solubility limit reported by Rhines and Mathewson were obtained by exposing copper sheets "to free access of air at a definite temperature in a Hevi Duty muffle furnace for periods varying between one day and three weeks." Data on the dissociation pressures of the oxides of copper are given in Table 11.5.

The observations on the solubility of *oxygen in iron* have been summarized by Smithells as follows:[66]

> The solubility of oxygen in *liquid iron* increases from 147 cm³/100 g (0.21 per cent oxygen by weight) at the melting point to 387 cm³/100 g (0.552 per cent by weight) at 1734° C. Oxygen is probably more soluble in $\gamma$ than in $\alpha$ or $\delta$ iron, and estimates of the solubility in the range between 800° C and 1000° C vary from 24 to 28 cm³/100 g.

Herty and Gaines[67] have shown that oxygen must dissolve in liquid iron as FeO, and it probably dissolves in the solid metal in the same form. This view is in accord with the observations on the solubility of oxygen in copper and also in cobalt.

The saturation solubility data obtained by Seybolt and Mathewson[68] for *oxygen in cobalt* are shown in Table 8.31. These results were observed, as in the case of copper, by heating the metal "in an oxygen atmosphere at the temperature chosen, for a sufficient time to obtain equilibrium by diffusion of oxygen through it."

**TABLE 8.31**

**Solid Solubility of Oxygen in Cobalt**

| °C | mg/100 g | Atomic Per Cent CoO | $cm^3/100$ g |
|---|---|---|---|
| 600 | 6 | 0.0221 | 4.2 |
| 700 | 9 | 0.0332 | 6.3 |
| 810 | 16 | 0.0588 | 11.2 |
| 875 | 10 | 0.0369 | 7.0 |
| 945 | 7 | 0.0258 | 4.9 |
| 1000 | 8 | 0.0295 | 5.6 |
| 1200 | 13 | 0.0479 | 9.1 |

Figure 8.17 taken from the original paper, shows a plot of $1/T$ versus log (oxygen content). The abrupt decrease in solubility at about 875° C (the point indicated by two concentric circles) leads to the conclusion that a crystallographic transformation of cobalt in the cobalt-oxygen system occurs at this temperature. In pure cobalt this transformation occurs at 850° C.

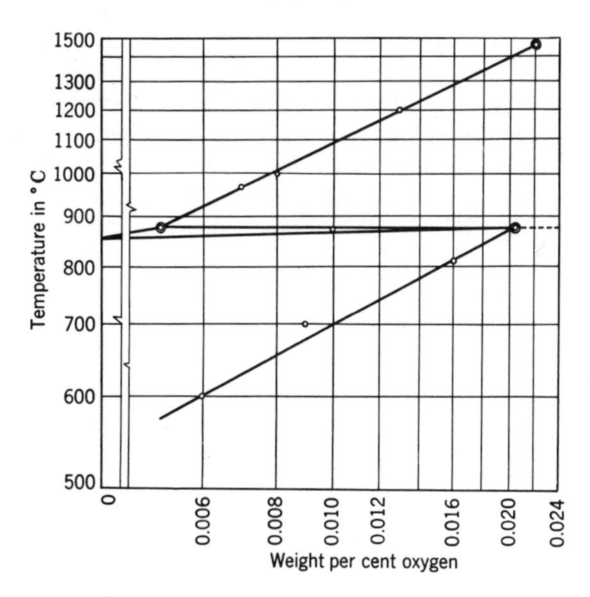


Fig. 8.17. Phase diagram for cobalt-cobaltous oxide system in which $1/T$ is plotted against log of percentage by weight of CoO, but temperature is shown in degrees Centigrade and the weight per cent oxygen is given. At the point indicated by two concentric circles (875° C) there is an abrupt decrease in solubility. (Seybolt and Mathewson.)

For the upper range of temperatures the linear plot yields the value $Q_S = 8650$ cal/mole oxygen, for the heat of solution.

In the case of *nickel*, the observation has been recorded by Merica and Waltenberg[69] that "nickel melted in the presence of oxygen yields an oxide (NiO) with which it forms a eutectic. This eutectic contains about 0.24 per cent oxygen, corresponding to 1.1 per cent NiO, and melts at 1438° C."

In Chapter 11 related topics are discussed, as dissociation pressures of the oxides of the metals.

The system *silver-oxygen* has been investigated by Steacie and Johnson[70] and also by Simons.[71] The solubility of oxygen in solid silver and the rate of solution are proportional to $(P)^{1/2}$. Table 8.32 gives the data obtained by

**TABLE 8.32**

**Solubility ($cm^3$ per 100 g) of Oxygen at 800 mm in Silver**

| $t°$ C: | 200 | 300 | 400 | 500 | 600 | 700 | 800 |
|---|---|---|---|---|---|---|---|
| $s$: | 1.3 | 0.924 | 0.828 | 0.905 | 1.26 | 1.84 | 3.37 |
| $10^2 s/(P_{mm})^{1/2}$: | 4.78 | 3.27 | 2.93 | 3.20 | 4.45 | 6.50 | 11.92 |

Steacie and Johnson for a pressure of 800 mm. The third row gives values of $s/(P_{mm})^{1/2}$, and Fig. 8.18 shows plots of four isobars giving values of $V_g$ ($= v_0\rho$) as a function of the temperature at constant pressure.

The most interesting feature about these data is the occurrence of a minimum at about 400° C. Steacie and Johnson have accounted for this minimum on the assumption that, below 400° C, the dissolved oxygen exists in the form of $Ag_2O$, and above this temperature as oxygen atoms, since $Ag_2O$ is a relatively unstable compound.

Simons has argued that the minimum is due to a balance between (1) increase in solubility of $Ag_2O$ in silver with increase in temperature, and (2) increase in dissociation pressure of the oxide with increase in temperature. The dissociation pressures of $Ag_2O$ for a series of temperatures are given in Table 11.3, and from these data it will be observed that at 400° C the dissociation pressure is about 115 atm.

According to Simons the solubility of $Ag_2O$ in silver is given by the relation

$$\log z = 0.0034(t - 300), \tag{8.23}$$

where $z$ = grams of $Ag_2O$ per 1000 g Ag, and $t$ = ° C.

The solubility of oxygen in *molten silver* was investigated by Sieverts and Hagenacker,[72] and Table 8.33 gives the results of their determinations. The third row gives the values of the ratio of oxygen atoms to silver atoms. As will be observed, there is about a fortyfold increase in the solubility at the melting point.

TABLE 8.33

Solubility of Oxygen at 1 atm in Molten Silver

| $t$° C: | 923 | 960.5 | 973 | 1024 | 1075 | 1125 |
|---|---|---|---|---|---|---|
| mg/100 g: | | mp | 305 | 295 | 277 | 264 |
| O: $10^4$ Ag: | | | 206 | 199 | 187 | 178 |
| Vol. $O_2$: vol. Ag: | | | 22.4 | 21.5 | 20.4 | 19.4 |
| $s$: | 5.43 | | 213.5 | 205.6 | 193.9 | 184.9 |

At 1075° C, 1 volume of molten silver in air (oxygen pressure about 150 mm) dissolves about 9 volumes of oxygen. As in the case of the solid metal $s$ varies as $P^{1/2}$.

The liberation of oxygen from silver during solidification has been studied by Allen[73] with the view of eliminating blowholes in the cast metal. He has made some interesting calculations on the internal pressures developed in molten silver in consequence of oxygen liberation and in the reduction of $Cu_2O$ in molten copper by hydrogen. By applying external pressure of sufficient magnitude, or by means of a deoxidant, sound

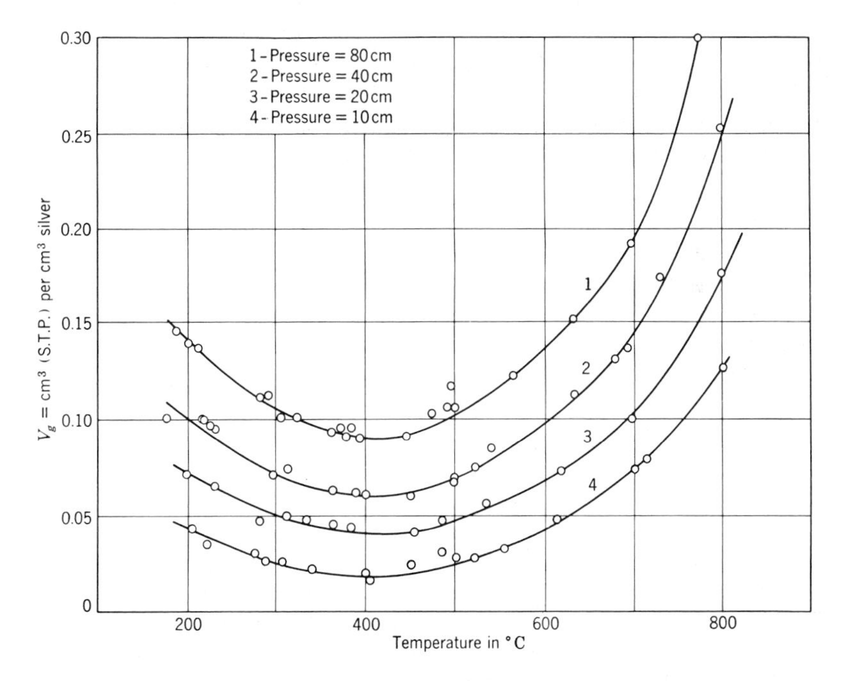


Fig. 8.18. Isobars at four pressures for the solubility of oxygen in silver (Steacie and Johnson).

castings can be obtained. Allen has also observed that the melting point of silver is lowered by oxygen in accordance with the equation

$$t_m = 961 - 22.31(P_{atm})^{1/2}, \tag{8.24}$$

where $t_m$ = melting point in degrees centigrade.

The solubility of *sulfur dioxide in molten copper* has been a very interesting topic of investigation on account of both its technical importance and its theoretical significance. The earlier workers on this problem, Sieverts and Krumbhaar,[74] and Sieverts and Bergner,[75] concluded that the solubility varies as $P^{1/3}$. However, such a relation could not be reconciled with any logical interpretation. It is therefore of interest that later work, carried out by Floe and Chipman,[76] cleared up the problem in a very satisfactory manner.

In *pure* molten copper, from which sulfur and oxygen are completely eliminated, $s$ varies as $P^{1/3}$. In the presence of an *excess of sulfur*, $s$ varies as $P^{1/2}$, and in both cases the linear plots pass through the origin. Excess of oxygen decreases the solubility. Figure 8.19 shows plots of $s$ (cm³/100 g) versus $(P_{mm})^{1/3}$.

Assuming that when $SO_2$ passes into solution the reaction is

$$6\text{Cu}\ (l) + \text{SO}_2\ (g) = \text{Cu}_2\text{S (in Cu)} + 2\text{Cu}_2\text{O},$$

there should exist an equilibrium constant

$$K = \frac{(\text{Cu}_2\text{S})(\text{Cu}_2\text{O})^2}{(\text{Cu})^6 P(\text{SO}_2)},$$

that is,

$$K' = \frac{(\%\text{S}) \times (\%\text{O})^2}{P(\text{SO}_2)}$$

$$= \frac{(\%\text{SO}_2)^3}{P(\text{SO}_2)},$$

since the percentage concentrations of sulfur and oxygen are the same and are proportional to $s$.

The values of $K'$ for $P_{mm}$ ($SO_2$) derived by Floe and Chipman for the three temperatures at which observations were made are as follows:

| | 1100°C | 1200°C | 1300°C |
|---|---|---|---|
| $K'$ = | $1.1 \cdot 10^{-5}$ | $2.4 \cdot 10^{-5}$ | $4.4 \cdot 10^{-5}$ |

The observation that, for each temperature, the value of the constant $K'$ exhibits a slight linear increase with pressure is interpreted by the investigators as due either to "deviations from the laws of dilute solutions or increased reaction with the fused-silica tube."

From the linear plot of log $K'$ versus $1/T$ the conclusion is drawn that, for the reaction

$$SO_2\,(g) = S\text{ (in Cu)} + 2O\text{ (in Cu)},$$
$$\Delta F = \text{increase in free energy}$$
$$= 30{,}700 + 0.38T.$$

That is, the heat of reaction is 30,700 cal/mole $SO_2$.

The solubility of oxygen and nitrogen in tantalum and the effect on the electrical resistance of the metal have been investigated in the General Electric Research Laboratory by Mrs. M. R. Andrews, and the topic is discussed in Chapter 9 in the section dealing with chemical clean-up.

The solubility of oxygen and of "oxide gases," such as $SO_2$, CO, $CO_2$, and $H_2O$, in metals and the chemical reactions involved have been discussed by Lepp in a series of papers.[77] In general, it can be stated that these gases react with metals in much the same manner as sulfur dioxide with copper, and at any temperature and pressure of gas an equilibrium state is established between oxidizing gas, metal, oxide, and reduction product.

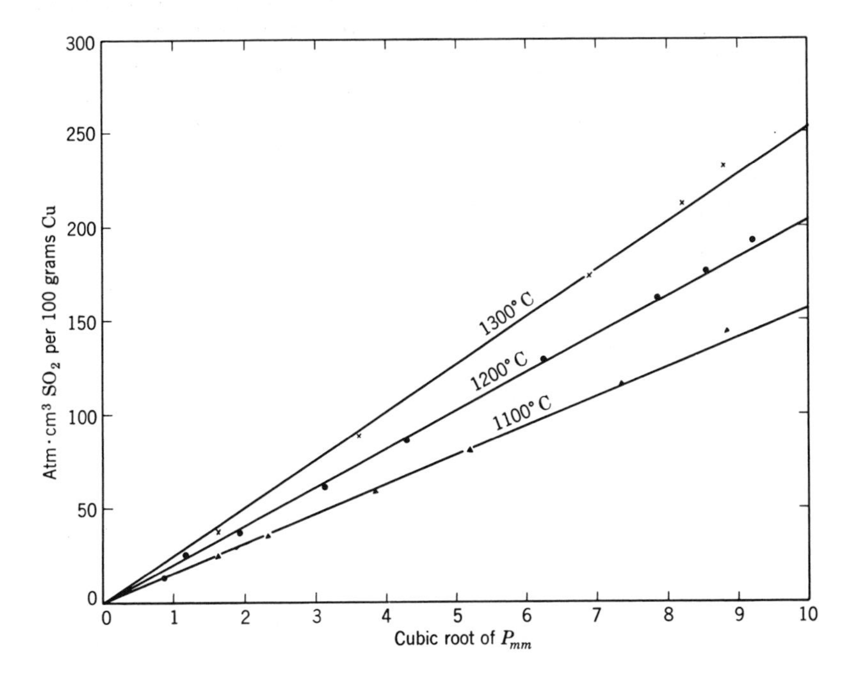


Fig. 8.19. Solubility of sulfur dioxide in molten copper at three temperatures plotted versus cube root of $P_{mm}$ (Floe and Chipman).

For an adequate discussion of this topic we must therefore consider these reactions (as is done in Chapter 11 from the thermodynamic point of view, that is, on the basis of free-energy data.

## 8.9. DIFFUSION OF GASES THROUGH METALS

As Steacie and Johnson have stated:[78]

> The phenomena of absorption of gases by metals, and diffusion of gases through metals, must be fundamentally connected. It is virtually impossible to conceive of a mechanism for diffusion other than that of solution on the high-pressure side of the metal and subsequent giving up of the gas on the low-pressure side which is supersaturated. Apparently, then, solution must precede diffusion. This conclusion is substantiated by the fact that there is no known case of a gas diffusing through a metal in which it is not apparently soluble.

It follows that, even where the solubility of a gas in a metal has not been determined, the fact that diffusion of this gas occurs through a certain metal is a valid reason for assuming that the gas is soluble to some extent in the metal.

That diffusion occurs by a motion of the atoms (or protons in the case of hydrogen) along interstitial sites follows from the observation that the rate of diffusion follows the $(P)^{1/2}$ law, and also that the rare gases and polyatomic molecules do not diffuse (at least noticeably) through metals. That in general diffusion of gases occurs through the crystal lattice and not along grain boundaries was shown by an experiment reported by Edwards[79] in which he found that "no difference could be detected between the rate at which hydrogen diffused through a single crystal of iron, and through the same material after it had been recrystallized into a mass of small crystals."

Evidently, rates of sorption and desorption of gases must depend on rates of diffusion of the gases through the metals, that is, on the nature of the concentration gradients at the two surfaces of the sample, and on the values of the diffusion constants.

Let $D$ = diffusion constant (square centimeters per second) and $C$ = concentration at any point inside the metal (cubic centimeters, STP) per cubic centimeter of metal.

Then the amount (atmosphere · centimeters$^3$ or equivalent micron · liters) passing through the area $A$ per unit time is given by the relation

$$q = -DA\frac{dC}{dx}, \tag{8.25}$$

where $dC/dx$ is the concentration gradient at the point $x$.

In the *stationary state* this gradient is constant throughout the thickness $d$ of the metal sheet. For $C$ we can substitute the values of $V_g$ (the volume of gas per volume of metal) at the high-pressure and low-pressure surfaces, and equation 8.25 assumes the form

$$q = \frac{DAK}{d}(V_{g1} - V_{g2}), \tag{8.26}$$

where, for a diatomic gas (such as $H_2$, $O_2$, or $N_2$),

$$V_{g1} = K(P_{mm1})^{1/2} \quad \text{and} \quad V_{g2} = K(P_{mm2})^{1/2}, \tag{8.27}$$

and $P_{mm1}$ and $P_{mm2}$ denote the pressures at the two surfaces.

Hence, equation 8.26 assumes the form

$$q = \frac{DAK}{d}[(P_{mm1})^{1/2} - (P_{mm2})^{1/2}]. \tag{8.28}$$

For a cylindrical tube of length $l$, inside diameter $a$, and outside diameter $b$,

$$q = DfK[(P_{mm1})^{1/2} - (P_{mm2})^{1/2}], \tag{8.29}$$

where

$$f = \frac{2\pi l}{\ln(b/a)} = \frac{2.73l}{\log(b/a)} \tag{8.30}$$

$$= \frac{\pi la}{d} \quad \text{for} \quad 2d \ll a \qquad (2d = b - a).$$

Since we shall consider diffusion through flat surfaces, and also assume $(P_{mm2})^{1/2} = 0$, the equation for diffusion becomes

$$q = DK\frac{A}{d}P^{1/2}, \tag{8.31}$$

where $P$ is the pressure at the surface into which the gas penetrates.

At constant pressure the rate of diffusion increases exponentially with the temperature, and from theoretical considerations Richardson, Nicol, and Parnell[80] derived the relation,

$$q = q_0 T^{1/2} \epsilon^{-b_0/T}, \tag{8.32}$$

where $q_0$ and $b_0$ are constants for a given gas-metal system.

As shown by Smithells, this equation can be replaced for all practical purposes by a relation which does not involve the $T^{1/2}$ factor, and taking into account equation 8.31, it is possible to express the rate at which gas diffuses in *atmosphere · centimeters*$^3$ *per square centimeter per second* by the relation

$$q = \frac{k_0}{d}(P_{mm})^{1/2}\epsilon^{-b_0/T}, \tag{8.33}$$

where $k_0$ is a characteristic of the gas-metal system, which involves the two constants designated above by $D$ and $K$. In the following discussion $d$ is expressed in *millimeters* and $P$ in *millimeters of mercury*.[81]

In terms of *micron · liters* ($Q_{\mu l}$) at 0° C, *per minute*,

$$Q_{\mu l} = 760 \times 60q = 45{,}600q. \tag{8.34}$$

The constant $b_0$ is defined by the relation

$$b_0 = \frac{E_0}{2R_0}, \tag{8.35}$$

where $E_0$ = heat of diffusion in gram-calories *per mole*, and $E_0/2$ = heat of diffusion in gram-calories *per gram-atom*.

In terms of micron · liters at $T = 298$,

$$Q_{\mu l}' = 760 \times 60q \times \frac{298}{273}$$

$$= 1.091 Q_{\mu l}. \tag{8.36}$$

For $P_{mm} = 760$, and $d$ = *1 mm*,

$$q = k_0(760)^{1/2}\epsilon^{-b_0/T};$$

that is,

$$\log q = \log k_0 + 1.4404 - \frac{b_0}{2.303T}$$

$$= \log k_0 + 1.4404 - \frac{E_0}{9.148T}, \tag{8.37}$$

and

$\log Q_{\mu l}$ = micron · liters (at 0° C) per square centimeter, *per minute*, per millimeter thickness, at $P_{mm} = 760$

$$= \log k_0 + 6.0994 - B/T \tag{8.38}$$

$$= C - B/T,$$

where

$$\left.\begin{aligned} C &= \log k_0 + 6.0994 \\ B &= E_0/9.148. \end{aligned}\right\} \tag{8.39}$$

Values of $k$ and $E_0$ are given for a number of gas-metal systems by Smithells.[82] From these, values of $C$ and $B$ were calculated which are given in Table 8.34. These constants were then used to calculate values of $t$ (°C) for which $Q_{\mu l}$ has values ranging from 0.1 to 100. In some cases, for which the value of $t$ thus calculated would exceed that of fusion, values of $t$ are given for lower values of $Q_{\mu l}$, which are indicated in parentheses following the value of $t$. The second column in the table gives the reference number for the source of the original data, as shown in the bibliography,

TABLE 8.34

**Constants for Rates of Diffusion of Gases through Metals and Values of $t°$ C for $Q_{\mu l}$ = micron · liters (0° C) per cm² per min per mm Thickness at 1 atm Pressure**

| System | Ref. | $C$ | $B$ | $t°$ C for $Q_{\mu l}$ = | | | |
|---|---|---|---|---|---|---|---|
| | | | | 0.1 | 1.0 | 10 | 100 |
| $H_2$-Ni | 1 | 4.4611 | 3371 | 344 | 483 | 701 | 1097 |
| | 2 | 4.0288 | 3031 | 330 | 479 | 727 | 1221 |
| | 3 | 4.1206 | 2929 | 299 | 438 | 666 | 1108 |
| | 4 | 4.2455 | 3017 | 302 | 438 | 657 | 1071 |
| | 5 | 4.2578 | 2898 | 278 | 408 | 616 | 1010 |
| $H_2$-Pt | 6 | 4.2486 | 4285 | 544 | 736 | 1016 | 1732 |
| | 3 | 4.1713 | 3936 | 488 | 671 | 968 | 1539 |
| | 7 | 5.010 | 4220 | 429 | 571 | 779 | 1130 |
| $H_2$-Pd | 9, 1 | 5.783 | 2318 | 69 | 128 | 212 | 340 |
| | 8 | 4.5765 | 2296 | 139 | 229 | 369 | 618 |
| $H_2$-Cu | 5 | 3.4611 | 3629 | 541 | 775 | 1041 (5) | |
| | 10 | 3.2755 | 4307 | 734 | 1042 | | |
| $H_2$-Fe | 5 | 3.3116 | 2100 | 214 | 361 | 735 | 1029 (50) |
| | 4 | 3.3035 | 2055 | 204 | 349 | 619 | 1303 |
| | 11 | 3.480 | 2405 | 264 | 418 | 697 | 1352 |
| $H_2$-Al | 5 | 5.723 | 6735 | $Q_{\mu l} = 1.03 \cdot 10^{-3}$ for 500° C; $= 1.02 \cdot 10^{-2}$ for 600° C | | | |
| $H_2$-Mo | 5 | 4.0679 | 4417 | 598 | 813 | 1167 | 1862 |
| $N_2$-Mo | 5 | 5.0185 | 9837 | 1362 | 1687 | 2174 | 2444 (25) |
| $N_2$-Fe | 11 | 3.7526 | 5204 | 822 | 1114 | 1431 (5) | |
| CO-Fe | 11 | 3.2133 | 4066 | 692 | 993 | 1344 (5) | |
| $O_2$-Ag | 12 | 4.413 | 4941 | 640 | 846 | | |
| | 13 | 4.673 | 4941 | 598 | 784 | 940 (4) | |
| | 13 | 5.121 | 4941 | 534 | 692 | 926 (4) | |

List III, which follows. In the following text, numbers in parentheses refer to items in this list.

*List III. References on Rates of Diffusion of Gases through Metals*

1. V. Lombard, *Compt. rend.*, **177,** 116 (1923).
2. H. G. Deming and B. C. Hendricks, *J. Am. Chem. Soc.*, **45,** 2857 (1923).
3. W. R. Ham, *J. Chem. Phys.*, **1,** 476 (1933); *Phys. Rev.*, **45,** 741 (1934).
4. G. Borelius and S. Lindblom, *Ann. Physik*, **82,** 201 (1926–1927); G. Borelius, *Ann. Physik*, **83,** 121 (1927).
5. C. J. Smithells and C. E. Ransley, *Proc. Roy. Soc. London*, **A, 150,** 172 (1935); **152,** 706 (1936).
6. O. W. Richardson, J. Nicol, and T. Parnell, *Phil. Mag.*, **8,** 1 (1904).
7. R. Jouan, *J. phys. radium*, **7,** 101 (1936).
8. R. M. Barrer, *Trans. Faraday Soc.*, **36,** 1235 (1940).
9. V. Lombard and C. Eichner, *Compt. rend.*, **194,** 1929 (1932); *Bull. soc. chim. France*, **53,** 1176 (1933); *Compt. rend.*, **196,** 1998 (1933).
10. E. O. Braaten and G. F. Clark, *Proc. Roy. Soc. London*, **A, 153,** 504 (1935–1936).
11. H. M. Ryder, *Elec. J.*, **17,** 161 (1920).
12. F. M. G. Johnson and P. Larose, *J. Am. Chem. Soc.*, **46,** 1377 (1924); **49,** 312 (1927).
13. L. Spencer, *J. Chem. Soc.*, **123,** 2124 (1923).

The values of $t$ given in Table 8.34 were derived by means of the relation

$$T = t + 273 = \frac{B}{C - \log Q_{\mu l}}. \tag{8.40}$$

In order to derive $t$ for $Q_{\mu l}' = 1.091\ Q_{\mu l}$ = rate in micron · liters at 25° C,

$$T = t + 273 = \frac{B}{C + 0.0380 - \log Q_{\mu l}'}. \tag{8.41}$$

A table of selected values of the constants $k_0$ and $b_0$ in equation 8.33 is shown in a publication by Jossem.[83] This table also gives values of $q$ for the different gas-metal systems for $T = 1200°$ K. Figure 8.20 shows a simple arrangement, recommended in this paper, for introducing hydrogen through a heated palladium tube into a system. The rate of penetration is controlled by the temperature of operation of the tungsten winding $H$.

The fact that hydrogen diffuses through palladium much more readily than through any other metal is utilized for the introduction of the pure gas into an evacuated system. Even at room temperature, the rate of diffusion may be fairly rapid, as has been observed by Dushman, especially if the palladium happens to be in a well-activated condition.

The $P^{1/2}$ relation for rate of diffusion is valid, in general, only at higher pressures (around atmospheric pressure and higher). At low pressures the rate decreases below the value predicted by this relation, and Smithells and Ransley (5) have deduced a modification of the relation which is based on

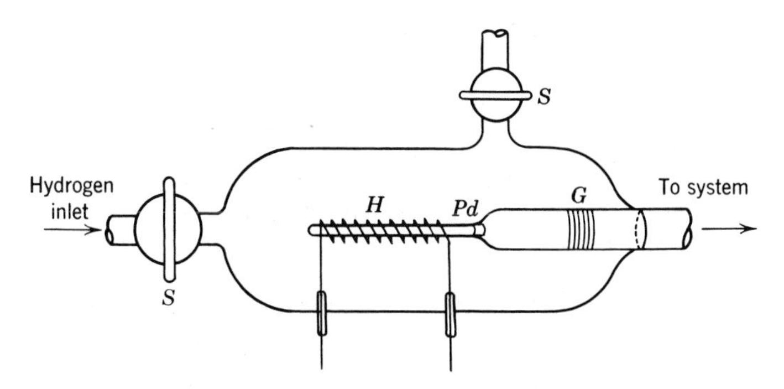


Fig. 8.20. A constant, controllable arrangement for the diffusion of hydrogen through palladium. $H$ = heater coil, $Pd$ = palladium tube, $G$ = graded seal, and $S$ = stopcock. (Jossem.)

the view that diffusion must be preceded by adsorption on the surface. They state, "Since at any instant diffusion can only occur through those parts of the surface which are covered by adsorbed gas, the *effective* area of the metal is $A\theta$, where $A$ is the total area of the surface and $\theta$ the fraction of the surface covered."

According to the Langmuir adsorption equation,

$$\theta = \frac{cP}{1 + cP},$$

where $c$ is a constant for any particular gas-metal system. Hence, the rate of diffusion should be given, at constant temperature (and for $d = 1$ mm), by

$$q = k(P)^{1/2}\left(\frac{cP}{1 + cP}\right), \tag{8.42}$$

where

$$k = k_0\epsilon^{-b_0/T}. \tag{8.43}$$

This conclusion has been confirmed by the observations made by Smithells and Ransley for a number of gas-metal systems, and also by Braaten and Clark (10). Illustrations of the validity of equation 8.42 will be given in the subsequent discussion.

There are two other sets of observations which are interpreted most satisfactorily by the adsorption theory.

Nitrogen will diffuse [as Smithells and Ransley observe] through iron, chromium, and molybdenum, with which, under suitable conditions, it can form nitrides, but not through copper and nickel, towards which it is chemically inert. We are therefore led to the conclusion that not only must adsorption occur as a preliminary to diffusion, but that activated adsorption with dissociation of the adsorbed molecules is necessary, and that purely physical adsorption with weak binding forces will not result in diffusion. This hypothesis explains the specific nature of diffusion.

In these respects diffusion of gases through metals is thus of quite a different nature from diffusion of gases through glasses, quartz, and similar materials. In such materials, as observed in Chapter 7, the rate of diffusion varies *linearly* with the pressure, thus indicating that diffusion in these cases is a *molecular*, and not an atomic, process.

The other set of observations which supports the adsorption theory includes those made on the *effects of surface treatment*. This is illustrated by the data in Table 8.35, taken from the publication by Smithells and

**TABLE 8.35**

**Effect of Surface Treatment of Metals on Rate of Diffusion of Hydrogen**

| Metal | Treatment | $t°$ C | $P_{mm}$ | $q$ |
|---|---|---|---|---|
| Nickel | Polished | 750 | 0.042 | $1.39 \cdot 10^{-6}$ |
| | Oxidized and reduced | 750 | 0.042 | 2.70 |
| | Polished | 750 | 0.091 | 2.91 |
| | Oxidized and reduced | 750 | 0.091 | 4.23 |
| Iron | Polished | 400 | 0.77 | $0.47 \cdot 10^{-7}$ |
| | Etched | 400 | 0.77 | 4.4 |
| | Polished | 590 | 0.073 | 1.28 |
| | Oxidized and reduced at 600° C | 590 | 0.073 | 0.76 |
| | Oxidized and reduced at 800° C | 590 | 0.073 | 1.54 |

Ransley, which give the results of observations on the rate of diffusion of hydrogen through nickel and iron. They state:

> Etching the surface with acid had a far more marked effect, increasing the rate of diffusion by 10 times. The increased adsorption is also shown by the much larger value of $c$ in equation 8.42 found for etched surfaces.

Such observations also probably account for the wide variation in the values of the diffusion constants and derived values of rates of diffusion in Table 8.34.

While, in general, $q$ (or $Q_{\mu l}$) varies with the pressure in accordance with equation 8.42, the rate of diffusion may become independent of the pressure if a compound is formed between the gas and the metal. Smithells and Ransley have shown that at 900° C and higher temperatures the rate of diffusion of oxygen through nickel is practically constant at pressures above 0.25 mm, because the diffusion is limited by the rate of formation of NiO.

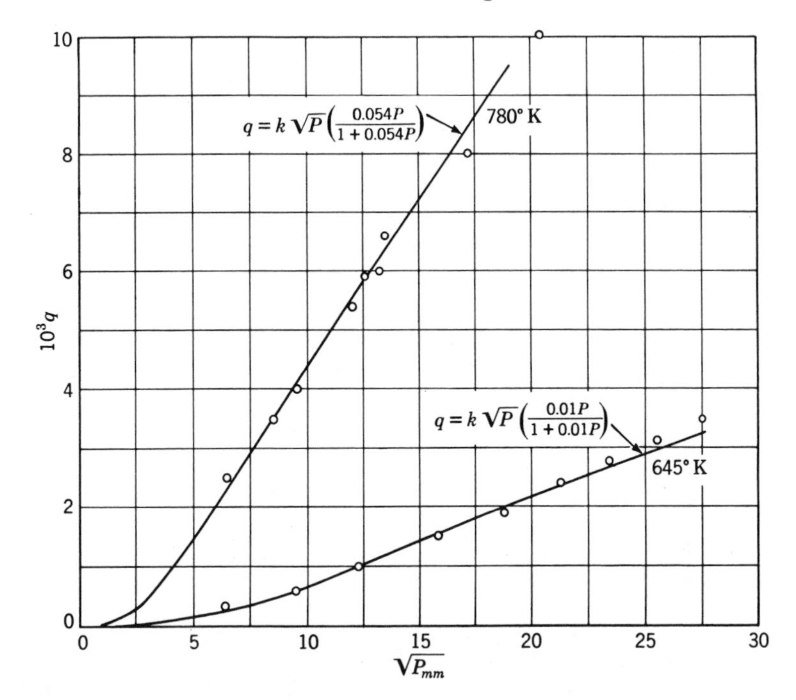


Fig. 8.21. Rate of diffusion ($q$) of hydrogen through palladium plotted versus $(P_{mm})^{1/2}$, illustrating application of equation 8.42. In the equations for each plot, $P$ corresponds to $P_{mm}$. (Smithells and Ransley.)

As will be observed from the data in Table 8.34, there is a very wide difference between the rates of diffusion obtained by at least two investigators for hydrogen through palladium. Smithells[84] states that this metal "appears to be particularly affected by preliminary treatment, and its capricious behavior has been noted by a number of workers."

The rate of diffusion for hydrogen in palladium is considerably greater than that observed for other gas-metal systems. Figure 8.21, from the paper by Smithells and Ransley, shows plots of observations reported by Lombard, which illustrate the application of equation 8.42. For the pressure in millimeters, $c$ has the value 0.054 for 780° K and 0.01 for 645° K. For $P_{mm} = 400$, the plots show $q = 10 \cdot 10^{-3}$ at 780° K and $q = 2.2 \cdot 10^{-3}$ at 645° K. (In Fig. 8.21 and subsequent figures, $10^n q$ signifies that the *unit of ordinates* is $10^{-n}$ atm · cm$^3$ per cm$^2$ per sec per mm thickness. Also, in the equations for the plots, $P = P_{mm}$.)

The data plotted in Fig. 8.21 lead to the values $C = 5.783$ and $B = 2318$, which are given in Table 8.34. On the other hand, the observations of Lombard and Eichner (9) lead to the values $C = 4.014$ and $B = 1063$, and

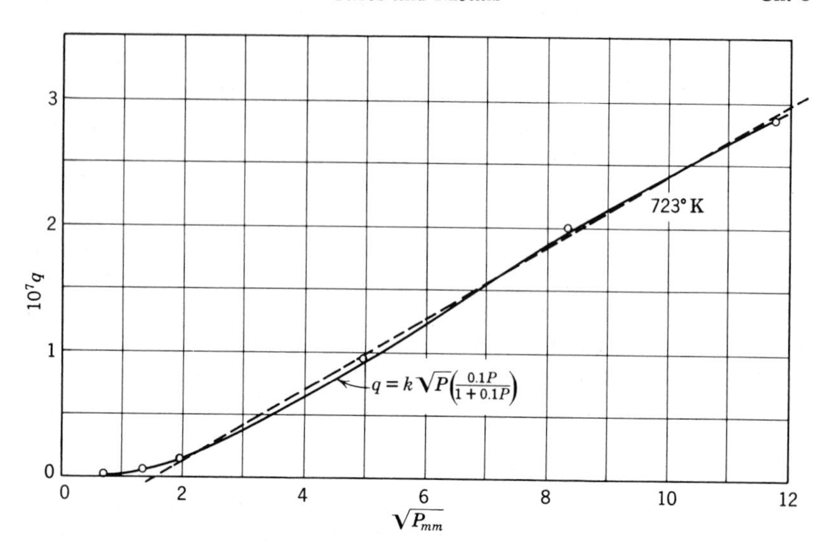


Fig. 8.22. Rate of diffusion ($q$) of hydrogen through copper plotted versus $(P_{mm})^{1/2}$. In the equation for the plot, $P$ corresponds to $P_{mm}$. (Smithells and Ransley.)

the constants given by Smithells and Ransley lead to the values $C = 4.712$ and $B = 918$. For a so-called "inactive" palladium, Barrer (8) deduced constants which correspond to $C = 4.577$ and $B = 2296$. Table 8.36 gives values of $Q_{\mu l}$ for a series of temperatures as derived by means of these four sets of values of $C$ and $B$.

**TABLE 8.36**

**Values of $Q_{\mu l}$ for $H_2$-Pd System**

| $T$ °K | $C = 4.712$ $B = 918$ | $C = 4.104$ $B = 1063$ | $C = 5.783$ $B = 2318$ | $C = 4.577$ $B = 2296$ |
|---|---|---|---|---|
| 600 | 1521 | 216 | 83 | 6 |
| 700 | 2518 | 386 | 294 | 20 |
| 800 | 3664 | 597 | 767 | 51 |
| 900 | 4920 | 838 | 1611 | 106 |
| 1000 | 6223 | 1100 | 2917 | 191 |

Figure 8.22 gives a plot of observations made by Smithells and Ransley on the diffusion of hydrogen through copper at 723° K. These observations are satisfactorily represented by equation 8.42 with the value $c = 0.1$ mm$^{-1}$. As shown in Table 8.34, the rates of diffusion observed in this investigation are much higher than those observed by Braaten and Clark.

Plots drawn by Smithells and Ransley of observations made by Borelius and Lindblom (4) on the diffusion of hydrogen through iron, at 571 and 633° K, are shown in Fig. 8.23, while Fig. 8.24 shows similar plots for 805 and 975° K. Evidently all the plots are in agreement with equation 8.42, with the value of $c$ increasing with increase in $T$ from 0.01 mm$^{-1}$ at 571° K to 0.1 mm$^{-1}$ at 975° K.

The effect of surface treatment on the diffusion of hydrogen through aluminum has been investigated by Smithells and Ransley. It was found that, by means of a special treatment for removal of the oxide film, the rate of diffusion could be increased about sixfold. But gradual poisoning of the surface by oxygen then caused the rate to decrease gradually. The constants in Table 8.34 were deduced from the plot for 558° C shown in Fig. 8.25.[85]

Finally, it is interesting to compare the rates of diffusion of gases through quartz with those observed for metals. A comparison of the data in Table 7.30 and those in Table 8.34 (allowing for the difference in the definition of $Q_{\mu l}$) shows that for hydrogen or nitrogen, at 1 atm, the rate of diffusion, for the same temperature and value of $d$, is much higher through many metals than through quartz.

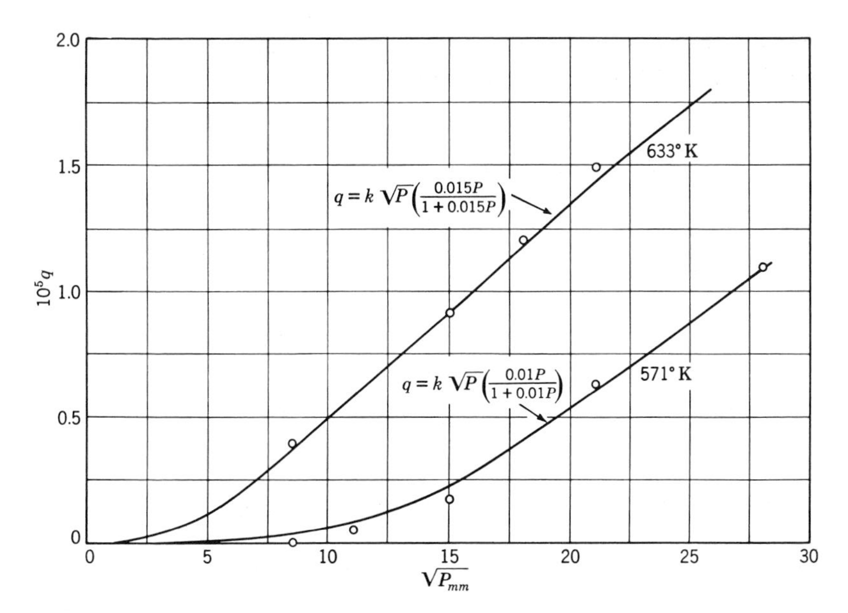


Fig. 8.23. Rate of diffusion ($q$) of hydrogen through iron, at lower temperatures, plotted versus $(P_{mm})^{1/2}$. In the equations for the two plots, $P$ corresponds to $P_{mm}$. (Borelius and Lindblom.)

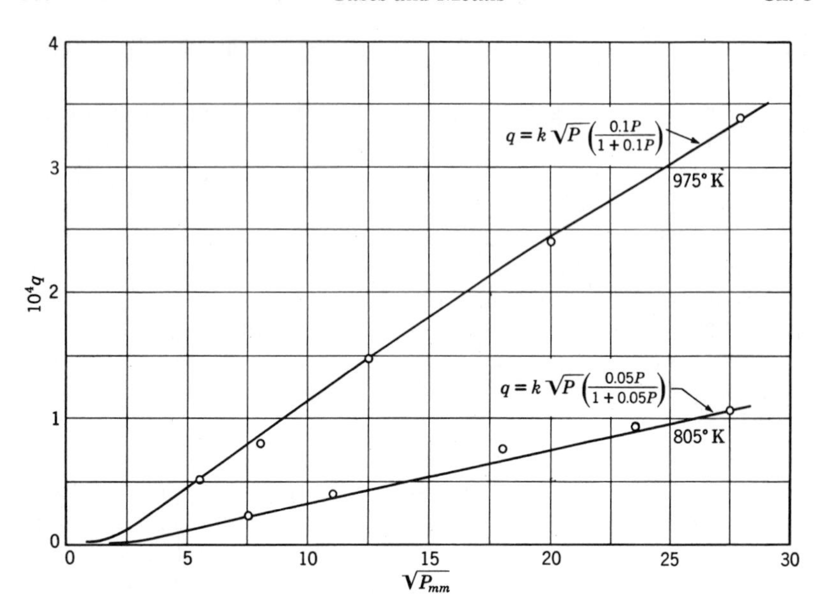


Fig. 8.24. Rate of diffusion ($q$) of hydrogen through iron, at higher temperatures, plotted versus $(P_{mm})^{1/2}$. In the equation for the two plots, $P$ corresponds to $P_{mm}$. (Borelius and Lindblom.)

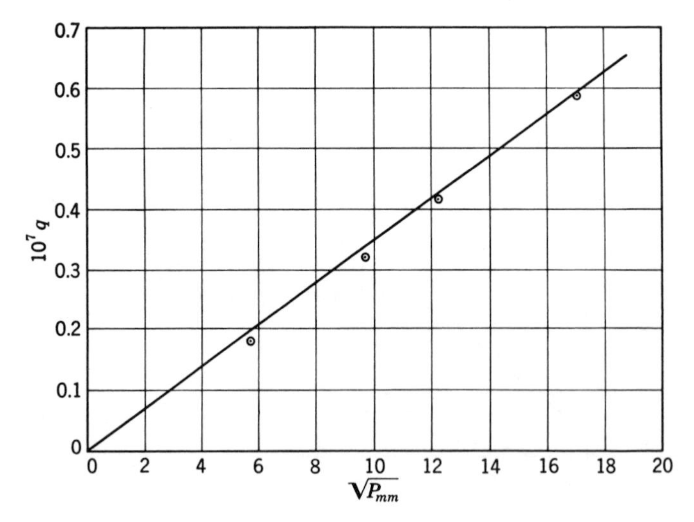


Fig. 8.25. Rate of diffusion ($q$) of hydrogen through aluminum at 558° C plotted versus $(P_{mm})^{1/2}$ (Smithells and Ransley).

The decade 1950–1960 saw diffusion studies in general greatly accelerated. This stimulus was due to the following developments:

1. Preparation and accessibility of radioactive and stable isotopes. Experiments using the stable isotopes have been greatly extended by widely increased use of mass spectrometers.
2. Nuclear magnetic resonance methods for studying self-diffusion.
3. Extension of the technique of Zener and others for measuring internal friction of solids.

There has been important work on hydrogen diffusion and permeation, particularly in steel.[86,87,88,89,90]

## 8.10. RATES OF SORPTION AND DESORPTION; CALCULATION OF DIFFUSION CONSTANTS

The rates at which these processes occur will, in general, depend upon rates of diffusion of gases into or out of plates or wires of metals in which the gases are soluble. These rates may therefore be deduced as particular solutions of Fick's law, which for diffusion in one direction is of the form[91]

$$\frac{\partial C}{\partial t} = \frac{\partial}{\partial x}\left(D \frac{\partial C}{\partial x}\right), \tag{8.44}$$

where $C$ denotes the concentration (grams per cubic centimeter) at time $t$, at a point distant $x$ from the origin, and $D$ is the diffusion constant (square centimeters per second).

This equation is applicable, for instance, to the desorption of gas from a thin plate. Let $2a$ denote the thickness of a sheet of metal for which the width and length are each very large compared to $2a$.

Let $C_0$ denote the concentration of dissolved gas, uniform throughout the range $x = 0$ to $x = 2a$, at $t = 0$. Then the fraction, $f$, of the total gas content which has diffused to the two surfaces at any time, $t$, is given by the relation

$$f = 1 - \frac{8}{\pi^2}\left(\epsilon^{-z} + \frac{\epsilon^{-9z}}{9} + \frac{\epsilon^{-25z}}{25} + \cdots\right), \tag{8.45}$$

where

$$z = \frac{\pi^2}{4} \cdot \frac{Dt}{a^2}.$$

From this equation, it follows that plots of $f$ versus $t/a^2$ should be similar for different samples, independently of the value of $D$. This conclusion, we shall find, is valid for all cases of diffusion.

A very convenient tabulation of the dimensionless number $Dt/a^2$ is given by Newman.[92] For equation 8.45, he gives tables and graphs for values of $Dt/a^2$ and $E = 1 - f$ for an infinite slab, a cylinder, and a sphere. He also gives detailed numerical examples to obtain these relations for shapes of various dimensions.

As is evident, the series in the parentheses in equation 8.45 converges rapidly. This is seen from the following values:

| | | | |
|---|---|---|---|
| $z$: | 0.20 | 0.25 | 0.30 |
| $\epsilon^{-z}$: | 0.8187 | 0.7788 | 0.7408 |
| $\epsilon^{-9z}/(9\epsilon^{-z})$: | 0.0225 | 0.0150 | 0.0102 |
| $Dt/a^2$: | 0.0811 | 0.1014 | 0.1216 |

For values of $Dt/a^2 > 0.0892$, the second term in parentheses in equation 8.45 is less than 2 per cent of the first term. Hence, to a first approximation, equation 8.45 can be written in the form

$$\log(1 - f) = \log\left(\frac{8}{\pi^2}\right) - \frac{\pi^2}{4 \times 2.303} \cdot \frac{Dt}{a^2}. \qquad (8.46)$$

That is, under these conditions a plot of $\log(1 - f)$ versus $t/a^2$ becomes linear, with a slope

$$\frac{\Delta \log(1 - f)}{\Delta(t/a^2)} = \frac{-\pi^2 D}{9.212}, \qquad (8.47)$$

from which the value of $D$ may be deduced.

By application of this method, the value deduced for the coefficient of diffusion of oxygen in steel at 1100° C was $D = 1 \cdot 10^{-9}$ cm$^2 \cdot$ sec$^{-1}$, which appears reasonable when compared with the value $D = 7.5 \cdot 10^{-10}$ cm $\cdot$ sec$^{-1}$ deduced by other investigators for 1000° C.

The preceding remarks may be illustrated by the following example. Let $a = 10^{-2}$ cm ($= 4 \cdot 10^{-3}$ in.). Then, for $D = 1 \cdot 10^{-9}$ cm$^2 \cdot$ sec$^{-1}$, and $Dt/a^2 = 0.0811$, $t = 2.25$ hr, and $f = 0.321$. To degas the plate to 10 per cent of its initial gas content ($f = 0.90$), the time required would be about 23.6 hr.

For the case in which the concentration gradient is uniform over a film of thickness $l$, equation 8.44 assumes the simple form

$$\frac{dC}{dt} = \frac{D(C_0 - C)}{l}$$

$$= k(C_0 - C). \qquad (8.48)$$

This relation for rate of absorption (and a similar one for rate of desorption) is observed quite frequently. Assuming that the different gases are adsorbed to the same extent on the surface of the metal, the values observed for $k$ are in the same ratio as those of $D$.

Van Liempt[93] has used a relation for calculating values of $D$ from rates of degassing, which for a metal plate of width very large compared with the thickness $2a$, has the form

$$\frac{C}{C_0} = 1 - \frac{2}{\pi^{1/2}} \int_0^y \epsilon^{-\beta^2}\, d\beta, \tag{8.49}$$

where $y = x/2(Dt)^{1/2}$; the integral is the "error function"; $C_0$ = initial concentration of gas, uniform throughout the thickness; and $C$ = concentration at distance $x$ from the central plane, at time $t$.

Equation 8.49 is identical with that deduced for the rate of diffusion of gas into a semi-infinite solid for the condition $C = C_0$ at the boundary ($x = 0$) for all values of $t$, and the initial condition $C = 0$ for all values of $x > 0$.

A discussion of the rather involved method used by van Liempt in applying this relation to rates of degassing is given by Barrer.[94]

It is of interest to note that from equation 8.49 it may be deduced that the rate at which gas diffuses into the solid, per unit area of surface, is given by

$$\frac{dQ}{dt} = C_0\left(\frac{D}{\pi t}\right)^{1/2}, \tag{8.50}$$

and hence the *total amount* which has diffused into the solid per unit area of surface at time $t$ is

$$Q = 2C_0\left(\frac{Dt}{\pi}\right)^{1/2}. \tag{8.51}$$

For the sorption and desorption of gas in the case of a *wire* heated by the passage of current, where diffusion is radial, equation 8.44 is replaced by the equation

$$\frac{\partial C}{\partial t} = \frac{D}{r} \cdot \frac{\partial}{\partial r}\left(r\frac{\partial C}{\partial r}\right). \tag{8.52}$$

The solution of this equation, as applied to the rate of degassing of a wire of radius $r$, has been given by Euringer.[95]

For *large values* of $Dt/r^2$ ($D$ in square centimeters per second, $t$ in seconds, and $r$ in centimeters), the solution has the form

$$\log G = \log\frac{2C_0 D}{r} - \frac{2.519\, Dt}{r^2}, \tag{8.53}$$

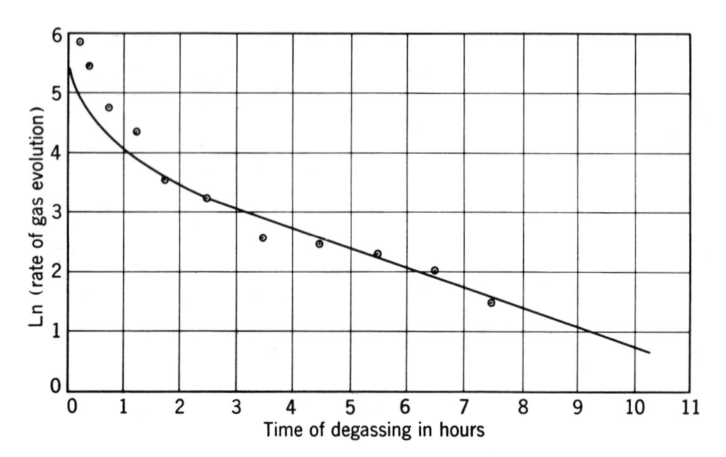


Fig. 8.26. Plot of natural log of rate of degassing versus time in hours. The full line shows the theoretical curve; the linear portion of the plot follows the relation log $G = A - Bt$, where $G$ denotes the rate of evolution of carbon monoxide from a nickel wire at 950° C. (Smithells and Ransley.)

where $G$ = rate of evolution of gas in atmosphere · centimeters$^3$ per square centimeter per second, and $C_0$ = initial concentration of gas in cubic centimeters (STP) per cubic centimeter (that is, $C_0 = V_g$ for $t = 0$).

Since it may be assumed that, at constant temperature, $D$ is a constant, it follows from equation 8.53 that a plot of log $G$ versus $t$ should yield a straight line (at least for sufficiently large values of $t$), the slope of which is given by $-2.519D/r^2$. Such a plot, obtained by Smithells and Ransley,[96] is shown in Fig. 8.26. This is a plot of the natural logarithm of $G$ versus $t$ ($G$ = rate of evolution of carbon monoxide) during the degassing of a nickel wire 1 mm in diameter at 950° C.

The absolute values of $D$ for carbon monoxide in nickel thus derived were

$$D = 4.0 \cdot 10^{-8}\ \text{cm}^2 \cdot \text{sec}^{-1} \quad \text{at } 950°\ \text{C}$$
$$= 14.0 \cdot 10^{-8}\ \text{cm}^2 \cdot \text{sec}^{-1} \quad \text{at } 1050°\ \text{C}.$$

The same investigators obtained, for the rate of diffusion of carbon monoxide through nickel at 900° C and *1 mm pressure*, the average value $q_1 = 2.05 \cdot 10^{-10}$ cm$^3$ (STP) per sec per cm$^2$ per mm thickness. From the values given in Table 8.34 for the constants $C$ and $B$ ($C = 4.2578$, $B = 2898$ for $H_2$-Ni; $C = 3.2133$, $B = 4066$ for CO-Fe) we derive the following values for $t = 900°$ C:

$$\text{CO-Fe,} \quad q_1 = 4.45 \cdot 10^{-7},$$
$$H_2\text{-Ni,} \quad q_1 = 4.87 \cdot 10^{-5}.$$

The value of $q_1$ for carbon monoxide in nickel is thus considerably smaller than the values for carbon monoxide in iron and for hydrogen in nickel. From a comparison of this value for carbon monoxide in nickel with the rate of evolution of carbon monoxide corresponding to the values of $G$ observed by them, Smithells and Ransley concluded that the gas could not be present in the nickel as carbon monoxide.

"The most reasonable explanation," according to them, "of the desorption of carbon monoxide from nickel is derived from a reaction between dissolved oxygen and carbon, probably present in the metal as nickel oxide and nickel carbide." This explanation was confirmed by determinations of rates of diffusion of oxygen and carbon through nickel. Table 8.37 gives the values of $D$ (square centimeters per second) obtained

**TABLE 8.37**

**Values of Diffusion Constants for Carbon and Oxygen in Nickel and Steel**

| Metal | $t°$ C | $10^8 D$ for Carbon | $10^{10} D$ for Oxygen |
|---|---|---|---|
| Nickel | 900 | 1.8 | 1.5 |
| | 950 | 3.7 | 6.8 |
| | 1000 | 7.2 | 23.6 |
| | 1050 | 13.3 | 78.0 |
| Steel | 900 | 3.8 | ... |
| | 950 | 8.7 | ... |
| | 1000 | 20.0 | 7.5 |

for these elements in nickel by Smithells and Ransley, and, for comparison, values obtained for steel by a group of English investigators.

At saturation, nickel contains about 1.1 per cent (by weight) NiO and 0.4 per cent carbon. These were the values of $C_0$ used in calculating the above values of $D$.

As Smithells and Ransley point out, "Commercial nickel always contains a small percentage of carbon, usually of the order of 0.03 per cent, which is much in excess of that required to account for the carbon monoxide desorbed."

From observations on the rate of evolution of hydrogen from a nickel wire, Euringer[95] obtained the following values of $D$ (square centimeters per second) for the system $H_2$-Ni:

| $t°$ C: | 165 | 125 | 85 |
|---|---|---|---|
| $10^8 D$: | 10.5 | 3.4 | 0.86 |

These results can be represented by the relation

$$D = 2.04 \cdot 10^{-3} \epsilon^{-8700/R_0 T}.$$

Barrer[97] has quoted the following results, obtained by A. Bramley and his associates, on the diffusion of nitrogen in steel:

| $t°$ C: | 800 | 850 | 900 | 950 | 1000 | 1050 | 1100 |
|---|---|---|---|---|---|---|---|
| $10^8 D$: | 1.2 | 3.0 | 6.0 | 10.8 | 13.5 | 25.0 | 40.0 |

These results are in agreement with the relation

$$D = 1.07 \cdot 10^{-1} \epsilon^{-34{,}000/R_0 T}.$$

Thus the values of $D$ for any gas-metal system can be represented as a function of $T$ by a relation similar to that used for $q$ (or $Q_{\mu l}$) above, that is, by a relation of the form

$$\log D = C' - \frac{B'}{T}. \tag{8.54}$$

Table 8.38 gives values of $B'$ derived from the values of $D$ given in

**TABLE 8.38**

**Values of $B'$ for Diffusion Constants Compared with Those of $B$ for Permeability**

| System | $B'$ | $B$ (Table 8.34) |
|---|---|---|
| $H_2$-Ni | 1,900 | 2898–3371 |
| CO-Ni | 8,100 | ... |
| C-Ni | 8,950 | ... |
| $O_2$-Ni | 17,150 | ... |
| $N_2$-Fe | 7,430 | 5204 |
| $N_2$-Mo | 5,816 | 9837 |
| $N_2$-W | 11,040 | ... |

Table 8.37 and, for comparison, the values of $B$ given in Table 8.34. The values for $N_2$-Mo and $N_2$-W were deduced by Norton and Marshall[98] from observations on the rate of gas evolution from samples which had been saturated previously with nitrogen.

That the values of $B'$ need not necessarily agree with those of $B$ follows from the fact that, according to equation 8.25, the variation with temperature of $q$ includes the variations of both $D$ and $C$.

The flow of heat is very similar to the diffusion process, and books on heat flow are of great help in solving diffusion problems. Fourier's law for linear heat flow, of which Fick's law (equation 8.44) is a special case, is:

$$\frac{\partial T}{\partial t} = h\frac{\partial T}{\partial x^2} \tag{8.55}$$

with $T$ = temperature, $t$ = time, and $x$ = distance. $h$ is the thermal diffusivity and is related to the thermal conductivity $K$ by the equation

$$h = \frac{K}{cd}. \tag{8.56}$$

Here $c$ = specific heat and $d$ = density; hence $cd$ is the heat capacity per unit volume of the solid. In the cgs system, the units of $h$ are square centimeters per second, as for the diffusion constant $D$. In solutions of the Fourier equation, $ht/a^2$ is called the Fourier number. This may well be applied in the diffusion case for the dimensionless number $Dt/a^2$. There is also the similar relationship of the diffusion constant $D$ to the permeability $P$ of gases through solids:

$$D = \frac{P}{S} \tag{8.57}$$

where $S$ is the solubility of the diffusing material in unit volume of the solid.

Carslaw and Jaeger[99] give curves and equations to enable the spatial distribution of temperature (or dissolved gas, in the diffusion case) to be determined.

## 8.11. EARLIER INVESTIGATIONS ON DEGASSING OF METALS

In the preceding sections it has been shown that gases are adsorbed on the surface and are also present in the interior of metal parts either in a state of solution or in the form of chemical compounds. Gases may also be present in microscopic blowholes or pores. The surface of certain metals, especially copper and iron, may be covered with a film of oxide which has a thickness varying from $10^{-4}$ to $10^{-6}$ cm.

If such metal parts are not previously degassed they will gradually evolve gas in an evacuated volume, especially if they are heated to even a slight extent, because of increased rate of diffusion from the inside towards the surface. Consequently, one of the most important problems in vacuum technique is the elimination of all sorbed gases from metal parts.

Metal parts may be preheated in pure, dry hydrogen whenever the nature of the metal permits.[98] A preliminary heating in vacuum by means of high frequency is extremely worth while, since, after such a treatment, metals may be stored for at least a few days in a dry atmosphere at 100–200° C without reabsorption of more than an almost negligible amount of gas.[100]

The literature on the amount and composition of the gases which are evolved from metals when degassed in vacuum at higher temperatures is very extensive. In this section and those that follow only the more important investigations can be reviewed.[101]

The vacuum-fusion furnace designed by Guldner[102,103] has come into wide use for quantitative studies of gas evolution.

In connection with his investigations on the causes of blackening of evacuated tungsten-filament lamps, Langmuir developed a method for analyzing the small amounts of gas contained in metals and glass.[104] He observed[105] that, when care is taken to denude the glass walls of occluded water vapor and carbon dioxide, the amount of gas evolved from a tungsten filament does not exceed ten times the volume of the metal. Most of the gas is eliminated at a temperature of 1500° C, and it consists of about 70–80 per cent by volume of carbon monoxide, the remainder being mostly hydrogen and carbon dioxide.

During the course of this investigation the amounts and composition of the gases evolved from other metals used in the construction of incandescent-filament lamps were determined. The metals were in the form of filaments having a total volume of about 30 cu mm, so that they could be heated by passing current through them. Different samples of so-called "untreated" nickel wires gave off amounts of gas varying from 5 to 15 cu mm (STP), consisting of about 75–90 per cent carbon monoxide and 20–10 per cent carbon dioxide with small amounts of hydrogen. Similar filaments of Monel metal, copper, and copper-coated nickel-iron alloy (used for leads in lead-glass stems) evolved amounts of gas varying from 3 to 20 cu mm of gas. The composition of the gas was found to be about the same as that evolved from nickel wires. It will be observed that in all these cases the volume of gas did not exceed that of the metal; but, since 1 cu mm (STP) corresponds to about 0.83 micron · liter at room temperature, it is evidently important to eliminate this gas before sealing off an evacuated device containing parts made of these metals.

Ryder,[106] by a modification of Langmuir's method of analysis, determined the nature and composition of the gases evolved on heating untreated commercial copper in vacuum. It was observed that the gases removed, in order of decreasing amounts, were carbon dioxide, carbon monoxide, water vapor, and nitrogen. The total amount of gas evolved at temperatures below 750° C was about 0.153 $cm^3$ per $cm^3$ of metal.

The gases in steel were investigated by Alleman and Darlington.[107] Some typical results obtained in this work are shown in Table 8.39.

TABLE 8.39

**Amounts and Compositions of Gas Evolved from Steel**

| Volume of Gas per Gram of Metal ($cm^3$) | Volume of Gas per $cm^3$ of Metal ($cm^3$) | Maximum Temperature (°C) | Composition of Gas (volume per cent) | | | | |
|---|---|---|---|---|---|---|---|
| | | | $CO_2$ | $O_2$ | CO | $H_2$ | $N_2$ |
| (1) 25.2 | 197 | 1468 | 0.13 | 2.08 | 59.8 | 18.18 | 19.81 |
| (2) 18.6 | 146 | 1500 | 1.20 | 0 | 79.8 | 11.65 | 7.35 |
| (3) 8.5 | 67 | 1100 | 0.68 | 1.57 | 26.15 | 43.40 | 28.20 |

The analyses of the three metal samples were as follows:

| Sample | (1) and (2) | (3) |
|---|---|---|
| Per cent carbon | 1.049 | 0.084 |
| Per cent silicon | 0.153 | 0.005 |
| Per cent phosphorus | 0.045 | 0.094 |
| Per cent sulfur | 0.028 | 0.082 |
| Per cent manganese | 0.405 | 0.536 |

Table 8.40 shows results of an analysis of the gases obtained from a

TABLE 8.40

**Compositions in Volume Per Cent of Gas Evolved from Bessemer Steel at a Series of Temperatures**

| Gas | At 1000° C | At 1250° C | At 1500° C | At 1675° C |
|---|---|---|---|---|
| $CO_2$ | 1.08 | 0.62 | 0.00 | 0.00 |
| $O_2$ | 2.4 | 3.07 | 4.28 | 6.25 |
| CO | 48.9 | 56.10 | 18.75 | 8.42 |
| $H_2$ | 21.16 | 15.08 | 4.20 | 1.10 |
| $N_2$ | 26.46 | 25.13 | 72.77 | 84.23 |

Bessemer steel (0.1 per cent carbon) at different temperatures. The total amount of gas evolved was 28.1 $cm^3$ per gram or 220 $cm^3$ per $cm^3$ of metal.

It will be observed that the hydrogen and carbon dioxide were evolved at 1250° C and lower, whereas the removal of nitrogen and oxygen required much higher temperatures. At the highest temperatures the metals were in a fused condition. Another interesting feature of these results is the relatively large total volume of gas evolved as compared with the results for tungsten, nickel, and copper already mentioned.

By 1929 the large-scale production of vacuum-fused metals had attained considerable importance, especially in Germany, and a paper by Rohn[108] deals with the technique and properties of vacuum-fused metals. According to this paper, the volume of gas obtained in this process is about 16–120 times that of the metal. The gas consists mostly of hydrogen and carbon monoxide and is ascribed to chemical reactions which have not been completed during reduction of the oxide and proceed to completion during the process of vacuum fusion. Thus carbon monoxide and dioxide are due to a reaction between carbon in the metal and residual oxides, while hydrogen is due to reaction between water vapor and the metals. (We shall find that this explanation of the origin of these gases is confirmed by many investigators.)

Rohn's observations on a chrome-nickel alloy showed that *most of the gas is evolved after the metal is fused*, and that this gas contains a relatively large proportion of carbon monoxide with smaller percentages of nitrogen, oxygen, and hydrogen. However, most of the hydrogen is evolved at temperatures below 1000° C.

As Rohn has pointed out, metals and alloys fused in vacuum are softer and more ductile than the same materials fused in air. For instance, copper containing a small amount of $Cu_2O$ does not exhibit embrittlement after fusion in vacuum, and furthermore the electrical conductivity is higher (by about 1 or 2 per cent) than that of ordinary commercial copper.

Naturally, the problem of developing a technique for the fusion of metals in vacuum and subsequent analysis of the gas evolved has received considerable attention. Some of the laboratory methods which have been developed will be described in a subsequent section. In a comprehensive review of the methods used previously, Hessenbruch[109] has also given data on the amount and composition of gases evolved from heated samples of different metals. Tables 8.41 and 8.42 show results observed for copper and nickel, respectively.

**TABLE 8.41**

**Amounts and Compositions of Gas in Copper Treated at 1250° C**

| Sample | Volume of Gas per 100 g Metal ($cm^3$) | Volume of Gas per $cm^3$ Metal ($cm^3$) | Composition of Gas (volume per cent) | | | |
|---|---|---|---|---|---|---|
| | | | $SO_2$ | $H_2$ | CO | $N_2$ |
| Refined Cu (1) | 2.72 | 0.24 | 58.2 | 7.90 | 19.50 | 14.40 |
| Refined Cu (2) | 1.46 | 0.13 | 85.7 | 7.43 | ... | 6.87 |
| Refined Cu (3) | 1.79 | 0.16 | 33.9 | 28.55 | 37.55 | ... |
| Refined Cu (4) | 0.73 | 0.07 | 61.1 | 11.11 | 22.22 | 5.57 |
| Electrolytic Cu | 7.97 | 0.71 | 10.9 | 39.52 | 49.70 | ... |

TABLE 8.42

Amounts and Compositions of Gas in Nickel Treated at 1470° C

| Sample | Volume of Gas per 100 g Metal ($cm^3$) | Volume of Gas per $cm^3$ Metal ($cm^3$) | Composition of Gas (volume per cent) | | | |
|---|---|---|---|---|---|---|
| | | | $CO_2$ | CO | $H_2$ | $N_2$ |
| Cube Ni | 482 | 42.9 | 2.30 | 90.0 | 3.50 | 4.20 |
| "Mond" Ni | 113 | 10.1 | 2.92 | 72.4 | 24.70 | ... |
| Electrolytic Ni | 7.88 | 0.7 | 0.00 | 21.1 | 78.9 | ... |

The $SO_2$ obtained in the case of copper is evidently due to the reaction

$$Cu_2S + 2Cu_2O = SO_2 + 6Cu,$$

which has been discussed in section 8.8 in connection with the determinations of the solubility of $SO_2$ in molten copper.[110]

In the case of electrolytic nickel, the large amount of hydrogen must be due to gas occluded by the metal during deposition.

Tables 8.43 and 8.44, also taken from Hessenbruch, show results obtained on samples of commercial aluminum evacuated at 1200° C and iron evacuated at 1550° C (both metals being in the fused condition). The analysis of the metal gives percentage by weight of the different constituents. The total volumes of gas in cubic centimeters per 100 $cm^3$ and in cubic centimeters per kilogram in Table 8.44, were derived by means of the conversion factors in Table 8.45.

Villachon and Chaudron[111] investigated the evolution of gases from metals in a vacuum furnace in which the pressure could be maintained at 0.02 mm at about 1700° C and at 0.002 mm at 1000° C. The fusion was carried out in magnesite crucibles. Then the samples were rolled to thin sheets, the nickel sheets being heated in quartz tubes to 800° C and the copper sheets and iron sheets being heated in Pyrex tubes to 600° C. The gas evolved was found to be composed of hydrogen and carbon monoxide, and the total amount was approximately equal in volume to that of the metal. This would correspond to about 11 $cm^3$ per 100 g of metal.

The oxide film on the surface of aluminum hinders the elimination of gas, and Chaudron[112] used an electric discharge to clean the surface. The surface was made the cathode in a high-voltage (130-kv) low current discharge (of a few milliamperes). The improvement obtained in removal of gas was certainly remarkable. Whereas, by heating the molten metal in a pressure of 10 microns, 20.0 $cm^3$ gas was obtained per 100 g of metal,

**TABLE 8.43**

**Amounts and Compositions of Gas in Aluminum Evacuated at 1200° C**

| Sample No. | Analysis of Metal | | | Total Volume of Gas | | Composition of Gas (volume per cent) | | | |
|---|---|---|---|---|---|---|---|---|---|
| | Al | Si | Fe | $cm^3/kg$ | $cm^3/100\ cm^3$ | $H_2$ | CO | $CO_2$ | $N_2$ |
| 1 | 97.61 | 1.24 | 1.15 | 46.1 | 12.5 | 71.60 | 11.90 | 5.65 | 10.85 |
| 2 | 99.52 | 0.22 | 0.26 | 39.4 | 10.6 | 74.70 | 11.16 | ... | 14.20 |
| 3 | 98.40 | 0.45 | 1.15 | 87.8 | 23.7 | 83.26 | 8.54 | 3.98 | 4.22 |
| 4 | 98.62 | 0.98 | 0.40 | 63.8 | 17.2 | 84.31 | 12.08 | 2.98 | 0.63 |
| 5 | 99.50 | 0.24 | 0.26 | 70.3 | 19.0 | 71.70 | 15.38 | 2.56 | 10.36 |
| 6 | 99.01 | 0.08 | 0.65 | 172.0 | 46.4 | 88.13 | 8.48 | 3.39 | ... |
| 7 | 99.01 | 0.06 | 0.65 | 124.5 | 33.6 | 77.33 | 17.20 | 5.47 | ... |

**TABLE 8.44**

**Amounts and Compositions of Gas in Iron Evacuated at 1550° C**

| Sample | Analysis of Metal | | | | Total Volume of Gas | | Gas Composition (per cent by weight) | | |
|---|---|---|---|---|---|---|---|---|---|
| | C | Si | Mn | S | $cm^3/kg$ | $cm^3/100\ cm^3$ | $O_2$ | $H_2$ | $N_2$ |
| Electrolytic | 0.02 | 0.02 | 0.02 | 0.0028 | 211.6 | 166.5 | 0.010 | 0.0010 | 0.0037 |
| Armco | 0.02 | ... | 0.05 | ... | 540.4 | 425.3 | 0.062 | 0.0007 | 0.0035 |
| Swedish | 0.04 | ... | ... | ... | 1221.6 | 961.6 | 0.152 | 0.0012 | 0.0029 |

**TABLE 8.45**

**Conversion Factors for Gas Analysis**

| 0.10 Per Cent by Weight | = $cm^3/kg$ | = $cm^3/100\ cm^3$ Fe |
|---|---|---|
| Hydrogen | 11,200 | 8853.0 |
| Oxygen | 700 | 553.4 |
| Nitrogen | 800 | 632.4 |
| Carbon monoxide | 800 | 632.4 |
| Carbon dioxide | 509 | 402.4 |

as much as 192 $cm^3$ per 100 g aluminum was extracted by use of the gas discharge.

For many purposes, a knowledge of the exact composition of the gas is not nearly as important as information about the total volume that can be removed by heating to a temperature just below fusion. On the basis of such data it is possible to compare the results obtained by different procedures for the preliminary treatment of metal parts. In order to obtain such data, Mrs Andrews,[113] working in the General Electric Research Laboratory, developed the arrangement shown diagrammatically in Fig. 8.27.

Samples of the metals to be investigated were heated in a molybdenum cup $M$ contained in a 7-in. bulb. The evolved gases were removed by means of a mercury pump into a reservoir to which was connected a McLeod gauge. Thus the total amount of gas could be measured. By

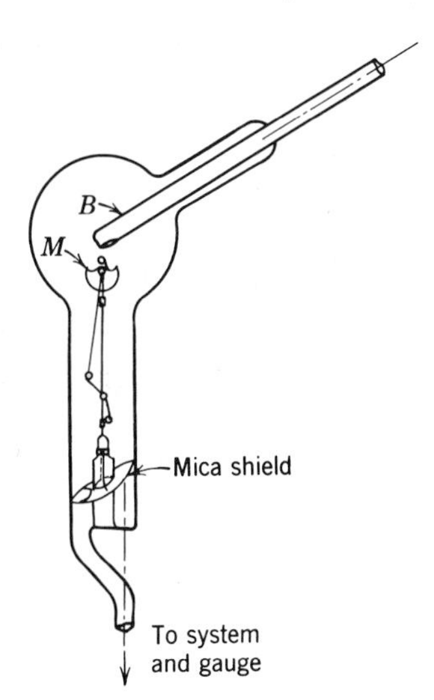


Fig. 8.27. Arrangement for heating samples in molybdenum cup for determination of gases evolved from metals (Andrews).

means of a liquid-air trap below the reservoir a determination could also be made of the amount of gas not condensable at about −190° C, which includes hydrogen, nitrogen, and carbon monoxide, while the condensable gas was mostly carbon dioxide.

The procedure adopted for making an analysis consisted in placing the sample in the tube *B*. The whole system was baked at 450° C for 1 hr, and the molybdenum cup was heated with the high frequency to degas it. The sample was then shaken into the cup by tapping the tube *B* gently, and the reservoir for collecting evolved gas was closed off from the final pump. The high frequency was applied again, the temperature of the cup being determined by an optical pyrometer, and the heating continued until the gas evolution had become negligible at the maximum temperature.

**TABLE 8.46**

**Amounts of Noncondensable and Condensable Gases Evolved by Heating Samples of Metals in Vacuum**

| Material | Noncondensable Gas ($cm^3$/kg) | Condensable Gas ($cm^3$/kg) |
|---|---|---|
| Ni wire, 0.030 in., $H_2$-annealed | 49.0 | 3.1 |
| | 76.0 | 3.9 |
| Ni strip, 8 mil | 60.2 | 8.2 |
| Ni strip, vacuum-treated (1 hr at 1000° C) | 40.5 | 5.5 |
| Mo sheet | 52.9 | 8.3 |
| Mo sheet, treated in $H_2$ (15 min at 1150° C) | 10.8 | 4.0 |
| Mo sheet heated in vacuum (1 hr at 1050° C) | 9.8 | 3.7 |
| Ni, $H_2$-annealed, diamond drawn | 96.0 | 6.5 |
| Cu, pure, deoxidized | 12.1 | 2.5 |
| | 6.3 | 2.1 |
| | 2.5 | 1.6 |
| Cu, heated to 1600° C in vacuum | 22.0 | ... |
| | 6.3 | 3.6 |
| Cu, boiled in vacuum in Mo cup | 5.4 | 3.3 |

The molybdenum cup was fastened on a pivot so that it could be emptied after the gas determination. The mica shield was inserted to prevent cracking of the seal. In this manner several samples could be inserted in tube *B* and a gas determination made on each, without opening the connection to the evacuating system.

Table 8.46 shows some typical results obtained by this method. Since the densities of both copper and nickel are approximately the same (8.9), 100 $cm^3$ per kg metal corresponds to 0.89 volume of gas per volume of metal. Thus the data show that the volume of gas rarely exceeded that of the metal, and by treatment at high temperature in vacuum the gas content could be reduced to less than 10 per cent of the volume of the metal.

An interesting point observed during this investigation, which was confirmed during the course of the work described in the next section, was that vacuum-fired samples exposed to a dry atmosphere take up extremely small amounts of gas unless touched with the fingers.

The effect of grease on the surface of samples to be tested for gas content has been emphasized by Steinhäuser.[114] He has shown that insufficient care in removing grease from the surface of aluminum leads to erratic results in the subsequent determinations of gases evolved on heating the samples in vacuum. The procedure recommended for preliminary treatment is as follows. After the samples are cleaned with a suitable volatile solvent (ether, alcohol, or benzene), they are heated to 350° C in a gas flame. Samples of aluminum treated in this manner, then heated to 1050° C in vacuum and allowed to solidify in vacuum, evolved consistently about 0.5 $cm^3$ per 100 g of metal (1.35 $cm^3$ per 100 $cm^3$ of aluminum). The gas was composed of about 70 per cent by volume hydrogen and 30 per cent methane. (Since methane could have come only from carbon introduced during the heating in a gas flame, the question arises whether even better results could not have been obtained by heating the samples to about 400° C in dry nitrogen.)

In the General Electric Research Laboratory, Marshall and Norton[115] obtained the data for copper shown in Table 8.47. The "surface gas" corresponds to that evolved at temperatures of 1000° C and lower; the major portion of the "interior gas" was evolved during volatilization of the metal. The gase volved consisted largely of carbon monoxide with smaller amounts of hydrogen and carbon dioxide, except in the case of electrolytic copper, where most of the surface gas was found to consist of hydrogen.

**TABLE 8.47**

**Volumes of Gas Evolved from Copper Heated in Vacuum**

| Nature of Sample | Volume of Gas ($cm^3$/100 $cm^3$) | | Total Volume ($cm^3$/kg) |
|---|---|---|---|
| | Surface Gas | Interior Gas | |
| 1. Electrolytic | 0.51 | 7.1 | 8.57 |
| 2. Commercial | 0.67 | 4.0 | 5.22 |
| 3. Vacuum-fused | 0.91 | 3.68 | 5.13 |
| 4. Vacuum-fused | 0.69 | 5.2 | 6.58 |

A summary of the range of values obtained in the different investigations reviewed in this section, for the amount of gas evolved, is given in Table 8.48. The results on steel given in Table 8.39 are omitted from this summary, since they are obviously very much higher than any values obtained subsequently. The conclusion that can be drawn from this summary is that the total volume of gas evolved rarely exceeds that of the metal.

**TABLE 8.48**

**Summary of Gas-Evolution Data**

| Table | Metal | $cm^3/kg$ Metal | $cm^3/100\ cm^3$ Metal |
|---|---|---|---|
| 8.41 | Copper | 7–80 | 6–72 |
| 8.42 | Nickel | 79–4800 | 70–4300 |
| 8.43 | Aluminum | 40–172 | 11–46 |
| 8.44 | Iron | 11–110 | 9–87 |
| 8.46 | Nickel | 46–102 | 41–92 |
| 8.46 | Copper | 4–22 | 3.6–20 |

## 8.12. INVESTIGATION ON DEGASSING OF METALS BY NORTON AND MARSHALL

A very comprehensive series of determinations of the amounts and composition of the gases evolved from molybdenum, tungsten, iron, nickel, and graphite was carried out by Norton and Marshall.[116] The analytical method used, which is described fully in the original paper, was a modification of the methods of Langmuir and Ryder.[117] Because of their importance in vacuum-tube technique, the results obtained are reviewed in the following sections at some length.

### Degassing of Molybdenum

The samples, thoroughly cleaned by a chemical procedure described in the paper, were heated to successively higher temperatures, and the gas evolved at each stage was collected and analyzed. The following quotation from the original paper describes the initial observations:

Hydrogen comes off most readily at the lower temperatures, as Alleman and Darlington[107] found with ferrous alloys. The amount of this is relatively small.

The next gas to appear is carbon monoxide, which comes off readily at 1000° C and lower. It persists, in small amounts, to the higher temperatures.

Nitrogen is more difficult to remove and requires a temperature of 1200° and up before it comes off.

In view of these observations, all subsequent degassing experiments were carried out at 1760° C. The total volume of gas evolved varied from 0.86 to 8.2 cm³ (STP) per kg. The composition of the gas showed the following variations in percentage by volume:

$N_2$: 22–98 (most frequently, over 60 per cent)
CO: 0–76 (most frequently, 15–36 per cent)
$CO_2$: 0–9 (most frequently, 3–6 per cent)
$H_2$: 0–17 (most frequently, 0 per cent)

The volume and composition of the gas evolved were found to be affected by the method of cleaning in the preparation of the sample for vacuum treatment. According to the authors,

> One of the most striking things found in this investigation of molybdenum was that a sample once degassed stays degassed, if proper precautions are taken. The most important requirement is to avoid handling the sample. . . . If precautions are taken, and the sample is handled with tweezers and kept wrapped in condenser paper, the surface contamination is reduced to a minimum.

Degassed samples were exposed to air and then degassed again. The volume of gas obtained in that case was found to correspond to an approximately unimolecular layer (physically adsorbed). This effect of storing on degassed samples, which, as mentioned previously, was also observed by Mrs Andrews, is evidently of extreme importance in high-vacuum technique, where it is frequently necessary or convenient to degas metal parts in one operation and then store them for subsequent use.

Another series of experiments showed that "samples of molybdenum fired in a tube hydrogen furnace at atmospheric pressure and 2000° C gave off less gas on subsequent degassing than was present in unfired samples." A subsequent degassing at 1760° C for $\frac{1}{2}$–1 hr was sufficient for complete removal of residual occluded gases. (All the degassing at high temperature was carried out by means of high-frequency heating.) It was observed that *the volume of this residual gas decreased with increase in purity of the hydrogen used*, as is illustrated by the observations recorded in Table 8.49.

The "percentage of total gas removed" was calculated by assuming 4.5 cm³ per kg as the normal amount of gas in a sample that had been given a similar cleaning treatment before heating in vacuum.

The effect, on the volume of gas evolved, of using extremely pure hydrogen such as is obtained by diffusion through heated palladium is very significant, and still more important is the observation that degassing may be accomplished by heating samples in purified hydrogen instead of in a vacuum furnace. As has, however, been mentioned previously,

TABLE 8.49

**Effect of Treatment of Molybdenum on Gas Content**

| Method | Gas Left in Mo ($cm^3/kg$) | | Average Percentage of Total Gas Removed |
|---|---|---|---|
| | Average | Lowest | |
| Hydrogen firing, tube furnace at 1 atm | 1.15 | 0.86 | 74.8 |
| Arsem vacuum furnace | 0.95 | 0.58 | 75.9 |
| Tungsten vacuum furnace | 0.82 | 0.44 | 81.5 |
| Purified line hydrogen, 1 atm | 0.47 | 0.39 | 89.6 |
| Line hydrogen diffused through heated palladium | 0.16 | 0.07 | 96.4 |

hydrogen firing is not always a safe procedure because of so-called embrittlement effects.

Observations made by Norton and Marshall on rates of degassing have been discussed in section 8.10.

**Degassing of Tungsten and Nickel**

To degas tungsten at a rate comparable with that observed for molybdenum at 1800° C, it was found necessary to heat the metal to 2300–2430° C. The amount of gas evolved from tungsten was observed to be about 15–25 per cent of that evolved from a similar sample of molybdenum and averaged about 0.3 $cm^3$ per kg. The average composition of the gas was about 67 per cent nitrogen, 30 per cent carbon monoxide, and 3 per cent hydrogen.

Since nickel has the same vapor pressure at 1030° C as molybdenum at 1760° C (about 3 microns), this fixed an upper limit to the temperature at which the nickel could be heated for any length of time in vacuum. A sample of vacuum-melted nickel heated to 1090° C for about 7 hr gave off 2.8 $cm^3$ per kg, while, from a sample of "Electro" nickel heated to 1100° C for 15 hr and then to 1200° C for 4 hr, 19.3 $cm^3$ per kg was obtained. In this case the gas evolved consisted of 60–90 per cent carbon dioxide, 2–40 per cent hydrogen, and 6–11 per cent nitrogen.

**Degassing of Iron**

Norton and Marshall describe the different types of iron investigated as follows:

1. Ordinary low-carbon stamping steel.
2. Svea iron. This is described[118] as a special type of exceptionally pure Swedish iron made by a secret process to approach a chemically pure iron.

3. Cast steel. A manganese molybdenum steel with the following analysis: carbon, 0.24 per cent; silicon, 0.34; sulfur, 0.010; phosphorus, 0.017; manganese, 1.29; molybdenum, 0.49.
4. Bain cast steel of analysis: carbon, 0.17 per cent; silicon, 0.37; sulfur, 0.010; phosphorus, 0.019; manganese, 0.47; chromium, 1.74; molybdenum, 0.69.
5. "New iron" made by the General Reduction Corporation, Detroit. . . . The composition was: carbon, 0.05; silicon, 0.008; sulfur, 0.006; phosphorus, 0.008; manganese, 0.18 per cent.

The samples were heated in vacuum to as high as 1370° C. The total amounts of gas (cubic centimeters per kilogram) evolved from each of the samples, designated by the numbers given above, were as follows:

1, 700; 2, 288; 2, 230; 3, 174; 4, 155; 5, 82; 5, 70.

The last sample, which was "New iron," previously fused by high-frequency heating, evidently evolved the lowest amount of gas. The gas evolved from the different samples of iron consisted mainly of nitrogen and carbon monoxide, with smaller amounts of carbon dioxide and hydrogen.

### Degassing of Graphite

Samples such as are used for anodes in Tungar rectifiers were given a preliminary firing in a hydrogen furnace at 1500° C and then heated in an Arsem vacuum furnace at 1800° C.

> At 2150° C [according to Norton and Marshall] it is possible to degas graphite so that subsequent heating at a higher temperature gives no further gas. It is very interesting to note that the gas evolved in the range 1700° C to 2200° C is predominantly nitrogen. A piece of graphite, which has been degassed completely, on subsequent exposure to air, absorbs oxygen, and some of this is so firmly bound that it is only desorbed as carbon monoxide on subsequent heating to 2150° C.

The volume evolved varied from 1.5 to 12.3 $cm^3$ per kg, and consisted mainly of hydrogen, carbon monoxide, and nitrogen with a much smaller percentage of carbon dioxide.

Table 8.50 presents a summary of the results obtained in this investigation. The second and third columns give the range of values observed for the amount of gas evolved in the different sets of experiments.

Only with iron and unprocessed graphite did the volume of gas evolved exceed that of the metal. Tungsten, since it is sintered in hydrogen at a temperature of about 3655° K (melting point), contains the smallest amount of gas per unit volume, and molybdenum has about 3–5 times as much.

**TABLE 8.50**

**Range of Values for Total Gas Evolved from Materials Investigated by Norton and Marshall**

| Material | Total Gas Evolved ($cm^3$, STP) | |
|---|---|---|
| | Per kg | Per 100 $cm^3$ |
| Tungsten | 0.2–0.4 | 0.4–0.8 |
| Molybdenum | 1–10 | 1–10 |
| Nickel | 3–20 | 3–18 |
| Iron (and steel) | 70–700 | 55–550 |
| Graphite, processed at 1800° C | 1.5–12 | 0.33–2.7 |
| Graphite, unprocessed | 270 and higher | 60 and higher |

The important conclusions that may be drawn from the results obtained by Norton and Marshall are the following:

1. A preliminary firing in *very pure hydrogen* at as high a temperature as possible yields a much more gas-free metal and should be used whenever hydrogen produces no deleterious effects on the mechanical or electrical properties of the metal.
2. Since the rate of evolution of gas increases rapidly with increase in temperature, the sample should be evacuated at as high a temperature as practicable.
3. The total amount of gas evolved per unit mass or unit volume depends to a great extent on the nature of the preliminary treatment, such as method of cleaning, on the purity of hydrogen used for annealing or for the elimination of oxides, and even on the metallurgical procedures followed in the production of the sample.

## 8.13. DETERMINATION OF OXYGEN CONTENT OF METALS FROM COMPOSITION OF GASES EVOLVED

According to Smithells,[119] "the total quantity of gas that can be extracted from commercial metals by heating *in vacuo* is usually between 1 and 20 $cm^3$ per 100 g of metal. With copper and aluminum the gas is largely hydrogen, but carbon monoxide is the chief constituent in iron and nickel." That the carbon monoxide is due to a reaction between carbon (usually present as carbide) and oxygen (or oxides) in the metal has been demonstrated by several investigators, including Smithells and Ransley, as noted in section 8.10.

In an excellent review dealing with the application of nickel in the radio-tube industry, Wise[120] has pointed out the importance of the oxygen content in the metal on account of its bad effect on the life of oxide-coated cathodes. The fact that iron contains about 10–50 times as much oxygen as nickel does is due to the inherent difference in ease of oxidation.

Furthermore, as Wise states,

> Nickel is melted by special methods which insure a minimum oxygen content, and finally an addition of about 0.1 per cent of magnesium is made which may further lower the available oxygen and which confers several other desirable properties on the nickel. The methods employed in annealing nickel are such as to reduce the oxygen content further.

Since, as will be shown in Chapter 11, many metal oxides posesss extremely low dissociation pressures even at 1500–1800° C, the "vacuum-fusion method" is frequently used for the purpose of determining the oxygen in the form of carbon monoxide. This procedure, developed by Vacher and Jordon,[121] consists in fusing the metal in vacuum in a graphite crucible.[122] Improvements in the method have been described by Reeve;[123] results obtained by him on the temperature of reduction of oxides by carbon are shown in Table 8.51. "Very slow reduction was observed at temperatures about 50° C below those quoted in the table."

**TABLE 8.51**

**Degree of Reduction of Oxides by Carbon**

| Oxide | Temperature of Reduction (°C) | Percentage Yield of Oxygen Collected as Carbon Monoxide |
|---|---|---|
| $FeO$, iron oxide | 1000–1050 | 97.3 |
| $MnO$, manganous oxide | 1150 | 86.9 |
| $SiO_2$, silica | 1300 | 100.0 |
| $Al_2O_3$, alumina | 1550–1600 | 75.0 |

It should be observed that, in terms of volume at STP,

$$1 \text{ per cent by weight of oxygen (O)} = 14 \text{ cm}^3 \text{ CO per g.}$$

$$1 \text{ per cent by weight of carbon (C)} = 18.67 \text{ cm}^3 \text{ CO per g.}$$

Thus, from a measurement of the carbon monoxide evolution at different temperatures, it is possible to determine the nature and amount of each oxide present in the metal. As an illustration of the application of the vacuum-fusion method, Reeve gives the analytical results shown in Table 8.52 of a bare weld wire which contained 0.021 per cent carbon, 0.09 per cent manganese, and 0.014 per cent silicon. It is of interest to

TABLE 8.52

**Percentage by Weight of Oxygen, Nitrogen, and Hydrogen Evolved from Bare Weld Iron at Different Temperatures**

| Temperature (° C) | Oxygen as | Per Cent | Per Cent $N_2$ | Per Cent $H_2$ |
|---|---|---|---|---|
| 1050 | FeO | 0.282 | 0.159 | 0.00025 |
| 1300 | $SiO_2$ | 0.0003 | 0.0077 | ... |
| 1500 | $Al_2O_3$ | 0.022 | 0.0141 | ... |

note that, as observed previously by other investigators, the greater part of the hydrogen in steel is evolved at about 1150° C and comparatively smaller amounts are evolved at 1300° C and higher.

A second method for the determination of oxygen in metals is that originally suggested by A. Ledebur, in which *carefully purified hydrogen* is passed over the metal at a high temperature and the oxygen is collected and weighed as $H_2O$. The method has been described very fully by Larsen and Brower,[124] and an improved technique has been described by the same investigators and Shenk.[125]

TABLE 8.53

**Percentage of Oxygen as FeO in Samples of Different Metals**

| Description of Metal Used | Per Cent Oxygen as FeO | Equivalent Oxygen as CO ($cm^3$/kg) |
|---|---|---|
| Ingot iron (low-carbon, high purity) | 0.040–0.060 | 560–840 |
| Electrolytic iron | 0.018–0.024 | 250–340 |
| Pure iron, vacuum-melted | 0.003–0.006 | 42–84 |
| Ingot iron heated in $H_2$ at 1500° C | 0.002 | 28 |
| Low- and medium-carbon milled steels | 0.012–0.033 | 168–462 |
| Commercial "A" nickel (four samples) | 0.001–0.002 | 14–28 |
| Iron-aluminum alloy, 0.25% Al* | 0.002 | 28 |
| Iron-aluminum alloy, 0.50% Al* | Trace | ... |
| Carbonyl iron, pressed from powder and rolled into sheet* | 0.012 | 168 |

* Values not given in Wise's table but taken from the paper by Larsen, Brower, and Shenk (Ref. 125).

Wise[120] has given, in Table 8.53, a summary of oxygen determinations, made by the last-mentioned investigators, on samples of iron, and also data for several heats of commercial "A" nickel. When the gas evolution from iron is compared with that from nickel, it is evident that nickel is much more suitable for vacuum devices than iron. With regard to other impurities Wise has made the following comments:

Phosphorus and sulfur while not gaseous at room temperature volatilize rather easily and in some respects behave like gases. The phosphorus content of nickel is very low, about 0.003 per cent, while that of iron may range from 0.018 to about 0.065 per cent. Phosphorus seems to be tolerated by thoriated cathodes but experiments indicate that it is detrimental to oxide-coated cathodes.

Sulfur in all tube parts is definitely detrimental, and this element is carefully controlled in nickel, the amount present being normally 0.005 per cent or less. The sulfur content of irons is usually 0.025 to 0.45 per cent. Some methods of steel manufacture lead to values as high as 0.060 per cent sulfur.

Sulfur is extremely detrimental to oxide-coated cathodes, and the use of material containing appreciable quantities of this element in plates and other tube parts is known to be extremely undesirable.

Considerable work has been done recently on vacuum-fusion analysis as a general method, and for oxygen determination. This is covered in three publications.[126,127,128]

The problem of the source of nitrogen in iron and its alloys has been discussed by Reeve. In view of the observations on the dissociation pressures of the nitrides, which are discussed in Chapter 11 (also see comments at the end of section 8.7), he has suggested that the nitrogen evolved at 1300° C is associated with silicon and that at 1570° C with aluminum.

The topic discussed in this section is intimately related to the problems, dealt with in Chapter 11, of the conditions for the reduction of oxides by carbon and hydrogen, and the reader is therefore referred to that chapter for further discussion of the subject.

## 8.14. EMBRITTLEMENT OF METALS BY HYDROGEN

In section 8.12 attention was drawn to the observation made by Norton and Marshall on the beneficial effect in reducing the gas content in metals of a preliminary heating in extremely pure hydrogen. However, it was also mentioned that in the case of tantalum such a treatment leads to injurious effects on the mechanical properties. Similar effects are also observed in the case of copper.

As stated in section 8.4, a tantalum filament, heated in hydrogen to a high temperature and then cooled in the gas to room temperature, takes up about 740 times its volume of hydrogen. It was first observed by Pirani[129] that such a filament, although originally quite ductile, becomes very brittle when saturated with hydrogen. It can be crushed easily, because of the formation of very small crystals. Also the electrical resistance increases about 100 per cent. However, if the filament is raised to a very high temperature in a good vacuum, the ductibility as well as the original resistance is restored. Pirani observed that in order to achieve this result the temperature must be high enough to sinter the filament and that the gas is not completely eliminated unless the pressure is maintained at as low a value as possible (much less than 100 microns). In fact, Pirani concluded that the only method of eliminating the gas completely was to fuse the metal and maintain it in that condition for a long time in a very good vacuum.

Pirani also observed that a tantalum filament is oxidized to $Ta_2O_5$ when heated to a high temperature in a stream of hydrogen, even when the gas has been passed over freshly reduced copper turnings (heated in an oven to 500–600° C) in order to remove oxygen and water vapor. Only by passing the gas over heated calcium or molten sodium was it found possible to purify it to such an extent as to prevent surface oxidation of the tantalum filament.

Indeed, tantalum filaments heated to a high temperature may be used to remove residual traces of water vapor and other oxidizing gases from hydrogen. In this respect tantalum may be employed instead of chromium powder, which is also an efficient reagent for the purification of hydrogen. (The reason for this characteristic behavior of both tantalum and chromium will be evident from the discussion in Chapter 11 of the equilibria between the oxides and hydrogen.)

The so-called "embrittlement" of copper which results from treatment of the metal in a hydrogen atmosphere is an interesting phenomenon, similar to that observed in the case of tantalum, which has received a great deal of attention from metallurgists. The following summary is based upon the comprehensive discussion of this topic by Wyman.[130]

Wyman has shown that annealing copper alternately in an oxidizing and in a reducing atmosphere is specially effective in causing embrittlement. During the oxidizing cycle, oxygen diffuses into the metal and brings about formation of $Cu_2O$; during the reducing cycle, hydrogen diffuses into the metal and the water vapor formed causes ruptures along grain boundaries.

Wyman's investigation led to the following conclusions:

1. Neither pure copper when oxidized or deoxidized nor rough-pitch copper up to 0.06 per cent oxygen is embrittled by hydrogen at temperatures of 400° C or below, at 4 hours.

2(*a*). The rate of oxygen penetration into "pure" copper decreases markedly below 800° C.

(*b*). The rate of hydrogen penetration into oxygen-bearing copper decreases markedly below 700° C. Consequently, up to about 800° C, the coppers containing no oxygen are superior to tough-pitch coppers for resistance to the embrittling action of oxidation-reduction cycles.

The elimination of oxygen by purifying process or by means of deoxidation of the copper is advantageous in increasing the resistance to the embrittling action of reducing gases.

Rhines and Anderson[131] have observed embrittlement not only when $Cu_2O$ is present but also when readily oxidizable alloying elements are present in the copper. The explanation is the same as that suggested by Wyman.

> Hydrogen [they state] diffuses through copper much faster than the water vapor can escape by diffusion and a pressure sufficient to rupture the material is built up. The cracks and holes apparently occur at the grain boundaries, because of inherent weakness there, because of more favorable conditions for nuclear formation there, because the oxide is more plentiful there, or because hydrogen enters more readily at the grain boundaries.

While Rhines and Anderson conclude that hydrogen is the only gas that will produce embrittlement of the type discussed above, other investigators have observed the same phenomenon with illuminating gas. This, however, involves no contradication with the previous statement, since this gas contains hydrogen as well as methane and both gases react with oxides.

In a review entitled "The Influence of Gas-Metal Diffusion in Fabricating Processes,"[132] Rhines has emphasized the important fact that hydrogen may diffuse into metals as a result of a reaction between the heated metal and water vapor, or as a result of chemical action, as in pickling or electrolysis. This leads to embrittlement in the case of other metals such as iron and its alloys and in nickels.

As has been pointed out in section 8.4, the densities of the "hydrides" of palladium and of other metals of Group B are all lower than those of the pure metals. That is, the lattice expands as a result of the presence of hydrogen atoms (or rather protons) in interstitial sites. Although the solutions of hydrogen in metals of Group A, such as iron and nickel, exhibit considerablly less expansion, there is, nevertheless, a slight expansion which is probably effective, as Rhines points out, in producing stresses which ultimately lead to cracking, flaking, and similar phenomena.

> Electroplates [Rhines states], particularly those of iron, nickel, and chromium, if formed under conditions where a considerable quantity of hydrogen is released, are often very brittle and give trouble in buffing. . . .
>
> When ferrous materials are subjected to stress at elevated temperatures, the presence of dissolved hydrogen promotes the development of cracks. Thus hot formed products contaminated with hydrogen are slowly cooled to avoid the development of shatter cracks. Flakes in wrought steel are believed to result from the development of fine cracks in the metal when it is worked without first being freed of hydrogen by sufficiently prolonged heating.
>
> Some other metals are similarly sensitive to hydrogen embrittlement. Electrolytic nickel, as produced, contains a considerable quantity of hydrogen,

which is partly retained through subsequent melting and casting operations if special precautions are not taken. When such nickel is rolled or forged, cracks develop. It is found possible to eliminate the major part of the hydrogen in melting by oxidation or by a carbon boil.

In order to recover ductility in samples of ferrous alloys which have been pickled, the usual procedure consists in allowing the samples to stand in the air for some time or baking them at a low temperature in order to let the hydrogen diffuse out.[133]

Uhlig[134] has observed that low-carbon manganese-iron alloys, treated at 1000° C in hydrogen and quenched, are embrittled by retained hydrogen, whereas the alloys treated in vacuum at 1000° C and quenched in vacuum are ductile, since the hydrogen is eliminated in the vacuum treatment. The important conclusion drawn from his investigation is that *the alloys are brittle only if hydrogen is retained.* Thus, if a treatment in hydrogen at high temperature is followed by a *rapid* quench, the alloy is brittle. If, however, the quench is sufficiently slow to permit the hydrogen to diffuse out as the alloy cools, then the ductility is retained.

In agreement with Wyman, Rhines emphasized the part played by oxides in the metal. He states:

> Many commercial metals and alloys contain oxygen either in solid solution or in the form of an oxide. When such materials are heated in the presence of hydrogen, carbon monoxide, or similar reducing gases, the hydrogen or carbon may diffuse into the metal and there reduce the oxides or react with the oxygen to form insoluble water vapor or one of the gaseous oxides of carbon. If this occurs when the products of reaction are confined, a considerable pressure may be generated and blisters or cracks will form.
>
> Repeated cycles [he continues] of oxidizing and reducing conditions promote the disintegration of materials that are not particularly sensitive to either conditions alone. Apparently minute cracks which are caused by grain-boundary oxidation and which would grow only slowly under constant oxidation are freed of the obstructing filling of oxide during the reduction phases, and so admit gas more readily on subsequent oxidation.

Such a "water cycle," as it has been designated, also occurs in evacuated devices if precautions are not taken to eliminate water vapor as completely as possible. The phenomenon was first observed by Langmuir in connection with his investigations on the causes of blackening in vacuum tungsten-filament lamps. If these lamps are not exhausted with sufficient care to remove all traces of water vapor or possible sources of evolution of the vapor, the vapor is dissociated at the incandescent filament with formation of oxygen and hydrogen. In contact with the filament at temperatures of about 1500° C and higher, atomic hydrogen is formed, which is an even more effective reducing agent than molecular hydrogen. The cycle which then occurs is as follows. The oxygen forms $WO_2$ and $WO_3$, which volatilize and condense on the glass walls. The atomic hydrogen reduces the oxides to tungsten, and the resulting water vapor then goes through the same cycle as before, thus causing a very rapid blackening of the glass bulb. Similar reaction may also occur with other metals. In a hot-cathode device, it results not only in rapid disintegration of the cathode but also in a considerable decrease in electron emissivity, because of the "poisoning" effect of oxygen.

To the vacuum-tube engineer interested in the fabrication of devices having metal containers, or to the designer of vacuum systems made of metal for large-scale operations, the degassing characteristics of the metals used are obviously of primary importance. It is for this reason that the rather lengthy discussion has been given in this and the preceding sections of topics which, at first glance, may appear to be more pertinent to metallurgy than to vacuum technique. Also, for the same reason, it has been considered worth while to add the remarks made in the following section, which deals with a closely allied topic.

## 8.15. INFLUENCE OF GASES ON MECHANICAL PROPERTIES

The influence of gases, particularly hydrogen, on the mechanical properties of steels and other metals has been widely studied. Several books and reviews on this subject are available.[135,136,137,138,140]

The articles by Perlmutter and Dodge[139,140] involve especially hydrogen at high pressures, with extensive references. Other books are by Robertson[141] and Queneau and Ratz.[142]

During the refining or melting operations, metals are quite likely to take up large volumes of gas. This is due primarily to the fact, emphasized in section 8.3, that the solubility of gases, especially that of hydrogen, in metals of Group A increases with the temperature and is greater in the fused metal than in the solid. That the increase in solubility at the melting point is considerable is shown by the data in Table 8.54, which are taken from the tables in section 8.3, indicated in the first column. As usual, $s$ = cubic centimeters (STP) per 100 g metal.

**TABLE 8.54**

**Increase in Solubility (at 1 atm) of Hydrogen in Metals at Melting Point**

| | | | | | | |
|---|---|---|---|---|---|---|
| Table 8.1 | Iron | $t°$ C: | 1500 | 1535 (mp) | 1550 | 1650 |
| | | $s$: | 12.75 | | 27.75 | 30.97 |
| Table 8.4 | Nickel | $t°$ C: | 1400 | 1452 (mp) | 1500 | 1600 |
| | | $s$: | 16.52 | | 41.6 | 43.1 |
| Table 8.7 | Copper | $t°$ C: | 1000 | 1083 (mp) | 1100 | 1200 |
| | | $s$: | 1.58 | 2.10 (solid)<br>6.10 (liquid) | 6.3 | 8.1 |

For hydrogen in aluminum, *s* is negligible for the solid but increases rapidly with the temperature above the melting point. For oxygen in silver at 1 atm, *s* increases from 5.4 at 923° C to 213.5 at 973° C (melting point = 960.5° C).

On solidification the large excess of gas held in solution is liberated and is likely to be trapped in the metal, as it crystallizes, in the form of minute cavities or blowholes, which result in unsound castings. As observed in the discussion of the data in Table 8.33 on the solubility of oxygen in molten silver, Allen[143] has shown that the trapped gases exert enormous pressures.

In general the gas that most frequently causes defects in the casting of metals such as iron, nickel, copper and aluminum is hydrogen. This may come from decomposition by the fused metal of water vapor or hydrocarbon vapor present in furnace gases, or from reaction with moisture in the sand molds.

In a melt which contains, in addition to hydrogen, a small amount of dissolved oxygen, the gases recombine on solidification to form steam, thus giving rise to what has been designated "reaction unsoundness" or "steam unsoundness." This was first observed by Allen in his investigation on gases in molten copper. He concluded that, as a result of the reaction between hydrogen and oxygen, "the blowholes are much smaller, more numerous, and tend to congregate in those parts of the crystal grain last to solidify and in the crystal boundaries."[144] This type of porosity is observed when the oxygen content exceeds a certain critical value; according to observations by Allen and Street,[145] this critical oxygen content is apparently connected with the solid solubility of the oxygen in the metal.

It should be observed that, in a paper by Allen and Hewitt,[146] it was shown that the behavior of the system consisting of molten copper in equilibrium with hydrogen, oxygen, and steam is "consistent with the ordinary mass-action laws, and that from the data obtained the dissociation pressure of cuprous oxide could be calculated."[147]

A review of the investigations carried out by himself and previous workers on gases in molten and solid copper has been given by Ellis.[148] Many of the observations reported in this paper, such as those on the solubility in copper of carbon dioxide and water vapor, are, however, not confirmed by other investigators.

The results obtained by Allen and his associates, as well as methods for the degassing of metals, have been discussed fully by Chaston.[149] The different methods which have been proposed for the degassing of metals are summarized by Chaston as follows:

1. Vacuum fusion.[150]
2. Presolidification.[151]

3. Bubbling a stream of nitrogen through the molten metal before casting.[152]
4. Use of oxidizing fluxes.[153]

The relative merits of the first three methods were also discussed in an earlier paper by Prytherch.[154] In a series of experiments on the effects of oxygen, hydrogen, and sulfur dioxide on the soundness of ingots he observed that the nature of the unsoundness due to dissolved gases "depends on the conditions of temperature, on rate of solidification, and possibly on shape of ingot obtained during casting." Furthermore he concluded that, "Under conditions of the experiments [carried out in this investigation] it has been found that the method of removing gases from copper by oxidation of the metal and its subsequent removal is extremely slow and incomplete."

Molten aluminum can absorb 10 times as much hydrogen as corresponds to equilibrium solubility; and, according to Claus and his associates,[152] this gas is due to a reaction between the metal and water vapor. This dissolved excess gas is likely to be emitted during solidification with intense frothing, and as a remedy it is recommended that the metal should be kept for some time at a temperature just above the melting point in a furnace atmosphere as free as possible from water vapor. These workers have also used the third method mentioned above, but for aluminum they have obtained good results by using chlorine or mixtures of this gas and nitrogen instead of pure nitrogen.

The second method, as described by Chaston, "consists simply in allowing the melt to cool slowly in a crucible so as to expel most of the dissolved hydrogen and then to remelt the metal as rapidly as possible."

Undoubtedly the most effective method of degassing metals is fusion in vacuum. Evidence for this statement has been presented in the previous sections of this chapter, and confirmation is given in a publication by Morse.[155] The gas content of vacuum-melted copper obtained in the laboratory of the National Research Corporation, as compared with oxygen-free commercial metal, is stated to be as follows:

| Gas | Gas Content (per cent by weight) | |
|---|---|---|
| | Vacuum-melted | Oxygen-free Commercial |
| Hydrogen | 0.00001 | 0.00012 |
| Oxygen | 0.00039 | 0.00045 |
| Other gases | 0.00002 | 0.00040 |
| Sulfur | 0.00006 | 0.00230 |

It should be noted that, according to the conversion factors in Table 8.45, the hydrogen content of the vacuum-melted metal corresponds to 1.12 atm · cm$^3$ per kilogram, and the oxygen content to 2.73 atm · cm$^3$ per kilogram.

The application of the methods discussed above to the degassing of aluminum has also been discussed by Erickson.[156]

Of special interest in this connection are the observations reported by Johnson, Arner, and Schwartz[157] on the slow evolution of gas from cast steel at *room temperature.* The gas consists of carbon monoxide, nitrogen, and hydrogen, and the rate of evolution increases both with increase in temperature and with decrease in the partial pressure of the evolved gases in the surrounding atmosphere. Furthermore, the rate varies with time in the same manner as that of a first-order reaction, so that after a period of several hundred hours the volume of gas evolved reaches a constant value which is of the order of a few cubic centimeters per 100 g of metal. Since 1 atm · cm$^3$ corresponds to approximately 760 micron · liters, it is evident that such a rate of gas evolution would make a material practically useless for application in the fabrication of vacuum devices.

One important conclusion that can be drawn from the discussion in this section is that in attempting to use cast metals in vacuum devices extremely great care should be exercised to make certain that the castings are vacuumtight. Though certain palliative procedures may be used for "vacuum-proofing" a porous container, they are obviously not always reliable and care should be taken, if it is necessary to use a casting, to check its soundness by preliminary tests.

Large-scale vacuum melting has come into wide use, and several publications giving many details are available.[158,159,160,161,162,163]

## 8.16. ANALYSIS OF GASES IN METALS

The so-called "low-pressure" method for the determination of gases in metals, glass, etc., was first developed by Langmuir.[165] It was further refined by Ryder[166] and has been modified in certain details and utilized extensively by subsequent investigators—Prescott,[167] Dalton,[168] Norton and Marshall,[169] and others.[164]

In its simplest form, the apparatus consists of a McLeod gauge for measuring the pressure of the gas (in a known total volume) and a Toepler pump for transferring the gas into other parts of the system in which each of the constituents can be determined by suitable chemical reactions. Mercury cut-offs are used to separate each individual reaction, and solid carbon dioxide and liquid air to condense out the different products. Without entering into a detailed description of the arrangements

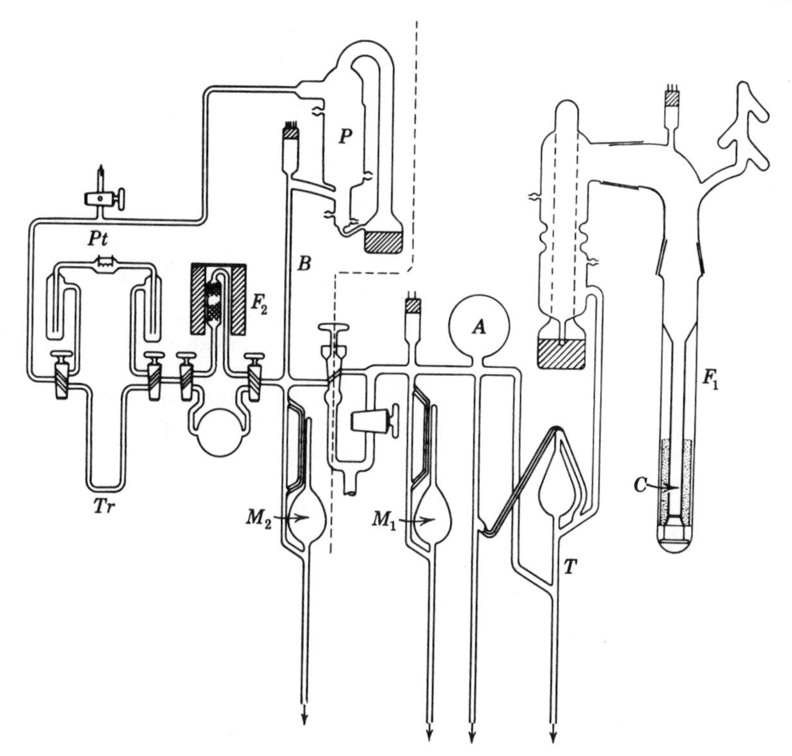


Fig. 8.28. System for determination of amount and composition of gas evolved from metals (Naughton).

employed by the different investigators mentioned above, the general principles involved may be illustrated by means of the system shown in Fig. 8.28, which was developed by J. J. Naughton of the General Electric Research Laboratory for the determination of gases in metals.

The equipment consists of two parts: (1) the gas-extraction and measurement section, and (2) the gas-analysis section. The *extraction section* (shown at the right in Fig. 8.28) consists, according to Naughton's description, of a

> ... quartz tube furnace, $F_1$, which contains the graphite crucible, $C$, and suitable shields. The gases are extracted by melting the metal in this furnace.[170] They are quickly swept away by a mercury-diffusion pump and concentrated by a Toepler pump, $T$, in a known volume, $A$. The pressure of gas in this known volume, measured with a McLeod gauge, $M_1$, gives the quantity of gas in the metal sample.
>
> The *gas-analysis section* is a circulatory system comprising a known volume, $B$, in which the gas being analyzed can be concentrated and measured.[171] A McLeod gauge, $M_2$, which is part of $B$, is provided for making this measurement.

A furnace containing a tube filled with copper oxide, $F_2$, a freeze-out trap, *Tr*, and a mercury-diffusion pump, *P*, for circulating the gas are the other essential parts of the system.

For analysis, a portion of the gas extracted from a metal is admitted to the known volume. Its pressure is measured and it is then circulated over frozen acetone to remove water vapor. After removal has been effected the remaining gas is concentrated in the known volume, *B*, and its pressure measured. The difference between the pressure before and after the removal of a component gives the pressure of the component gas in the mixture. Carbon dioxide is removed by passing through the freeze-out trap surrounded by liquid nitrogen. Hydrogen is first oxidized to water vapor by passing through the copper oxide furnace, and the water is removed as before. Similarly, the carbon monoxide is oxidized to carbon dioxide which is then removed as already noted. These are the gases usually found in the gas extracted from a metal. At times methane is also a component of the gas mixture. This gas is removed by admitting a known amount of pure oxygen and burning the mixture over a hot platinum filament, Pt. Traps are provided before and after this filament to prevent oxidation of mercury vapor and consequent error.

With this gas-analysis equipment [Naughton states] it is possible to analyze a sample of gas whose total volume is as small as 0.005 $cm^3$ (STP).

One further precaution that has to be taken in connection with the use of the equipment is to ensure that good thermal contact is maintained between the metal and the crucible. This is sometimes accomplished by means of a metal flux, such as iron, in which carbon is readily soluble.

In the case of iron-manganese alloys the manganese evaporates at relatively low temperatures,[172] and the deposit adsorbs considerable amounts of hydrogen. Hence in a determination of hydrogen in such alloys by the vacuum-extraction method, Naughton has used tin as a flux[173] in order to extract the gas at a temperature at which evaporation of manganese is negligible.

The determination of hydrogen in ferrous materials by vacuum extraction at 800° C has been investigated by Holm and Thompson.[174] They observed that as much as 90 per cent of the hydrogen introduced into iron alloys by cathodic charging or by heating in hydrogen at high temperatures is evolved from the metal during a few days' storage at room temperature. The final values for the hydrogen content are, in general, of the same order as the solubility of the gas at room temperature.

A great many investigators have dealt with the problem of oxygen determination, and the topic has been discussed at length in a paper by Thompson, Vacher, and Bright.[175]

A method for the analysis of gases which was first described by Campbell[176] and has been applied successfully by many subsequent investigators is designated by Farkas and Melville[177] as "low-temperature distillation." The principle of the method can be illustrated by the technique, developed in the General Electric Research Laboratory by F. J. Norton, for the determination of $H_2O$ and $CO_2$ in the gases evolved in the evacuation of devices containing cellulose materials.

The gas to be analyzed is collected in a known volume, $V$ liters, which includes the volume of the gauge and that of an appendix. Thus, if the pressure is $P_{\mu}$, the total quantity is $P_{\mu}V$ micron · liters. On immersing the appendix in liquid air, the $H_2O$ and $CO_2$ are frozen out and the residual pressure $P_{\mu 0}$ corresponds to the amount of noncondensable constituents ($N_2$, $O_2$, CO, and $H_2$). On removing the liquid air and letting the temperature of the appendix rise at a measurable rate (−195° C to 25° C in 20 min), the $CO_2$ begins to evaporate, and, as the temperature increases, the pressure versus temperature plot obtained (shown in curve $AA'$, Fig. 8.29) corresponds to the vapor-pressure curve of $CO_2$. (According to the data in Tables 10.12(*a*) and (*b*), the vapor pressure of solid $CO_2$ is 0.5 micron at −169° C and increases to 10 microns at −158° C.) After all the $CO_2$ has evaporated, the pressure indicated is $P_{\mu 1}$. As the temperature increases beyond that corresponding to this pressure, there is only a negligible increase in pressure (as shown by the approximately horizontal portion $A'B$ in Fig. 8.29) until a temperature is reached at which the vapor pressure of $H_2O$ becomes significant. (As shown in Table 10.11, the vapor pressure is 0.4 micron at −80° C and increases to about 100 microns at −40° C.) Thus a plot of pressure versus temperature is obtained, such as $BB'$ in Fig. 8.29.

The approximately horizontal portion, $B'C$, indicates the pressure $P_{\mu 2}$ obtained on complete evaporation of the water vapor. Thus the composition of the gas (by volume) is as follows:

$$\text{Noncondensable gas: } \frac{100P_{\mu 0}}{P_{\mu 0} + P_{\mu 1} + P_{\mu 2}} \text{ per cent,}$$

$$\text{Carbon dioxide: } \frac{100P_{\mu 1}}{P_{\mu 0} + P_{\mu 1} + P_{\mu 2}} \text{ per cent,}$$

$$\text{Water vapor: } \frac{100P_{\mu 2}}{P_{\mu 0} + P_{\mu 1} + P_{\mu 2}} \text{ per cent.}$$

The technique as developed by Norton consists in connecting a photoelectric recorder to a thermocouple which measures the temperature of the appendix and a second recorder to a thermocouple gauge which measures the pressure in the system. This procedure yields two simultaneous records, as shown in Fig. 8.29, in which $AA'BB'C$ gives a record of pressure versus time, while the linear plot, $DD$, gives the record of temperatures versus time.

As Norton has pointed out, the same procedure can be used for the analysis of gases that are evolved during exhaust. A trap to which a thermocouple is attached is inserted between the system and the pump, and, during pumping, liquid air is applied to the trap for a definite period. The liquid air is then removed and the trap warmed slowly. The temperature and pressure recorders are started. As the temperature rises to a value at which a particular gas begins to evaporate, a peak will be observed on the pressure recorder, and the position of this peak as well as its height will indicate both the nature of the gas and its relative amount. In this manner a succession of peaks is observed from which the nature of the constituents and their relative amounts may be determined. Farkas and Melville have described the application of the method to the analysis of a mixture of paraffin hydrocarbons and also refer to the refinement of the methods used by Sebastian and Howard.[178]

Kenty and Reuter[179] used a modification of the same method for the analysis of gases evolved during exhaust. After identification of the condensable gases by their condensation points, a known amount of oxygen is added to the noncondensable residue. The mixture is compressed into a small volume and ignited by a heated platinum filament. The $H_2O$ and $CO_2$ formed are then determined as described above. Any residual oxygen is determined by noting the decrease in pressure in the presence of a heated tungsten filament, and the residue is then assumed to be nitrogen.

Noncondensable gas can also be analyzed by comparing the absolute pressure reading, as read by a McLeod gauge, with the indication on a thermocouple (or Pirani) gauge which has been calibrated for the different possible constituent gases and mixtures of these gases. This method is

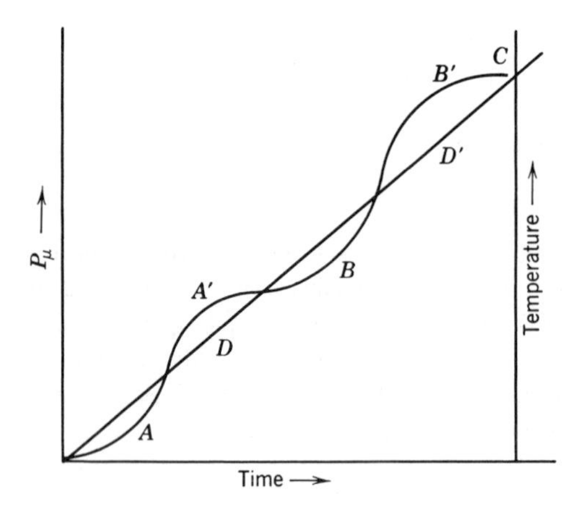


Fig. 8.29. Method of "low-temperature distillation" for analysis of gases evolved. A thermocouple gauge gives the record ($AA'$ $BB'$ $C$) of the vapor pressure and amount of each constituent; the linear plot $DD'$ shows the variation in temperature with time. (Norton.)

similar to that used by Dickinson and Mitchell[180] for the analysis of hydrogen-nitrogen mixtures.

Kenty and Reuter have also taken the speed of exhaust through a standard capillary as an additional criterion for the analysis of the evolved gas. This depends upon the fact that speed of exhaust varies inversely as $M^{1/2}$.

Kenty[181] has used the fractional distillation method to determine small amounts of oxygen, nitrogen, and hydrogen in krypton and xenon.

A powerful tool, the mass spectrometer, has come into wide use for the study and analysis of gases from metals and other materials. Norton[182] has used it for gas-permeation and analysis studies. Two extensive papers by Hintenberger and Börnenburg[183] on applications to vacuum practice give 150 references, and there have been many other studies.[184,185]

## 8.17. RARE GASES IN METALS

Under purely thermal conditions, the rare gases do not dissolve in, or permeate, true metals. In the borderline semiconducting metals, silicon and germanium, van Wieringer[186] has measured the permeation of helium by the mass spectrometer.

If rare-gas ions are formed, as in a gas discharge, they can be driven into a metal and later released when the metal is heated. This was done by Koch,[187] and there has been considerable work relating to this process.[188,189]

The diffusion of radioactive xenon and krypton in silver has been studied by Tobin.[190] He pointed out that, if a rare-gas atom must be activated to enter a metal lattice, this may be a very high energy barrier. For example, if helium must be ionized to enter, over 400 kcal per mole are represented.

Besides the method of ion impact to put rare gases into metals, there are nuclear-disintegration processes which may form these rare gas atoms *in situ*. This method has become very important in atomic energy generation. It is discussed by R. S. Barnes *et al.*[191] Other workers also have been active in this field.[192,193,194,195,196,197]

## REFERENCES AND NOTES

1. C. J. Smithells, *Gases and Metals*, Chapman and Hall, London, 1937.

2. R. M. Barrer, *Diffusion in and through Solids*, Cambridge University Press, 1951.

3. W. Jost, *Diffusion in Solids, Liquids, Gases*, Academic Press, New York, 1952.

4. W. Jost, *Diffusion: Measurement and Methods*, German monograph, *Fortschr. physik. Chem.*, Darmstadt, Germany, 1957.

5. D. P. Smith, L. W. Eastwood, D. J. Carney, and C. E. Sims, *Gases in Metals*, American Society for Metals, Cleveland, 1953.

6. A. Nikuradse and R. Ulbrich, *Das Zweistoffsystem Gas-Metall*, R. Oldenbourg, Munich, 1950.

7. W. Seith, *Diffusion in Metallen*, Berlin, 1955, section 21.

8. C. R. Cupp, "Gases in Metals," Chapter 3 in Vol. IV, *Progress in Metal Physics*, London, 1953.

9. G. Ziliani, *I gas nei metalli*, Milan, 1952, 82 pp. (Italian).

10. *Tabellen der Electronphysik*, M. von Ardenne, Berlin, 1956, Vol. II, section 3.

11. *Handbuch der Physik*, S. Flügge, Editor, Springer Verlag, Berlin, 1958, Vol. 12; "Vacuum Physics," R. Jaekel, pp. 515–607, "Ultrahigh Vacua," D. Alpert, pp. 609–662.

12. J. Schweitzer, Le Vide, **14,** 165 (1959) (ultrahigh vacuum).

13. J. Yarwood, *J. Sci. Instr.*, **34,** 297 (1957).

14. G. J. van Amerongen, *J. Appl. Phys.*, **17,** 972 (1946).

15. F. J. Norton, *J. Appl. Phys.*, **28,** 34 (1957).

16. C. J. Smithells, *Metals Reference Book*, Butterworths, London, 1955, Vol. II, pp. 553–560.

17. *Metals Handbook*, 1939 edition, p. 427.

18. N. P. Allen, *J. Inst. Metals*, **43,** 81 (1930).

19. M. B. Bever and C. F. Floe, *Trans. A.I.M.E.*, **156,** 149 (1944).

20. Plots of $s$ versus molybdenum content for different temperatures are shown in Smithells (Ref. 1), pp. 185–186.

21. W. D. Robertson, thesis, M.I.T., 1942.

22. G. I. Azarian and E. Epremian, thesis, M.I.T., 1943, a copy of which was kindly given to Dushman by Mr. Epremian.

23. Plots of the data for a series of temperatures are shown in Smithells (Ref. 1), p. 182.

24. J. Erickson, *Light Metal Age*, July, 1944, p. 19.

25. C. J. Smithells (Ref. 1), p. 161.

26. For some of the metals, the values given by C. J. Smithells (Ref. 1), p. 161, were used.

27. A. Sieverts and A. Gotta, *Z. anorg. Chem.*, **172,** 1 (1928); **187,** 155 (1930).

28. For review of the earlier literature see J. W. McBain, *The Sorption of Gases and Vapours by Solids*, George Routledge and Sons, London, 1932, pp. 288 *et seq*.

29. A. Sieverts and G. Zapf, *Z. physik. Chem.*, **A, 174,** 359 (1935).

30. A. Sieverts and W. Danz, *Z. physik. Chem.*, **B, 34,** 158 (1936).

31. J. R. Lacher, *Proc. Roy. Soc. London*, **A, 161,** 525 (1937).

32. The significance of the subscripts $\alpha$ and $\beta$ will appear from the subsequent discussion.

33. This is discussed more fully in the book entitled *The Theory of the Properties of Metals and Alloys*, by N. F. Mott and H. Jones, The Clarendon Press, Oxford, 1936, pp. 189 *et seq*.

34. C. J. Smithells (Ref. 1).

35. A. S. Darling, *Platinum Metals Rev.*, **2,** 16 (1958).

36. R. E. Norberg, *Phys. Rev.*, **86,** 745 (1952).

37. F. A. Lewis, *Platinum Metals Rev.*, **4,** 132 (1960).

38. J. B. Hunter, *Platinum Metals Rev.*, **4,** 130 (1960).

39. G. Hägg, *Z. physik. Chem.*, **B, 11,** 433 (1931).

40. Alloys of lanthanum and cerium are designated "misch metal."

41. M. Pirani, *Z. Elektrochem.*, **11,** 555 (1905).
42. A. Sieverts and H. Brüning, *Z. physik. Chem.*, **A, 174,** 365 (1935).
43. A. Sieverts and H. Brüning, *Z. physik. Chem.*, **A, 163,** 409 (1932).
44. H. Hagen and A. Sieverts, *Z. physik. Chem.*, **A, 165,** 1 (1933).
45. H. Hagen and A. Sieverts, *Z. physik. Chem.*, **A, 174,** 247 (1935).
46. A brief résumé is given by C. J. Smithells (Ref. 1), pp. 191–198.
47. R. H. Fowler and C. J. Smithells, *Proc. Roy. Soc. London*, **A, 160,** 37 (1937). See also R. M. Barrer (Ref. 2), pp. 152–155.
48. E. Müller and K. Schwabe, *Z. Elektrochem.*, **35,** 165 (1929).
49. L. Mond, W. Ramsay, and J. Shields, *Phil. Trans. Roy. Soc. London*, **A, 191,** 105 (1898).
50. A. Gutbier and W. Schieferdecker, *Z. anorg. Chem.*, **184,** 305 (1929).
51. A. F. Benton, *J. Am. Chem. Soc.*, **48,** 1850 (1926).
52. S. Valentiner, *Ber. deut. physik. Ges.*, **13,** 1003 (1911).
53. A very comprehensive review of the earlier observations made by this investigator and others on the sorption characteristics of palladium and the other metals of this group is given in McBain (Ref. 28), pp. 286–305.
54. E. B. Maxted, *J. Chem. Soc.*, **115,** 1050 (1919).
55. C. J. Smithells (Ref. 1), p. 170.
56. A. Sieverts, *Z. physik. Chem.*, **A, 155,** 299 (1931).
57. A. Sieverts, G. Zapf, and H. Moritz, *Z. physik. Chem.*, **A, 183,** 19 (1938–1939).
58. J. Chipman and D. W. Murphy, *Trans. A.I.M.E.*, **116,** 179 (1935).
59. A. Sieverts and H. Brüning, *Arch. Eisenhüttenw.*, **7,** 641 (1933).
60. F. J. Norton and A. L. Marshall, *Trans. A.I.M.E.*, **156,** 351 (1944), and personal communication.
61. E. Martin, *Arch. Eisenhüttenw.*, **3,** 407 (1929–1930).
62. J. H. de Boer and J. D. Fast, *Rec. trav. chim.* **55,** 459 (1936).
63. C. J. Smithells, (Ref. 1), p. 175.
64. F. N. Rhines and C. H. Mathewson, *Trans. A.I.M.E.*, **111,** 337 (1934).
65. A. Phillips and E. N. Skinner, *Trans. A.I.M.E.*, **143,** 301 (1941).
66. C. J. Smithells (Ref. 1), p. 177.
67. C. H. Herty, Jr., and J. M. Gaines, *Trans. A.I.M.E.*, **80,** 142, *Iron and Steel* (1928).
68. A. U. Seybolt and C. H. Mathewson, *Trans. A.I.M.E.*, **117,** 156 (1935).
69. P. D. Merica and R. G. Waltenberg, *Technol. Paper, Natl. Bur. Standards*, **19,** No. 281, p. 155 (1924–1925).
70. E. W. R. Steacie and F. M. G. Johnson, *Proc. Roy. Soc. London*, **A, 112,** 542 (1926).
71. J. H. Simons, *J. Phys. Chem.*, **36,** 652 (1932).
72. A. Sieverts and J. Hagenacker, *Z. physik. Chem.*, **68,** 115 (1909).
73. N. P. Allen, *J. Inst. Metals*, **49,** 317 (1932).
74. A. Sieverts and W. Krumbhaar, *Z. physik. Chem.*, **74,** 295 (1910).
75. A. Sieverts and E. Bergner, *Z. physik. Chem.*, **82,** 257 (1913).
76. C. F. Floe and J. Chipman, *Trans. A.I.M.E.*, **143,** 287 (1941); **147,** 28 (1942).
77. H. Lepp, *Metal Ind.*, **47,** 315 (1935); **53,** 27, 59, 79, 103, 131 (1938).
78. E. W. R. Steacie and F. M. G. Johnson, *Proc. Roy. Soc. London*, **A, 112,** 542 (1926).
79. C. A. Edwards, *Trans. A.I.M.E.*, **117,** 13 (1935).
80. O. W. Richardson, J. Nicol, and T. Parnell, *Phil. Mag.*, **8,** 1 (1904).

81. These are the units used by Smithells, and $q$ is identical with the quantity designated $D$ in his treatise. The constant $k_0$ is identical with the constant designated $K$ by Smithells and also with the constant $P_0$ used by Barrer (Ref. 2, p. 168).

82. C. J. Smithells (Ref. 1), p. 95, Table 9; also by Smithells and C. E. Ransley, *Proc. Roy. Soc. London*, **A, 150,** 172 (1935), in Table V, p. 182.

83. E. L. Jossem, *Rev. Sci. Instr.*, **11,** 164 (1940). This contains additional references to those given in List III on p. 574.

84. C. J. Smithells (Ref. 1), p. 113.

85. C. J. Smithells and C. E. Ransley, *Proc. Roy. Soc. London*, **A, 152,** 706 (1936).

86. P. L. Chang and W. D. G. Bennett, *J. Iron Steel Inst.*, **170,** 205 (1952).

87. A. Demarez, A. G. Hock, and F. A. Meunier, *Acta Met.*, **2,** 214 (1954).

88. T. M. Stross and F. C. Tompkins, *J. Chem. Soc.*, **1956,** 230.

89. P. Bastien, *Métaux and corrosion*, **25,** 248 (1950) (46 refs.).

90. J. D. Hobson, *J. Iron Steel Inst.*, **189,** 315 (1958).

91. Solutions of this equation and that for diffusion in a wire are discussed by Barrer (Ref. 2), Chapter 1, and in a number of mathematical treatises on Fourier's series and Bessel functions.

92. A. B. Newman, *Trans. Am. Inst. Chem. Engrs.*, **27,** 203, 310 (1931); *Chem. and Met. Eng.*, **38,** 710 (1931).

93. J. A. M. van Liempt, *Rec. travaux chim.*, **57,** 871 (1938).

94. R. M. Barrer (Ref. 2), pp. 215–217.

95. G. Euringer, *Z. physik*, **96,** 37 (1935).

96. C. J. Smithells and C. E. Ransley, *Proc. Roy. Soc. London*, **A, 155,** 195 (1936), Fig. 4.

97. R. M. Barrer (Ref. 2), p. 224.

98. F. J. Norton and A. L. Marshall, *Trans. Am. Inst. Met. Eng.*, **156,** 351 (1944).

99. H. S. Carslaw and J. C. Jaeger, *Conduction of Heat in Solids*, Oxford University Press, 1947 edition, pp. 83–84; 1959 edition, pp. 101–102.

100. See subsequent remarks in this section.

101. The following remarks are taken from Dushman's review of this topic in *J. Franklin Inst.*, **211,** 728–736 (1931).

102. W. G. Guldner, *Bell Labs. Record*, **29,** 18 (1951).

103. A. L. Beach and W. G. Guldner, *Am. Soc. Testing Materials Spec. Tech. Pub. No.* 222, pp. 15–23 (1957).

104. I. Langmuir, *J. Am. Chem. Soc.*, **35,** 105 (1912). Also see section 8.16.

105. I. Langmuir, *Trans. Am. Inst. Elec. Eng.*, **32,** 1921 (1913).

106. H. M. Ryder, *J. Am. Chem. Soc.*, **40,** 1656 (1918); *J. Franklin Inst.*, **187,** 508 (1919).

107. G. Alleman and C. J. Darlington, *J. Franklin Inst.*, **185,** 161 333, 461 (1918). This paper contains a comprehensive review of previous investigations on gases in ferrous alloys.

108. W. Rohn, *Z. Metallkunde*, **21,** 12 (1929).

109. W. Hessenbruch, *Z. Metallkunde*, **21,** 46 (1929).

110. See also the remarks by E. H. Schulz, *Z. Metallkunde*, **21,** 7 (1929).

111. A. Villachon and G. Chaudron, *Rev. mét.*, **27,** 368 (1930).

112. G. Chaudron, *Aluminum*, **19,** 594 (1937); O. Kubaschewski, *Z. Elektrochem.*, **44,** 152 (1938).

113. Mrs. M. R. Andrews, *J. Franklin Inst.*, **211,** 689 (1931).

114. K. Steinhäuser, *Z. Metallkunde*, **26,** 136 (1934).

115. Reported by Dushman, *J. Franklin Inst.*, **211,** 735 (1931).

116. F. J. Norton and A. L. Marshall, *Trans. Am. Inst. Mining Met. Engrs.*, **156,** 351 (1944).
117. See section 8.16 for description of this method.
118. H. C. Todd, *Radio Eng.*, **12,** 18 (1932).
119. C. J. Smithells (Ref. 1), p. 204.
120. E. M. Wise, *Proc. Inst. Radio Engrs.*, **25,** 714 (1937).
121. H. C. Vacher and L. Jordan, *J. Research Natl. Bur. Standards*, **7,** 375 (1931).
122. This method is discussed at greater length in section 8.16.
123. L. Reeve, *Trans. Am. Inst. Mining Met. Engrs.*, **113,** 82 (1934).
124. B. M. Larsen and T. E. Brower, *Trans. A.I.M.E.*, **100,** 196 (1932).
125. B. M. Larsen, T. E. Brower, and W. E. Shenk, *Trans. A.I.M.E.*, **113,** 61 (1934). These publications give references to the extensive literature on this topic.
126. R. S. McDonald, J. E. Fagel, and E. W. Balis, *Anal. Chem.*, **27,** 1632 (1955).
127. "Determination of Gases in Metals," *ASTM Symposium STP No.* 222 (1957).
128. "Determination of $N_2$ in Steel," *Spec. Rept. No.* 62, Iron and Steel Institute, London, 1958.
129. M. Pirani, *Z. Elektrochem.*, **11,** 555 (1905).
130. L. L. Wyman, *Trans. A.I.M.E.*, **104,** 141 (1933); **111,** 305 (1934) and **137,** 291 (1940); *Gen. Elec. Rev.*, **37,** 120 (1934).
131. F. N. Rhines and W. A. Anderson, *Trans. A.I.M.E.*, **143,** 312 (1941).
132. F. N. Rhines, *Trans. A.I.M.E.*, **156,** 335 (1944).
133. See remarks in section 8.16 in connection with the discussion of the determination of hydrogen in iron alloys.
134. H. H. Uhlig, *Trans. Am. Inst. Mining Met. Engrs.*, **158,** (*Iron and Steel*), 183 (1944).
135. D. P. Smith, D. J. Carney, L. W. Eastwood, and C. E. Simons, *Gases in Metals,* American Society for Metals, Cleveland, 1953.
136. D. P. Smith, *Hydrogen in Metals*, Chicago, 1948.
137. R. W. Buzzard and H. E. Cleaves, "Hydrogen Embrittlement of Steel," *Natl. Bur. Standards Circ.* 511 (1200 references, 1800–1949).
138. *Los Alamos Sci. Lab. Rept.* 2283, 1958; Bibliography, "Hydrogen Embrittlement," 1952–1958.
139. D. D. Perlmutter and B. F. Dodge, *Ind. Eng. Chem.*, **48,** 885 (1956).
140. B. F. Dodge, *Trans. Am. Soc. Mech. Engrs.*, **75,** 331 (1953).
141. W. D. Robertson, Editor, *Stress Corrosion Cracking and Embrittlement*, John Wiley & Sons, New York, 1956.
142. B. R. Queneau and G. A. Ratz, *The Embrittlement of Metals*, American Society for Metals, Cleveland, 1956.
143. N. P. Allen, *J. Inst. Metals*, **49,** 317 (1932).
144. N. P. Allen, *J. Inst. Metals*, **43,** 81 (1930).
145. N. P. Allen and A. C. Street, *J. Inst. Metals*, **51,** 233 (1933). This investigation reports observations on "the effects of hydrogen and oxygen on the unsoundness of copper-nickel alloys."
146. N. P. Allen and T. Hewitt, *J. Inst. Metals*, **51,** 257 (1933).
147. This is discussed in Chapter 11.
148. O. W. Ellis, *Trans. Am. Inst. Mining Met. Engrs.*, **106,** 487 (1933).
149. J. C. Chaston, *Metal Treatment*, **11,** 212, 213 (1944–1945). Chaston also gives the following references on degassing of metals: D. R. Tullis, *J. Inst. Metals*, **40,** 55 (1928); W. Rosenhain, J. D. Grogan, and T. H. Schofield, *J. Inst. Metals*, **44,** 305

(1930); J. D. Grogan and T. H. Schofield, *J. Inst. Metals*, **49,** 301 (1932); D. Hanson and I. G. Slater, *J. Inst. Metals*, **46,** 187, 216 (1931).

150. W. J. P. Rohn, *J. Inst. Metals*, **42,** 203 (1929). See also remarks in section 8.11.

151. S. L. Archbutt, *J. Inst. Metals*, **33,** 227 (1925).

152. W. Claus, S. Briesemeister, and E. Kalähne, *Z. Metallkunde*, **21,** 268 (1929).

153. W. T. Pell-Walpole, *J. Inst. Metals*, **70,** 127 (1944).

154. W. E. Prytherch, *J. Inst. Metals*, **43,** 73 (1930).

155. R. S. Morse, *Ind. Eng. Chem.*, **30,** 1064 (1947). Furnaces for the vacuum fusion of metals have been described in the following publications: A. U. Seybolt, *Metals Progress*, **50,** 1102 (1946), and A. R. Kaufmann and E. Gordon, *Metals Progress*, **52,** 387 (1947).

156. J. L. Erickson, *Light Metal Age*, **2,** No. 4, 19 (1944).

157. H. H. Johnson, L. H. Arner, and H. A. Schwartz, *Trans. Am. Soc. Metals*, **37,** 309 (1946).

158. W. A. Fischer and H. Engelbrecht, *Stahl u. Eisen*, **74,** 1515 (1959).

159. *Vacuum Metallurgy*, a Symposium, Electrochemical Society, Boston, 1955.

160. T. J. Kopkin and V. De Pierre, "Vacuum Melting Bibliography," *Iron Age*, January, 1956.

161. "Vacuum Processes," *J. Iron Steel Inst.*, **190,** 312, 414, 432 (1958).

162. G. M. Gill, E. Meson, and G. W. Austin, "Thermodynamics of Vacuum Melting," *J. Iron Steel Inst.*, **191,** 172 (1959).

163. J. Suiter, "Thermodynamics," *J. Australian Inst. Met.*, **4,** 95 (1959).

164. An excellent review of methods of "micro-gas analysis" is given in *Experimental Methods in Gas Reactions*, by A. Farkas and H. W. Melville, Cambridge University Press, 1939, pp. 174–201.

165. I. Langmuir, *J. Am. Chem. Soc.*, **34,** 1310 (1912).

166. H. M. Ryder, *J. Am. Chem. Soc.*, **40,** 1656 (1918).

167. C. H. Prescott, Jr., *J. Am. Chem. Soc.*, **50,** 3237 (1928).

168. R. H. Dalton, *J. Am. Chem. Soc.*, **57,** 2150 (1935).

169. F. J. Norton and A. L. Marshall, *Trans. Am. Inst. Mining Met. Engrs.*, **102,** 287 (1932); **104,** 136 (1933).

170. The heating is carried out by means of an induction furnace.

171. In order to increase the accuracy of the determination of this volume, Naughton and Uhlig have used a modified design of mercury-vapor pump which is described in *Ind. Eng. Chem., Anal. Ed.*, **15,** 750 (1943).

172. See vapor-pressure data in Table 11.2.

173. J. Naughton, *Trans. Am. Inst. Mining Met. Engrs.*, **162,** 385 (1945). This is one of a number of papers given at a symposium on the determination of hydrogen in steel and published in the same volume of the *Transactions*, pp. 353–412.

174. V. C. F. Holm and J. G. Thomson, *J. Research Natl. Bur. Standards*, **26,** 245 (1941), also J. G. Thompson, *Trans. Am. Inst. Mining Met. Engrs.*, **162,** 369 (1945).

175. J. G. Thompson, H. C. Vacher, and H. A. Bright, *J. Research Natl. Bur. Standards*, **18,** 259 (1937). This and the previous paper contain further references to the literature on these topics. See also remarks in section 8.13.

176. N. R. Campbell, *Proc. Phys. Soc.*, **33,** 287 (1920–1921).

177. A. Farkas and H. W. Melville (Ref. 164), p. 187.

178. J. J. Sebastian and H. C. Howard, *Ind. Eng. Chem., Anal. Ed.*, **6,** 172 (1934).

179. C. Kenty and F. W. Reuter, Jr., *Rev. Sci. Instr.*, **18,** 918 (1947).

180. R. G. Dickinson and A. C. G. Mitchell, *Proc. Natl. Acad. Sci.*, **12,** 692 (1926).

181. This method has been used by C. Kenty, *Rev. Sci. Instr.*, **17,** 158 (1946), for the determination of a small amount of impurity in xenon.

182. F. J. Norton, *J. Am. Ceram. Soc.*, **36,** 90–96 (1953); *J. Appl. Phys.*, **28,** 34–39 (1957).

183. H. Hinterberger and E. Börnenburg, *Vacuum Technique*, **7,** 121–130, 159–170 (1958).

184. S. A. Landeen, H. E. Farnsworth, and R. K. Sherburne, *Phys. Rev.*, **78,** 351 (1959).

185. J. S. Wagener and P. T. Marth, *J. Appl. Phys.*, **28,** 1027 (1957).

186. J. D. Fast, H. G. Van Bueren, and J. Philibert, Editor, "Diffusion in Metals Symposium," Bibliotheque Technique Phillip, Ed. Eindhoven, 1957.

187. J. Koch, *Nature*, **161,** 566 (1948).

188. D. E. Rimmer and A. H. Cottrell, *Phil. Mag.*, **2,** 1345 (1957).

189. L. H. Varnerin and J. H. Carmichael, *J. Appl. Phys.*, **28,** 913 (1957).

190. J. M. Tobin, *Acta Met.*, **5,** 398 (1957).

191. "Swelling and Gas Diffusion in Irradiated Uranium," *Proc. 2nd Intern. Conf. Peaceful Uses of Atomic Energy*, Geneva, 1958, Vol. 5, pp. 481, 543.

192. K. E. Zimen and P. Schmelling, *Z. Electro. chem.*, **58,** 599 (1954).

193. Z. and L. Dahl, *Z. Naturforsch.*, **12a,** 167 (1957).

194. D. W. Lillie, "Effects of Gases in Metallic Lattices," *Proc. 2nd Intern. Conf. Peaceful Uses of Atomic Energy*, Geneva, 1958, Vol. 6, p. 221.

195. M. B. Reynolds, *Nuclear Sci. and Eng.*, **3,** 428 (1958).

196. *Solid State Diffusion*, a Symposium, Centre d'Etudes Nucleaire de Saclay, 1958, p. 57, R. S. Barnes, "Diffusion and Precipitation of Helium Injected into Metals"; p. 65, J. Blin, "Energy of Solution of Rare Gases in Metals"; p. 73, J. A. Stohr and L. Alfille, "Apparatus for Studying the Diffusion of Rare Gases in Stainless Steel."

197. C. W. Tucker and F. J. Norton, "On the Location and Motion of Rare Gas Atoms in Metals" and "Krypton Evolution from Metallic Uranium," *J. Nuclear Materials*, **2,** 329–334 and 350–352 (1960).

# 9 Chemical and electrical clean-up and ultrahigh vacuum

Revised by V. L. Stout and T. A. Vanderslice

## 9.1. GENERAL REMARKS

During the last decade of the nineteenth century when commercial production of carbon-filament lamps was initiated, it was recognized that the degree of vacuum attainable by means of the pumps available at that time was not sufficiently good for long-life operation of the lamps. In order to improve the vacuum in a sealed-off lamp, Malignani, in 1894, introduced the use of a suspension, in a viscous liquid, of red phosphorus which was painted on the inside of the exhaust tube of the lamp. After the tubulation from the pump was closed off, the filament was flashed at a higher voltage than normal and the result was a considerable decrease in pressure of residual gases. Subsequently other chemicals were introduced with the same purpose, and these came to be designated as "getters," that is, clean-up reagents for residual gases. As will be evident from the discussion in the following sections, the mechanism of clean-up by getters is different from that involved in the use of coconut charcoal, palladium black, and similar adsorbents mentioned in Chapter 7.

In 1907 F. Soddy proposed the volatilization of calcium as a method for reducing the pressure in a sealed-off tube below that obtained by the use of a pump. During the course of his early investigations on vacuum tungsten-filament lamps Langmuir observed that the pressure of residual gases in a lamp was decreased as a result of chemical reactions between the tungsten and residual gases; that is, evaporated tungsten functioned as a getter.

The greatest impetus to further investigations on getters came as a result of the increasing applications of vacuum technique for the industrial production of hot-cathode, high-vacuum devices. Even though it is possible by means of vapor pumps to obtain pressures sufficiently low for the operation of such electronic devices, the use of getters not only permits a reduction of the time of exhaust but also serves to maintain, in the sealed-off tube, the degree of vacuum necessary for efficient operation and long life. Thus the getters not only "get" the residual gases, but also "keep" them in much the same manner as, for instance, sodium retains the oxygen with which it has combined to form the oxide. In fact, the present-day large-scale production of the vast numbers of electronic devices would be well-nigh impossible if it were not for the aid rendered by getters.

In hot-cathode devices the residual gases may act deleteriously in two ways. First, if the mean free path for an electron is of the same order of magnitude as, or less than, the distance between anode and cathode, gas molecules are ionized by collision with electrons; and, since one positive ion can neutralize the space-charge effect due to hundreds of electrons, it requires only a relatively small concentration of ions to eliminate the control of the space current by means of the anode or grid voltage. Second, the electron emission is decreased in general either by chemical reaction at the cathode surface or by positive ion bombardment with the consequent removal by sputtering of the material that is active in the emission of electrons.[1]

The action of the getter is often quite complex and frequently is not well understood. In practical devices the situation is commonly complicated by the presence of more than one gas, electric fields, ions, and unknown metal vapors. Four important processes limiting gettering rates are: (1) the formation of adsorbed monolayers, (2) the confinement of reaction products to surfaces, (3) the solution of gases in metal, and (4) the vaporization of reaction products. In the process of gettering a particular gas by a single metal it is common that several physical mechanisms occur in sequence to form the complete gettering act.

Many studies have been made in order to understand better the capacity and rate limitations of getters. The vacuum systems used for these studies have been varied; and, although most of the earlier studies were performed with static systems, at present investigations are most frequently made with dynamic systems.

Metals in many shapes and forms can serve as getters. The two most useful types are vaporized metallic films and bulk metal. Commonly vaporized metallic films, flashed getters, operate at the ambient temperature, and bulk-metal getters operate at temperatures above the ambient.

Another type of clean-up occurs when voltage is applied between the anode and hot cathode. This has been designated *electrical clean-up*; it is due to interaction between electrons emitted from the cathode and the molecules in the gas. The latter may be dissociated to form atoms that are extremely active chemically. Such dissociation may also be produced, as in the case of hydrogen, by contact of the hydrogen with a tungsten filament operated at temperatures above 2000° K. Furthermore, as a result of collisions between electrons and molecules, excited molecules and positive ions are formed, as mentioned previously. Under the action of the applied voltage the ions are driven with high velocities into the negatively charged walls and electrodes. The retention by the latter of the gas thus cleaned up depends in a very complex manner on the temperature, on the nature of getter deposit and of electrodes, and on the pressure of gas.

## PART A. CHEMICAL CLEAN-UP

Revised by V. L. Stout

### 9.2. APPARATUS AND TECHNIQUES FOR MEASUREMENT OF GETTERING RATES

Reports of getter studies are frequently in conflict. It is not uncommon for investigators to draw different conclusions from measurements that have been gleaned from what appear to be similar systems. Points of conflict may arise because of slight differences in experimental technique. Certain salient features in experimental technique must be closely observed in order to obtain reproducible results.

The system in which the measurement is to be made must be clean. To assure cleanliness the system should be baked to avoid water-vapor contamination. Diffusion pumps should be trapped to prevent back diffusion of the pumping vapor. Yet traps must be at such a temperature and physical location that they do not condense the gas being measured. When other metals form a part of the system, they should not collect or evolve gas.

A clean sample of gas is important. Water and oil vapors can be easily carried into a clean system with the gas sample. The composition can change during the gettering measurement. It is almost impossible to contain pure oxygen near a hot filament. Carbon associated with the filament as an impurity is attacked by the oxygen to form CO and $CO_2$.[2,3]

The properties of the getter metal should be well defined. Oxidation and gettering studies have shown that several per cent of an additional metal can produce a marked effect on the gas attack of the primary metal. Surface preparation is critical. Under many conditions of gettering the rate of gas sorption is limited by surface reaction, making the rate directly dependent upon the surface area. The getter material itself can contain gas that may evolve or contaminate the surface during the gettering measurement. This gas should be removed, in so far as possible, before measurements are begun.

The temperature of the getter and of the gas with which it reacts should be well controlled. Atoms and molecules of the same gas react differently with a metal, since the initial energies are different. The temperature of the getter should be constant over its surface. The temperature-measuring device should not alter the temperature of the getter and should not combine with the getter metal or the gas.

A simple continuous method of pressure measurement is desirable. It is undesirable for the measuring device to react with the gas. For example, the ions generated in an ionization gauge can be pumped; the mercury of a McLeod gauge can alloy with the getter; oil from a manometer can contaminate a getter's surface.

It is desirable to know how much gas has been cleaned up by a getter. Generally, as a getter becomes saturated with gas its sorption rate is decreased. Mixtures of gases cause varied results. In particular, gases which cause a surface-limited reaction can inhibit clean-up of other gases.

Although gettering experiments are difficult, many significant flashed and bulk-getter investigations have been performed in both static and dynamic systems. Each of these systems has certain advantages and disadvantages.

The static system consists of a fixed and known volume enclosure that is filled with gas at a measured pressure. The gettering rate is determined by the change in pressure per unit time. The static system is hard to use, since it requires filling with gas for each set of gettering measurements. The pressure in the enclosure may vary three decades during the process of measurement and is difficult to measure with a high degree of precision. A McLeod gauge is capable of covering the high-pressure range but, as it cannot be read continuously, is difficult to use. An ionization gauge may be employed at low pressures.

Static systems for measurement of getters have been used by many investigators.[4,5,6] Measurements are made by trapping a fixed volume of gas at a known pressure and observing the change of pressure as a function of time. From the gas law we know that $dP/dt = \dfrac{KT\,dN/V}{dt}$, so that the rate of change of pressure is proportional to the rate at which gas molecules are being gettered. A static system perhaps is somewhat easier to put

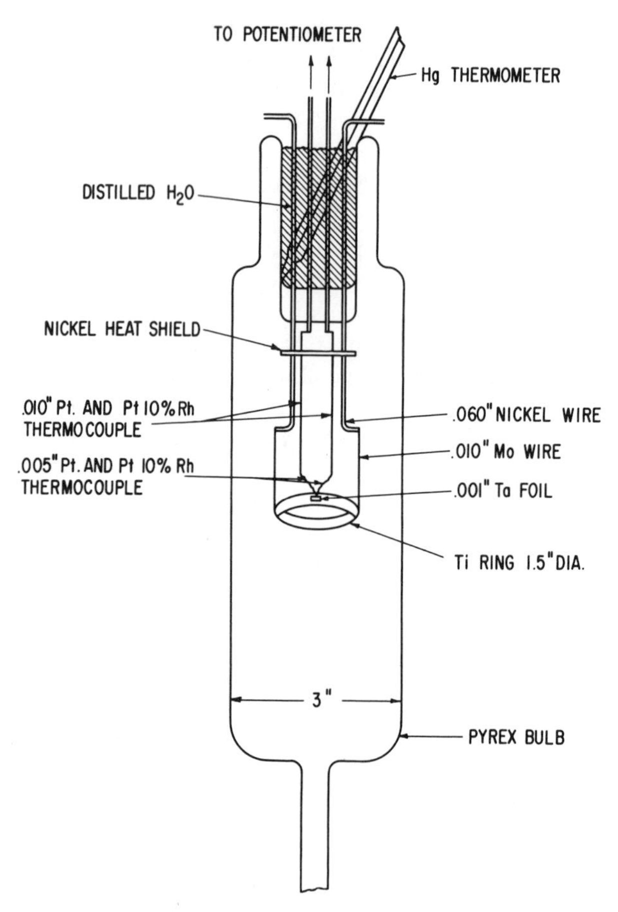


Fig. 9.1. Getter bulb used for gas-sorption studies.

together than a dynamic system and for this reason finds its greatest use in making quick qualitative measurements of gettering rates. A good static system for getter evaluation should (1) be pumped by a diffusion or ion-type pump, (2) be completely bakable to the pump connection, and (3) be provided with cold traps.

One physical arrangement for the getter chamber, in this case for the study of a bulk getter, is shown in Fig. 9.1. Several features are shown. Temperature measurement by a 0.005-in.-diameter thermocouple wire helps prevent heat-conduction loss from the junction. For the same reason the support is limited to two 0.010-in. molybdenum wires. A thin foil of tantalum placed between the thermocouple junction and the getter prevents alloying. The getter is heated by induced current. The r-f field is capable of producing ions that can becloud results. In this case there was no optical evidence of a discharge.

The dynamic system introduced by Wagener[7] has two principal advantages over the static system. Gettering rates can be measured continuously, and measurements can be made at constant or varying gas pressure above the getter.

An experimental system for the type of measurement that was designed by Wagener is shown in Fig. 9.2. For such a system the gas flow being sorbed by the getter is given by

$$I = GP_s = F(P_p - P_s),$$

where $G$, defined by Wagener, is the rate of sorption (cubic centimeters per second), $F$ is the conductance of the capillary, and $P_p$ and $P_s$ are the pressure of the gas at the inlet and outlet ends of the capillary, respectively. Della Porta and his collaborators[8] have indicated that $G$, as found by their measurements with a similar system, is not a constant, and thus they prefer to interpret their experimental data in terms of $I$, the gas current.

The physical dimensions for such a system were first given by Wagener.[7] The volume of the getter bulb in his initial work was about 50 cm$^3$; it was connected to the manifold through a capillary tube with an inside diameter of 2 mm and a length of 40 mm. The manifold was pumped by a diffusion pump, and the pressure $P_p$ was measured by an ionization gauge that was connected into the system. Gas admitted to the system through a needle valve permitted adjustment of the pressure $P_p$ from the ultimate of the system upward.

Both Wagener and della Porta have used systems similar to the one shown in Fig. 9.2. Gettering results that are obtained from a dynamic system must be interpreted with caution. The observed gettering rate may be determined by the capillary; and, since small quantities of gas can be measured, sorption by other surfaces and electrical clean-up become more critical. The close proximity of the ion gauge to the getter increases the latter difficulty. It has been pointed out by della Porta, Origlio, and Argano[8] that it is necessary to distinguish between sorption caused by the ions, activated gas molecules, the hot filament, and the usual gettering.

They have reported that in the case of nitrogen the pumping effect of the hot filament (in the absence of ionizing electron current) becomes noticeable only at very high temperatures (about 2700° K). This pumping effect does not change the velocity of sorption of the getter but produces an additive effect.

They further state:

On the other hand, the electron current produces an increase in the gettering velocity of barium films which is proportional to the ionization produced.

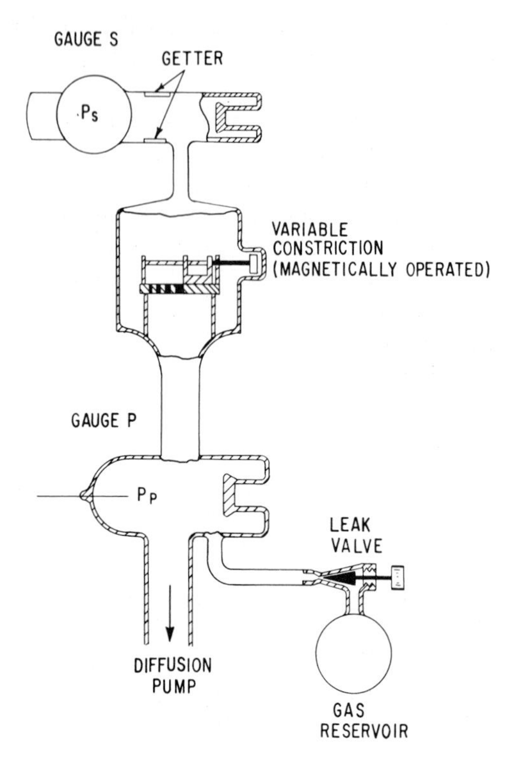


Fig. 9.2. An experimental arrangement for measuring getter rates and capacities at very low pressures.

For hydrogen we have, as is already well known, a relatively large increase in the pumping action of the hot filament. However, this occurs at a much lower temperature. This increase is once again independent of the presence of the barium film, but is partially dependent on the temperature of the surface.

The increase produced by the ionizing electron current is practically independent of the evaporated barium film, as opposed to what is found for nitrogen.

In the case of carbon monoxide, for which the effects of the hot filament only become apparent at temperatures above 2600° K, a very small influence of the electron current is noticeable in the presence of the getter, but it is almost negligible when the getter is not present.

Gas-metal reactions can be studied by observing the change in weight of the getter while the gettering reaction takes place. Such a technique has been used by Gulbransen[9] in studying the oxidation of metals. One characteristic of the microbalance reported by him is a sensitivity of $2.5 \cdot 10^{-7}$ g. The pressure coefficient was found to be less than $0.3 \cdot 10^{-6}$ g for a 1-atm change, and the temperature coefficient is about $0.8 \cdot 10^{-6}$ g/° C. The microbalance was operated in a vacuum of $10^{-5}$ mm of Hg as described in the initial studies but is undoubtedly capable of ultrahigh vacuum. It had a stability of better than $1 \cdot 10^{-6}$ g during a period of 24 hr, and is used in a static system.

The sensitivities of the measurement systems described above are different. As mentioned, the microbalance has a sensitivity of about $2.5 \cdot 10^{-7}$ g. This is equivalent to 0.145 micron · liter of oxygen. The sensitivity of the dynamic method of getter study can easily be $10^{-5}$ micron · liter, which for oxygen is equivalent to $1.72 \cdot 10^{-11}$ g. The static system, using pressure measurement, has a sensitivity that depends upon the nature of the actual experiment but can easily be made to fall in the range of 0.1–1.0 micron · liter, which again, for oxygen, is equivalent to $1.72 \cdot 10^{-5}$ to $1.72 \cdot 10^{-6}$ g.

## 9.3. MECHANISMS OF SORPTION BY FLASHED GETTERS

In the gettering act, metal and gas react chemically. Thus it is possible to predict the amount of gas that can be gettered by knowing the chemical formula of the reaction product.

Let $MX_n$ represent the formula, where $M$ stands for the atomic weight of the metal, and $X$ that of the nonmetallic element (the molecular weight of which is assumed to be $X_2$). Then,

$$M = \frac{n}{2} \times \frac{22{,}415 \times 298}{273} \quad \text{atm} \cdot \text{cm}^3 \text{ of gas at } 25^\circ \text{ C.}$$

Hence

$$1 \text{ mg} = n \times \frac{9296}{M} \text{ micron} \cdot \text{liters at } 25^\circ \text{ C.}$$

Table 9.1 gives values of the ratio (micron · liters per milligram) for a number of metals used as getters and for the three gases hydrogen, nitrogen, and oxygen. The figures in parentheses are based on the assumption that the hydrides obey stoichiometric relations, which, as discussed in Chapter 8, is not actually true for many of these metals.

The amounts of the gases required to form these chemical compounds are seldom taken up in practice. The rate of gettering depends upon the amount of gas that has been previously sorbed, so that as the getter approaches the saturation limit the gettering rate becomes impractically slow. For this reason from an operational point of view the amount of gas that could be taken up if a chemical compound were to form is of little interest. The capacity of a getter is determined by the amount of gas that can be gettered before pressure build-up causes the device to malfunction. Ehrke and Slack[5] obtained comparable results for various getters. These results are shown in Table 9.2.

TABLE 9.1
**Volume of Gas per Unit Mass of Metal for Formation of Chemical Compound**

| Metal | $M$ | $9296/M$ | Micron · liters/mg for Oxide | Nitride | Hydride |
|---|---|---|---|---|---|
| Magnesium | 24:32 | 382.2 | 382.2 | 254.8 | 764.4 |
| Aluminum | 26.97 | 344.6 | 516.9 | 344.6 | ... |
| Silicon | 28.06 | 331.3 | 662.6 | ... | 1325.2 |
| Calcium | 40.08 | 231.9 | 231.9 | 154.6 | 463.8 |
| Titanium | 47.90 | 194.1 | 388.2 | 194.1 | (388.2) |
| Strontium | 87.63 | 106.1 | 106.1 | 70.8 | 212.3 |
| Zirconium | 91.22 | 101.9 | 203.8 | 101.9 | (203.8) |
| Barium | 137.36 | 67.67 | 67.7 | 45.1 | 135.4 |
| Lanthanum | 138.92 | 66.91 | 100.4 | ... | (200.8) |
| Cerium | 140.13 | 66.36 | 132.7 | 66.4 | (265.4) |
| Thorium | 232.12 | 40.05 | 80.1 | 40.1 | (120.2) |
| Uranium | 238.07 | 39.04 | 53.6 | ... | ... |

The theoretical values in the last column are taken from Table 9.1. The superiority of gettering by diffuse deposits is evident, as is also the fact that barium and misch metal (active constituents mostly cerium and lanthanum) were the most active.

With regard to the nature of each of the getters the following remarks are of interest. Both the *aluminum* and *magnesium* were in the form of wire twisted around the turns of a tungsten coil. Difficulty was experienced in degassing these metals without causing some evaporation. With neither getter was it possible to clean up nitrogen, hydrogen, or carbon dioxide, either at room temperature or by heating the deposit to 200° C or even higher. The clean-up of oxygen by diffuse magnesium appeared to be permanent, since no gas evolution occurred even on heating the deposit to 265° C. No clean-up was observed in the case of hydrogen and nitrogen.

*Barium* was found to be the best clean-up reagent. It was obtained from two commercial products the batalum and "KIC" getter. "The batalum getter," according to Ehrke and Slack, "was found to exert a very powerful clean-up action when vaporized in the presence of air. On the other hand, when flashed in a high vacuum after careful degassing, its gettering powers were much diminished." For this reason the experiments were carried out by evaporating the getter in the presence of argon, to yield a diffuse deposit. The quantitative determinations mentioned in Table 9.2 were made with the KIC getter, and they refer to the amounts cleaned up at

TABLE 9.2

**Volume of Gas Cleaned Up per Unit Mass of Metal**

(Ehrke and Slack)

| Getter | Gas | Micron · liters/mg Cleaned Up: Bright Deposit | Diffuse Deposit | Theoretical Value |
|---|---|---|---|---|
| Aluminum | $O_2$ | 7.5 | 38.6 | 516.9 |
| | $N_2$, $H_2$, $CO_2$ | None | None | ... |
| Magnesium | $O_2$ | 20 | 202 | 382.2 |
| | $CO_2$ | None | Slight | ... |
| | $N_2$, $H_2$ | None at room temperature | None at room temperature | ... |
| Thorium | $O_2$ | 7.45 | 33.15 | 80.1 |
| | $H_2$ | 19.45 | 53.7 | 120.2 |
| Uranium | $O_2$ | 10.56 | 9.26 | 53.6 |
| | $H_2$ | 8.9 | 21.5 | ... |
| Misch metal | $O_2$ | 21.2 | 50.9 | 100–133 |
| | $H_2$ | 46.1 | 63.9 | 200–265 |
| | $N_2$ | 3.18 | 16.1 | 66 |
| | $CO_2$ | 2.2 | 44.8 | ... |
| Barium | $O_2$ | 15.2 | 45 | 67.7 |
| | $H_2$ | 87.5 | 73.0 | 135.4 |
| | $N_2$ | 9.5 | 36.1 | 45.1 |
| | $CO_2$ | 5.21 | 59.5 | ... |

room temperature. Heating the deposits to 200° C caused additional clean-up, but further heating to 300° C resulted in evolution of nitrogen and hydrogen which had been taken up at the lower temperature. In the presence of mercury vapor barium is ineffective as a getter.

As is evident from the data in Table 9.2, *misch metal* was found to have about the same gettering power as barium. Small slugs of the misch metal were inserted into tungsten coils and evaporated by passing current through the coil.

The *thorium* and *uranium*, in the form of small pieces of wire, were used in the same manner as the misch metal, and both diffuse and bright deposits were obtained by evaporation. Though these metals proved to be effective getters for hydrogen and oxygen, they are rather inconvenient to use on account of the high temperature required for flashing, and the presence of oxides on the surface.

From their observations Ehrke and Slack concluded that the clean-up by deposits formed as a result of evaporation of the metals is of the "contact gettering" type. This is in accord with the statement that normally the area covered by the deposit was about 75 cm$^2$. Now to form a monolayer of oxygen or hydrogen about $2.5 \cdot 10^{-2}$ micron · liter/cm$^2$ is required; and, though no indication is given in the paper regarding actual quantity of metal per square centimeter of deposit, it may be assumed that the amounts of gas cleaned up were of the same order of magnitude as that required to form monolayers.

Subsequent investigation of gas clean-up carried out at reduced pressures more nearly equivalent to those required for vacuum tubes gave somewhat different results for barium.[10] It is also reasonable to assume that a heavier, thicker deposit of barium was used in these measurements. The results are given in Table 9.3.

**TABLE 9.3**

**Absorption Capacity of a Barium Getter**

(della Porta)

| Gas | Quantity Absorbed |
|---|---|
| Oxygen | 50 micron · liters/mg |
| Moisture | 61 micron · liters/mg |
| Hydrogen | 13 micron · liters/cm$^2$ |
| Carbon dioxide | 1.8 micron · liters/cm$^2$ |
| Carbon monoxide | 3.7 micron · liters/cm$^2$ |
| Nitrogen | 2.25 micron · liters/cm$^2$ |
| Dry air | 2.6 micron · liters/cm$^2$ |

The oxygen- and moisture-absorption capacity is independent of the area of the getter but does depend upon the weight of the getter deposit. These data are difficult to compare with those of Ehrke and Slack (see Table 9.2). However, in general the absorption capacities reported by della Porta are greater than those of Ehrke and Slack. In particular, the absorption capacities per square centimeter of deposit are nearly a factor of 10 greater.

Wagener[11] defines the capacity as that quantity of gas which can be sorbed before the gettering rate is reduced to the value which is equivalent to the gettering rate of a 0.5-cm$^2$ oxide cathode.[12] Values of getter capacities based on this definition are given in Table 9.4.

TABLE 9.4

**Getter Capacities for Gases and Metals**

(Wagener)

(At room temperature for a deposit of 0.04 mg/cm$^2$, in micron · liters per square centimeter)

| Getter Material | Gas | | | | | |
|---|---|---|---|---|---|---|
| | $CH_4$ | CO | $CO_2$ | $H_2$ | $N_2$ | $O_2$ |
| Ba | 0 | 0.2 | 0.2 | 0 | 0 | 1 |
| Sr | ... | 0.1 | ... | ... | ... | ... |
| Mg | ... | 0 | ... | ... | ... | ... |
| Mo | 0 | 0.15 | ... | 0.05 | 0.05 | 0.5 |
| Ta | 0 | 0.08 | ... | 0.5 | 0.1 | 1 |
| Ti | 0 | 0.15 | 0.13 | 2 | 0.05 | 1 |

These data lead to the conclusion that 25 micron · liters of $O_2$ per mg of barium, tantalum, or titanium is the practical getter-capacity limit for a vacuum device that uses an oxide cathode.

There are several other significant features about these data: (1) the sorption capacity for oxygen is greatest, (2) methane is sorbed rather poorly, and (3) carbon monoxide is sorbed by all materials except magnesium. The amount of carbon monoxide sorbed would form a layer of between 5 and 20 monolayers on the geometric surface. This, Wagener feels, may indicate sorption on internal surfaces which are believed to be correspondingly larger.

Such an explanation of the limitation of gettering capacity is consistent with ideas of della Porta, who states that one can define getter capacity in a number of ways:

*a.* a superficial adsorption capacity, which is strictly related to the surface area of the film and to its roughness factor;
*b.* a capacity related to the real surface of the film dependent on the film structure and the mass of the evaporated barium;
*c.* a total capacity which is only dependent on the mass of metal evaporated.

At the gas pressures and temperatures usually found in electron tubes della Porta concludes that the actual capacity will lie between the total capacity and the capacity of adsorption of the real surface.

The mechanisms of the gettering processes can be inferred from gettering-rate measurements, and the actual capacity depends upon saturation of one or more modes of sorption. Mechanisms for sorption of nitrogen by barium were suggested at the Fifth National Vacuum Symposium by della Porta and Ricca.[13] Their data, as shown in Fig. 9.3,

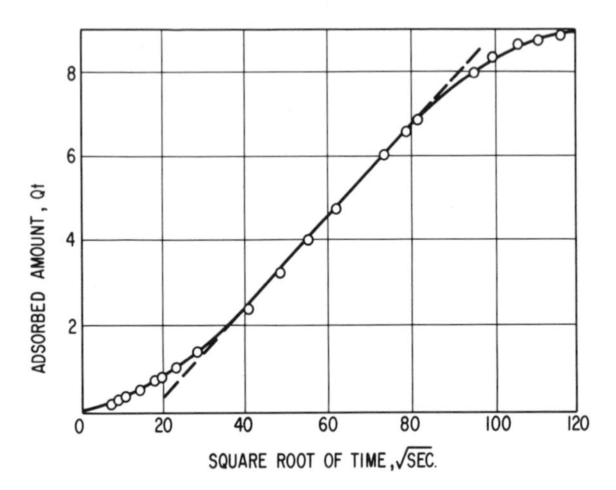


Fig. 9.3. Time dependence of adsorbed amount of nitrogen by barium film. Weight of evaporated barium, 2 mg; surface area, 100 cm²; adsorption temperature 25° C; pressure, $3 \cdot 10^{-4}$.

indicate two mechanisms of gettering. Initially adsorption takes place. During the time interval $t^{1/2} = 40$ to $t^{1/2} = 80$ the getter is diffusion limited. This is indicated by the sorbed amount $Qt$ being proportional to the square root of time. Examination of the initial velocity of sorption as a function of temperature leads these authors to conclude that the activation energy for chemisorption of nitrogen on barium films is about 3000 cal/mole. Diffusion-controlled "getting-rate" data were collected and plotted as a function of temperature in Fig. 9.4. This shows log $d(Qt)/d(t)^{1/2}$, where $Qt$ is the quantity of gas sorbed, plotted as the ordinate, while $T^{-1}$ is plotted as the abscissa. Two different diffusion mechanisms control the rate of gettering after the initial adsorption of nitrogen on barium. For temperatures lower than 140° C the activation energy for diffusion is 2300 cal/mole, and for temperatures higher than 140° C the activation energy is 11,000 cal/mole.

Similar conclusions are drawn about the processes of carbon monoxide and hydrogen sorption by barium.[8]

The activation energies of chemisorption are believed to be 1000 and 1500 cal/mole, respectively.[8] Origlio and della Porta feel that hydrogen and nitrogen are sorbed in the atomic state, while with carbon monoxide the sorption involves molecules. However, in the preceding paragraphs, gettering rates for all gases, including hydrogen and nitrogen, are discussed in terms of calories per mole. Table 9.5 summarizes the results of della Porta *et al.*

TABLE 9.5

**Results from Gettering Studies with Barium**

(della Porta *et al.*)

| | $N_2$ | $H_2$ | CO |
|---|---|---|---|
| Initial velocity of sorption (number of molecules sticking) ($cm^2 \cdot$ sec/mm at 298° K) | $1.93 \cdot 10^{16}$ | $1.04 \cdot 10^{17}$ | $1.32 \cdot 10^{18}$ |
| Activation energy for initial adsorption (cal/mole) | 3,000 | 1500 | ... |
| Critical temperature, $T_c$ (° C) | 140 | ... | 80 |
| Activation energy for $T < T_c$ (cal/mole) | 2,300 | ... | $<1000$ |
| Activation energy for $T > T_c$ (cal/mole) | 11,000 | 3000 | 9000 |

The actual mechanism of adsorption in the initial gettering stage is open to a great deal of question. By comparing the number of molecules sorbed per second with those arriving per second, one finds that a very small fraction are sorbed when they first strike the getter.

The processes of gettering for other flashed getters are similar to those for barium. The reaction between gases and metals at room temperature

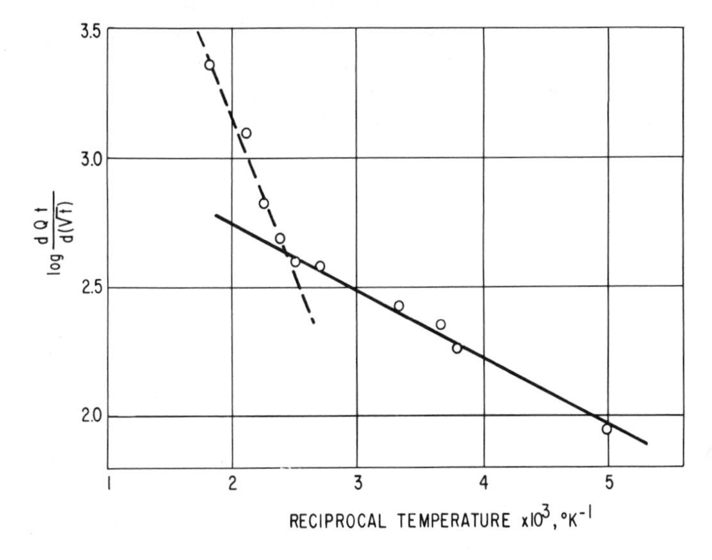


Fig. 9.4. Temperature dependence of nitrogen diffusion in barium films. Weight of evaporated barium, 2 mg; surface area, 100 $cm^2$; pressure, $5 \cdot 10^{-4}$.

TABLE 9.6

Initial Gettering Rates for Different Getters and Gases

(Wagener)

(High = >20 per cent of theoretical maximum; low = <0.01 per cent of theoretical maximum, measured at room temperature)

| Getter Material | Gas | | | | | |
|---|---|---|---|---|---|---|
| | $CH_4$ | CO | $CO_2$ | $H_2$ | N | $O_2$ |
| Ba | Low | High | High | Low | Low | High |
| Sr | ... | High | ... | ... | ... | ... |
| Mg | ... | Low | ... | ... | ... | ... |
| Mo | Low | High | ... | High | High | High |
| Ta | Low | High | ... | High | High | High |
| Ti | Low | High | High | High | High | High |

... = not measured.

seems to be most favorable for barium and titanium. The capacities described by Wagener are greatest for these two metals, and initial gettering rates are favorable, as shown in Table 9.6, taken from Wagener.[11]

Barium and titanium flashed getters are outstanding. The fact that the flashing temperature of titanium is higher than that of barium, being about 2000° C compared to 1000° C, would be a consideration in their use. Since titanium does not alloy with mercury, it may be valuable as a getter in mercury-vapor devices. Another advantage is that it can operate at high temperatures without revaporizing or migrating. This should make it useful in certain power-tube applications where barium may have undesirable side effects.

## 9.4. FLASHED GETTERS IN APPLICATIONS

As indicated above, the ability to sorb gas is not the sole requirement that is placed on metals to be used as flashed getters. Other requirements have been summarized by Espe, Knoll, and Wilder[14] as follows:

1. During vacuum bake-out of the device and system at temperatures up to at least 400° C the vapor pressure of the getter metal should not exceed $10^{-2}$ mm mercury.

2. The getter should be readily vaporized at a temperature which is not too low to prevent adequate bake-out and outgassing, and should not be so high that it causes vaporization of other parts of the device, especially of the getter leads. For nickel leads the upper temperature is about 1000° C and the lower temperature, consistent with previously stated considerations, would be about 600° C.

3. After flashing, the getter deposit on the wall of the device should have a negligibly low pressure $p < 10^{-7}$ mm Hg at the operating temperature of the envelope.

4. Between the ambient temperature and the operating temperature the getter should be active for all oxidizing gases. The reaction-product film on the getter should be porous or incoherent so that the gas attack is not limited to a surface layer.

5. The chemical compound formed should be stable, so that gases gettered at the ambient temperature are not released during operation. In gas-filled tubes, i.e., thyratrons, the getter should not react or alloy with the fill gas.

It has also been pointed out by Espe, Knoll, and Wilder how several metals fulfill the requirements set forth. The first requirement is satisfied by all alkaline-earth metals. Magnesium has a somewhat higher vapor pressure and thus cannot be used in high-temperature devices. Requirement 2 is met by all alkaline-earth metals but not by aluminum. Requirement 3 is met by all the alkaline-earth metals but not by magnesium. The fourth requirement is met by all the alkaline-earth metals but not by magnesium and aluminum. Requirement 5 is met by all alkaline-earth metals. Thus we see that the alkaline-earth metals are valuable. These metals alloy with mercury; and, since magnesium does not, magnesium has been widely used as a getter for mercury-vapor tubes.

Five methods for the production of the active metal films for getters have been described by Lederer and Wamsley.[15] These all depend on changes produced by the addition of heat.

1. A change of chemical form. Phosphorus is an example.

2. Evaporation of an active metal from the metal which is initially protected by an oxide film. Aluminum is an example of this case.

3. Evaporation from an alloy which is inactive in the atmosphere at room temperature. Barium, which is evaporated from a barium-aluminum alloy, is an example.

4. Evaporation of an active metal from a protective metal case. Barium in a nickel case is an example of this type of getter.

5. Production of a getter by means of a chemical reaction. The batalum getter produces barium by chemical reaction between barium berylliate and tantalum.

Although the last three methods of producing active getter metals are all used in commercial practice, method 3 seems to be favored.

Certain precautions should be observed when using flashed getters in electron tubes so as to avoid deleterious side effects. Some of these effects that can be produced are as follows.

1. Electrical leakage may result when evaporated getter material deposits on the press or tube base.

2. The work function of the grid is lowered if getter material deposits there. This can cause (*a*) grid emission, and (*b*) changes in the control potentials.

3. A mirror-like getter deposit reflects radiant energy, thereby reducing the desired radiation loss from tubes.

4. Getter metal landing on positive electrodes can alter the secondary electron-emission properties of surfaces, which may (*a*) change the electrical impedance of the device, or (*b*) cause the potential of insulating surfaces to shift.

These deleterious effects can, in general, be avoided partially by the proper choice and location of the getter.

Some difficulties that arise with the getters themselves are: (1) preflashing of the getter, (2) ineffective getter, (3) getter peel, (4) deterioration of the getter in storage, (5) deterioration of the getter during seal-in, and (5) violent evaporation of the getter. The causes and remedies for most of these difficulties with barium getters are fairly well known.

Preflashing occurs most frequently with magnesium-alloy getters. Either the getter should be moved to a cooler zone, or relocated so that less r-f energy is coupled into it during the processing of other tube parts, or a barium-aluminum getter should be substituted for the magnesium alloy.

An ineffective getter deposit of a dark color can result if the getter is too close to the glass wall on which the film is being evaporated. In this case heat from the getter during flashing liberates, from the glass, gas which reacts with the metal being deposited. A shiny getter that is ineffective can be produced by overfiring a clad getter. In this case nickel, copper, or iron, the cladding metal, is evaporated over the inner surface of the barium metal, thus sealing the barium from gas attack. A getter can become ineffective during the operation of the tube it protects. When this happens, the proper remedy may require better processing (vacuum firing of metal parts or a better bake of glassware) rather than the use of a larger getter.

The peeling of a getter film away from the glass on which it is evaporated may have at least two causes. Peeling will occur when the glass surface under the deposit is not clean. Induced current in the getter film will cause heating of the getter and glass, thus producing a chemical reaction that liberates vapor from the glass. The vapor causes the getter to peel. Air directed externally onto the glass in the area of the getter deposit during flashing can be helpful in preventing peel. Barium deposited on a cool wall also seems to have a greater absorbing capacity than barium deposited on a warm wall.

Deterioration of getters in improper storage occurs through atmospheric gas attack. Water vapor can be particularly harmful. Storage in vacuum or in a noble gas at low pressures is practiced by manufacturers in order to assure an active product.

A getter may be destroyed upon reaching a high temperature in air, which may occur during seal-in. This problem is most easily solved by moving the getter to a lower temperature zone. It may also be possible to partially protect the getter by surrounding it with a noble gas during the time of seal-in. In any event, it is desirable to choose a getter with a high flashing temperature if the getter must be located close to a glass seal.

Violent vaporization of a getter can be caused by overheating or by defective material. Overheating can be prevented frequently by flashing the getter more slowly. It is stated that this difficulty arises most often with magnesium-alloy assemblies.

The selection of the proper getter is dealt with in literature available from getter manufacturers. Some of the factors governing this selection are (1) the quantity and types of gas to be gettered, (2) the presence of a filling gas, (3) the ambient temperature at which the device must operate, (4) the temperatures required for manufacture, (5) the space available, and (6) the energy required to produce flashing.

Barium is the favorite material for low-voltage vacuum devices. When properly alloyed with aluminum, it produces a relatively inert metal that can stand some exposure to the atmosphere. This alloy is a good source of barium.

The quality of a getter depends upon the selection of the metals forming the getter, the processing that the metals receive during fabrication, the handling, and the storage for shipment. These factors are, of course, out of the hands of the user.

## 9.5. BULK GETTERS

Bulk getters are distinguished from flashed getters in that the limiting capacity of the bulk getter depends usually upon the mass of the getter and not on the surface area. However, the surface area is an important factor. A bulk getter may have any physical shape. It may be a part of the device structure or may perform only as the getter; it may be coated upon structural members of the device.

Several metals, such as titanium, zirconium, tantalum, thorium, columbium, and alloys such as cerium-misch metal-aluminum-thorium,[16] zirconium-aluminum or zirconium-aluminum-nickel,[17] and zirconium-titanium[18] are useful bulk getters. Although all these materials perform differently, the gas mechanisms of sorption are somewhat similar. At room temperature the gas reaction with a bulk getter is usually limited to the surface. Once adsorption is complete, the rate of formation of a surface-reaction product may be the process that controls the rate of gettering. In this case the rate limit is established by the diffusion of an atom or ion through the reaction product. This causes the gettering process to have a parabolic time dependence. In the high-temperature region, which varies for different getters, the rate of gettering depends upon the rate of gas diffusion into the bulk of the metal. Again the rate of gettering has a parabolic time dependence. Small quantities of gas that are dissolved by the metal reside as atoms or ions at interstitial positions in metal lattice. The lattice expands somewhat to accommodate the foreign element. When large amounts of gas have been gettered, a new crystal phase develops, the reaction product, which usually appears along the grain boundaries within the metal.

When a second phase has developed, the getter becomes brittle and subject to fracture. The quantity of gas that can be sorbed before a second crystal phase develops is given by a constitutional diagram. Figure 9.5, the oxygen-titanium diagram,[19] shows that the second crystalline phase, TiO, develops when about 33 atomic per cent oxygen has been dissolved. Thus $N_t$ titanium atoms are capable of reacting with $N_o$ oxygen atoms before a new crystal phase develops where $N_t$ and $N_o$ are related by the expression $N_o/(N_t + N_o) = 0.33$. From existing phase diagrams for other gases and metals it is possible to learn about bulk-getter performance as indicated by the development of a second phase.

Before a bulk getter can be used successfully in a vacuum device it should be processed to clean the surface and to drive out gas. In theory a good bulk getter should not release any gas once gas has been sorbed, but in practice all the getters do liberate gas.

Titanium and zirconium release principally hydrogen, but water vapor and methane may also be released during the outgassing of these materials.[3] The methane and water vapor are probably formed when hydrogen from the interior of the getter diffuses to the surface and combines there with carbon and oxygen. Titanium can be outgassed at 1100° C, while zirconium should be heated to about 1300° C.

Tantalum, according to measurements of Wagener,[3] evolves water vapor, carbon monoxide, nitrogen, and hydrogen when heated to approximately 2200° C. The fact that nitrogen is evolved from tantalum is consistent with earlier findings by Andrews.[20] The equilibrium nitrogen pressure above tantalum for a constant amount of gas sorbed was an exponential function of the temperature. Measurements by Zubler confirm these results.[21]

Columbium requires a temperature of 1650° C for outgassing. At this temperature occluded and absorbed gases are expelled, and more importantly columbium oxide is volatilized. Columbium would be expected to react with nitrogen as tantalum does. Thorium, like zirconium and titanium, sorbs hydrogen at low temperatures and evolves the sorbed hydrogen at elevated temperatures. For this reason it would be expected that some water vapor and methane might be evolved during the outgassing of thorium. This will depend upon its previous surface treatment. A desorption study by Wagener[12] showed that less than 1 per cent of the oxygen and carbon dioxide gettered by thorium can be desorbed by heating.

Methane and water vapor can be sorbed by bulk getters, but the sorption is not a simple process. For example, titanium at 1200° C causes the dissociation of methane and dissolves carbon, while at this temperature

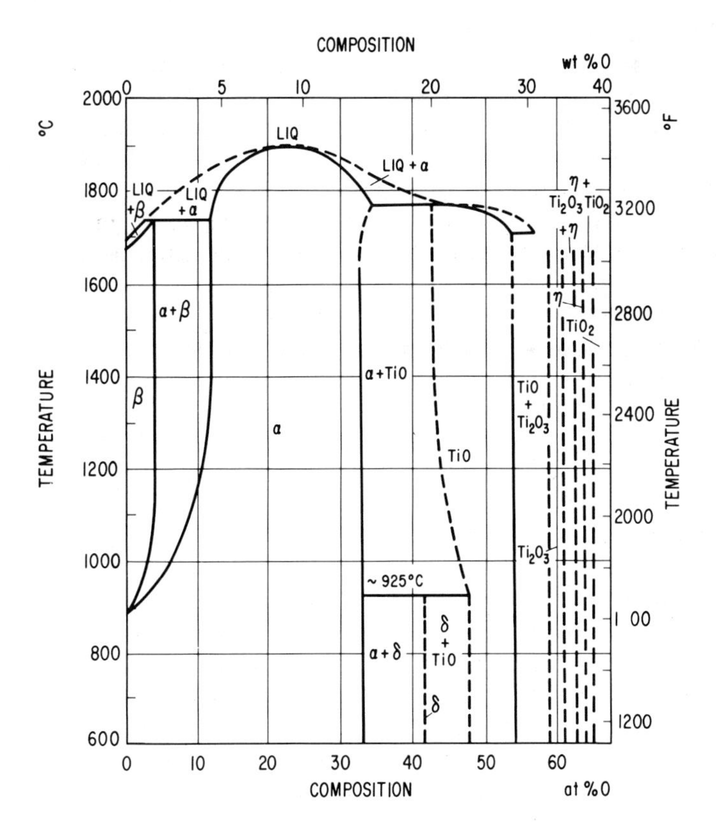


Fig. 9.5. Constitutional diagram of titanium-oxygen system (Bumps, Kessler, and Hansen).

little hydrogen is dissolved. When the methane is completely dissociated, and the temperature of the titanium is reduced, hydrogen is dissolved.

Water vapor and methane react with titanium in a similar manner. At low temperatures, that is, less than 400° C, the water vapor-titanium reaction is limited at the surface by the formation of an oxide layer. Sorption of hydrogen is blocked by a surface oxide. At higher temperatures, greater than 500° C, the surface oxide is dissolved.

The sorption of hydrogen by bulk getters takes place at a low temperature, and hydrogen is released by these getters when they are heated. Since hydrogen and other gases are common to vacuum tubes, it became useful to create a getter for sorption of gas combinations. This has led to the study and use of alloy getters.

One alloy getter, the Ceto getter, was developed in Germany during World War II. It is prepared by alloying cerium-misch metal with 28 per cent aluminum. A powder is prepared from this alloy, which is mixed with thorium powder in a ratio of 1 to 4, and the mixture is pressed into bars. The bars are sintered in vacuum at 1000° C and then are ground up and made into a paste with amylacetate and a nitrocellulose binder. The paste is applied to tube parts by painting. Coating thicknesses vary between 15 and 25 mg/cm$^2$. The Ceto getter is activated by heating to 1000° C and then sorbs gas in the temperature range of 200–500° C. This temperature is such that large quantities of all gases are sorbed and held.

Titanium and zirconium alloys have been studied.[22] Alloys are readily prepared by inert arc melting and can be purchased commercially.

By using the fixed volume-changing pressure method of measurement, parabolic rate constants were determined for various titanium-zirconium compositions. Plots of these rate constants versus temperature are shown in Fig. 9.6. All alloys sorbed oxygen more readily than the individual elements, and an alloy containing about 87.5 atomic per cent zirconium–12.5 atomic per cent titanium proved to be the best oxygen getter. At a temperature of 375° C this composition getters about 1000 times more rapidly than titanium and about 100 times more rapidly than zirconium. Measurements of gettering rates made with other gases showed that the alloys gettered as well as or better than the single elements. The composition which sorbed air most rapidly fell in the 50–63 atomic per cent zirconium range. In the temperature range of 900–1100° K the 75–25 atomic per cent titanium alloy was found to sorb $CO_2$ most readily, the sorption-rate constant varying from 0.6 to 4.5 micron · liters · cm$^2$ · min$^{1/2}$ respectively. The 75 atomic per cent zirconium alloy sorbed hydrogen and oxygen simultaneously at 300° C.

The superior gettering properties of alloys are probably the result of reduced diffusion barriers within the metal grains. It is interesting to note that the axial ratio, $c/a$, for the titanium-zirconium lattice has a maximum at about 80 atomic per cent zirconium. As there is further evidence that the easy direction of diffusion for oxygen is along the close-packed directions, it would be expected that the diffusion barriers for gettering would be least for the larger $c/a$ ratios. Although several alloys getter oxygen more readily, the sorption of carbon and nitrogen may not be greatly enhanced.

Temperature ranges for operation of the various getters are given in degrees Centigrade as follows: tantalum, 700–1200; zirconium, 500–1600; titanium, 500–1100; zirconium-titanium alloys, 300–1100; and Ceto, 200–700. The low temperature limit is generally set by the ability of the getter to get and hold hydrogen in the presence of small amounts of other gas. The upper temperature limit is set by sublimation of the getter or by dissociation of reaction products other than the metal hydrides. Just as there are precautions to be observed with film or flashed getters, so certain precautions are necessary in using bulk getters.

1. The getter should not be located close to a glass wall. If the glass becomes too hot, it will liberate water vapor. A bulk getter operating improperly in this manner can cause the pressure in a closed-off volume to rise. The increased quantity of gas (hydrogen) is produced by the

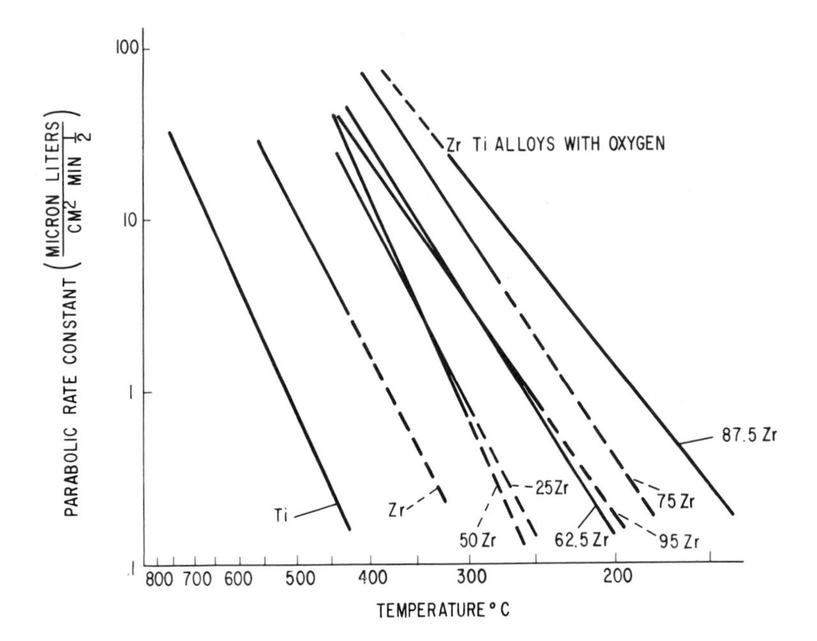


Fig. 9.6. Values of the rate constants for oxygen sorption by titanium-zirconium alloys.

reduction of water vapor. The existence of such a condition can be detected by reducing the temperature of the getter. If this is accompanied by a decrease of pressure, it is fairly certain that hydrogen gas is present.

2. A getter should not be used for a structural member of a tube (*a*) when it may be required to getter a large enough quantity of gas to cause embrittlement, and (*b*) when the bulk getter may be irradiated by electrons. Electron irradiation of many metals that contain gases can cause the release of positive gas ions. The ions are accelerated to the cathode, where they do considerable damage. For this reason the use of tantalum for grids and anodes is very much open to question.[2]

3. As bulk getters operate at elevated temperatures only, certain other metals should be joined to them in order to avoid evaporation and alloying.

4. No bulk getter should be heated to its operating temperature during high-pressure processing. A thin strip of zirconium-titanium alloy can be ignited with a match.

5. When a getter is activated it should be in a good vacuum, and the activating process should cease when the pressure over the bulk getter reaches or comes close to the ultimate of the system.

6. The getter should be kept within previously specified temperature limits. A getter that is operated at too low a temperature is ineffective, and one that is operated too hot may vaporize or will not be effective in the sorption of one or more gases. At the upper temperature limit all the bulk getters discussed will have a limited effectiveness for the sorption of hydrogen.

## 9.6. INCANDESCENT FILAMENTS

Another mechanism of gettering takes place at the surface of a hot metal when a gas-metal reaction product is vaporized. At temperatures above about 1250° K and at pressures below about 75 microns, $WO_3$ distills from tungsten as fast as formed, leaving a clean surface.[23] The rate of clean-up of oxygen at any given temperature $T$ of the filament is found to be proportional to the pressure, that is, proportional to the rate at which oxygen molecules strike the tungsten surface. However, only a fraction $\beta$ of the molecules incident on the surface react with the tungsten.

Let $V$ = volume of the system (cubic centimeters),

$P_\mu$ = pressure at any instant,

$A$ = area of filament (square centimeters),

and we shall assume that the molecules possess a velocity corresponding to 25° C.

Then the rate of clean-up is given by

$$-n_1 V \frac{dP_\mu}{dt} = \nu_1 \beta A P_\mu, \tag{9.1}$$

where $n_1$ = number of molecules per cubic centimeter at 1 micron and 25° C
$= 3.240 \cdot 10^{13}$,
$v_1$ = number of molecules incident on a surface per second per square centimeter at 1 micron and 25° C
$= 3.60 \cdot 10^{17}$.

Hence equation 9.1 becomes

$$-\frac{dP_\mu}{dt} = \frac{1.111 \cdot 10^4 \beta A P_\mu}{V}. \tag{9.2}$$

It follows that the pressure decreases with time $t$ (in seconds) in accordance with the relation

$$\log P_{\mu 0} - \log P_\mu = \frac{1.111 \cdot 10^4 \beta A t}{2.303 V}, \tag{9.3}$$

where $P_{\mu 0}$ = pressure at $t = 0$.

Hence the interval of time, $\Delta t_0$, required to reduce the pressure to 0.1 of its initial value is

$$\Delta t_0 = \frac{2.073 \cdot 10^{-4} V}{\beta A}. \tag{9.4}$$

The values of $\beta$ actually observed by Langmuir for a series of temperatures are shown in Table 9.7.

**TABLE 9.7**

**Values of the "Accommodation" Coefficient for the Clean-up of Oxygen by Tungsten***

| $T$(° K) | $\beta$ | $T$(° K) | $\beta$ |
|---|---|---|---|
| 1070 | 0.00033 | 2020 | 0.049 |
| 1270 | 0.0011 | 2290 | 0.095 |
| 1470 | 0.0053 | 2520 | 0.12 |
| 1570 | 0.0094 | 2770 | 0.15 |
| 1770 | 0.0255 | ... | ... |

* Values given in *J. Am. Chem. Soc.*, **31,** 1139 (1915).

Thus, for $A = 0.1$, $V = 1000$, and $T = 1470$, it follows from equation 9.4 that $\Delta t_0 = 6.52$ min, and at $T = 2290$, $\Delta t_0 = 21.8$ sec.

For the range $T = 1070$–$1770$, the values of $\beta$ can be represented as a function of $T$ by a relation of the form

$$\log \beta = A - \frac{B}{T},$$

where the activation energy, $Q = 2.303R_0B$, is about 26,500 cal/mole oxygen.

The clean-up of oxygen by molybdenum is similar to that of tungsten. The vapor pressure of $MoO_3$ is given as[24]

$$\log P \text{ (mm Hg)} = 16{,}140T^{-1} - 5.53 \log T + 30.69 (25\text{–}795^\circ \text{ C}) \quad (9.5)$$

and

$$\log P \text{ (mm Hg)} = 14{,}560T^{-1} - 7.04 \log T + 34.07 (795\text{–}1460^\circ \text{ C}). \quad (9.6)$$

Nitrogen and carbon monoxide do not attack molybdenum or tungsten as does oxygen, but they can be gettered by the vaporized metals.

Another method of clean-up of incandescent filaments occurs through the dissociation of gas molecules into atoms. When a wire of tungsten, palladium, or platinum is heated to a temperature above about 1500° K, in *hydrogen* at a low pressure, some of the molecules striking the filament are dissociated into atoms. At temperatures above 2000° K, this dissociation of the gas in contact with the filament causes the absorption of a very large quantity of heat. The atomic hydrogen produced diffuses out and recombines at the walls to form molecules with liberation of heat. This explains the observation that the energy loss from a tungsten wire heated in hydrogen is many times greater than it should be on the assumption that the heat loss is due exclusively to thermal conduction by the gas. According to most recent determinations, the heat of dissociation of molecular hydrogen is 102,700 cal/mole. Thus the energy liberated in recombination of atoms is considerable, and this fact has been applied by Langmuir to the development of atomic hydrogen welding.

In a series of papers Langmuir[25] discussed the mechanism of dissociation and calculated the degree of dissociation $\alpha$ as a function of the temperature and pressure. Table 9.8 gives values of this quantity for a series of temperatures and two pressures.

This atomic hydrogen [Langmuir states] has remarkable properties. It is readily adsorbed by glass surfaces at room temperatures, although more strongly at liquid-air temperatures; but only a very small amount (a few cubic millimeters) can be so retained because the atoms evidently react together to form

TABLE 9.8

**Degree of Dissociation ($\alpha$) of Hydrogen for the Range $T = 1600$–$3000°$ K**

| $T$ | $P_{mm} = 760$ | $P_{mm} = 1$ | $T$ | $P_{mm} = 760$ | $P_{mm} = 1$ |
|---|---|---|---|---|---|
| 1600 | 0.00020 | 0.0055 | 2400 | 0.0216 | 0.447 |
| 1800 | 0.00093 | 0.025 | 2600 | 0.044 | 0.692 |
| 2000 | 0.0033 | 0.087 | 2800 | 0.081 | 0.86 |
| 2200 | 0.0092 | 0.224 | 3000 | 0.13 | 0.94 |

molecular hydrogen as soon as they come in contact, even at liquid-air temperatures. The atomic hydrogen reacts instantly at room temperature with oxygen, phosphorus, and many reducible substances such as $WO_3$, $PtO_2$, etc.

Thus the clean-up of hydrogen in the presence of an incandescent tungsten filament is due to dissociation of molecules into atoms, which is followed by an adsorption of the atoms on the glass walls.

Of special interest in this connection is the reaction with water vapor. It was observed very early in the production of vacuum incandescent tungsten-filament lamps that lamps exhausted at low temperatures (say 100–200° C) blacken rapidly during life.[26] Langmuir has remarked,

A lamp made up with a side tube containing a little water which is kept cooled by a freezing mixture of solid carbon dioxide and acetone (−78.5° C) will blacken very rapidly when running at normal efficiency, although the vapor pressure of water at this temperature is only about $4 \cdot 10^{-4}$ mm.

The explanation [Langmuir states] of the behavior of water vapor seems to be as follows:

The water vapor coming into contact with the filament is decomposed, the oxygen combining with the tungsten and the hydrogen being evolved. The oxide distils to the bulb, where it is subsequently reduced to metallic tungsten by atomic hydrogen given off by the filament, water vapor being simultaneously produced. The action can thus repeat itself indefinitely with a limited quantity of water vapor.

Several experiments indicated that the amount of tungsten that was carried from the filament to the bulb was often many times greater than the chemical equivalent of the hydrogen produced, so the deposit on the bulb could not well be formed by the simple attack of the filament by water vapor.

Another experiment demonstrated that even the yellow oxide, $WO_3$, could be reduced at room temperature by atomic hydrogen. A filament was heated in a well-exhausted bulb containing a low pressure of oxygen; this gave an invisible deposit of the yellow oxide on the bulb. The remaining oxygen was pumped out and dry hydrogen was admitted. The filament was now lighted to a temperature (2000° K) so low that it could not possibly produce blackening under ordinary conditions. In a short time the bulb became distinctly dark, thus indicating a reduction of the oxide by the active hydrogen. Further treatment in hydrogen failed to produce any further darkening, showing that the oxide could only be reduced superficially.

Even in the absence of water vapor, the "water-vapor cycle" may occur in the presence of a low pressure of hydrogen if traces of oxides are present on the filament leads.

*Chlorine* molecules in contact with a heated tungsten filament are dissociated into atoms in the same manner as hydrogen molecules. Langmuir has described the following typical observations made with chlorine in the presence of a heated tungsten filament:

By heating a tungsten filament for a short time to 3000° K in a high vacuum a sufficient quantity volatilizes to form a black deposit on the bulb. If, now, a low pressure of chlorine be admitted to the bulb, this does not perceptibly attack the deposit on the bulb nor the filament, even if the bulb is heated to 200° C. However, if the filament is now heated to a high temperature, while the bulb is kept cool, the tungsten deposit on the bulb soon disappears. The chlorine has evidently been activated or dissociated by the filament and the atoms formed travel at these low pressures directly from the filament to the bulb without having any chance to recombine on the way.

The experiment is more striking if two filaments be placed side by side in the same bulb containing a very low pressure of chlorine. If one of the filaments be heated to a high temperature it is found that the other one, which remains cold, is gradually eaten away on the side facing the hot filament until it finally disappears completely. The hot one does not lose at all in weight, but may even become heavier by having tungsten deposited on it by the decomposition of the chloride formed by the attack of the cold filament.

Langmuir also obtained evidence, in connection with his investigations on thermionic emission, that oxygen molecules incident on a tungsten filament at very high temperatures are dissociated into atoms. Only, since the heat of dissociation of molecular oxygen is higher than that of hydrogen (about 117,000 cal/mole),[27] the degree of dissociation at any given temperature and pressure is considerably less than that of hydrogen under the same conditions. Thus the degree of dissociation of oxygen at 1 atm pressure is 1 per cent between 2500 and 2600° K, and about 6 per cent at 3000° K.

The observations on the dissociation of chlorine and oxygen at the surface of an incandescent filament account for the operation of the so-called "regenerative" getters[28] used formerly in the commercial production of vacuum lamps. Such getters as tungsten oxychloride and potassium thallium chloride were introduced into the lamp at some point where the temperature would be sufficiently high to obtain a low pressure of chlorine. The atomic chlorine would then combine with the tungsten deposit on the glass walls and thus maintain a clear glass surface for a much longer time than in the absence of these getters.

Other materials, such as barium chlorate, were used to produce, by dissociation of the compound, a low pressure of oxygen, which would then prevent blackening by a mechanism similar to that involved in the case of chlorine compounds.

# PART B. ELECTRICAL CLEAN-UP OF GASES

Revised by T. A. Vanderslice

## 9.7. PHENOMENA IN COLD-CATHODE DISCHARGE TUBES

The gradual disappearance of gas in a low-pressure gaseous discharge tube was first observed by Plücker in 1858. As the pressure decreases, the voltage required to pass current through the tube increases, and finally the tube becomes "nonconducting"—at least with the voltage available. In the gas-filled Roentgen tube this phenomenon was known as "hardening," since the X-rays emitted increase in hardness, or penetrating power, with increase in the voltage required to operate the tube.

A large number of investigators have dealt with this phenomenon and a great variety of explanations has been offered.[29] The observations made by these investigators and the different interpretations given form the subject of the following sections.

However, before we can discuss all this work, it is necessary to consider briefly the mechanism of electrical conduction in a gas discharge such as that which occurs in a Geissler tube. The carriers of electric charges consist of both electrons and positive ions. Because of their greater mobility and smaller mass, the electrons contribute by far the larger fraction of the current. Of special interest is the characteristic manner in which the voltage is distributed along a long tube wherein a positive column discharge occurs. Figure 9.7, based on observations made by Found and Forney,[30] shows the potential distribution in a discharge in nitrogen at 0.5 mm Hg pressure. The tube diameter was 6 cm, and the current, 5 ma. The distribution of light intensity along the tube is shown at the bottom. While that region in the discharge, known as the positive column, produces most of the light, a considerable fraction of the overall voltage is consumed in the narrow region adjacent to the cathode which is designated the "cathode dark space." This *cathode drop* varies from about 50 to 400 volts or higher, depending on the nature of both the electrode and the gas, the pressure, the area of the cathode, and the current. There is a slight gradient in the Faraday dark space, but under certain conditions this may become zero or even negative. Throughout the positive column there

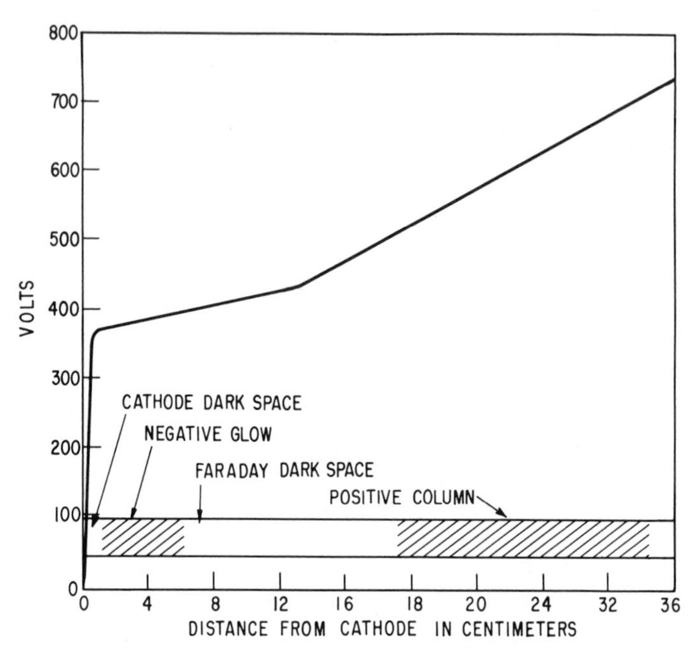


Fig. 9.7. Distribution of potential in positive column discharge in nitrogen (Found and Forney).

is a practically constant positive voltage gradient. The voltage drop at the anode may be positive, zero, or negative, depending upon the current density and other factors. However, it is always very small compared to the cathode drop.

In the cathode fall region the ions are accelerated in accordance with the relation

$$\tfrac{1}{2}m_p v_p^2 = Ve, \tag{9.7}$$

that is,

$$v_p = 1.390 \cdot 10^6 \left(\frac{V}{M}\right)^{1/2} \quad \text{cm} \cdot \text{sec}^{-1}, \tag{9.8}$$

where $m_p$ and $v_p$ denote the mass and velocity, respectively, of the positive ion, $e$ denotes the charge on an electron or singly charged positive ion, $V$ denotes the voltage drop, and $M$ denotes the molecular weight of the gas in which the positive ions are produced.

Thus, for $H_2^+$, $v_p = 9.790 \cdot 10^5 V^{1/2}$ cm · sec$^{-1}$, and for $N_2^+$, $v_p = 2.626 \cdot 10^5 V^{1/2}$ cm · sec$^{-1}$. Comparing equation 9.8 with the relation deduced in Chapter 1 for the root-mean-square velocity,

$$v_r = 15{,}794 \left(\frac{T}{M}\right)^{1/2} \quad \text{cm} \cdot \text{sec}^{-1},$$

it follows that for any ion, independently of the value of $M$, the equivalent absolute temperature is given by the relation

$$T = 7.734 \cdot 10^3 V. \tag{9.9}$$

That is, the energy of the positive ion corresponds to that of the normal molecule at a very high temperature. This fact is of fundamental importance in most of the explanations which have been given of electrical clean-up phenomena.

The ions are produced as a result of collisions between electrons emitted from the cathode and gas molecules. For every gas there is a minimum value of the energy of the electrons at which such ionization occurs; this is designated the *ionization potential* $V_i$. The values of $V_i$ vary from 3.87 volts for cesium vapor to 24.48 for helium. The velocity, $v_e$, acquired by an electron accelerated through a potential drop of $V$ volts is given by the equation, similar to 9.7 and 9.8,

$$\tfrac{1}{2} m_0 v_e^2 = Ve, \tag{9.10}$$

where $m_0$ denotes the mass of the electron. That is,

$$v_e = 5.9314 \cdot 10^7 V^{1/2} \quad \text{cm} \cdot \text{sec}^{-1}. \tag{9.11}$$

At voltages below $V_i$, electrons colliding with the gas molecules are able to excite them, if they have the proper energy, to higher energy states (or levels), and it is by spontaneous transitions from these levels to lower levels (including the normal state of the atom or molecule) that spectral lines are emitted.[31]

Figure 9.8 shows such an energy-level diagram for *mercury*. When electrons emitted from the cathode collide with mercury atoms, no interchange of energy occurs until the electrons have been accelerated through a potential of 4.9 volts. At this energy two excited states are produced, designated $2^3P_0$ and $2^3P_1$, of which the first one is metastable since no spontaneous transitions can occur to the lowest ($^1S_0$) or normal state. On the other hand, a transition from the $2^3P_1$ to the normal state is accompanied by emission of the ultraviolet line $\lambda 2537$. As the accelerating voltage on the electrons is increased beyond 4.9 volts, excited states of higher energy content are produced, thus giving rise to the different spectral lines (indicated in the diagrams by lines connecting the energy levels). At 10.4 volts one of the electrons in the outer "shell" of the atom is completely removed, leaving a mercury ion ($Hg^+$).

The frequency of the line emitted as a result of any one transition is given by the relation,

$$h\nu = \Delta V \cdot e, \tag{9.12}$$

that is,

$$\nu = 10^8 \, c\lambda^{-1} = 2.4185 \cdot 10^{14} \, \Delta V \, \text{sec}^{-1} = c\Delta\tilde{\nu} \tag{9.13}$$

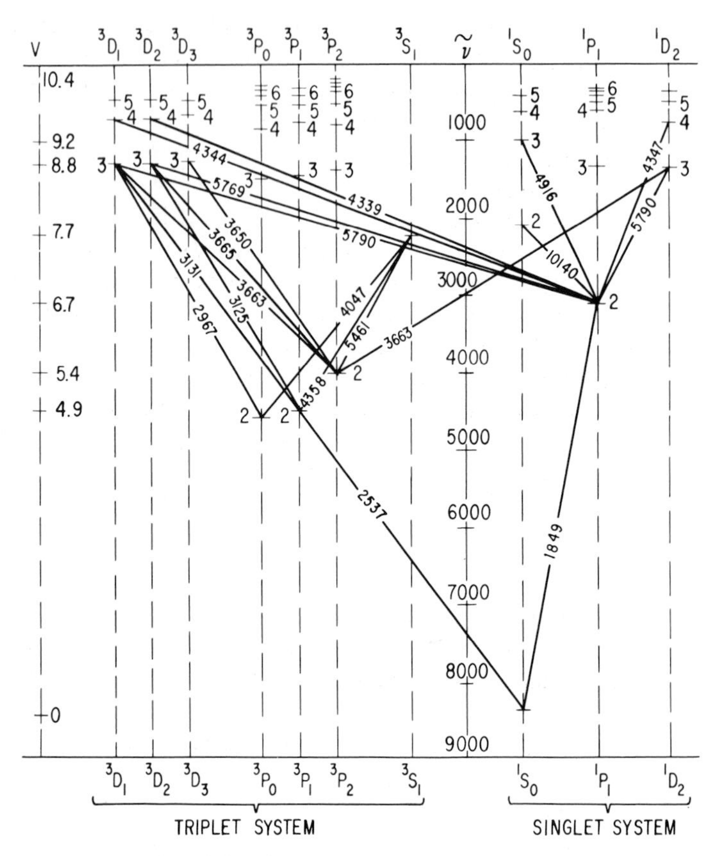


Fig. 9.8. Energy levels and lines in arc spectrum of mercury. Levels are designated by wavenumbers, $\tilde{\nu}$, or electron volts, V, required for their excitation.

where $h = 6.624 \cdot 10^{-27}$ erg sec, $c$ = velocity of light (centimeters per second), $\Delta V$ is the difference in the voltages required to excite the higher- and lower-energy levels between which the transition occurs that gives rise to the spectral line (wavelength = $\lambda$ in angstroms), and $\Delta\tilde{\nu}$ is the difference in wavenumbers ($\tilde{\nu}$) of the levels.

As mentioned above, positive ions, when accelerated towards a negatively charged electrode or other surface, acquire considerable energy. It would be correct to assume that such high-speed particles are to some degree driven into the surfaces they strike. Furthermore, because of the high energy acquired in passing through the cathode fall, the positive ions cause a disintegration of the cathode. This phenomenon is commonly called "sputtering." Wehner's data[32] for sputtering by mercury ions at various voltages are shown in Fig. 9.9 for iron and copper. As can be seen from the figure, at low voltages the sputtering yield is low, while at higher voltages and consequently higher kinetic energies of the bombarding species the sputtering yield increases. In general, the heavier the mass of the bombarding species, the higher the sputtering yield. Uncharged high-velocity neutral species also cause sputtering. In Fig. 9.10 is shown the sputtering yield for hydrogen atoms.[33]

Consequently, in a cold-cathode discharge there are ions with velocities corresponding to millions of degrees Kelvin as well as high-energy neutral atoms and metastable atoms. Sputtered metal is also being deposited in various regions of the tube. The chemical activity of the dissociated molecules is much higher than that of the diatomic species.[34] Atomic nitrogen, for example, reacts with various hydrocarbons to form HCN.[35] Blodgett[36] has shown that evaporated copper deposits do not adsorb molecular hydrogen but do adsorb considerable quantities of atomic hydrogen. These types of phenomena present in gas discharges are sufficient to make the clean-up mechanism quite complex.

Brodetsky and Hodgson[37] have given the following list of possible explanations for gas clean-up:

1. Chemical action between the gas and glass, cathode and/or anode.
2. Mechanical action between the gas and glass, cathode and/or anode.
3. Chemical action due to active species such as atomic nitrogen.

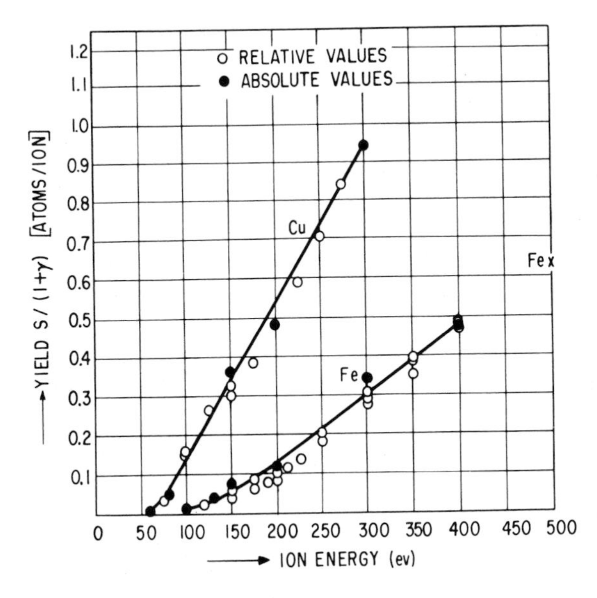


Fig. 9.9. Sputtering yields of copper and iron (Wehner).

4. Mechanical occlusion of the gas in the glass, cathode, and/or anode.
5. Mechanical occlusion of the gas in the sputtered deposit.

In addition, all possible combinations of the above may occur.

According to Langmuir[38] both the chemical and mechanical types of mechanisms are responsible for the clean-up of nitrogen in the presence of a hot tungsten cathode when an electric discharge passes. At higher pressures, the reaction is "electrochemical," the nitrogen combining with the tungsten to form the nitride $WN_2$. At very low pressures and high anode voltages, the action is apparently purely mechanical (Langmuir designates this the "electrical" clean-up). The nitrogen is driven into the glass in such a form that it can be recovered by heating. The action is thus apparently reversible, and only limited quantities of nitrogen can be cleaned up in this manner. The electrical clean-up, as distinguished from the electrochemical, also exhibits distinct fatigue effects.

Willows[39] observed that the amount of clean-up of gas in a discharge tube varied according to the nature of the glass from which the tube was made. The absorption was least in Jena glass, more in lead glass, and greatest in soda glass. At constant current, the amount of gas occluded

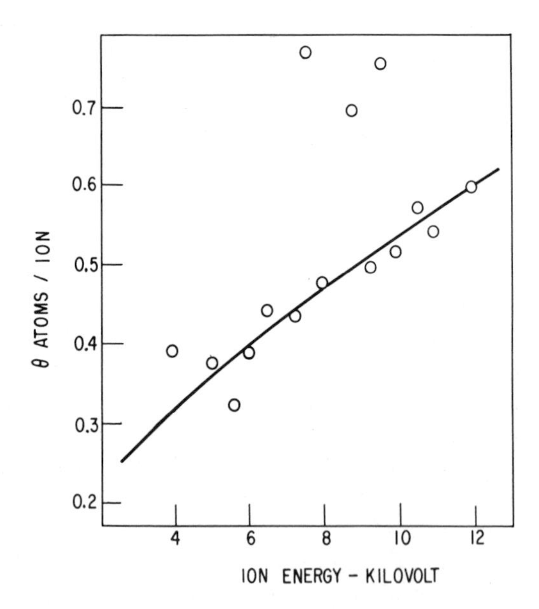


Fig. 9.10. Sputtering coefficients for unanalyzed hydrogen beams normally incident on silver (the relation of the numbers and energies of neutrals to the ion energy is not known) (Moore).

was found to increase with decrease in pressure. The conclusion arrived at was that the clean-up is due to chemical reactions between the gases and the glass walls. In support of this view, Hill[40] has shown that absorption is observed with air in an electrodeless discharge. His experiments were carried out in the range of pressures varying from 0.4 to 0.04 mm of mercury. During these experiments the bulbs became covered with a dark deposit on the walls, presumably due to oxidation reactions. On exhausting these bulbs and filling them with hydrogen at about 1 mm pressure, the discharge caused a rapid disappearance of the gas and the deposit became lighter colored, which Hill accounted for by chemical reduction. Willows repeated Hill's experiments[41] with quartz vacuum tubes and observed that "a new quartz bulb does not absorb air, but if fed with repeated doses of hydrogen—which are absorbed when an electrodeless discharge is passed—it becomes very active. If discharges in hydrogen are alternated with those in air the bulb can be made to absorb large quantities of either gas and the activity with each gradually increases." Willows accounted for these results by assuming, as Hill did, alternate oxidation and reduction.

Leck[42] measured the amount of inert gases adsorbed on glass and metal surface under high-vacuum conditions. In the case of the metals adsorption was brought about by bombarding the surfaces with ions of the inert gases with energies up to 5000 ev. The surfaces were degassed by heating, and the total quantity of gas desorbed was measured by using a mass spectrometer. The mass spectrometer enabled the actual gas of interest to be detected in the presence of a large background.

Much of the previous work done in this area depended on pressure measurements alone to measure recovery, and the experimenter was never sure what the measurement really meant. Leck's results with tungsten, nickel, and platinum showed that the total quantity of gas cleaned up increased with increasing energy of the bombarding ions to 5000 volts. Recovery occurred at temperatures which would indicate a higher binding energy than that for physical adsorption. Ions could also dislodge, "and take the place of molecules already adsorbed at the surface." No quantitative measurements of the intensities or energies of the incident ions on glass surfaces were made. Stable absorption took place, and, to release all the inert gas atoms from the surface, a temperature of 500–600° K had to be used.

Chemically active gases clean up generally at much faster rates than inert gases. Mey[43] showed that, when potassium and sodium amalgams are used as electrodes, compounds of these metals with hydrogen and nitrogen are formed during the discharge, and Gehlhoff[44] utilized this observation to purify rare gases (argon, helium, neon, etc.). In the presence of a glow discharge with a heated alkali metal as cathode, all the chemically active gases are removed from a mixture of chemically active and rare gases. Nitrogen is completely absorbed even with the alkali metal at ordinary temperatures. Complete absorption of hydrogen occurs with sodium at 290° C and with potassium at 175° C, and rubidium and cesium are effective at even lower temperatures. Gehlhoff assumed that chemical combination occurs between the vapors of the metals and the residual gases in an active state.

In this connection it is interesting to refer briefly to the experiments carried out by Strutt[45] on the formation and properties of a chemically active modification of nitrogen. He observed that, on passing a condenser discharge through nitrogen at low pressures, a form of nitrogen is obtained which shows an intense yellow glow and is extremely active chemically. As is well known, nitrogen in the ordinary state is very inert chemically. It combines with other elements with difficulty and only under special conditions, such as high temperature or high pressure. On the other hand, Strutt found that the nitrogen passed through the discharge tube under the above conditions was very active chemically. "Drawing it by the pump over a small pellet of phosphorus, a violent reaction occurs, red phosphorus is formed, and the yellow glow is quenched. At the same time the gas is absorbed." Similarly, active nitrogen combines readily with iodine, sulfur, and arsenic. There is no doubt, therefore, that the formation of active nitrogen must be taken into account in explaining clean-up effects in electrical discharges.

Winkler[35] and his students have made extensive investigations of the reactions of active nitrogen. Kistiakowsky and Volpi,[46] for example, have demonstrated that active nitrogen is in reality atomic nitrogen.

In a similar manner Newman accounted for the absorption of hydrogen by phosphorus, sulfur, and iodine, by assuming the formation of an active modification of the gas. This is probably the same form as that produced by Wendt and Landauer[47] in subjecting hydrogen to alpha rays from radium emanation or to high-potential corona discharges.

Observations made by Vegard[48] confirmed the generally held view that clean-up is due to sorption of gas by the metal sputtered from the cathode. He found that sorption is small as long as the cathode drop is below a certain "threshold" value. The rate of sorption apparently increased with increase in cathode drop, and the same parallelism held for the amount of cathode sputtering. Thus, in oxygen, gold electrodes showed more sputtering than platinum electrodes. Vegard also observed that with helium there was a measurable clean-up, which was, however, less than was obtained under similar conditions with either nitrogen or oxygen. With hydrogen both adsorption and evolution were found to occur.

The *clean-up of rare gases* in a cold-cathode discharge tube has been investigated by Alterthum, Lompe, and Seeliger.[49] The discharge tube used was 2 cm in diameter and had hollow cylindrical electrodes made of thin sheet iron, which could be degassed with high-frequency heating. The cylinders were covered at the end nearest the seals by flat circular disks. Pressures were measured by means of a Pirani gauge, and the source of current was a transformer operated at 50 cycles and having an open-circuit voltage of 4500 volts. The pressures used were varied from several millimeters to 0.2 mm Hg, and the currents, from 20 to 100 ma. In a tube containing neon at 7 mm Hg and operated at 100 ma, the voltage remained fairly constant at about 250 volts for several hundred hours. Then, as rate of clean-up increased, the voltage increased to 500 volts at 1000 hr and the pressure decreased to less than 1 mm Hg.

It was observed that in neon, at constant current, the rate of clean-up is low at higher pressures but increases rapidly with decrease in pressure and finally becomes approximately independent of the pressure when it has decreased to about 3 mm Hg. The rate of clean-up was found to be proportional to the current and independent of the form and length of the discharge tube.

It was shown that the gas is cleaned up for the most part by the electrodes and, furthermore, that this clean-up occurs at the cathode. The fact that the clean-up did not occur on the glass walls and was independent of the amount of sputtering leads to the conclusion that the neon was *cleaned up as positive ions* driven into the cathode by the high cathode fall. This is in agreement with the observation that rate of clean-up increased with increase in cathode drop.

Similar observations were made with discharges in argon and helium. The rates of clean-up in the different gases at 2 mm Hg and 100 ma were found to be in the ratios 0.4 : 1 : 10 for argon, neon, and helium, respectively. "This also," Alterthum, Lompe, and Seeliger state, "is in agreement with the above hypothesis, since, the smaller the atoms (or ions), the more easily they will be able to penetrate into the metal and diffuse throughout it." It should also be added that the cathodic sputtering is greatest for argon and least for helium, which is another argument for the conclusions drawn from this investigation regarding the mechanism of clean-up.

In two subsequent papers,[50] Alterthum and Lompe investigated the effect of other factors on the clean-up of rare gases in a discharge tube such as that used in the previous work. In order to compare the data obtained under different conditions they defined a factor $Z$ by means of the relation

$$t = \frac{ZPV}{i}, \tag{9.14}$$

where $t$ denotes the period (in hours) required to clean up $PV$ mm $\cdot$ cm$^3$

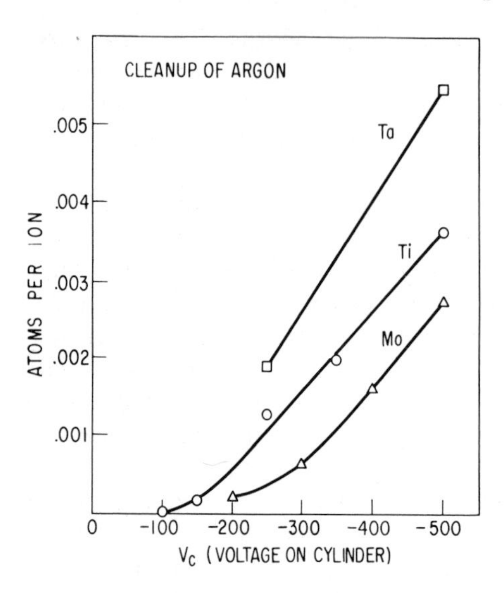


Fig. 9.11. Number of argon atoms cleaned up per argon ion striking a cylinder, as a function of voltage.

with a discharge of $i$ ma. Thus $Z$ is a proportionality factor which corresponds to the period in hours required to decrease the quantity of gas by 1 mm · cm³ with a discharge of 1 ma. The quantity $Z$ involves all the factors that affect rate of clean-up, that is, nature of gas, nature of electrodes, variation in cathode drop with electrode shape, temperature of electrodes, etc.

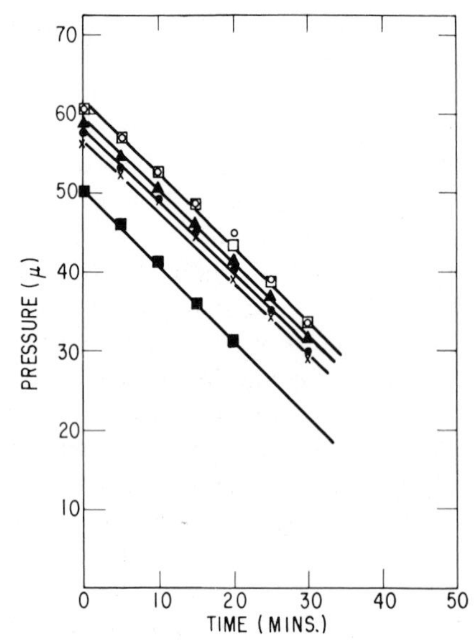


Fig. 9.12. Pressure versus time data for clean-up of argon at a constant discharge current of 200 ma. Different lines indicate separate runs after recharging system with argon.

From these observations and a series of later papers[51,52,53,54] by Alterthum, Bartholomeyczyk, Seeliger, and their collaborators it may be concluded that electrical clean-up in the rare gases takes place by impact of positive ions on the electrodes. Reddan and Rouse[55] observed clean-up of helium with onset of sputtering.

Blodgett and Vanderslice[56] have shown that the clean-up of neon, argon, and krypton cannot be explained by simple ion penetration. Clean-up is governed largely by two factors: first the rate at which metal is being sputtered and, second, the potential of the surface on which the metal lands. At small negative or positive potentials on the surface collecting sputtered metal, there is a slow clean-up rate caused by uncharged species being buried by sputtered metal. At more negative potentials burial of ions becomes important, and clean-up is much more rapid. The clean-up is faster with the more easily sputtered metal (Fig. 9.11).

The clean-up of inert gases by sputtered metal does not saturate; that is, upon recharging the tube to the same initial pressure it was found that the clean-up occurred at the same rate (Fig. 9.12). Hundreds of equivalent monolayers of argon have been cleaned up without any detectable change in rate. Recovery was affected by heating to the evaporation temperatures of the metal. However, an electrode collecting ions at a uniform current density over the surface will have a net clean-up rate of zero after the initial disappearance of a small amount of gas. Resputtering occurs, liberating the gas as fast as it is being cleaned up.

## 9.8. CLEAN-UP IN ELECTRODELESS DISCHARGES

In order to eliminate the effect of electrodes a number of investigators have carried out experiments on clean-up in electrodeless discharges. Such discharges are obtained by coupling glass bulbs or rings containing the gas with coils carrying a high-frequency current, the frequency of which varies from $10^5$ sec$^{-1}$ to higher values.

Mierdel[57] used bulbs of glass and quartz to which were attached appendixes for cooling purposes. The clean-up was observed to be greatest for oxygen. With the appendix at liquid-air temperature the pressure decreased in less than a minute from 85 to 2 microns Hg, and in order to maintain the discharge fresh oxygen had to be fed into the bulb. On warming the appendix to room temperature, most of the gas was re-evolved. Mierdel considers the clean-up due to formation of ozone.

With hydrogen, nitrogen, and air similar clean-up effects were observed, the amount of clean-up being least with hydrogen. Again, a considerable fraction of the absorbed gas was re-evolved on raising the temperature of the appendix. Mierdel accounts for the clean-up in hydrogen as due to formation of atomic hydrogen. This is condensed on the walls at liquid-air temperature and re-evaporates in the molecular form at room temperature.

With regard to the clean-up of hydrogen, Hiedemann,[58] working in Mierdel's laboratory, showed that the extent and rate of disappearance of the gas depend to a large extent on the nature of the preliminary treatment of the glass. Furthermore he observed a normal and "abnormal" clean-up effect. The normal is accounted for by the condensation of atomic hydrogen at low temperature; to explain the abnormal clean-up, it is suggested (on the basis of other investigations) that compounds of silicon and hydrogen (silanes) are formed at low temperatures.

The clean-up of hydrogen *at room temperature* was investigated by Johnson.[59] The pressure was measured by means of a very sensitive mercury manometer. The range of the manometer was about 0.5 mm Hg, with a detectable sensitivity of $10^{-4}$ mm.

It was observed that the amount of hydrogen cleaned up under the influence of a high-frequency discharge was a maximum for any given tube immediately after the tube had been baked out at 300° C and thoroughly degassed. On repeating the bake-out and evacuation and letting in fresh hydrogen to the same initial pressure (about 0.1 mm) the amount of gas cleaned up was less and decreased still more with successive repetitions of bake-out and evacuation. Taking into account the area available for adsorption, it was found that the maximum possible values of the fraction of the glass surface covered with a monolayer of atomic hydrogen were 0.66 and 0.36 for the two tubes used in the investigation.

Siegler and Dieke[60] studied the clean-up of argon and xenon in electrodeless discharges at 15,200 and 2500 Mc at pressures from 1 to 75 microns Hg. Clean-up occurred at the walls of the glass vessel whenever the discharge plasma covered the surface. No clean-up was found beyond the plasma region. The rate of clean-up was large when the walls were outgassed and decreased monotonically as the density of the cleaned-up gas increased. The rate was independent of discharge frequency but proportional to the intensity. The rate decreased with increasing pressure. The cleaned-up gas was liberated by heating the discharge tube, and complete recovery was effected. Siegler and Dieke conclude that clean-up occurs by penetration of ions possessing considerable kinetic energy into the walls.

Experiments by Wheeldon[61] on the absorption of sodium and argon by glass are in general agreement with the above work. A significant difference in absorption was measured for various types of glass, as shown in Fig. 9.13. Wheeldon concludes that argon is absorbed only when ionized in a discharge, and that the rate of absorption is more or less independent of temperature below 300° C. At higher temperature a partial evolution

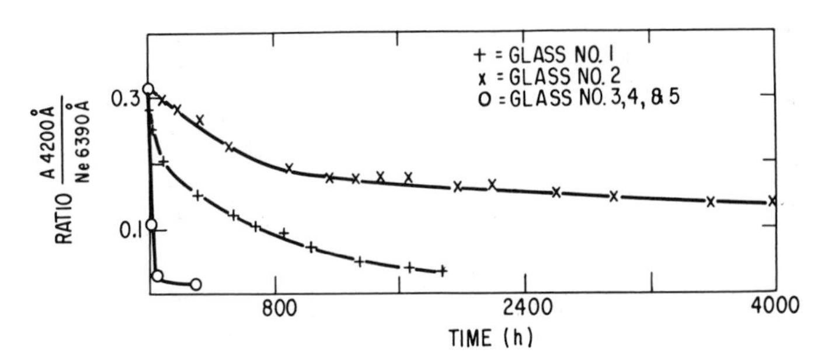


Fig. 9.13. Argon absorption by various glasses (Wheeldon).

occurs. Wheeldon observed no diffusion of the argon through the walls of the glass tube.

## 9.9. ELECTRON EMISSION AND SPACE-CHARGE PHENOMENA IN HOT-CATHODE LOW-PRESSURE DEVICES

In a cold-cathode discharge tube the electrons are generated at the cathode by means of the positive ions which acquire a high velocity in the cathode-fall region. The ions cause sputtering of the cathode, and the sputtering is accompanied by a clean-up of the gas. If, however, a hot cathode is used, electrons are emitted by a mechanism analogous to the evaporation of atoms from a solid, and the cathode drop is reduced to a value which is, in general, only slightly higher than the ionizing potential $V_i$ of the gas. Under certain conditions the cathode fall may be even less than $V_i$. As shown by Hull,[62] no sputtering occurs if the cathode drop is maintained below a certain critical value which depends upon the nature of the cathode and the composition of the gas. For instance, for thoriated tungsten cathodes this "disintegration voltage" is about 27 for neon ions and 17 for mercury ions.

Before considering voltage clean-up phenomena in hot-cathode devices it is essential to review briefly certain fundamental characteristics of hot cathodes.[63] The electron emission from a heated cathode varies with the absolute temperature in accordance with the relation

$$I = AT^2\epsilon^{-b_0/T}, \tag{9.15}$$

where $I$ = emission per unit area, at $T$ degrees $K$, and $A$ and $b_0$ are constants. This relation is more conveniently written in the form

$$\log I = \log A + 2 \log T - \frac{b_0}{2.303T}. \tag{9.16}$$

**TABLE 9.9**

**Electron Emission**

($I$ = amperes per square centimeter; $W$ = watts per square centimeter)

| | Tungsten | | Molybdenum | | Tantalum | | Columbium | | ThW |
|---|---|---|---|---|---|---|---|---|---|
| $T$ °K | $I$ | $W$ | $I$ | $W$ | $I$ | $W$ | $I$ | $W$ | $I$ |
| 1000 | | | | | | | | | $1.73 \cdot 10^{-7}$ |
| 1200 | | | | | | | | | $3.95 \cdot 10^{-5}$ |
| 1400 | | | | | | | | | $2.03 \cdot 10^{-3}$ |
| 1600 | $9.27 \cdot 10^{-7}$ | 7.74 | $2.39 \cdot 10^{-6}$ | 6.30 | $9.1 \cdot 10^{-6}$ | 7.36 | $2.19 \cdot 10^{-5}$ | 6.40 | $4.06 \cdot 10^{-2}$ |
| 1800 | $4.47 \cdot 10^{-5}$ | 14.2 | $1.05 \cdot 10^{-4}$ | 11.3 | $3.32 \cdot 10^{-4}$ | 13.3 | $6.95 \cdot 10^{-4}$ | 11.4 | 0.428 |
| 2000 | $1.00 \cdot 10^{-3}$ | 24.0 | $2.15 \cdot 10^{-3}$ | 19.2 | $6.21 \cdot 10^{-3}$ | 21.6 | $1.16 \cdot 10^{-2}$ | 18.5 | 2.864 |
| 2200 | $1.33 \cdot 10^{-2}$ | 38.2 | $2.59 \cdot 10^{-2}$ | 30.7 | $6.78 \cdot 10^{-2}$ | 34.2 | 0.115 | 29.9 | |
| 2400 | 0.116 | 57.7 | 0.215 | 47.0 | 0.509 | 51.3 | 0.800 | 45.3 | |
| 2600 | 0.716 | 83.8 | 1.29 | 69.5 | 2.25 | 75.4 | 5.28 | 67.0 | |
| 2800 | 3.54 | 117.6 | 6.04 | 98.0 | 12.53 | 105.5 | 60.67 | 130.6 | |
| 3000 | 14.15 | 160.5 | 23.28 | 116.0 | 45.60 | 144.4 | | | |
| log $A$: | 1.780 | | 1.780 | | 1.740 | | 1.571 | | 0.477 |
| $b_0/2.303$: | 22,750 | | 22,100 | | 21,110 | | 20,226 | | 13,680 |

Table 9.9 shows values of $I$, the emission in amperes per square centimeter, as well as $W$, the watts radiated per square centimeter, at a series of temperatures, for four types of cathodes.[64] Values of log $A$ and of $b_0/2.303$ used in evaluating the emission data are given in the two lowest rows.

The thoriated tungsten is obtained by flashing a tungsten filament containing 0.5–1.5 per cent $ThO_2$ at a high temperature (above 2700° K), which reduces some of the oxide to metallic thorium. This is then followed by heating at 2100–2200° K, thus causing diffusion of thorium to the surface, where it forms a monatomic layer which possesses much higher electron emission than pure tungsten. At temperatures above about 2000° K the thorium evaporates at a rate which exceeds that of diffusion to the surface with a resultant decrease in electron emissivity.

In many electronic devices, an oxide-coated cathode is used, which consists of a nickel or nickel-alloy filament coated with a mixture of the carbonates of barium, strontium, and calcium. The filament must be "activated" by heat treatment in order to produce a monatomic film of barium which is the source of electrons. The values of $A$ and $b_0$ for the activated emitter vary with composition of the coating, method of preparation, and nature of metal support.[65] The emission at 1000° K is usually found to lie, within a factor of 10, at the value of 10 ma · $cm^{-2}$ for BaO, 1 ma · $cm^{-2}$ for SrO, and 0.1 ma · $cm^{-2}$ for CaO. The values of $b_0/2.303$ for BaO lie around 5500, and the values for SrO and CaO are higher. At temperatures above about 1200° K, the rate of evaporation of barium atoms increases rapidly with increase in temperature and the surface loses its activity.

The value of the space current passing from the hot cathode to the anode, in a good vacuum, is limited either by temperature saturation or by "space charge." The former may be calculated for a given area and temperature of cathode by means of equation 9.15 or 9.16. This emission is obtained, however, only if the anode voltage is sufficient to overcome the repulsion between electrons. As first shown by Langmuir (1913), the space-charge current is given by a relation of the form

$$i_s = sV^{3/2}, \tag{9.17}$$

where $s$ is a constant, the value of which depends upon the geometry of the cathode and anode.

For instance, for parallel plane surfaces,

$$i_s = \frac{2.325 \cdot 10^{-6}}{d^2} V^{3/2} \quad \text{amp} \cdot \text{cm}^{-2}, \tag{9.18}$$

where $d$ = distance (in centimeters) between the two electrodes. For a

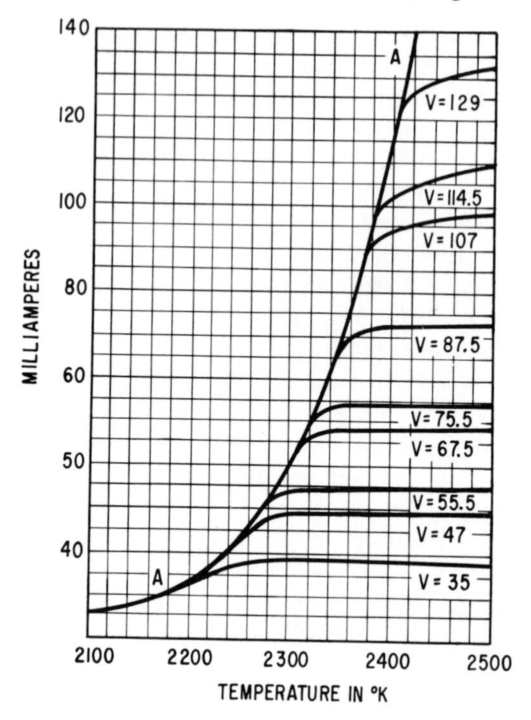


Fig. 9.14. The limitation of electron current in a hot-cathode diode by (*a*) temperature of the cathode, as shown by the curve *AA*, or (*b*) anode volts, as shown by the horizontal lines for different values of the anode potential.

filamentary cathode of radius $r_0$ along the axis of a cylindrical anode of radius $r$ (in centimeters), the current (in amperes) per centimeter length is given by the relation

$$i_s = 14.65 \times 10^{-6} \frac{V^{3/2}}{r\beta^2} \quad \text{amp/cm length,} \tag{9.19}$$

where $\beta^2$ is a function of $r/r_0$, which has the value 1 for $r/r_0 = \infty$.

The manner in which the electron current between the cathode and the anode may be limited either by temperature of the cathode or by voltage between the two electrodes is illustrated by the plots in Fig. 9.14. These plots give the variation in electron current observed from a tungsten filament 0.25 mm in diameter located along the axis of a cylinder 2.54 cm in diameter and 7.62 cm long. The maximum available electron current as a function of the temperature is given by the exponential curve. The horizontal plots give the space-charge-limited currents for different anode voltages. Curves such as shown in Fig. 9.14 are obtained only at extremely low pressure (usually of the order of $10^{-3}$ micron Hg or less).

## 9.10. CLEAN-UP IN HOT-CATHODE DISCHARGES

When, however, the gas pressure exceeds a certain value (depending principally upon the nature of the gas and the geometrical dimensions of the device) some of the electrons no longer travel directly from the cathode to the anode. Collisions with gas molecules occur, and if the anode voltage exceeds the ionizing potential positive ions are produced which tend to neutralize a part or the whole of the negative space charge produced by the electrons; consequently, the electron current reaches the saturation value, corresponding to the temperature of the filament, at much lower voltages than when gas is not present.

These phenomena are illustrated by the following observations with a small hot-cathode high-vacuum rectifier (diode) in which a tungsten spiral was used as cathode and a molybdenum cylinder (enclosing the spiral) as anode. To the bulb containing these electrodes was attached an appendix containing a small amount of mercury, and the whole arrangement was exhausted to a very high vacuum. By immersing the appendix in liquid air the pressure of the mercury vapor could be reduced to such a low value that no ionizing effects occurred. Figure 9.15 shows the characteristics of the tube, under these conditions, at two different filament currents. It will be observed that at a filament current of 1.35 amp (curve *A*) the electron current varied with the anode voltage in accordance with the $\frac{3}{2}$ power relation. At a filament current of 1.25 amp, corresponding to a lower temperature, the electron current increased at first in accordance with the same voltage law and then tended to reach the value 38 ma, corresponding to the emission for that particular temperature (curve *B*).

On removing the liquid air from the appendix, so that the mercury attained room temperature and the pressure in the device reached the value of about $2 \cdot 10^{-3}$ mm (corresponding to the vapor pressure of mercury at this temperature), the characteristic obtained with the filament at 1.25 amp was that shown in curve *C*. At 18 volts a distinct blue glow appeared, and at 26 volts this glow extended practically all the way down the appendix. It will be observed that, simultaneously with this appearance of blue glow, the electron current increased rapidly until it reached practically the same saturation value as that obtained under good vacuum conditions. That is, even at such a low pressure of mercury vapor, the space-charge phenomena practically disappeared, and the saturation electron emission corresponding to 1.25-amp filament current was obtained at relatively low voltages.

Similar phenomena are observed in hot-cathode devices with all other gases. At very low pressures the electron current varies with the voltage according to the $\frac{3}{2}$ power law; but at higher pressures (ordinarily about

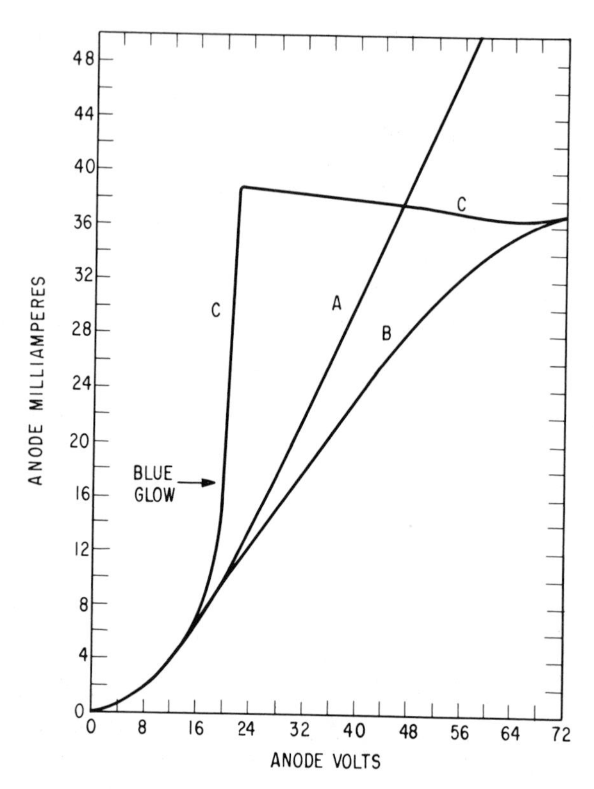


Fig. 9.15. The effect of mercury vapor on the characteristics of hot-cathode diode. Curve *A*, space-charge limitation in good vacuum. Filament current, 1.35 amp. Curve *B*, space-charge limitation at lower values of anode potential and temperature limitation at higher values of anode potential. Filament current, 1.25 amp. Curve *C*, same filament temperature as curve *B*, but in the presence of mercury vapor at $P_{\mu} = 2.0$.

1–2 microns) blue glow appears as the voltage is raised above the ionizing potential and simultaneously with this blue glow the current increases rapidly, as shown in curve *C*, Fig. 9.15, to the saturation emission at the given filament temperature. In the case of most gases this blue glow does not ordinarily persist very long, owing to clean-up effects that accompany its appearance, and consequently the electron current decreases again gradually until it attains the limiting value corresponding to space charge.

The clean-up phenomena that are observed in hot-cathode devices are extremely interesting and are of great technical importance. The experimental evidence obtained in the General Electric Research Laboratory and by other investigators points to the conclusion that these effects are primarily due to the formation of ions by collision between electrons and gas molecules. There is little or no evidence for any electrical clean-up below ionizing potentials.

A description of some observations made during earlier experiments by Mrs. M. R. Andrews, H. Huthsteiner, and S. Dushman are of interest in this connection. In these experiments a tube containing two adjacent tungsten filaments was used, so that either filament could be made cathode, and the rate of clean-up was investigated for argon and nitrogen. Gas from a large reservoir was allowed to flow continuously through this tube at practically constant pressure, and the rate of clean-up was determined by collecting the gas after it left the tube. The pressures used in most of the experiments were so low that no blue glow effects were observed, and during any one set of observations both the electron current and the anode voltage were maintained constant. The magnitude of the current was varied in different runs from a few microamperes to several milliamperes, and the anode potential was varied from 25 to 250 volts. The rate of clean-up was observed to increase almost linearly with increase in pressure. With electron currents below a milliampere the rate of clean-up also varied linearly with the current, but it showed a tendency to increase much less rapidly with electron currents exceeding this value. At anode voltages in excess of 25, the rate of clean-up was observed to be practically constant and independent of the voltage. With freshly baked-out glass bulbs, the fatigue effects observed were quite pronounced. On covering the glass surface inside with a tungsten deposit (by evaporation of one of the filaments) the rate of clean-up was found to increase considerably, and much more gas could be cleaned up before fatigue effects occurred. This effect is undoubtedly connected with later observations of Alpert and coworkers on clean-up in ionization gauges discussed in section 9.10. The rate of clean-up of argon was found to be about half of that of nitrogen under otherwise similar conditions.

Very interesting phenomena were observed on passing a discharge in mercury vapor. In an otherwise thoroughly exhausted glass bulb, it was observed that under these conditions *hydrogen is liberated* from the walls. The fact that the discharge might evolve gas from the walls instead of causing it to disappear had already been noticed by Campbell[66] as occurring to some extent in nitrogen, carbon monoxide, and argon. In the presence of mercury vapor this phenomenon was extremely marked.

It should be noted that the gas thus liberated cannot be liberated by mere heating of the walls to their softening point; gas can be attached to the walls in some such way that it can be liberated by the discharge but not by heating. Of course, the attachment may consist of chemical combination; it is possible that glass contains hydrogen chemically combined, probably as water. But it should be observed that the hydrogen liberated, if piled up on the glass, would form a layer at least 25 molecules deep. Since the potential driving the discharge in these experiments was often as low as 50 volts, it is hardly to be expected that the electrons or ions could penetrate so far into the glass simply in virtue of the energy which they receive from the discharge. It seems easier to believe that a layer on the surface, subject to the action of these particles, is constantly renewed by diffusion from within.

That the clean-up effects are more pronounced in the presence of blue glow has been confirmed by work in the General Electric Research Laboratory. However, the rate of voltage clean-up in hot-cathode devices is quite appreciable even at very low pressures, as will be pointed out in a later section. The actual values of the glow potentials are dependent on the geometry of the electrodes and the size of the bulb. For instance, blue glow in mercury vapor has been observed at as low as 18 volts, whereas the value obtained by Campbell at a pressure of 2 microns was 32.5 volts.

In general, however, it would appear that the clean-up mechanisms occurring in hot-cathode discharge devices are not fundamentally different from those that occur in cold-cathode discharges. The important distinctions are connected with differences in the type of discharge. With hot-cathode devices the arc drop is in most cases lower than in cold-cathode devices, and consequently there tends to be less sputtering. Also, the hot-cathode surface discourages adsorption of ions and excited species.

## 9.11. ELECTRICAL CLEAN-UP IN HOT-CATHODE TUBES AT VERY LOW PRESSURES

In all the work discussed in the previous sections the clean-up was investigated at pressures in the range of 1–100 microns Hg or even higher. With the advent of ultrahigh-vacuum techniques the pressure range down to at least $10^{-12}$ mm Hg has received much attention in the literature.

The arrangement used by von Meyern[67] is shown in Fig. 9.16. It consists of a tube with a tungsten-filament cathode *K*, and a platinum ring-shaped anode *A*, 2 cm in diameter. The discharge tube is set inside a solenoid *S* by means of which a magnetic field can be applied. Pressures are measured by an ionization gauge, and the voltage on the anode can be varied from 0 to 1000 volts.

In the presence of a magnetic field acting parallel to the axis of the tube the electron paths are elongated spirals, and consequently the electrons pass through the ring for a certain distance, then reverse their paths, and repeat this a number of times before they are collected by the anode. Hence the ionizing distance is very much greater than it would be in the absence of a magnetic field, and, as would be expected, the rate of clean-up is much more rapid in the presence of the magnetic field.

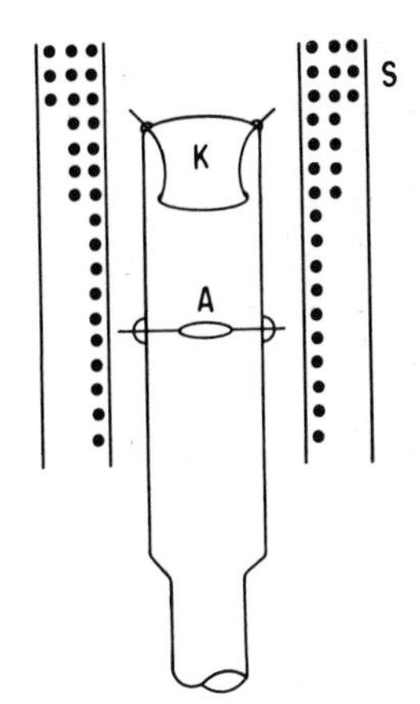


Fig. 9.16. Tube with hot cathode $K$ and ring anode $A$ in magnetic field produced by the solenoid $S$. Clean-up observed in absence and also in presence of magnetic field (von Meyern).

Figure 9.17 illustrates the results obtained for air and hydrogen in the absence of a magnetic field (curves 1 and 2) and in the presence of a field of 280 gauss (curves 3 and 4). The initial pressure in all four cases was $1 \cdot 10^{-4}$ mm Hg, and, as will be observed, the rate of clean-up in the presence of the field was approximately exponential during the first minute. The initial value of the electron current was 1.5 ma, and the anode potential was 820 volts for hydrogen and 900 volts for air. After the discharge was stopped the residual gas pressure remained at the low value for a while and then gradually increased to about $1.7 \cdot 10^{-5}$ mm Hg in a period of several hours.

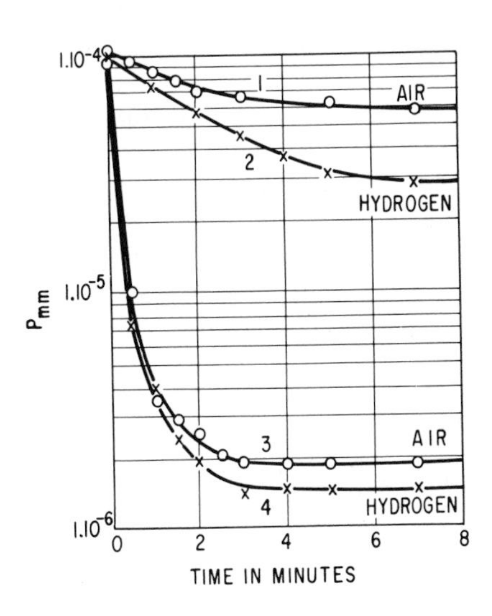


Fig. 9.17. Plot of $P_{mm}$ (on log scale) versus time during clean-up of air and hydrogen in absence of magnetic field (curves 1 and 2) and in presence of magnetic field of 280 gauss (curves 3 and 4) (von Meyern).

On heating the bulb to 200° C, all the absorbed gas was re-evolved; there was no appreciable absorption by the anode. A large fraction of the gas cleaned up in the presence of a magnetic field was evolved at room temperature if, after an equilibrium state was reached, the magnetic field was cut off. Von Meyern found that, as the clean-up proceeded, the rate decreased. This was ascribed to re-emission of gas from the surfaces. The rate of clean-up increased with increasing ion velocity between 400 and 800 volts. It is likely, in view of the later work, that metal films were present and played an important role.

Schwarz[68] showed that clean-up is dependent on negative charges on the glass walls. The walls can become charged negatively by electron impact provided that the secondary-electron yield for electrons is less than unity. The positive ions are driven into the glass walls by the field, and in some cases the walls adsorb as much as 1.7 times the number required to form a monolayer. Not only may positive ions be driven into the walls, but also these positive ions can drive out gases already sorbed in the walls, thus changing the composition of residual gases. The "interchange" phenomenon was recognized by the change in the value of the ratio of positive-ion current to electron current ($i_p/i_e$) at constant pressure. In one experiment nitrogen was cleaned up by successive introduction of fresh samples until complete saturation was attained at a pressure of a few tenths of a micron Hg. The gauge was then thoroughly evacuated, and argon introduced. Although the pressure as measured by a Gaede molecular manometer was constant, the value of the ratio $i_p/i_e$ changed continuously, in such a manner as to indicate a substitution of argon for nitrogen in the walls. Finally most of the gas in the gauge consisted of nitrogen. In another experiment, nitrogen was found to displace argon from the walls of the tube.

Bloomer and Haine[69] found that the cleaned-up gas in an ionization gauge is held by the glass walls of the tube, particularly in the vicinity of the electrode structure. Varnerin and Carmichael[70] showed that a Bayard-Alpert gauge does not clean up helium efficiently unless there is a thin film of metal on the walls of the gauge. This film is sputtered or evaporated in ordinary use. Young[71] found similar results for helium but discovered that nitrogen was effectively collected on various surfaces. He found that the ion current striking the wall of an Alpert gauge is 5–10 times that striking the collector. The maximum pumping speed was 0.1 liter · sec$^{-1}$ for nitrogen. On the basis of ion current to walls it is possible to account for a pumping speed of 0.1–0.2 liter · sec$^{-1}$. The clean-up for helium is shown in Fig. 9.18. The departure of the curve from linearity is ascribed to a depletion of available sites and to spontaneous re-emission of gas at a rate proportional to the number of trapped

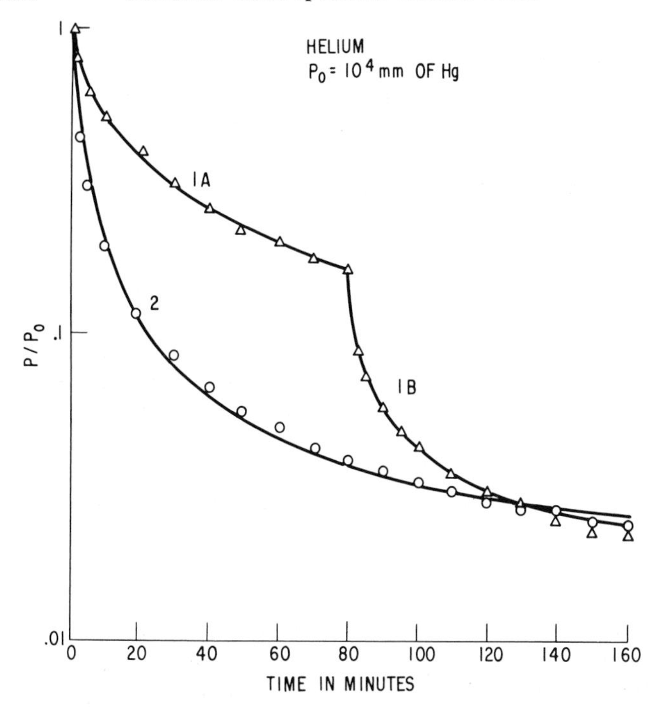


Fig. 9.18. Clean-up of helium in ionization gauge. Curve 1*A*, bare glass opposite electrode structure; curves 1*B* and 2, slight metal deposit opposite electrode structure (Young).

gas atoms. That is, the rate of clean-up is the difference in rate between the number striking the surface and the number leaving. Initially, the concentration on the surface is low and consequently the number leaving is low. As the concentration on the surface increases, the number leaving increases, since the rate of departure is proportional to the number of trapped gas atoms. Eventually, an equilibrium is reached where the number striking and sticking is balanced by the number leaving, and the rate of clean-up is zero. The curve through the points in the figure represents the best fit of Young's derived expression.

The exponential decrease in pressure with time in a conventional ion gauge is expected on the basis of a simple argument put forth by Varnerin and Carmichael.[72] The rate of loss of gas atoms is proportional to the number of ions, which in an ionization gauge is proportional to the gas density. Therefore, an exponential decrease is expected. Varnerin and Carmichael have made a study of clean-up of helium on a bombarded molybdenum cylinder. They report an increase in trapping efficiency with ion energy, as shown in Fig. 9.19. The saturation of clean-up is explained

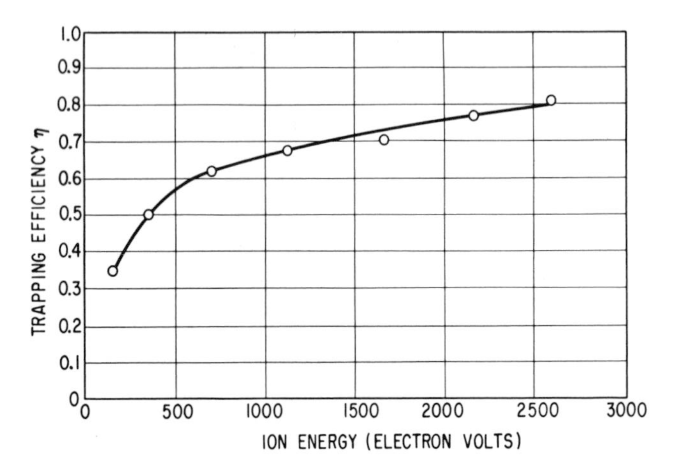


Fig. 9.19. Trapping efficiency as a function of ion energy (Varnerin and Carmichael).

by a phenomenological theory. The rate of clean-up can be expressed by the difference in rate between the number of ions striking the surface, multiplied by a "sticking probability," and the rate of re-emission. However, from a study of the re-emission process Varnerin and Carmichael show that the rate of re-emission is not exponentially decreasing, as would be expected on Young's model, where it was assumed that each re-emission event occurred with equal probability. The rate of re-emission is a function not only of the number of atoms trapped but also of the time at which the trapping occurs.

Brown and Leck[73] used a mass spectrometer to study the desorption of gas in a Penning discharge. The gases were accelerated to a cathode with energies up to 4000 volts. The gas could be recovered by bombardment with a different gas. Their work confirmed the observations of Schwarz, discussed previously, on gas interchange. In addition Brown and Leck report that the interchange adsorption and interchange processes were independent of the chemical nature of the gas or metal surface and conclude from this that the molecules are probably bound to the cathode surface by van der Waals forces. The gases were evolved upon bake-out between 100 and 200° C. As is pointed out by Brown and Leck, it is difficult to understand on the basis of van der Waals binding why gas adsorbed from the discharge is held more strongly than that taken up as atoms. Physical adsorption of noble gases, for example, occurs only well below room temperature, whereas the surface layer formed by ion bombardment is stable up to 100° C. The authors report that the maximum quantity of gas recovered corresponds approximately to that needed to form a complete surface layer on the cathode. If it is assumed that the adsorption is due entirely to molecules held at the cathode, the chance of an ion "sticking" approaches 100 per cent for many gases. In Brown and Leck's work, the glass walls were shielded from the discharge.

Carmichael and Trendelenburg[74] have used the mass-spectrometer technique of Brown and Leck to study the re-emission of inert gases from a metal surface. Helium, neon, argon, and krypton were pumped ionically at about 100 ev into a nickel target and subsequently released by a similar bombardment, using a different noble gas. The re-emission data are described in terms of a model where the atoms are trapped in the target in such a fashion that the probability for release by a bombarding ion is not the same for each atom. The difference in probability for release is due to a certain depth distribution of the trapped atoms in the target material. Saturation of the depth distribution was observed for neon, argon, and krypton at concentrations below one equivalent monolayer.

It is clear from the foregoing discussion that much is still to be learned concerning the mechanism of clean-up in a hot-filament discharge at low pressure.

## 9.12. ULTRAHIGH VACUUM

In certain types of investigations, such as those dealing with thermionic and photoelectric emission, contact potentials, secondary electron emission, and surface chemistry, it is absolutely essential to obtain surfaces that are as clean as possible. It is also necessary to maintain such a good vacuum that these surfaces shall not be contaminated by traces of residual gases, at least for a sufficiently long period to make certain that the observed data are characteristic of the clean surface.

Even at as low a pressure as $10^{-6}$ micron Hg, the rate at which oxygen molecules at room temperature strike a surface is $3.6 \cdot 10^{11}$ molecules per $cm^2$ per sec. Assuming that the number of molecules required to form a monolayer is $8.26 \cdot 10^{14}$ per $cm^2$, and that every molecule incident on the surface is adsorbed, 10 per cent of the surface would become covered with a layer of oxygen molecules in about 3.8 min. A contamination of this magnitude would decrease the photoelectric emission by a factor of the order of 10, at least, and thus lead to quite erroneous observations.

It should be borne in mind that the ultimate pressure obtained in any device represents an equilibrium state between the rate at which gas is evolved from glass walls and metal parts, and the rate at which these gases are removed, either by adsorption or by pumping. Hence it is of extreme importance to take every precaution in investigations such as these under discussion to treat all parts of a system or device, while under exhaust, in such a manner that subsequent evolution of gases will be reduced to the minimum.

The story of the development of a vacuum technique adequate for the purpose mentioned begins with the investigations of Langmuir and his associates, in 1912–1913. At that time a considerable number of physicists believed that electron emission from an incandescent metal surface at low pressures was due to the presence of residual traces of gas and that consequently this thermionic emission should disappear in a very good vacuum.

In order to demonstrate the existence of a pure thermionic emission and the validity of the space-charge relation, Langmuir[75] used a bulb containing two hairpin tungsten filaments either of which could serve as electron emitter and the other as anode. During exhaust (by means of a Gaede rotary mercury pump) the bulb was heated for 1 hr at 360° C, which was as high a temperature as was practical without deformation of the bulb by atmospheric pressure. A liquid-air trap was placed between bulb and pump. The filaments were heated to 2700–2800° K in order to degas them, and the bulb was sealed off. In order to improve the vacuum, the filaments were aged at 2400° K for about 24 hr, which served to clean up the residual oxygen and nitrogen. "To still further improve the vacuum in some cases," Langmuir states, "the entire bulb was immersed in liquid air and the filaments again heated to high temperature for a short time." As shown subsequently by Dushman,[76] observations taken with the molecular gauge indicated that the lowest pressure obtained in Langmuir's experiments was certainly well below about $4 \cdot 10^{-4}$ micron Hg.[114]

This technique, which involves heating the glass during exhaust for 1 or more hours at about 360° C for soft glass (and at 500° C for Pyrex), heating the electrodes either by passage of current in the case of filaments or by electron bombardment to as high a temperature as practicable, sealing off the pump, and improving the vacuum in the sealed-off device by chemical or electrochemical clean-up—this procedure for obtaining a good vacuum became recognized practice.

Even with careful processing such as discussed above, a direct measurement of the residual pressure was not possible below certain values, and one finds in the older literature such statements as "a sticking vacuum (McLeod gauge) was obtained." Nottingham had recognized the X-ray limit of the conventional ionization gauge as discussed in Chapter 5. Apker employed the flash-filament technique of Langmuir to measure residual pressures below this X-ray limit, or measurement of $10^{-8}$ mm Hg. The technique is a simple one and consists merely of heating a filament to a high temperature for a sufficient interval to desorb the adsorbed gas from the surface. The filament is cooled for a known time and then heated rapidly to a temperature where all the adsorbed gas is desorbed. From the change in pressure in the known volume the number of molecules (atoms) adsorbed on the filament per square centimeter can be calculated. The rate at which molecules arrive at the filament surface at a given pressure can be calculated from kinetic theory, and therefore from the measured number of atoms adsorbed per square centimeter per second the residual pressure of adsorbable gas atoms can be calculated. It was recognized by Apker that the sticking probability of various gases is in general less than unity and also that the flash-filament technique does not measure non-adsorbable gases.

Anderson[77] and Nottingham[78,79] expressed the opinion that the constancy of emission from a clean tungsten surface is an even more sensitive indication of vacuum conditions in the system than the ionization gauge. At the time these opinions were expressed the statement was undoubtedly true.

In 1950 Bayard and Alpert[80] described their ionization gauge, which made it possible to measure vacuum below $10^{-10}$ mm Hg. In a series of publications from the Westinghouse laboratories the techniques of ultrahigh vacuum were described, and today in most laboratories these practices are standard. "With the ultrahigh vacuum technique," Alpert reported, "we achieve vacuums between $10^{-9}$ and $10^{-10}$ mm Hg in a manner which is in no case more difficult than the procedures conventionally used to attain the usual 'good' vacuum of $10^{-7}$ mm Hg. . . ." Once a reliable pressure-measuring device was available, the techniques of obtaining low pressure were relatively simple.

### Vacuum Systems

For any operating vacuum system there will be established a stationary pressure $P_e$, which corresponds to the equilibrium between the flux of gas into the system and the rate of exhaust. This rate of flow of gas $Q$ into the system may arise from desorption from the walls of the system, back diffusion from the pumps, or passage of atmospheric gases through the walls. For a system of volume $V$ and pumping speed $S$, the rate of reduction of pressure is given by

$$\frac{dP}{dt} = -\frac{S}{V}P + \frac{Q}{V}. \tag{9.20}$$

For the simple case in which $S$ and $Q$ are constants, independent of pressure, the solution of this equation is

$$P = \frac{Q}{S} + P_0 - \frac{Q}{S}\exp - \frac{t}{V/S}, \tag{9.21}$$

where $V/S$ is now the pumping-time constant or characteristic pumping time of the system, that is, the time required to pump the system down to 0.368 of its initial pressure $P_0$. The equilibrium pressure finally attained in the system will be

$$P_e = \frac{Q}{S}. \tag{9.23}$$

Obviously, then, to achieve ultrahigh vacuum the rate of gas leakage into the system, divided by the pumping speed of the system, must be small. In order to keep the influx of gas low, it is necessary to keep back diffusion from pumps at a minimum by using effective traps with low backstreaming pumps.

Desorption from walls of the system, which is a major source of gas, is generally decreased by baking the system to temperatures up to 500° C or by immersing the system in a refrigerant such as liquid helium.[81] The required high-temperature bake-out makes the use of stopcocks, glass taper fittings, wax, etc., impractical.

The leak rate of gas through the walls of the system must be small. Alpert and Buritz[82] have shown that the ultimate limit to the achievement of very low pressures in glass systems is fixed by the diffusion of atmospheric helium through the walls of the system. They measured this rate for borosilicate glass as approximately $10^{-15}$ mm · liter · $sec^{-1}$ · $cm^{-2}$. A 1-liter system with a surface area of 1000 $cm^2$ would require a pumping speed of only 0.01 liter · $sec^{-1}$ to maintain a pressure of $10^{-10}$ mm Hg. Norton[83] has made measurements of the permeation rate of helium through various glasses, and the results are shown in Fig. 9.20. The rate for borosilicate glass is in substantial agreement with the value given by Alpert and Buritz.

The permeation of atmospheric helium can be decreased by a selection of the glass used in the vacuum system. Corning 1720, for example, has a permeation rate $10^{-5}$ lower than ordinary borosilicate glass (Corning 7740). Since the diffusion rate of helium is much lower through crystalline materials than through glass, it is expected that an all ceramic-metal system would have a low permeation rate for helium and that the ultimate pressure would be limited only by the outgassing of the metal and ceramic.

Another limit to the achievement of low pressures is the finite vapor pressure of the tungsten filament in the ionization gauge. At 2300° K, for example,[84] the vapor pressure is $10^{-12}$ mm Hg. This limits the temperature to which a tungsten filament may be heated to achieve sufficient emission in an ultrahigh-vacuum ionization gauge. However, lanthanum boride[85] produces 25 ma · $cm^2$ emission at 1425° K, and the vapor pressure is about $10^{-4}$ lower than tungsten at the same electron emission current density.

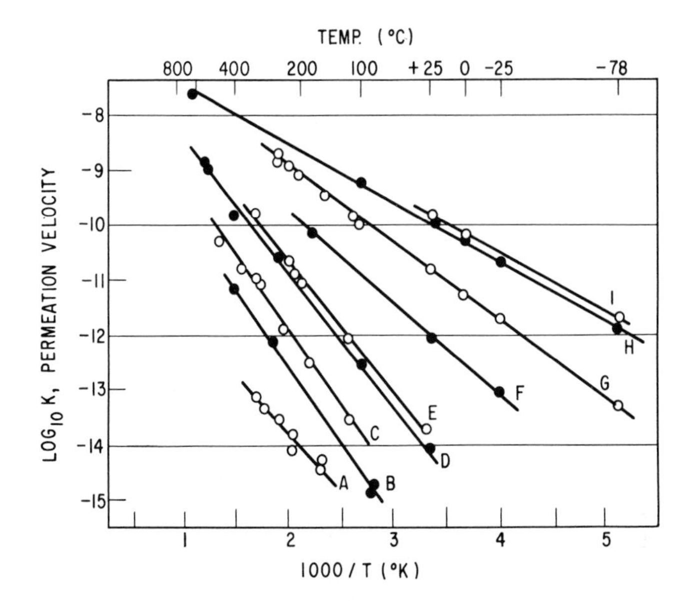


Fig. 9.20. Permeation velocity $K$ of helium diffusing through various glasses. The permeation velocity, $\log_{10} K$, is plotted against the reciprocal of the absolute temperature, $1000/T$. The permeation velocity $K$ is in units of $cm^3$ gas (NTP)/sec/$cm^2$ area/mm thickness/cm Hg (gas pressure difference). Curve *A*, lead borate glass; *B*, X-ray shield glass; *C*, combustion tubing No. 1720; *D*, soda-lime glass No. 0080; *E*, phosphate glass; *F*, borosilicate glass No. 650; *G*, chemical Pyrex brand glass No. 7740; *H*, fused silica; and *I*, Vycor brand glass (Norton).

Figure 9.21 shows a diagrammatic sketch of an exhaust system which has been designed and used by L. Apker, in the General Electric Research Laboratory, in connection with an investigation of the work functions of semiconductors. The dashed rectangles indicate ovens that can be arranged to bake different parts of the system. The following exhaust procedure, used by Apker, is based on the investigations carried out by Nottingham on high-vacuum technique.

Initially the tube *A*, the gauge *G*, and the liquid-air traps *Tr*-2 and *Tr*-1 are given a prolonged bake-out at 500° C. Then the two lower ovens surrounding *Tr*-1 are removed, liquid air is put around it, and the bake-out is continued for a longer period on the rest of the system. At the end of this treatment, the upper oven is removed, all metal parts are heated by high-frequency and electron bombardment, and the tube is again heated at a moderate temperature.

If the tube is to be operated on the exhaust system, liquid air is applied to *Tr*-2 immediately after the last bake-out. If, however, the tube is to be

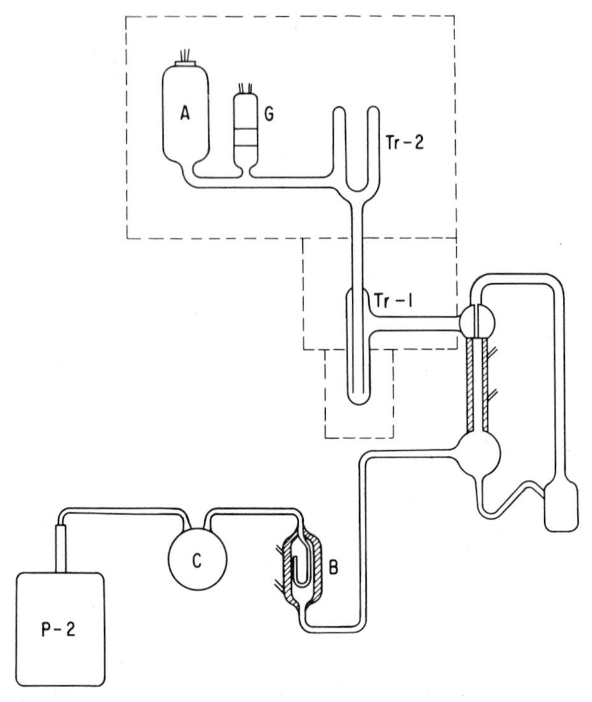


Fig. 9.21. Diagrammatic sketch of exhaust system for production of extremely low pressure (Nottingham, Apker). *A*, tube to be evacuated; *G*, ionization gauge; *Tr*-1 and *Tr*-2, liquid-air traps; *P*-1, mercury-vapor pump; and *P*-2, rotary oil pump. The bulbs *B* and *C* prevent diffusion of oil vapor into the mercury-vapor pump.

sealed off, the constriction is heated to near the softening point of the glass for about 15 min, and pumping is continued for a couple of hours, with the tube hot. The getter is then flashed, liquid air is applied to *Tr*-2, and the tube is sealed off.

In the past few years a variety of components specifically designed for ultrahigh-vacuum systems has become available. A discussion of these follow.

## Pumps for Ultrahigh Vacuum

*Diffusion pumps.* The operation of a diffusion pump depends on the effusion of gas molecules from the region being evacuated into the directed vapor stream of the pump.[86] Some of the directed momentum of the vapor molecules is imparted to the gas molecules in the system, thus creating a pressure gradient causing the gas to flow through the pump. This process is independent of pressure once the pressure has been lowered by a backing pump to a value at which molecular flow or effusion can occur, and the pumping speed will rise to a plateau value and remain there for all lower pressures. The rate of gas evolution from the pump itself becomes significant at lower pressures and eventually exceeds the measured rate of introduction of gases into the pump from the vacuum system. Eventually, then, the net pumping speed of the diffusion pump falls to zero because of the gas evolution and/or backstreaming of the pump fluid from the pump itself.

Hagstrum[87] and Becker and Hartman[88] at the Bell Telephone Laboratories have used mercury-diffusion pumps with liquid nitrogen-cooled traps to attain ultrahigh vacuum. Experiments by Venema and Bandringa[89] have demonstrated that a properly designed mercury pump and a series of traps will attain a vacuum of at least $10^{-12}$ mm Hg. Vanderslice at the General Electric Research Laboratory has achieved a pressure of $3 \times 10^{-12}$ mm Hg at room temperature, using standard mercury diffusion pumps.

Oil-diffusion pumps have very high pumping speeds, but until recently it was felt that their pressure limit was in the vicinity of $10^{-7}$ mm Hg. Even with a high pumping speed, the ultimate pressure is limited by the high value of backstreaming. Alpert[90] has shown that at some pressure between $10^{-7}$ and $10^{-8}$ mm Hg the conventional oil-diffusion pump becomes a source of contamination for a "clean" vacuum system. The oil pumps were at first used to evacuate the system down to $10^{-7}$ mm Hg; after the system was closed off from the oil pumps, the pumping action of the ionization gauge was utilized to pump to lower pressures. Alpert[90,91] designed a copper-foil trap, shown in Fig. 9.22, that would prevent the

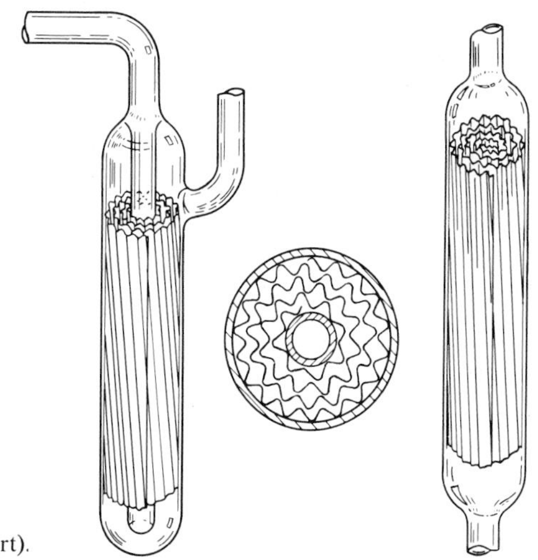

Fig. 9.22. Copper-foil trap (Alpert).

backstreaming of oil vapor but would not seriously restrict the conductance and pumping speed of the system. With the gettering action of the copper eliminating backstreaming, the oil pump evacuated the system to pressures below $10^{-10}$ mm Hg. The copper-foil trap was effective even at room temperature and has been effective for periods of several months. The trap was reactivated by baking with the rest of the system. The exact mechanism whereby the trap adsorbs backstreaming oil vapor at room temperature is not fully understood, but with proper traps an oil-diffusion pump may be used in ultrahigh-vacuum experiments.

The work of Biondi,[92] utilizing nonrefrigerated molecular sieve traps, has had and will continue to have a profound impact on the use of oil pumps for ultrahigh-vacuum experiments. With these traps it is now possible to achieve, for reasonably long times, ultrahigh vacuum with oil pumps without the use of refrigerants. The most popular materials at the moment are zeolite and alumina with some advantages being claimed for alumina. L. A. Harris[93] has demonstrated that the activity of oxide cathodes may be maintained for a period of several days, and perhaps longer, with the use of alumina traps. A comparative test without alumina and with conventional cold trapping showed that a similar oxide cathode loses much of its activity in a comparable time.

*Ion pumps.* The ionization gauge itself may be used for pumping, once the pressure has been brought to a low value. The pumping speed of ionization gauges is low, of the order of hundredths of a liter per second, but with a low influx of gas, $Q$, the speed is sufficient for a well baked-out system. Alpert[94] has demonstrated that a Bayard-Alpert gauge with titanium continually evaporated to the wall has a pumping speed of at least 20 liters $\cdot$ sec$^{-1}$ at low pressures.

The most important development in recent years has been the advent of commercially available ion pumps with relatively high pumping speeds down to at least $10^{-9}$ mm Hg. Hall[95] reports a pressure of $10^{-11}$ mm Hg, using a baked metal system and an ion pump.

The cold-cathode ion pump is basically a variation of a Penning-type ionization gauge consisting of a ring anode and two cathode plates in a

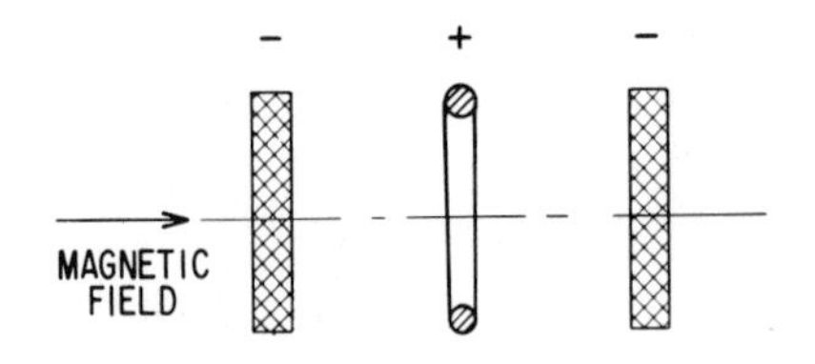


Fig. 9.23. Penning pump.

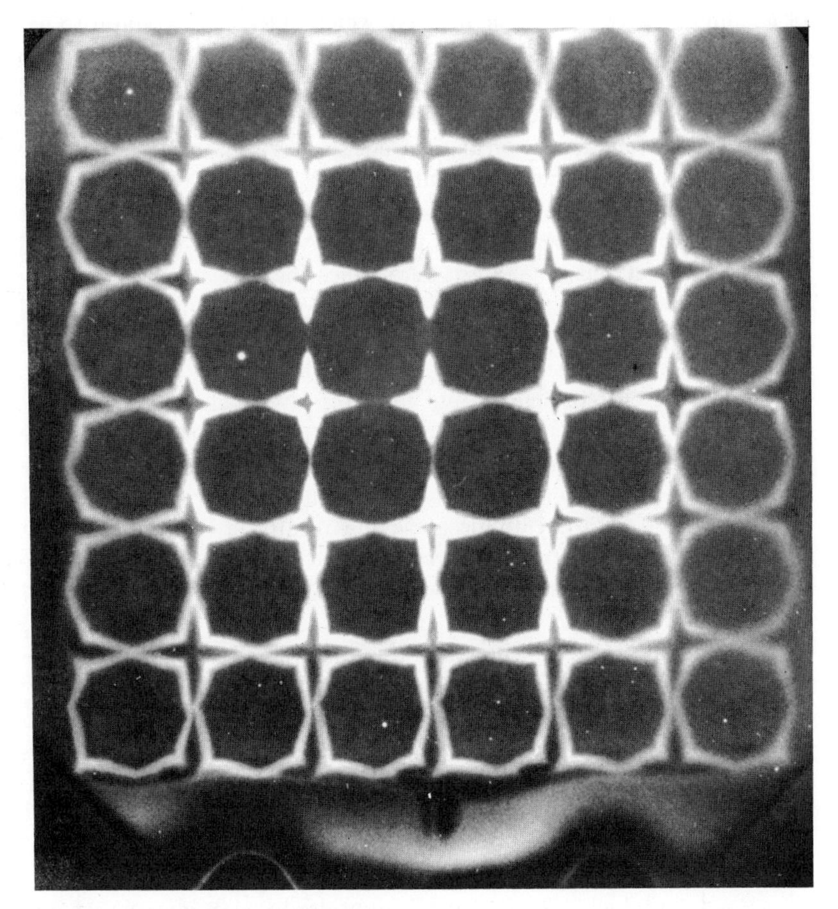

Fig. 9.24. Radiograph showing the location of radioactive krypton 85 in the sputtered deposit on the cathode of an ion pump (Vanderslice).

magnetic field as shown in Fig. 9.23. The anode is operated at high positive potential with respect to the cathode. Electrons emitted by the cold cathode are forced into a spiral path by the presence of a strong magnetic field. The increased electron path results in a high probability for collision and ionization between electrons and gas molecules. The positive ions then bombard the cathode and sputter metal from the cathode. The sputtered metal deposits in various regions of the tube, forming stable compounds with the chemically active gas molecules. Chemically inert gas atoms such as argon are also pumped, but by a different mechanism. In an ion pump much of the inert gas is initially found in the sputtered deposit on the cathode. Figure 9.24 shows a radiograph of the cathode of a commercial two-electrode pump in which radioactive krypton 85 has been pumped. The pump was disassembled and the cathode placed in contact with a photographic film, showing that the krypton was located chiefly in the sputtered deposit on the cathode.

If the ion-current density on an electrode surface is uniform, the clean-up of inert gas on that surface is zero after a short period of time. Initially a small number of ions will be embedded in the surface, but as soon as sputtering occurs, embedded gas is released. A condition is soon reached in which the net clean-up rate is zero, since the rate at which ions are driven into the newly exposed surface is just equal to the rate of release of gas from the surface by sputtering. If an ion-current density gradient exists on the electrode surface, there will be a build-up of sputtered metal in areas of low current density, and consequently gas clean-up will occur. The Penning-type discharge in a two-electrode pump is somewhat unstable during operation, producing changes in the distribution of the ion-current density at the cathode surface. This results in the resputtering of previously sputtered metal on the cathode and consequently in the release of gas previously cleaned up. This is frequently referred to as the ion-pump memory effect. This effect is reported to be lessened by grooving the cathode.[96]

Inert gases are also cleaned up at the anode to some extent.[96,97] The gas is found in the metal which has been sputtered from the cathode. There is the question of how the gas sticks to the anode, since unexcited inert gas atoms are not adsorbed on the electrode surfaces for long enough times at room temperature to be covered by sputtered metal, and the ions normally do not have sufficient kinetic energy to reach the anode. There are several plausible explanations: (1) neutral inert gas atoms excited by the discharge stick to the anode and are covered by sputtered metal from the cathode; (2) high velocity neutral atoms produced by charge exchange are driven into the anode and then covered by sputtered metal from the cathode; (3) the gas ions gain sufficient kinetic energy to reach the anode through oscillations in the discharge; and (4) the gas is swept out of the space between the electrodes by the high flux of the metal vapor. The first two explanations appear more likely. Although the anode mechanism is less efficient than coverage of ions at the cathode, the clean-up is permanent because resputtering of material from the anode cannot occur.

Three electrode ion pumps have been studied in the General Electric Research Laboratory[97] and elsewhere.[98] In the arrangement shown in Fig. 9.25, the third electrode is placed in a favorable position to receive sputtered metal from the cathode. The voltage applied to this electrode is intermediate between the cathode and the anode, so that the inert gas

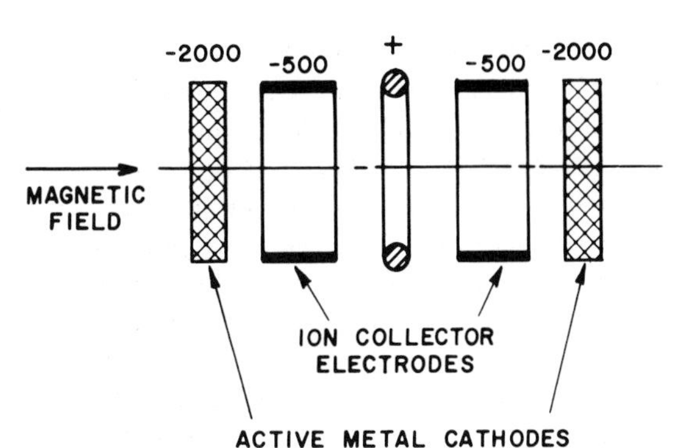


Fig. 9.25. Three-electrode pump.

ions are driven into it with sufficient energy to be embedded but not to cause appreciable sputtering. Thus gas cleaned up by this electrode is permanent.

Ion pumps offer great promise of replacing many conventional diffusion pumps. Their principal advantages are that the system in never exposed to contamination from diffusion-pump fluids and no refrigerants are necessary. Their current disadvantages are the difficulty of pumping organic vapors and their inability to handle large amounts of gas at high pressure.

*Cryogenic pumps.* Immersion of the whole or part of a vacuum system in liquid air or immersion of a trap containing charcoal in liquid air was a convenient way of producing low pressures in the days before the advent of diffusion pumps. In recent years with the need for large systems for space simulation and routine experiments at ultrahigh vacuum, cryogenic pumping is again becoming popular. At liquid-helium temperatures, only helium and hydrogen have appreciable vapor pressures. All other gases condense out on the surface of the cooled container. The sticking probability for most gases approaches unity as the temperature is lowered, and the total pressure can be reduced to the ultrahigh-vacuum range in the absence of hydrogen or helium. If hydrogen or helium is present in quantities greater than monolayer amounts, it can be removed by other means, such as ion pumping. The speed of a cryogenic pump is determined by the area of the cold surface and the conductance of the connecting tubing. For the efficient design of a cryogenic pump, it is necessary to reduce the radiation incident on the trap from the walls of the vessel. The helium trap is normally jacketed by liquid nitrogen.

Cryogenic pumping shows much promise for large specialized system applications, such as space-simulation chambers.

## Ultrahigh-Vacuum Components

Since mercury cutoffs, standard stopcocks, and rubber gaskets are not bakable, their use in ultrahigh-vacuum systems is prohibited. Stopcock grease, mercury, or other fluids used in cut-offs tend to contain large amounts of occluded gas. This has led to the development of a new line of hardware components for ultrahigh-vacuum systems.

*Vacuum valves.* Alpert[99] described a valve (Fig. 9.26) which can withstand bake-out at ~450° C and has a sufficiently large conductance for

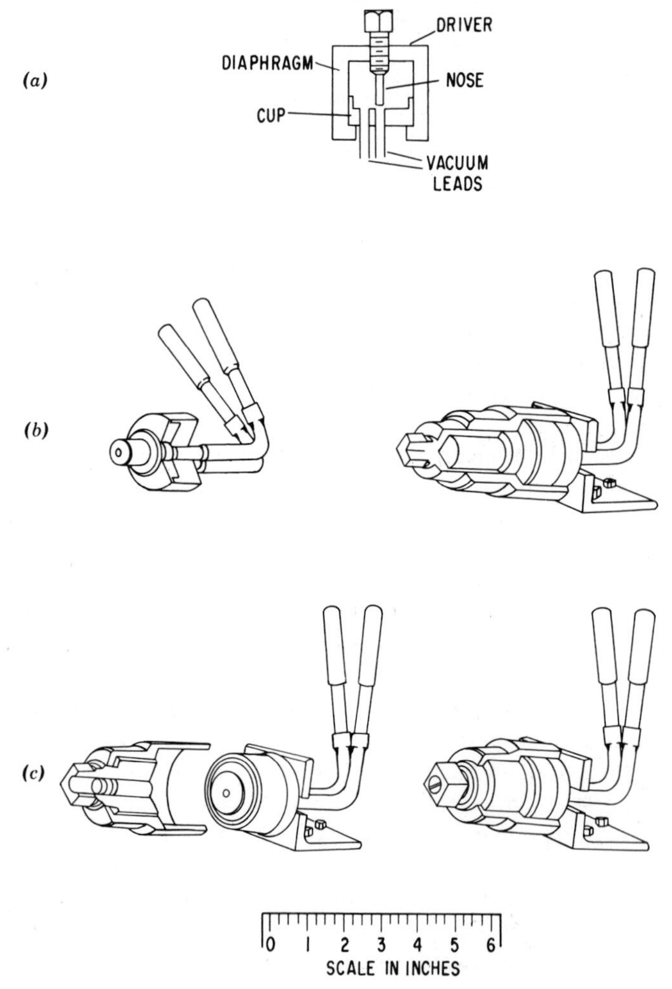


Fig. 9.26. Vacuum valve. (*a*) Schematic diagram. (*b*) Valve with differential screw driving mechanism. (*c*) Valve with driver using simple screw (Alpert).

Fig. 9.27. Granville-Phillips metal valve.

ultrahigh-vacuum systems. There have been many improvements in this general type of valve.[100] A popular commercially available valve made by the Granville-Phillips Corporation of Pullman, Wash., is shown in Fig. 9.27. This particular valve has a conductance variable from 1 liter $sec^{-1}$ to less than $10^{-14}$ liter $sec.^{-1}$

Molten-metal valves are becoming more popular.[99] The advantage of this type of valve over ordinary mercury cut-offs is that at room temperature the metal is a solid having a low vapor pressure. The metal is raised into the seal-off position while the valve is hot and then is allowed to solidify by cooling; this completes the seal. These valves lack some of the convenience of other types but have the advantage of simplicity.

*Demountable seals.* The convenience of a demountable seal is often desired even in an ultrahigh-vacuum device. The usual device consists of a metal gasket crushed between two flanges.[90,101] The gasket metal is selected for its ductility as well as its low vapor pressure at bake-out temperatures. Pattee[102] described a demountable joint which was used in a field emission apparatus. The seal (Fig. 9.28) consists of a knife-edge which seats in an annealed copper gasket. The knife-edge is machined on heavy-wall stainless steel tubing. It has a 0.005-in. radius at the tip and a 45-degree wedge angle. Adjustment is permitted by deformation of the copper by the clamping screws. Gaskets as long as 1 cm have been used successfully.

**Modern Vacuum Systems**

A major requirement of an ultrahigh vacuum system is that it be capable of being baked at high temperatures. This is most conveniently done by

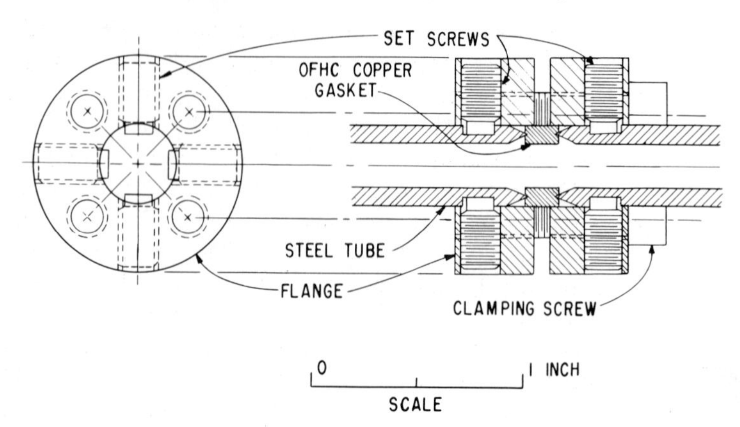


Fig. 9.28. Demountable vacuum seal (Pattee).

having a furnace placed around the entire vacuum system. There are many possible arrangements, and one which has been found convenient in the General Electric Research Laboratory is shown in Fig. 9.29. Two walls of the oven are fixed permanently in place. These can be made of insulated

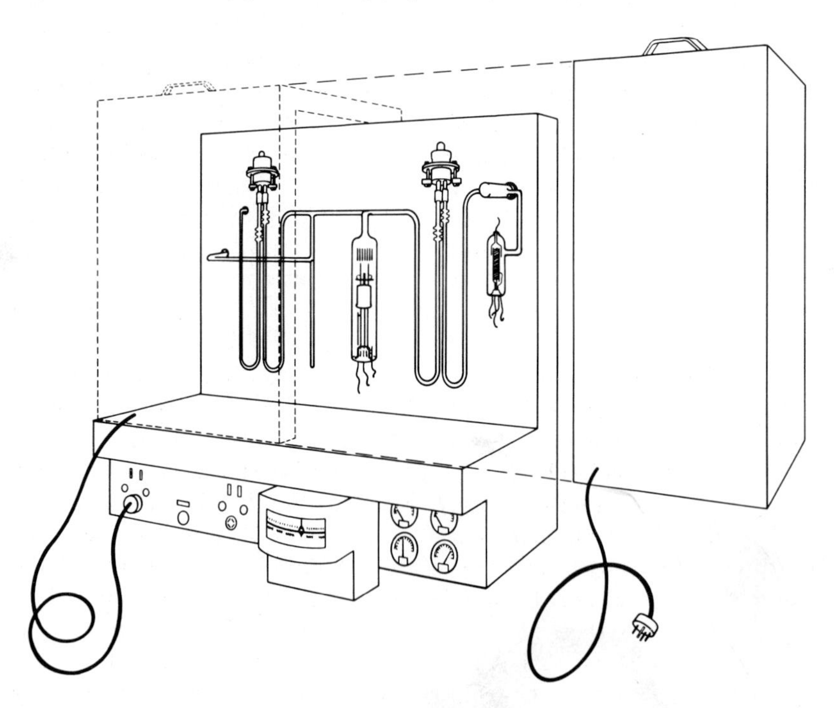

Fig. 9.29. Typical laboratory ultrahigh-vacuum system (Vanderslice).

stainless steel or asbestos board. The vacuum components, such as valves, are attached directly to the surface with flexible glass bellows or long glass U-bends inserted between each fixed point. Light-weight insulated aluminum modular ovens are placed over the system during bake-out, and the temperature is controlled by any convenient pyrometer. Only part of the heating units (Calrods) is controlled by the pyrometer; the others are controlled manually. If it is desired to bring the system quickly to temperature, the manually controlled Calrods are turned on. When the bake-out temperature is approached, these heating units are turned off and the automatically controlled Calrods maintain the temperature in the vicinity of the desired value. The automatically controlled heating units do not have enough power to heat the oven more than 50° C above the desired temperature. For this reason, if the controller fails to function, there is less danger of melting the glassware. In practice, however, the automatically controlled heaters are used to bring the furnace to temperature in a few hours.

The procedure is much the same whether mercury, oil, or ion pumps are used. The system is first pumped to $\sim 10^{-6}$ mm Hg, and a cursory check for leaks is made. The ovens are usually placed over the system, and the automatic controller is set to the desired temperature. The entire system, including the traps, back to the pumps, is baked at 425° C. After 4 or 5 hr at 425° C, the furnace on the trap closest to the pump is shut off. With mercury pumps, the trap is cooled by a Dewar flask filled with liquid nitrogen. When alumina traps are used in conjunction with oil-diffusion pumps, the trap is simply allowed to cool to room temperature. With ion pumps no trap is needed. A common procedure is to use two ion pumps in series, one of which is baked out and the other serves to pump the system during bake-out. When mercury or oil pumps are used, the system is kept hot for approximately 4 hr after the traps have cooled before being allowed to cool slowly to room temperature.

The ovens are removed, the pressure of the system is checked with an ionization gauge, and a check is made for small leaks. The detection of leaks is quite simple in the pressure range below $10^{-8}$ mm Hg. Acetone and carbon tetrachloride, for example, applied to a small leak cause the positive ion-current reading on the ionization gauge to increase suddenly, or they clog the leak temporarily and cause the current to decrease. Helium and hydrogen can also be used with success. Measurement of the rate of rise of pressure also will detect a leak, and by shutting off various sections of the system, the leak can be isolated. Final pinpointing is usually done with the ionization gauge. If no leaks are found, the metal parts are outgassed at high temperatures by electron bombardment or by r-f heating. In some cases a second bake-out is done after the outgassing, and pressures of $10^{-10}$ mm Hg are routinely obtained. Pressures of $10^{-12}$ mm Hg have been achieved in some cases.

## REFERENCES AND NOTES

1. A detailed discussion of the effects of gases on emission from cathodes is given by G. Hermann and S. Wagener, *The Oxide-Coated Cathode*, Vol. 2, Chapman and Hall, London, 1951.

2. J. R. Young, *J. Appl. Phys.*, **30,** 1671 (1959).

3. J. S. Wagener and P. T. Marth, *J. Appl. Phys.*, **28,** 1027 (1957).

4. A. L. Reimann, *Phil. Mag.*, **16,** 673 (1933); **18,** 1117 (1934).

5. L. F. Ehrke and C. M. Slack, *J. Appl. Phys.*, **11,** 129 (1940).

6. V. L. Stout and M. D. Gibbons, *J. Appl. Phys.*, **26,** 1488 (1955).

7. S. Wagener, *Brit. J. Appl. Phys.*, **1,** 225 (1950).

8. P. della Porta, *Vacuum*, **10,** 181 (1960).

9. E. A. Gulbransen, *Rev. Sci. Instr.*, **15,** 201 (1944).

10. *Getters*, Technical Publication of S.A.E.S., Milano, Italy, 1958.

11. S. Wagener, *Proceedings of the National Conference on Tube Techniques*, 1958, p. 1.

12. S. Wagener, *Proc. Phys. Soc.*, **67,** 369 (1954).

13. P. della Porta and F. Ricca, *Transactions of the 5th National Symposium on Vacuum Technology*, Pergamon Press, 1958, p. 25.

14. W. Espe, M. Knoll, and M. P. Wilder, *Electronics*, **23,** 80 (1950).

15. E. A. Lederer and D. H. Wamsley, *RCA Rev.*, **2,** 117 (1937).

16. W. Espe, *Powder Met. Bull.*, **3,** 100 (1956).

17. British Patent Specification No. 755804.

18. V. L. Stout and M. D. Gibbons (to be published).

19. A. D. McQuillian and M. K. McQuillian, *Titanium*, Academic Press, New York, 1956.

20. M. R. Andrews, *J. Am. Chem. Soc.*, **54,** 1845 (1932).

21. Private communication from E. Zubler.

22. V. L. Stout and M. D. Gibbons, U.S. Patent No. 2,926, 981 (1957).

23. I. Langmuir, *J. Am. Chem. Soc.*, **35,** 105 (1913).

24. O. Kubaschewski and B. E. Hopkins, *Oxidation of Metals and Alloys*, Butterworths Scientific Publications, London, 1953.

25. I. Langmuir, *J. Am. Chem. Soc.*, **37,** 417 (1915), and subsequent papers.

26. I. Langmuir, *Trans. A.I.E.E.*, **32,** 1893 (1913).

27. H. L. Johnston and M. K. Walker, *J. Am. Chem. Soc.*, **55,** 187 (1933).

28. J. W. Howell and H. Schroeder, *History of the Incandescent Lamp*, The Maqua Company, Schenectady, N.Y., 1927, p. 131.

29. A very comprehensive report of the earlier work on this topic has been given by E. Pietsch, *Ergeb. der exakt. Naturw.*, **5,** 213 (1926).

30. C. G. Found and J. D. Forney, *A.I.E.E. Trans.*, **47,** 855 (1928).

31. A discussion of the Bohr theory of the origin of spectral lines is given in a large number of treatises, for instance, H. Semat, *Introduction to Atomic Physics*, Rinehart and Co., New York, 1946, and S. Dushman, Chapter II. *Treatise on Physical Chemistry*, Vol. I, H. S. Taylor and S. Glasstone, Editors, D. Van Nostrand Company, Princeton, N.J., 1942.

32. G. K. Wehner, *Phys. Rev.*, **108,** 35 (1957).

33. A. Weiss, L. Heldt, and W. J. Moore, *J. Chem. Phys.*, **29**, 7 (1958).
34. F. O. Rice and T. A. Vanderslice, *J. Am. Chem. Soc.*, **80**, 291 (1958).
35. D. A. Armstrong and C. A. Winkler, *J. Phys. Chem.*, **60**, 1100 (1956).
36. K. B. Blodgett, *J. Chem. Phys.*, **29**, 39 (1958).
37. S. Brodetsky and B. Hodgson, *Phil. Mag.*, **31**, 478 (1916).
38. I. Langmuir, *J. Am. Chem. Soc.*, **35**, 931 (1913).
39. R. S. Willows, *Phil. Mag.*, **1**, 503 (1901).
40. S. E. Hill, *Proc. Phys. Soc. London*, **25**, 35 (1912).
41. R. S. Willows, *Proc. Phys. Soc. London*, **28**, 124 (1915–16).
42. J. H. Leck, *Brit. J. Appl. Phys.*, **6**, 161 (1955).
43. K. Mey, *Ann. Physik*, **11**, 127 (1903).
44. G. Gehlhoff, *Ber. deut. physik. Ges.*, **13**, 271 (1911).
45. R. J. Strutt, *Proc. Roy. Soc. London*, **A, 85**, 219 (1911).
46. G. B. Kistiakowsky and G. G. Volpi, *J. Chem. Phys.*, **27**, 1141 (1957).
47. G. L. Wendt and R. S. Landauer, *J. Am. Chem. Soc.*, **49**, 930 (1920).
48. L. Vegard, *Ann. Physik*, **50**, 769 (1916).
49. H. Alterthum, A. Lompe, and R. Seeliger, *Z. tech. Physik*, **17**, 407 (1936).
50. H. Alterthum and A. Lompe, *Z. tech. Physik*, **19**, 113 (1938).
51. W. Bartholomeyczyk, W. Funk, and R. Seeliger, *Z. Physik*, **117**, 651 (1940/41).
52. W. Bartholomeyczyk, *Ann. Physik*, **42**, 534 (1942/43).
53. R. Seeliger, *Naturwissenschaften*, **30/31**, 461 ( 1942).
54. W. Lumpe and R. Seeliger, *Z. Physik*, **121**, 546 (1943).
55. M. J. Reddan and G. F. Rouse, *Trans. Am. Inst. Elec. Engrs.*, **70**, 1924 (1951).
56. K. B. Blodgett and T. A. Vanderslice, *J. Appl. Phys.*, **31**, 1017 (1960).
57. G. Mierdel, *Ann. Physik*, **85**, 612 (1928).
58. E. Hiedemann, *Ann. Physik*, **8**, 456 (1931).
59. M. C. Johnson, *Proc. Roy. Soc. London*, **A, 123**, 603 (1929).
60. E. H. Siegler and G. H. Dieke, "Clean-up of Rare Gases in Electrodeless Discharges," ONR Contract NONR 248, Task Order VIII, *Tech. Report No.* VIII.
61. J. W. Wheeldon, *Brit. J. Appl. Phys.*, **10**, 295 (1959).
62. A. W. Hull, *Gen. Elec. Rev.*, **32**, 213 (1929).
63. See, for example, Karl R. Spangenberg, *Vacuum Tubes*, McGraw-Hill Book Company, New York, 1948.
64. The data were taken from the following sources: tungsten: H. A. Jones and I. Langmuir, *Gen. Elec. Rev.*, **30**, 310, 354, 408 (1927); molybdenum: A. G. Worthing, *Phys. Rev.*, **28**, 190 (1946); tantalum: M. D. Fiske, *Phys. Rev.*, **61**, 513 (1942), and L. Malter and D. B. Langmuir, *Phys. Rev.*, **55**, 743 (1939); columbium: A. L. Riemann and C. K. Grant, *Phil. Mag.*, **22**, 34 (1936); thoriated tungsten (ThW): S. Dushman.
65. See, for example, J. P. Blewett, *J. Appl. Phys.*, **10**, 668, 831 (1939).
66. N. R. Campbell, *Phil. Mag.*, **41**, 685 (1921); N. R. Campbell and J. W. H. Ryde, *Phil. Mag.*, **40**, 585 (1920). Observations on the clean-up effect in the presence of blue glow have also been reported by C. F. Hagenow, *Phys. Rev.*, **13**, 415 (1919).
67. W. von Meyern, *Z. Physik*, **84**, 531 (1933).
68. Von Helmut Schwarz, *Z. Physik*, **117**, 23 (1940); **122**, 437 (1944).
69. R. N. Bloomer and M. E. Haine, *Vacuum*, **111** (1953).
70. L. J. Varnerin and J. H. Carmichael, *J. Appl. Phys.*, **26**, 782 (1955).
71. J. R. Young, *J. Appl. Phys.*, **27**, 926 (1956).
72. L. J. Varnerin and J. H. Carmichael, *J. Appl. Phys.*, **28**, 913 (1957).
73. E. Brown and J. H. Leck, *Brit. J. Appl. Phys.*, **6**, 161 (1955).
74. L. J. Carmichael and E. A. Trendelenburg, *J. Appl. Phys.*, **29**, 1570 (1958).

75. I. Langmuir, *Phys. Rev.*, **2,** 450 (1913).
76. S. Dushman, *Phys. Rev.*, **5,** 212 (1915).
77. P. A. Anderson, *Phys. Rev.*, **57,** 122 (1940).
78. W. B. Nottingham, *J. Appl. Phys.*, **8,** 762 (1937).
79. W. B. Nottingham, *Phys. Rev.*, **55,** 203 (1939).
80. R. T. Bayard and D. Alpert, *Rev. Sci. Instr.*, **21,** 571 (1950).
81. P. A. Redhead and E. V. Kornelsen, *Vakuum-Technik* (in press).
82. D. Alpert and R. S. Buritz, *J. Appl. Phys.*, **25,** 202 (1954).
83. F. J. Norton, *J. Am. Ceram. Soc.*, **36,** 90 (1953).
84. D. Alpert and R. S. Buritz, *J. Appl. Phys.*, **25,** 202 (1954).
85. J. M. Lafferty, *J. Appl. Phys.*, **22,** 299 (1951).
86. G. W. Sears, *Rev. Sci. Instr.*, **20,** 458 (1949).
87. H. D. Hagstrum, *Phys. Rev.*, **89,** 244 (1953).
88. J. A. Becker and C. D. Hartman, *J. Phys. Chem.*, **57,** 153 (1953).
89. A. Venema and M. Bandringa, *Philips Tech. Rev.*, **20,** 6 (1958).
90. D. Alpert, *J. Appl. Phys.*, **24,** 860 (1953).
91. D. Alpert, *Rev. Sci. Instr.*, **22,** 536 (1951).
92. M. A. Biondi, *Rev. Sci. Instr.*, **30,** 831 (1959).
93. L. A. Harris, *Rev. Sci. Instr.*, **31,** 903 (1960).
94. D. Alpert, *Vacuum*, **9,** 89 (1959).
95. L. D. Hall, *Science*, **128,** 279 (1958).
96. Robert L. Jespen, Arthur B. Francis, Sherman L. Rutherford, and Burdell E. Kietzmann, *Transactions of the 7th National Symposium on Vacuum Technology*, Pergamon Press, p. 45 (1960).
97. K. B. Blodgett and T. A. Vanderslice, Gaseous Electronics Conference, October 1958. [See *Bull. Am. Phys. Soc.*, **4,** 111 (1959).]
98. W. M. Brubaker, *Transactions of the 6th National Symposium on Vacuum Technology*, Pergamon Press, p. 302 (1959).
99. D. Alpert, *Handbuch der Physik*, *Vol. XII*, Springer-Verlag (1958).
100. For example, see Norm N. Axelrod, *Rev. Sci. Instr.*, **30,** 944 (1959).
101. Henrich Hintenberger, *Z. Naturforsch.*, **6a,** 459 (1959).
102. H. H. Pattee, *Rev. Sci. Instr.*, **25,** 1132 (1954).

# 10 Vapor pressures and rates of evaporation

Revised by L. R. Koller

## 10.1. THERMODYNAMICAL RELATIONS

In connection with investigations and industrial developments in the field of high vacua, a knowledge of the vapor pressures of the materials used is often of great importance. Whether the problem concerns metals operated at high temperatures or gases and vapors at very low temperatures, information on the vapor pressures and rates of evaporation is frequently essential in order to be able to interpret results observed and also to guide the experimenter in the choice of metals for specific temperature requirements. This information is also of great importance for the increasing number of technological applications of evaporated films of metals and compounds.

In discussing the available data on materials of interest in high-vacuum technique it has appeared to be most convenient to present this information in several sections dealing, respectively, with metals and carbon, alloys, inorganic compounds, and condensable and noncondensable gases at low temperatures. Thus, the data presented in the first part of this chapter are of interest in vacuum metallurgy, and in those investigations where it is desirable either to raise the temperature to as high a value as possible without obtaining appreciable evaporation or to actually evaporate some of the metal to form a thin deposit. On the other hand, it is of interest in using low-temperature traps to know the vapor pressures, at these temperatures, of gases which it is desired to eliminate from the vacuum system. Such data are presented in the latter part of this chapter.

The methods that have been followed for the determination of vapor pressures have been described fully both in treatises on the subject of physical measurements and in a large number of publications. These methods differ according to the range of pressures under investigation and according to the volatility of the material in the temperature range under investigation. Of special interest in high-vacuum work are data on the temperatures corresponding to vapor pressures below 1 mm of mercury. For this reason most of the data given in this chapter are applicable to this range of pressures.

The methods for the determination of vapor pressures of metals have been reviewed by Ditchburn and Gilmour.[1] Most of them are based upon the application of Knudsen's relation for the rate at which molecules strike a surface, and some of these applications have been discussed in Chapter 1. The methods for the determination of low vapor pressures of oils have been discussed in a paper by Verhoek and Marshall.[2]

For highly refractory metals, such as tungsten, molybdenum, tantalum, and columbium, the vapor pressures have been deduced from determinations of the rate of loss in weight of filaments at constant temperatures. For these metals the principal problem has been the accurate determination of the temperature. An illustration of the extreme precautions that must be taken to secure such accuracy is given in the paper by Marshall, Dornte, and Norton[3] on the evaporation of iron and copper.

As will be discussed more fully in Chapter 11, the most reliable method for correlating experimental observations on vapor-pressure data is based on the calculation of a function known as the *free energy*. Denoting this function by $\Delta F$, the vapor pressure in *atmospheres* may be derived from values of $\Delta F$ (calories per gram-mole) by means of the relation

$$\log P_{atm} = -\frac{\Delta F}{4.575T}. \tag{10.1}$$

In terms of other units of pressure, equation 10.1 takes the forms

$$\log P_{mm} = 2.881 - \frac{\Delta F}{4.575T}, \tag{10.2a}$$

$$\log P_{\mu} = 5.881 - \frac{\Delta F}{4.575T}, \tag{10.2b}$$

and

$$\log P_{\mu b} = 6.0057 - \frac{\Delta F}{4.575T}. \tag{10.2c}$$

In the range of vapor pressures below about 1 mm of mercury it has been observed that, for most materials, the vapor pressure can be represented as a function of $T$ by the relation

$$\log P = A - \frac{B}{T}, \tag{10.3}$$

where $A$ and $B$ are constants.

Comparing this relation with the first three equations, it follows that, in the low-pressure range,

$$\Delta F = L_0 - IT, \tag{10.4}$$

where $L_0$, the latent heat of evaporation at $T = 0$, is determined by the relation

$$L_0 = 4.575B, \tag{10.5}$$

and $I$ is an integration constant. Since $I$ is usually expressed in terms of the pressure in atmospheres,

$$A = \log x + \frac{I}{4.575}, \tag{10.6}$$

where $x$ = conversion factor from atmospheres to the particular unit of pressure in which it is desired to express $P$. Thus, in equation 10.2*a*, log 760 = 2.881.

The great advantage of equation 10.3 is that it makes it possible to calculate very readily values of $T$ corresponding to given values of $P$ by means of the relation

$$T = \frac{B}{A - \log P}. \tag{10.7}$$

From the values of $P$, the rate of evaporation can be derived by means of the following equations:

Let $W$ = rate of evaporation in grams per square centimeter per second, and $M$ = gram-molecular weight. Then,

$$\log W = \bar{5}.6410 + 0.5 \log M + \log P_{\mu b} - 0.5 \log T \tag{10.8a}$$

$$= \bar{5}.7660 + 0.5 \log M + \log P_{\mu} - 0.5 \log T \tag{10.8b}$$

$$= C - 0.5 \log T - \frac{B}{T}, \tag{10.9}$$

where

$$C = A\ (\text{microbars}) + 0.5 \log M + \bar{5}.6410 \tag{10.10a}$$

$$= A\ (\text{microns}) + 0.5 \log M + \bar{5}.7660 \tag{10.10b}$$

Let $\mu$ = rate of evaporation in number of molecules or atoms per square centimeter per second. Then

$$\log \mu = 19.4208 + \log P_{\mu b} - 0.5 \log (MT) \tag{10.11a}$$

$$= 19.5458 + \log P_{\mu} - 0.5 \log (MT) \tag{10.11b}$$

$$= 23.7798 + \log W - \log M. \tag{10.12}$$

For metals for which *rates* of evaporation have been determined directly and can be represented by a relation of the form

$$\log W = A_1 - \frac{B_1}{T}, \tag{10.13}$$

the values of the vapor pressure, in microns, have been calculated by means of equation 10.8*b*, that is,

$$\log P_\mu = 4.2340 + \log W + 0.5 \log \frac{T}{M}. \tag{10.14}$$

From a plot of values of log $P$ versus $1/T$ it was then possible to calculate values for the constants $A$, $B$, and $C$ in equations 10.3 and 10.10*b*.

Since $P_{\mu b} = 1.333 P_\mu$,

$$\log P_{\mu b} = 0.1250 + \log P_\mu. \tag{10.15}$$

In the tables in the following section *values of P are given in terms of microns*.

The latest compilation of vapor-pressure data is that of Honig,[4] and most of the data in Tables 10.1 and 10.2, except as otherwise noted, are based on this compilation. Honig's review is largely based on the three compilations of Kubaschewski and Evans,[5] Stull and Sinke,[6] and Hultgren,[7] respectively. Earlier collections of vapor-pressure data are those of Law,[8] based on prewar data, and of Brewer,[9] which contains additional information obtained during World War II. Kelley[10] gives calculations of the free energy as a function of the temperature for all the metals for which data were available in 1935.

## 10.2. VAPOR-PRESSURE DATA AND RATES OF EVAPORATION FOR METALS AND CARBON

Any compilation of this kind rests on the data of many workers who made measurements by different techniques under widely varying experimental conditions. Accordingly the values given are in no sense final but represent, in the opinion of the compiler, the most reliable ones.

Table 10.1 gives values of the temperature in degrees Centigrade corresponding to pressures of $10^{-2}$, $10^{-1}$, 1, 10, 100, and 1000 microns, as well as the temperatures at the melting and boiling points and the corresponding pressure for 58 elements. On the second line for each element is given $W$, calculated by means of equation 10.8*b*.

A large amount of experimental work has been carried out in the past, with discordant results, on the vaporization of graphite. Some of the uncertainty results from disagreement concerning the state of carbon in the vapor. Species from $C_1$ up to $C_8$[11,12,13] have been found by mass spectrometer, and various investigators list $C_1$, $C_3$, $C_5$, and $C_8$ as dominant. Marshall and Norton[14] estimate that monatomic carbon is 99.4 per cent of the total and that diatomic carbon is less than 0.6 per cent. Their values for vapor pressure, which fall approximately in the middle of later determinations, are about an order of magnitude below those of Honig's shown in Table 10.1.

Values of the constants $A$, $B$ and $C^*$, based on Norton and Marshall's data, are as follows:

$$A = 14.03, \quad B = 38.4 \cdot 10^3, \quad C = 10.336.$$

The value now generally accepted for the heat of sublimation is 170 kcal/mole. A good historical summary of the controversy over the determination of the heat of sublimation of carbon is given by Kern.[15]

A determination of the vapor pressure of titanium by J. Morrison gives values considerably below those in Honig's tabulation. The constants (see Table 10.2), based on Morrison's data, are:

$$A = 13.4728, \quad B = 24.659 \cdot 10^3, \quad C = 10.08.$$

Antimony evaporates largely in the form of molecules, $Sb_2$, and the values of $W$ given for this metal in Table 10.1 are based on this conclusion, that is, $M = 243.52$.

From the values of $T$ given by Honig for the series of values of $P$ in Table 10.1, values were calculated of the constants $A$ and $B$ in equation 10.3 which were in most satisfactory agreement with the data. (It is a pleasure to acknowledge the assistance of Mr. D. S. Story, who made these calculations.) In so far as possible these values were calculated for values of the data within the experimentally determined range. Values of $A$ (for the pressure in *microns*), $B$, and the constant $C$, defined by means of equation 10.10*b*, are given in Table 10.2.

Brewer[9] has calculated values of $T_B$, the boiling point (degrees Kelvin), and of $L$, the heat of vaporization (calories per mole) at $T_B$. It will be observed that these values given in Table 10.3 are uniformly lower than the values of $L_0$ calculated from those of $B$ in Table 10.2, since Brewer deduced his values on the basis of more rigorous arguments. Table 10.3 also gives values of $T_1$ and $T_{10}$, the temperatures (in degrees Kelvin) corresponding to $P_\mu = 1$ and $P_\mu = 10$, respectively. From these data, values have been derived of $T_1/T_B$, $T_{10}/T_B$, and $L/T_B$. The last column in the table gives values of $Z$, the atomic number. The value of $L/T_B$ is designated the Trouton constant. As will be observed, the values of this constant are low for the alkali metals and considerably higher for carbon and the transition elements. For about half the total number of metals $L/T_B = 25 \pm 3$. It is also evident that the values of $T_1/T_B$ and $T_{10}/T_B$ increase with increase in value of $L/T_B$. That this is in accord with the Clapeyron relation may be demonstrated as shown on p. 701.

**TABLE 10.1**

**Temperatures for Given Values of $P_\mu$ and Corresponding Rates of Evaporation**

| Metal | Ref. | Data Temp. Range (° C) | $t$(° C) and $W$ | $P_\mu =$ | | | | | | $t_m$ | $P_\mu$ | $t_B$ |
|---|---|---|---|---|---|---|---|---|---|---|---|---|
| | | | | $10^{-2}$ | $10^{-1}$ | 1 | 10 | 100 | 1000 | | | |
| Li | 4 | 459–1080 | $t$: | 348 | 399 | 460 | 534 | 623 | 737 | 181 | | 1331 |
| | | | $W$: | $6.17 \cdot 10^{-8}$ | $5.93 \cdot 10^{-7}$ | $5.68 \cdot 10^{-6}$ | $5.41 \cdot 10^{-5}$ | $5.13 \cdot 10^{-4}$ | $4.84 \cdot 10^{-3}$ | | | |
| Na | 4 | 264–928 | $t$: | 158 | 195 | 238 | 290 | 355 | 437 | 98 | $10^{-4}$ | 890 |
| | | | $W$: | $1.35 \cdot 10^{-7}$ | $1.29 \cdot 10^{-6}$ | $1.24 \cdot 10^{-5}$ | $1.18 \cdot 10^{-4}$ | $1.12 \cdot 10^{-3}$ | $1.05 \cdot 10^{-2}$ | | | |
| K | 4 | 100–760 | $t$: | 91 | 123 | 162 | 208 | 266 | 341 | 64 | $9 \cdot 10^{-4}$ | 766 |
| | | | $W$: | $1.91 \cdot 10^{-7}$ | $1.83 \cdot 10^{-6}$ | $1.75 \cdot 10^{-5}$ | $1.66 \cdot 10^{-4}$ | $1.57 \cdot 10^{-3}$ | $1.47 \cdot 10^{-2}$ | | | |
| Rb | 4 | ... | $t$: | 64 | 95 | 133 | 176 | 228 | 300 | 39 | $2.6 \cdot 10^{-3}$ | 701 |
| | | | $W$: | $2.94 \cdot 10^{-7}$ | $2.81 \cdot 10^{-6}$ | $2.68 \cdot 10^{-5}$ | $2.55 \cdot 10^{-4}$ | $2.41 \cdot 10^{-3}$ | $2.22 \cdot 10^{-2}$ | | | |
| Cs | 4 | ... | $t$: | 46 | 75 | 110 | 152 | 206 | 277 | 30 | $2.6 \cdot 10^{-3}$ | 685 |
| | | | $W$: | $3.77 \cdot 10^{-7}$ | $3.61 \cdot 10^{-6}$ | $3.44 \cdot 10^{-5}$ | $3.26 \cdot 10^{-4}$ | $3.07 \cdot 10^{-3}$ | $2.87 \cdot 10^{-2}$ | | | |
| Cu | 4 | 969–1606 | $t$: | 942 | 1032 | 1142 | 1272 | 1427 | 1622 | 1084 | 0.26 | 2578 |
| | | | $W$: | $1.33 \cdot 10^{-7}$ | $1.29 \cdot 10^{-6}$ | $1.24 \cdot 10^{-5}$ | $1.18 \cdot 10^{-4}$ | $1.13 \cdot 10^{-3}$ | $1.07 \cdot 10^{-2}$ | | | |
| Ag | 4 | 721–1000 | $t$: | 757 | 832 | 922 | 1032 | 1167 | 1337 | 961 | 2.5 | 2162 |
| | | | $W$: | $1.89 \cdot 10^{-7}$ | $1.82 \cdot 10^{-6}$ | $1.75 \cdot 10^{-5}$ | $1.68 \cdot 10^{-4}$ | $1.60 \cdot 10^{-3}$ | $1.51 \cdot 10^{-2}$ | | | |
| Au | 4 | 727–987 | $t$: | 987 | 1082 | 1197 | 1332 | 1507 | 1707 | 1063 | $5 \cdot 10^{-2}$ | 2709 |
| | | | $W$: | $2.31 \cdot 10^{-7}$ | $2.26 \cdot 10^{-6}$ | $2.14 \cdot 10^{-5}$ | $2.05 \cdot 10^{-4}$ | $1.94 \cdot 10^{-3}$ | $1.84 \cdot 10^{-2}$ | | | |
| Be | 4 | 899–1279 | $t$: | 902 | 987 | 1092 | 1212 | 1367 | 1567 | 1283 | 28 | 2097 |
| | | | $W$: | $5.11 \cdot 10^{-8}$ | $4.93 \cdot 10^{-7}$ | $4.74 \cdot 10^{-6}$ | $4.55 \cdot 10^{-5}$ | $4.33 \cdot 10^{-4}$ | $4.08 \cdot 10^{-3}$ | | | |
| Mg | 4 | 736–1020 | $t$: | 287 | 330 | 382 | 442 | 517 | 612 | 650 | $2.5 \cdot 10^{-3}$ | 1104 |
| | | | $W$: | $1.22 \cdot 10^{-7}$ | $1.17 \cdot 10^{-6}$ | $1.12 \cdot 10^{-5}$ | $1.08 \cdot 10^{-4}$ | $1.02 \cdot 10^{-3}$ | $0.97 \cdot 10^{-2}$ | | | |
| Ca | 4 | 527–647 | $t$: | 402 | 452 | 517 | 592 | 687 | 817 | 850 | $1.9 \cdot 10^{3}$ | 1492 |
| | | | $W$: | $1.42 \cdot 10^{-7}$ | $1.37 \cdot 10^{-6}$ | $1.31 \cdot 10^{-5}$ | $1.26 \cdot 10^{-4}$ | $1.19 \cdot 10^{-3}$ | $1.12 \cdot 10^{-2}$ | | | |
| Sr | 4 | ... | $t$: | 342 | 394 | 456 | 531 | 623 | 742 | 770 | $1.8 \cdot 10^{3}$ | 1367 |
| | | | $W$: | $2.20 \cdot 10^{-7}$ | $2.11 \cdot 10^{-6}$ | $2.02 \cdot 10^{-5}$ | $1.93 \cdot 10^{-4}$ | $1.82 \cdot 10^{-3}$ | $1.71 \cdot 10^{-2}$ | | | |
| Ba | 4 | 1060–1138 | $t$: | 417 | 467 | 537 | 617 | 727 | 867 | 710 | $7 \cdot 10$ | 1637 |
| | | | $W$: | $2.60 \cdot 10^{-7}$ | $2.51 \cdot 10^{-6}$ | $2.40 \cdot 10^{-5}$ | $2.28 \cdot 10^{-4}$ | $2.16 \cdot 10^{-3}$ | $2.03 \cdot 10^{-2}$ | | | |

| | | | | | | | | | | | | |
|---|---|---|---|---|---|---|---|---|---|---|---|---|
| | | | W: | $2.15 \cdot 10^{-7}$ | $2.07 \cdot 10^{-6}$ | $1.99 \cdot 10^{-5}$ | $1.90 \cdot 10^{-4}$ | $1.81 \cdot 10^{-3}$ | $1.71 \cdot 10^{-2}$ | | | |
| Cd | 4 | 200–260 | t: | 149 | 182 | 221 | 267 | 321 | 392 | 321 | $10^{2}$ | 765 |
| | | | W: | $3.01 \cdot 10^{-7}$ | $2.90 \cdot 10^{-6}$ | $2.78 \cdot 10^{-5}$ | $2.66 \cdot 10^{-4}$ | $2.54 \cdot 10^{-3}$ | $2.44 \cdot 10^{-2}$ | | | |
| Hg | 4 | ... | t: | −28 | −8 | 16 | 45 | 81 | 125 | −39 | $2.6 \cdot 10^{-3}$ | 357 |
| | | | W: | $5.28 \cdot 10^{-7}$ | $5.08 \cdot 10^{-6}$ | $4.86 \cdot 10^{-5}$ | $4.63 \cdot 10^{-4}$ | $4.39 \cdot 10^{-3}$ | $4.14 \cdot 10^{-2}$ | | | |
| B | 4 | ... | t: | 1687 | 1827 | 1977 | 2157 | 2377 | 2657 | 2027 | 2 | 3927 |
| | | | W: | $4.33 \cdot 10^{-8}$ | $4.19 \cdot 10^{-7}$ | $4.05 \cdot 10^{-6}$ | $3.89 \cdot 10^{-5}$ | $3.73 \cdot 10^{-4}$ | $3.55 \cdot 10^{-3}$ | | | |
| Al | 4 | 1137–1195 | t: | 882 | 972 | 1082 | 1207 | 1347 | 1547 | 659 | ... | 2447 |
| | | | W: | $8.92 \cdot 10^{-8}$ | $8.59 \cdot 10^{-7}$ | $8.23 \cdot 10^{-6}$ | $7.88 \cdot 10^{-5}$ | $7.53 \cdot 10^{-4}$ | $7.10 \cdot 10^{-3}$ | | | |
| Sc | SD* | ... | t: | 1058 | 1161 | 1282 | 1423 | 1595 | 1804 | 1397 | 6.6 | 2400† |
| Y | SD* | ... | t: | 1249 | 1362 | 1494 | 1649 | 1833 | 2056 | 1477 | 0.76 | |
| La | 4 | 1327–1627 | t: | 1262 | 1377 | 1527 | 1697 | 1897 | 2147 | 920 | ... | 3367 |
| | | | W: | $1.56 \cdot 10^{-7}$ | $1.69 \cdot 10^{-6}$ | $1.62 \cdot 10^{-5}$ | $1.55 \cdot 10^{-4}$ | $1.48 \cdot 10^{-3}$ | $1.40 \cdot 10^{-2}$ | | | |
| Ce | SD* | ... | t: | 1004 | 1091 | 1190 | 1305 | 1439 | 1599 | 785 | $5.5 \cdot 10^{-6}$ | 1400† |
| Nd | 4 | 899–1279 | t: | 957 | 1062 | 1192 | 1342 | 1537 | 1777 | 1024 | $4.2 \cdot 10^{-2}$ | 3087 |
| | | | W: | $2.00 \cdot 10^{-7}$ | $1.92 \cdot 10^{-6}$ | $1.83 \cdot 10^{-5}$ | $1.74 \cdot 10^{-4}$ | $1.65 \cdot 10^{-3}$ | $1.55 \cdot 10^{-2}$ | | | |
| Ga | 4 | 957–1245 | t: | 757 | 842 | 937 | 1057 | 1197 | 1372 | 37 | ... | 2237 |
| | | | W: | $1.52 \cdot 10^{-7}$ | $1.46 \cdot 10^{-6}$ | $1.40 \cdot 10^{-5}$ | $1.34 \cdot 10^{-4}$ | $1.27 \cdot 10^{-3}$ | $1.22 \cdot 10^{-2}$ | | | |
| In | 4 | 727–1075 | t: | 670 | 747 | 837 | 947 | 1077 | 1242 | 156 | ... | 2091 |
| | | | W: | $2.04 \cdot 10^{-7}$ | $1.96 \cdot 10^{-6}$ | $1.88 \cdot 10^{-5}$ | $1.79 \cdot 10^{-4}$ | $1.70 \cdot 10^{-3}$ | $1.61 \cdot 10^{-2}$ | | | |
| Tl | 4 | ... | t: | 412 | 468 | 535 | 615 | 713 | 837 | 304 | $3 \cdot 10^{-5}$ | 1467 |
| | | | W: | $3.19 \cdot 10^{-7}$ | $3.06 \cdot 10^{-6}$ | $2.93 \cdot 10^{-5}$ | $2.80 \cdot 10^{-4}$ | $2.66 \cdot 10^{-3}$ | $2.50 \cdot 10^{-2}$ | | | |
| C | 4 | 2084–2597 | t: | 1977 | 2107 | 2247 | 2427 | 2627 | 2867 | ... | ... | (3927) |
| | | | W: | $4.27 \cdot 10^{-8}$ | $4.14 \cdot 10^{-7}$ | $4.03 \cdot 10^{-6}$ | $3.89 \cdot 10^{-5}$ | $3.76 \cdot 10^{-4}$ | $3.61 \cdot 10^{-3}$ | | | |
| Si | 4 | ... | t: | 1177 | 1282 | 1357 | 1547 | 1717 | 1927 | 1415 | 1 | 2787 |
| | | | W: | $8.12 \cdot 10^{-8}$ | $7.84 \cdot 10^{-7}$ | $7.54 \cdot 10^{-6}$ | $7.24 \cdot 10^{-5}$ | $6.93 \cdot 10^{-4}$ | $6.59 \cdot 10^{-3}$ | | | |
| Ti | Morrison | 1111–1323 | t: | 1321 | 1431 | 1558 | 1703 | 1877 | 2083 | 1800 | ... | >3000† |
| | | | W: | $1.01 \cdot 10^{-7}$ | $0.98 \cdot 10^{-6}$ | $0.94 \cdot 10^{-5}$ | $0.90 \cdot 10^{-4}$ | $0.86 \cdot 10^{-3}$ | $0.82 \cdot 10^{-2}$ | | | |
| Zr | 4 | 1676–1781 | t: | 1837 | 2002 | 2187 | 2397 | 2647 | 2977 | 1852 | $10^{-2}$ | 4405 |
| | | | W: | $1.21 \cdot 10^{-7}$ | $1.17 \cdot 10^{-6}$ | $1.12 \cdot 10^{-5}$ | $1.08 \cdot 10^{-4}$ | $1.03 \cdot 10^{-3}$ | $0.98 \cdot 10^{-2}$ | | | |
| Th | SD* | ... | t: | 1686 | 1831 | 1999 | 2196 | 2431 | 2715 | 1827 | $9.3 \cdot 10^{-2}$ | 4500† |
| | | | W: | $2.01 \cdot 10^{-7}$ | $1.94 \cdot 10^{-6}$ | $1.86 \cdot 10^{-5}$ | $1.79 \cdot 10^{-4}$ | $1.71 \cdot 10^{-3}$ | $1.63 \cdot 10^{-2}$ | | | |
| Ge | 4 | 1237–1612 | t: | 1037 | 1142 | 1262 | 1407 | 1582 | 1797 | 937 | $7 \cdot 10^{-4}$ | 2827 |
| | | | W: | $1.37 \cdot 10^{-7}$ | $1.32 \cdot 10^{-6}$ | $1.27 \cdot 10^{-5}$ | $1.21 \cdot 10^{-4}$ | $1.15 \cdot 10^{-3}$ | $1.09 \cdot 10^{-2}$ | | | |

**TABLE 10.1** *(Continued)*

| Metal | Ref. | Data Temp. Range (° C) | $t$(° C) and $W$ | $P_\mu =$ $10^{-2}$ | $10^{-1}$ | 1 | 10 | 100 | 1000 | $t_m$ | $P_\mu$ | $t_B$ |
|---|---|---|---|---|---|---|---|---|---|---|---|---|
| Sn | 4 | 1151–1415 | $t$: | 882 | 977 | 1092 | 1227 | 1397 | 1612 | 232 | ... | 2679 |
| | | | $W$: | $1.87 \cdot 10^{-7}$ | $1.80 \cdot 10^{-6}$ | $1.72 \cdot 10^{-5}$ | $1.64 \cdot 10^{-4}$ | $1.56 \cdot 10^{-3}$ | $1.46 \cdot 10^{-2}$ | | | |
| Pb | 4 | ... | $t$: | 487 | 551 | 627 | 719 | 832 | 977 | 328 | ... | 1751 |
| | | | $W$: | $3.05 \cdot 10^{-7}$ | $2.93 \cdot 10^{-6}$ | $2.80 \cdot 10^{-5}$ | $2.67 \cdot 10^{-4}$ | $2.53 \cdot 10^{-3}$ | $2.38 \cdot 10^{-2}$ | | | |
| V | 4 | 1389–1609 | $t$: | 1432 | 1551 | 1687 | 1847 | 2037 | 2287 | 1857 | 10 | 3377 |
| | | | $W$: | $1.01 \cdot 10^{-7}$ | $0.98 \cdot 10^{-6}$ | $0.94 \cdot 10^{-5}$ | $0.90 \cdot 10^{-4}$ | $0.87 \cdot 10^{-3}$ | $0.82 \cdot 10^{-2}$ | | | |
| Cb | SD* | ... | $t$: | 2194 | 2355 | 2539 | ... | ... | ... | 2500* | 0.6 | 3700† |
| | | | $W$: | $1.16 \cdot 10^{-7}$ | $1.08 \cdot 10^{-6}$ | $1.06 \cdot 10^{-5}$ | ... | ... | ... | | | |
| Ta | 4 | 1727–2997 | $t$: | 2397 | 2587 | 2807 | 3067 | 3372 | 3737 | 2997 | 6 | 5427 |
| | | | $W$: | $1.52 \cdot 10^{-7}$ | $1.47 \cdot 10^{-6}$ | $1.41 \cdot 10^{-5}$ | $1.36 \cdot 10^{-4}$ | $1.30 \cdot 10^{-3}$ | $1.24 \cdot 10^{-2}$ | | | |
| $P_4$ | 4 | ... | $t$: | 107 | 130 | 157 | 187 | 222 | 262 | 597 | ... | 431 |
| | | | $W$: | $3.3 \cdot 10^{-7}$ | $3.24 \cdot 10^{-6}$ | $3.13 \cdot 10^{-5}$ | $3.03 \cdot 10^{-4}$ | $2.92 \cdot 10^{-3}$ | $2.81 \cdot 10^{-2}$ | | | |
| $Sb_2$ | 4 | ... | $t$: | 382 | 427 | 477 | 542 | 617 | 757 | 630 | $10^2$ | 1637 |
| | | | $W$: | $2.52 \cdot 10^{-7}$ | $2.43 \cdot 10^{-6}$ | $2.35 \cdot 10^{-5}$ | $2.26 \cdot 10^{-4}$ | $2.16 \cdot 10^{-3}$ | $2.01 \cdot 10^{-2}$ | | | |
| Bi | 4 | 409–497 | $t$: | 450 | 508 | 578 | 661 | 762 | 892 | 271 | ... | 1559 |
| | | | $W$: | $3.14 \cdot 10^{-7}$ | $3.02 \cdot 10^{-6}$ | $2.89 \cdot 10^{-5}$ | $2.76 \cdot 10^{-4}$ | $2.62 \cdot 10^{-3}$ | $2.47 \cdot 10^{-2}$ | | | |
| Cr | 4 | 889–1288 | $t$: | 1062 | 1162 | 1267 | 1392 | 1557 | 1737 | 1903 | $5.5 \cdot 10^{-3}$ | 2665 |
| | | | $W$: | $1.15 \cdot 10^{-7}$ | $1.11 \cdot 10^{-6}$ | $1.07 \cdot 10^{-5}$ | $1.03 \cdot 10^{-4}$ | $0.98 \cdot 10^{-3}$ | $0.94 \cdot 10^{-2}$ | | | |
| Mo | 4 | 1797–2231 | $t$: | 1987 | 2167 | 2377 | 2627 | 2927 | 3297 | 2577 | 8 | 4827 |
| | | | $W$: | $1.20 \cdot 10^{-7}$ | $1.16 \cdot 10^{-6}$ | $1.11 \cdot 10^{-5}$ | $1.06 \cdot 10^{-4}$ | $1.01 \cdot 10^{-3}$ | $0.95 \cdot 10^{-2}$ | | | |
| W | 4 | ... | $t$: | 2547 | 2757 | 3007 | 3297 | 3647 | ... | 3377 | $2 \cdot 10$ | 5527 |
| | | | $W$: | $1.49 \cdot 10^{-7}$ | $1.44 \cdot 10^{-6}$ | $1.38 \cdot 10^{-5}$ | $1.32 \cdot 10^{-4}$ | $1.26 \cdot 10^{-3}$ | ... | | | |
| U | 4 | 1357–1697 | $t$: | 1442 | 1582 | 1737 | 1927 | 2157 | 2447 | 1130 | $10^{-5}$ | 3927 |
| | | | $W$: | $2.17 \cdot 10^{-7}$ | $2.09 \cdot 10^{-6}$ | $2.01 \cdot 10^{-5}$ | $1.92 \cdot 10^{-4}$ | $1.83 \cdot 10^{-3}$ | $1.73 \cdot 10^{-2}$ | | | |

| | | | | | | | | | | | | |
|---|---|---|---|---|---|---|---|---|---|---|---|---|
| Se | 4 | ... | $t$: | 144 | 167 | 197 | 232 | 277 | 347 | 217 | 3.2 | 685 |
| | | | $W$: | $2.54 \cdot 10^{-7}$ | $2.47 \cdot 10^{-6}$ | $2.39 \cdot 10^{-5}$ | $2.31 \cdot 10^{-4}$ | $2.21 \cdot 10^{-3}$ | $2.08 \cdot 10^{-2}$ | | | |
| $Te_2$ | 4 | ... | $t$: | 261 | 296 | 336 | 383 | 438 | 520 | 450 | $1.6 \cdot 10^{-2}$ | 987 |
| | | | $W$: | $4.03 \cdot 10^{-7}$ | $3.91 \cdot 10^{-6}$ | $3.78 \cdot 10^{-5}$ | $3.64 \cdot 10^{-4}$ | $3.50 \cdot 10^{-3}$ | $3.31 \cdot 10^{-2}$ | | | |
| Po | 4 | 438–745 | $t$: | 187 | 220 | 263 | 314 | 382 | 472 | 254 | $8 \cdot 10^{-1}$ | 962 |
| | | | $W$: | $3.94 \cdot 10^{-7}$ | $3.81 \cdot 10^{-6}$ | $3.65 \cdot 10^{-5}$ | $3.49 \cdot 10^{-4}$ | $3.30 \cdot 10^{-3}$ | $3.10 \cdot 10^{-2}$ | | | |
| Mn | 4 | ... | $t$: | 697 | 767 | 852 | 947 | 1067 | 1227 | 1244 | $1.2 \cdot 10^{-3}$ | 2051 |
| | | | $W$: | $1.39 \cdot 10^{-7}$ | $1.34 \cdot 10^{-6}$ | $1.29 \cdot 10^{-5}$ | $1.24 \cdot 10^{-4}$ | $1.18 \cdot 10^{-3}$ | $1.12 \cdot 10^{-2}$ | | | |
| Re | 4 | ... | $t$: | 2367 | 2557 | 2787 | 3057 | 3397 | | 3180 | 25 | 5627 |
| | | | $W$: | $1.55 \cdot 10^{-7}$ | $1.50 \cdot 10^{-6}$ | $1.44 \cdot 10^{-5}$ | $1.38 \cdot 10^{-4}$ | $1.31 \cdot 10^{-3}$ | ... | | | |
| Fe | 4 | 1092–1246 | $t$: | 1107 | 1207 | 1322 | 1467 | 1637 | 1847 | 1539 | 30 | 2857 |
| | | | $W$: | $1.17 \cdot 10^{-7}$ | $1.13 \cdot 10^{-6}$ | $1.09 \cdot 10^{-5}$ | $1.05 \cdot 10^{-4}$ | $0.99 \cdot 10^{-3}$ | $0.95 \cdot 10^{-2}$ | | | |
| Co | 4 | 1090–1249 | $t$: | 1162 | 1262 | 1377 | 1517 | 1697 | 1907 | 1495 | 67 | 2877 |
| | | | $W$: | $1.18 \cdot 10^{-7}$ | $1.14 \cdot 10^{-6}$ | $1.10 \cdot 10^{-5}$ | $1.06 \cdot 10^{-4}$ | $1.01 \cdot 10^{-3}$ | $0.96 \cdot 10^{-2}$ | | | |
| Ni | 4 | 1034–1310 | $t$: | 1142 | 1247 | 1357 | 1497 | 1667 | 1877 | 1452 | 5 | 2839 |
| | | | $W$: | $1.19 \cdot 10^{-7}$ | $1.15 \cdot 10^{-6}$ | $1.11 \cdot 10^{-5}$ | $1.06 \cdot 10^{-4}$ | $1.01 \cdot 10^{-3}$ | $0.96 \cdot 10^{-2}$ | | | |
| Ru | SD* | ... | $t$: | 1913 | 2058 | 2230 | 2431 | 2666 | 2946 | 2427 | 9.8 | ... |
| | | | $W$: | $1.26 \cdot 10^{-7}$ | $1.22 \cdot 10^{-6}$ | $1.18 \cdot 10^{-5}$ | $1.13 \cdot 10^{-4}$ | $1.08 \cdot 10^{-3}$ | $1.04 \cdot 10^{-2}$ | | | |
| Rh | 4 | ... | $t$: | 1587 | 1707 | 1857 | 2027 | 2247 | 2527 | 1966 | 5.5 | (3727) |
| | | | $W$: | $1.37 \cdot 10^{-7}$ | $1.33 \cdot 10^{-6}$ | $1.28 \cdot 10^{-5}$ | $1.23 \cdot 10^{-4}$ | $1.18 \cdot 10^{-3}$ | $1.12 \cdot 10^{-2}$ | | | |
| Pd | 4 | ... | $t$: | 1157 | 1262 | 1387 | 1547 | 1727 | 1967 | 1550 | 11 | 3127 |
| | | | $W$: | $1.59 \cdot 10^{-7}$ | $1.54 \cdot 10^{-6}$ | $1.48 \cdot 10^{-5}$ | $1.41 \cdot 10^{-4}$ | $1.35 \cdot 10^{-3}$ | $1.27 \cdot 10^{-2}$ | | | |
| Os | SD | ... | $t$: | 2101 | 2264 | 2451 | 2667 | 2920 | 3221 | 2697 | ... | >5300† |
| | | | $W$: | $1.65 \cdot 10^{-7}$ | $1.60 \cdot 10^{-6}$ | $1.54 \cdot 10^{-5}$ | $1.48 \cdot 10^{-4}$ | $1.42 \cdot 10^{-3}$ | $1.36 \cdot 10^{-2}$ | | | |
| Ir | 4 | ... | $t$: | 1797 | 1947 | 2107 | 2307 | 2527 | 2827 | 2454 | 50 | (4127) |
| | | | $W$: | $1.78 \cdot 10^{-7}$ | $1.72 \cdot 10^{-6}$ | $1.66 \cdot 10^{-5}$ | $1.60 \cdot 10^{-4}$ | $1.53 \cdot 10^{-3}$ | $1.46 \cdot 10^{-2}$ | | | |
| Pt | 4 | ... | $t$: | 1602 | 1742 | 1907 | 2077 | 2317 | 2587 | 1770 | $1.7 \cdot 10^{-1}$ | 3827 |
| | | | $W$: | $1.88 \cdot 10^{-7}$ | $1.82 \cdot 10^{-6}$ | $1.75 \cdot 10^{-5}$ | $1.68 \cdot 10^{-4}$ | $1.60 \cdot 10^{-3}$ | $1.52 \cdot 10^{-2}$ | | | |

* SD = Dushman, 1st edition of this book.
† *Handbook of Chemistry and Physics*, 39th edition.

TABLE 10.2

Constants in Relations for Evaporation of Metals*

| Metal | $A$† | $10^{-3} \cdot B$ | $C$ |
|---|---|---|---|
| Li | 10.99 | 8.07 | 7.18 |
| Na | 10.72 | 5.49 | 7.17 |
| K | 10.28 | 4.48 | 6.84 |
| Rb | 10.11 | 4.08 | 6.84 |
| Cs | 9.91 | 3.80 | 6.74 |
| Cu | 11.96 | 16.98 | 8.63 |
| Ag | 11.85 | 14.27 | 8.63 |
| Au | 11.89 | 17.58 | 8.80 |
| Be | 12.01 | 16.47 | 8.25 |
| Mg | 11.64 | 7.65 | 8.10 |
| Ca | 11.22 | 8.94 | 7.79 |
| Sr | 10.71 | 7.83 | 7.45 |
| Ba | 10.70 | 8.76 | 7.54 |
| Zn | 11.63 | 6.56 | 8.30 |
| Cd | 11.56 | 5.72 | 8.35 |
| B | 13.07 | 29.62 | 9.36 |
| Al | 11.79 | 15.94 | 8.27 |
| Sc* | 11.94 | 18.57 | 8.53 |
| Y* | 12.43 | 21.97 | 9.17 |
| La | 11.60 | 20.85 | 8.44 |
| Ce* | 13.74 | 20.10 | 10.58 |
| Ga | 11.41 | 13.84 | 8.09 |
| In | 11.23 | 12.48 | 8.03 |
| Tl | 11.07 | 8.96 | 7.99 |
| C | 15.73 | 40.03 | 12.04 |
| Si | 12.72 | 21.30 | 9.21 |
| Ti | 12.50 | 23.23 | 9.11 |
| Zr | 12.33 | 30.26 | 9.08 |
| Th* | 12.52 | 28.44 | 9.47 |
| Ge | 11.71 | 18.03 | 8.40 |
| Sn | 10.88 | 14.87 | 7.69 |
| Pb | 10.77 | 9.71 | 7.69 |
| V | 13.07 | 25.72 | 9.69 |
| Cb* | 14.37 | 40.40 | 11.12 |
| Ta | 13.04 | 40.21 | 9.93 |

TABLE 10.2 (*Continued*)

| Metal | $A$† | $10^{-3} \cdot B$ | $C$ |
|---|---|---|---|
| $Sb_2$ | 11.15 | 8.63 | 7.96 |
| Bi | 11.18 | 9.53 | 8.10 |
| Cr | 12.94 | 20.00 | 9.56 |
| Mo | 11.64 | 30.85 | 8.40 |
| W | 12.40 | 40.68 | 9.30 |
| U | 11.59 | 23.31 | 8.54 |
| Mn | 12.14 | 13.74 | 8.77 |
| Fe | 12.44 | 19.97 | 9.08 |
| Co | 12.70 | 21.11 | 9.35 |
| Ni | 12.75 | 20.96 | 9.40 |
| Ru* | 13.50 | 33.80 | 10.27 |
| Rh | 12.94 | 27.72 | 9.71 |
| Pd | 11.78 | 19.71 | 8.56 |
| Os* | 13.59 | 37.00 | 10.50 |
| Ir | 13.07 | 31.23 | 9.98 |
| Pt | 12.53 | 27.28 | 9.44 |

* Dushman, 1st edition of this book.
† Values of $A$ for the pressure in microns.

From the second law of thermodynamics it follows that

$$\frac{d \ln P}{d\,(1/T)} = -\frac{L}{R_0}. \tag{10.16}$$

Assuming that $L$ is a constant, this relation becomes

$$\frac{\log P_{\mu B} - \log P_{\mu}}{1/T - 1/T_B} = \frac{L}{4.574}, \tag{10.17}$$

where $P_{\mu B} = 7.60 \cdot 10^5$, and $\log P_{\mu B} = 5.881$.

Hence, for $P_{\mu} = 1$,

$$\frac{L/T_B}{26.71 + L/T_B} = \frac{T_1}{T_B}, \tag{10.18}$$

and for $P_{\mu} = 10$,

$$\frac{L/T_B}{22.14 + L/T_B} = \frac{T_{10}}{T_B}. \tag{10.19}$$

For $L/T_B = 25$, $T_1/T_B = 0.483$ and $T_{10}/T_B = 0.530$.

Table 10.4 gives a number of illustrations of the application of equations 10.18 and 10.19; and, as will be observed, the agreement between the values thus calculated and those given in Table 10.3 is good, especially in

## TABLE 10.3

**Absolute Temperatures Corresponding to Three Values of the Vapor Pressure, also Values of the Latent Heat of Evaporation, and of the Trouton Constant**

| Element | $T_1$ | $T_{10}$ | $T_B$ | $T_1/T_B$ | $T_{10}/T_B$ | $L$ | $L/T_B$ | $Z$ |
|---|---|---|---|---|---|---|---|---|
| Li | 712 | 787 | 1640 | 0.4342 | 0.4799 | 32,480 | 19.81 | 3 |
| Na | 511 | 564 | 1187 | 0.4304 | 0.4751 | 23,120 | 19.47 | 11 |
| K | 434 | 480 | 1052 | 0.4128 | 0.4562 | 18,880 | 17.95 | 19 |
| Rb | 396 | 438 | 952 | 0.4160 | 0.4601 | 18,110 | 19.02 | 37 |
| Cs | 383 | 426 | 963 | 0.3977 | 0.4424 | 16,320 | 16.94 | 55 |
| Cu | 1414 | 1546 | 2868 | 0.4929 | 0.5391 | 72,810 | 25.39 | 29 |
| Ag | 1209 | 1320 | 2485 | 0.4863 | 0.5311 | 60,720 | 24.42 | 47 |
| Au | 1589 | 1738 | 3239 | 0.4907 | 0.5365 | 81,800 | 25.25 | 79 |
| Be | 1403 | 1519 | 2780 | 0.5047 | 0.5464 | 73,900 | 26.59 | 4 |
| Mg | 656 | 716 | 1399 | 0.4688 | 0.5117 | 31,470 | 22.49 | 12 |
| Ca | 801 | 878 | 1755 | 0.4564 | 0.5003 | 36,740 | 20.94 | 20 |
| Sr | 748 | 822 | 1657 | 0.4514 | 0.4962 | 33,610 | 20.28 | 38 |
| Ba | 819 | 902 | 1911 | 0.4286 | 0.4721 | 35,665 | 18.66 | 56 |
| Zn | 565 | 616 | 1180 | 0.4787 | 0.5220 | 27,880 | 23.62 | 30 |
| Cd | 493 | 537 | 1038 | 0.4750 | 0.5175 | 23,870 | 23.00 | 48 |
| Hg | 291 | 321 | 634 | 0.4590 | 0.5063 | 13,980 | 22.06 | 0 |
| B | 1512 | 1628 | 2800 | 0.5400 | 0.5813 | 75,000 | 26.79 | 85 |
| Al | 1162 | 1269 | 2600 | 0.4470 | 0.4881 | 67,950 | 26.13 | 13 |
| Sc | 1555 | 1696 | 3000 | 0.5183 | 0.5654 | 80,000 | 26.67 | 21 |
| Y | 1767 | 1922 | 3500 | 0.5048 | 0.5618 | 90,000 | 25.71 | 39 |
| La | 1515 | 1654 | 3000 | 0.5050 | 0.5513 | 80,000 | 26.67 | 57 |
| Ce | 1463 | 1578 | 2800 | 0.5225 | 0.5635 | 73,000 | 26.07 | 58 |
| Ga | 1238 | 1366 | 2700 | 0.4584 | 0.5058 | 60,000 | 22.22 | 31 |
| In | 1113 | 1225 | 2440 | 0.4560 | 0.5020 | 53,800 | 22.05 | 49 |
| Tl | 800 | 879 | 1730 | 0.4625 | 0.5082 | 38,810 | 22.44 | 81 |
| C | 2744 | 2954 | 4775 | 0.5746 | 0.6186 | 180,000 | 37.70 | 6 |
| Si | 1496 | 1616 | 2750 | 0.5434 | 0.5876 | 71,000 | 25.82 | 14 |
| Ti | 1657 | 1819 | 3400 | 0.4873 | 0.5350 | 84,000 | 24.71 | 22 |
| Zr | 2089 | 2274 | 3850 | 0.5426 | 0.5906 | 120,000 | 31.17 | 40 |
| Th | 2272 | 2469 | 4500 | 0.5049 | 0.5487 | 130,000 | 28.89 | 90 |
| Ge | 1385 | 1524 | 2980 | 0.4648 | 0.5113 | 68,000 | 22.82 | 32 |
| Sn | 1315 | 1462 | 3000 | 0.4383 | 0.4874 | 70,000 | 23.33 | 50 |
| Pb | 898 | 991 | 2010 | 0.4468 | 0.4930 | 42,400 | 21.10 | 82 |
| V | 1998 | 2161 | 3800 | 0.5259 | 0.5687 | 106,000 | 27.89 | 23 |
| Cb | 2812 | ... | 5400 | 0.5207 | ... | 155,000 | 28.77 | 41 |
| Ta | 3093 | ... | 6300 | 0.4910 | ... | 183,000 | 29.05 | 73 |
| $Sb_2$ | 868 | 951 | 1890 | 0.4592 | 0.5031 | 40,000 | 21.16 | 51 |
| Bi | 882 | 971 | 1900 | 0.4641 | 0.5110 | 42,600 | 22.42 | 83 |
| Cr | 1363 | 1478 | 2495 | 0.5463 | 0.5924 | 72,970 | 29.24 | 24 |

TABLE 10.3 (*Continued*)

| Element | $T_1$ | $T_{10}$ | $T_B$ | $T_1/T_B$ | $T_{10}/T_B$ | $L$ | $L/T_B$ | $Z$ |
|---|---|---|---|---|---|---|---|---|
| Mo | 2568 | 2806 | 5077 | 0.5058 | 0.5527 | 128,420 | 25.29 | 42 |
| W | 3289 | 3582 | 5950 | 0.5529 | 0.6020 | 184,580 | 30.12 | 74 |
| U | 2003 | 2171 | 3800 | 0.5270 | 0.5714 | 110,000 | 28.95 | 92 |
| Mn | 1151 | 1253 | 2370 | 0.4857 | 0.5286 | 54,380 | 22.95 | 25 |
| Fe | 1583 | 1720 | 3008 | 0.5264 | 0.5719 | 84,620 | 30.94 | 26 |
| Co | 1767 | 1922 | 3370 | 0.5243 | 0.5703 | 93,000 | 27.60 | 27 |
| Ni | 1644 | 1783 | 3110 | 0.5285 | 0.5732 | 90,480 | 29.09 | 28 |
| Ru | 2503 | 2704 | 4500 | 0.5562 | 0.6011 | 148,000 | 32.90 | 44 |
| Rh | 2244 | 2422 | 4150 | 0.5408 | 0.5837 | 127,000 | 30.61 | 45 |
| Pd | 1678 | 1839 | 3440 | 0.4877 | 0.5346 | 89,000 | 25.87 | 46 |
| Os | 2724 | 2940 | 4900 | 0.5559 | 0.5999 | 162,000 | 33.06 | 76 |
| Ir | 2613 | 2829 | 4800 | 0.5444 | 0.5893 | 152,000 | 31.66 | 77 |
| Pt | 2177 | 2363 | 4100 | 0.5310 | 0.5761 | 122,000 | 29.76 | 78 |

view of the fact that, actually, $L$ decreases with increase in $T$. However, the decrease in value in the range $T = T_1$ to $T = T_B$ does not, in general, exceed more than a few per cent.

TABLE 10.4

**Relation between the Value of $L/T_B$ and Values of $T_1/T_B$ and $T_{10}/T_B$**

| Element | $L/T_B$ | $T_1/T_B$ | | $T_{10}/T_B$ | |
|---|---|---|---|---|---|
| | | Eq. 10.18 | Table 10.3 | Eq. 10.19 | Table 10.3 |
| Cs | 16.94 | 0.388 | 0.398 | 0.434 | 0.442 |
| Li | 19.81 | 0.426 | 0.434 | 0.472 | 0.480 |
| Pb | 21.10 | 0.441 | 0.447 | 0.488 | 0.493 |
| Sn | 23.33 | 0.466 | 0.438 | 0.513 | 0.487 |
| Cu | 25.39 | 0.487 | 0.493 | 0.534 | 0.539 |
| V | 27.89 | 0.511 | 0.526 | 0.557 | 0.569 |
| W | 30.12 | 0.530 | 0.553 | 0.576 | 0.602 |
| Ru | 32.90 | 0.552 | 0.556 | 0.598 | 0.601 |
| C | 37.70 | 0.585 | 0.575 | 0.630 | 0.619 |

## 10.3. GENERAL REMARKS ON VACUUM DISTILLATION OF METALS AND DEPOSITION OF FILMS

For the distillation of a metal in a vacuum, in order to purify it, Kroll[16] considers 1.8 mm of mercury as suitable for commercial operation. The temperature required for this purpose may be obtained from Table 10.1; it will be observed that for some metals this pressure is attained below the melting temperature (or triple point)—that is, the solid sublimes—whereas other metals must be in the liquid state, at a temperature considerably above that of fusion.

Under these conditions, as will be discussed in the next section, nonvolatile metals can be purified of more-volatile constituents. For instance, tin can be freed from lead, since the temperature at which the pressure is 1 mm is about 977° C for lead and about 1612° C for tin.

Also, during the process of distillation (or sublimation) dissolved gases are evolved, and some compounds such as hydrides, nitrides, or oxides may be dissociated. (See the discussion of dissociation pressures in Chapter 11.)

The deposition of thin films by vacuum evaporation has received considerable attention from a large number of investigators. Important early contributions towards the development of a satisfactory technique, especially for the coating of glass with aluminum, were made by Strong[17] and Cartwright.[18] Since then, methods have been developed for evaporating films of most of the metallic elements which are below their melting points at room temperature.

The deposition of nonmetals and compounds has become increasingly important both in science and in technology. The deposition of films by evaporation in a good vacuum is now utilized for a wide variety of purposes. These include the formation of highly reflecting metal mirrors,[19] inorganic[20] protective coatings, interference filters,[21] luminescent films,[22] photoconductors,[23] and semiconductors.[24]

Some of these applications are carried out on a large scale commercially. Examples are metal coating of capacitor tissue, ornamental metal coatings, and optical nonreflecting coatings. Evaporation of films has also become an important step in the fabrication of many electronic devices.

The literature on the subject is quite extensive. It has been treated very comprehensively in the book by Holland.[25] The formation and general physical properties of thin films have been discussed by Mayer,[26] while the optical properties have been covered by Heavens.[27]

In general, deposition is carried out in a vacuum of 0.1 micron or better and at temperatures $t_{10}$ (or $T_{10}$), corresponding to a *vapor pressure of about 10 microns*. Values of $t_{10}$ (or $T_{10}$) have been given in Tables 10.1 and 10.3.

If the temperature used for evaporation is much in excess of the value $t_{10}$, there is danger of excessive sputtering (due, frequently, to too rapid evolution of sorbed gases) and the deposit may be both somewhat spongy and nonadherent to the surface of the glass, quartz, or other material on which condensation occurs.

In the presence of residual traces of gases having a pressure of a few microns or more, the deposit is also diffuse and extremely nonadherent, so that it can be rubbed off very readily. As mentioned in Chapter 9, such a diffuse deposit has many advantages for use as a "getter."

The rate of evaporation is decreased in the presence of a gas at pressures exceeding 1 mm; and at higher pressures the rate of evaporation decreases to very low values. As is well known, this fact has been utilized in the development of the gas-filled tungsten-filament lamp.

An important feature of the deposition of films is the preparation of the substrate. This must be rigorously cleaned,[25] the method of course depending on the nature of the material. After the substrate is placed in the vacuum chamber, additional cleaning and degassing are desirable. This may be accomplished by simply heating *in vacuo* or by a low-pressure high-voltage glow discharge.[28] A movable shield is usually placed in front of the substrate so that the evaporator may be well degassed before any material is deposited on the substrate.

A wide variety of evaporators are used, depending on the nature of the material evaporated. Some metals can be fused in the form of a bead onto a refractory-metal heater wire and evaporated from this by passing a current through the wire. This method can be used only with metals which adhere well to the filament, have sufficient vapor pressure to evaporate at temperatures below the melting point of the filament, and do not form alloys with melting points below the temperature used for the evaporation. Caldwell[29] investigated combinations of nine filament materials and twenty-seven metals for evaporation which satisfy these conditions and has presented the results in convenient tabular form.

Materials which do not satisfy these requirements may be evaporated from a conical wire basket, where the surface tension prevents the metal from flowing out between the closely spaced turns. Alternatively a boat of refractory metal may be used. The boat is heated by the passage of current. Boats are particularly well suited for inorganic compounds which cannot conveniently be fused onto a filament. These are often compressed into a pellet rather than used in the powder form. Ceramic or graphite crucibles may also be used. They may be heated by radiation from a filament, by electron bombardment, or by r-f induction.

Olsen, Smith, and Crittenden[30] prepared crucibles by painting closely wound conical spirals of tungsten with a water suspension of alumina and drying in air. By repeated applications and drying, a thick ceramic shell is built up. This is sintered in vacuum at about 1500° C. The coating may also be applied by spraying. Beryllium oxide has been used in a similar manner but is not recommended on account of its toxic properties.

Figures 10.1 and 10.2 show a number of different kinds of evaporators. Table 10.5, taken from Holland,[25] is an excellent compilation of the various methods and materials for evaporating metals.

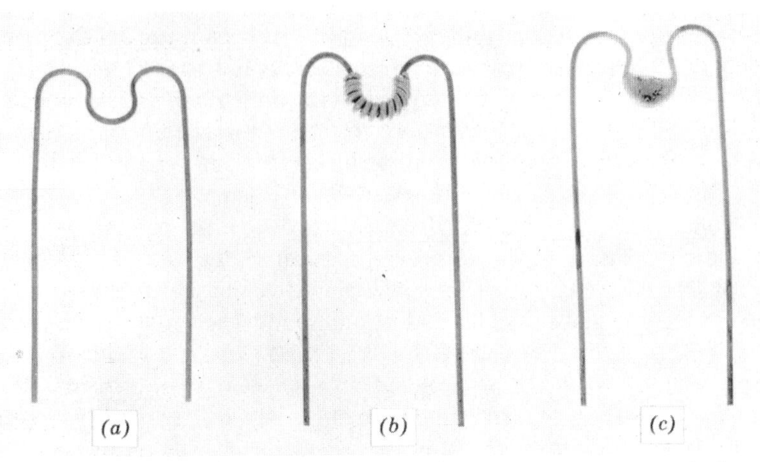


Fig. 10.1. Tungsten filament for evaporating aluminum. (*a*) Heater wire; (*b*) heater wire overwound with aluminum; (*c*) after melting aluminum *in vacuo*.

The geometrical arrangement of vapor sources and substrate in the vacuum in order to achieve various distributions of coating is discussed by Holland and by Bond.[31] Complete equipment for vacuum coating is now available from several manufacturers of high-vacuum equipment.

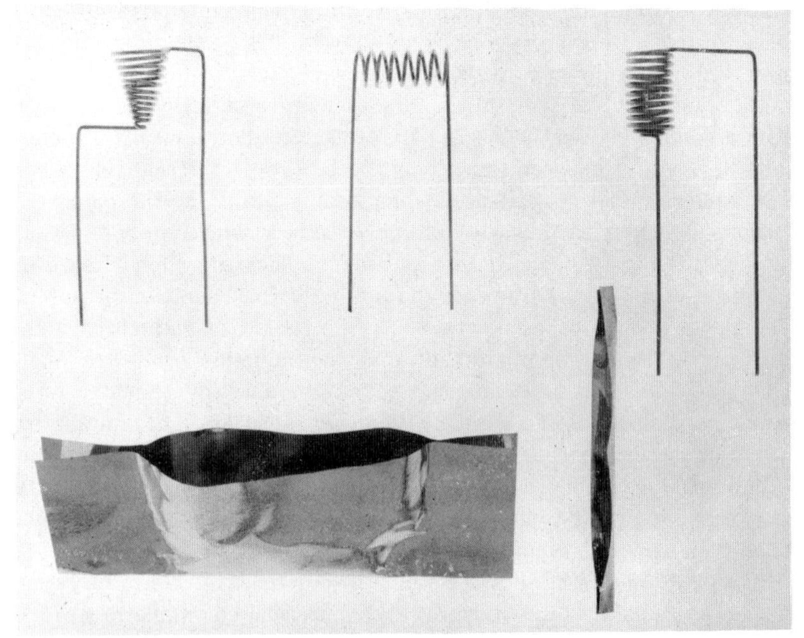

Fig. 10.2. Several kinds of evaporator filaments and boats.

## TABLE 10.5

## Techniques and Refractory Support Materials for Evaporating Metals, Together with Some Useful Vapor-Pressure Data

| Metal | Melting Point (° C) | Evaporation Temp. (° C), Vapor Pressure = 10 microns Hg* | Resistance Heated Sources in Order of Merit | | Other Evaporation Techniques | | Remarks |
|---|---|---|---|---|---|---|---|
| | | | Filaments† | Boats | Source‡ | Method of Heating | |
| Aluminium | 660 | 996(?)<br>1148(?) | W, Ta, helical coil | ... | ... | ... | Alloys freely with refractory metals and reacts with carbon and oxide crucibles. |
| Antimony ($Sb_2$) | 630 | 678 | Chromel, Ta, conical basket | Mo<br>Ta | Alumina crucible | External W heater | Wets Chromel. |
| Arsenic | ... | ... | ... | ... | Alumina crucible | External W heater | ... |
| Barium | 717 | 629 | W, Ta, Mo, Cb, Ni, Fe, Chromel, conical basket | Ta, Mo | ...<br>... | —<br>— | Freely wets without alloying with the heater metals quoted. Reacts with alumina. |
| Beryllium | 1284 | 1246 | Ta, W, Mo, conical basket | ... | Carbon crucible<br>BeO crucible | Electron bombardment<br>High frequency | Wets heater metals quoted. |
| Bismuth | 271 | 698 | Chromel, Ta, W, Mo, Cb, conical basket | Fe, Mo, Ta | Alumina crucible | External W heater | Wets Chromel. |
| Boron | 2000<br>−2080 | 1355 | — | ... | Carbon | Resistance heated | Deposits from carbon probably impure. |
| Calcium | 810 | 605 | W, conical basket | ... | Alumina crucible | External W heater | |
| Cadmium | 321 | 264<br>(subl.) | Chromel, Cb, Ta, Mo, W, Ni, Fe, conical basket | Mo, Ta | Alumina crucible<br>Iron crucible | External W heater<br>Nichrome heater | Freely wets Chromel and Cb. |
| Carbon | 3700<br>±100 | 2681 | ... | ... | Pointed carbon rods pressed together, forming high-resistance contact | Resistance heating | ... |
| Cerium | 785 | 1305 | ... | ... | ... | ... | ... |
| Cesium | 29 | 153 | ... | ... | ... | ... | ... |
| Chromium | 1900 | 1205 | W, conical basket | ... | Electro-deposited Cr on W helical coil | Sublimation | |
| Cobalt | 1478 | 1649 | Cb, W | ... | Alumina or BeO crucible<br>Electro-plated W spiral | Embedded W heater<br>Resistance heater | Alloys with W, Ta, Mo, Cb, Pt. Evaporant weight must not exceed 35 per cent of that of W spiral. |
| Columbium | 2500 | (VP at MP = 1 micron Hg) | W, helical coil | ... | ... | ... | Refractory metal used as a source material. |
| Copper | 1083 | 1273 | Pt, helical coil; Cb, Mo, Ta, W, conical basket | Mo, Ta | Alumina crucible | Embedded W heater | Copper alloys with Ni, Fe, Chromel. Does not wet readily Mo, W, Ta. |
| Gallium | 30 | 1093 | ... | ... | BeO, $SiO_2$, $Al_2O_3$ | ... | Alloys with metals, oxides quoted resist attack up to 1000° C. |

## TABLE 10.5 (*Continued*)

| Metal | Melting Point (° C) | Evaporation Temp. (° C), Vapor Pressure = 10 microns Hg* | Resistance Heated Sources in Order of Merit | | Other Evaporation Techniques | | Remarks |
|---|---|---|---|---|---|---|---|
| | | | Filaments† | Boats | Source‡ | Method of Heating | |
| Germanium | 959 | 1251 | Ta, Mo, W, conical basket | Mo, Ta | Alumina crucible<br>Carbon crucible | Embedded W heater<br>Resistance heater | Wets Ta and Mo. |
| Gold | 1063 | 1465 | W, Mo, conical basket | Mo | ... | ... | Wets but reacts with Ta, possible alloy formation. Partially wets W, Mo. |
| Indium | 157 | 952 | W, Fe, conical basket | Mo | ... | ... | ... |
| Iridium | 2454 | 2556 | ... | ... | ... | ... | ... |
| Iron | 1535 | 1447 | W, helical coil | ... | Alumina or BeO crucible | Embedded W heater | Alloys with W, Ta, Mo, Cb. Evaporant must not exceed 35 per cent of the weight of W filament. |
| Lead | 328 | 718 | Fe, Ni, Chromel, conical basket | Mo<br>... | Alumina or iron crucible<br>Carbon (?) | External Nichrome heater<br>Resistance heater | Does not wet W, Ta, Mo, and Cb. |
| Lithium | 179 | 514 | ... | ... | Mild steel (?) crucible | Nichrome heater | ... |
| Magnesium | 651 | 443 (subl) | W, Ni, Fe, Ta, Mo, Cb, Chromel, conical basket | Mo, Ta | Iron crucible<br>Carbon | External Nichrome heater<br>Resistance heater | Does not melt when volatilized from open spirals and boats. |
| Manganese | 1244 | 980 | W, Ta, Mo, Cb, conical basket | ... | Alumina crucible | Embedded W heater | Freely wets heater metals quoted. |
| Molybdenum | 2622 | 2533 | ... | ... | ... | ... | Refractory metal used for filament and boat-type vapor sources. Evaporates rapidly if oxidized to form $MoO_3$. |
| Nickel | 1455 | 1510 | W, heavy-gauge helical coil | ... | Alumina or BeO crucible | Embedded W heater | Alloys with Mo, Ta, and W. Evaporant must not exceed 30 per cent of the weight of W filament. |
| Palladium | 1555 | 1566 | W, helical coil | ... | ... | ... | ... |
| Platinum | 1774 | 2090 | Multistrand W filament with Pt wire twisted together | ...<br>* | Electro-deposited Pt on W spiral | Resistance heater | Alloys with Ta and partially with W. Platinum may be used as a source heater for metal oxides to prevent decomposition of charge. |
| Rhodium | 1967 | 2149 | ... | ... | Electro-deposited Rh on W spiral<br>Resistance heated Rh foil | Sublimation<br>Sublimation | Requires very low pressure for deposition of neutral transmitting films. |
| Selenium | 217 | 234 | Chromel, Fe, Mo, Cb, conical basket | Mo, Ta | Alumina crucible | Nichrome external heater or radiant heater | Very volatile, may contaminate plant. Wets filament metals quoted. |
| Silicon | 1410 | 1343 | ... | ... | BeO crucible | Embedded W heater | Difficult to prepare Si films free from SiO contamination. |

TABLE 10.5 (*Continued*)

| Metal | Melting Point (° C) | Evaporation Temp. (° C), Vapor Pressure = 10 microns Hg* | Resistance Heated Sources in Order of Merit | | Other Evaporation Techniques | | Remarks |
|---|---|---|---|---|---|---|---|
| | | | Filaments† | Boats | Source‡ | Method of Heating | |
| Silver | 961 | 1047 | Ta, Mo, Cb, Fe, Ni, Chromel, helical coil or W conical basket | Mo, Ta | Electrodeposited Ag on Mo helical coil | Resistance heater | Ag does not wet W. Can be kept in basket by binding fine platinum wire on outside. |
| Strontium | 771 | 549 | W, Ta, Mo, Cb, conical basket | ... | Carbon | Resistance heated | Freely wets without alloying with all filament metals quoted. |
| Tantalum | 2996 | (VP = 1 micron Hg at 2820°) | ... | ... | ... | — | Refractory metal used for source heaters. |
| Tellurium | 452 | (VP = 760 mm Hg at 1390°) | W, Ta, Mo, Cb, Ni, Fe, Chromel, conical basket | ... | Alumina crucible | External Nichrome heater | Very volatile, may lead to plant contamination. Wets without alloying all metal heaters quoted. |
| Thallium | 304 | 606 | Ni, Fe, Cb, Ta, conical basket | ... | Alumina crucible | External Nichrome heater | Freely wets metal heaters quoted without alloying. Partially wets W, Ta, but not Mo. |
| Thorium | 1827 | 2196 | W, conical basket | ... | ... | — | Wets W heater. |
| Tin | 232 | 1189 | Chromel, helical coil; Mo, Ta, conical basket | Mo, Ta | Alumina crucible Carbon | Embedded W heater Resistance heater | Wets Chromel and Mo. |
| Titanium | 1727 | 1546 | W, Ta, conical basket or helical coil | ... | Carbon | Resistance heated | Ti reacts with W spiral, deposit contains trace of W. Ti does not react with Ta but filament may burn out during premelting of Ti. |
| Tungsten | 3382 | 3309 | ... | ... | ... | ... | Refractory metal used for source heaters. Evaporates more readily if surface oxidized to form volatile $WO_3$ or $WO_2$. |
| Uranium | 1132 | 1898 | W, conical basket | ... | ... | ... | Forms oxidized deposits at lowest gas pressures. |
| Vanadium | 1697 | 1888 | W, Mo, conical basket | ... | ... | ... | Alloys with Ta and partially with W. Wets but does not alloy with Mo. |
| Yttrium | 1477 | 1649 | ... | ... | ... | ... | ... |
| Zinc | 419 | 343 (subl) | W, Ta, Mo, Cb, conical basket | ... | Alumina or iron crucible Carbon | External Nichrome heater Resistance heater | Wets without alloy formation all filament metals quoted. |
| Zirconium | 2127 | 2001 | W, conical basket or helical coil | ... | ... | ... | Requires low pressures <0.1 micron Hg to prevent film oxidation. Evap. characteristics similar to Ti. |

* Vapor-pressure data are mainly those reported by Dushman. In many cases rapid evaporation will commence only at temperatures well above the values quoted, because of the existence of surface oxide films.

† Suitable filament materials taken from the reports of Cartwright and Strong, Countryman, Caldwell, and Holland.

‡ The preparation of alumina and beryllium oxide crucibles with embedded heaters is described by Olsen, Smith, and Crittenden.

According to Strong, a good reflecting surface of aluminum is obtained with a thickness of about 1000 A, and presumably the thickness required for surfaces of other metals is of the same order of magnitude. From the values for $W$ given in Table 10.1 it should, therefore, be possible, for a given area of deposit and temperature of evaporation, to calculate the time required to obtain the necessary thickness of the metal desired as a deposit.

In evaporating films from filaments of very refractory metals, like molybdenum, columbium, tungsten, and tantalum, in a vacuum, it is often necessary to keep the temperature constant. This is done by maintaining a constant value of the product $V(I)^{1/3}$, where $V$ denotes the voltage drop along the filament, and $I$ the heating current. For the metals mentioned, data are available for calculating both the current and the volts per centimeter length, for any given value of the diameter, as a function of $T$.[32]

For a considerable number of metals, radiation emissivities have been determined, by means of which the temperature can be calculated either from the value of the watts per square centimeter (in vacuum) or from "brightness temperatures" as measured by an optical pyrometer.[32,33]

## 10.4. VAPOR PRESSURES AND EVAPORATION PHENOMENA OF ALLOYS

According to the law of Raoult, which applies to dilute solutions, the vapor pressure of the solution is lower than that of the pure solvent by an amount which is proportional to the concentration of the solute. Let $P_A$ and $P_{AS}$ denote the vapor pressures of the pure solvent and solution, respectively, at any given temperature, and let $x_B$ denote the mole-fraction, of the solute present in the solution. Then, according to the law of Raoult,

$$\frac{P_A - P_{AS}}{P_A} = x_B = \frac{n_B}{n_A + n_B}, \tag{10.20}$$

where $n_A$ and $n_B$ denote the number of moles of solvent and solute, respectively.

Since

$$x_A = \frac{n_A}{n_A + n_B}, \tag{10.21}$$

it follows from equation 10.20 that the following relation should be valid:

$$a_A = \frac{P_{AS}}{P_A} - x_A, \tag{10.22}$$

where $a_A$ is designated the *activity* of $A$ in the solution.

Applying these relations to a *binary* alloy in which one constituent $A$, which is present in larger proportion, is by far more volatile than $B$, we may regard $B$ as solute, and consequently the decrease in vapor pressure of $A$ should be in accordance with equation 10.20.

Furthermore, it follows, as a consequence of the application of the second law of thermodynamics, that, corresponding to the decrease in vapor pressure of magnitude $P_A - P_{AS} = \Delta P_A$, there is an *increase* ($\Delta T_A$) in the temperature at which the vapor pressure of the "solution" has a value of magnitude $P_A$. This increase is given by the relation

$$\frac{\Delta T_A}{\Delta P_A/P_A} = \frac{R_0 T_A^{\,2}}{L_A}, \tag{10.23}$$

where $R_0$ = molar gas constant (= 1.9865 cal/mole) and $L_A$ = heat of vaporization of $A$, in calories per mole.

From equations 10.23 and 10.20 it follows that

$$\Delta T_A = \frac{R_0 T_A^{\,2} \cdot x_B}{L_A} = C_A T_A^{\,2} x_B, \tag{10.24}$$

where $C_A$ is a constant characteristic of the "solvent" $A$.

These relations should be applicable to both the solid and the liquid states of the alloys, provided that the value of $n_B$ is small compared with that of $n_A$.

If, furthermore, both $A$ and $B$ in the pure states possess vapor pressures which are of the same order of magnitude, we would expect to observe a decrease in vapor pressure of $B$ due to addition of $A$, as well as a decrease in vapor pressure of $A$ due to addition of $B$.

Before proceeding with the discussion of particular illustrations of the above remarks, it is well to emphasize the fact that in this section the composition of an alloy is given in terms of *mole-fractions* rather than in terms of per cent by weight. Let $q_A$ denote the per cent by weight of $A$ and $q_B$, that of $B$, in a binary alloy. Then $q_A + q_B = 100$ and

$$q_A = \frac{100 x_A M_A}{x_A M_A + (1 - x_A) M_B} = \frac{100 x_A M_A}{x_A (M_A - M_B) + M_B}, \tag{10.25}$$

where $M_A$ and $M_B$ denote the gram-molar (or gram-atomic) masses.

Also, given the per cent by weight of each constituent, the mole-fractions can be deduced by means of the relation

$$x_A = \frac{q_A/M_A}{q_A/M_A + (100 - q_A)/M_B} = \frac{q_A}{q_A + (100 - q_A) M_A/M_B}. \tag{10.26}$$

An illustration of the application of equation 10.24 is found in the observations made by Schneider and Esch[34] on the rise in the boiling point of magnesium obtained by alloying it with different metals. The boiling point of pure magnesium at 760 mm pressure is 1380° K, and the latent heat of evaporation per gram-atom ($M = 24.32$) is 32,570 cal.

In this case, the composition was expressed in terms of $n_B$, the number of gram-atoms of the nonvolatile constituent $B$ per 1000 g of magnesium. Hence,

$$x_B = \frac{n_B}{n_B + 41.12}, \tag{10.27}$$

and, consequently, the rise in boiling point according to equation 10.24 should be given by the relation

$$\Delta T_A = \frac{1.9865 \times (1380)^2}{32{,}570} \cdot x_B \tag{10.28}$$

$$= \frac{116.2 n_B}{n_B + 41.12}.$$

Figure 10.3 shows a plot of $\Delta T_A$ versus $n_B$, calculated by means of this relation, and also values actually observed for a number of alloys of magnesium with silver, aluminum, tin, lead, and copper in the liquid

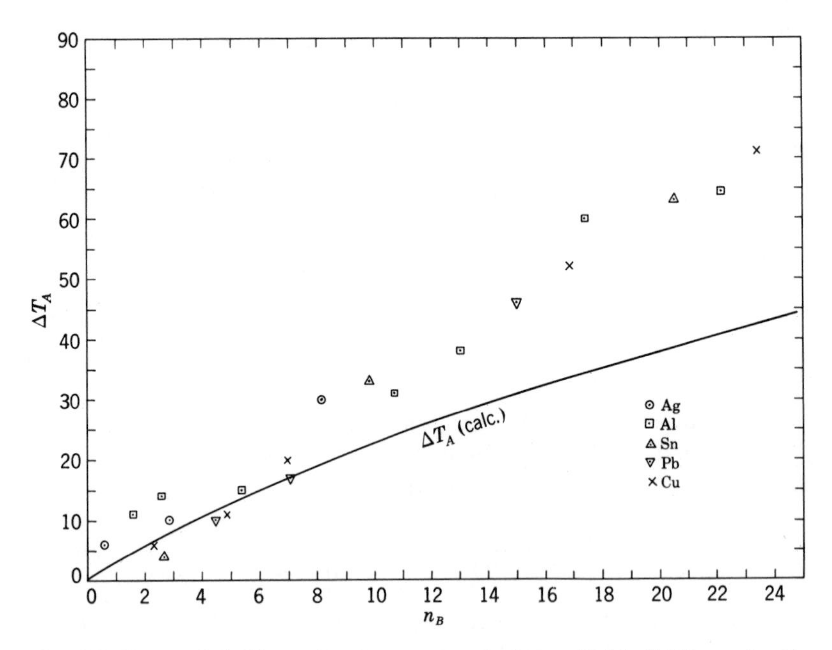


Fig. 10.3. Increase in boiling point of magnesium, in degrees Kelvin ($\Delta T_A$), as a function of gram-atoms ($n_B$) of alloying metal per 1000 grams magnesium (Schneider and Esch).

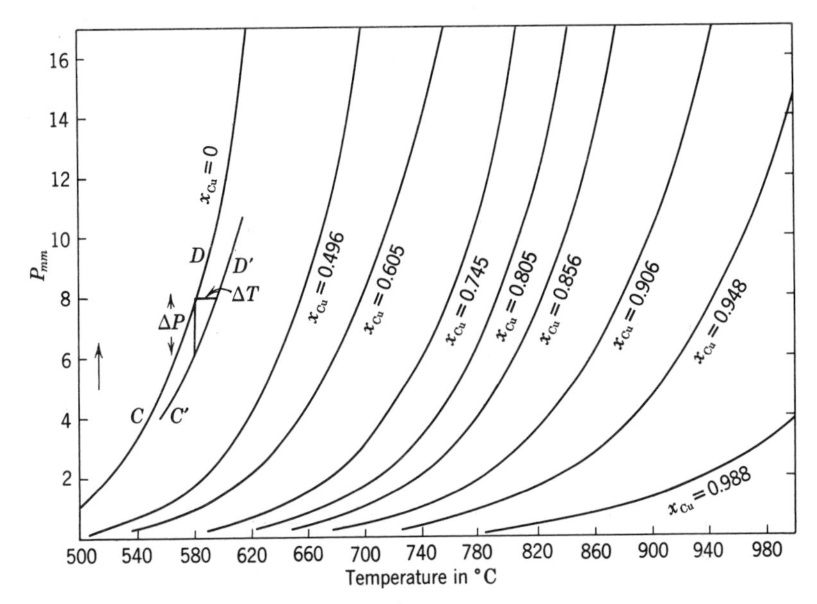


Fig. 10.4. Vapor pressure versus temperature plots for zinc-copper alloys for different values of $x_{Cu}$, the mole fraction of copper. Curve *CD* is the vapor-pressure plot for pure zinc. For an alloy the vapor pressure for a given temperature is decreased by $\Delta P$, and the temperature for a given value of the vapor pressure is increased by $\Delta T$.

state. It will be observed that, with increase in the value of $n_B$, there is an increasing difference between the values of $\Delta T_A$ observed and those calculated. The discrepancy is such as to indicate a higher effective value of $x_B$ than that calculated from the value of $n_B$.

From the observations made by Hargreaves[35] on the vapor pressure of zinc in brasses, the curves shown in Fig. 10.4 have been plotted for a series of alloys in which the mole-fraction of copper was varied from $x_{Cu} = 0.988$ to $x_{Cu} = 0.496$. The curve designated *CD* is a vapor-pressure plot for pure zinc (based on the vapor-pressure values deduced by Kelley[10]). For an alloy containing a very small mole-fraction of copper, the vapor-pressure plot would be represented by some curve such as *C'D'*, and the corresponding values of $\Delta T$ and $\Delta P$ have been indicated.

As will be observed, the vapor pressure of zinc, at any given temperature, decreases with increase in $x_{Cu}$. Table 10.6 gives values of the activity of zinc for the series of brasses at 727° C, as defined by the relation

$$a_{Zn} = \frac{P_{Zn}}{90.3}, \tag{10.29}$$

where $P_{Zn}$ is the vapor pressure (in millimeters) as observed for the alloys

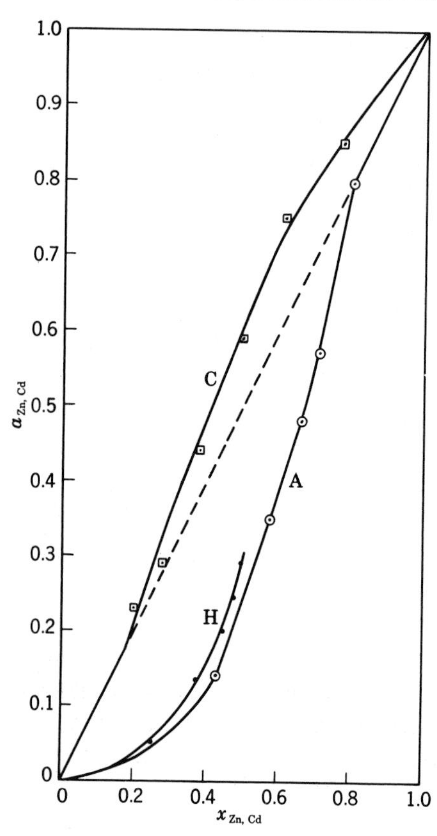


Fig. 10.5. Plots of the activities ($a$) of zinc and cadmium versus mole fraction ($x$) for alloys with copper. Curve $A$, plot for zinc-copper alloys at 700° C; curve $H$, plot for zinc-copper alloys at 727° C (from data in Table 10.6), curve $C$, plot for cadmium-copper alloys in the range 700–850° C. The dashed line represents the values deduced on the basis of Raoult's law. (Curves $A$ and $C$ from Schneider and Schmid.)

and 90.3 mm is the vapor pressure, at this temperature, of pure zinc. The ratio

$$f_{Zn} = \frac{a_{Zn}}{x_{Zn}}, \tag{10.30}$$

is designated the *activity coefficient* and should be equal to 1 for an ideal solution.

The values of $a_{Zn}$ versus $x_{Zn}$ have been plotted in curve $H$, in Fig. 10.5. Curve $A$ shows a similar plot of data deduced from values of $P_{Zn}$ observed at 700° C.[36] As will be observed, the values of $a_{Zn}$ for zinc-copper alloys containing 0.8 mole-fraction of zinc or higher follow the relation predicted by equation 10.22. On the other hand, for values of $x_{Zn}$ less than 0.8, the vapor pressures are lower than those calculated on the basis of equation 10.22, as shown by the observation that $a_{Zn}$ is less than $x_{Zn}$.

Figure 10.6 shows the constitution of copper-zinc alloys, and, as is seen from the diagram, all the alloys containing up to about 69 per cent zinc

TABLE 10.6

**Values of $q_{Zn}$, $x_{Zn}$, $a_{Zn}$, and $f_{Zn}$ for Brasses at 727° C**

| $q_{Zn}$ | $x_{Zn}$ | $a_{Zn}$ | $f_{Zn}$ | |
|---|---|---|---|---|
| 5.3 | 0.052 | 0.0045 | 0.0865 | |
| 9.7 | 0.094 | 0.0086 | 0.0915 | |
| 14.8 | 0.144 | 0.0185 | 0.128 | |
| 20.0 | 0.195 | 0.0301 | 0.154 | |
| 26.0 | 0.255 | 0.0545 | 0.214 | |
| 38.0 | 0.373 | 0.134 | 0.48 | for $\beta$ phase |
| | | | 0.33 | for $\alpha$ phase |
| 46.0 | 0.453 | 0.200 | 0.46 | |
| 48.8 | 0.481 | 0.246 | 0.51 | |
| 51.1 | 0.504 | 0.291 | 0.58 | |

are solid at 727° C. At 697° C, the alloys containing 30 to 19.5 per cent copper exhibit a transition from the solid to the liquid state, and all alloys containing less than 19.5 per cent copper are fused at this temperature. Thus, according to curve *A* in Fig. 10.5, $f_{Zn} = 1$ for the alloys that are in the fused state at 700° C.

If log $P_{Zn}$ for a given copper-zinc alloy is plotted versus $1/T$, a straight line is obtained, the slope of which yields a value of *L*, the heat of sublimation. Such a series of plots of the data shown in Fig. 10.4 leads, as shown by Hargreaves, to values of *L* ranging from 36,000 cal/g-atom for the

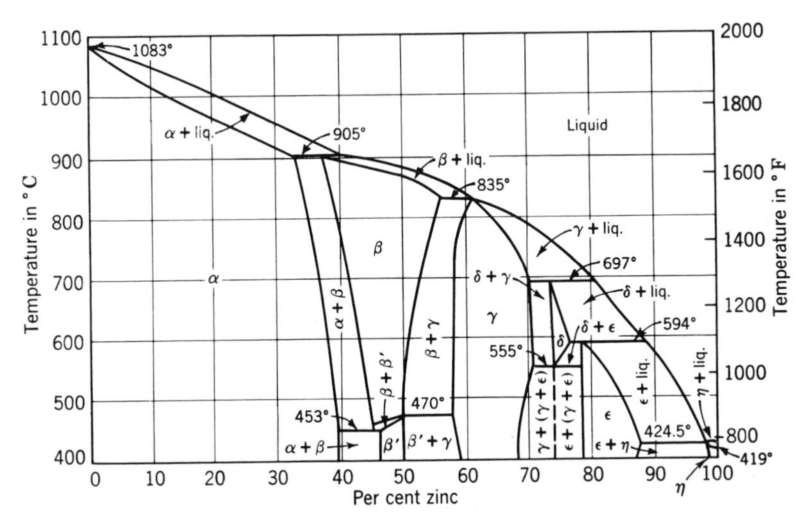


Fig. 10.6. Constitution diagram of copper-zinc alloys (*Metals Handbook*).

alloy containing 1.2 per cent by weight of zinc to 31,500 cal/g-atom for the alloy containing 51.1 per cent zinc. The value for liquid zinc is 28,600 cal/g-atom.

One important conclusion from these observations, which is valid for most alloys, is that in order to maintain the vapor pressure of zinc in a brass at, for instance, 1 mm, the temperature of evaporation must be increased continuously as the zinc content decreases. At 960° C, the vapor pressure of pure copper is about $10^{-5}$ mm (see Table 10.1), and, in the presence of even a small content of zinc, the vapor pressure of copper must be considerably smaller. Hence the elimination of the zinc becomes progressively more and more difficult with decreasing content in the alloy.

Curve $C$ in Fig. 10.5 shows a plot of $a_{Cd}$ for cadmium-copper alloys[36] in the range 700–850° C. At these temperatures all the alloys are in the fused state, and as will be observed $f_{Cd}$ is greater than 1 for all alloys containing more than 0.2 mole-fraction of cadmium. That is, the vapor pressure of cadmium for these alloys is greater than that corresponding to equation 10.22.

The brasses are typical of binary alloys in which one constituent is practically nonvolatile as compared with the other. An interesting illustration of a binary alloy in which this is not the case is furnished by the zinc-cadmium system, which has been studied by Burmeister and Jellinek.[38]

**TABLE 10.7**

**Vapor Pressures (in Millimeters) for Zinc-Cadmium Alloys at 682° C**

(Burmeister and Jellinek)

| $x_{Cd}$ | $P_{Cd}$ | $a_{Cd}$ | $x_{Zn}$ | $P_{Zn}$ | $a_{Zn}$ | $P_{Zn} + P_{Cd}$ |
|---|---|---|---|---|---|---|
| 1 | 252 | 1 | 0 | 0 | 0 | 252 |
| 0.749 | 205 | 0.815 | 0.251 | 20.5 | 0.472 | 225.5 |
| 0.459 | 153 | 0.633 | 0.541 | 30.7 | 0.706 | 183.7 |
| 0.238 | 112 | 0.444 | 0.762 | 37 | 0.850 | 149 |
| 0 | ... | ... | 1 | 44 | 1 | 44 |

Table 10.7 gives the vapor pressures of zinc and cadmium at 682° C for the pure metals and three alloys. At this temperature the metals and alloys are above their melting points.

The values of the vapor pressures of the individual constituents and the total pressures are plotted in Fig. 10.7; Fig. 10.8 shows plots of the activities. Similar plots of vapor pressures of fused tin-zinc alloys and tin-cadmium alloys[38] (at 700 and 540° C, respectively) are shown in

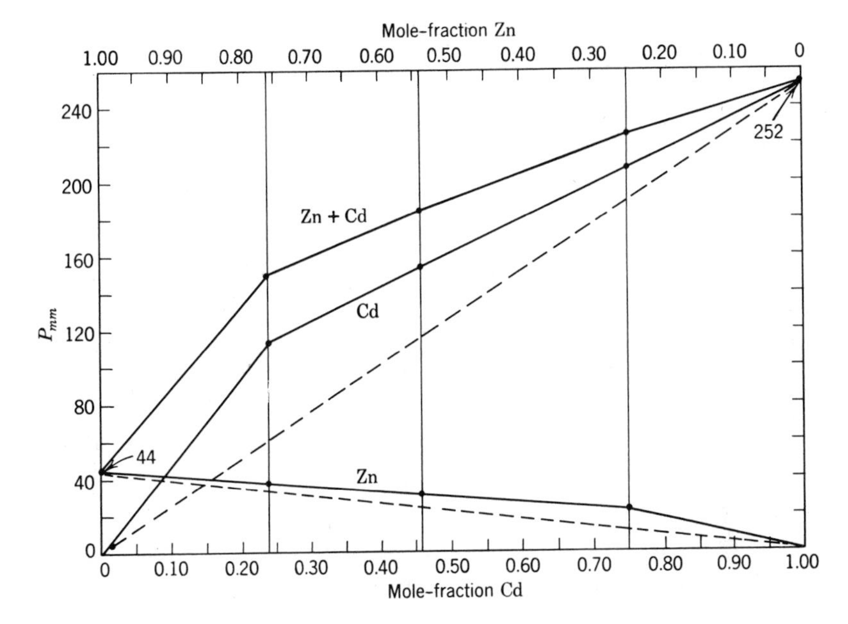


Fig. 10.7. Vapor pressures of zinc and cadmium and total vapor pressure for zinc-cadmium alloys at 682° C, plotted versus mole fraction of each constituent (Burmeister and Jellinek).

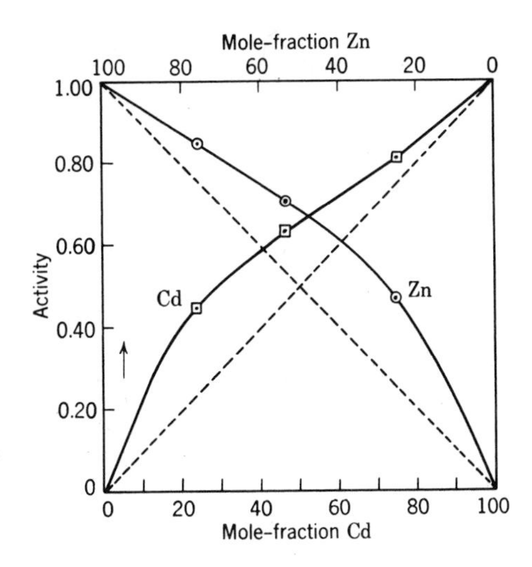


Fig. 10.8. Plots of activity versus mole fraction for zinc and cadmium in zinc-cadmium alloys, corresponding to vapor-pressure plots in Fig. 10.7.

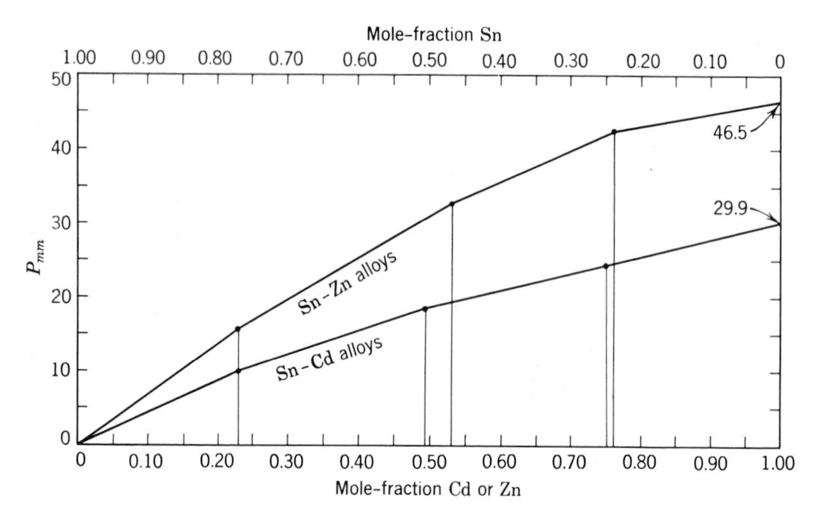


Fig. 10.9. Vapor pressure of zinc in tin-zinc alloys at 700° C, and vapor pressure of cadmium in tin-cadmium alloys at 540° C, plotted versus mole fraction of zinc and cadmium respectively (Burmeister and Jellinek).

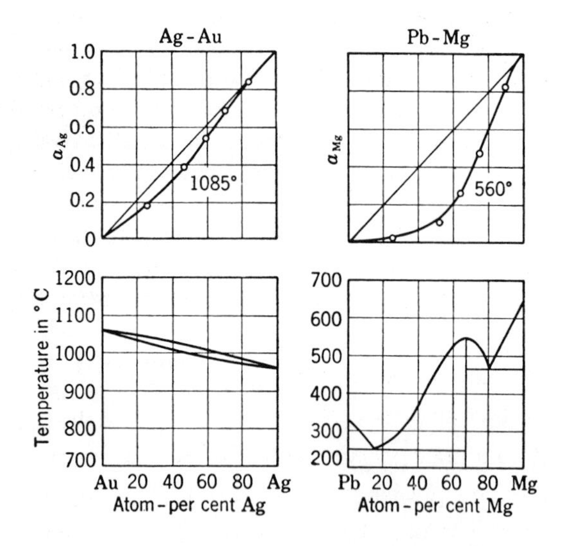


Fig. 10.10. Plots of the activity ($a$) versus atom-per cent for silver in fused silver-gold alloys at 1085° C, and for magnesium in fused lead-magnesium alloys at 560° C. Corresponding phase diagrams are shown below (from Kubaschewski).

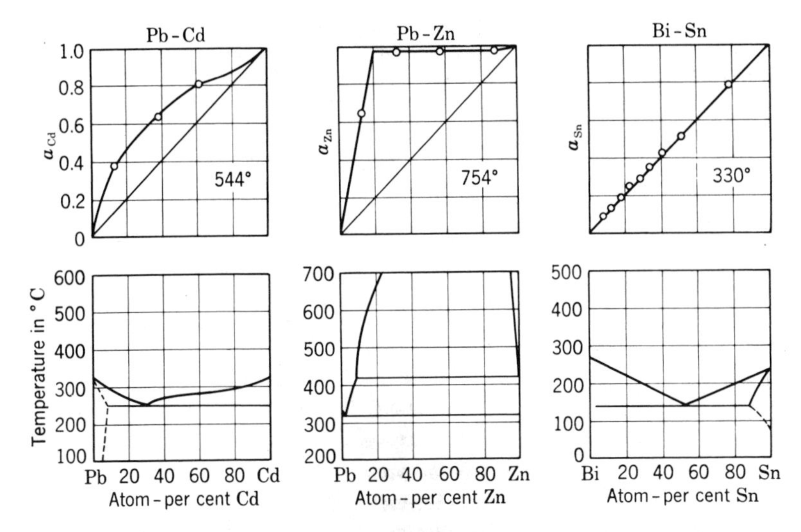


Fig. 10.11. Plots of the activity versus atom-per cent of cadmium in cadmium-lead alloys at 544° C, of zinc in zinc-lead alloys at 754° C, and of tin in bismuth-tin alloys at 330° C, with corresponding phase diagrams (from Kubaschewski).

Fig. 10.9. In both these cases the values of $f_{Zn}$ and $f_{Cd}$ are greater than 1 for all the compositions.

Figure 10.10[39] shows plots of the activities of silver and magnesium in fused alloys and, for comparison, phase diagrams for these alloys. Similar plots are shown in Fig. 10.11[39] for the volatile constituents in fused alloys of lead-cadmium, lead-zinc, and bismuth-tin. The system lead-zinc is of special interest because of the limited solubility of zinc in lead.

The vapor pressures of aluminum-zinc alloys over the range of temperatures 600–1100° C have been investigated by Schneider and Stoll.[40] Figure 10.12 shows plots of log $P_{mm}$ versus $1/T$ for the three alloys having the compositions:

$$\begin{aligned} 2.12 \text{ atomic per cent Zn} &= 5.03 \text{ per cent by weight,} \\ 15.02 \text{ atomic per cent Zn} &= 30.0 \text{ per cent by weight,} \\ 49.04 \text{ atomic per cent Zn} &= 70.0 \text{ per cent by weight.} \end{aligned}$$

According to the constitution diagram shown in Fig. 10.13,[41] the alloy containing 5.03 per cent by weight zinc melts at about 608° C, and equilibrium exists between the $\gamma$ phase and the liquid to about 647° C. Above this temperature the alloy is completely liquid. The plot in Fig. 10.12 shows that in this range of temperatures the vapor pressures are abnormally higher. A similar phenomenon is observed for the alloy containing 30 per cent by weight zinc in the range 513–615° C, which is

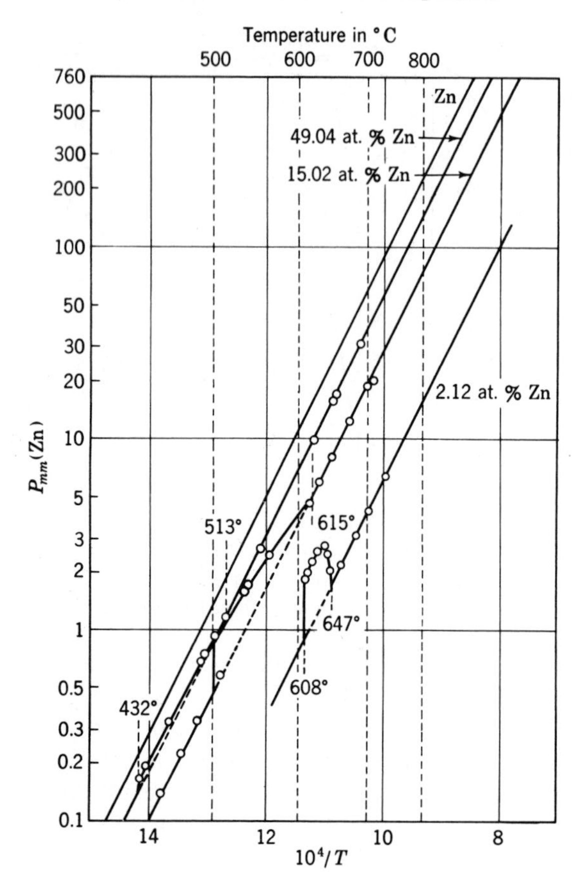


Fig. 10.12. Plots of $P_{mm}$ (zinc) versus $1/T$ on semilog scale for three zinc-aluminum alloys and for pure zinc (Schneider and Stoll).

the range of existence of $\gamma$ + liquid. For the alloy containing 70 per cent by weight zinc the $\gamma$ + liquid range is from about 443° C to slightly above 500° C, and the log $P_{mm}$ versus $1/T$ plot shows a deviation from the linear plot over this range of temperatures. The log $P_{mm}$ versus $1/T$ plot for pure zinc shown in Fig. 10.12 is that for the liquid phase, since the melting point is 419.4° C.

Plots of $a_{Zn}$ versus $x_{Zn}$ for the aluminum-zinc system for 650 and 800° C are shown in Fig. 10.14. As will be observed, the activities are higher than those predicted by Raoult's law.

Data on the vapor pressures of mercury over amalgams with a number of metals (for the range 300–324° C) are given in *International Critical Tables*.[42] As shown by Poindexter[43] and Hughes and Poindexter,[44] "a

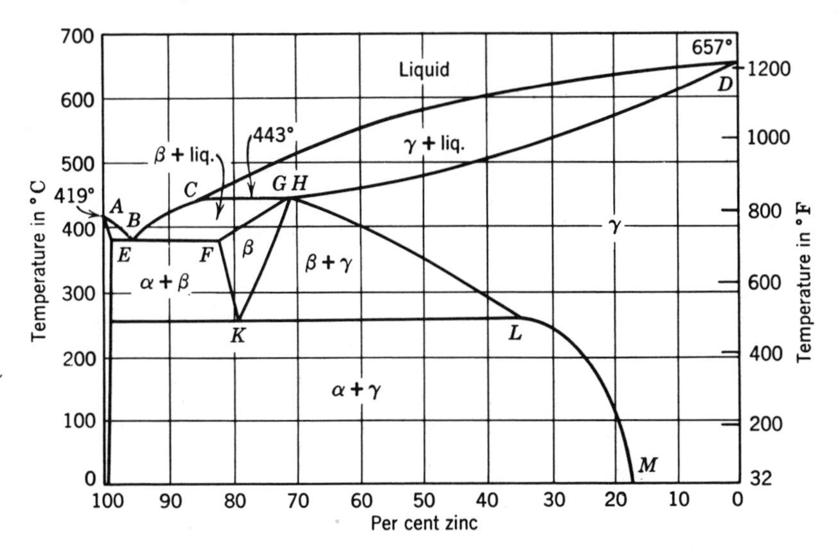


Fig. 10.13. Constitution diagram for zinc-aluminum alloys (*Metals Handbook*).

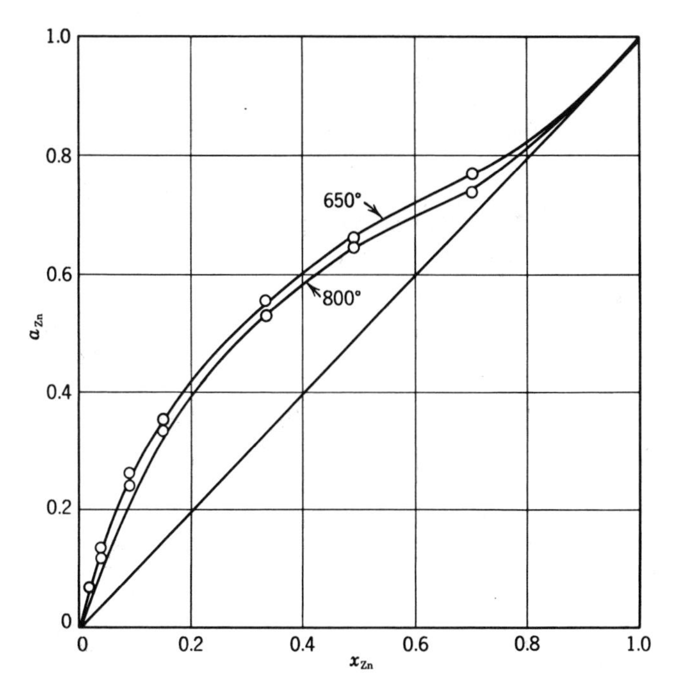


Fig. 10.14. Plots of activity of zinc ($a_{Zn}$) versus mole fraction ($x_{Zn}$) in zinc-aluminum alloys.

sodium or potassium trap is almost, if not quite, as effective as liquid air for stopping mercury vapor." Poindexter[45] has concluded, from plots of log $P$ versus $1/T$ for a sodium amalgam containing a 50 per cent mole-fraction of sodium, that at 20° C the vapor pressure is $3 \cdot 10^{-8}$ mm mercury. However, in the case of amalgams with other metals than those of the alkali group, the deviations from Raoult's law are not at all as great. The plots of $a_{Hg}$ versus $x_{Hg}$ are very similar to those shown for alloys of zinc-cadmium and some of the other metals.[46]

Frequently, in degassing parts made of high-melting-point alloys it is of considerable interest to be able to predict the relative volatilities of different constituents in a given alloy. From the preceding discussion it follows that at any given temperature the vapor pressure of each constituent in the alloy must be much lower than that for the pure metal. As a rough approximation the validity of Raoult's law may be assumed, that is, equation 10.22.

For instance, let us assume that we have an iron-manganese-carbon alloy containing 13 per cent by weight manganese and 0.20 per cent carbon.[47] Since the atomic weights of iron and manganese are 55.85 and 54.93, respectively, the mole-fraction of manganese is approximately 0.13. Hence, we may make the following assumptions:

1. Vapor pressure of Mn in alloy = 0.13 × (Vapor pressure of pure Mn).
2. Vapor pressure of Fe in alloy = 0.87 × (Vapor pressure of pure Fe).

From Table 10.1 it is seen that, at 1000° C, the vapor pressure of iron is about $10^{-3}$ micron, while that of manganese is about 10 microns. Therefore, in the alloy, the vapor pressures of manganese and iron will be approximately 1 micron and $9 \cdot 10^{-4}$ micron, respectively, and hence the manganese will evaporate about 1000 times more rapidly than the iron. The effect of the carbon may be neglected under these conditions.

As another illustration, let us consider an iron-nickel alloy containing 25 atomic per cent nickel. At 1200° C, the vapor pressures of iron and nickel are about 0.1 and 0.03 micron, respectively. Assuming the validity of Raoult's law, the vapor pressures of iron and nickel in the alloy would be about 0.075 and 0.008 micron, respectively. Consequently, we would conclude that the iron would evaporate about ten times as fast as the nickel.

In view of the fact that, as far as Dushman was aware, there are no experimental data available on the rates of evaporation of alloys of the more refractory metals, the application of Raoult's law represents probably the only method by which any semiquantitative conclusion can be deduced in such cases.

## 10.5. VAPOR PRESSURES OF OXIDES AND OTHER INORGANIC COMPOUNDS

As in the case of metals, vapor pressures of a number of inorganic compounds of interest in vacuum technique can be calculated from free-energy data by means of equation 10.2 *a* or *b*. Some calculated and some observed values for a number of compounds are given in Table 10.8. The data from the different sources cover different pressure ranges.

A compilation by Stull[48] gives the vapor pressures of 1200 organic and 300 inorganic compounds, as well as many elements.

From their observations on the vapor pressures of four *solid* halides, in the range $T = 945°$ K to $T = 600°$ K, Zimm and Mayer[49] deduced the following relations for vapor pressures:

$$\text{KCl: } \log P_{mm} = 17.006 - 11{,}879/T - 2 \log T.$$
$$\text{KBr: } \log P_{mm} = 17.141 - 11{,}543/T - 2 \log T.$$
$$\text{KI: } \log P_{mm} = 17.277 - 11{,}263/T - 2 \log T.$$
$$\text{NaCl: } \log P_{mm} = 17.093 - 12{,}288/T - 2 \log T.$$

From these relations two-term expressions were derived of the form

$$\log P_{mm} = A - \frac{B}{T}, \tag{10.31}$$

from which, in turn, values of $t°$ C were calculated for values of the pressure ranging from $P_\mu = 10^{-2}$ to $P_\mu = 100$, which are included in Table 10.8 in the rows designated by reference 2.

Kangro and Wieking[50] found that their determination of the vapor pressures of the *fused* halides could be expressed in the form of equation 10.31 with the values of $A$ (for the pressure in millimeters) and $B$ given in Table 10.9. These were used to derive values of the pressure at the melting point. As will be observed these values are higher than those calculated from Kelley's values of $\Delta F$. The table also gives the values of $P_{mm}$ observed at 860, 950, and 990° C.

Binary mixtures of compounds which are mutually soluble exhibit a decrease in the vapor pressure of each compound which is similar to that exhibited by binary alloys and which was discussed in the previous section. As an illustration of this phenomenon Table 10.10 gives vapor-pressure data for fused mixtures of KCl and NaCl as observed by Kangro and Wieking.[50]

## 10.6. VAPOR PRESSURES AT LOW TEMPERATURES

In high-vacuum technique refrigerants are used to condense out a number of gases and vapors. Table 10.11 contains vapor-pressure data for

**TABLE 10.8**

**Vapor Pressure of Oxides, Halides, Etc.**

| Compound and Ref. | Melting Point (°C) | $P_{mm}$ at Melting Point | $t°$ C for $P_{mm}$ = | | | $t°$ C for $P_{\mu}$ = | | | | |
|---|---|---|---|---|---|---|---|---|---|---|
| | | | 0.076 | 0.76 | 7.6 | $10^{-2}$ | $10^{-1}$ | 1 | 10 | 100 |
| $Al_2O_3$ (10) | 2050 | 0.347 | ... | 2122 | 2351 | ... | ... | ... | ... | ... |
| BaO (69) | 1923 | ... | ... | ... | ... | 1150 | 1263 | 1395 | 1552 | 1742 |
| CaO (10) | 2572 | ... | ... | ... | ... | 1583 | 1717 | 1873 | 2055 | 2271 |
| $MoO_3$ (10) | 795 | 7.50 | 661 | 726 | 801 | ... | ... | ... | ... | ... |
| $SiO_2$ (10) | 1710 | 7.60 | ... | ... | 1710 | ... | ... | ... | ... | ... |
| SrO (10) | 2430 | ... | ... | ... | ... | 1420 | 1258 | 1631 | 1757 | 1901 |
| $P_4O_{10}$ (10) | 358 (subl) | ... | ... | ... | ... | 42 | 63 | 87 | 116 | 149 |
| AgCl (10) | 455 | ... | ... | ... | ... | 479 | 536 | 607 | 690 | 790 |
| LiCl (10) | 613 | 0.021 | 663 | 769 | 911 | ... | ... | ... | ... | ... |
| NaCl (10) | 801 | 0.31 | 743 | 850 | 996 | 479 (49) | 532 (49) | 592 (49) | 662 (49) | 745 (49) |
| KCl (10) | 776 | 0.44 | 704 | 806 | 948 | 457 (49) | 508 (49) | 567 (49) | 635 (49) | 716 (49) |
| RbCl (10) | 718 | 0.23 | ... | 777 | 917 | ... | ... | ... | ... | ... |
| CsCl (10) | 646 | 0.14 | ... | 730 | 864 | ... | ... | ... | ... | ... |

TABLE 10.9

Vapor Pressure of Fused Halides

| Compound and Ref. | $A$ | $B$ | $P_{mm}$ at Melting Point | $P_{mm}$ at 860° C | $P_{mm}$ at 950° C | $P_{mm}$ at 990° C |
|---|---|---|---|---|---|---|
| LiCl (50) | 7.172 | 7187 | 0.12 | 6.7 | 10.9 | 30.2 |
| NaCl (50) | 7.906 | 8475 | 1.03 | 2.8 | 9.2 | 16.2 |
| KCl (50) | 7.245 | 7450 | 1.38 | 5.0 | 14.0 | 22.4 |
| RbCl (50) | 7.326 | 7809 | 0.28 | 2.7 | 8.6 | 14.0 |
| CsCl (50) | 8.302 | 8400 | 0.18 | 7.7 | 25.6 | 46.0 |

TABLE 10.10

Vapor Pressures of Fused Mixtures of KCl ($A$) and NaCl ($B$)

| $t$° C | $P_{mm}(A)$ | $P_{mm}(B)$ | Mole-Fraction (per cent) $A$ | Mole-Fraction (per cent) $B$ |
|---|---|---|---|---|
| 860 | 5.0 | ... | 100 | 0 |
| | 3.7 | 0.7 | 74.5 | 25.5 |
| | 2.2 | 1.0 | 49.3 | 50.7 |
| | 1.1 | 1.9 | 25.6 | 74.4 |
| | ... | 2.8 | 0 | 100 |
| 950 | 14.0 | ... | 100 | 0 |
| | 8.9 | 1.9 | 74.4 | 25.6 |
| | 6.2 | 3.5 | 49.6 | 50.4 |
| | 3.1 | 6.5 | 25.8 | 74.2 |
| | ... | 9.2 | 0 | 100 |
| 990 | 22.4 | ... | 100 | 0 |
| | 13.0 | 4.9 | 75.9 | 24.1 |
| | 9.2 | 7.3 | 49.5 | 50.5 |
| | 6.7 | 9.8 | 24.3 | 75.7 |
| | ... | 16.2 | 0 | 100 |

ice and mercury. In each case two sets of values are given, one taken from *International Critical Tables*,[51] and the other extending the values to lower temperatures derived from the expressions for $\Delta F$ calculated by Kelley.[10] According to the latter, the vapor pressure of ice is given by the relation

$$\log P_{mm} = 1.207 + 3.857 \log T + 3.41 \cdot 10^{-3} T + 4.875 \cdot 10^{-8} T^2 - \frac{2461}{T}. \tag{10.32}$$

**TABLE 10.11**

**Vapor Pressures of Ice and Mercury at Low Temperatures**

| $t$° C | $T$ °K | $P_\mu$ (KK) | $P_\mu$ (ICT) | $t$° C | $T$° K | $P_\mu$ (KK) | $P_\mu$ (ICT) |
|---|---|---|---|---|---|---|---|
| | | | Ice ($H_2O$) | | | | |
| 0 | 273.2 | ... | 4579 | −80 | 193.2 | $4.13 \cdot 10^{-1}$ | $4.0 \cdot 10^{-1}$ |
| −10 | 263.2 | 1960 | 1950 | −90 | 183.2 | $7.45 \cdot 10^{-2}$ | $7.0 \cdot 10^{-2}$ |
| −20 | 253.2 | 779 | 776 | −100 | 173.2 | $1.10 \cdot 10^{-2}$ | ... |
| −30 | 243.2 | 293.1 | 285.9 | −110 | 163.2 | $1.25 \cdot 10^{-3}$ | ... |
| −40 | 233.2 | 99.4 | 96.6 | −120 | 153.2 | $1.13 \cdot 10^{-4}$ | ... |
| −50 | 223.2 | 29.85 | 29.6 | −130 | 143.2 | $6.98 \cdot 10^{-6}$ | ... |
| −60 | 213.2 | 8.40 | 8.08 | −140 | 133.2 | $2.93 \cdot 10^{-7}$ | ... |
| −70 | 203.2 | 1.98 | 1.94 | −150 | 123.2 | $7.4 \cdot 10^{-12}$ | ... |
| −78.5 | 194.7 | 0.53 | ... | −183 | 90.2 | $1.4 \cdot 10^{-19}$ | ... |
| | | | Mercury (Hg) | | | | |
| 30 | 303.2 | 3.13 | 2.78 | −38.9 | 234.3 | $2.51 \cdot 10^{-3}$ | ... |
| 25 | 298.2 | 2.05 | 1.84 | −45 | 228.2 | $1.05 \cdot 10^{-3}$ | ... |
| 20 | 293.2 | 1.38 | 1.20 | −50 | 223.2 | $4.94 \cdot 10^{-4}$ | ... |
| 10 | 283.2 | $5.62 \cdot 10^{-1}$ | $4.90 \cdot 10^{-1}$ | −60 | 213.2 | $9.89 \cdot 10^{-5}$ | ... |
| 0 | 273.2 | $2.22 \cdot 10^{-1}$ | $1.85 \cdot 10^{-1}$ | −70 | 203.2 | $1.68 \cdot 10^{-5}$ | ... |
| −10 | 263.2 | $7.87 \cdot 10^{-2}$ | $6.06 \cdot 10^{-2}$ | −78.6 | 194.6 | $3.13 \cdot 10^{-6}$ | ... |
| −20 | 253.2 | $2.64 \cdot 10^{-2}$ | $1.81 \cdot 10^{-2}$ | −80 | 193.2 | $2.38 \cdot 10^{-6}$ | ... |
| −30 | 243.2 | $7.85 \cdot 10^{-3}$ | $4.78 \cdot 10^{-3}$ | −100 | 173.2 | $2.39 \cdot 10^{-8}$ | ... |
| −38.9 | 234.3 | $2.51 \cdot 10^{-3}$ | $1.24 \cdot 10^{-3}$ | −183 | 90.2 | $3.48 \cdot 10^{-29}$ | ... |

The relations derived for the vapor pressure of mercury from Kelley's calculated expressions for $\Delta F$ are as follows:

$$\text{Liquid} \rightarrow \text{Gas:} \quad \log P_{mm} = 10.377 - 0.8254 \log T - \frac{3285}{T}. \tag{10.33}$$

$$\text{Solid} \rightarrow \text{Gas:} \log P_{mm} = 9.453 - 0.2011 \log T - 6.558 \cdot 10^{-4} T - \frac{3379}{T}. \tag{10.34}$$

Table 10.12($a$) gives vapor-pressure data for solid carbon dioxide according to Meyers and Van Dusen.[52] The sublimation temperature of the solid at atmospheric pressure is −78.5° C.

The relation derived from Kelley's expression for $\Delta F$ is

$$\log P_{mm} = 8.882 + 0.8702 \log T - 3.891 \cdot 10^{-3} T - \frac{1408}{T}. \tag{10.35}$$

**TABLE 10.12(*a*)**

**Vapor Pressure of Solid Carbon Dioxide**

(Pressure in microns of mercury)

| $t$° C | 0 | 1 | 2 | 3 | 4 | 5 | 6 | 7 | 8 | 9 |
|---|---|---|---|---|---|---|---|---|---|---|
| −180 | 0.013 | 0.008 | 0.006 | 0.004 | 0.003 | 0.0017 | 0.0011 | 0.0007 | 0.0005 | 0.0003 |
| −170 | 0.37 | 0.27 | 0.20 | 0.14 | 0.10 | 0.074 | 0.052 | 0.037 | 0.026 | 0.018 |
| −160 | 5.9 | 4.6 | 3.6 | 2.7 | 2.1 | 1.58 | 1.19 | 0.90 | 0.67 | 0.50 |
| −150 | 60.5 | 48.8 | 39.2 | 31.4 | 25.1 | 19.9 | 15.8 | 12.4 | 9.8 | 7.6 |
| −140 | 431 | 359 | 298 | 247 | 204 | 168 | 138 | 113 | 92 | 75 |

(Pressure in millimeters of mercury)

| $t$° C | 0 | 1 | 2 | 3 | 4 | 5 | 6 | 7 | 8 | 9 |
|---|---|---|---|---|---|---|---|---|---|---|
| −130 | 2.31 | 1.97 | 1.68 | 1.43 | 1.22 | 1.03 | 0.87 | 0.73 | 0.61 | 0.51 |
| −120 | 9.81 | 8.57 | 7.46 | 6.49 | 5.63 | 4.88 | 4.22 | 3.64 | 3.13 | 2.69 |
| −110 | 34.63 | 30.76 | 27.27 | 24.14 | 21.34 | 18.83 | 16.58 | 14.58 | 12.80 | 11.22 |
| −100 | 104.81 | 94.40 | 84.91 | 76.27 | 68.43 | 61.30 | 54.84 | 48.99 | 43.71 | 38.94 |
| −90 | 279.5 | 254.7 | 231.8 | 210.8 | 191.4 | 173.6 | 157.3 | 142.4 | 128.7 | 116.2 |
| −80 | 672.2 | 618.3 | 568.2 | 521.7 | 478.5 | 438.6 | 401.6 | 367.4 | 335.7 | 306.5 |
| −70 | 1486.1 | 1377.3 | 1275.6 | 1180.5 | 1091.7 | 1008.9 | 931.7 | 859.7 | 792.7 | 730.3 |
| −60 | 3073.1 | 2865.1 | 2669.7 | 2486.3 | 2314.2 | 2152.8 | 2001.5 | 1859.7 | 1726.9 | 1602.5 |
| −50 | ... | ... | ... | ... | ... | ... | ... | 3780.9 | 3530.2 | 3294.6 |

Triple point, −56.602 ± 0.005° C; 3885.2 ± 0.4 mm.

This has been used to calculate values of $P$ for solid carbon dioxide in the range below $-150°$ C, and Table 10.12(*b*) gives a comparison between these values (KK) and those given by Meyers and Van Dusen (MVD).

**TABLE 10.12(*b*)**

**Vapor Pressures of Solid Carbon Dioxide at Very Low Temperatures**

| $t°$ C | $T°$ K | $P_\mu$ (KK) | $P_\mu$ (MVD) | $t°$ C | $T°$ K | $P_\mu$ (KK) | $P_\mu$ (MVD) |
|---|---|---|---|---|---|---|---|
| −150 | 123.1 | 61.94 | 60.5 | −183 | 90.1 | $4.2 \cdot 10^{-3}$ | $4.0 \cdot 10^{-3}$ |
| −160 | 113.1 | 7.30 | 5.9 | −187 | 86.1 | $7.9 \cdot 10^{-4}$ | $7.0 \cdot 10^{-4}$ |
| −170 | 103.1 | $3.82 \cdot 10^{-1}$ | $3.8 \cdot 10^{-1}$ | −189 | 84.1 | ... | $3.0 \cdot 10^{-4}$ |
| −175 | 98.1 | ... | $7.4 \cdot 10^{-2}$ | −193 | 80.1 | $5.85 \cdot 10^{-5}$ | ... |
| −180 | 93.1 | ... | $1.3 \cdot 10^{-2}$ | −196 | 77.1 | $1.05 \cdot 10^{-5}$ | ... |

Table 10.13(*a*) gives vapor pressures for the more common "noncondensable" gases.[53] Triple points are shown by asterisks.

Kelley has given the following vapor-pressure relations for the *solid* state:

$$\text{Argon:} \quad \log P_{mm} = 7.620 - \frac{411.1}{T}. \tag{10.36}$$

$$\text{Krypton:} \quad \log P_{mm} = 7.478 - \frac{555.2}{T}. \tag{10.37}$$

$$\text{Xenon:} \quad \log P_{mm} = 8.00 - \frac{842}{T}. \tag{10.38}$$

A chart showing plots of the vapor pressure versus temperature for a number of different gases in the range 100 to $-200°$ C has been published by Berl.[54]

Table 10.13(*b*) gives the temperature in degrees Centigrade for a series of values of the pressure in microns, as derived from Kelley's calculated values of $\Delta F$. The values of $t$ for 7600, 760, and 76 microns were taken from his tables; those for lower pressures were extrapolated from a plot of $\log P_u$ versus $1/T$ for the above values of $P_u$. For argon, krypton, and xenon the two-term relations given above for the solid state were used. The last column in the table gives the boiling points according to Kelley for comparison with those given in the previous table.

In addition to the vapor pressures of the stable phases (solid or liquid) it is sometimes necessary to have values for vapor pressures in equilibrium with another phase. Thus, for example, in measurements of surface area by adsorption of krypton at liquid-air temperature the adsorbed film behaves more nearly as a liquid than as a solid. For this reason liquid-phase vapor pressures are used, rather than the values for the solid, which

**TABLE 10.13(*a*)**

**Vapor Pressures of Gases at Low Temperatures**

| | | | | | | | | | |
|---|---|---|---|---|---|---|---|---|---|
| He | $t°$ C | −271.1 | −269.9 | −268.9 | | | | | |
| | $P_{mm}$ | 3 | 197 | 760 | | | | | |
| $H_2$ | $t°$ C | −259.14 | −258.46 | −257.99 | −257.21 | −256.19 | −254.73 | −252.74 | |
| | $P_{mm}$ | 51.4* | 79.9 | 100 | 149.7 | 232.5 | 397.6 | 760 | |
| Ne | $t°$ C | −257.62 | −257.48 | −254.92 | −253.16 | −251.24 | −250.22 | −248.5 | −245.92 |
| | $P_{mm}$ | 0.55 | 2.4 | 7.8 | 28.2 | 91 | 148 (*s*) | 325 (*l*) | 760 |
| $N_2$ | $t°$ C | −216.13 | −213.18 | −211.11 | −210.90 | −206.20 | −202.16 | −198.30 | −195.82 |
| | $P_{mm}$ | 21.8 | 47.0 | 71.9 | 96.4* | 177.6 | 329.4 | 561.3 | 760 |
| CO | $t°$ C | −220.6 | −209.1 | −205.7 | −192.0 | (Mp = −205.0) | | | |
| | $P_{mm}$ | 4 | 50 | 111.3 | 760 | | | | |
| Ar | $t°$ C | −205.3 | −200 | −195.6 | −189.3 | −186.1 | −185.9 | | |
| | $P_{mm}$ | 38.3 | 84.6 | 208.8 | 515.6* | 672.7 | 760 | | |
| $O_2$ | $t°$ C | −215.73 | −210.76 | −205.82 | −201.42 | −196.39 | −192.73 | −187.61 | −182.95 |
| | $P_{mm}$ | 2.68 (*l*) | 9.59 | 28.07 | 64.01 | 129.5 | 239.5 | 457.6 | 760 |
| $CH_4$ | $t°$ C | −183.2 | −182 | −176 | −170 | −167 | −165 | −163 | −161.5 |
| | $P_{mm}$ | 70 (*s*) | 94.0 (*l*) | 190.5 | 353.2 | 468.2 | 559.3 | 663.8 | 760 |
| Kr | $t°$ C | −192 | −188.5 | −182.0 | −169* | −165.3 | −157.0 | −151.8 | −152.9 |
| | $P_{mm}$ | 4.8 | 8.5 | 22.6 | 132.5 | 207.9 | 522.2* | 760 | 760 |
| Xe | $t°$ C | −178.0 | −162.0 | −152.3 | −138.0 | −127.0 | −118.1 | −112.0 | −109.1 |
| | $P_{mm}$ | 0.3 | 3.0 | 10.0 | 60.0 | 185.0 | 373.5 | 615.5* | 760 |

* Asterisks denote triple points.

TABLE 10.13(*b*)

**Temperatures in Degrees Centigrade for Pressures in Microns**

(Kelley)

| Gas | 0.076 | 0.76 | 7.6 | 76 | 760 | 7600 | BP (° C) |
|---|---|---|---|---|---|---|---|
| Ne | | | | −259.3 | −257.4 | −254.9 | −245.95 |
| $N_2$ | −242.1 | −239.2 | −236.0 | −231.9 | −226.8 | −220.0 | −195.78 |
| CO | −238.5 | −235.5 | −232.0 | −227.8 | −222.7 | −216.1 | −191.5 |
| Ar | −238.1 | −234.8 | −230.9 | −226.0 | −220.1 | −212.1 | −185.8 |
| $O_2$ | −237.3 | −234.1 | −230.3 | −225.6 | −219.7 | −211.4 | −183.0 |
| $CH_4$ | −229.1 | −225.1 | −220.1 | −214.1 | −206.5 | −196.7 | −161.4 |
| Kr | −225.2 | −220.7 | −215.2 | −208.4 | −200.1 | −188.9 | −152.1 |
| Xe | −203.6 | −197.4 | −189.9 | −180.8 | −169.6 | −155.0 | −108.0 |

is the stable phase at these temperatures. Thus equations such as those given in the *International Critical Tables*[55] (argon) or by Meihuizen and Crommelin[56] (krypton) are appropriate. For further discussion see the work of Beebe *et al.*,[57] of Gaines,[58] and of Gaines and Cannon.[59]

Charts showing plots of vapor pressure versus the temperature in degrees Centigrade for a large number of gases and vapors of inorganic and organic compounds have been published by Berl.[54] Since these plots cover the range from −190 to 200° C and from pressures above atmospheric to very low values, they should prove of value for information on the vapor pressures of elements and compounds not discussed in the above review.

Also R. V. Smith has published a vapor-pressure chart for volatile hydrocarbons[60] which shows values down to $10^{-2}$ mm plotted in accordance with the relation

$$\log P_{mm} = A - \frac{B}{T}. \tag{10.39}$$

## 10.7. BOILING POINT OF LIQUID AIR

Ordinary air consists of a mixture of 78.02 per cent nitrogen, 20.99 per cent oxygen, 0.94 per cent argon, and traces of hydrogen, carbon dioxide, and the other rare gases.[61] Hence the molecular percentage of oxygen is 23.5.

Since the boiling point of nitrogen is −195.78° C and that of oxygen −182.95° C, the "boiling point" of liquid air lies between these two values, depending on the composition. The earliest investigation on this topic was carried out by Baly.[62] Later, Dodge and Dunbar[63] repeated the measurements and extended the range to higher and lower pressures.

Table 10.14 gives the equilibrium compositions, as obtained in this investigation, for the range of temperatures from that of pure nitrogen to that of pure oxygen. Figure 10.15 shows a plot of these data.

**TABLE 10.14**

**Compositions of Vapor and Liquid Phases for Nitrogen-Oxygen Mixtures at 1 atm**

| $T°$ K | $T - 273.13$ | Mole Fraction Oxygen in Liquid | Mole Fraction Oxygen in Vapor |
|---|---|---|---|
| 77.35 | −195.78 | 0 | 0 |
| 77.86 | −195.27 | 8.10 | 2.30 |
| 78.78 | −194.35 | 21.60 | 6.50 |
| 79.72 | −193.41 | 33.35 | 11.05 |
| 80.65 | −192.48 | 43.38 | 15.65 |
| 81.60 | −191.53 | 52.17 | 21.05 |
| 82.48 | −190.65 | 59.53 | 26.26 |
| 83.38 | −189.75 | 66.20 | 32.05 |
| 84.28 | −188.85 | 72.27 | 38.40 |
| 85.22 | −187.91 | 77.80 | 45.80 |
| 86.18 | −186.95 | 82.95 | 54.00 |
| 87.15 | −185.98 | 87.60 | 63.40 |
| 88.14 | −184.99 | 91.98 | 73.90 |
| 89.18 | −183.95 | 96.15 | 86.03 |
| 90.16 | −182.97 | 100.00 | 100.00 |

At any given temperature the compositions of the vapor and liquid phases are given by the two points, such as $A$ and $B$, on the line parallel to the axis of abscissas. For freshly produced liquid air, the composition of the liquid in equilibrium with air is 56 moles of oxygen per 100 moles of the binary mixture; and the boiling point[64] is (82 − 273.13 =) −191.1° C. However, as the liquid continues to evaporate it loses nitrogen much more rapidly than oxygen; the boiling point rises along the lower curve and ultimately attains the value −183° C, corresponding to the boiling point of pure oxygen. It is, therefore, strictly speaking, not correct to speak of the "temperature of liquid air," since this has no well-defined value.

Furthermore, where liquid air is rectified,[65] as occurs in many industries, for the purpose of producing only nitrogen or oxygen alone, the liquid phase may be either of these, in a more or less purified state.

If the temperature of the liquid air surrounding a trap is below about −191° C and the exhaust system is opened to the atmosphere, air will condense in the trap and interfere with subsequent exhaust operation. *It is therefore always a safe rule to remove the refrigerant and permit the*

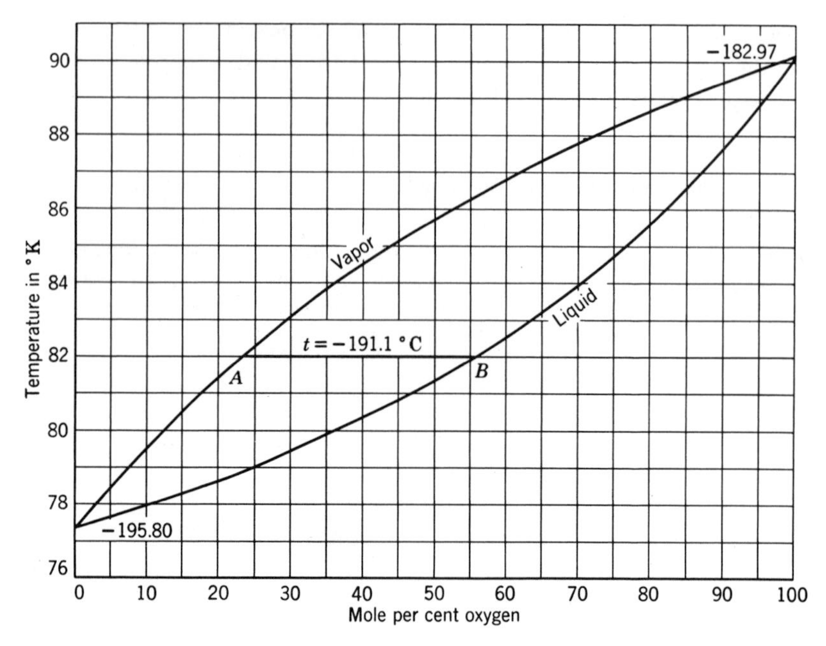


Fig. 10.15. Compositions of vapor and liquid phases for oxygen-nitrogen mixtures at 1 atm, showing corresponding boiling points (Dodge and Dunbar). For air the composition of the gas corresponds to *A*, and that of the liquid phase in equilibrium corresponds to *B*.

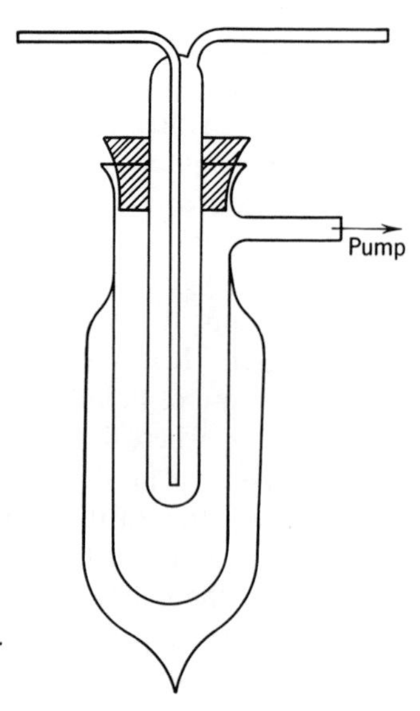


Fig. 10.16. Arrangement for lowering the temperature of liquid air (Klemenc).

*trap to reach room temperature before air or preferably dry nitrogen is let into the system.* In fact it may be of considerable advantage to warm the trap moderately before letting in gas, in order to remove any traces of water, mercury, carbon dioxide, etc., that may be condensed or adsorbed on the walls. According to Klemenc,[66] fresh liquid air is colorless and turns blue on evaporation of the nitrogen.

By reducing the pressure over the liquid air, the temperature may be lowered to −220° C. Figure 10.16, taken from the book by Klemenc, shows a simple arrangement for this purpose. The vapor pressures of pure oxygen and nitrogen at temperatures below their boiling points at 1 atm are given in Tables 13(*a*) and 13(*b*).

## 10.8. VAPOR PRESSURES OF OILS, GREASES, AND WAXES

### Oils

Data on the vapor pressures of oils used in vapor pumps have been given in Chapter 3. The oils customarily used in rotary mechanical pumps, such as those described in Chapter 3, have vapor pressures of the order of 0.1 micron or less at room temperature.

Values obtained by H. Huthsteiner in the General Electric Research Laboratory on a commercial transformer oil are given in Table 10.15.

**TABLE 10.15**

**Approximate Vapor Pressures of Commercial Oil**

| | | | | | |
|---|---|---|---|---|---|
| (*a*) Oil preheated to eliminate dissolved air and water: | | | | | |
| $t°$ C: | 100 | 74 | 57 | 34 | |
| $P_\mu$: | 614 | 70–56 | 6 | 1 | (extrapolated) |
| (*b*) Distilled out about $\frac{1}{3}$ of the oil: | | | | | |
| $t°$ C: | 73 | 66 | 56 | 49 | |
| $P_\mu$: | 25 | 8.5 | 2.6 | 1 | (extrapolated) |
| (*c*) Distilled out about $\frac{3}{4}$ of the oil: | | | | | |
| $t°$ C: | 84 | 77 | 60 | | |
| $P_\mu$: | 2.8 | 1.0 | 0.1 (extrapolated) | | |

The Knudsen effusion method was used, and the vapor pressure was calculated from the loss in weight of the oil by assuming an average molecular weight of 142.

The solubilities $\alpha$ of some of the more common gases in transformer oil, according to Loomis,[67] are shown in Table 10.16. By definition,

$$\alpha_t = \left(\frac{273.1}{t + 273.1}\right) \times \left(\frac{\text{g} \cdot \text{cm}^{-3} \text{ in liquid}}{\text{g} \cdot \text{cm}^{-3} \text{ in gas}}\right). \tag{10.40}$$

TABLE 10.16

**Solubilities of Gases in Transformer Oil***

| Gas | Values of $10^3 \alpha_t$ | | |
|---|---|---|---|
| | $t = 25°$ C | $t = 50°$ C | $t = 80°$ C |
| $H_2$ | 51 | ... | 69 |
| $N_2$ | 84.8 | ... | 91.6 |
| $O_2$ | 156.2 | ... | 148.5 |
| $CO_2$ | 990 | 770 | 570 |

* A mineral oil of Pennsylvania base, 96 per cent saturated hydrocarbons distilling between 300 and 400° C. At 25° C, density is 0.840, and at 80° C, 0.800.

Assuming that the pressure is 1 atm, the quantity $q$ of gas in solution is $7.6 \cdot 10^2 \alpha_t$ micron · liters · cm$^{-3}$ of liquid. Thus, for $\alpha_t = 50 \cdot 10^{-3}$, $q = 38$ micron · liters · cm$^{-3}$. Also, all such blended oils always contain traces of more-volatile hydrocarbons. Consequently, any oil to be used in a vacuum pump must be "conditioned" for a certain period before attaining minimum vapor pressure.

## Waxes and Greases

The waxes and greases used in vacuum technology are necessarily materials with low vapor pressures. The true vapor pressure of such a

TABLE 10.17

**Vapor Pressure of Apiezon Compounds**

| Compound | Vapor Pressure after Evolution of Dissolved Air (mm at $t°$ C) | Melting Point (° C) | Temp. to Which Sealing Media Should Be Raised to Apply | Safe Max. Temp. for Use (° C) |
|---|---|---|---|---|
| Apiezon Grease L | $10^{-3}$ at 300<br>$10^{-10}$ to $10^{-11}$ at 20 | 47 | Room temp. | 30 |
| Apiezon Grease M | $10^{-3}$ at 200<br>$10^{-7}$ to $10^{-8}$ at 20 | 44 | Room temp. | 30 |
| Apiezon Grease N | $10^{-3}$ at 250<br>$10^{-8}$ to $10^{-9}$ at 20 | 43 | Room temp. | 30 |
| Apiezon Grease T | $10^{-8}$ (approx) at 20 | 125 | Room temp. | 110 |
| Apiezon Sealing Compound Q | $10^{-4}$ at 20 | 45 | Room temp. | 30 |
| Apiezon Wax W | $10^{-3}$ at 180 | 85 | 100° C | 80 |
| Apiezon Wax W100 | $10^{-3}$ at 180 | 55 | 80° C | 50 |
| Apiezon Wax W40 | $10^{-3}$ at 180 | 45 | 40–50° C | 30 |

material is determined with considerable difficulty, since it can be determined only after the removal of the more volatile constituents as well as the occluded gases. Hence the results reported by different observers vary according to the test methods used. The data in Table 10.17 are taken from information supplied by the manufacturers.

**TABLE 10.18**

**Comparison of Vacuum Greases[63]**

| Trade Name | Melting Point (° C) | Vapor Pressure at 20° C (mm Hg) |
|---|---|---|
| Lubriseal | 40 | $<10^{-5}$ |
| Vacuseal Light | 50 | $10^{-5}$ |
| Vacuseal Heavy | 60 | $10^{-5}$ |
| Celvacene Light | 90 | $10^{-6}$ |
| Celvacene Medium | 120 | $<10^{-6}$ |
| Celvacene Heavy | 130 | $<10^{-6}$ |
| Stopcock grease | 215 | $<10^{-5}$ |
| Apiezon L | 47 | $10^{-11}$ ($10^{-3}$ at 300° C) |
| Apiezon M | 44 | $6 \cdot 10^{-7}$ ($10^{-3}$ at 200° C) |

Several different greases have been compared by Normand, as shown in Table 10.18.[68]

## REFERENCES AND NOTES


1. R. W. Ditchburn and J. C. Gilmour, *Revs. Modern Phys.*, **13,** 310 (1941).
2. F. H. Verhoek and A. L. Marshall, *J. Am. Chem. Soc.*, **61,** 2737 (1939) (the vapor pressures and accommodation coefficients of four nonvolatile compounds).
3. A. L. Marshall, R. W. Dornte, and F. J. Norton, *J. Am. Chem. Soc.*, **59,** 1161 (1937) (the vapor pressures of copper and iron).
4. R. E. Honig, "Vapor Pressure Data for the More Common Elements," *RCA Rev.*, **18,** 195 (1957).
5. O. Kubaschewski and E. L. Evans, *Metallurgical Thermochemistry*, John Wiley & Sons, New York, 2nd edition, 1956.
6. D. R. Stull and G. C. Sinke, "Thermodynamic Properties of the Elements," *Advances in Chemistry*, Series 18, American Chemical Society, Washington, D.C., 1956.
7. R. R. Hultgren, "Selected Values for the Thermodynamic Properties of Metals and Alloys," Report of Minerals Research Laboratory, University of California, Berkeley, Calif., 1956.
8. R. R. Law, *Rev. Sci. Instr.*, **19,** 920 (1948) (vapor-pressure data for various substances).
9. L. Brewer, *The Chemistry and Metallurgy of Miscellaneous Materials: Thermodynamics*, McGraw-Hill Book Company, New York, 1950, p. 13.

10. K. K. Kelley, *U.S. Bur. Mines Bull. No.* 383 (1935) (reissued in 1960 as *U.S. Bur. Mines Bull. No.* 585).
11. R. E. Honig, *J. Chem. Phys.*, **22,** 126 (1954).
12. R. J. Thorn and G. H. Winslow, *J. Chem. Phys.*, **26,** 186 (1957).
13. G. Glockler, *J. Chem. Phys.*, **22,** 159 (1954).
14. A. L. Marshall and F. J. Norton, *J. Am. Chem. Soc.*, **72,** 2166 (1950).
15. D. M. Kern, *J. Chem. Educ.*, **33,** 272 (1956).
16. W. J. Kroll, *Trans. Am. Electrochem. Soc.*, **87,** 571 (1956).
17. J. Strong, *Procedures in Experimental Physics*, Prentice-Hall, Englewood Cliffs, N.J., 1938.
18. C. H. Cartwright and J. Strong, *Rev. Sci. Instr.*, **2,** 189 (1931).
19. G. Hass, *J. Opt. Soc. Am.*, **49,** 593 (1959).
20. G. Hass and N. W. Scott, *J. Opt. Soc. Am.*, **39,** 179 (1949).
21. H. D. Polster, *J. Opt. Soc. Am.*, **42,** 21 (1952).
22. C. Feldman and M. O'Hara, *J. Opt. Soc. Am.*, **47,** 300 (1957).
23. G. G. Kretschmar and L. E. Schilberg, *J. Appl. Phys.*, **28,** 865 (1957).
24. K. Günther, *Naturwissenschaften*, **45,** 415 (1958).
25. L. Holland, *The Vacuum Deposition of Thin Films*, John Wiley & Sons, New York, 1956.
26. H. Mayer, *Physik Dünner Schichten*, Wissenschaftliche Verlagsgesellschaft M.B.H., Stuttgart, Vol. I, 1950; Vol. II, 1955.
27. O. Heavens, *Optical Properties of Thin Solid Films*, Academic Press, New York, 1955.
28. N. C. Benson, G. Hass, and N. W. Scott, *J. Opt. Soc. Am.*, **40,** 687 (1950).
29. W. C. Caldwell, *J. Appl. Phys.*, **12,** 779 (1941).
30. L. O. Olsen, C. S. Smith, and E. C. Crittenden, Jr., *J. Appl. Phys.*, **16,** 425 (1945).
31. W. L. Bond, *J. Opt. Soc. Am.*, **44,** 429 (1954) (notes on solution of problems in odd-job vapor coating).
32. A. G. Worthing, *Astrophys. J.*, **61,** 146 (1925); H. A. Jones and I. Langmuir, *Gen. Elec. Rev.*, 354, 408 (1927); W. E. Forsythe and E. Q. Adams, *J. Opt. Soc. Am.*, **35,** 108 (1945).
33. For tungsten: *Smithsonian Tables*, 9th edition, p. 102. For tantalum: D. B. Langmuir and L. Malter, *Phys. Rev.*, **55,** 743 (1939); Baldwin, Shilts, and Coomes, *Tech. Report on ONR contract No.* Nonr 1623(00), NR 372-731; Notre Dame Physical Electronics Group, August 1955. For molybdenum: *Temperature, Its Measurement and Control in Science and Industry*, Reinhold Publishing Corporation, New York, 1941.
34. A. Schneider and U. Esch, *Z. Elektrochem.*, **45,** 888 (1939).
35. R. Hargreaves, *J. Inst. Metals*, **64,** 115 (1939).
36. A. Schneider and H. Schmid, *Z. Elektrochem.*, **48,** 627 (1942).
37. *Metals Handbook*, 1939 edition, p. 1367.
38. E. Burmeister and K. Jellinek, *Z. physik. Chem.*, **A, 165,** 121 (1933).
39. O. Kubaschewski, *Z. Elektrochem.*, **48,** 559, 646 (1942).
40. E. K. Stoll, *Z. Elektrochem.*, **47,** 527 (1941).
41. *Metals Handbook*, 1939 edition, p. 1738.
42. *International Critical Tables*, 1928, Vol. 3, p. 284.
43. F. E. Poindexter, *J. Opt. Soc. Am.*, **9,** 629 (1924).
44. A. L. Hughes and F. E. Poindexter, *Phil. Mag.*, **50,** 423 (1925) (see Chapter 4, section 9).
45. F. E. Poindexter, *Phys. Rev.*, **28,** 208 (1926).

46. See also the following references: Ma-Hg (375° C), K. Hauffe, *Z. Elektrochem.*, **46,** 348 (1940); K-Hg (300° C), Cd-Hg (284° C), J. S. Pedder and S. Barratt, *J. Chem. Soc.*, **1933,** 537; K-Hg (184–445° C), R. W. Millar, *J. Am. Chem. Soc.*, **49,** 3003 (1927).
47. Constitution diagram given in *Metals Handbook*, 1939 edition, p. 411.
48. D. R. Stull, *Ind. Eng. Chem.*, **39,** 517, 1947.
49. B. Zimm and J. E. Mayer, *J. Chem. Phys.*, **12,** 362 (1944).
50. W. Kangro and H. Wieking, *Z. physik. Chem.*, **A,183,** 199 (1938–1939).
51. *International Critical Tables*, 1928, Vol. 3, pp. 206 and 210. Also *Handbook of Chemistry and Physics*, Chemical Rubber Publishing Company, Cleveland, Ohio, 1958, and *Handbook of Chemistry* by N. A. Lange, Handbook Publishers, Sandusky, Ohio, 1956.
52. C. H. Meyers and M. S. van Dusen, *J. Research Natl. Bur. Standards*, **10,** 381 (1933).
53. Values for helium and neon from *Handbook of Chemistry and Physics* (see note 51). Values for hydrogen, nitrogen, carbon monoxide, argon, oxygen, and methane from *Handbook of Chemistry*, (see note 51). For boiling points of these gases see F. J. Allen and R. B. Moore, *J. Am. Chem. Soc.*, **53,** 2522 (1931).
54. E. Berl, *Chem. Met. Eng.*, **51,** 132 (1944). See also *Chem. Met. Eng.*, **53,** 130 (1946) for table of boiling points of a number of elements and inorganic compounds.
55. *International Critical Tables*, 1928, Vol. 3, p. 203.
56. J. J. Meihuizen and C. A. Crommelin, *Physica*, **4,** 1 (1937).
57. R. A. Beebe *et al.*, *J. Am. Chem. Soc.*, **67,** 1554 (1945).
58. G. L. Gaines, *J. Phys. Chem.*, **62,** 1526 (1958).
59. G. L. Gaines and P. Cannon, *J. Phys. Chem.*, **64,** 997 (1960).
60. *U.S. Bur. Mines Inform. Circ.* No. 7215 (August 1942).
61. *International Critical Tables*, Vol. 1, p. 393. The density at 0° C and 1 atm is 1.2930 $g \cdot liter^{-1}$.
62. E. C. C. Baly, *Phil. Mag.*, **49,** 517 (1900).
63. B. F. Dodge and A. K. Dunbar, *J. Am. Chem. Soc.*, **49,** 591 (1927).
64. This is the value of the absolute temperature for 0° C used by Dodge and Dunbar.
65. The subject of rectification of mixtures of $O_2$–$N_2$, $O_2$–Ar, CO–$N_2$, Ar–$N_2$, $N_2$–$H_2$, $CO_2$–$H_2$ and hydrocarbons is discussed at length by M. Ruhemann, *The Separation of Gases*, Clarendon Press, Oxford, 1940.
66. A. Klemenc, *Die Behandlung und Reindarstellung von Gasen*, Leipzig, 1938.
67. *International Critical Tables*, 1928, Vol. 3, pp. 261–270.
68. *U.S. Atomic Energy Comm. TID*-5210.
69. J. P. Blewett, H. A. Liebhafsky, E. F. Hennelly, *J. Chem. Phys.*, **7,** 480 (1939).

# 11 Dissociation pressures of oxides, hydrides, and nitrides

Revised by Edward L. Simons

## 11.1. FREE ENERGY

As mentioned in Chapter 8, many metals form compounds by reaction with oxygen, hydrogen, and nitrogen. In general, these reactions are exothermic, and the dissociation of the resulting compounds increases with increasing temperature. Some of the oxides volatilize at temperatures so low that dissociation occurs only in the vapor phase; water is one of the most important illustrations of this type of oxide. However, for an oxide, nitride, or hydride that exists as a solid over a considerable range of temperatures there exists at any given temperature a dissociation pressure, which represents an equilibrium between the compound, the gas, and the metal. The dissociation pressure of an oxide is a direct measure of its thermal stability and is one of the factors that define the conditions under which it can be converted to the pure metal by chemical reducing agents.

Dissociation and reduction reactions can best be discussed in terms of certain concepts of thermodynamics. Any chemical (or physical) reaction is characterized by a change in what is known as the *enthalpy* or *heat content* of the system. This change, denoted by the symbol $\Delta H$, can often be measured calorimetrically and is numerically equal to the heat absorbed or liberated when the reaction is carried out under conditions of constant pressure. For endothermic reactions, which involve an absorption of heat by the reacting system, $\Delta H$ is positive. For exothermic reactions, $\Delta H$ is negative.

In the latter part of the nineteenth century, Thomsen and Berthelot suggested that the heat evolved in a reaction be used as a measure of the chemical affinity of the reacting substances. Indeed, the reactions that take place most readily at room temperature are exothermic. However, from considerations based on the second law of thermodynamics, it has been shown that the direction and extent of a reaction are governed by the sign and magnitude of the change in *free energy*, designated by $\Delta F$. The conclusion deduced may be stated as follows: A reaction can occur at constant temperature and pressure *only if the value of* $\Delta F$ *is negative.* That is, *the free energy must decrease.*[1] The values of $\Delta H$ and $\Delta F$ are ordinarily given in calories per mole.

From the second law of thermodynamics it follows that, at any given temperature and constant pressure,

$$\left[\frac{\partial(\Delta F)}{\partial T}\right]_p = \frac{\Delta F - \Delta H}{T}; \tag{11.1}$$

that is,

$$\Delta F = \Delta H + T\left[\frac{\partial(\Delta F)}{\partial T}\right]_p. \tag{11.2}$$

Dividing both sides of this equation by $T^2$, we obtain the relation

$$\frac{\Delta F}{T^2} - \frac{1}{T}\left[\frac{\partial(\Delta F)}{\partial T}\right]_p = \frac{\Delta H}{T^2}; \tag{11.3}$$

that is,

$$\left[\frac{\partial(\Delta F/T)}{\partial T}\right]_p = -\frac{\Delta H}{T^2} \tag{11.4}$$

or

$$\left[\frac{\partial(\Delta F/T)}{\partial(1/T)}\right]_p = \Delta H. \tag{11.5}$$

Equations 11.1 and 11.2 constitute the Gibbs-Helmholtz relation, which states the relation between the change in free energy, the change in enthalpy, and the temperature coefficient of the free energy. At any temperature above absolute zero $\Delta F$ may be negative, in spite of a positive value of $\Delta H$, if the second term on the right-hand side of equation 11.2 is negative and numerically greater than the enthalpy change.[2]

The variation of $\Delta H$ with temperature is given by

$$\left[\frac{\partial(\Delta H)}{\partial T}\right]_p = \Delta Cp, \tag{11.6}$$

in which $\Delta Cp$ represents the difference between the heat capacities of the products and of the reactants. From specific heat measurements on the products and reactants, $\Delta Cp$ can be expressed as an empirical function of the temperature. Equation 11.6 can then be integrated to give

$$\Delta H = \Delta H_0 - aT - b \cdot 10^{-3}T^2 + 2c \cdot 10^5 T^{-1}, \tag{11.7}$$

where $a$, $b$, and $c$ are the empirical constants derived from specific heat measurements, and $\Delta H_0$ is an integration constant.

Equations 11.4 and 11.5 also represent the variation with temperature of the standard free-energy change,[3] denoted by $\Delta F^\circ$. By substituting equation 11.7 in equation 11.4 and integrating, we obtain

$$\Delta F^\circ = \Delta H_0 + aT \ln T + b \cdot 10^{-3}T^2 + c \cdot 10^5 T^{-1} + IT. \tag{11.8}$$

Thus, from a knowledge of the constants $\Delta H_0$, $a$, $b$, $c$, and $I$, it is possible to calculate $\Delta F^\circ$ for a given reaction. In principle, one can deduce the value of $\Delta F^\circ$ at any desired temperature from a single determination of $\Delta F^\circ$ at some other temperature and from calorimetrically measured values of the constants in equation 11.7.

In practice, the standard free-energy change for a process is often determined from measured values of the equilibrium constant, to which it is related by the equation

$$\Delta F^\circ = -R_0 T \ln K. \tag{11.9}$$

We shall now illustrate the application of the above relations to certain typical physical and chemical reactions.

## Evaporation of Solids or Liquids

The equilibrium constant for the evaporation of a solid or a liquid is simply the vapor pressure of the system, which is uniquely determined by the temperature.[4] Hence, equation 11.9 can be written as

$$\frac{\Delta F}{T} = -R_0 \ln P_{atm}. \tag{11.10}$$

Substitution of equation 11.10 into equation 11.4 yields

$$\frac{d \ln P_{atm}}{dT} = \frac{\Delta H_v}{R_0 T^2}, \tag{11.11}$$

in which $\Delta H_v$ is the molar heat of vaporization. General integration of equation 11.11, assuming $\Delta H_v$ constant, gives

$$\log P_{atm} = \frac{-\Delta H_v}{4.575T} + \text{constant}, \tag{11.12}$$

in which logarithms have been converted to the base 10, $R_0 = 1.9865$

$\text{cal} \cdot \text{deg}^{-1} \cdot \text{mole}^{-1}$, and $2.303R_0 = 4.575$. Equation 11.12 can be written in the more general form

$$\log P_{atm} = A - \frac{B}{T}, \tag{11.13}$$

which has been used in Chapter 10.

### Dissociation of an Oxide of a Nonvolatile Metal

We can express the isothermal equilibrium between a metal, its oxide, and oxygen in the form

$$M_mO_{2n}\,(s) = mM\,(s) + nO_2\,(g). \tag{11.14}$$

The standard free energy of this *dissociation* reaction is $-\Delta F^\circ$, where $\Delta F^\circ$ is the standard free energy of *formation* of a mole of $M_mO_{2n}$ from its elements at $T^\circ$ K. If we assume that oxygen is the only gaseous species, the equilibrium constant for the above reaction is given by

$$K = P^n_{atm}\,(O_2). \tag{11.15}$$

It thus follows from equation 11.9 that

$$\log P_{atm}\,(O_2) = \frac{1}{n}\frac{\Delta F^\circ}{4.575T} \tag{11.16}$$

and

$$\log P_{mm}\,(O_2) = 2.881 + \frac{1}{n}\frac{\Delta F^\circ}{4.575T}. \tag{11.17}$$

For instance, for the reaction

$$M\,(s) + 0.5O_2\,(g) = MO\,(s), \tag{11.18}$$

where M represents a divalent metal,

$$\log P_{atm}\,(O_2) = \frac{2\Delta F^\circ}{4.575T}. \tag{11.19}$$

Equations 11.16 and 11.17 apply equally well to the equilibrium between two nonvolatile oxides of a given metal and oxygen. In this case $\Delta F^\circ$ is obtained by appropriate combination of the standard free energies of formation of the two oxides. For example, the standard free energy for the formation of hematite from magnetite,

$$2Fe_3O_4\,(s) + 0.5O_2\,(g) = 3Fe_2O_3\,(s), \tag{11.20}$$

is given by

$$\Delta F^\circ = 3\Delta F^\circ_{Fe_2O_3} - 2\Delta F^\circ_{Fe_3O_4}. \tag{11.21}$$

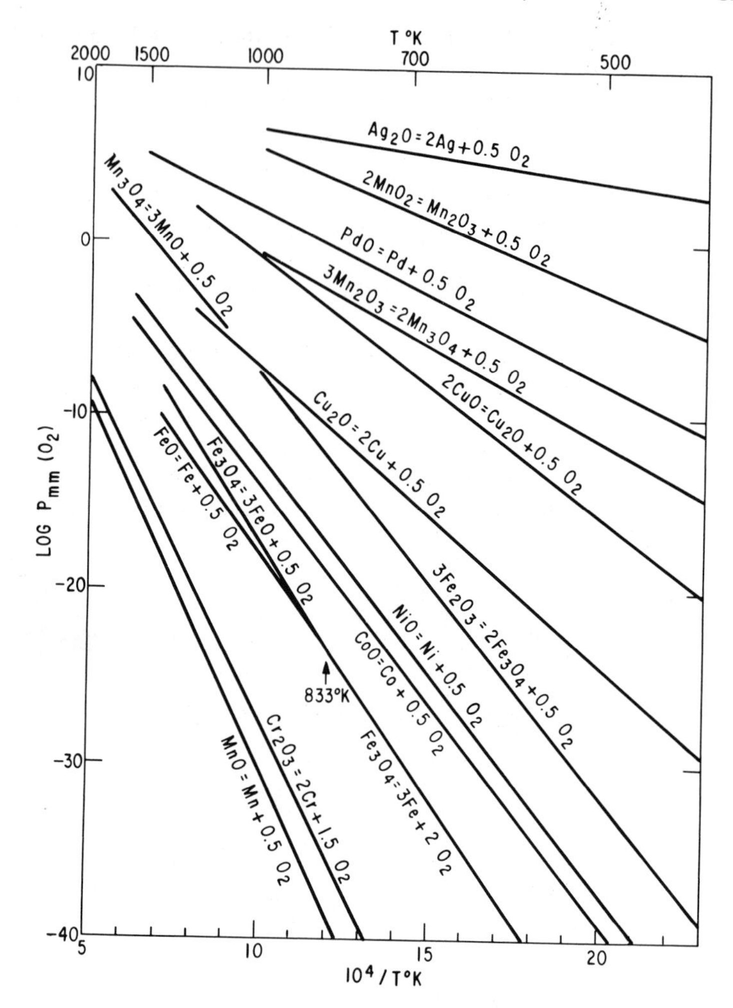


Fig. 11.1. Dissociation pressures of some oxides.

Coughlin[5] has prepared a convenient compilation of heats and free energies of formation of inorganic oxides. These data can be represented by equations 11.7 and 11.8 or, in the case of the free energy, by the following linear equation, which is satisfactory to within ±600 cal:

$$\Delta F^\circ = A + BT. \tag{11.22}$$

In Table 11.1 are listed the constants deduced by Coughlin for equations 11.7, 11.8, and 11.22 for all but one of the oxides to be discussed in this chapter. Richardson and Jeffes[6] have also prepared a compilation of linear free-energy equations for the oxides of importance in the metallurgy of iron and steel.

In the sections that follow, the dissociation pressures have been calculated from Coughlin's standard free-energy values by the use of equation 11.17. A comparison of equations 11.17 and 11.22 shows that the dissociation pressures, like vapor pressures, can be expressed as a linear function of $1/T$, using the appropriate values of $A$ and $B$ from Table 11.1. Thus

$$\log P_{mm}(O_2) = 2.881 + \frac{B}{4.575n} + \frac{A}{4.575nT}. \tag{11.23}$$

Figure 11.1 is a graphical representation of this linear relationship for some of the oxides discussed in this chapter. The calculated pressures represent an intensive thermodynamic property of the equilibrium system and provide a convenient way of comparing the relative stabilities of the metal oxides for which free energies of formation are available. Direct experimental determination of pressures becomes extremely difficult below $10^{-10}$ mm.

## Dissociation of an Oxide of a Volatile Metal

If the metal has an appreciable vapor pressure at the temperature of the dissociation reaction, equation 11.14 can be written as

$$M_mO_{2n}(s) = mM(g) + nO_2(g). \tag{11.24}$$

The free-energy change for reaction 11.24 is given by

$$\Delta F_{24} = -\Delta F^\circ + m\,\Delta F_v, \tag{11.25}$$

where $\Delta F^\circ$ is the standard molar free energy of formation of the oxide, and $\Delta F_v$ is the free energy of vaporization per gram-atom of metal. The equilibrium constant for reaction 11.24 is given by

$$K_{24} = P^m_{atm}(M) \cdot P^n_{atm}(O_2), \tag{11.26}$$

and equations 11.25 and 11.26 are related by

$$\Delta F_{24} = -R_0T \ln K_{24}. \tag{11.27}$$

If the oxide is the only source of metal vapor and oxygen gas,

$$P\,(\text{total}) = P\,(M) + P\,(O_2) \tag{11.28}$$

and

$$P\,(M) = \frac{m}{n} P\,(O_2). \tag{11.29}$$

Values of $\Delta F_v$ have been tabulated by Stull and Sinke.[7]

At temperatures above the normal boiling point of the metal, the vapor phase is the standard state of the metal to which the standard free energy

## TABLE 11.1

**Coefficients in Equations for Standard Free Energy of Formation of Oxides (Calories per Mole)**

(Coughlin)

$\Delta F^\circ = \Delta H_0 + 2.303aT \log T + b \cdot 10^{-3}T^2 + c \cdot 10^5 T^{-1} + IT$ $\quad\quad$ $\Delta F^\circ = A + BT$

| Reaction and Temperature Range (° K) of Validity | | $\Delta H_0$ | 2.303$a$ | $b$ | $c$ | $I$ | Temperature Range (° K) of Validity | $A$ | $B$ |
|---|---|---|---|---|---|---|---|---|---|
| C (*graphite*) + $\frac{1}{2}O_2$ (*g*) = CO (*g*) | (298.16–2000) | −25,400 | +2.05 | +0.27 | −1.095 | −28·79 | ... | −26,760 | −20.98 |
| C (*graphite*) + $O_2$ (*g*) = $CO_2$ (*g*) | (298.16–2000) | −93,690 | +1.63 | −0.07 | −0.23 | −5.64 | ... | −94,260 | −0.27 |
| 2Cr (*c*) + $\frac{3}{2}O_2$ (*g*) = $Cr_2O_3$ (*β*) | (298.16–1823) | −274,670 | −14.07 | +2.01 | +0.69 | +105.65 | ... | −271,300 | +61.82 |
| 2Cr (*l*) + $\frac{3}{2}O_2$ (*g*) = $Cr_2O_3$ (*β*) | (1823–2000) | −278,030 | +2.33 | −0.35 | +1.57 | +58.29 | ... | −279,190 | +66.00 |
| Co (*α*, *β*) + $\frac{1}{2}O_2$ (*g*) = CoO (*c*) | (298.16–1400) | −56,910 | +0.69 | ... | ... | +16.03 | ... | ... | ... |
| Co (*γ*) + $\frac{1}{2}O_2$ (*g*) = CoO (*c*) | (1400–1763) | −58,160 | −1.15 | ... | ... | +22.71 | (298.16–2000) | −57,380 | +18.65 |
| Co (*l*) + $\frac{1}{2}O_2$ (*g*) = CoO (*c*) | (1763–2000) | −65,680 | −6.22 | ... | ... | +43.43 | ... | ... | ... |
| 3Co (*α*, *β*, *γ*) + $2O_2$ (*g*) = $Co_3O_4$ (*c*) | (298.16–1500) | −207,300 | −2.30 | ... | ... | +90.56 | (298.16–1500) | −206,590 | +82.86 |
| 2Cu (*c*) + $\frac{1}{2}O_2$ (*g*) = $Cu_2O$ (*c*) | (298.16–1357) | −40,550 | −1.15 | −1.10 | −0.10 | +21.92 | (298.16–1357) | −39,630 | +16.15 |
| 2Cu (*l*) + $\frac{1}{2}O_2$ (*g*) = $Cu_2O$ (*c*) | (1357–1502) | −43,880 | +8.47 | −2.60 | −0.10 | −3.72 | (1357–1502) | −43,200 | +18.66 |
| 2Cu (*l*) + $\frac{1}{2}O_2$ (*g*) = $Cu_2O$ (*l*) | (1502–2000) | −37,710 | −12.48 | +0.25 | −0.10 | +54.44 | (1502–2000) | −28,510 | +8.91 |
| Cu (*c*) + $\frac{1}{2}O_2$ (*g*) = CuO (*c*) | (298.16–1357) | −37.740 | −0.64 | −1.40 | −0.10 | +24.87 | (298.16–1357) | −36,850 | +20.48 |
| Cu (*l*) + $\frac{1}{2}O_2$ (*g*) = CuO (*c*) | (1357–1720) | −39,410 | +4.17 | −2.15 | −0.10 | +12.05 | (1357–1720) | −37,300 | +20.66 |
| Cu (*l*) + $\frac{1}{2}O_2$ (*g*) = CuO (*l*) | (1720–2000) | −41,060 | −11.35 | +0.25 | −0.10 | +59.09 | (1720–2000) | −33,110 | +18.21 |
| $H_2$ (*g*) + $\frac{1}{2}O_2$ (*g*) = $H_2O$ (*l*) | (298.16–373.16) | −70,600 | −18.26 | +0.64 | −0.04 | +91.67 | (298.16–425) | −44,140 | +53.80 |
| $H_2$ (*g*) + $\frac{1}{2}O_2$ (*g*) = $H_2O$ (*g*) | (298.16–2000) | −56,930 | +6.75 | −0.64 | −0.08 | −8.74 | (298.16–1500) | −33,120 | +27.90 |
| 0.947Fe (*α*) + $\frac{1}{2}O_2$ (*g*) = $Fe_{0.947}O$ (*c*) | (298.16–1033) | −65,320 | −11.26 | +2.61 | +0.44 | +48.60 | ... | ... | ... |
| 0.947Fe (*β*) + $\frac{1}{2}O_2$ (*g*) = $Fe_{0.947}O$ (*c*) | (1033–1179) | −62,380 | +4.08 | −0.75 | +0.235 | +3.00 | ... | ... | ... |
| 0.947Fe (*γ*) + $\frac{1}{2}O_2$ (*g*) = $Fe_{0.947}O$ (*c*) | (1179–1650) | −66,750 | −8.04 | +0.67 | −0.10 | +42.28 | (298.16–1650) | −63,200 | +15.47 |
| 0.947Fe (*γ*) + $\frac{1}{2}O_2$ (*g*) = $Fe_{0.947}O$ (*l*) | (1650–1674) | −64,200 | −18.72 | +1.67 | −0.10 | +73.45 | (1650–2000) | −57,070 | +11.60 |
| 0.947Fe (*δ*) + $\frac{1}{2}O_2$ (*g*) = $Fe_{0.947}O$ (*l*) | (1674–1803) | −59,650 | −6.84 | +0.25 | −0.10 | +34.81 | ... | ... | ... |
| 0.947Fe (*l*) + $\frac{1}{2}O_2$ (*g*) = $Fe_{0.947}O$ (*l*) | (1803–2000) | −63,660 | −7.48 | +0.25 | −0.10 | +39.12 | ... | ... | ... |
| 3Fe (*α*) + $2O_2$ (*g*) = $Fe_3O_4$ (*magnetite*) | (298.16–900) | −268,310 | +5.87 | −12.45 | +0.245 | +73.11 | ... | ... | ... |
| 3Fe (*α*) + $2O_2$ (*g*) = $Fe_3O_4$ (*β*) | (900–1033) | −272,300 | −54.27 | +11.65 | +0.245 | +233.52 | ... | ... | ... |
| 3Fe (*β*) + $2O_2$ (*g*) = $Fe_3O_4$ (*β*) | (1033–1179) | −262,990 | −5.71 | +1.00 | −0.40 | +89.19 | (298.16–900) | −265,660 | +76.81 |
| 3Fe (*γ*) + $2O_2$ (*g*) = $Fe_3O_4$ (*β*) | (1179–1674) | −276,990 | −44.05 | +5.50 | −0.40 | +213.52 | (900–1803) | −261,200 | +71.36 |
| 3Fe (*δ*) + $2O_2$ (*g*) = $Fe_3O_4$ (*β*) | (1674–1803) | −262,560 | −6.40 | +1.00 | −0.40 | +91.05 | (1803–2000) | −248,240 | +64.42 |

| | | | | | | | | | |
|---|---|---|---|---|---|---|---|---|---|
| $3Fe\ (l) + 2O_2\ (g) = Fe_3O_4\ (\beta)$ | (1803–1874) | −275,280 | −8.74 | +1.00 | −0.40 | +104.84 | … | … | … |
| $3Fe\ (l) + 2O_2\ (g) = Fe_3O_4\ (l)$ | (1874–2000) | −257,240 | −26.89 | +1.00 | −0.40 | +155.46 | … | … | … |
| $2Fe\ (\alpha) + \frac{3}{2}O_2\ (g) = Fe_2O_3$ (*hematite*) | (298.16–950) | −200,000 | −13.84 | −1.45 | +1.905 | +108.26 | … | … | … |
| $2Fe\ (\alpha) + \frac{3}{2}O_2\ (g) = Fe_2O_3\ (\beta)$ | (950–1033) | −202,960 | −42.64 | +7.85 | +0.13 | +188.48 | (298.16–950) | −195,450 | +61.38 |
| $2Fe\ (\beta) + \frac{3}{2}O_2\ (g) = Fe_2O_3\ (\beta)$ | (1033–1050) | −196,740 | −10.27 | +0.75 | −0.30 | +92.26 | (950–1179) | −192,800 | +58.30 |
| $2Fe\ (\beta) + \frac{3}{2}O_2\ (g) = Fe_2O_3\ (\gamma)$ | (1050–1179) | −193,200 | −0.39 | −0.13 | −0.30 | +59.96 | (1179–1800) | −192,550 | +58.12 |
| $2Fe\ (\gamma) + \frac{3}{2}O_2\ (g) = Fe_2O_3\ (\gamma)$ | (1179–1674) | −202,540 | −25.95 | +2.87 | −0.30 | +142.85 | … | … | … |
| $2Fe\ (\delta) + \frac{3}{2}O_2\ (g) = Fe_2O_3\ (\gamma)$ | (1674–1800) | −192,920 | −0.85 | −0.13 | −0.30 | +61.21 | … | … | … |
| $Mn\ (\alpha) + \frac{1}{2}O_2\ (g) = MnO\ (c)$ | (298.16–1000) | −92,600 | −4.21 | +0.97 | +0.155 | +29.66 | … | … | … |
| $Mn\ (\beta) + \frac{1}{2}O_2\ (g) = MnO\ (c)$ | (1000–1374) | −91,900 | +1.84 | −0.39 | +0.34 | +12.15 | (298.16–1374) | −91,980 | +17.48 |
| $Mn\ (\gamma) + \frac{1}{2}O_2\ (g) = MnO\ (c)$ | (1374–1410) | −89,810 | +7.30 | −0.72 | +0.34 | −6.05 | (1374–1517) | −93,000 | +18.25 |
| $Mn\ (\delta) + \frac{1}{2}O_2\ (g) = MnO\ (c)$ | (1410–1517) | −89,390 | +8.68 | −0.72 | +0.34 | −10.70 | (1517–2000) | −97,360 | +21.12 |
| $Mn\ (l) + \frac{1}{2}O_2\ (g) = MnO\ (c)$ | (1517–2000) | −93,350 | +7.99 | −0.72 | +0.34 | −5.90 | … | … | … |
| $3Mn\ (\alpha) + 2O_2\ (g) = Mn_3O_4\ (\alpha)$ | (298.16–1000) | −332,400 | −7.41 | +0.66 | +0.145 | +106.62 | … | … | … |
| $3Mn\ (\beta) + 2O_2\ (g) = Mn_3O_4\ (\alpha)$ | (1000–1374) | −330,310 | +10.75 | −3.42 | +0.70 | +54.07 | (298.16–1374) | −330,900 | +83.72 |
| $3Mn\ (\gamma) + 2O_2\ (g) = Mn_3O_4\ (\alpha)$ | (1374–1410) | −324,050 | +27.12 | −4.41 | +0.70 | −0.50 | (1374–1517) | −330,600 | +83.50 |
| $3Mn\ (\delta) + 2O_2\ (g) = Mn_3O_4\ (\alpha)$ | (1410–1445) | −322,800 | +31.27 | −4.41 | +0.70 | −14.46 | (1517–1800) | −339,000 | +89.03 |
| $3Mn\ (\delta) + 2O_2\ (g) = Mn_3O_4\ (\beta)$ | (1445–1517) | −328,870 | −4.56 | +1.00 | −0.40 | +95.20 | … | … | … |
| $3Mn\ (l) + 2O_2\ (g) = Mn_3O_4\ (\beta)$ | (1517–1800) | −340,730 | −6.63 | +1.00 | −0.40 | +109.60 | … | … | … |
| $2Mn\ (\alpha) + \frac{3}{2}O_2\ (g) = Mn_2O_3\ (c)$ | (298.16–1000) | −230,610 | −5.96 | −0.06 | +0.945 | +80.74 | … | … | … |
| $2Mn\ (\beta) + \frac{3}{2}O_2\ (g) = Mn_2O_3\ (c)$ | (1000–1374) | −229,210 | +6.15 | −2.78 | +1.315 | +45.70 | (298.16–1517) | −228,840 | +61.08 |
| $2Mn\ (\gamma) + \frac{3}{2}O_2\ (g) = Mn_2O_3\ (c)$ | (1374–1410) | −225,030 | +17.06 | −3.44 | +1.315 | +9.33 | (1517–1700) | −236,270 | +65.96 |
| $2Mn\ (\delta) + \frac{3}{2}O_2\ (g) = Mn_2O_3\ (c)$ | (1410–1517) | −224,200 | +19.82 | −3.44 | +1.315 | +0.05 | … | … | … |
| $2Mn\ (l) + \frac{3}{2}O_2\ (g) = Mn_2O_3\ (c)$ | (1517–1700) | −232,110 | +18.44 | −3.44 | +1.315 | +9.65 | … | … | … |
| $Mn\ (\alpha) + O_2\ (g) = MnO_2\ (c)$ | (298.16–1000) | −126,400 | −8.61 | +0.97 | +1.555 | +70.14 | (298.16–1000) | −124,140 | +43.14 |
| $Hg\ (l) + \frac{1}{2}O_2\ (g) = HgO$ (*red*) | (298.16–629.88) | −21,760 | +0.85 | −2.47 | −0.10 | +24.81 | (298.16–629.88) | −21,470 | +25.18 |
| $Hg\ (g) + \frac{1}{2}O_2\ (g) = HgO$ (*red*) | (629.88–1500) | −36,920 | −2.92 | −2.47 | −0.10 | +59.42 | (629.88–1500) | −33,140 | +44.12 |
| $Mo\ (c) + O_2\ (g) = MoO_2\ (c)$ | (298.16–2000) | −132,910 | −3.91 | … | … | +47.42 | (298.16–2000) | −131,530 | +33.95 |
| $Ni\ (\alpha) + \frac{1}{2}O_2\ (g) = NiO\ (c)$ | (298.16–633) | −57,640 | −4.61 | +2.16 | −0.10 | +34.41 | … | … | … |
| $Ni\ (\beta) + \frac{1}{2}O_2\ (g) = NiO\ (c)$ | (633–1725) | −57,460 | −0.14 | −0.46 | −0.10 | +23.27 | (298.16–2000) | −57,370 | +22.09 |
| $Ni\ (l) + \frac{1}{2}O_2\ (g) = NiO\ (c)$ | (1725–2000) | −58,830 | +7.23 | −1.36 | −0.10 | +1.76 | … | … | … |
| $2Ag\ (c) + \frac{1}{2}O_2\ (g) = Ag_2O\ (c)$ | (298.16–1000) | −7,740 | −4.14 | … | … | +27.84 | (298.16–1000) | −6,720 | +44.49 |
| $W\ (c) + O_2\ (g) = WO_2\ (c)$ | (298.16–1500) | −137,180 | −1.38 | … | … | +45.56 | … | −136,750 | +40.93 |

of formation values refer. Consequently the $\Delta F_v$ term in equation 11.25 is zero.

### Reduction of an Oxide by Hydrogen

Let us now consider the reaction

$$2n\text{H}_2\,(g) + \text{M}_m\text{O}_{2n}\,(s) = 2n\text{H}_2\text{O}\,(g) + m\text{M}\,(s). \tag{11.30}$$

The standard free energy for this reduction is given by

$$\Delta F^\circ_R = 2n\,\Delta F^\circ_{\text{H}_2\text{O}} - \Delta F^\circ_{\text{M}_m\text{O}_{2n}}, \tag{11.31}$$

and its equilibrium constant is

$$K_R = \left[\frac{P\,(\text{H}_2\text{O})}{P\,(\text{H}_2)}\right]^{2n}. \tag{11.32}$$

Because the ratio $P\,(H_2O)/P\,(H_2)$ is independent of the units used for the measurement of pressure, the notation $(H_2O)/(H_2)$ has frequently been used to characterize the equilibrium involved in the reduction of an oxide by hydrogen. If carbon monoxide is the reducing agent, the equilibrium is characterized by the ratio $(CO_2)/(CO)$. The higher the ratio, the lower the equilibrium oxygen pressure in the system and the more complete the reduction. In general, a reducing environment is one in which the oxygen pressure is maintained at a value less than the dissociation pressure of the oxide at the same temperature.

Because $\Delta F^\circ_R$ and $K_R$ are related as shown in equation 11.9, values of $(H_2O)/(H_2)$ can be obtained from standard free-energy values for water and for the oxide.

## 11.2. FREE ENERGY OF FORMATION OF WATER VAPOR

The preceding section has shown that equilibrium calculations for the reduction of metal oxides by hydrogen require a knowledge of the free energy of formation of water vapor. Values of $\Delta F^\circ_W$ from Coughlin[5] are listed in Table 11.2; the appropriate constants for equations 11.7, 11.8, and 11.22 are in Table 11.1.

The formation reaction may be written as

$$\text{H}_2\,(g) + 0.5\text{O}_2\,(g) = \text{H}_2\text{O}\,(g). \tag{11.33}$$

Since all the species in this equilibrium are gaseous, the equilibrium constant is given by

$$K_W = \frac{P_{atm}\,(\text{H}_2\text{O})}{P_{atm}\,(\text{H}_2)\,P^{1/2}_{atm}(\text{O}_2)}. \tag{11.34}$$

TABLE 11.2

**Values of $\Delta F^\circ_W$, $\log K_W$, and $100\alpha$ at 760 mm and 0.76 mm**

| $T^\circ$ K | $10^3/T$ | $\Delta F^\circ_W$ | $\log K_W$ | $100\alpha$ |  |
|---|---|---|---|---|---|
|  |  |  |  | $P_{mm} = 760$ | $P_{mm} = 0.76$ |
| 298 | 3.356 | −54,635 | 40.0742 | $2.42 \cdot 10^{-25}$ | $2.42 \cdot 10^{-24}$ |
| 600 | 1.667 | −51,150 | 18.6339 | $4.76 \cdot 10^{-11}$ | $4.76 \cdot 10^{-10}$ |
| 800 | 1.250 | −48,640 | 13.2899 | $1.74 \cdot 10^{-7}$ | $1.74 \cdot 10^{-6}$ |
| 1000 | 1.000 | −46,030 | 10.0612 | $2.47 \cdot 10^{-5}$ | $2.47 \cdot 10^{-4}$ |
| 1200 | 0.8333 | −43,360 | 7.8980 | $6.84 \cdot 10^{-4}$ | $6.84 \cdot 10^{-3}$ |
| 1400 | 0.7143 | −40,640 | 6.3451 | $7.42 \cdot 10^{-3}$ | $7.42 \cdot 10^{-2}$ |
| 1600 | 0.6250 | −37,880 | 5.1748 | $4.47 \cdot 10^{-2}$ | $4.47 \cdot 10^{-1}$ |
| 1800 | 0.5556 | −35,110 | 4.2634 | $1.81 \cdot 10^{-1}$ | 1.81 |
| 2000 | 0.5000 | −32,310 | 3.5312 | $5.57 \cdot 10^{-1}$ | 5.57 |

Values of $\log K_W$, calculated by use of equation 11.9, are also tabulated in Table 11.2, along with values of the degree of dissociation $\alpha$, derived as follows.

From 1 mole $H_2O$ vapor, at pressure $P_{atm}$, there are produced at equilibrium $(1 - \alpha)$ moles $H_2O$, $\alpha$ moles $H_2$, and $0.5\alpha$ mole $O_2$. Hence, the total number of moles at equilibrium is $(1 + 0.5\alpha)$, and it follows from equation 11.34 that

$$K_W = \frac{(1 - \alpha)(2 + \alpha)^{1/2}}{\alpha^{3/2} P_{atm}^{1/2}}, \tag{11.35}$$

that is,

$$\alpha^3 = \frac{(1 - \alpha)^2(2 + \alpha)}{K_W{}^2 P_{atm}}. \tag{11.36}$$

For $\alpha \ll 0.01$,

$$\alpha^3 = \frac{2}{K_W{}^2 P_{atm}} = \frac{1520}{K_W{}^2 \cdot P_{mm}}. \tag{11.37}$$

From these relations it follows that, at constant temperature, $\alpha$ varies inversely as $P^{1/3}$.

The last two columns in Table 11.2 give values of $100\alpha$, the per cent dissociation at $P_{mm} = 760$ mm and $P_{mm} = 0.76$ mm, respectively, calculated by means of the relations

$$\log \alpha = 0.1003 - \tfrac{2}{3} \log K_W \text{ (for } P_{mm} = 760),$$

and

$$\log \alpha = 1.1003 - \tfrac{2}{3} \log K_W \text{ (for } P_{mm} = 0.76).$$

The value of $\alpha$ for $T = 2000$ and $P_{mm} = 0.76$ is only approximate, since equation 11.37 is no longer valid for such large values of $\alpha$.

In the presence of hydrogen or oxygen, the degree of dissociation of the water vapor is depressed, as may be deduced from the following considerations.

Let $\beta$ = number of moles added $O_2$ per mole $H_2O$, and let $\alpha_1$ denote the degree of dissociation obtained as a result of the presence of additional oxygen. Then,

$$K_W = \frac{(1-\alpha_1)}{\alpha_1 P_{atm}^{1/2}}\left(\frac{2+\alpha_1+2\beta}{\alpha_1+2\beta}\right)^{1/2}. \tag{11.38}$$

For $\alpha_1 \ll 0.01$, and $\beta$ large compared to $\alpha_1$, this equation leads to the relation

$$\alpha_1{}^2 = \frac{1}{K_W{}^2 P_{atm}}\left(\frac{1+\beta}{\beta}\right). \tag{11.39}$$

Comparing this with equation 11.36, we find that, for the same value of $P$,

$$\left(\frac{\alpha_1}{\alpha_0}\right)^2 = \frac{\alpha_0}{2}\left(\frac{1+\beta}{\beta}\right), \tag{11.40}$$

which shows that $\alpha_1$ must always be considerably less than $\alpha_0$, where $\alpha_0$ denotes the degree of dissociation in the absence of excess of either oxygen or hydrogen.

## 11.3. DISSOCIATION PRESSURES OF OXIDES OF SILVER, PALLADIUM, AND MERCURY

Three relatively unstable metal oxides are those of silver, palladium, and mercury.

Pitzer and Smith[8] have reviewed the different observations that have been made on the dissociation pressures of silver oxide, $Ag_2O$, in the temperature range 446–773° K. Selected values are listed in Table 11.3, along with pressures calculated from free-energy values below 446° K.

The dissociation pressures of palladium oxide, PdO, also given in Table 11.3, include experimental values obtained by Schenck and Kurzen[9] in the range 963–1063° K.

The dissociation of mercuric oxide, HgO, affords a good example of an equilibrium involving the oxide of a volatile metal. The normal boiling point of mercury is 629.88° K.

As seen by the equation

$$HgO\,(s) = Hg\,(g) + 0.5O_2\,(g), \tag{11.41}$$

TABLE 11.3

Dissociation Pressures of Oxides of Silver and Palladium

| $Ag_2O = 2Ag + 0.5O_2$ | | $PdO = Pd + 0.5O_2$ | |
|---|---|---|---|
| $T$ (° K) | $P_{mm}(O_2)$ | $T$ (° K) | $P_{mm}(O_2)$ |
| 298 | 0.164 | 700 | 0.003 |
| 400 | 79.0 | 800 | 0.146 |
| 446 | 422 | 900 | 3.54 |
| 575 | $1.56 \cdot 10^4$ | 963 | 10.9 |
| 676 | $8.70 \cdot 10^4$ | 1013 | 41.3 |
| 773 | $2.95 \cdot 10^5$ | 1063 | 138 |

the pressure of mercury, $P$(Hg), is twice $P(O_2)$, the pressure of oxygen, provided that no liquid mercury is present. This relationship is illustrated in columns 2 and 3 of Table 11.4, which is based upon the results obtained

TABLE 11.4

Dissociation Pressures of Mercuric Oxide

$HgO = Hg(g) + 0.5O_2$

| No Liquid Hg Present | | | | Liquid Hg Present | |
|---|---|---|---|---|---|
| $T$ (° K) | $P_{mm}$(Hg) | $P_{mm}(O_2)$ | $K^*$ | $P_{mm}$(Hg) | $P_{mm}(O_2)$ |
| 500 | 0.446 | 0.223 | 0.210 | 39.64 | $2.82 \cdot 10^{-5}$ |
| 550 | 3.54 | 1.77 | 4.74 | 146.2 | $1.05 \cdot 10^{-3}$ |
| 600 | 20.0 | 10.0 | 63.5 | 433.3 | $2.15 \cdot 10^{-2}$ |
| 650 | 86.8 | 43.4 | 571 | 1082.7 | $2.78 \cdot 10^{-1}$ |
| 700 | 304 | 152 | 3750 | (2340)† | 2.57 |

† Extrapolated.

by Taylor and Hulett[10] in their investigation of this equilibrium over the range 633–753° K. Their complete results may be represented by the equation

$$\log P_{mm}(O_2) = 9.268 - \frac{4960}{T}, \tag{11.42}$$

which has been used to obtain the pressure values listed in column 3, including those extrapolated below 633° K.

The equilibrium constant[11] for this dissociation is

$$K^* = P_{mm}(\mathrm{Hg})\, P_{mm}^{1/2}(O_2). \tag{11.43}$$

Equation 11.43 is valid whether liquid mercury is present or not. In the absence of liquid mercury, it can be simplified to

$$K^* = 2P_{mm}^{3/2}\,(O_2). \tag{11.44}$$

By combining equations 11.42 and 11.44, we can use Taylor and Hulett's data to express $K^*$ as a function of temperature:

$$\log K^* = 14.203 - \frac{7440}{T}. \tag{11.45}$$

If liquid mercury is present in the equilibrium chamber, the pressure of mercury in the gas phase will be the vapor pressure of mercury, values of which, from standard tables, are given in column 5 of Table 11.4. Since the value of $K^*$ must remain constant at any given temperature, the oxygen pressures in the presence of both solid mercuric oxide and liquid mercury must drop to the values given in column 6. In other words, the degree of dissociation of mercuric oxide is decreased in the presence of liquid mercury. A corollary is that the dissociation of mercuric oxide gives rise, at a given temperature, to a pressure of mercury vapor that is less than the vapor pressure of liquid mercury at the same temperature.

The same results can be obtained by using free-energy data as described for oxides of volatile metals in section 11.1.

## 11.4. DISSOCIATION AND REDUCTION OF OXIDES OF COPPER

The important dissociation reactions for copper oxides are

$$2CuO = Cu_2O + 0.5O_2 \tag{11.46}$$

and

$$Cu_2O = 2Cu + 0.5O_2. \tag{11.47}$$

The experimental data on these reactions have been reviewed by Randall, Nielsen, and West.[12] Calculated values of the dissociation pressures and of $(H_2O)/(H_2)$ are given in Table 11.5 for temperatures below the melting point of copper, 1357° K.

Dissociation pressures for reaction 11.47 are too low for direct measurement and must be calculated from free-energy data, which have been determined experimentally from measurements of $(H_2O)/(H_2)$ in the reduction of $Cu_2O$ to copper by hydrogen[13] or, more reliably, from measurements of electrical potentials in copper-cuprous oxide cells.[14]

Direct measurements have been made above 1000° K for reaction 11.46 by Smyth and Roberts,[15] whose data are in satisfactory agreement with those of Table 11.5. By use of a mass spectrometer, which can measure accurately oxygen pressures as low as $10^{-6}$ mm, Norton[16] has extended the range of direct measurement of the equilibria of reaction 11.46 down

TABLE 11.5

**Dissociation Pressures for Oxides of Copper and Equilibrium Constants for Their Reduction by Hydrogen**

| | $2CuO = Cu_2O + 0.5O_2$ | $2CuO + H_2 = Cu_2O + H_2O$ | $Cu_2O = 2Cu + 0.5O_2$ | $Cu_2O + H_2 = 2Cu + H_2O$ |
|---|---|---|---|---|
| $T(° K)$ | $P_{mm}(O_2)$ | $(H_2O)/(H_2)$ | $P_{mm}(O_2)$ | $(H_2O)/(H_2)$ |
| 600 | $1.09 \cdot 10^{-11}$ | $5.16 \cdot 10^{11}$ | $1.48 \cdot 10^{-19}$ | $6.00 \cdot 10^{7}$ |
| 700 | $4.36 \cdot 10^{-8}$ | $2.91 \cdot 10^{10}$ | $1.86 \cdot 10^{-15}$ | $6.01 \cdot 10^{6}$ |
| 800 | $1.93 \cdot 10^{-5}$ | $3.10 \cdot 10^{9}$ | $2.51 \cdot 10^{-12}$ | $1.12 \cdot 10^{6}$ |
| 900 | $1.97 \cdot 10^{-3}$ | $5.09 \cdot 10^{8}$ | $6.11 \cdot 10^{-10}$ | $2.83 \cdot 10^{5}$ |
| 1000 | $8.00 \cdot 10^{-2}$ | $1.18 \cdot 10^{8}$ | $4.96 \cdot 10^{-8}$ | $9.30 \cdot 10^{4}$ |
| 1100 | 1.81 | $3.72 \cdot 10^{7}$ | $1.65 \cdot 10^{-6}$ | $3.55 \cdot 10^{4}$ |
| 1200 | $1.90 \cdot 10^{1}$ | $1.25 \cdot 10^{7}$ | $3.33 \cdot 10^{-5}$ | $1.65 \cdot 10^{4}$ |
| 1300 | $1.50 \cdot 10^{2}$ | $5.11 \cdot 10^{6}$ | $3.91 \cdot 10^{-4}$ | $8.27 \cdot 10^{3}$ |

to 750° K. His results fall on the extrapolated extension of Smyth and Roberts' high-temperature data and may be represented by

$$\log P_{mm}(O_2) = \frac{-13{,}236}{T} + 12.36. \tag{11.48}$$

The dissociation of CuO to $Cu_2O$ is the only one of the above reactions for which the oxygen pressures are sufficiently high at temperatures of 900° K and above to be of interest in actual exhaust operations. On the other hand, because the values of $(H_2O)/(H_2)$ are very large for the reductions corresponding to reactions 11.46 and 11.47, even at 1300° K, the reduction of the oxides to metal occurs readily, even when the hydrogen contains considerable water vapor. If, however, the ratio $(H_2O)/(H_2)$ exceeds that listed in Table 11.5, oxidation rather than reduction occurs. For instance, at 1200° K, a steam-hydrogen mixture will oxidize copper if the value of this ratio is greater than $1.65 \cdot 10^4$.

Although the reduction of CuO by hydrogen begins at 200° C, a reaction rate rapid enough for industrial practice is not reached until about 500° C. Thermodynamic calculations, such as those made above, supply *no criterion whatever regarding the velocity of a reaction.* For example, Tatievskaya and coworkers[17,18] have shown that, although the dissociation pressure for the CuO–$Cu_2O$ equilibrium is about $10^8$ times higher than that for the $Cu_2O$–Cu equilibrium, the rates of reduction of both oxides by hydrogen are approximately equal at 473–623° K. The rate of reduction is not related to the dissociation pressure of the oxide but is governed by the adsorption of hydrogen on the oxide surface.

When hydrogen is used to eliminate oxide impurities from copper, precautions must be taken to avoid the phenomenon of hydrogen "embrittlement," described in section 8.14.

## 11.5. DISSOCIATION OF OXIDES OF IRON AND THEIR REDUCTION BY HYDROGEN

Below 833° K only two oxides of iron, hematite ($Fe_2O_3$) and magnetite ($Fe_3O_4$), are thermodynamically stable. Their dissociation reactions may be written as

$$3Fe_2O_3 = 2Fe_3O_4 + 0.5O_2 \tag{11.49}$$

and

$$Fe_3O_4 = 3Fe + 2O_2. \tag{11.50}$$

Above 833° K, wüstite (FeO) is a thermodynamically stable phase, and the following dissociation reactions are possible:

$$Fe_3O_4 = 3FeO + 0.5O_2 \tag{11.51}$$

and

$$FeO = Fe + 0.5O_2. \tag{11.52}$$

These relationships are illustrated in Figure 11.1 by the intersection at 833° K of the lines for reactions 11.50, 11.51, and 11.52.

Below 1400° K dissociation pressures for all but reaction 11.49 are too low for direct measurement. The free-energy values from which they have been calculated are generally based upon measurements of reduction equilibria involving carbon monoxide or hydrogen,[19,20] and upon heat-capacity measurements.[21] By use of a mass-spectrometer technique Norton[22] has made direct measurements of the $Fe_2O_3$–$Fe_3O_4$ equilibrium down to 1100° K, where the oxygen pressure is of the order of $10^{-6}$ mm. His data are in satisfactory agreement with the calculated values in Table 11.6 and may be represented by

$$\log P_{mm}(O_2) = \frac{-24{,}912}{T} + 17{,}281. \tag{11.53}$$

Although the values of dissociation pressure and $(H_2O)/(H_2)$ in Table 11.6 have been calculated in accordance with the stoichiometry of the above equations, none of these oxides is a true stoichiometric compound. This is particularly so for wüstite, which is a phase of variable composition, whose atomic ratio of iron to oxygen is less than one.[19] The free-energy values in Table 11.1 are actually for wüstite of a composition $Fe_{0.947}O$.

Smiltens[23] has calculated the corrections necessary to account for the nonstoichiometric compositions of the oxides in reaction 11.49. The corrected free-energy values, however, do not differ significantly from those given by Coughlin in Table 11.1.

TABLE 11.6
Dissociation Pressures for Oxides of Iron and Equilibrium Constants for Their Reduction by Hydrogen

| $T$ (° K) | $3Fe_2O_3 = 2Fe_3O_4 + 0.5O_2$ | $3Fe_2O_3 + H_2 = 2Fe_3O_4 + H_2O$ | $Fe_3O_4 = 3Fe + 2O_2$ | $Fe_3O_4 + 4H_2 = 3Fe + 4H_2O$ | $Fe_3O_4 = 3FeO + 0.5O_2$ | $Fe_3O_4 + H_2 = 3FeO + H_2O$ | $FeO = Fe + 0.5O_2$ | $FeO + H_2 = Fe + H_2O$ |
|---|---|---|---|---|---|---|---|---|
| | $P_{mm}$ ($O_2$) | ($H_2O$)/($H_2$) | $P_{mm}$ ($O_2$) | ($H_2O$)/($H_2$) | $P_{mm}$ ($O_2$) | ($H_2O$)/($H_2$) | $P_{mm}$ ($O_2$) | ($H_2O$)/($H_2$) |
| 400 | $2.11 \cdot 10^{-44}$ | $9.27 \cdot 10^{5}$ | $5.02 \cdot 10^{-62}$ | $1.43 \cdot 10^{-3}$ | | | | |
| 500 | $1.48 \cdot 10^{-32}$ | $3.42 \cdot 10^{5}$ | $1.93 \cdot 10^{-47}$ | $1.24 \cdot 10^{-2}$ | | | | |
| 600 | $1.39 \cdot 10^{-24}$ | $1.84 \cdot 10^{5}$ | $9.37 \cdot 10^{-38}$ | $4.78 \cdot 10^{-2}$ | | | | |
| 700 | $7.89 \cdot 10^{-19}$ | $1.24 \cdot 10^{5}$ | $7.24 \cdot 10^{-31}$ | $1.19 \cdot 10^{-1}$ | | | | |
| 800 | $1.86 \cdot 10^{-14}$ | $9.63 \cdot 10^{4}$ | $9.96 \cdot 10^{-26}$ | $2.23 \cdot 10^{-1}$ | $6.22 \cdot 10^{-26}$ | 0.176 | $1.17 \cdot 10^{-25}$ | 0.241 |
| 900 | $6.53 \cdot 10^{-11}$ | $9.26 \cdot 10^{4}$ | $9.09 \cdot 10^{-22}$ | $3.46 \cdot 10^{-1}$ | $1.50 \cdot 10^{-21}$ | 0.444 | $4.39 \cdot 10^{-22}$ | .318 |
| 1000 | $3.31 \cdot 10^{-8}$ | $7.60 \cdot 10^{4}$ | | | $4.16 \cdot 10^{-18}$ | 0.851 | $9.18 \cdot 10^{-19}$ | .400 |
| 1100 | $5.42 \cdot 10^{-6}$ | $6.44 \cdot 10^{4}$ | | | $3.12 \cdot 10^{-15}$ | 1.54 | $2.89 \cdot 10^{-16}$ | .470 |
| 1200 | $4.12 \cdot 10^{-4}$ | $5.82 \cdot 10^{4}$ | | | $7.44 \cdot 10^{-13}$ | 2.47 | $3.63 \cdot 10^{-14}$ | .547 |
| 1300 | $1.28 \cdot 10^{-2}$ | $4.72 \cdot 10^{4}$ | | | $6.82 \cdot 10^{-11}$ | 3.45 | $2.26 \cdot 10^{-12}$ | .628 |
| 1400 | $3.00 \cdot 10^{-1}$ | $4.40 \cdot 10^{4}$ | | | $4.06 \cdot 10^{-9}$ | 5.11 | $7.24 \cdot 10^{-11}$ | .683 |

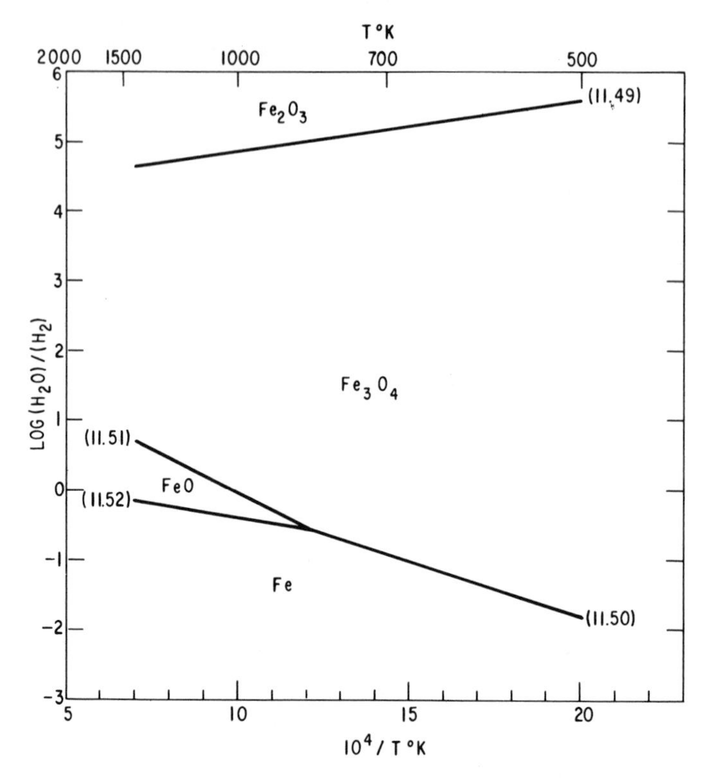


Fig. 11.2. Plots of the log of the equilibrium-pressure ratios $(H_2O)/(H_2)$ versus $1/T$ for the oxidation of iron by water vapor and the reduction of oxides of iron by hydrogen.

Figure 11.2 shows the hydrogen-reduction equilibria that correspond to equations 11.49, 11.50, 11.51, and 11.52. Below 833° K, the reduction of $Fe_3O_4$ to Fe can be prevented by maintaining a ratio of $(H_2O)/(H_2)$ that lies above curve 11.50. Above 833° K the reduction may yield FeO if the ratio lies between curves 11.51 and 11.52. The ratio required to prevent the reduction of $Fe_2O_3$ to $Fe_3O_4$ is so high that this reaction can be considered as practically irreversible.

## 11.6. DISSOCIATION AND REDUCTION OF OXIDES OF NICKEL AND COBALT

In Table 11.7 are listed calculated values of $P_{mm}(O_2)$ and $(H_2O)/(H_2)$ for the dissociation and reduction equilibria involving the following pairs of solid phases: NiO-Ni, CoO-Co, and $Co_3O_4$-CoO.

TABLE 11.7

**Dissociation Pressures for Oxides of Nickel and Cobalt and Equilibrium Constants for Their Reduction by Hydrogen**

| | $NiO = Ni + 0.5O_2$ | $NiO + H_2 = Ni + H_2O$ | $CoO = Co + 0.5O_2$ | $CoO + H_2 = Co + H_2O$ | $Co_3O_4 = 3CoO + 0.5O_2$ | $Co_3O_4 + H_2 = 3CoO + H_2O$ |
|---|---|---|---|---|---|---|
| $T$ (° K) | $P_{mm}$ ($O_2$) | ($H_2O$)/($H_2$) | $P_{mm}$ ($O_2$) | ($H_2O$)/($H_2$) | $P_{mm}$ ($O_2$) | ($H_2O$)/($H_2$) |
| 600 | | | | | $2.32 \cdot 10^{-11}$ | $7.52 \cdot 10^{11}$ |
| 800 | | | | | $1.12 \cdot 10^{-4}$ | $7.49 \cdot 10^{9}$ |
| 1000 | $3.51 \cdot 10^{-13}$ | 248 | $8.30 \cdot 10^{-15}$ | 38.1 | $5.99 \cdot 10^{-1}$ | $3.23 \cdot 10^{8}$ |
| 1200 | $4.89 \cdot 10^{-9}$ | 201 | $1.35 \cdot 10^{-10}$ | 32.0 | $1.83 \cdot 10^{2}$ | $3.88 \cdot 10^{7}$ |
| 1400 | $4.31 \cdot 10^{-6}$ | 167 | $1.15 \cdot 10^{-7}$ | 27.2 | | |
| 1600 | $6.83 \cdot 10^{-4}$ | 142 | $2.05 \cdot 10^{-5}$ | 24.6 | | |

The large values of ($H_2O$)/($H_2$) (= $K_R$), which increase with decreasing temperature, lead to the conclusion that these oxides are readily reduced by hydrogen. As in the case of the copper oxides, however, the equilibrium constants yield no information about the relative rates of reaction. Despite the enormous difference between the equilibrium dissociation pressures of $Co_3O_4$ and CoO, the rate of reduction of the former is only about twice that of the latter.[24]

## 11.7. DISSOCIATION AND REDUCTION OF OXIDES OF MANGANESE, CHROMIUM, MOLYBDENUM, AND TUNGSTEN

There are at least four oxides of manganese, which may be regarded as successive stages of oxidation of manganese or successive stages in the reduction of $MnO_2$. These reactions and the temperature ranges in which dissociation has been observed are as follows:[25]

$$2MnO_2 \rightarrow Mn_2O_3 + 0.5O_2 \quad (500\text{–}900^\circ\ C) \tag{11.54}$$

$$3Mn_2O_3 \rightarrow 2Mn_3O_4 + 0.5O_2 \quad (900\text{–}1000^\circ\ C) \tag{11.55}$$

$$Mn_3O_4 \rightarrow 3MnO + 0.5O_2 \quad (\text{above } 1300^\circ\ C) \tag{11.56}$$

$$MnO \rightarrow Mn + 0.5O_2 \quad (\text{very high temperatures}) \tag{11.57}$$

Thus, according to Krüll, dissociation of $MnO_2$ is already perceptible (but occurs at a very slow rate) at 500° C and occurs very rapidly at 900° C.

The values of $P_{mm}$ ($O_2$) and ($H_2O$)/($H_2$) given in Table 11.8 were derived from the appropriate free-energy equations of Table 11.1, many of which were calculated from high-temperature reduction equilibria.

The important stable oxide of chromium is $Cr_2O_3$. Values of $P_{mm}$ ($O_2$) and ($H_2O$)/($H_2$), derived from Table 11.1, are given in Table 11.9.

TABLE 11.8

**Dissociation Pressures for Oxides of Manganese and Equilibrium Constants for Their Reduction by Hydrogen**

| | $2MnO_2 = Mn_2O_3 + 0.5O_2$<br>$2MnO_2 + H_2 = Mn_2O_3 + H_2O$ | | $3Mn_2O_3 = 2Mn_3O_4 + 0.5O_2$<br>$3Mn_2O_3 + H_2 = 2Mn_3O_4 + H_2O$ | | $Mn_3O_4 = 3MnO + 0.5O_2$<br>$Mn_3O_4 + H_2 = 3MnO + H_2O$ | | $MnO = Mn + 0.5O_2$<br>$MnO + H_2 = Mn + H_2O$ | |
|---|---|---|---|---|---|---|---|---|
| $T$ (° K) | $P_{mm}$ ($O_2$) | ($H_2O$)/($H_2$) | $P_{mm}$ ($O_2$) | ($H_2O$)/($H_2$) | $P_{mm}$ ($O_2$) | ($H_2O$)/($H_2$) | $P_{mm}$ ($O_2$) | ($H_2O$)/($H_2$) |
| 500 | $7.05 \cdot 10^{-4}$ | $7.47 \cdot 10^{19}$ | | | | | | |
| 600 | $4.73 \cdot 10^{-1}$ | $1.07 \cdot 10^{17}$ | | | | | | |
| 700 | $4.29 \cdot 10^{1}$ | $9.13 \cdot 10^{14}$ | | | | | | |
| 800 | $1.62 \cdot 10^{3}$ | $2.84 \cdot 10^{13}$ | $3.49 \cdot 10^{-5}$ | $1.32 \cdot 10^{10}$ | | | | |
| 900 | $2.18 \cdot 10^{4}$ | $1.69 \cdot 10^{12}$ | $1.18 \cdot 10^{-2}$ | $1.25 \cdot 10^{9}$ | | | | |
| 1000 | | | $1.62 \cdot 10^{-1}$ | $1.68 \cdot 10^{8}$ | | | | |
| 1100 | | | 1.26 | $3.10 \cdot 10^{7}$ | | | | |
| 1200 | | | 7.54 | $7.87 \cdot 10^{6}$ | $3.63 \cdot 10^{-4}$ | $5.47 \cdot 10^{4}$ | | |
| 1400 | | | | | $2.17 \cdot 10^{-1}$ | $3.74 \cdot 10^{4}$ | | |
| 1600 | | | | | $1.86 \cdot 10^{1}$ | $2.34 \cdot 10^{4}$ | $3.29 \cdot 10^{-15}$ | $3.11 \cdot 10^{-4}$ |
| 1800 | | | | | $5.14 \cdot 10^{2}$ | $1.51 \cdot 10^{4}$ | $2.93 \cdot 10^{-12}$ | $1.14 \cdot 10^{-3}$ |
| 2000 | | | | | | | $6.88 \cdot 10^{-10}$ | $3.23 \cdot 10^{-3}$ |

TABLE 11.9

**Dissociation Pressures for Chromium Oxide and Equilibrium Constants for Its Reduction by Hydrogen**

| | $Cr_2O_3 = 2Cr + 1.5O_2$ | $Cr_2O_3 + 3H_2 = 2Cr + 3H_2O$ |
|---|---|---|
| $T$ (° K) | $P_{mm}$ ($O_2$) | ($H_2O$)/($H_2$) |
| 1000 | $2.53 \cdot 10^{-28}$ | $6.65 \cdot 10^{-6}$ |
| 1200 | $9.35 \cdot 10^{-22}$ | $8.77 \cdot 10^{-5}$ |
| 1400 | $4.53 \cdot 10^{-17}$ | $5.40 \cdot 10^{-4}$ |
| 1600 | $1.50 \cdot 10^{-13}$ | $2.10 \cdot 10^{-3}$ |
| 1800 | $8.09 \cdot 10^{-11}$ | $5.98 \cdot 10^{-3}$ |

The reduction of $Cr_2O_3$ by hydrogen is characterized by a value of ($H_2O$)/($H_2$) at about 1000° K that corresponds to an equilibrium concentration of only about 1 part of water vapor in $1.7 \cdot 10^5$ parts of hydrogen, by volume. Consequently, any moisture present in hydrogen in excess of this concentration will be reduced by metallic chromium according to the reaction

$$2Cr\,(s) + 3H_2O\,(g) = Cr_2O_3\,(s) + 3H_2\,(g). \qquad (11.58)$$

This reaction, therefore, has been used to obtain the extremely dry hydrogen required for many annealing and firing processes.

Ramsey, Caplan, and Burr[26] have determined the composition of a hydrogen-water vapor mixture that would just cause an oxide-coated strip of chromium to lose weight. Their data may be represented by

$$\log \frac{(H_2O)}{(H_2)} = \frac{-6660}{T} + 1.68. \qquad (11.59)$$

Free-energy and dissociation-pressure values calculated from these measurements agree more closely with the free-energy equation of Richardson and Jeffes[6] than with that of Table 11.1. The latter was deduced by combining appropriate heat-capacity and entropy data with the value obtained experimentally by Mah[27] for the heat of formation of $Cr_2O_3$, that is, $\Delta H^\circ_{298} = -272.7$ kcal/mole. Ramsey, Caplan, and Burr have combined the same heat-capacity and entropy data with their equilibrium-constant measurements to arrive at $\Delta H^\circ_{298} = -268.5$ kcal/mole. They feel that the difference is probably within the combined experimental uncertainty of both values.

Molybdenum forms two oxides, $MoO_3$ and $MoO_2$. The former has a relatively low melting point of 1068° K, at which its vapor pressure is $10^{-2}$ atm.[28] We shall consider only the reduction equilibria involving $MoO_2$ and Mo. The values of ($H_2O$)/($H_2$) in Table 11.10 were calculated

TABLE 11.10

**Equilibrium Constants for the Reduction of Oxides of Molybdenum and Tungsten by Hydrogen**

| $T$ (° K) | $(H_2O)/(H_2)$ $MoO_2 + 2H_2 = Mo + 2H_2O$ |
|---|---|
| 1000 | 0.317 |
| 1200 | 0.524 |
| 1400 | 0.745 |

| $T$ (° K) | $WO_3 \rightarrow WO_{2.90}$ | $WO_{2.90} \rightarrow WO_{2.72}$ | $WO_{2.72} \rightarrow WO_2$ | $WO_2 \rightarrow W$ |
|---|---|---|---|---|
| 873 | 2.29 | 1.21 | 0.875 | 0.172 |
| 973 | 5.12 | 3.21 | 1.09 | 0.266 |
| 1073 | 9.26 | 7.18 | 1.31 | 0.391 |
| 1173 | | | 1.49 | 0.534 |
| 1273 | | | | 0.677 |

from the measurements of Gocken[29] in the range 949–1334° K, which may be represented by

$$\log \frac{(H_2O)}{(H_2)} = \frac{-1297}{T} + 0.798. \tag{11.60}$$

Tungsten forms four oxides, $WO_3$, $WO_{2.90}$ ($W_{20}O_{58}$), $WO_{2.72}$ ($W_{18}O_{49}$), and $WO_2$. Griffis[30] has studied the equilibrium reduction by hydrogen of each of these oxides into the next lower member of the series, ending with the reduction of $WO_2$ to W. His results are given in Table 11.10. The relatively low values of $(H_2O)/(H_2)$ for the last two equilibria indicate that at these temperatures water vapor readily oxidizes not only tungsten but also $WO_2$ to the next higher oxide.

Reduction equilibrium values for the $MoO_2$-Mo and $WO_2$-W systems can also be obtained from the free-energy equations of Table 11.1. Coughlin, however, gives no data for $WO_{2.90}$, and considers the other intermediate oxide as $WO_{2.75}$ instead of $WO_{2.72}$.

The vapor species formed when $MoO_3$, $MoO_2$, and $WO_3$ are heated in vacuum are listed in Table 11.11, which is discussed in the following section.

## 11.8. OTHER MODES OF DISSOCIATION OF OXIDES

In the previous sections we have considered dissociations that involve just the solid oxide, the free metal, and oxygen. It is only for this type of equilibrium that the dissociation pressures calculated from the standard free energies of formation of the oxide correspond to the experimental values. When some oxides are heated to high temperatures in vacuum, other modes of dissociation and vaporization lead to a vapor pressure greater than that calculated on the assumption of dissociation directly to the elements.[31]

The gaseous oxide molecules produced under these conditions usually have an atomic ratio different from that of the solid oxide. For example, gaseous $Al_2O$ and AlO have been observed when $Al_2O_3$ was heated to 2200° K in vacuum.[32] On the other hand, no change in atomic ratio was produced when $MoO_3$ was heated to 900° K in vacuum, but the principal gaseous species were the trimer, tetramer, and pentamer of the starting material.[33] Identification of the volatile species has been made by use of the mass spectrograph and by examination of emission or absorption spectra. Summaries of the results of investigations in this branch of high-temperature chemistry have been prepared by Brewer,[34] Porter,[35] Ackerman and Thorn,[36] and Inghram and Drowart.[37]

The information in Table 11.11 has been taken from the paper by Inghram and Drowart[37] and from the original sources cited in the table. The volatile species listed were identified in the molecular beam effusing from an equilibrium system in a Knudsen cell. Some were obviously produced by reaction of the oxide with the crucible material (see section 11.10). The review by Inghram and Drowart[37] describes the use of a mass spectrometer to determine the composition of the gaseous phase, the pressure of each of the gaseous species, and the variation of each pressure with temperature. From these data some thermodynamic properties of the gaseous oxide molecules can be calculated.

## 11.9. REDUCTION OF OXIDES BY CARBON MONOXIDE, OTHER REDUCING ATMOSPHERES, AND CARBON

In industrial metallurgical practice, carbon monoxide or some other reducing atmosphere is often used instead of hydrogen for the reduction of oxides.

Consider the reaction

$$2n\mathrm{CO}\,(g) + \mathrm{M}_m\mathrm{O}_{2n}\,(s) = 2n\mathrm{CO}_2\,(g) + m\mathrm{M}\,(s), \qquad (11.61)$$

whose equilibrium constant, analogous to that for hydrogen reductions, can be written as $(CO_2)/(CO)$.

It is convenient to consider reaction 11.61 as the sum of

$$\mathrm{M}_m\mathrm{O}_{2n}\,(s) = m\mathrm{M}\,(s) + n\mathrm{O}_2\,(g) \qquad (11.62)$$

TABLE 11.11

**Volatile Species Observed during Evaporation of Oxides in Knudsen Effusion Cell**

| System | Crucible | Temperature (° K) | Species in Order of Decreasing Pressure | Reference |
|---|---|---|---|---|
| $Al_2O_3$ | W | 2100–2500 | Al, O, AlO, $Al_2O$, $WO_2$, $WO_3$, WO, $Al_2O_2$, $O_2$ | 1, 2 |
| $Al_2O_3$ | Mo | 2100–2500 | Al, O, $Al_2O$, AlO, $MoO_2$, MoO, $MoO_3$, $Al_2O_2$ | 3 |
| $Al_2O_3$-Al | $ZrO_2$ | 1750 | $Al_2O$, Al | 4 |
| BaO | $Al_2O_3$ | 1500–1800 | Ba, BaO, $Ba_2O$, $Ba_2O_2$, $Ba_2O_3$, $O_2$ | 5 |
| BeO | W | 2242 | Be, O, $WO_2$, $WO_3$, $O_2$, $(BeO)_{1-6}$, $WO_2 \cdot (BeO)_{1-4}$ $WO_3 \cdot (BeO)_{1-3}$ | 6 |
| $B_2O_3$-B | $Al_2O_3$ | 1300–1500 | $B_2O_2$, $B_2O_3$ | 7 |
| $La_2O_3$-La | $ThO_2$, Ta | 1700–1900 | LaO, La | 8 |
| $Li_2O$ | Pt | 1400 | Li, $O_2$, $Li_2O$, LiO | 9 |
| MgO | $Al_2O_3$ | 1950 | Mg, $O_2$, MgO | 10 |
| $MoO_3$ | Pt | 800–1000 | $(MoO_3)_{x=3,4,5}$ | 11 |
| $MoO_2$ | $Al_2O_3$, Mo | 1600–1800 | $MoO_3$, $(MoO_3)_2$, $MoO_2$, $(MoO_3)_3$ | 12 |
| $SiO_2$ | $Al_2O_3$ | 1900 | SiO, $SiO_2$, $O_2$, O | 13 |
| $SiO_2$-Si | $Al_2O_3$ | 1463 | SiO, $Si_2O_2$ | 13 |
| SrO | $Al_2O_3$ | 2100 | Sr, $O_2$, SrO | 14 |
| $Ta_2O_5$-Ta | Ta | 2000–2300 | $TaO_2$, TaO | 15 |
| $TiO_2$-Ti | Ta, Mo or $ThO_2$ | 1840 | TiO, Ti | 16 |
| $Ti_2O_3$ | Ta, Mo | 2194 | TiO, Ti, $TiO_2$ | 16 |
| VO | W | 1945 | VO, V, $VO_2$ | 17 |
| $V_2O_5$ | Pt | ... | $V_4O_{10}$, $V_4O_8$, $V_6O_{12}$, $V_6O_{14}$, $V_2O_4$ | 17 |
| $WO_3$ | Pt | 1492 | $(WO_3)_{x=3,4,5}$ | 18 |
| $ZrO_2$ | Ta | 2000 | ZrO, TaO | 19 |

*References for Table 11.11*


1. G. DeMaria, J. Drowart, and M. G. Inghram, *J. Chem. Phys.*, **30,** 318 (1959).
2. J. Drowart, G. DeMaria, R. P. Burns, and M. G. Inghram, *J. Chem. Phys.*, **32,** 1366–1372 (1960).
3. G. DeMaria, R. P. Burns, J. Drowart, and M. G. Inghram, *J. Chem. Phys.*, **32,** 1373–1377 (1960).
4. R. F. Porter, P. Schussel, and M. G. Inghram, *J. Chem. Phys.*, **23,** 339 (1955).
5. M. G. Inghram, W. A. Chupka, and R. F. Porter, *J. Chem. Phys.*, **23,** 2159 (1955).
6. W. A. Chupka, J. Berkowitz, and C. F. Giese, *J. Chem. Phys.*, **30,** 827 (1959).

7. M. G. Inghram, R. F. Porter, and W. A. Chupka, *J. Chem. Phys.*, **25,** 498 (1956).
8. W. A. Chupka, M. G. Inghram, and R. F. Porter, *J. Chem. Phys.*, **24,** 792 (1956).
9. J. Berkowitz, W. A. Chupka, G. D. Blue, and J. L. Margrave, *J. Phys. Chem.*, **63,** 644 (1959).
10. R. F. Porter, W. A. Chupka, and M. G. Inghram, *J. Chem. Phys.*, **23,** 1347 (1955).
11. J. Berkowitz, M. G. Inghram, and W. A. Chupka, *J. Chem. Phys.*, **26,** 842 (1957).
12. R. P. Burns, G. DeMaria, J. Drowart, and R. T. Grimley, *J. Chem. Phys.*, **32,** 1363–1366 (1960).
13. R. F. Porter, W. A. Chupka, and M. G. Inghram, *J. Chem. Phys.*, **23,** 216 (1955).
14. R. F. Porter, W. A. Chupka, and M. G. Inghram, *J. Chem. Phys.*, **23,** 1347 (1955).
15. M. G. Inghram, W. A. Chupka and J. Berkowitz, *J. Chem. Phys.*, **27,** 569 (1957).
16. J. Berkowitz, W. A. Chupka, and M. G. Inghram, *J. Phys. Chem.*, **61,** 1569 (1957).
17. J. Berkowitz, W. A. Chupka, and M. G. Inghram, *J. Chem. Phys.*, **27,** 87 (1957).
18. J. Berkowitz, W. A. Chupka, and M. G. Inghram, *J. Chem. Phys.*, **27,** 85 (1957).
19. W. A. Chupka, J. Berkowitz, and M. G. Inghram, *J. Chem. Phys.*, **26,** 1207 (1957).

---

and

$$2n[\mathrm{CO}\,(g) + 0.5\mathrm{O}_2\,(g) = \mathrm{CO}_2\,(g)], \tag{11.63}$$

so that reduction equilibria can be calculated from the appropriate combination of free energies of formation from Table 11.1 with free-energy values for reaction 11.63, given by

$$\Delta F^\circ{}_{\mathrm{C}} = \Delta F^\circ{}_{\mathrm{CO}_2} - \Delta F^\circ{}_{\mathrm{CO}} \tag{11.64}$$

Standard free energies of formation for CO, $CO_2$, and for reaction 11.63 are given in Table 11.12, along with values of the equilibrium constant for reaction 11.63, which may be expressed as

$$K_C = \frac{P_{atm}(\mathrm{CO}_2)}{P_{atm}(\mathrm{CO})\,P_{atm}^{1/2}(\mathrm{O}_2)} \tag{11.65}$$

The values of $\Delta F^\circ{}_{\mathrm{C}}$ in column 4 of Table 11.12 are of the same order of magnitude as the values of $\Delta F^\circ{}_W$ in Table 11.2. As a consequence, the equilibrium ratio $(CO_2)/(CO)$ for the isothermal reduction of an oxide with carbon monoxide will be of the same order of magnitude as the equilibrium ratio $(H_2O)/(H_2)$ for the isothermal reduction of the same oxide with hydrogen. This is shown, for the reduction of FeO, by the ratios listed in the second and third columns of Table 11.13, which is taken from a publication by Austin and Day.[38] The temperature coefficients of these equilibrium constants have opposite signs because the standard free energy of formation of FeO lies between the values of $\Delta F^\circ{}_W$ and $\Delta F^\circ{}_{\mathrm{C}}$.

Besides hydrogen and carbon monoxide, other types of reducing atmospheres are often used in metallurgical practice. The equilibrium constants for three of these reducing gases[38] are shown in the last three

TABLE 11.12

Free Energies of Formation of CO and $CO_2$ and Equilibrium Constants for Oxidation of CO to $CO_2$

| $T$ (° K) | $\Delta F°$: $C+O_2=CO_2$ | $C+0.5O_2=CO$ | $CO+0.5O_2=CO_2$ | $K_C=\frac{P(CO_2)}{P(CO)\cdot P^{1/2}(O_2)}$ |
|---|---|---|---|---|
| 298 | −94,260 | −32,808 | −61,452 | $1.19 \cdot 10^{45}$ |
| 600 | −94,440 | −39,360 | −55,080 | $1.16 \cdot 10^{20}$ |
| 800 | −94,540 | −43,680 | −50,860 | $7.88 \cdot 10^{13}$ |
| 1000 | −94,610 | −47,940 | −46,670 | $1.59 \cdot 10^{10}$ |
| 1200 | −94,660 | −52,150 | −42,510 | $5.54 \cdot 10^{7}$ |
| 1400 | −94,690 | −56,310 | −38,380 | $9.82 \cdot 10^{5}$ |
| 1600 | −94,730 | −60,430 | −34,300 | $4.85 \cdot 10^{4}$ |
| 1800 | −94,720 | −64,480 | −30,240 | $4.70 \cdot 10^{3}$ |
| 2000 | −94,720 | −68,510 | −26,210 | $7.32 \cdot 10^{2}$ |

columns of Table 11.13. In the water gas reaction the atmosphere will be oxidizing or reducing according as the values of the ratios $P_{CO}/P_{CO_2}$ and $P_{H_2}/P_{H_2O}$ are less than, or greater than the individual values of $K_1$ and $K_2$ shown in the table.

Both the producer-gas reaction (leading to formation of carbon monoxide) and the decomposition of methane are evidently favored by increase in temperature.

Table 11.14, taken from the treatise by Stansel,[39] gives the composition of a number of gaseous mixtures used in industrial metallurgical operations. The composition of commercial producer gas varies according to whether the coal used is bituminous or anthracite. The active components are carbon monoxide and hydrogen, while about half the mixture is nitrogen.

Coke-oven gas consists mainly of hydrogen and methane, and the carbon monoxide content ranges from 5 to 9 per cent. The composition of commercial water-gas mixtures is very close to the theoretical, which is equal parts of hydrogen and carbon monoxide. Carburetted water gas contains, in addition to the constituents mentioned in the table, 2 per cent benzene.

The selection of an atmosphere for any particular service involves a number of considerations of a practical nature, which are discussed by Stansel. In using such gas mixtures for the high-temperature treatment of metal parts for application in high-vacuum devices or systems, every precaution must be taken to avoid deleterious chemical and physical effects.

In a series of papers Lepp has discussed the reactions of oxides with hydrogen, carbon monoxide, and other reducing gases.[40] Although hydrogen reacts only with the oxide, carbon monoxide can also react with the metal obtained by reduction of the oxide.

### TABLE 11.13

**Equilibrium Constants for the Reduction of FeO by Hydrogen and Carbon Monoxide and for Three Other Reactions**

| $T$ (° K) | $K_1{}^* = (H_2)/(H_2O)$ | $K_2{}^* = (CO)/(CO_2)$ | $K_3 = \frac{(H_2)(CO_2)}{(H_2O)(CO)}$ | $K_4 = \frac{P_{atm}^2\,(CO)}{P_{atm}\,(CO_2)}$ | $K_5 = \frac{P_{atm}^2\,(H_2)}{P_{atm}\,(CH_4)}$ |
|---|---|---|---|---|---|
| 673 | 9.12** | 0.74** | 12.3 | $9 \cdot 10^{-5}$ | $5.66 \cdot 10^{-2}$ |
| 773 | 4.68** | 0.96** | 4.88 | $4.7 \cdot 10^{-3}$ | $4.22 \cdot 10^{-1}$ |
| 873 | 2.99 | 1.17 | 2.55 | $9.6 \cdot 10^{-2}$ | $2.09 \cdot 10^{0}$ |
| 973 | 2.38 | 1.53 | 1.56 | $1.06 \cdot 10^{0}$ | $7.16 \cdot 10^{0}$ |
| 1073 | 2.00 | 1.90 | 1.05 | $7.48 \cdot 10^{0}$ | $2.01 \cdot 10$ |
| 1173 | 1.72 | 2.24 | 0.765 | $3.767 \cdot 10$ | $4.83 \cdot 10$ |
| 1273 | 1.51 | 2.57 | 0.589 | $1.465 \cdot 10^{2}$ | $1.024 \cdot 10^{2}$ |
| 1373 | 1.37 | 2.88 | 0.474 | $4.634 \cdot 10^{2}$ | $1.920 \cdot 10^{2}$ |
| 1473 | 1.26 | 3.21 | 0.395 | $1.244 \cdot 10^{3}$ | $3.35 \cdot 10^{2}$ |
| 1573 | 1.18 | 3.49 | 0.339 | $2.951 \cdot 10^{3}$ | $5.47 \cdot 10^{2}$ |

Reactions:
(1) $Fe + H_2O \rightarrow FeO + H_2$.
(2) $Fe + CO_2 \rightarrow FeO + CO$.
(3) $CO + H_2O \rightarrow CO_2 + H_2$ Water-gas reaction.
(4) $C + CO_2 \rightarrow 2CO$ Producer-gas reaction.
(5) $CH_4 = C + 2H_2$ Decomposition of methane.

* The ratios $K_1$ and $K_2$ given in the table are the *reciprocals* of the ratios defined previously in section 11.1. The data designated with two asterisks are probably not very accurate, since the stable phase at temperatures of 673° K and 773° K is $Fe_3O_4$, not FeO.

TABLE 11.14

**Commercial Reducing Gases and Their Analysis**

| Gas | Constituents (per cent by volume) | | | | | | | |
|---|---|---|---|---|---|---|---|---|
| | $CH_4$ | $C_2H_6$ | $C_2H_4$ | $CO_2$ | CO | $H_2$ | $O_2$ | $N_2$ |
| Natural gas, Ohio | 80.5 | 18.2 | ... | ... | ... | ... | ... | 1.3 |
| Natural gas, Penna. | 67.6 | 31.3 | ... | ... | ... | ... | ... | 1.1 |
| Manufactured coal gas | 40 | ... | 4.2 | 0.50 | 6.0 | 46.0 | 0.50 | 1.5 |
| Water gas | 2 | ... | ... | 4 | 45.0 | 45.0 | .50 | 2 |
| Blue water gas | ... | ... | ... | 3.5 | 43.4 | 51.8 | ... | 1.30 |
| Carburetted water gas | 20 | ... | 12.05 | 0.22 | 25.9 | 36.50 | .03 | 3.24 |
| Producer gas (bit.) | 2.5 | ... | 0.4 | 2.5 | 27 | 12.0 | .3 | 55.3 |
| Producer gas (anth.) | 0.31 | ... | 0.31 | 6.57 | 25.07 | 18.73 | ... | 49.01 |
| Oil gas | 58.3 | ... | 17.4 | ... | ... | 24.3 | ... | ... |
| Blast-furnace gas | ... | ... | ... | 14.48 | 22.81 | 5.07 | ... | 57.6 |

For example, carbon monoxide reacts at about 200° C with iron and nickel to form the volatile and relatively unstable carbonyls, $Fe(CO)_5$ and $Ni(CO)_4$. At about 500–600° C, carbon monoxide may react with metals to form carbides. For example, we have the reactions

$$3Fe + 2CO \rightarrow Fe_3C + CO_2 \tag{11.66}$$

and

$$4Ni + CO \rightarrow Ni_3C + NiO, \tag{11.67}$$

as well as the oxidation reaction,

$$Ni + CO_2 \rightarrow NiO + CO. \tag{11.68}$$

At the melting point of iron, carbon monoxide can oxidize as well as carburize the metal,

$$4Fe + CO \rightarrow FeO + Fe_3C. \tag{11.69}$$

Hence, from the point of view of eliminating oxides from metal parts that are intended for use in high-vacuum devices, carbon monoxide should be used with considerable caution. The same remark applies to other reducing atmospheres obtained by cracking hydrocarbon gases. For example, methane reacts with iron according to the equation

$$3Fe + CH_4 \rightarrow Fe_3C + 2H_2 \tag{11.70}$$

to form the carbide. This is the well-known process of cementation.

In a similar manner the reaction between certain oxides and carbon leads to the formation of carbides. For instance, Prescott has studied the two reactions

$$ZrO_2 + 3C \rightarrow ZrC + 2CO \text{[41]} \tag{11.71}$$

and

$$ThO_2 + 4C \rightarrow ThC_2 + 2CO. \text{[42]} \tag{11.72}$$

For the first reaction, he obtained the following relation for the pressure of carbon monoxide in atmospheres:

$$\log P_{atm} = 8.592 - \frac{16{,}580}{T}, \tag{11.73}$$

and, for the second, the relation

$$\log P_{atm} = 8.069 - \frac{19{,}325}{T}. \tag{11.74}$$

From these relations the following values for the pressure of carbon monoxide are deduced for a series of values of $T$. The second row gives the values for the reduction of $ZrO_2$; the third row, those for the reduction of $ThO_2$:

| $T$ (° K): | 1600 | 1800 | 2000 | 2200 | 2400 |
|---|---|---|---|---|---|
| (Reduction of $ZrO_2$), $P_{atm}$ (CO): | $1.7 \cdot 10^{-2}$ | $2.4 \cdot 10^{-1}$ | 2.00 | 11.35 | ... |
| (Reduction of $ThO_2$), $P_{atm}$ (CO): | $1 \cdot 10^{-4}$ | $4.6 \cdot 10^{-2}$ | $2.55 \cdot 10^{-2}$ | 0.191 | 1.04 |

Thus $ZrO_2$ is much more readily reduced by carbon than $ThO_2$. The values of $\Delta F$ corresponding to relations 11.73 and 11.74, as derived by Prescott, are:

$$\Delta F = 151{,}800 - 78.68T \tag{11.75}$$

and

$$\Delta F = 176{,}970 - 73.892T, \tag{11.76}$$

respectively.

## 11.10. REDUCTION OF OXIDES BY METALS

Oxides of nonvolatile metals can be reduced by pure metals whose own oxides are more stable than the starting material. Thus oxides such as those of chromium, iron, titanium, and zirconium can be reduced to the pure metal by reaction with active metals such as aluminum and calcium. The reactions of barium oxide with various metals are particularly important to vacuum technology because this oxide is widely used as the electron-emitting coating for cathodes of vacuum and gas-discharge tubes. The activation of these cathodes involves the formation of free barium, and an understanding of the process requires a knowledge of the equilibrium pressure of barium vapor produced by the dissociation or reduction of barium oxide. It must be emphasized, however, that a process as complex as the activation of oxide-coated cathodes cannot be understood solely in terms of a single thermodynamic parameter. A detailed discussion of the many other pertinent factors has been presented by Rittner[43] and Herrmann[44] among others.

As part of his broad treatment of the subject, Rittner has used thermochemical data on various metal oxides to calculate the equilibrium pressures of barium produced by reactions of the type

$$BaO\,(s) + M\,(s) = MO\,(s) + Ba\,(g), \tag{11.77}$$

where M represents a reducing metal. Following the procedures of section 11.1, we can write the standard free-energy change for reaction 11.77 as

$$\Delta F_{77} = \Delta F^{\circ}{}_{MO} - \Delta F^{\circ}{}_{BaO} + \Delta F_v, \tag{11.78}$$

where $\Delta F^{\circ}$ designates the standard free energy of formation of the solid oxides, and $\Delta F_v$ is the free energy of vaporization per gram-atom of barium.

Some of Rittner's results are shown in Table 11.15. The upper limit for $P$ (Ba) is set by the vapor pressure of pure barium, while the lower limit is set by the thermal dissociation of barium oxide.

**TABLE 11.15**

**Equilibrium Pressure of Barium Produced by Reduction of BaO**

$$\log P_{mm}\,(Ba) = A + \frac{B}{T}$$

| Reaction | $A$ | $B$ | $P_{mm}$ (Ba) at 1000° K |
|---|---|---|---|
| $BaO = Ba\,(g) + \frac{1}{2}O_2$ | +9.76 | −24,900 | $8 \cdot 10^{-16}$ |
| $BaO + Fe = Ba\,(g) + FeO$ | +9.08 | −23,200 | $8 \cdot 10^{-15}$ |
| $\frac{4}{3}BaO + \frac{1}{3}Mo = Ba\,(g) + \frac{1}{3}BaMoO_4$ | +8.56 | −19,800 | $6 \cdot 10^{-12}$ |
| $BaO + \frac{2}{3}Cr = Ba\,(g) + \frac{1}{3}Cr_2O_3$ | +8.28 | −17,700 | $4 \cdot 10^{-10}$ |
| $\frac{4}{3}BaO + \frac{1}{3}W = Ba\,(g) + \frac{1}{3}BaWO_4$ | +8.56 | −17,400 | $2 \cdot 10^{-9}$ |
| $BaO + C\,(graphite) = Ba\,(g) + CO$ | +10.31 | −15,800 | $4 \cdot 10^{-6}$ |
| $BaO + \frac{1}{2}Ce = Ba\,(g) + \frac{1}{2}CeO_2$ | +8.13 | −11,800 | $2 \cdot 10^{-4}$ |
| $Ba\,(s) = Ba\,(g)$ | +7.83 | −9,730 | $1 \cdot 10^{-2}$ |

Rittner has also considered the reduction of barium oxide by carbon monoxide or hydrogen, shown as

$$BaO\,(s) + CO\,(g) = Ba\,(g) + CO_2\,(g), \tag{11.79}$$

$$BaO\,(s) + H_2\,(g) = Ba\,(g) + H_2O\,(g). \tag{11.80}$$

By assuming no outside source of carbon dioxide or water vapor, Rittner computed the following expressions for the barium pressure:

$$\log P_{mm}\,(Ba) = \frac{-11{,}200}{T} + 5.71 + 0.5 \log P_{atm}\,(CO), \tag{11.81}$$

$$\log P_{mm}\,(Ba) = \frac{-12{,}300}{T} + 6.81 + 0.5 \log P_{atm}\,(H_2). \tag{11.82}$$

These equations indicate that carbon monoxide and hydrogen are moderately active reducing agents only at pressures of about 1 atm, where they give rise to a barium pressure of about $3 \cdot 10^{-6}$ mm at 1000° K.

The reaction between a metal and an oxide at high temperatures may also produce gaseous oxide species. Some of the species listed in Table 11.11 arose from reaction between the solid oxide and the molybdenum, tantalum, or tungsten crucible. For example, Chupka, Berkowitz, and Inghram[45] have shown by mass spectrometry that the following reaction occurred when $ZrO_2$ was heated in vacuum in a tantalum Knudsen effusion cell at 2000° K:

$$Ta\,(s) + ZrO_2\,(s) = TaO\,(g) + ZrO\,(g). \qquad (11.83)$$

Both the oxide sample and the container were transformed into volatile species. The ratio of oxide to metal transported out of the cell may vary considerably with the cell material. Ackerman and Thorn[46] have shown that for the evaporation of aluminum oxide from a tungsten cell at 2587° K the ratio was 1.17, and the gaseous products were $Al_2O$, Al, O, and WO. When a tantalum cell was used, the ratio of oxide to tantalum transported was only 0.19, and the gaseous products were Al and TaO.

Reaction between aluminum oxide and tantalum to produce aluminum vapor can occur in vacuum at 1873° K, even when the two materials are not in contact with each other. Navias[47] has observed the formation of an aluminum mirror on the walls of an evacuated glass envelope that contained a sapphire rod surrounded by, but not in contact with, a concentric tantalum heater tube. No sign of reaction was observed in several hours at 2173° K when a tungsten heater tube was used.

In the light of this evidence, the rate of weight loss alone should not be used to determine the vapor pressure of a substance by the Knudsen effusion technique. Such measurements should be supplemented by analysis of the effusing species in order to avoid the error that might otherwise be introduced if there is an unsuspected reaction between sample and container.

## 11.11. DISSOCIATION OF HYDRIDES OF THE ALKALI AND ALKALINE-EARTH METALS

In Chapter 8 a comparison was made between the heats of formation of the so-called hydrides of the titanium group of metals and the hydrides of the alkali and alkaline-earth metals. These hydrides are formed when hydrogen is passed over the heated metals, so that the metal functions as a "getter" or chemical clean-up reagent. (Getters were discussed in Chapter 9.) At higher temperatures these hydrides dissociate, and at each temperature there exists an equilibrium pressure.

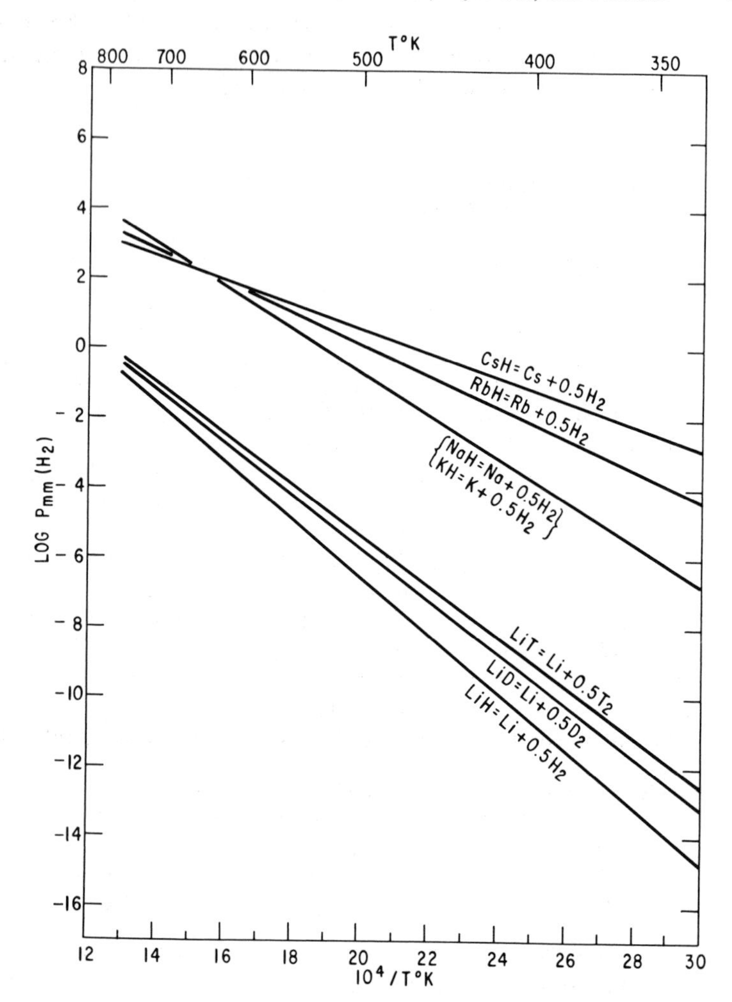


Fig. 11.3. Dissociation pressures of some alkali metal hydrides.

The experimental measurements of these equilibrium pressures of hydrogen, deuterium, or tritium can be represented by the general equation

$$\log P_{mm}\,(H_2) = A + \frac{B}{T}. \tag{11.84}$$

Values of $A$ and $B$ and the vapor pressures at three temperatures are given in Table 11.16. Some of the pressure-temperature relationships are plotted in Fig. 11.3. They show that lithium hydride is considerably more stable than the other alkali metal hydrides. Its dissociation-pressure line is almost coincident with that for calcium hydride, which is not shown in Fig. 11.3. The substitution of deuterium or tritium for hydrogen increases the dissociation pressure of the alkali metal compound.

**TABLE 11.16**

**Dissociation Pressures of Alkali and Alkaline-Earth Metal Hydrides and Deuterides**

| | $\log P_{mm}(H_2) = A + \frac{B}{T}$ | | $P_{mm}(H_2)$ at $T°$ K | | | |
|---|---|---|---|---|---|---|
| Compound | $A$ | $B$ | 500 | 600 | 700 | Ref. |
| LiH | +9.9258 | −8244 | $2.74 \cdot 10^{-7}$ | $1.53 \cdot 10^{-4}$ | $1.41 \cdot 10^{-2}$ | 1 |
| LiD | +9.4836 | −7626 | $1.71 \cdot 10^{-6}$ | $5.94 \cdot 10^{-4}$ | $3.88 \cdot 10^{-2}$ | 1 |
| LiT | +9.2456 | −7325 | $3.94 \cdot 10^{-6}$ | $1.09 \cdot 10^{-3}$ | $6.04 \cdot 10^{-2}$ | 1 |
| NaH | +11.9250 | −6318 | $1.95 \cdot 10^{-1}$ | $2.48 \cdot 10^{1}$ | $7.93 \cdot 10^{2}$ | 2 |
| NaD | +13.1994 | −6915 | $2.34 \cdot 10^{-1}$ | $4.73 \cdot 10^{1}$ | $2.09 \cdot 10^{3}$ | 2 |
| KH | +11.6535 | −6186 | $1.91 \cdot 10^{-1}$ | $2.21 \cdot 10^{1}$ | $6.55 \cdot 10^{2}$ | 3 |
| KD | +12.1977 | −6319 | $3.65 \cdot 10^{-1}$ | $4.63 \cdot 10^{1}$ | $1.48 \cdot 10^{3}$ | 3 |
| RbH | +9.20 | −4534 | 1.36 | $4.40 \cdot 10^{1}$ | $5.28 \cdot 10^{2}$ | 4 |
| CsH | +7.50 | −3476 | 3.53 | $5.09 \cdot 10^{1}$ | $3.42 \cdot 10^{2}$ | 4 |
| $CaH_2$ | +9.073 | −7782 | $3.23 \cdot 10^{-7}$ | $1.27 \cdot 10^{-4}$ | $9.03 \cdot 10^{-3}$ | 5 |

*References for Table 11.16*


1. F. K. Heumann and O. N. Salmon, *U.S. Atomic Energy Comm. KAPL*-1667, (1956), 54 pages.
2. E. F. Sollers and J. L. Crenshaw, *J. Am. Chem. Soc.*, **59**, 2724 (1937).
3. E. F. Sollers and J. L. Crenshaw, *J. Am. Chem. Soc.*, **59**, 2015 (1937).
4. L. Hackspill and A. Borocco, *Bull. soc. chim.*, **6**, 91 (1939).
5. W. C. Johnson, M. F. Stubbs, A. E. Sidwell, and A. Pechukas, *J. Am. Chem. Soc.*, **61**, 318 (1939).


No dissociation-pressure equations are available for strontium and barium hydrides, although Schumb, Sewell, and Eisenstein[48] report a dissociation pressure of about 0.24 mm for $BaH_2$ at 873° K. The considerable literature on the dissociation of calcium hydride has been reviewed by Kilb,[49] and the general field of hydride chemistry has been covered by Hurd.[50]

## 11.12. DISSOCIATION PRESSURES OF NITRIDES

Metal nitrides are formally analogous to metal oxides, and their dissociation reactions can be treated as shown in section 11.1. Thus, for the reaction

$$M_mN_{2n}(s) = mM(s) + nN_2(g) \qquad (11.85)$$

the dissociation pressure is given by

$$\log P_{mm}\,(N_2) = 2.881 + \frac{1}{n}\frac{\Delta F^\circ}{4.575T}, \tag{11.86}$$

where $\Delta F^\circ$ is the standard free energy of formation per mole of $M_mN_{2n}$.

Thermodynamic data on nitrides have been compiled by Kelley,[51] by Brewer, Bromley, Giles, and Lofgren,[52] and by Pearson and Ende.[53] For many nitrides the free energy of formation can be represented by equation 11.22, and the dissociation pressure, therefore, can be calculated by an equation like 11.23:

$$\log P_{mm}\,(N_2) = 2.881 + \frac{B}{4.575n} + \frac{A}{4.575nT}. \tag{11.87}$$

Table 11.17 lists values of $A$ and $B$ for some nitrides, and the standard free energies of formation at 298° K for others.[54] The use of quotation marks around a formula, for example, "VN," designates a homogeneous phase of variable composition within which the nominal composition lies. In these cases the condensed phases in equilibrium with gaseous nitrogen are the nonstoichiometric nitride phase and often a solution of nitrogen in the metal. For these equilibria the free-energy values calculated by means of equation 11.22 are in error by at least $\pm 10$ kcal/mole. The $A$ and $B$ values of Table 11.17 have been used in equation 11.87 to construct Fig. 11.4, which shows some nitride dissociation pressures as a function of temperature.

Extrapolation of the lines in Fig. 11.4 shows that at 2000° K the dissociation pressures of even the most stable nitrides are greater than $10^{-6}$ mm. Thus, despite their high melting points, the nitrides are of limited usefulness as refractory materials in high-vacuum applications.

The binary system chromium-nitrogen affords a good example of equilibria that involve a solid solution, a nonstoichiometric nitride phase, and a stoichiometric nitride phase. Seybolt and Oriani[55] have studied the pressure-temperature-composition relations between the chromium-nitrogen terminal solid solution and the first nitride phase, "$Cr_2N$." Smith[56] has extended these measurements to the CrN composition and has established the composition limits for the "$Cr_2N$" phase. The temperature-composition relationships are represented in Fig. 11.5, which shows that at 1673° K solid chromium metal can dissolve 0.3 weight per cent nitrogen, and that the phase "$Cr_2N$" actually covers a range of compositions from $Cr_2N_{0.760}$ (9.3 per cent N) to $Cr_2N$ (11.9 per cent N). These composition limits are virtually independent of temperature in the range 1073–1473° K.

The dissociation equilibria in this binary system are

$$Cr_2N_{0.760} = 2Cr\ (\text{sat. with } N_2) + 0.380N_2 \tag{11.88}$$

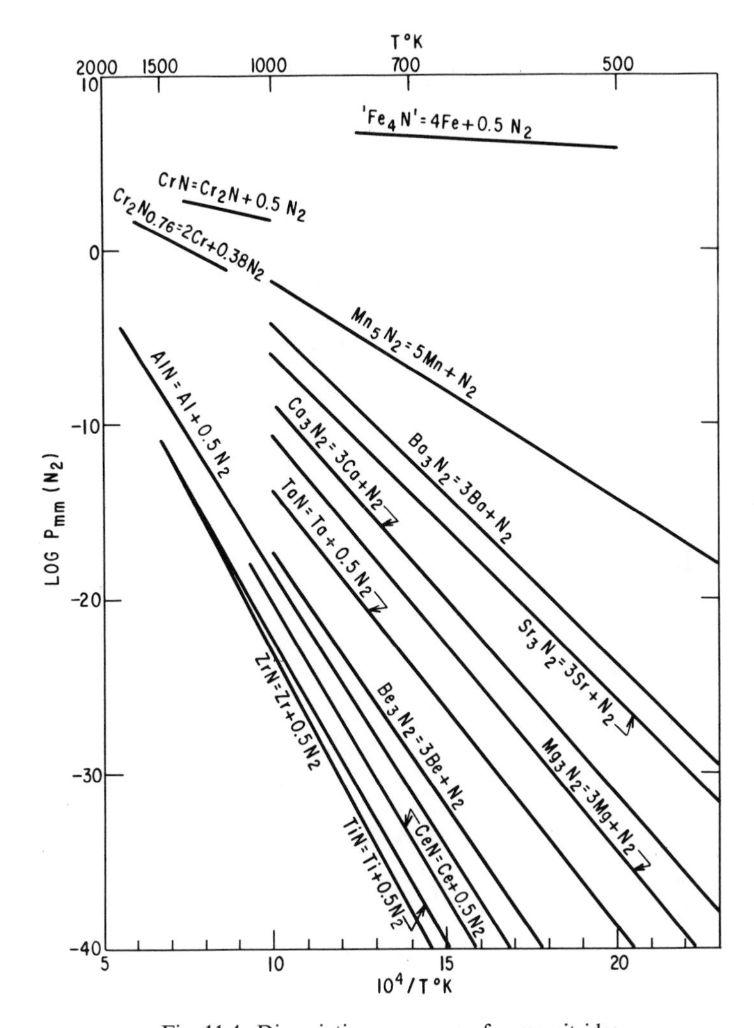


Fig. 11.4. Dissociation pressures of some nitrides.

and

$$CrN = Cr_2N + 0.5N_2. \tag{11.89}$$

The measured values of the equilibrium nitrogen pressures for these reactions are given in Table 11.18 and included in Fig. 11.4. They are consistently lower than the pressure values obtained by earlier investigators and summarized in the free-energy equations given by Pearson and Ende.[53]

Because the solubility of nitrogen in solid chromium is so low, the activity of chromium in the solid-solution phase can be considered as

**TABLE 11.17(*a*)**

**Coefficients in Equations for Standard Free Energies of Formation of Nitrides (Calories per Mole)**

$$\Delta F^\circ = A + BT$$

| Nitride | $A$ | $B$ | Reference |
|---|---|---|---|
| AlN | −72,150 | +23.3 | 53 |
| $Ba_3N_2$ | −89,900 | +57.4 | 51 |
| $Be_3N_2$ | −133,500 | +40.6 | 51 |
| BN | −26,250* | +9.7* | 53 |
| $Ca_3N_2$ | −103,200 | +50.2 | 51 |
| CeN | −78,000 | +25.0 | 51 |
| "$Cr_2N$" | | See Table 11.18 | |
| CrN | | See Table 11.18 | |
| "$Fe_4N$" | −2,900 | +12.3 | 53 |
| LaN | −72,100 | +25.0 | 51 |
| $Li_3N$ | −47,470 | +34.0 | 51 |
| $Mg_3N_2$ | −109,600 | +47.41 | 53 |
| $Mn_5N_2$ | −57,770 | +36.4 | 51 |
| $Si_3N_4$ | −180,000 | +80.4 | 53 |
| $Sr_3N_2$ | −91,400 | +50.9 | 51 |
| TaN | −57,625 | +19.35 | 53 |
| "TiN" | −80,250 | +22.20 | 53 |
| $U_3N_4$ | −275,000 | +81.9 | 51 |
| "VN" | −41,650 | +19.9 | 53 |
| ZrN | −81,900 | +22.31 | 53 |

* Valid only in range 1200–2300° K; applies to hexagonal BN.

**TABLE 11.17(*b*)**

**Standard Free Energies of Formation at 298° K (Kilocalories per Mole) and Melting Points of Some Refractory Nitrides**

| Nitride | Melting Point (° K) | $\Delta F^\circ_{298}$ | Reference |
|---|---|---|---|
| HfN | 3580 | −81.4 | 54 |
| NbN | 2573 | −50.0 | 54 |
| ScN | 2923 | −60.5 | 52 |
| TaN | 3360 | −53.2 | 54 |
| "TiN" | 3200 | −73.7 | 54 |
| ZrN | 3255 | −80.5 | 54 |

TABLE 11.18

Dissociation Pressures for Chromium Nitrides [$P_{mm}$ ($N_2$)]

| $T$ (° K) | $Cr_2N_{0.760}$ = $2Cr_{(ss)} + 0.380N_2$ | $T$ (° K) | $CrN$ = $Cr_2N + 0.5N_2$ |
|---|---|---|---|
| 1273 | 0.50 | 1182 | 227 |
| 1373 | 1.96 | 1219 | 284 |
| 1473 | 6.25 | 1263 | 390 |
| 1573 | 17.6 | 1359 | 740 |
| 1673 | 42.1 | | |

unity. The free energies corresponding to reactions 11.88 and 11.89 can, therefore, be calculated by applying equation 11.17 to the data of Table 11.18, using values[57] of 0.38 and 0.50 for $n$.

Within the homogeneous "$Cr_2N$" range in Fig. 11.5 the system is bivariant, and its dissociation pressure depends upon both its temperature and its composition. At 1330° K Smith[56] observed that the relationship between the equilibrium nitrogen pressure and the mole-fraction ($x$) of nitrogen in the "$Cr_2N$" phase could be expressed by the equation

$$\log P_{mm}(N_2) = 43.1x - 11.69. \qquad (11.90)$$

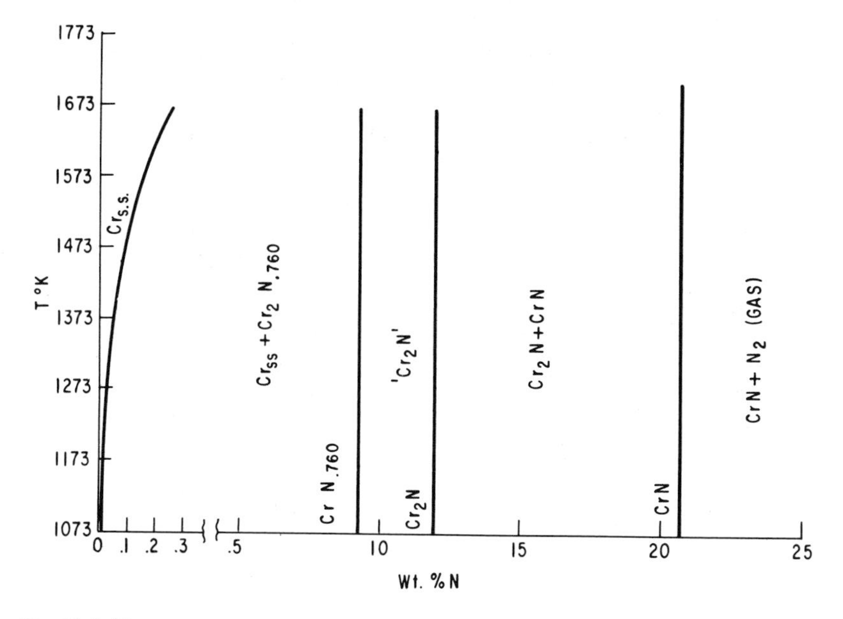


Fig. 11.5. Temperature-composition diagram for the Cr-N system from 1073 to 1673° K.

## REFERENCES AND NOTES

1. These concepts are discussed comprehensively in different treatises on thermodynamics, such as the following:

*a.* L. E. Steiner, *Introduction to Chemical Thermodynamics*, McGraw-Hill Book Company, New York, 1948.

*b.* F. D. Rossini, *Chemical Thermodynamics*, John Wiley & Sons, New York, 1950.

*c.* I. M. Klotz, *Chemical Thermodynamics*, Prentice-Hall, Englewood Cliffs, N.J., 1950.

*d.* M. A. Paul, *Principles of Chemical Thermodynamics*, McGraw-Hill Book Company, New York, 1951.

2. This actually occurs in the case of some batteries.

3. For a discussion of standard states, see any of the treatises cited in Ref. 1.

4. The thermodynamic equilibrium constant is the activity of the vapor. In this chapter the vapor phase is assumed to be ideal, so that its activity is numerically equal to its pressure. The standard state of unit activity is the ideal gas at 1 atmosphere.

5. J. P. Coughlin, "Contributions to the Data on Theoretical Metallurgy, XII. Heats and Free Energies of Formation of Inorganic Oxides," *U.S. Bureau Mines Bull. No.* 542, Washington, 1954.

6. F. D. Richardson and J. H. E. Jeffes, *J. Iron Steel Inst.*, **160,** 261 (1948); **161,** 229 (1949); **163,** 147, 397 (1949); **166,** 213 (1950).

7. D. R. Stull and G. C. Sinke, "Thermodynamic Properties of the Elements," *Advances in Chemistry*, Series No. 18, American Chemical Society, 1956.

8. K. S. Pitzer and W. V. Smith, *J. Am. Chem. Soc.*, **59,** 2633 (1937).

9. R. Schenck and F. Kurzen, *Z. anorg. allgem. Chem.*, **220,** 97 (1934).

10. G. B. Taylor and G. A. Hulett, *J. Phys. Chem.*, **17,** 565 (1913).

11. For convenience, the equilibrium constant, $K^*$, is expressed in pressure units of mm Hg. The thermodynamic $K$ is expressed in pressure units of atmospheres.

12. M. Randall, R. F. Nielsen, and G. H. West, *J. Ind. Eng. Chem.*, **23,** 388 (1931).

13. C. G. Maier, *U.S. Bur. Mines Bull. No.* 2926, Washington, D.C., 1929.

14. C. G. Maier, *J. Am. Chem. Soc.*, **51,** 194 (1929).

15. F. H. Smyth and H. S. Roberts, *J. Am. Chem. Soc.*, **42,** 2582 (1920); H. S. Roberts and F. H. Smyth, *J. Am. Chem. Soc.*, **43,** 1061 (1921).

16. F. J. Norton, unpublished results from the General Electric Research Laboratory.

17. E. P. Tatievskaya and G. I. Chufarov, *Bull. acad. sci. U.R.S.S., Classe sci. tech.*, **1946,** 1005, as reported in *Chem. Abst.*, **41,** 1538h (1947).

18. E. P. Tatievskaya, M. G. Zhuravleva, and G. I. Chufarov, *Izvest. Akad. Nauk S.S.S.R., Otdel. Tekh. Nauk*, **1949,** 1235, as reported in *Chem. Abst.*, **45,** 4537b (1951).

19. L. S. Darken and R. W. Gurry, *J. Am. Chem. Soc.*, **67,** 1398 (1945), and **68,** 799 (1946).

20. H. Peters and G. Mann, *Z. Elektrochem.*, **63,** 244 (1959).

21. J. P. Coughlin, E. G. King, and K. R. Bonnickson, *J. Am. Chem. Soc.*, **73,** 3891 (1951).

22. F. J. Norton, unpublished results from the General Electric Research Laboratory.

23. J. Smiltens, *J. Am. Chem. Soc.*, **79,** 4877 (1957).

24. G. I. Chufarov, M. G. Zhuravleva, and E. P. Tatievskaya, *Doklady Akad. Nauk S.S.S.R.*, **73,** 1209 (1950), as reported in *Chem. Abst.*, **45,** 426i (1951).

25. F. Krüll, *Z. anorg. Chem.*, **208,** 134 (1932); also A. Simon and F. Fehér, *Z. Electrochem.*, **38,** 137 (1932).

26. J. N. Ramsey, D. Caplan, and A. A. Burr, *J. Electrochem. Soc.*, **103,** 135 (1956).

27. A. D. Mah, *J. Am. Chem. Soc.*, **76,** 3363 (1954).

28. R. J. Ackerman, R. J. Thorn, C. Alexander, and M. Tetenbaum, *J. Phys. Chem.*, **64,** 350 (1960).
29. N. A. Gocken, *Trans. AIME,* **197,** 1019 (1953).
30. R. C. Griffis, *J. Electrochem. Soc.*, **106,** 418 (1959).
31. D. T. Livey, *J. Less-Common Metals*, **1,** 145 (1959).
32. G. DeMaria, J. Drowart, and M. G. Inghram, *J. Chem. Phys.*, **30,** 318 (1959).
33. J. Berkowitz, M. Inghram, and W. A. Chupka, *J. Chem. Phys.*, **26,** 842 (1957).
34. L. Brewer, *Chem. Revs.*, **52,** 1 (1953).
35. R. F. Porter, *Proceedings of the Symposium on High Temperature—A Tool for the Future*, Stanford Research Institute, 182 (1956).
36. R. J. Ackerman and R. J. Thorn, *U.S. Atomic Energy Comm. TID*-7530 (Pt. 1), 14 (1957).
37. M. G. Inghram and J. Drowart, *International Symposium on High Temperature Technology*, Stanford Research Institute, 219 (1959).
38. J. B. Austin and M. J. Day, *Symposium on Controlled Atmospheres*, American Society for Metals, Cleveland, Ohio, 1942, page 20.
39. N. R. Stansel, *Industrial Electric Heating*, John Wiley & Sons, New York, 1933.
40. H. Lepp, *Metal Ind. London*, **53,** 27, 59, 79, 103, 131 (1938).
41. C. H. Prescott, *J. Am. Chem. Soc.*, **48,** 2534 (1926).
42. C. H. Prescott and W. B. Hincke, *J. Am. Chem. Soc.*, **49,** 2744 (1927).
43. E. S. Rittner, *Philips Research Repts.*, **8,** 184 (1953).
44. G. Herrmann, *The Oxide-Coated Cathode*, Chapman and Hall, London, 1951 (translated by S. Wagener), Vols. 1 and 2.
45. W. A. Chupka, J. Berkowitz, and M. G. Inghram, *J. Chem. Phys.*, **26,** 1207 (1957). See also H. W. Goldstein, P. N. Walsh, and D. White, *J. Phys. Chem.*, **64,** 1087 (1960).
46. R. J. Ackerman and R. J. Thorn, *J. Am. Chem. Soc.*, **78,** 4169 (1956). For discussion, see J. Drowart, G. DeMaria, R. P. Burns, and M. G. Inghram, *J. Chem. Phys.*, **32,** 1366 (1960).
47. L. Navias, *Am. Ceram. Soc. Bull.*, **38,** 256 (1959).
48. W. C. Schumb, E. F. Sewell, and A. S. Eisenstein, *J. Am. Chem. Soc.*, **69,** 2029 (1947).
49. E. P. Kilb, *U.S. Atomic Energy Comm. APEX* 485, 1959, 57 pages.
50. D. T. Hurd, *An Introduction to the Chemistry of the Hydrides*, John Wiley & Sons, New York, 1952.
51. K. K. Kelley, *U.S. Bur. Mines Bull. No.* 407, Washington, D.C., 1937.
52. L. Brewer, L. A. Bromley, P. W. Giles, and N. L. Lofgren, *The Chemistry and Metallurgy of Miscellaneous Materials: Thermodynamics*, L. L. Quill, Editor, McGraw-Hill Book Company, New York, 1950.
53. J. Pearson and U. J. C. Ende, *J. Iron Steel Inst.*, **175,** 52 (1953).
54. A. D. Mah and N. L. Gelbert, *J. Am. Chem. Soc.*, **78,** 3261 (1956).
55. A. U. Seybolt and R. A. Oriani, *Trans. AIME,* **206,** 556 (1956).
56. W. H. Smith, Ph.D. thesis, Rensselaer Polytechnic Institute, June, 1958.
57. In Ref. 55, the free-energy values for reaction 11.88 were calculated with $n = 0.50$ instead of 0.38.

# Name Index

# Subject Index